Multiples and Submultiples for SI Units

Prefix	Symbol	Multiplier	Use
tera	T	$1{,}000{,}000{,}000{,}000 = 10^{12}$	1 terameter (Tm)
giga	G	$1{,}000{,}000{,}000 = 10^{9}$	1 gigameter (Gm)
mcga	M	$100{,}000 = 10^{6}$	1 megameter (Mm)
kilo	k	$1{,}000 = 10^{3}$	1 kilometer (km)
centi*	c	$0.01 = 10^{-2}$	1 centimeter (cm)
milli	m	$0.001 = 10^{-3}$	1 millimeter (mm)
micro	μ	$0.000001 = 10^{-6}$	1 micrometer (μm)
nano	n	$0.000000001 = 10^{-9}$	1 nanometer
—	Å	$0.0000000001 = 10^{-10}$	1 angstrom (Å)*
pico	p	$0.000000000001 = 10^{-12}$	1 picometer (pm)

*The use of the centimeter and the angstrom is discouraged but they are still widely used.

The SI Base Units for Seven Fundamental Quantities and Two Supplemental Quantities

Quantity	Unit	Symbol
Base Units		
Length	meter	m
Mass	kilogram	kg
Time	second	s
Electric current	ampere	A
Temperature	kelvin	K
Luminous intensity	candela	cd
Amount of substance	mole	mol
Supplemental Units		
Plane angle	radian	rad
Solid angle	steradian	Sr

D0169338

PHYSICS

SEVENTH EDITION

PHYSICS

Paul E. Tippens

Professor Emeritus
Southern Polytechnic State University

Higher Education

Boston   Burr Ridge, IL   Dubuque, IA   Madison, WI   New York   San Francisco   St. Louis
Bangkok   Bogotá   Caracas   Kuala Lumpur   Lisbon   London   Madrid   Mexico City
Milan   Montreal   New Delhi   Santiago   Seoul   Singapore   Sydney   Taipei   Toronto

Higher Education

PHYSICS, SEVENTH EDITION

Published by McGraw-Hill, a business unit of The McGraw-Hill Companies, Inc., 1221 Avenue of the Americas, New York, NY 10020. Copyright © 2007 by The McGraw-Hill Companies, Inc. All rights reserved. No part of this publication may be reproduced or distributed in any form or by any means, or stored in a database or retrieval system, without the prior written consent of The McGraw-Hill Companies, Inc., including, but not limited to, in any network or other electronic storage or transmission, or broadcast for distance learning.

Some ancillaries, including electronic and print components, may not be available to customers outside the United States.

This book is printed on acid-free paper.

1 2 3 4 5 6 7 8 9 0 VNH/VNH 0 9 8 7 6 5

ISBN-13   978–0–07–301267–4
ISBN-10   0–07–301267–X

Publisher: *Margaret J. Kemp*
Senior Sponsoring Editor: *Daryl Bruflodt*
Developmental Editor: *Liz Recker*
Marketing Manager: *Todd L. Turner*
Senior Project Manager: *Gloria G. Schiesl*
Senior Production Supervisor: *Laura Fuller*
Lead Media Project Manager: *Judi David*
Senior Media Producer: *Jeffry Schmitt*
Senior Designer: *David W. Hash*
Cover/Interior Designer: *Rokusek Design*
(USE) Cover Image: *©Karl Weatherly/CORBIS*
Lead Photo Research Coordinator: *Carrie K. Burger*
Supplement Producer: *Melissa M. Leick*
Compositor: *TechBooks/GTS, York, PA*
Typeface: *10.5/12 Times Roman*
Printer: *Von Hoffmann Corporation*

Library of Congress Cataloging-in-Publication Data

Tippens, Paul E.
    Physics / Paul E. Tippens. — 7th ed.
        p. cm.
    Includes index.
        ISBN 978–0–07–301267–4 — ISBN 0–07–301267–X
    1. Physics—Textbooks. I. Title.

QC21.3.T57   2007
530—dc22   2005054018
                                                CIP

About the Author

Paul E. Tippens has authored two successful textbooks for McGraw-Hill Companies and all of the ancillary materials associated with those books. His most successful book, *PHYSICS,* won the prestigious McGuffey Award, a longevity award for textbooks whose excellence has been demonstrated over time. Additionally, he is the author of *BASIC TECHNICAL PHYSICS, Second Edition.* These textbooks have been translated into Spanish, French, Chinese, and Japanese editions. Other works include four volumes of computer tutorials and numerous papers in major journals. Dr. Tippens is an active member of the Text and Academic Authors Association (TAA) and a strong advocate of its mission to provide information, advice, and networking for creators of intellectual property. He served for two years as the VP/President elect of TAA, because he sees TAA as a major player in guaranteeing a voice for authors in securing their intellectual property rights. He holds a EdD in Educational Administration from Auburn University, a Masters degree in Physics from the University of Georgia, and has completed numerous short courses in at least four other major universities. Currently, he is an Emeritus Professor at Southern Polytechnic State University in Marietta, Georgia, where he has taught college physics for 30 years.

To Jared Andrew Tippens and Elizabeth Marie Tippens
And to the supporting cast: Sarah, Travis, and Ryan Tippens
Grandchildren are great—all the joy and no responsibility

Brief Contents

Contents

Preface

The seventh edition of *Physics* is written for a one-year, transferable course in introductory physics. The emphasis on applications and the broad range of topics makes it suitable for students majoring in science and technology as well as for those majoring in biology, the health professions, and the environmental sciences. It may also be used for introductory courses in a variety of trade and industrial institutions, where the need is for an applied course that will not limit the future educational choices of such students. The mathematics, which has been reviewed extensively, assumes some familiarity with algebra, geometry, and trigonometry, but not calculus.

Physics began in earlier editions as an extensive project to address the need for a textbook that presents the fundamental concepts of physics in ways that can be understood and applied by students with a variety of backgrounds and preparation. The goal was to develop a textbook that is readable and easy to follow, but also one that provides a strong and rigorous preparation. The generous input from many dedicated users of the first six editions has helped to perpetuate that goal, and the work has received national recognition in the form of the prestigious McGuffey Award presented by the Text and Academic Authors Association (TAA) for excellence and longevity.

Three trends are significantly affecting modern instruction in college physics—the basic foundation for advanced study in almost any area:

1. Science and technology are growing exponentially.

2. Available jobs and career choices require more and more understanding of physics fundamentals.

3. The secondary-level preparation in math and science (for a variety of reasons) is not improving rapidly enough.

The focus of the seventh edition of *Physics* is to attack both ends of the problems caused by these trends. We provide the necessary mathematics background, and we do not compromise the educational outcomes.

Organization

The text consists of 39 chapters covering the entire spectrum of physics: *Mechanics, Thermal Physics, Wave Motion, Sound, Electricity, Magnetism, Light and Optics,* and *Atomic and Nuclear Physics.* This standard sequence can fit the requirements of a two-semester sequence or it can be used in a three-quarter program with a slight rearrangement of the topics. Shorter courses are also possible with a judicious selection of topics. Where possible, the coverage is designed so the order of topics can be changed.

There are a few areas where the coverage differs from that presented in most standard textbooks. A major distinction is the recognition that many students enter their first course in physics unable to apply basic skills in algebra and trigonometry. They have had the prerequisite courses, but for a variety of reasons seem unable to apply the concepts to solve problems. The dilemma is how to succeed with students without sacrificing standards. In *Physics,* we devote an entire chapter to a review of the mathematics and trigonometry required for solving physics problems. Other textbooks, if they provide such a review at all, usually do so in an appendix or ancillary product. Our approach enables students to recognize the *importance* of mathematics and to assess rather quickly their needs and deficiencies. It can easily be eliminated depending on the preparation of students or at the discretion of an individual instructor; however, it cannot be ignored as a critical requirement for problem solving.

Next, we address the need for meeting standards by covering statics before dynamics. Newton's first, second, and third laws are covered early to provide a qualitative understanding of force, but the full treatment of the second law is delayed until the concepts of free-body diagrams and static equilibrium are understood. This allows students to develop their understanding in a logical and continuous manner, while the skills in mathematics are slowly reinforced. In other books, the treatment of statics in later chapters often requires a review of forces and vectors. With the approach of this text, it is possible to provide more detailed examples of important applications of Newton's second law.

We also include a chapter on *Simple Machines* to give instructors the option of emphasizing many real world examples involving concepts of force, torque, work, energy, and efficiency. This chapter can easily be omitted if there are time constraints, but it has been very popular with some colleges where applications are paramount.

Modern Physics is treated as a survey course in the principles of *relativity, atomic physics,* and *nuclear physics.* Here the coverage is traditional and the topics have been selected in a way that helps students understand and apply the basic theories underlying many modern applications of atomic and nuclear physics.

New to the Seventh Edition

Content Changes

- **Treatment of Vectors.** The traditional component method of vector addition is emphasized, but an option that permits the use of unit vectors has been added.

- **Newton's Second Law.** The relationship between acceleration and force is introduced earlier to provide a qualitative understanding of force, but the more detailed treatment continues after considerable practice with free-body diagrams.

- **Rotational Kinetic Energy.** A significant addition extends the treatment of rotation in conservation of energy problems by covering the problem of objects that are both translating and rotating.

- **Electromagnetic Waves.** A more extensive discussion of electromagnetic waves precedes the treatment of light and optics.

- **Examples.** New examples have been added and all have been reworked to simplify the discussion and to clarify the problem-solving process.

- The sections on electrochemistry and the chapter on electronics have been eliminated based on input from past users and reviewers.

Improved Art Program

- **Opening Chapter Photos.** An effort has been made to make physics more visual by including introductory photographs for each chapter with brief annotations. These pictures were carefully selected to demonstrate concepts and applications covered in each chapter.

- **Figures.** All of the figures have been revised and/or redrawn. In many cases, photo objects have been inserted to enhance the line art, and a more extensive use of color adds contrast for emphasis.

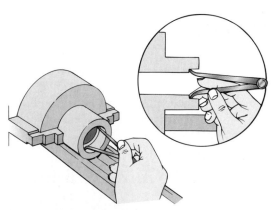

Figure 3.3 Using calipers to measure an inside diameter.

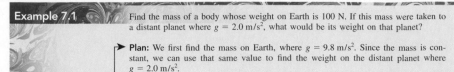

Example 7.1 Find the mass of a body whose weight on Earth is 100 N. If this mass were taken to a distant planet where $g = 2.0$ m/s^2, what would be its weight on that planet?

Plan: We first find the mass on Earth, where $g = 9.8$ m/s^2. Since the mass is constant, we can use that same value to find the weight on the distant planet where $g = 2.0$ m/s^2.

Solution:

$$m = \frac{W}{g} = \frac{100 \text{ N}}{9.8 \text{ m/s}^2} = 10.2 \text{ kg}$$

The weight on the planet is

$$W = mg = (10.2 \text{ kg})(2 \text{ m/s}^2); \qquad W = 20.4 \text{ N}$$

Planning Statements

A frequent comment from beginning students is "I just don't know how to get started." To address this concern we have included an extra step for many of the worked examples given in the text. The **_Plan_** statement bridges the gap between reading a problem and applying a learning strategy.

Everyday Physics Topics

Marginal notes are located throughout the text to generate interest and stimulate further study.

Online Learning Center

www.mhhe.com/tippens McGraw-Hill offers a wealth of online features and study aids that greatly enhance the physics teaching and learning experiences.

Digital Content Manager

This CD-ROM contains every illustration from the text. Instructors can use this artwork to create customized classroom presentations and other course tools.

PHYSICS TODAY

Do you know how much time satellites have in the sunlight to charge their batteries? For low Earth orbit, they have 60 min of sunlight and 35 min of darkness. Geosynchronous Earth orbit (GEO) satellites, which are much farther out, spend less time in Earth's shadow. They spend 22.8 hours in sunlight and 1.2 hours in darkness. The power to run the satellites must come entirely from batteries during the dark period.

Interactives

A total of 16 interactives are now available on the Digital Content Manager CD-ROM and online through the Online Learning Center. These interactives offer a fresh and dynamic method to teach and learn the physics basics by providing applets that are completely accurate and work with real data.

Retained Features

Several features retained from previous editions will capture and maintain the attention of students. These include:

Mathematics Preparation

Chapter 2 is devoted entirely to a review of the mathematics and trigonometry required for solving physics problems.

Chapter Objectives

To address the problem of educational outcomes, each chapter begins with a clear statement of objectives. The student knows from the beginning which topics are important and what outcomes are expected.

Problem-Solving Strategies

Throughout the text, we have included highlighted sections detailing the step-by-step procedures for solving difficult physics problems. Students may use this as a guide until they become familiar with the reasoning processes needed to apply the fundamental concepts presented in the text. The strategies are reinforced through many examples in the text.

Informative Writing Style

A hallmark of previous editions and a continued factor in the seventh edition is the presentation of physics in a friendly and informative manner.

Use of Color

Color is used to highlight pedagogical features in the text. Examples, learning strategies, key equations are shaded, and contrast is given to the important portions of figures.

Text Examples

Throughout each chapter are numerous worked examples. These serve as models for students on how to use the concepts covered in the text. Students learn to first visualize the situation, devise a plan for solving the problem, and then implement what they have learned to solve the problem.

End of Chapter Material

A carefully devised set of learning aids at the end of each chapter helps students review the chapter content, evaluate their grasp of key concepts, and utilize what they have learned.

- **Summaries.** A detailed summary is given for all of the essential concepts. Important equations are also highlighted in the text and summarized at the conclusion of each chapter.

- **Key Terms.** The key terms listed at the end of each chapter are highlighted in bold italic when they first appear in the text. These include all the main terms covered in the chapter so that students can verify their understanding of the concepts behind each term.

- **Review Questions.** More than 500 thought-provoking questions have been provided to stimulate thought and enhance conceptual thinking.

- **Problems and Additional Problems.** More than 1,750 carefully selected problems, ranging from simple, to moderate, to complex in difficulty are provided. In the seventh edition, considerable effort has been made to

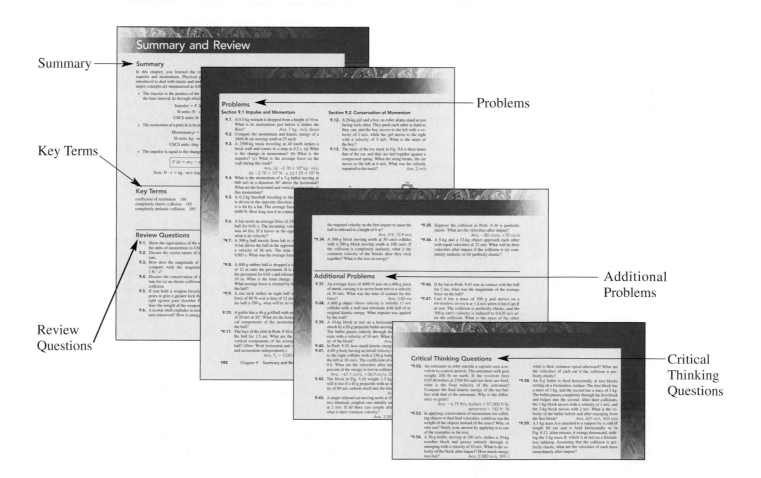

Summary — Key Terms — Review Questions — Problems — Additional Problems — Critical Thinking Questions

verify the accuracy of all problems and the answers given to the odd-numbered problems in the textbook. *Note:* Simple Problems do not have an asterisk next to them. Moderate Problems have one asterisk and complex problems have two asterisks.

- **Critical Thinking Questions.** Approximately 250 problems require moderate or greater thought than other problems in the text. They serve as learning examples that guide the students and build problem-solving skills. *Note:* Depending on the nature of the question, some answers are provided for the even-numbered questions and some for odd-numbered questions.

Supplements

Online Learning Center
www.mhhe.com/tippens
Student Online Resources include:

- **Study Questions.** True-false, multiple-choice, and completion questions are included.
- **Tutorials.** The author has prepared a comprehensive, web-based set of instructional PowerPoint modules for each chapter in the text. These tutorials are excellent for review prior to lectures, after lectures, before examinations, and before the final examination. They are also very useful to students who miss classes or who desire additional practice and discussion of physical concepts.
- **Interactives.** McGraw-Hill is proud to bring you an assortment of outstanding Interactive Applets like no other. These "Interactives" offer a fresh and dynamic method for teaching the physics basics by providing students with applets that are completely accurate

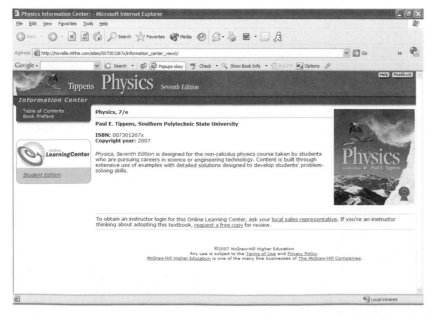

and work with real data. Interactives allow students to manipulate parameters and gain better understanding of 16 of the more difficult physics topics by watching the effect of these manipulations. Each Interactive includes an analysis tool (interactive model), a tutorial describing its function, and content describing its principle themes. Users can jump between these exercises and analysis tools with just the click of the mouse.

Instructor Online Resources include *all of the above,* plus:

- An Instructor's Manual, which includes the solutions to all the end of chapter problems and notes for laboratory experiments.
- The Online Learning Center can be easily loaded into course management systems such as Blackboard, WebCT, eCollege, and PageOut.

Digital Content Manager
This CD-ROM contains every illustration, photograph, and table from the text, and 16 interactives. The software makes customizing your multimedia presentation easy. You can organize figures in any order you want; add labels, lines, and your own artwork; integrate material from other sources; edit and annotate lecture notes; and have the option of placing your multimedia lecture into another presentation program such as PowerPoint.

Instructor's Testing and Resource CD-ROM
The accompanying electronic testing program is flexible and easy to use. The program allows instructors to create tests from book specific items. It accommodates a wide range of question types, and instructors may add their own questions. Multiple versions of the test can be created, and any test can be exported for use with course management systems such as WebCT, BlackBoard or PageOut. The program is available for Windows and Macintosh environments.

Instructor's Manual
The Instructor's Manual is found on the Tippens Online Learning Center and on the Instructor's Testing and Resource CD, and can be accessed only by instructors.

Custom Publishing

Did you know that you can design your own text or laboratory manual using any McGraw-Hill text and your personal materials to create a custom product that correlates specifically to your syllabus and course goals? Contact your McGraw-Hill sales representative to learn more about this option.

Acknowledgments

Reviewers of the Seventh Edition

Special thanks and appreciation go out to reviewers of the seventh edition. Their contributions, constructive suggestions, new ideas, and invaluable advice played an important role in the development of this seventh edition and its supplements. These reviewers include:

Abraham C. Falsafi   *National Institute of Technology*
Baher Hanna   *Owens Community College*
Kevin Hulke   *Chippewa Valley Technical College*
Benjamin C. Markham   *Ivy Tech State College*
James L. Meeks   *West Kentucky Community & Technical College*
John S. Nedel   *Columbus State Community College*
Russell Patrick   *Southern Polytechnic State University*
Sulakshana Plumley   *Community College of Allegheny County*
August Ruggiero   *Essex County College*
Erwin Selleck   *SUNY College of Technology at Canton*
Rich Vento   *Columbus State Community College*
Carey Witkov   *Broward Community College*
Todd Zimmerman   *Madison Area Technical College*

Special Acknowledgments

The author and McGraw-Hill would like to thank Rich Vento, professor at Columbus State Community College, for performing a complete accuracy check on the manuscript for the seventh edition. Rich's feedback proved invaluable to this edition.

A special thank you also goes to Russell Patrick, professor at Southern Polytechnic State University, for updating the testbank that accompanies *Physics.*

Reviewers of Previous Editions

The following reviewed previous editions of this book. Their comments and advice greatly improved the readability, accuracy, and currency of the book.

Shaikh Ali   *City College of Fort Lauderdale*
Fred Einstein   *County College of Morris*
Miles Kirkhuff   *Lincoln Technical Institute*
Henry Merrill   *Fox Valley Technical College*
Sam Nalley   *Chattanooga State Technical Community College*
Ajay Raychaudhuri   *Seneca College of Arts and Technology*
Charles A. Schuler   *California State University of Pennsylvania*
Scott J. Tippens   *Southern Polytechnic State University*
Bob Tyndall   *Forsyth Technical Community College*
Ron Uhey   *ITT Tech Institute*
Cliff Wurst   *Motlow State Community College*

The McGraw-Hill Book Team

The author wishes to express his enormous respect and gratitude for the efforts of the fine team of professionals at McGraw-Hill who have given countless hours of their time and expertise to the development and production of the seventh edition of *Physics.* Special thanks goes to my developmental editor, Liz Recker, far and away the best editor I've worked with in my many years with McGraw-Hill. Gloria Schiesl, the senior project manager, worked long and hard to make the production process run as smoothly as possible. Daryl Bruflodt (Sponsoring Editor), Todd Turner (Marketing Manager), Jeffry Schmitt (Media Producer), Judi David (Media Project Manager), Carrie Burger (Lead Photo Research Coordinator), Laura Fuller (Production Supervision), and Shirley Oberbroeckling (Managing Developmental Editor) also played key roles in this revision.

1

Introduction

Kennedy Space Center, Florida. In the Payload Hazardous Servicing Facility, workers watch as the Mars Exploration Rover-2 (MER-2) rolls over ramps to test its mobility and maneuverability. Scientists and engineers use the scientific method to verify that the vehicle can perform tasks similar to those needed for Mars exploration. (*Photo by NASA.*)

Knowledge of physics is essential to an understanding of our world. No other science has been as active in revealing the causes and effects of natural events. A casual glance at our past demonstrates a continuum of experimentation and discovery ranging from early measurements of gravity to later conquests of space. By studying objects at rest or in motion, scientists have been able to derive fundamental laws for many applications in mechanical engineering. The investigation of the principles that govern the production of heat, light, and sound has added countless applications that have served to make us more comfortable and more able to cope with our environment. Research and development in the areas of electricity, magnetism, atomic physics, and nuclear physics have led to a modern world that would have been unthinkable a mere 50 years ago (Fig. 1.1).

It is difficult to imagine a single product available today that does not require an application of some physical principle. This means that regardless of your career choice, you will need to understand physics in some way. Granted, there are some occupations and professions that do not require the depth of understanding necessary for engineering applications, but all fields of work utilize and apply these concepts. With a thorough understanding of mechanics, heat, sound, and electromagnetism, you carry with you the building blocks for almost any career. If you find it necessary or desirable to change careers either before or after graduation, you will be able to draw from a general base of science and mathematics. By taking this course seriously and by devoting an unusual amount of time and energy to it, you will have less trouble in the future. In your later coursework and on the job, you will be riding the crest of the wave instead of merely staying afloat in an angry sea.

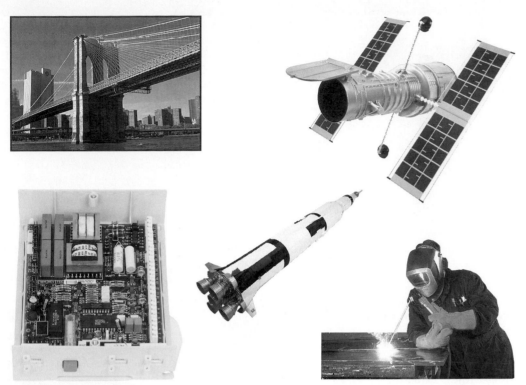

Figure 1.1   Applications of the principles of physics are found in many occupations. (*Photos by Hemera, Inc.*)

What Is Physics?

Even if you have previously taken courses in high-school physics, you probably still have a rather "fuzzy" idea of what *physics* really means and how it might differ, for example, from *science*. For our purpose, the sciences can be divided between *biological* and *physical*. The biological sciences deal with living things. The physical sciences deal primarily with the nonliving sides of nature.

> **Physics** can be defined as the science that investigates the fundamental concepts of matter, energy, and space and the relationships among them.

In terms of this broad definition, there are no clear boundaries among the physical sciences. This is evident from the overlapping fields of biophysics, chemical physics, astrophysics, geophysics, electrochemistry, and so forth.

The goal of this textbook is to provide a basic introduction to the world of physics. The emphasis is on applications, and the broad field of physics will be narrowed to the essential concepts that underlie all technical knowledge. You will study mechanics, heat, light, sound, electricity, and atomic structure. The most basic of these topics, and probably the most important for beginning students, is mechanics.

Mechanics is concerned with the position (statics) and motion (dynamics) of matter in space. *Statics* represents the study of physics associated with bodies at rest. *Dynamics* is concerned with a description of motion and its causes. In each case, the engineer or technician is concerned with measuring and describing physical quantities in terms of their cause and effect.

An engineer, for example, uses physical principles to determine which type of bridge structure will be the most efficient for a given situation. The concern is for the *effect* of forces. If the completed bridge fails, the *cause* of failure must be analyzed to apply this knowledge to future construction. It is important to note that by *cause* the scientist means the sequence of physical events leading to an *effect*.

What Part Is Played by Mathematics?

Mathematics serves many purposes. It is philosophy, art, metaphysics, and logic. These values are subordinate, however, to its main value as a tool for the scientist, engineer, or technician. One of the rewards of a first course in physics is the growing awareness of the relevance of mathematics. A study of physics reveals specific applications of basic mathematics.

Suppose you wish to predict how long it takes to stop a car traveling at a given speed. First, you would attempt to control as many of the variables as possible. In trial runs, you would want the braking in each trial to be uniform so that the average speed would be close to half of the initial speed. Using symbols, we might write

$$v_{\text{avg}} = \frac{v_i}{2}$$

The road conditions, slope of the road, the weather, and other parameters would also be controlled. For each run, you would record the initial speed v_i, the stopping distance x, and the time t. You might record the initial speed, the change in speed, and the distance and time required to stop the car. When all these facts are recorded, the data might be used to establish a tentative relationship. We cannot do this without the tools of mathematics.

From the definition of speed as distance traveled per unit of time, we recognize that the stopping distance x for our example might be the product of the average velocity $v_i/2$ and the time t. Our tentative relationship might be

$$x = \frac{v_i}{2}t \qquad \text{or} \qquad x = \frac{v_i t}{2}$$

Note that we have used symbols to represent the important parameters and mathematics to express their relationship.

This statement is a *workable hypothesis*. From this equation, we can predict the stopping distance for any car given its initial speed and stopping time. When it has been used long enough for us to be reasonably sure that it is true, we call it a *scientific theory*. In other words, any scientific theory is a workable hypothesis that has withstood the test of time.

Thus, we can see that mathematics is useful in deriving formulas that describe physical events accurately. Mathematics plays an even larger role in solving such formulas for specific quantities.

For example, in the previous formula, it would be relatively simple to find values for x, v_i, or t when the other quantities are known. However, many physical relationships involve a much more specialized knowledge of algebra, trigonometry, or even calculus. How easily you derive or solve a theoretical relationship depends on your background in mathematics.

A review of the mathematics required for this text is presented in Chapter 2. If you are unfamiliar with any of the topics discussed, you should study this chapter carefully. Pay particular attention to the sections on powers of 10, literal equations, and trigonometry. Skill in applying the tools of mathematics will largely determine your success in any physics course.

How Should I Study Physics?

Reading technical material is different from other reading. Attention to specific meanings of words must be given if the material is to be understood. Graphs, drawings, charts, and photographs are often included in technical literature. They are always helpful and may even be essential to the description of physical events. You should study them thoroughly so that you understand the principles clearly.

Much of what you learn will be from classroom lectures and experiments. The beginning student often asks, How can I concentrate fully on the lecture and at the same time take accurate notes? Of course, it may not be possible to understand fully all the concepts presented and still take complete notes. You must learn to note only the significant portions of each lesson. Make sure you listen attentively to the explanation of various topics. Learn to recognize key words such as *work, force, energy,* and *momentum.*

Adequate preparation before class should give you a good idea of which portions of the lecture are covered in the text and which are not. If a problem or definition is in the text, it is usually better to jot down a key word and concentrate fully on what the instructor is saying. These notes can be expanded later.

Each student who enters a beginning physics class should have the prerequisites and ability to pass the course. There may be other reasons for failure: perhaps motivation, heavy class loads, outside jobs, sickness, or personal problems. The following advice comes from seasoned instructors who have a history of success with beginning physics students.

- *The ultimate responsibility for learning rests with the student.* The instructor is only a facilitator, the college is only a campus, and the text is only a book. Come to class each day on time and prepared for the scheduled instruction. Study the material in advance and make a note of questions to ask the instructor.

- *Timely learning is efficient learning.* It is better to study an hour each day of the week, than to study for 20 hours every weekend. After each lecture or recitation, use the very next free period to reinforce your understanding of the topics covered. Work a few example problems. The longer you wait, the more you forget, and the more you waste your time. If you wait until the weekend, it often requires an hour or more just to review and reconstruct the lecture from notes. *Cramming for tests will not work.* Instead, review problems that you have already worked and attempt to work similar problems in the textbook.

- *Learning is rarely completed in the classroom.* In order to retain and build on what you learn in the classroom, solving problems on your own is essential. Seek help from others, including your instructor, after you make your best attempt to solve assigned problems. There is no substitute for active participation in the thinking and procedures needed to solve problems.

- *Brush up on basic skills.* Chapter 2 on technical mathematics stresses those skills that may be rusty or weak. Make sure that you understand these topics.

- *Study the syllabus.* Know what will be covered on exams, when they will be given, and how they will be used to determine your grade.

- *Find a class partner and obtain his/her phone number.* Use a *buddy system* so that each can notify the other of class/lab activities that are unavoidably missed. Have this person pick up extra handouts and instructions given in your absence.

- *Organization is the key to effective learning.* Keep a loose-leaf filler notebook with tabbed sections: Handouts, Notes, Problems, Graded Exams, Graded Labs.

- *If you get into trouble, seek help immediately.* A multitude of learning resources are available for students today that years ago were just a dream. There are computer-assisted tutorials, the Internet, solution guides, problem-solving manuals, and even other textbooks covering the same topics. Your instructor or your library personnel can give you the direction you need, but you must take the responsibility.

After many years of teaching college physics, I have noted the most common reason some students get into trouble with beginning physics is due to ineffective planning and organization. Today, a typical student may be taking as many as two or three other subjects concurrently with physics—some even more. Additionally, they may be working in a part-time job; they may be married with children; they may be loaded with campus activities; and they may also be taking physics BEFORE completing the necessary preparatory mathematics courses. It soon becomes obvious that there is not sufficient time for significant immersion into any single area of study. Therefore, you must set a rigorous schedule with firm goals and priorities. To help you with such planning, I offer the following additional points for your consideration:

- As far as preparation for college and for your future in a modern technical world, physics is your *most important* beginning course. *(I will gladly debate this statement with anyone, and I often do.)*

- Don't expect to fully understand physics principles in the same way as you might learn about other, nontechnical subjects. The real understanding of physics comes through *application* and *problem solving*. You must apply a concept *very soon* after it is presented, or you will waste hours of time trying to reconstruct your thoughts later. Try to schedule a free period immediately after your physics course and attempt to work example problems while the lecture is fresh.

- Plan your study habits around the nature of the subjects you are taking. Many core subjects requiring extensive reading and reporting can be treated differently from mathematics and physics. Each is important, but mathematics and physics cannot be efficiently learned in cram sessions. When each successive topic *requires* understanding of the preceding topics, one can get hopelessly behind rather quickly.

- I have never taught a course in physics without someone complaining that "test anxiety" is a major reason for their poor performance. Granted, this is a very real problem, and it is much worse for some than for others. I think the best way to deal with the pressure of physics exams is with thorough and adequate preparation. You should work as many example problems as possible before taking examinations. In basketball, the game may be on the line with a final free throw. The successful player is the one who has shot so many foul shots that his/her reflexes are conditioned to respond even under pressure.

2

Technical Mathematics

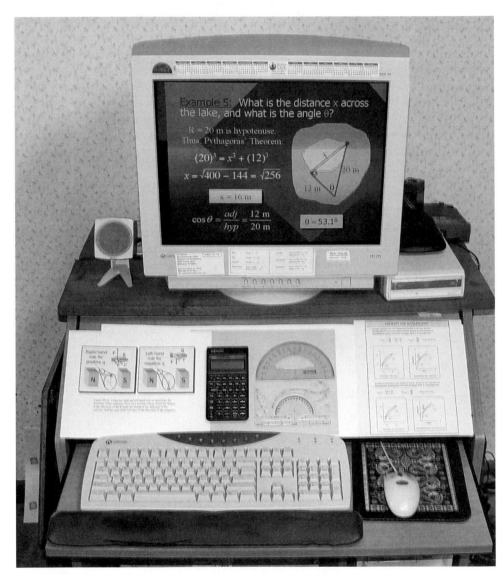

Mathematics is the fundamental tool for all science. The graph on the computer screen shows an application of trigonometry. (*Photo by Paul E. Tippens.*)

Objectives

After completing this chapter, you should be able to

1. Demonstrate your ability to add, subtract, multiply, and divide technical measurements.

2. Solve simple formulas for any quantity appearing in the formula and evaluate by substitution.

3. Solve simple problems that require operations with exponents and radicals.

4. Perform common mathematical operations in scientific notation.

5. Construct a graph from given technical data and interpret new information from the graph.

6. Apply the elementary rules of geometry to determine unknown sides and angles in given situations.

It is often frustrating to open a physics book and see that it begins with mathematics. Naturally, you want to learn only those things for which you see a definite need. You want to make measurements, to work with machines or engines, to get your hands on something, or, at the very least, to know that your time is not wasted. Depending on your previous experience, you may omit much or all of this chapter at the discretion of your instructor. Keep in mind that fundamentals are important and that some skills in mathematics are essential. You may have a complete understanding of force, mass, energy, and electricity as concepts. But you may not be able to apply the concepts to your work because of weaknesses in basic mathematics. Mathematics is the language of physics. In this book, an effort has been made to make that language as simple and as relevant as possible.

In any industrial or technical occupation, we are concerned with physical measurements of some kind. It might be the length of a board, the area of a sheet of metal, the number of bolts to be ordered, the stress on an aircraft wing, or the pressure in an oil tank. The only way we can make sense out of such data is through the use of numbers and symbols. Mathematics provides the necessary tools for organizing the data and for predicting outcomes. For example, the formula $F = ma$ expresses the relationship between a net force F and the acceleration a that the force produces. The quantity m is a symbol that represents the mass of an object (a measure of the amount of matter it contains). Through appropriate mathematical steps, we can use formulas such as this to predict future events. However, a general knowledge of algebra and geometry is required in many instances. This chapter provides you with a review of a few essential concepts in mathematics. Chapter sections may be assigned or omitted at the discretion of your instructor.

2.1 Signed Numbers

Frequently, it is necessary to work with negative numbers as well as positive numbers. For example, a temperature of −10°C means 10 degrees "below" a zero reference point, and 24°C refers to a temperature 24 degrees "above" zero (see Fig. 2.1). The numbers refer to the *magnitude* of the temperature, and the plus or minus sign refers to the *direction* from zero. The minus sign in −10°C does not indicate a lack of temperature; it means that the temperature is less than zero. The number 10 in −10°C describes how far the temperature is from zero; the minus sign is necessary to indicate the direction from zero.

The value of a number without its sign is called its *absolute value*. In other words, if we disregard the signs of +7 and −7, the value is the same. Each number is 7 units from zero. The absolute value of a number is indicated by vertical bar symbols. The number +7 does not equal the number −7, but |+7| does equal |−7|. When arithmetic operations are performed with signed numbers, the absolute values are used.

Plus and minus signs are also used to indicate arithmetic operations; for example,

7 + 5 means "add the number +5 to the number +7"

7 − 5 means "subtract the number +5 from the number +7"

If we wish to indicate the addition or subtraction of negative numbers, parentheses are helpful:

(+7) + (−5) means "add the number −5 to the number +7"

(+7) − (−5) means "subtract the number −5 from the number +7"

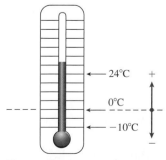

Figure 2.1

When signed numbers are added, it is useful to recall the following rule:

Addition Rule: To add two numbers of like sign, we add the absolute values of the numbers and give the sum the common sign. To add two numbers of unlike sign, we find the difference of their absolute values and give the result the sign of the number of larger value.

Consider the following examples:

$$(+6) + (+2) = +(6 + 2) = +8$$
$$(-6) + (-2) = -(6 + 2) = -8$$
$$(+6) + (-2) = +(6 - 2) = +4$$
$$(-6) + (+2) = -(6 - 2) = -4$$

Now, let's examine the procedure for subtraction. Whenever a number has another subtracted from it, we change the sign of the second number and then add it to the first, using the addition rule. In the expression $7 - 5$, the number $+5$ is to be subtracted from the number $+7$. The subtraction is accomplished by changing $+5$ to -5 and then adding the two numbers of unlike sign: $(+7) + (-5) = +(7 - 5) = +2$.

Subtraction Rule: To subtract one signed number b from another signed number a, we change the sign of b and then add it to a, using the addition rule.

Consider the following examples:

$$(+8) - (+5) = 8 - 5 = 3$$
$$(+8) - (-5) = 8 + 5 = 13$$
$$(-8) - (+5) = -8 - 5 = -13$$
$$(-8) - (-5) = -8 + 5 = -3$$

Example 2.1

The velocity of an object is considered positive when it is moving upward and negative when it is moving downward. What is the change in velocity of a ball if it strikes the floor at 12 meters per second (m/s) and rebounds upward at 7 m/s? Refer to Fig. 2.2.

Plan: We first choose the upward direction as positive so that we can use consistent signs for velocity. The initial velocity is -12 m/s because the ball is moving *downward*. Later its velocity is $+7$ m/s because it is moving *upward*. The *change* in velocity will be the final velocity less the initial velocity.

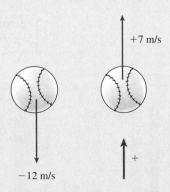

Figure 2.2

Solution:

$$\text{Change in velocity} = \text{final velocity} - \text{initial velocity}$$
$$= (+7 \text{ m/s}) - (-12 \text{ m/s})$$
$$= 7 \text{ m/s} + 12 \text{ m/s} = 19 \text{ m/s}$$

Without understanding signed numbers, we might have guessed that the change in speed was only 5 m/s $(12 - 7)$. A moment's thought, however, makes us realize that the speed must first decrease to zero (a change of 12 m/s) and then attain a speed of 7 m/s in the opposite direction (an additional change of 7 m/s).

When two or more numbers are to be multiplied, each number is called a *factor,* and the result is the *product.* We can now state the multiplication rule for signed numbers:

Multiplication Rule: If two factors have like signs, their product is positive. If two factors have unlike signs, their product is negative.

Examples are

$$(+2)(+3) = +6 \qquad (-3)(-4) = +12$$
$$(-2)(+3) = -6 \qquad (-3)(+4) = -12$$

An extension of the multiplication rule is often helpful for products resulting from several factors. Rather than multiplying a series of factors two at a time, we might recall that

The product will be positive if all factors are positive or if there is an even number of negative factors. The product will be negative if there is an odd number of negative factors.

Consider the following examples:

$$(-2)(+2)(-3) = +12 \qquad \text{(two negative factors—even)}$$
$$(-2)(+4)(-3)(-2) = -48 \qquad \text{(three negative factors—odd)}$$
$$(-3)^3 = (-3)(-3)(-3) = -27 \qquad \text{(three negative factors—odd)}$$

Notice that in the last example a superscript 3 was used to indicate the number of times the number -3 was to be taken as a factor. The superscript 3 written in this manner is called an *exponent.*

When two numbers are to be divided, the number being divided is called the *dividend.* The number divided into the dividend is called the *divisor.* The result of division is called the *quotient.* The rule for dividing signed numbers is as follows:

Division Rule: The quotient of two numbers of like sign is positive, and the quotient of two numbers of unlike sign is negative.

For example,

$$(+2) \div (+2) = +1 \qquad (-4) \div (-2) = +2$$
$$\frac{+4}{-2} = -2 \qquad\qquad \frac{-4}{+2} = -2$$

If either the numerator or the denominator of a fraction contains two or more factors, the following rule is also useful:

The quotient is negative if the total number of negative factors is odd; otherwise, the quotient is positive.

For example,

$$\frac{(-4)(3)}{2} = -6 \qquad \text{odd}$$

$$\frac{(-2)(-2)(-3)}{(2)(-3)} = +2 \qquad \text{even}$$

You should practice the application of all the rules in this entire section. It is a serious mistake to assume that you understand these concepts without adequate proof. A major source of errors in physics problems can be traced to signed numbers.

2.2 Algebra Review

Algebra is really a generalization of arithmetic in which letters are used to replace numbers. For example, we will learn that the space occupied by some objects (their volume) can be calculated by multiplying the length by the width by the height. By assigning letters to each of these measurements, we can establish a general *formula,* such as

$$\text{Volume} = \text{length} \times \text{width} \times \text{height}$$
$$V = l \cdot w \cdot h \tag{2.1}$$

The advantage of formulas is that they work for any number of situations. Given the length, width, and height of any rectangular solid, we can use Eq. (2.1) to calculate its volume. If we wish to determine the volume of a rectangular metal block, we need only *substitute* the proper numbers into the formula.

Example 2.2

Calculate the volume of a solid whose length is 6 centimeters (cm), whose width is 4 cm, and whose height is 2 cm.

Plan: Recall or locate the formula for finding the volume, and then substitute the given quantities.

Solution: Substitution yields

$$V = lwh$$
$$= (6 \text{ cm})(4 \text{ cm})(2 \text{ cm})$$
$$= 48 (\text{cm} \times \text{cm} \times \text{cm}) = 48 \text{ cm}^3$$

The treatment of units that results in volume being expressed as cubic centimeters is discussed later. For now, you should concentrate on the substitution of numbers.

When numbers are substituted for letters in a formula, it is very important to insert the proper sign of the number. Consider the following formula:

$$P = c^2 - ab$$

Suppose that $c = +2$, $a = -3$, and $b = +4$. Remember that plus and minus signs in formulas do not apply to any of the numbers that might be substituted. In this example, we have

$$P = (c)^2 - (a)(b)$$
$$= (+2)^2 - (-3)(+4)$$
$$= 4 + 12 = 16$$

It is easy to see how confusing a sign in the formula with the sign of a substituted number can result in an error.

It is frequently necessary to solve a formula or an equation for some letter that is only a part of the formula. For example, suppose we wanted to give a formula for the length of a rectangular solid in terms of its volume, height, and width. The letters in $V = lwh$ would need to be rearranged so that the letter l would appear by itself on the left side. Formula rearrangement is not difficult if we recall several rules for working with equations.

Basically, an equation is a mathematical statement that two expressions are equal. For example,

$$2b + 4 = 3b - 1$$

is an equation. In this case, it is evident that the letter b represents the *unknown* quantity. If we substitute $b = 5$ into both sides of this equation, we obtain $14 = 14$. Thus, $b = 5$ is the *solution* to the equation.

We can obtain solutions for equalities by performing the same operations on each side of an equation. Consider the equality $4 = 4$. If we add, subtract, multiply, or divide the number 2 into both sides, we do not change the fact that the two sides are equal. We *do* increase or decrease the magnitude of each side, but the equality remains. (You should verify this statement for the equality $4 = 4$.) Note also that taking the square or square root of each side does not disturb the equality. By performing a series of identical operations to each side of an equation, we can eventually obtain an equality with one letter by itself on the left side. In this case, we have *solved* the equation for that letter.

Example 2.3

Solve the following equation for m:

$$3m - 5 = m + 3$$

Plan: The point is to get m by itself on one side of the equals sign and a number by itself on the other side. As long as we add or subtract the *same* quantity from each side, the equation will remain a true statement.

Solution: First, add $+5$ to both sides, and then subtract m from both sides:

$$3m - 5 + 5 = m + 3 + 5$$
$$3m = m + 8$$
$$3m - m = m + 8 - m$$
$$2m = 8$$

Finally, we divide both sides by 2:

$$\frac{2m}{2} = \frac{8}{2}$$
$$m = 4$$

To check this answer, we substitute $m = 4$ into the original equation and obtain $7 = 7$. This shows that $m = 4$ is the solution.

In formulas, the solution to an equation may also be expressed in terms of letters. For example, the literal equation

$$ax - 5b = c$$

might be solved for x in terms of a, b, and c. In cases such as this one, we decide in advance which letter is to be the "unknown." In our example, we will choose x. The remaining letters are treated as though they were known numbers. Adding $5b$ to both sides, we have

$$ax - 5b + 5b = c + 5b$$
$$ax = c + 5b$$

Now, we can divide both sides by a to obtain

$$\frac{ax}{a} = \frac{c + 5b}{a}$$

$$x = \frac{c + 5b}{a}$$

which is the solution for x. The values for a, b, and c in a given situation can be substituted to find a particular value for x.

Example 2.4

The volume of a right circular cone is expressed by the formula

$$V = \frac{\pi r^2 h}{3} \tag{2.2}$$

What is the height of such a cone of radius $r = 3$ cm and $V = 81$ cubic centimeters (cm^3)?
Assume that $\pi = 3.14$.

Plan: We must first solve the formula for h in terms of r and V. Then we will need to substitute given values for V, π, and r.

Solution: Multiplying both sides by the number 3 gives

$$3V = \pi r^2 h$$

Dividing both sides by πr^2 yields

$$\frac{3V}{\pi r^2} = \frac{\pi r^2 h}{\pi r^2} \qquad \text{or} \qquad \frac{3V}{\pi r^2} = \frac{h}{1}$$

Thus, the height h is given by

$$h = \frac{3V}{\pi r^2}$$

Substitution of the known values for V, π, and r gives

$$h = \frac{3(81 \text{ cm}^3)}{(3.14)(3 \text{ cm})^2} = \frac{243 \text{ cm}^3}{28.26 \text{ cm}^2} = 8.60 \text{ cm}$$

The height of the cone is 8.60 cm.

2.3   Exponents and Radicals (Optional)

It is often necessary to multiply the same quantity a number of times. A shorthand method of indicating the number of times a quantity is taken as a factor uses a superscript number called an **exponent**. This notation works according to the following scheme:

For any number a:	For the number 2:
$a = a^1$	$2 = 2^1$
$a \times a = a^2$	$2 \times 2 = 2^2$
$a \times a \times a = a^3$	$2 \times 2 \times 2 = 2^3$
$a \times a \times a \times a = a^4$	$2 \times 2 \times 2 \times 2 = 2^4$

The powers of the number a are read as follows: a^2 is read "a squared," a^3 is read "a cubed," and a^4 is read "a to the fourth power." More generally, we speak of a^n as "a to the nth power." In these examples, the letter a is called the **base,** and the superscript numbers 1, 2, 3, 4, and n are called the *exponents.*

We will review several rules to follow in working with exponents.

Rule 1: When two quantities of the same base are multiplied, their product is obtained by adding the exponents algebraically:

$$(a^m)(a^n) = a^{m+n} \qquad \textit{Multiplication Rule} \quad \textbf{(2.3)}$$

Examples:

$$(2^4)(2^3) = 2^{4+3} = 2^7$$
$$y^8 y^6 = y^{14}$$
$$x^2 x^5 y^3 x^3 = x^{2+5+3} y^3 = x^{10} y^3$$

Rule 2: When a is not zero, a negative exponent may be defined by either of the following expressions:

$$a^{-n} = \frac{1}{a^n} \quad \text{and} \quad a^n = \frac{1}{a^{-n}} \qquad \textit{Negative Exponent} \quad \textbf{(2.4)}$$

Examples:

$$3^{-4} = \frac{1}{3^4} = \frac{1}{81} \qquad 10^2 = \frac{1}{10^{-2}}$$
$$a^{-5} = \frac{1}{a^5} \qquad \frac{x^{-3} y^2}{a^{-4} b^3} = \frac{a^4 y^2}{x^3 b^3}$$

Rule 3: Any quantity raised to the power of zero is equal to 1:

$$a^0 = 1 \qquad \textit{Zero Exponent} \quad \textbf{(2.5)}$$

Examples:

$$x^3 y^0 = x^3 \qquad (x^3 y^2)^0 = 1$$

Rule 4: The quotient of two nonzero quantities of the same base is found by taking the algebraic difference of their exponents:

$$\frac{a^m}{a^n} = a^{m-n} \qquad \textit{Division} \quad \textbf{(2.6)}$$

Examples:

$$\frac{2^3}{2} = 2^{3-1} = 2^2 \qquad \frac{2^5}{2^7} = 2^{5-7} = 2^{-2} = \frac{1}{2^2}$$
$$\frac{a^{-3}}{a^{-5}} = a^{-3-(-5)} = a^{-3+5} = a^2$$

Rule 5: When a quantity a^m is raised to the power n, the exponents are multiplied:

$$(a^m)^n = a^{mn} \qquad \textit{Power of a Power} \quad \textbf{(2.7)}$$

Examples:

$$(2^2)^3 = 2^{2 \cdot 3} = 2^6 \qquad (2^{-3})^2 = 2^{-6} = \frac{1}{2^6}$$

$$(a^2)^4 = a^8 \qquad (a^2)^{-4} = a^{-8} = \frac{1}{a^8}$$

Rule 6: The power of a product and of a quotient is obtained by applying the exponent to each of the factors:

$$(ab)^n = a^n b^n \qquad \left(\frac{a}{b}\right)^n = \frac{a^n}{b^n} \qquad\qquad \textbf{(2.8)}$$

Examples:

$$(2 \cdot 3)^2 = 2^2 \cdot 3^2 = 4 \cdot 9 = 36$$
$$(ab)^3 = a^3 b^3$$
$$(ab^2)^3 = a^3 (b^2)^3 = a^3 b^6$$
$$\left(\frac{ax^3}{y^2}\right)^4 = \frac{a^4 x^{12}}{y^8}$$

If $a^n = b$, then not only is b equal to the *n*th power of a, but also by definition, a is said to be the *n*th *root* of b. In general, this fact is expressed by using a ***radical*** ($\sqrt{\ }$):

$$\sqrt[n]{b} \qquad \text{*n*th root of } b$$

Consider the following statements:

$$2^2 = 4 \text{ means that 2 is the *square root* of 4, or } \sqrt{4} = 2$$
$$2^3 = 8 \text{ means that 2 is the *cube root* of 8, or } \sqrt[3]{8} = 2$$
$$2^5 = 32 \text{ means that 2 is the *fifth root* of 32, or } \sqrt[5]{32} = 2$$

A radical may also be expressed by using a fractional exponent. In general, we may write

$$\sqrt[n]{b} = b^{1/n}$$

For example,

$$\sqrt[3]{8} = 8^{1/3} \qquad \text{or} \qquad \sqrt{10} = 10^{1/2}$$

There are two additional rules to follow when you are working with radicals.

Rule 7: The *n*th root of a product is equal to the product of the *n*th roots of each factor:

$$\sqrt[n]{ab} = \sqrt[n]{a}\,\sqrt[n]{b} \qquad\qquad \textit{Roots of a Product} \quad \textbf{(2.9)}$$

Examples:

$$\sqrt{4 \cdot 16} = \sqrt{4}\,\sqrt{16} = 2 \cdot 4 = 8$$
$$\sqrt[5]{ab} = \sqrt[5]{a}\,\sqrt[5]{b}$$

Rule 8: The roots of a power are found by using the definition of fractional exponents:

$$\sqrt[n]{a^m} = a^{m/n} \qquad\qquad \textit{Roots of Powers} \quad \textbf{(2.10)}$$

Examples:

$$\sqrt[3]{2^9} = 2^{9/3} = 2^3 = 8$$

$$\sqrt{10^{-4}} = 10^{-4/2} = 10^{-2} = \frac{1}{10^2}$$

$$\sqrt{4 \times 10^8} = \sqrt{4}\,\sqrt{10^8} = 2(10)^{8/2} = 2 \times 10^4$$

$$\sqrt[3]{8 \times 10^{-6}} = \sqrt[3]{8}(10)^{-6/3} = 2 \times 10^{-2}$$

Most of the problems in this textbook will require only a limited understanding of these rules. Squares, cubes, square roots, and cube roots are encountered most often. However, a good understanding of the rules for exponents and radicals is helpful.

2.4 Solution to Quadratic Equations

Often in solving physics problems, it is necessary to obtain the solution to a second-degree equation that has an unknown raised to the second power. For example, in kinematics, the position of a particle in a gravitational field varies with time according to the relation:

$$x = v_0 t + \tfrac{1}{2}at^2$$

where x is the displacement, v_0 is the intitial velocity, a is the acceleration, and t is the time. Note that the appearance of t^2 means that there are two times at which the displacement may be the same. Such equations are called **quadratic equations.** Although there are several methods for solving these equations, perhaps the most useful technique for physics problems is to apply the quadratic formula.

Given a quadratic equation of the form

$$ax^2 + bx + c = 0$$

with a not equal to zero, the solutions are found from the quadratic formula:

$$x = \frac{-b \pm \sqrt{b^2 - 4ac}}{2a}$$

Example 2.5

Solve the following equation for x: $3x^2 = 12 + 5x$

Plan: The highest power of the unknown x is 2, and the quadratic formula may be applied. We must write the equation in quadratic form, determine the constants a, b, and c, and then solve for x using the formula.

Solution: The quadratic form is $ax^2 + bx + c = 0$, so we may write

$$3x^2 - 5x - 12 = 0$$

By inspection, we see that $a = 3$, $b = -5$, and $c = -12$. Now, we solve for x by substitution into the quadratic formula:

$$x = \frac{b - \sqrt{b^2 - 4ac}}{2a}$$

$$= \frac{-(-5) \pm \sqrt{(-5)^2 - 4(3)(-12)}}{2(3)}$$

$$= \frac{+5 \pm \sqrt{(25) + (144)}}{2(3)} = \frac{+5 \pm \sqrt{169}}{6} = \frac{5 \pm 13}{6}$$

To find the two solutions for x, use first the plus sign and then the minus:

First solution: $x = \dfrac{5 + 13}{6} = \dfrac{18}{6}$ or $x = +3$

Second solution: $x = \dfrac{5 - 13}{6} = \dfrac{-8}{6}$ or $x = -1.33$

The two answers are $x = +3$ and $x = -1.33$. Depending on the conditions of the problem, one of the solutions may be true mathematically but not possible physically. You should always try to interpret your results in light of the stated conditions.

Example 2.6

A ball is thrown upward with an initial speed of $v_0 = 20$ m/s. The acceleration due to gravity is $g = -9.80$ m/s^2. Given that the displacement $y = v_0 t + \frac{1}{2}gt^2$, find the two times at which the displacement $y = 12$ m above the release point.

Plan: We must insert the given values for g, y, and v_0 to obtain a quadratic equation with time t as our unknown. Next, we will write the equation in quadratic form and solve for t using the quadratic formula.

Solution: Substitution yields

$$y = v_0 t + \tfrac{1}{2}gt^2 \qquad \text{or} \qquad (12) = 20t + \tfrac{1}{2}(-9.8)t^2$$

Here, we have left out the units so that the unknown t is clearly indicated. Writing this expression in quadratic form, we obtain

$$4.9t^2 - 20t + 80 = 0$$

Now, we apply the quadratic formula to find the two solutions for a.

$$t = \frac{b - \sqrt{b^2 - 4ac}}{2a}$$

$$= \frac{-(-20) \pm \sqrt{(-20)^2 - 4(4.9)(12)}}{2(4.9)}$$

$$= \frac{+20 \pm \sqrt{400 - 235}}{9.8} = \frac{-20 \pm 12.8}{9.8}$$

Again, we find the two solutions by using first the plus and then the minus sign:

First solution: $t = \dfrac{20 + 12.8}{6} = \dfrac{32.8}{6}$ or $t = +3.35$ s

Second solution: $t = \dfrac{20 - 12.8}{9.8} = \dfrac{7.17}{9.8}$ or $t = +0.732$ s

The ball first reaches the height of 12 m at a time $t = 0.732$ s after release. It later returns to that same displacement at the time $t = 3.35$ s.

2.5   Scientific Notation

In scientific work, one frequently encounters very large or very small numbers. For example, a machinist may measure the thickness of a thin metal sheet to be 0.00021 in. Similarly, an engineer may encounter an area of 130,000 m^2 for an airport runway. It is convenient to be able to express these numbers, respectively, as 2.1×10^{-4} in. and 1.3×10^5 m^2. Powers of

10 are used to keep track of the decimal point, and we are not forced to carry a large number of zeros along with our calculations. The expression of any number as a number between 1 and 10 times an integral power of 10 is called *scientific notation.*

Electronic calculators are often equipped with buttons that allow even beginning students to use scientific notation in many computations. You are virtually certain to encounter scientific notation, even if your job does not require frequent use of numbers expressed in this form. Review the manual that comes with your calculator to learn how to work with powers of 10 on the calculator.

Consider the following multiples of 10 and examples of their use in scientific notation:

$$0.0001 = 10^{-4} \qquad 2.34 \times 10^{-4} = 0.000234$$
$$0.001 = 10^{-3} \qquad 2.34 \times 10^{-3} = 0.00234$$
$$0.01 = 10^{-2} \qquad 2.34 \times 10^{-2} = 0.0234$$
$$0.1 = 10^{-1} \qquad 2.34 \times 10^{-1} = 0.234$$
$$1 = 10^{0} \qquad 2.34 \times 10^{0} = 2.34$$
$$10 = 10^{1} \qquad 2.34 \times 10^{1} = 23.4$$
$$100 = 10^{2} \qquad 2.34 \times 10^{2} = 234.0$$
$$1000 = 10^{3} \qquad 2.34 \times 10^{3} = 2340.0$$
$$10,000 = 10^{4} \qquad 2.34 \times 10^{4} = 23,400.0$$

To write a number larger than 1 in scientific notation, you must determine the number of times the decimal point must be moved to the left in order to arrive at the shorthand notation. Examples are

$$467 = 4\,6\,7. = 4.67 \times 10^2$$
$$30 = 3\,0. = 3.0 \times 10^1$$
$$35,700 = 3\,5\,7\,0\,0. = 3.57 \times 10^4$$

Any decimal number less than 1 can be written as a number between 1 and 10 times a *negative* power of 10. The negative exponent in this case is the number of times the decimal point is moved to the right. This is always one more than the number of zeros that separate the first digit from the decimal. Examples are

$$0.24 = 0.2\,4 = 2.4 \times 10^{-1}$$
$$0.00327 = 0.0\,0\,3\,2\,7 = 3.27 \times 10^{-3}$$
$$0.0000469 = 0.0\,0\,0\,0\,4\,6\,9 = 4.69 \times 10^{-5}$$

To transfer from scientific notation to decimal notation, the procedure is simply reversed.

By recalling the laws of exponents, scientific notation can be used in multiplication and division of very small or very large numbers. When two numbers are multiplied, the exponents of 10 are added. For example, 200×4000 may be written $(2 \times 10^2)(4 \times 10^3) = (2)(4) \times (10^2)(10^3) = 8 \times 10^5$. Other examples are

$$2200 \times 40 = (2.2 \times 10^3)(4 \times 10^1) = 8.8 \times 10^4$$
$$0.0002 \times 900 = (2.0 \times 10^{-4})(9.0 \times 10^2) = 1.8 \times 10^{-1}$$
$$1002 \times 3 = (1.002 \times 10^3)(3 \times 10^0) = 3.006 \times 10^3$$

Similarly, when one number is divided by another number, the exponent of 10 in the denominator is subtracted from the exponent of 10 in the numerator. Examples are

$$\frac{7000}{35} = \frac{7 \times 10^3}{3.5 \times 10^1} = \frac{7.0}{3.5} \times 10^{3-1} = 2.0 \times 10^2$$

$$\frac{1200}{0.003} = \frac{1.2 \times 10^3}{3.0 \times 10^{-3}} = \frac{1.2}{3.0} \times 10^{3-(-3)} = 4.0 \times 10^5$$

$$\frac{0.008}{400} = \frac{8 \times 10^{-3}}{4 \times 10^2} = \frac{8}{4} \times 10^{-3-2} = 2.0 \times 10^{-5}$$

When two numbers in scientific notation are added, care must be taken to adjust all numbers to be added so that they have identical powers of 10. Examples are

$$2000 + 400 = 2 \times 10^3 + 0.4 \times 10^3 = 2.4 \times 10^3$$
$$0.006 - 0.0008 = 6 \times 10^{-3} - 0.8 \times 10^{-3} = 5.2 \times 10^{-3}$$
$$4 \times 10^{-21} - 6 \times 10^{-20} = 0.4 \times 10^{-20} - 6 \times 10^{-20} = -5.6 \times 10^{-20}$$

Scientific calculators automatically make the necessary adjustments when adding and subtracting such numbers.

Scientific notation and powers of 10 take on an important and special significance when you work with metric units. In Chapter 3, you will see that multiples of 10 are used to define a number of units in the metric system. For example, 1 kilometer is defined as a thousand (1×10^3) meters, and 1 millimeter is defined as one-thousandth (1×10^{-3}) of a meter.

2.6   Graphs

Frequently, it is desirable to show the relationship between two quantities in the form of a graph. For example, we know that a car traveling at constant speed covers the same distances every minute (min) that it travels. We might record the distance traveled in feet at particular times as follows:

Distance, ft	200	400	600	800	1000
Time, min	1	2	3	4	5

Along the bottom of a sheet of graph paper, we might establish a time scale, maybe letting each division equal 1 min. On the left side of the paper, we can establish a distance scale. The scale selected should fill the graph paper. (This makes it easier to locate points on the graph.) Simple scale divisions are 1 division = 1, 2, or 5 times some power of 10. Good examples are 1 division $= 1 \times 10^3 = 1000$, or 1 division $= 2 \times 10^0 = 2$, or 1 division $= 5 \times 10^{-2} = 0.05$. Awkward scale divisions such as 3 divisions = 100 ft should be avoided because they hinder the location of points. In our example, we let each division represent 200 ft. The data may then be plotted on the graph, as shown in Fig. 2.3. Each point on the horizontal line has a corresponding point on the vertical line. For example, the distance traveled after 3 min is 600 ft. Notice that when the points are connected, a straight line results.

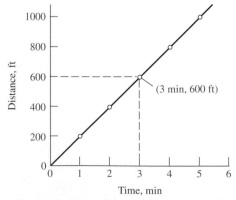

Figure 2.3   A graph of distance as a function of time (a *direct* relationship).

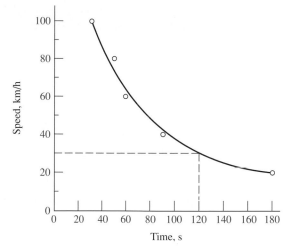

Figure 2.4   A graph of the time required to cover a distance of 1 km as a function of speed (an *inverse* relationship).

When a graph of one quantity versus another results in a straight line passing through the origin, there is a *direct relationship.* In this example, the distance traveled is directly proportional to the time. When one quantity changes, the other changes by the same proportion. Doubling the time elapsed doubles the distance traveled.

Indirect, or *inverse relationships,* in which an increase of one quantity results in a *proportional* decrease in another quantity, are also encountered. If we were to decrease the speed of an automobile, we would find that greater and greater time intervals would be required to cover the same distance. Suppose we measure the time in seconds (s) required to travel a distance of 1 kilometer (km) [0.621 mile (mi)] at speeds of 20, 40, 60, 80, and 100 kilometers per hour (km/h). The following data are recorded:

Speed, km/h	20	40	60	80	100
Time, s	180	90	60	45	36

A graph of these data is shown in Fig. 2.4. Notice that the graph of an inverse relationship is not a straight line but a curved one.

A graph can be used to obtain information that was not available before the graph was constructed. For example, in Fig. 2.4, we can learn that a time of 120 s would be required to cover the distance if our speed were 30 km/h.

2.7   Geometry

In this brief review, it is assumed that you already have an understanding of the concept of a point and a line. Other important concepts are reviewed only to the extent that they may be needed in physics problems. A lengthy review of the many possible theorems in this field is unnecessary. We will begin with angles and lines.

The **angle** between two straight lines is defined by drawing a circle with its center at the point of intersection (see Fig. 2.5a). The magnitude of angle A is proportional to the fraction of a complete circle that lies between the two lines. Angles are measured in **degrees,** as defined in Fig. 2.5b. One degree (°) is a portion of a circle equal to 1/360 of a full revolution (rev). Thus, there are 360° in 1 rev:

$$1° = \frac{1}{360} \text{ rev} \qquad 1 \text{ rev} = 360° \qquad \textbf{(2.11)}$$

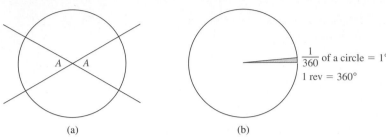

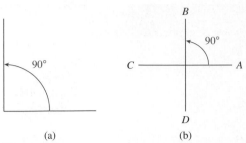

Figure 2.5   An angle is a fraction of a complete circle. One degree is a portion of a circle equal to 1/360 of a full revolution.

Figure 2.6   (a) A right angle is one-fourth of a circle. (b) Lines that cross at right angles are said to be perpendicular.

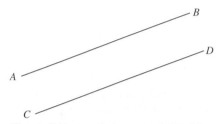

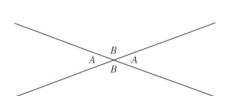

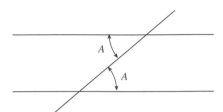

Figure 2.7   Parallel lines extended indefinitely will never intersect ($AB \parallel CD$).

Figure 2.8   When two straight lines intersect, the opposing angles are equal.

Figure 2.9   When a straight line intersects two parallel lines, the alternate interior angles are equal.

A special name is given to the angle that forms one-fourth of 1 rev, or 90°. It is called a ***right angle*** and is illustrated in Fig. 2.6a. When two straight lines meet so that the angle between them is a right angle, they are said to be ***perpendicular.*** Line *CA* in Fig. 2.6b is perpendicular to line *BD*. This may be written

$$CA \perp BD$$

where $\perp$ means "is perpendicular to."

Two straight lines are said to be ***parallel*** if the extension of their sides will never intersect. In Fig. 2.7, line *AB* is parallel to line *CD*. We write

$$AB \parallel CD$$

where $\parallel$ means "is parallel to."

The application of geometry usually requires only a few general rules. We will describe three of the most important.

Rule 1: When two straight lines intersect, they form opposing angles that are equal (Fig. 2.8).

Rule 2: When a straight line intersects (cuts across) two parallel lines, the alternate interior angles are equal (Fig. 2.9).

Note from Fig. 2.9 that the angles *A* are on alternate sides of the intersecting line and are inside the two parallel lines. According to Rule 2, these *alternate interior* angles are equal. (The remaining two interior angles are also equal.)

Example 2.7

Two studs at a construction site are braced by a cross member, as shown in Fig. 2.10. Determine the angle C, using geometry.

Plan: Assume that the two studs are parallel, and note that the cross member forms a straight line intersecting two parallel lines. Start with the given angle, then apply Rules 1 and 2 to find each of the other angles.

Solution: Angle A is 60° since opposite angles are equal by Rule 1. Angle B is 60° because alternate interior angles are equal by Rule 2 ($B = A$). Finally, apply Rule 1 again to see that angle C is equal to 60°. From this example, it is seen that *alternate exterior* angles are also equal, but no new rule is necessary.

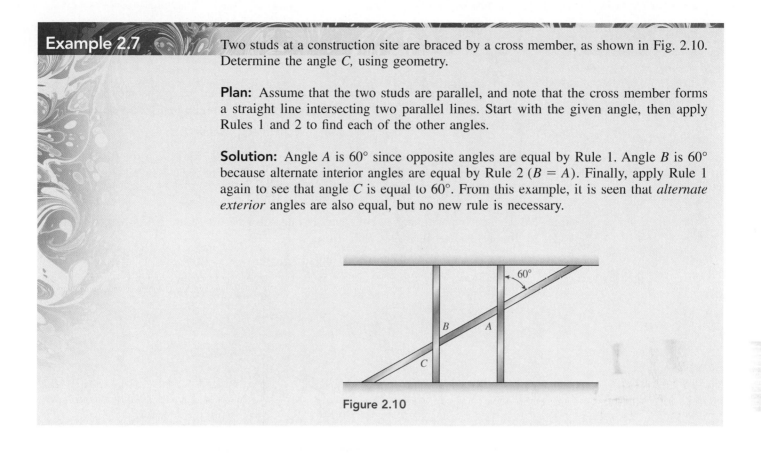

Figure 2.10

A *triangle* is a plane closed figure with three sides. Figure 2.11 shows a general triangle with sides a, b, and c and angles A, B, and C. Such a triangle in which no two sides are equal and no two angles are equal is called a *scalene triangle.*

A special triangle of interest to us is a *right triangle,* as shown in Fig. 2.12. A right triangle has one angle equal to 90° (two of the sides are perpendicular). The side opposite the 90° angle is called the *hypotenuse.*

Rule 3: For any triangle, the sum of the interior angles is equal to 180°.

$$A + B + C = 180°$$

Corollary: For any right triangle ($C = 90°$), the sum of the two smaller angles is equal to 90°.

$$A + B = 90°$$

In this case, the angles A and B are said to be *complimentary angles.*

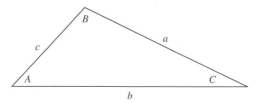

Figure 2.11   A scalene triangle.

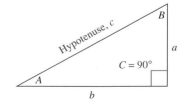

Figure 2.12   A right triangle in which one of the interior angles is a right angle.

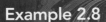

Example 2.8

Apply the rules of geometry to determine the unknown angles for the arrangement drawn in Fig. 2.13.

Plan: Look over the entire figure, searching for lines that are perpendicular (where right triangles might be formed). Using the given 30° angle, apply one of the rules of geometry to find other angles.

Solution: Since the line MC is perpendicular to line RQ, we have a right triangle in which the smallest angle is 30°. Applying the corollary to Rule 3 gives

$$30° + B = 90° \quad \text{or} \quad B = 60°$$

Now, since opposite angles are equal, D must also be 60°. Line NF is perpendicular to line RP so that $A + D = 90°$. Thus,

$$A + 60° = 90° \quad \text{and} \quad A = 30°$$

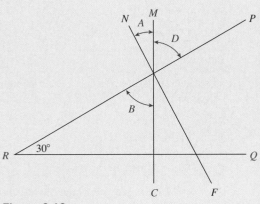

Figure 2.13

Another important rule for geometry involves the sides of a right triangle. The *Pythagorean theorem* will be discussed in Section 2.8.

2.8   # Right Triangle Trigonometry

Often it is necessary to determine lengths and angles from three-sided figures known as *triangles*. By learning a few principles that apply to all right triangles, you can significantly improve your ability to work with vectors. Moreover, handheld calculators make many of the calculations relatively simple.

First, let's review some of the things we already know about right triangles. We will follow the convention that uses Greek letters for angles and Roman letters for sides. Commonly used Greek symbols are

α alpha	β beta	γ gamma
θ theta	ϕ phi	δ delta

In the right triangle drawn as Fig. 2.14, the symbols R, x, and y refer to the side dimensions, and θ, ϕ, and 90° are the angles. Recall that the sum of the smaller angles in a right triangle is 90°:

$$\phi + \theta = 90° \qquad \textit{Right Triangle}$$

The angle ϕ is said to be the complement of θ, and vice versa.

The pythagorean theorem:
$$R^2 = x^2 + y^2$$
$$R = \sqrt{x^2 + y^2}$$

Figure 2.14

There is also a relationship between the sides, which is known as the **Pythagorean theorem:**

Pythagorean Theorem: The square of the hypotenuse is equal to the sum of the squares of the other two sides.

$$R^2 = x^2 + y^2 \qquad \textit{Pythagorean Theorem} \quad \textbf{(2.12)}$$

The *hypotenuse* is defined as the longest side. It is conveniently located as that side directly opposite the right angle—the line joining the two perpendicular sides.

Example 2.9

What length of guy wire is needed to stretch from the top of a 12-m telephone pole to a ground stake located 8 m from the foot of the pole?

Plan: Draw a rough sketch of the problem as in Fig. 2.15, noting that the guy wire forms a right triangle with the pole perpendicular to the ground. Label the figure and apply the *Pythagorean theorem* to find the length of the guy wire.

Solution: Identify the length R of the cable as the hypotenuse of a right triangle; then, from the Pythagorean theorem,

$$R^2 = (12 \text{ m})^2 + (8 \text{ m})^2$$
$$= 144 \text{ m}^2 + 64 \text{ m}^2 = 208 \text{ m}^2$$

Taking the square root of both sides gives

$$R = \sqrt{208 \text{ m}^2} = 14.4 \text{ m}$$

Remember to give your answer in three significant figures. In this text, we assume that all measurements have three significant digits. In other words, the height of the pole is 12.0 m, and the base of the triangle is 8.00 m, even though they are given as 12 m and 8 m for convenience.

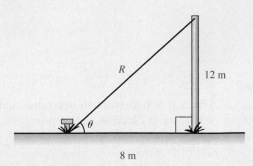

Figure 2.15

In general, to find the hypotenuse, we could express the Pythagorean theorem as

$$R = \sqrt{x^2 + y^2} \qquad \textit{Hypotenuse} \quad \textbf{(2.13)}$$

On some electronic calculators, the sequence of entries might be as follows:

$$x \quad \boxed{x^2} \quad \boxed{+} \quad y \quad \boxed{x^2} \quad \boxed{=} \quad \boxed{\sqrt{x}}$$

In this instance, x and y are the values of the shorter sides, and the boxed symbo[...] eration keys on the calculator. You should verify the solution to the previous [...] using $x = 8$ and $y = 12$. (The input procedure varies depending on the make of c[...]

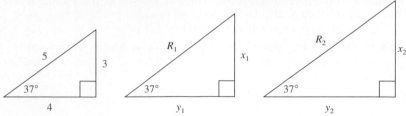

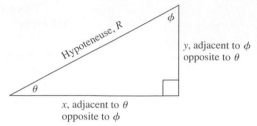

Figure 2.16   All right triangles with the same internal angles are similar; that is, their sides are proportional.

Figure 2.17

Of course, the Pythagorean theorem can also be used to find either of the shorter sides if the remaining sides are known. Solution for x or for y yields

$$x = \sqrt{R^2 - y^2} \qquad y = \sqrt{R^2 - x^2} \qquad (2.15)$$

Trigonometry is the branch of mathematics that takes advantage of the fact that similar triangles are proportional in size. In other words, for a given angle, the ratio of any two sides is the same regardless of the overall dimensions of the triangle. For the three triangles in Fig. 2.16, the ratios of corresponding sides are equal as long as the angle is 37°. From Fig. 2.16, it is seen that

$$\frac{3}{4} = \frac{x_1}{y_1} = \frac{x_2}{y_2}$$

and

$$\frac{4}{5} = \frac{y_1}{R_1} = \frac{y_2}{R_2}$$

Once an angle is identified in a right triangle, the sides *opposite* and *adjacent* to that angle may be labeled. The meanings of opposite, adjacent, and hypotenuse are given in Fig. 2.17. You should study this figure until you understand fully the meaning of these terms. Verify that the side opposite to θ is y and that the side adjacent to θ is x. Also notice that the sides described by "opposite" and "adjacent" change if we refer to angle ϕ.

In a right triangle, there are three side ratios that are important. They are the ***sine***, the ***cosine***, and the ***tangent***, defined as follows for angle θ:

$$\sin \theta = \frac{\text{opp } \theta}{\text{hyp}}$$

$$\cos \theta = \frac{\text{adj } \theta}{\text{hyp}} \qquad (2.16)$$

$$\tan \theta = \frac{\text{opp } \theta}{\text{adj}}$$

To make sure that you understand these definitions, you should verify the following for the triangles in Fig. 2.18:

$$\sin \theta = \frac{9}{15} \qquad \cos \gamma = \frac{m}{H} \qquad \tan \alpha = \frac{y}{x}$$

$$\sin \alpha = \frac{y}{R} \qquad \cos \beta = \frac{n}{H} \qquad \tan \phi = \frac{12}{9}$$

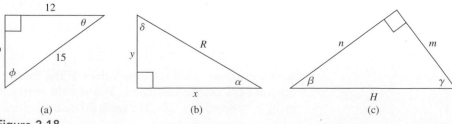

Figure 2.18

You should first identify the right angle and label the longest side (opposite the 90° angle) as the hypotenuse. Then, for a particular angle, the sides opposite and adjacent may be identified.

The constant values for the trigonometric functions are readily available from any scientific calculator. You should read the instructions that were included with your calculator to make sure that you know how to find the sine, cosine, or tangent of any angle and to find the angle whose sine, cosine, or tangent is some given ratio. The exact procedure differs for many calculators. Using your calculator, see if you can verify that

$$\cos 47° = 0.682$$

With most calculators, we would enter the number 47 and then strike the $\boxed{\cos}$ key to display the answer. You should verify the following:

$$\tan 38° = 0.781 \qquad \cos 31° = 0.857$$
$$\sin 22° = 0.375 \qquad \tan 65° = 2.144$$

To find the angle whose tangent is 1.34 or to find the angle whose sine is 0.45, we would reverse the above process. On a calculator, for example, we would enter the number 1.34, then we would look for one of the following sequences, depending on the calculator: $\boxed{\text{INV}}$ $\boxed{\tan}$, $\boxed{\text{ARC}}$ $\boxed{\tan}$, or $\boxed{\tan^{-1}}$. Any of these will give the angle whose tangent is the entered value. In the previous examples, we find that

$$\tan \theta = 1.34 \qquad \theta = 53.3°$$
$$\sin \theta = 0.45 \qquad \theta = 26.7°$$

You should now be able to apply trigonometry to find unknown angles or sides in a right triangle. The following problem-solving procedure will be helpful.

Problem-Solving Strategy

Application of Trigonometry

1. Draw the right triangle from the stated conditions of the problem. (Label all sides and angles with either their known values or a symbol for an unknown value.)

2. Isolate an angle for study; if an angle is known, choose that one.

3. Label each side according to whether its relation to the chosen angle is opp, adj, or hyp.

4. Decide which side or angle is to be found.

5. Recall the definitions of the trigonometric functions:

$$\sin \theta = \frac{opp}{hyp} \qquad \cos \theta = \frac{adj}{hyp} \qquad \tan \theta = \frac{opp}{adj}$$

6. Choose the trigonometric function that involves (a) the unknown quantity and (b) no other unknown quantity.

7. Write the trigonometric equation and solve for the unknown.

Example 2.10

What is the length of the rope segment x in Fig. 2.19?

Plan: Step 1 in the problem-solving strategy is already complete. We will proceed with the remaining steps until we determine the length of the rope segment x.

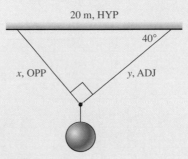

Figure 2.19

Solution: In steps 2 and 3, the 40° angle is selected for reference, and *opp, adj,* and *hyp* are labeled on the figure. In step 4, the decision is made to solve for *x* (the side opposite to the 40° angle). Next, the *sine* function is chosen because it involves *opp* and *hyp.*

$$\sin 40° = \frac{x}{20 \text{ m}}$$

Multiplying each side by 20 m allows us to solve for *x.*

$$x = (20 \text{ m}) \sin 40° \qquad \text{or} \qquad x = 12.9 \text{ m}$$

On some calculators, we might calculate *x* as follows:

$$(20 \text{ m}) \sin 40° = 20 \boxed{\text{X}} 40 \boxed{\text{sin}} \boxed{=} = 12.9 \text{ m}$$

The procedure differs with many calculators. You should verify this answer and also use your calculator to show that side *y* = 15.3 m.

Example 2.11

A car drives up the ramp in Fig. 2.20. If the base of the ramp is 20 m and its height is 4.3 m, what is the angle of the incline?

Plan: Draw and label a sketch (see Fig. 2.20) noting the given information and the relationships to the angle of incline. Then follow the problem-solving strategy.

Solution: Identify the sides *opp, adj,* and *hyp* for the angle θ and note that the *tangent* function is the only one that involves the two known sides. We write

$$\tan \theta = \frac{opp}{adj} = \frac{4.3 \text{ m}}{20 \text{ m}} \qquad \text{or} \qquad \tan \theta = 0.215$$

The angle θ is the *angle whose tangent is* 0.215. From our calculator, we obtain

$$\theta = 12.1°$$

The sequence on *some* calculators might be as follows:

$$4.3 \boxed{\div} 20 \boxed{=} \boxed{\tan^{-1}}$$

Some calculators use INV TAN, ATAN, ARCTAN, or other symbols instead of $\tan^{-1}$. Again, it will be necessary for you to study the manual that was included with your calculator.

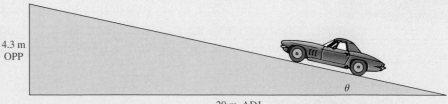

4.3 m
OPP

θ

20 m, ADJ

Figure 2.20

Summary and Review

Summary

This chapter was intended as a review of technical mathematics. Now that you have completed the chapter, go back and review the chapter objectives. If you feel uneasy about any of the objectives, perhaps another review would be helpful. Make sure that you understand the important topics in this chapter before applying the concepts of physics in later chapters. Remember the following points:

- When adding numbers of like sign, we add the absolute values of the numbers and give to the sum the common sign. To add numbers of unlike sign, we find the difference of their absolute values and affix the sign of the larger.

- To subtract a number b from a number a, we change the sign of number b and then add it to number a, using the addition rule.

- When we multiply or divide a group of signed numbers, the result is negative if the total number of negative factors is odd; otherwise, the result is positive.

- Formulas can be rearranged to solve for a particular unknown by performing similar operations (addition, subtraction, multiplication, division, etc.) to both sides of an equality.

- The following rules apply for exponents and radicals (optional):

$$a^m a^n = a^{m+n} \qquad a^{-n} = \frac{1}{a^n}$$

$$\frac{a^m}{a^n} = a^{m-n} \qquad (a^m)^n = a^{mn}$$

$$(ab)^n = a^n b^n \qquad \left(\frac{a}{b}\right)^n = \frac{a^n}{b^n}$$

$$\sqrt[n]{ab} = \sqrt[n]{a}\,\sqrt[n]{b} \qquad a^m = a^{m/n}$$

- Scientific notation uses positive or negative powers of 10 to express large or small numbers in shorthand notation.

- Graphs are used to give a continuous description of the relationship between two variables, based on observed data.

- When two straight lines intersect, they form opposing angles that are equal.

- When a straight line intersects two parallel lines, the alternate interior angles are equal.

- For any triangle, the sum of the interior angles is 180°; for a right triangle, the sum of the smaller angles is 90°.

- Application of the Pythagorean theorem and basic trigonometric functions is essential when studying physics.

$$R^2 = x^2 + y^2 \qquad \mathrm{Sin}\ \theta = \frac{\mathrm{Opp}}{\mathrm{Hyp}}$$

$$\mathrm{Cos}\ \theta = \frac{\mathrm{Adj}}{\mathrm{Hyp}} \qquad \mathrm{Tan}\ \theta = \frac{\mathrm{Opp}}{\mathrm{Adj}}$$

Key Terms

angle 19	hypotenuse 21	right triangle 21
base 13	parallel 20	scalene triangle 21
cosine 24	perpendicular 20	scientific notation 17
degree 19	product 9	sine 24
dividend 9	Pythagorean theorem 23	tangent 24
divisor 9	quadratic equation 15	triangle 21
exponent 9, 12	quotient 9	trigonometry 24
factor 9	radical 14	
formula 10	right angle 20	

Review Questions

2.1. The sum of two numbers is always greater than their difference. Is this statement true? Justify your answer by giving examples.

2.2. If you subtract the number (-8) from the number $(+4)$, what is the result? What if instead you subtracted the second number from the first? What is their sum?

2.3. On a cold winter day, the temperature changes from $-5°C$ to $+10°C$. What is the change in temperature? What is the change if the temperature cools back to $-5°C$? Explain the difference.

2.4. Is it true that a negative number raised to an odd power will always be negative?

2.5. Distinguish clearly between -9^2 and $(-9)^2$. Are they the same? Why or why not?

2.6. Discuss two ways in which positive and negative signs are used when working with formulas. When you substitute signed numbers into formula contain the operations of addition and subtraction, what precautions should you take?

2.7. When you transpose a term from one side of an equation to the other side, the sign is changed. Explain how this procedure works and why it works.

2.8. Cross-multiplication is sometimes used in formula rearrangement in which one fraction is equal to another fraction. For example,

$$\frac{a}{b} = \frac{c}{d} \quad \text{becomes} \quad ad = bc$$

Explain why this procedure works, and discuss the dangers involved.

2.9. A common error in formula rearrangement is canceling terms instead of factors. The following are *not* permitted:

$$\frac{x + y}{x} \neq y \qquad \frac{x^2 + y^2}{x + y} \neq \frac{x + y}{1}$$

2.10. If the graph of two variables (x, y) is a straight line passing through the origin, can we say that when x increases by 10 units, y must also increase by 10 units? Can we say that if the value of x doubles, the value of y must also double? Explain why or why not.

2.11. When two parallel lines are crossed by a transversal, the alternate interior angles are equal. Are the alternate exterior angles also equal?

2.12. One of the angles of a right triangle is 33°. What are the other two angles?

2.13. A window is 6 ft tall. A diagonal two-by-four, 9 ft in length, fits exactly from the top corner to the bottom corner. How wide is the window?

2.14. The complement ϕ of an angle θ is that angle ϕ such that $\phi + \theta = 90°$. Show that the sine of an angle is equal to the cosine of its complement.

2.15. If angles ϕ and θ are complementary, show that $\tan \theta$ is the reciprocal of $\tan \phi$.

Problems

Section 2.1 Signed Numbers

In items 2.1 through 2.26, perform the indicated operation.

2.1. $(+2) + (+5)$ Ans. +7

2.2. $(-2) + (6)$

2.3. $(-4) - (-6)$ Ans. +2

2.4. $(+6) - (+8)$

2.5. $(-3) - (+7)$ Ans. −10

2.6. $(-15) - (+18)$

2.7. $(-4) - (+3) - (-2)$ Ans. −5

2.8. $(-6) + (-7) - (+4)$

2.9. $(-2)(-3)$ Ans. +6

2.10. $(-16)(+2)$

2.11. $(-6)(-3)(-2)$ Ans. −36

2.12. $(-6)(+2)(-2)$

2.13. $(-3)(-4)(-2)(2)$ Ans. −48

2.14. $(-6)(2)(3)(-4)$

2.15. $(-6) \div (-3)$ Ans. +2

2.16. $(-14) \div (+7)$

2.17. $(+16) \div (-4)$ Ans. −4

2.18. $(+18) \div (-6)$

2.19. $\dfrac{-4}{-2}$ Ans. +2

2.20. $\dfrac{+16}{-4}$

2.21. $\dfrac{(-2)(-3)(-1)}{(-2)(-1)}$ Ans. −3

2.22. $\dfrac{(-6)(+4)}{(-2)}$

2.23. $\dfrac{(-16)(4)}{2(-4)}$ Ans. +8

2.24. $\dfrac{(-1)(-2)^2(12)}{(6)(2)}$

2.25. $(-2)(+4) - \dfrac{(-6)}{(+2)} - (-5)$ Ans. 0

2.26. $(-2)(-2)^2 + \dfrac{(-3)(-2)(-8)}{(-4)(1)} - (-6)^3$

In items 2.27 through 2.30, solve the given problem.

2.27. Distances above the ground are positive, and distances below the ground are negative. If an object is dropped from 20 ft above the ground into a 12-ft hole, what is the difference between the initial position and the final position?   Ans. 32 ft

2.28. In physics, work is measured in joules (J) and may be positive or negative depending on the direction of the force that does the work. What is the total work done if the works of the forces are 20 J, −40 J, and −12 J?

2.29. The temperature of a bolt is −12°C. (a) If the temperature increases by 6°C, what is the new temperature? (b) If the original temperature decreases by 5°C, what is the resultant temperature? (c) If the original temperature is multiplied by a factor of −3, what is the resultant temperature?
     Ans. (a) −6°C, (b) −17°C, (c) 36°C

2.30. A metal expands when heated and contracts when cooled. Assume that the length of a rod changes by 2 millimeters (mm) for each 1°C change in temperature. What is the total change in the length of this rod as the temperature changes from -5 to $-30°C$?

Section 2.2 Algebra Review

In items 2.31 through 2.46, determine the value for x when $a = 2$, $b = -3$, and $c = -2$.

2.31. $x = a + b + c$ Ans. -3
2.32. $x = a - b - c$
2.33. $x = b + c - a$ Ans. -7
2.34. $x = b(a - c)$
2.35. $x = \dfrac{b - c}{a}$ Ans. $-\dfrac{1}{2}$
2.36. $x = \dfrac{a + b}{c}$
2.37. $x = b^2 - c^2$ Ans. $+5$
2.38. $x = \dfrac{-b}{ac}$
2.39. $x = \dfrac{a}{bc}(a - c)$ Ans. $+\dfrac{4}{3}$
2.40. $x = a^2 + b^2 + c^3$
2.41. $x = \sqrt{a^2 + b^2 + c^2}$ Ans. $\sqrt{17}$
2.42. $x = ab(c - a)^2$
2.43. $2ax - b = c$ Ans. $-\dfrac{5}{4}$
2.44. $ax + bx = 4c$
2.45. $3ax = \dfrac{2ab}{c}$ Ans. $+1$
2.46. $\dfrac{4ac}{b} = \dfrac{2x}{b} - 16$

In items 2.47 through 2.56, solve the equations for the unknown letter.

2.47. $5m - 16 = 3m - 4$ Ans. $m = 6$
2.48. $3p = 7p - 16$
2.49. $4m = 2(m - 4)$ Ans. $m = -4$
2.50. $3(m - 6) = 6$
2.51. $\dfrac{x}{3} = (4)(3)$ Ans. $x = 36$
2.52. $\dfrac{p}{3} = \dfrac{2}{6}$
2.53. $\dfrac{96}{x} = 48$ Ans. $x = 2$
2.54. $14 = 2(b - 7)$
2.55. $R^2 = (4)^2 + (3)^2$ Ans. $R = +5$
2.56. $\dfrac{1}{2} = \dfrac{1}{P} + \dfrac{1}{6}$

In items 2.57 through 2.70, solve the formulas for the letter indicated.

2.57. $V = IR$, R Ans. $R = \dfrac{V}{I}$
2.58. $PV = nRT$, T
2.59. $F = ma$, a Ans. $a = \dfrac{F}{m}$
2.60. $s = vt + d$, d
2.61. $F = \dfrac{mv^2}{R}$, R Ans. $R = \dfrac{mv^2}{F}$
2.62. $s = \frac{1}{2}at^2$, a
2.63. $2as = v_f^2 - v_o^2$, a Ans. $a = \dfrac{v_f^2 - v_o^2}{2s}$
2.64. $C = \dfrac{Q^2}{2V}$, V
2.65. $\dfrac{1}{R} = \dfrac{1}{R_1} + \dfrac{1}{R_2}$, R Ans. $R = \dfrac{R_1 R_2}{R_1 + R_2}$
2.66. $MV = Ft$, t
2.67. $mv_2 - mv_1 = Ft$, v_2 Ans. $V_2 = \dfrac{Ft + mv_1}{m}$
2.68. $\dfrac{P_1 V_1}{T_1} = \dfrac{P_2 V_2}{T_2}$, T^2
2.69. $v = v_o + at$, a Ans. $a = \dfrac{v - v_o}{t}$
2.70. $c^2 = a^2 + b^2$, b

Section 2.3 Exponents and Radicals

In items 2.71 through 2.92, simplify the given expressions, using the laws of exponents and radicals.

2.71. $2^5 \cdot 2^7$ Ans. 2^{12}
2.72. $3^2 \cdot 2^3 \cdot 3^3$
2.73. $x^7 x^3$ Ans. x^{10}
2.74. $x^7 x^{-5} x^3$
2.75. $a^{-3} a^2$ Ans. $\dfrac{1}{a}$
2.76. $a^3 a^{-2} b^{-3} b$
2.77. $\dfrac{2^3}{2^5}$ Ans. $\dfrac{1}{2^2}$
2.78. $\dfrac{2a^3 b}{2ab^3}$
2.79. $\dfrac{2x^{17}}{x^{12}}$ Ans. $2x^5$
2.80. $(ab)^{-2}$
2.81. $(m^{-3})^{-2}$ Ans. m^6
2.82. $(n^3 c^{-2})^{-2}$
2.83. $(4 \times 10^2)^3$ Ans. 64×10^6
2.84. $(6 \times 10^{-2})^{-2}$
2.85. $\sqrt[3]{64}$ Ans. 4
2.86. $\sqrt[4]{81}$

2.87. $\sqrt[5]{x^{15}}$ 　　　　　Ans. x^3

2.88. $\sqrt{a^4 b^6}$

2.89. $\sqrt{4 \times 10^4}$ 　　　　Ans. 2×10^2

2.90. $\sqrt[3]{8 \times 10^{-27}}$

2.91. $\sqrt[5]{32a^{10}}$ 　　　　　Ans. $2a^2$

2.92. $\sqrt{(x + 2)^2}$

Section 2.5 Scientific Notation

In items 2.93 through 2.100, convert the decimal numbers to scientific notation.

2.93. 40,000 　　　　　　Ans. 4×10^4

2.94. 67

2.95. 480 　　　　　　Ans. 4.80×10^2

2.96. 497,000

2.97. 0.0021 　　　　　　Ans. 2.1×10^{-3}

2.98. 0.789

2.99. 0.087 　　　　　　Ans. 8.7×10^{-2}

2.100. 0.000967

In items 2.101 through 2.108, convert the numbers to decimal notation.

2.101. 4×10^6 　　　　　Ans. 4,000,000

2.102. 4.67×10^3

2.103. 3.7×10^1 　　　　　Ans. 37

2.104. 1.4×10^5

2.105. 3.67×10^{-2} 　　　　Ans. 0.0367

2.106. 4×10^{-1}

2.107. 6×10^{-3} 　　　　　Ans. 0.006

2.108. 4.17×10^{-5}

In items 2.109 through 2.132, simplify and express as a single number written in scientific notation.

2.109. $400 \times 20,000$ 　　　　Ans. 8×10^6

2.110. 37×2000

2.111. $(4 \times 10^{-3})(2 \times 10^5)$ 　　Ans. 8×10^2

2.112. $(3 \times 10^{-1})(6 \times 10^{-8})$

2.113. $(6.7 \times 10^3)(4.0 \times 10^5)$ 　Ans. 2.68×10^9

2.114. $(3.7 \times 10^{-5})(200)$

2.115. $(4 \times 10^{-3})^2$ 　　　　Ans. 1.60×10^{-5}

2.116. $(3 \times 10^6)^3$

2.117. $(6000)(3 \times 10^{-7})$ 　　Ans. 1.8×10^{-3}

2.118. $(4)(300)(2 \times 10^{-2})$

2.119. $7000 \div (3.5 \times 10^{-3})$ 　Ans. 2.00×10^6

2.120. $60 \div 30,000$

2.121. $(6 \times 10^{-5}) \div (3 \times 10^4)$ Ans. 2×10^{-9}

2.122. $(4 \times 10^{-7}) \div (7 \times 10^{-7})$

2.123. $\dfrac{4600}{0.02}$ 　　　　　Ans. 2.3×10^5

2.124. $\dfrac{(1600)(4 \times 10^{-3})}{1 \times 10^{-2}}$

2.125. $4.0 \times 10^2 + 2 \times 10^3$ 　Ans. 2.40×10^3

2.126. $6 \times 10^{-5} - 4 \times 10^{-6}$

2.127. $6 \times 10^{-3} - 0.075$ 　　Ans. -6.90×10^{-2}

2.128. $0.0007 - 4 \times 10^{-3}$

2.129. $\dfrac{4 \times 10^{-6} + 2 \times 10^{-5}}{4 \times 10^{-2}}$ 　Ans. 6×10^{-4}

2.130. $\dfrac{6 \times 10^3 + 4 \times 10^2}{1 \times 10^{-3}}$

2.131. $\dfrac{600 - 3000}{0.0003}$ 　　　Ans. -8×10^6

2.132. $(4 \times 10^{-3})^2 - 2 \times 10^{-5}$

Section 2.6 Graphs

2.133. Plot a graph for the following data recorded for an object falling freely from rest.

Speed, ft/s	32	63	97	129	159	192	225
Time, s	1	2	3	4	5	6	7

What do you expect the speed to be after 4.5 s? How much time is required for the object to attain a speed of 100 ft/s?　Ans. $v = 144$ ft/s, $t = 3.1$ s

2.134. The advance of a right-hand screw is directly proportional to the number of complete turns. For a particular screw, the following data are recorded:

Advance, in.	0.5	1.0	1.5	2.0	2.5	3.0
Number of turns	16	32	48	64	80	96

Plot a graph with the number of turns as the horizontal divisions and the advance in inches as the vertical divisions. What number of turns is required to advance the screw 2.75 in.?

2.135. Plot a graph showing the relationship between frequency and wavelength of electromagnetic waves. The following data are given:

Frequency, kilohertz (kHz)	150	200	300	500	600	900
Wavelength, meters (m)	2000	1500	1000	600	500	333

What are the wavelengths of electromagnetic waves with frequencies of 350 kHz and 800 kHz?　Ans. 857 m, 375 m

2.136. The electric power loss in a resistor varies directly with the square of the current. In a particular experiment, the following data are recorded:

Current, amperes (A)	1.0	2.5	4.0	5.0	7.0	8.5
Power, watts (W)	1.0	6.5	16.2	25.8	50.2	72.0

Plot a graph and from the curve determine the power loss when the current is (a) 3.2 A and (b) 8.0 A.

Section 2.7 Geometry

Note: If lines appear to be parallel or perpendicular, assume that they are.

2.137. What are the magnitudes of the angles in Fig. 2.21 by your estimation?

Ans. 90°, 180°, 270°, and 45°

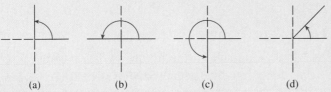

(a) (b) (c) (d)

Figure 2.21

2.138. Use a ruler and a protractor to measure lines and angles. Draw two parallel lines *AB* and *CD* that are separated by 2 cm. Now draw a third line *EF* that crosses each of the other lines at any angle other than 90°. Verify Rules 1 and 2 by measuring the angles formed by the transversal *EF*. Now draw another transversal line *GH* slanted in the other direction that intersects line *AB* at the same point as for line *EF*. Verify Rule 3 for the triangle you have just formed.

2.139. Determine angles *A* and *B* for each of the arrangements drawn in Fig. 2.22.

Ans. (a) *A* = 17°, *B* = 35°, (b) *A* = 50°, *B* = 40°

2.140. Determine angles *A* and *B* in Fig. 2.23.

Section 2.8 Right Triangle Trigonometry

In items 2.141 through 2.158, use your calculator to evaluate each example.

2.141. sin 67°   Ans. 0.921
2.142. cos 48°
2.143. tan 59°   Ans. 1.66
2.144. sin 34°
2.145. cos 29°   Ans. 0.875
2.146. tan 15°
2.147. 20 cos 15°   Ans. 19.3
2.148. 400 sin 21°
2.149. 600 tan 24°   Ans. 267
2.150. 170 cos 79°
2.151. 240 sin 78°   Ans. 235
2.152. 1400 tan 60°
2.153. $\dfrac{200}{\sin 17°}$ Ans. 684
2.154. $\dfrac{300}{\sin 60°}$
2.155. $\dfrac{167}{\cos 78°}$ Ans. 803

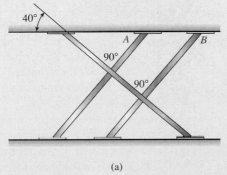

(a)

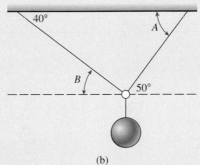

(b)

Figure 2.22

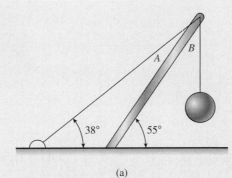

(a)

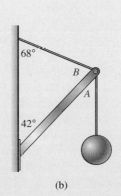

(b)

Figure 2.23

2.156. $\dfrac{256}{\cos 16°}$

2.157. $\dfrac{670}{\tan 17°}$ Ans. 2190

2.158. $\dfrac{2000}{\tan 51°}$

In items 2.159 through 2.167, determine the unknown angles.

2.159. $\sin \theta = 0.811$ Ans. 54.2°

2.160. $\sin \theta = 0.111$

2.161. $\tan \theta = 1.2$ Ans. 50.2°

2.162. $\tan \theta = 0.511$

2.163. $\cos \beta = 0.228$ Ans. 76.8°

2.164. $\cos \theta = 0.81$

2.165. $\cos \theta = \dfrac{400}{500}$ Ans. 36.9°

2.166. $\tan \theta = \dfrac{16}{4}$

2.167. $\sin \phi = \dfrac{140}{270}$ Ans. 31.2°

In items 2.168 through 2.175, solve the triangles for the unknown angles and sides.

2.168.

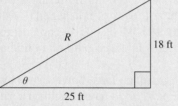

2.169.

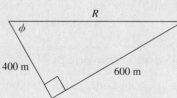

Ans. $R = 721$ m, $\phi = 56.3°$

2.170.

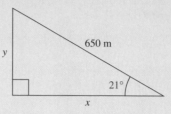

2.171.

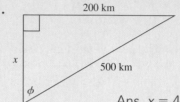

Ans. $x = 458$ km, $\phi = 23.6°$

2.172.

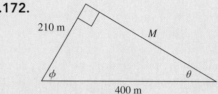

2.173.

Ans. $x = 164$ in., $y = 202$ in.

2.174.

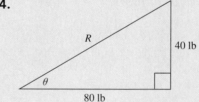

2.175.

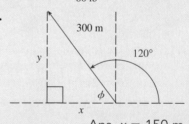

Ans. $x = 150$ m, left, $y = 260$ m

Additional Problems

2.176. Early one morning, a barometer reads 30.21. When an afternoon storm occurs, the reading decreases by 0.59 in. What is the later reading?

2.177. A thermometer reads 29.0°C. After some time in a freezer, it reads −15°C. What is the change in temperature?   Ans. −44C°

2.178. The change in temperature of an object is −34°C. If the original temperature was 20°C, what was the final temperature?

2.179. A length of wood is cut into six pieces, each 28 cm long. Each cut wastes 1 mm of wood. What was the original length of the board in inches?
     Ans. 66.3 in.

2.180. The volume V of a right circular cylinder is the area of the bottom (πr^2) times the height h. Given the radius, write a formula to find the height h.

2.181. The centripetal force F_C is found by multiplying the mass m by the square of its velocity v and then dividing by the radius R of the circle. Write the formula and solve it for the radius R.

$$\text{Ans. } F_C = \frac{mv^2}{R}, \ R = \frac{mv^2}{F_C}$$

2.182. Solve the equation $xb + cd = a(x + 2)$ for x and find the value of x when $a = 2$, $b = -2$, $c = 3$, and $d = -1$.

2.183. Solve for b when $c^2 = a^2 + b^2$ and find the value of b when $a = 50$ and $c = 20$.   Ans. 53.9

2.184. Newton's law of gravitation is written as $F = Gm_1 m_2 / R^2$. The following numerical values are given: $G = 6.67 \times 10^{-11}$, $m_1 = 4 \times 10^{-8}$, $m_2 = 3 \times 10^{-7}$, and $R = 4 \times 10^{-2}$. What is the value of F?

2.185. The length of a rod is initially $L_0 = 21.41$ cm at $t_0 = 20°C$. It is heated to a final temperature $t = 100°C$. The new length is given by

$$L = L_0 + \alpha L_0(t - t_0)$$

where $\alpha = 2 \times 10^{-3}/°C$. What is the value of L?   Ans. $L = 24.84$ cm

2.186. Plot a graph of the function $y = 2x$. From the graph, verify that $x = 3.5$ when $y = 7$.

2.187. Determine the unknown angles in Fig. 2.24a and b.
    Ans. (a) $A = 30°$, $B = 60°$, $C = 60°$;
    (b) $A = 60°$, $B = 30°$, $C = 120°$, $D = 60°$

2.188. Subtract -4 cm from -8 cm to get length A. Add -6 cm to $+14$ cm to get a length B. What is the length $C = A - B$? Is $A - B$ the same value as $B - A$? Consider a number line along the x axis. Draw a line from the origin to point C. Then draw a line from the origin to the point $B - A$. How far are these points apart?
    Ans. -12 cm, no, 24 cm

2.189. The period T of a pendulum is the time required to make a complete oscillation (out and back). The period is given by the following equation:

$$T = 2\pi \sqrt{\frac{L}{g}}$$

where L is the length of the pendulum and g is the acceleration due to gravity. (a) Solve for the length L. (b) If the length L of the pendulum is quadrupled, how much larger will the period be? (c) If the period of a pendulum on Earth is 2.0 s, what would be the period on the Moon, where g is one-sixth of its Earth value?

*2.190. The length of a tiny chip is 3.45×10^{-4} m and its width is 9.77×10^{-5} m. (a) Find the area and perimeter of the chip. (b) If the width is doubled and the length is cut in half, what is the change in the area, and what is the change in the perimeter?
    Ans. (a) 3.37×10^{-8} m^2, 8.85×10^{-4} m;
    (b) 0, -1.5×10^{-4} m

*2.191. The pressure of a gas in a storage tank is dependent on the temperature. The following measurements are recorded:

Temperature, K	300	350	400	450	500	550
Pressure, lb/in.^{2}	400	467	535	598	668	733

Plot a graph of these data. What is the slope of this graph? Can you write a description of the relationship between pressure and temperature on the basis of this information? What would you expect the pressure to be at the temperatures of 420 and 600 degrees kelvin (K)?

*2.192. The voltage (V) in volts and the electric current (I) in milliamperes is given for a particular resistor. The recorded data follows:

Voltage, V	10	20	30	40	50	60
Current, mA	145	289	435	581	724	870

Plot a graph of these data. What would you expect the current in the resistor to be for applied voltages of 26 and 48 V?   Ans. 377 mA, 696 mA

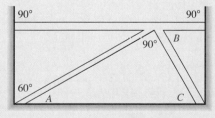

(a)

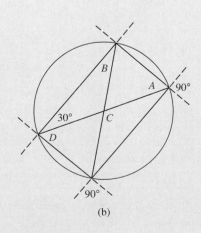

(b)

Figure 2.24

3

Technical Measurement and Vectors

PARCS is an atomic-clock mission scheduled to fly on the International Space Station (ISS) in 2008. The mission, funded by NASA, involves a laser-cooled cesium atomic clock and a time-transfer system using Global Positioning System (GPS) satellites. PARCS will fly concurrently with SUMO (Superconducting Microwave Oscillator), a different sort of clock that will be compared against the PARCS clock to test certain theories. The objectives of the mission are to test gravitational theory, study laser-cooled atoms in microgravity, and improve the accuracy of timekeeping on Earth. (*Photo by NASA.*)

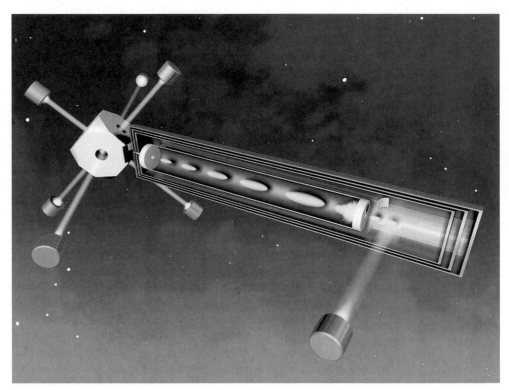

Objectives

After completing this chapter, you should be able to

1. Write the base units for mass, length, and time in SI and U.S. Customary System (USCS) units.
2. Define and apply the SI prefixes that indicate multiples of base units.
3. Convert from one unit to another unit for the same quantity when given the necessary definitions.
4. Define a vector quantity and a scalar quantity and give examples of each.
5. Determine the components of a given vector.
6. Find the resultant of two or more vectors.

The application of physics, whether in the shop or in a technical laboratory, always requires measurements of some kind. An automobile mechanic might be measuring the diameter, or *bore*, of an engine cylinder. Refrigeration technicians might be concerned with volume, pressure, and temperature measurements. Electricians use instruments that measure electric resistance and current, and mechanical engineers are concerned about the effects of forces whose magnitudes must be accurately determined. In fact, it is difficult to imagine any occupation that is not involved with the measurement of some physical quantity.

In the process of physical measurement, we are often concerned with the direction as well as the magnitude of a particular quantity. The length of a wooden rafter is determined by the angle it makes with the horizontal. The direction of an applied force determines its effectiveness in producing a change of position. The direction in which a conveyor belt moves is often just as important as the speed with which it moves. Such physical quantities as *displacement, force,* and *velocity* are often encountered in industry. In this chapter, the concept of *vectors* is introduced to permit the study of both the magnitude and the direction of physical quantities.

3.1 Physical Quantities

The language of physics and technology is universal. Facts and laws must be expressed in an accurate and consistent manner if everyone is to mean exactly the same thing by the same term. For example, suppose an engine is said to have a piston displacement of 3.28 liters (200 cubic inches). Two questions must be answered if this statement is to be understood: (1) How is the *piston displacement* measured, and (2) what is the *liter?*

Piston displacement is the volume that the piston displaces, or "sweeps out," as it moves from the bottom of the cylinder to the top. It is really not a displacement in the usual sense of the word; it is a volume. A standard measure for volume that is easily recognized throughout the world is the liter. Therefore, when an engine has a label on it that reads "piston displacement = 3.28 liters," all mechanics will give the same meaning to the label.

In the previous example, the piston displacement (volume) is an example of a *physical quantity.* Notice that this quantity was defined by describing the procedure for its measurement. In physics, all quantities are defined in this manner. Other examples of physical quantities are length, weight, time, speed, force, and mass.

A physical quantity is measured by comparison with some known standard. For example, we might need to know the length of a metal bar. With appropriate instruments, we might determine the length of the bar to be 4 meters. The bar does not contain four things called "meters"; it is merely compared with the length of some standard known as a "meter." The length could also be represented as 13.1 ft or 4.37 yards if we used other known measures.

The ***magnitude*** of a physical quantity is given by a *number* and a *unit* of measure. Both are necessary because either the number or the unit by itself is meaningless. Except for pure numbers and fractions, it is necessary to include the unit with the number when listing the magnitude of any quantity.

The *magnitude* of a physical quantity is specified completely by a number and a unit, for example, 20 meters or 40 liters.

Since there are many different measures for the same quantity, we need a way of keeping track of the exact size of particular units. To do this, it is necessary to establish standard measures for specific quantities. A ***standard*** is a permanent or easily determined physical record of the size of a unit of measurement. For example, the standard for measuring electrical resistance, the *ohm,* might be defined by comparison with a standard resistor whose resistance is accurately known. Thus, a resistance of 20 ohms would be 20 times as great as that of a standard 1-ohm resistor.

Remember that every physical quantity is defined by telling how it is measured. Depending on the measuring device, each quantity can be expressed in a number of different units. For example, some distance units are *meters, kilometers, miles,* and *feet,* and some speed units are *meters per second, kilometers per hour, miles per hour,* and *feet per second.* Regardless of the units chosen, however, distance must be a *length,* and speed must be a *length* divided by *time.* Thus, *length* and *length/time* are the *dimensions* of the physical quantities *distance* and *speed.*

Note that speed is defined in terms of two more fundamental quantities (length and time). It is convenient to establish a small number of base quantities such as length and time from which all other physical quantities can be derived. Thus, we might say that speed is a *derived* quantity and that length or time is a *base* quantity. If we reduce all physical measurements to a small number of quantities with standard base units, there will be less confusion in their application.

3.2   The International System

The international system of units is called *Système International d'Unités (SI)* and is essentially the same as what we have come to know as the *metric system.* The International Committee on Weights and Measures has established seven base quantities and has assigned official base units to each quantity. A summary of these quantities, their base units, and the symbols for the base units is given in Table 3.1.

Each of the units listed in Table 3.1 has a specific measurable definition that can be duplicated anywhere in the world. Of these base units only one, the *kilogram,* is currently defined in terms of a single physical sample. This standard specimen is kept at the International Bureau of Weights and Measures in France. Copies of the original specimen have been made for use in other nations. All other units are defined in terms of reproducible physical events and can be accurately determined at a number of locations throughout the world.

We can measure many quantities, such as volume, pressure, speed, and force, that are combinations of two or more *fundamental quantities.* However, no one has ever encountered a measurement that cannot be expressed in terms of length, mass, time, current, temperature, luminous intensity, or amount of substance. Combinations of these quantities are referred to as *derived* quantities, and they are measured in derived units. Several common derived units are listed in Table 3.2.

The SI units are not fully implemented in many industrial applications. The United States is making progress toward the adoption of SI units. However, wholesale conversions are costly, particularly in many mechanical and thermal applications, and total conversion to the international system will require some time. For this reason it is necessary to be familiar with older units for physical quantities. The U.S. Customary System (USCS) units for several important quantities are listed in Table 3.3.

Table 3.1

The SI Base Units for Seven Fundamental
Quantities and Two Supplemental Quantities

Quantity	Unit	Symbol
Base Units		
Length	meter	m
Mass	kilogram	kg
Time	second	s
Electric current	ampere	A
Temperature	kelvin	K
Luminous intensity	candela	cd
Amount of substance	mole	mol
Supplemental Units		
Plane angle	radian	rad
Solid angle	steradian	sr

Table 3.2

Derived Units for Common Physical Quantities

Quantity	Derived Units	Symbol	
Area	square meter	m^2	
Volume	cubic meter	m^3	
Frequency	hertz	Hz	s^{-1}
Mass density (density)	kilogram per cubic meter	kg/m^3	
Speed, velocity	meter per second	m/s	
Angular velocity	radian per second	rad/s	
Acceleration	meter per second squared	m/s^2	
Angular acceleration	radian per second squared	rad/s^2	
Force	newton	N	$kg \cdot m/s^2$
Pressure (mechanical stress)	pascal	Pa	N/m^2
Kinematic viscosity	square meter per second	m^2/s	
Dynamic viscosity	newton-second per square meter	$N \cdot s/m^2$	
Work, energy, quantity of heat	joule	J	$N \cdot m$
Power	watt	W	J/s
Quantity of electricity	coulomb	C	
Potential difference, electromotive force	volt	V	J/C
Electric field strength	volt per meter	V/m	
Electric resistance	ohm	Ω	V/A
Capacitance	farad	F	C/V
Magnetic flux	weber	Wb	$V \cdot s$
Inductance	henry	H	$V \cdot s/A$
Magnetic flux density	tesla	T	Wb/m^2
Magnetic field strength	ampere per meter	A/m	
Magnetomotive force	ampere	A	
Luminous flux	lumen	lm	$cd \cdot sr$
Luminance	candela per square meter	cd/m^2	
Illuminance	lux	lx	lm/m^2
Wave number	1 per meter	m^{-1}	
Entropy	joule per kelvin	J/K	
Specific heat capacity	joule per kilogram kelvin	$J/(kg \cdot K)$	
Thermal conductivity	watt per meter kelvin	$W/(m \cdot K)$	
Radiant intensity	watt per steradian	W/sr	
Activity (of a radioactive source)	1 per second	s^{-1}	

Table 3.3

U.S. Customary System Units

Quantity	SI Unit	USCS Unit
Length	meter (m)	foot (ft)
Mass	kilogram (kg)	slug (slug)
Time	second (s)	second (s)
Force (weight)	newton (N)	pound (lb)
Temperature	kelvin (K)	degree Rankine (R)

It should be noted that, even though the foot, pound, and other units are frequently used in the United States, they have been redefined in terms of the SI standard units. Thus, all measurements are currently based on the same standards.

3.3   Measurement of Length and Time

The standard SI unit of length, the ***meter (m),*** was originally defined as one ten-millionth of the distance from the North Pole to the equator. For practical reasons, this distance was marked off on a standard platinum-iridium bar. In 1960, the standard was changed to allow greater access to a more accurate measure of the meter based on an atomic standard. One meter was said to be equal to exactly 1,650,763.73 wavelengths of orange-red light from krypton 86. The number was chosen so that the new standard would be close to the older standard. However, even this standard was not without its problems. The wavelength of light emitted from krypton was uncertain because of the processes occurring within the atom during emission. Also, the development of stabilized lasers made it possible to measure a wavelength much more accurately in terms of time and the velocity of light. In 1983, the latest (and probably the final) standard for the meter was adopted:

> One *meter* is the length of path traveled by a light wave in a vacuum in a time interval of 1/299,792,458 second.

The new standard for the meter is more precise, but it also has other advantages. Its definition depends on the standard for time (s), and it hinges on a standard value for the velocity of light. The velocity of light is now exactly

$$c = 2.99792458 \times 10^8 \text{ m/s} \qquad \text{(exact by definition)}$$

A standard value for the speed of light makes sense because, according to Einstein's theory, the speed of light is a fundamental constant. Moreover, any future refinements to the standard for measuring time will automatically improve the standard for length. Of course, we do not usually need to know the precise definition of length to make accurate practical measurements. Many tools, such as simple metersticks and calipers, are calibrated to agree with the standard measure.

The original definition of time depended on the idea of a solar day as the time interval between two successive passages of the Sun over a given meridian on the Earth. One ***second*** was then 1/86,400 of a mean solar day. It's not hard to imagine the difficulties and the inconsistencies associated with such a standard. In 1976, the SI standard for time was defined as follows:

> One *second* is the time needed for 9,192,631,770 vibrations of a cesium atom.

Thus, the atomic standard for 1 second is the period of vibration of an atom of cesium. The best cesium clocks are so accurate that they gain or lose no more than 1 second in 300,000 years.

Because this measure of time tends to drift ahead of mean solar time, the National Bureau of Standards periodically inserts a *leap second,* usually once a year on December 31. Therefore, the last minute of each year often contains 61 seconds, rather than 60 seconds.

A distinct advantage of the metric system over other systems of units is the use of prefixes to indicate multiples of the base unit. Table 3.4 defines the accepted prefixes and demonstrates their use to indicate multiples and subdivisions of the meter. From the table, you can determine that

$$1 \text{ meter (m)} = 1000 \text{ millimeters (mm)}$$
$$1 \text{ meter (m)} = 100 \text{ centimeters (cm)}$$
$$1 \text{ kilometer (km)} = 1000 \text{ meters (m)}$$

Table 3.4

Multiples and Submultiples for SI Units

Prefix	Symbol	Multiplier	Use
tera	T	$1,000,000,000,000 = 10^{12}$	1 terameter (Tm)
giga	G	$1,000,000,000 = 10^{9}$	1 gigameter (Gm)
mega	M	$1,000,000 = 10^{6}$	1 megameter (Mm)
kilo	k	$1,000 = 10^{3}$	1 kilometer (km)
centi	c	$0.01 = 10^{-2}$	1 centimeter (cm)*
milli	m	$0.001 = 10^{-3}$	1 millimeter (mm)
micro	μ	$0.000001 = 10^{-6}$	1 micrometer (μm)
nano	n	$0.000000001 = 10^{-9}$	1 nanometer (nm)
—	Å	$0.0000000001 = 10^{-10}$	1 angstrom (Å)*
pico	p	$0.000000000001 = 10^{-12}$	1 picometer (pm)

*The use of the centimeter and the angstrom is discouraged, but they are still widely used.

The relationship between the centimeter and the *inch* can be seen in Fig. 3.1. By definition, 1 inch is equal to exactly 25.4 millimeters. This definition and other useful definitions are as follows (symbols for the units are in parentheses):

$$1 \text{ inch (in.)} = 25.4 \text{ millimeters (mm)}$$
$$1 \text{ foot (ft)} = 0.3048 \text{ meter (m)}$$
$$1 \text{ yard (yd)} = 0.914 \text{ meter (m)}$$
$$1 \text{ mile (mi)} = 1.61 \text{ kilometers (km)}$$

In reporting data, it is preferable to use the prefix that will allow the number to be expressed in the range from 0.1 to 1000. For example, 7,430,000 meters should be expressed as 7.43×10^{6} m, and then it should be reported as 7.43 megameters, abbreviated 7.43 Mm. It would not usually be desirable to write this measurement as 7430 kilometers (7430 km) unless the distance is being compared with other distances measured in kilometers. In the case of the quantity 0.00064 ampere, it is proper to write either 0.64 milliampere (0.64 mA) or 640 microamperes (640 μA). Normally, prefixes are chosen for multiples of a thousand.

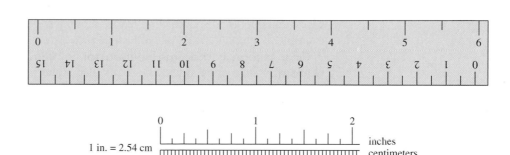

Figure 3.1 A comparison of the inch to the centimeter as a measure of length.

3.4 Significant Figures

Some numbers are exact, and others are approximate. If we select 20 bolts from a bin and use only one-fourth of them, the numbers 20 and ¼ are considered to be exact quantities. If we measured the length or the width of a rectangular plate, however, the accuracy of the measurement would be determined by the precision of the instrument and by the skill of the observer. Suppose we measure the width of such a plate with a vernier caliper and find it to be 3.42 cm. The last digit is estimated and is therefore susceptible to error. The actual width is between 3.40 cm and 3.50 cm. To write this width as 3.420 cm would imply greater accuracy than was justified. We say that the number 3.42 has three *significant figures,* and we are careful not to write more numbers or zeros than are meaningful.

All physical measurements are assumed to be approximate, with the last significant digit having been determined by an estimation of some kind. When writing such numbers, we often include some zeros to indicate properly the location of the decimal point. Except for these zeros, however, all other digits are considered significant digits. For example, the distance 76,000 m has only two significant digits. We assume that the three zeros following the 6 were needed to locate the decimal unless there is information to the contrary. Other examples are

4.003 cm	4 significant figures
0.34 cm	2 significant figures
60,400 cm	3 significant figures
0.0450 cm	3 significant figures

Zeros that are not specifically needed to locate the decimal are significant (as in the last two examples).

With the widespread use of calculators, students often report their results with a greater accuracy than is warranted. For example, suppose we measure the length of a rectangular plate to be 9.54 cm and its width to be 3.4 cm. The area of such a plate might be calculated as 32.436 cm^2 (five significant figures). But a chain is only as strong as its weakest link. Since the width was accurate to only two significant figures, we cannot report *more* than two significant figures in our result. The area should properly be written as 32 cm^2. The number in our calculator display gives a false indication of the accuracy of the reported area. This will be misleading to others who did not participate in the measurement. A significant figure is a *reliably known* digit.

> **Rule 1:** When approximate numbers are multiplied or divided, the number of significant digits in the reported result is the same as the number of significant digits in the *least* accurate of the factors. By "least accurate," we mean the factor having the lowest number of significant digits.

A different problem arises when approximate numbers are added or subtracted. In such cases, consideration should be given to the *precision* indicated by each measurement. For example, a length of 7.46 m is *precise* to the nearest hundredth of a meter, and a length of 9.345 m is precise to the nearest thousandth of a meter. The sum of a group of measurements *may* have more significant digits than some of the individual measurements, but it cannot be more *precise*. As an example, suppose we determine the perimeter of the rectangular plate described previously. We write

$$9.54 \text{ cm} + 3.4 \text{ cm} + 9.54 \text{ cm} + 3.4 \text{ cm} = 25.9 \text{ cm}$$

The least precise measurement was to the nearest tenth of a centimeter; therefore, the perimeter should be given to the nearest tenth of a centimeter even though it has *three* significant figures.

> **Rule 2:** When approximate numbers are added or subtracted, the number of decimal places in the result should equal the smallest number of decimal places of any term in the sum.

Classroom work and laboratory work are often treated differently. In the laboratory, we know the uncertainties in each measurement, and we must round our reported answers to the appropriate accuracy and precision. Since the limitations of each measure are not usually known in the classroom or with homework problems, we will follow the practice of assuming that all given data are accurate to three significant figures. Thus, an 8-cm rod would be treated as though it were 8.00 cm long. A speed of 51 km/h would be assumed to be 51.0 km/h. To avoid rounding errors, you must carry *at least* one more significant figure in your calculations than you ultimately plan to report. For example, if you wish to report three significant figures, you should carry at least four in all of your calculations.

3.5 Measuring Instruments

The choice of a measuring instrument is determined by the accuracy required and by the physical conditions surrounding the measurement. A basic choice for the mechanic or machinist is most often the steel rule, such as the one shown in Fig. 3.2. This rule is often accurate enough when you are measuring openly accessible lengths. Steel rules may be graduated as fine as thirty-seconds or even sixty-fourths of an inch. Metric rules are usually graduated in millimeters.

For the measurement of inside and outside diameters, calipers such as those shown in Fig. 3.3 may be used. The caliper itself cannot be read directly and therefore must be set to a steel rule or a standard size gauge.

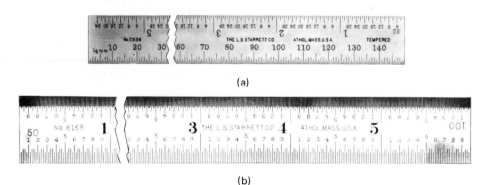

(a)

(b)

Figure 3.2   Some 6-in. (15-cm) steel scales. (a) Scales 1/32 in. and 0.5 mm. (b) Scales 1/100 and 1/50 in. (*The L. S. Starrett Company.*)

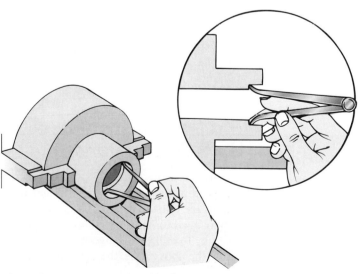

Figure 3.3   Using calipers to measure an inside diameter.

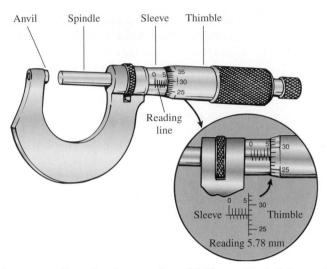

Figure 3.4 A micrometer caliper, showing a reading of 5.78 mm. (*The L. S. Starrett Company.*)

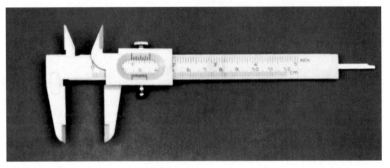

Figure 3.5 The vernier caliper. (*The L. S. Starrett Company.*)

Figure 3.6 Measuring the depth of a shoulder with a micrometer depth gauge.

The best accuracy possible with a steel rule is determined by the size of the smallest graduation and is on the order of 0.01 in. or 0.1 mm. For greater accuracy, the machinist often uses a standard micrometer caliper, as in Fig. 3.4, or a standard vernier caliper, as in Fig. 3.5. These instruments make use of sliding scales to record very accurate measurements. Micrometer calipers can measure to the nearest ten-thousandth of an inch (0.002 mm), and vernier calipers are used to measure within 0.001 in. or 0.02 mm.

The depth of blind holes, slots, and recesses is often measured with a micrometer depth gauge. Fig. 3.6 shows such a gauge used to measure the depth of a shoulder.

3.6 Unit Conversions

Because so many different units are required for a variety of jobs, it is often necessary to convert a measurement from one unit to another. For example, suppose that a machinist records the outside diameter of a pipe as $1\frac{3}{16}$ in. To order a fitting for the pipe, the machinist may need to know this diameter in millimeters. Such conversions can be accomplished easily by treating units algebraically and using the principle of cancellation.

In the above case, the machinist should first convert the fraction to a decimal.

$$1\tfrac{3}{16} \text{ in.} = 1.19 \text{ in.}$$

Next, the machinist should write down the quantity to be converted, giving both the number and the unit (1.19 in.). The definition that relates inches to millimeters is recalled:

$$1 \text{ in.} = 25.4 \text{ mm}$$

Since this statement is an equality, we can form two ratios, each equal to 1.

$$\frac{1 \text{ in.}}{25.4 \text{ mm}} = 1 \qquad \frac{25.4 \text{ mm}}{1 \text{ in.}} = 1$$

Note that the number 1 does not equal the number 25.4, but the *length* of 1 in. is equal to the *length* of 25.4 mm. Thus, if we multiply some other length by either of these ratios, we will get a new number, but we will not change the length. Such ratios are called **conversion factors.** Either of the above conversion factors may be multiplied by 1.19 in. without changing the length represented. Multiplication by the first ratio does not give a meaningful result. Note that units are treated as algebraic quantities.

$$(1.19 \text{ in.})\left(\frac{1 \text{ in.}}{25.4 \text{ mm}}\right) = \left(\frac{1.19}{25.4}\right)\left(\frac{\text{in.}^2}{\text{mm}}\right) \qquad \text{Wrong!}$$

Multiplication by the second ratio, however, gives the following results:

$$(1.19 \text{ in.})\left(\frac{25.4 \text{ mm}}{1 \text{ in.}}\right) = \frac{(1.19)(25.4)}{(1)} \text{ mm} = 30.2 \text{ mm}$$

Therefore, the outside diameter of the pipe is 30.2 mm.

Sometimes it is necessary to work with quantities that have multiple units. For example, *speed* is defined as *length* per unit *time* and may have units of *meters per second* (m/s), *feet per second* (ft/s), or other units. The same algebraic procedure can help with conversion of multiple units.

Problem-Solving Strategy

Procedure for Converting Units

1. Write down the quantity to be converted.
2. Define each unit appearing in the quantity to be converted in terms of the desired unit(s).
3. For each definition, form two conversion factors, one being the reciprocal of the other.
4. Multiply the quantity to be converted by those factors that will cancel all but the desired units.

Example 3.1

Convert a speed of 60 km/h to units of meters per second.

Plan: We recall two definitions that might result in four possible conversion factors that will cancel the nondesired units.

Solution: We must change kilometers into miles and hours into seconds.

$$1 \text{ km} = 1000 \text{ m} \qquad \nearrow \quad \frac{1 \text{ km}}{1000 \text{ m}}$$
$$\searrow \quad \frac{1000 \text{ m}}{1 \text{ km}}$$

$$1 \text{ h} = 3600 \text{ s} \begin{cases} \dfrac{1 \text{ h}}{3600 \text{ s}} \\ \dfrac{3600 \text{ s}}{1 \text{ h}} \end{cases}$$

We write down the quantity to be converted and then choose conversion factors that will cancel nondesired units.

$$60 \, \frac{\cancel{\text{km}}}{\cancel{\text{h}}} \left(\frac{1000 \text{ m}}{1 \, \cancel{\text{km}}} \right) \left(\frac{1 \, \cancel{\text{h}}}{3600 \text{ s}} \right) = 16.7 \, \frac{\text{m}}{\text{s}}$$

Additional examples of the procedure are as follows:

$$30 \, \frac{\cancel{\text{mi}}}{\cancel{\text{h}}} \left(\frac{5280 \text{ ft}}{1 \, \cancel{\text{mi}}} \right) \left(\frac{1 \, \cancel{\text{h}}}{60 \, \cancel{\text{min}}} \right) \left(\frac{1 \, \cancel{\text{min}}}{60 \text{ s}} \right) = 44 \text{ ft/s}$$

$$20 \, \frac{\cancel{\text{lb}}}{\cancel{\text{in.}^2}} \left(\frac{1550 \, \cancel{\text{in}^2}}{1 \text{ m}^2} \right) \left(\frac{4.448 \text{ N}}{1 \, \cancel{\text{lb}}} \right) = 1.38 \times 10^5 \text{ N/m}^2$$

If the required definitions are not available in this chapter, they can be found on the inside front cover of this book.

When you are working with technical formulas, it is always helpful to substitute units as well as numbers. For example, the formula for speed v is

$$v = \frac{x}{t}$$

where x is the distance traveled in a time t. Thus, if a car travels 400 m in 10 s, its speed will be

$$v = \frac{400 \text{ m}}{10 \text{ s}} = 40 \, \frac{\text{m}}{\text{s}}$$

Notice that the units of velocity are meters per second, written m/s.

Whenever velocity appears in a formula, it must always have units of *length* divided by *time*. These are said to be the **dimensions** of velocity. There may be many different units for a given physical quantity, but the dimensions result from a definition, and they do not change.

When you are working with formulas, it will be useful to remember two rules concerning dimensions:

Rule 1: If two quantities are to be added or subtracted, they must be of the same dimensions.

Rule 2: The quantities on both sides of an equals sign must be of the same dimensions.

Example 3.2   Assume that the distance x measured in meters (m) is a function of initial speed v_0 in meters per second (m/s), acceleration a in meters per second squared (m/s²), and time t in seconds (s). Show that the following formula is dimensionally correct:

$$x = v_0 t + \tfrac{1}{2} a t^2$$

Plan: Recall that the dimensions of each term must be the same and that the dimensions on each side of the equality must be the same. Since x is in meters, each term in the equation must reduce to meters if it is dimensionally correct.

Solution: Substituting units for the quantities in each term, we have

$$m = \frac{m}{s} \, (s) + \frac{m}{s^2} \, (s)^2 \qquad \text{gives} \qquad m = m + m$$

Notice that both Rule 1 and Rule 2 are satisfied. Therefore, the equation is dimensionally correct.

The fact that an equation is dimensionally correct is a valuable check. Such an equation still may not be a *true* equation, but at least it is consistent dimensionally.

3.7 Vector and Scalar Quantities

Some quantities can be described totally by a number and a unit. Only the *magnitudes* are of interest in an area of 12 m², a volume of 40 ft³, or a distance of 50 km. Such quantities are called **scalar quantities.**

> A *scalar quantity* is specified completely by its magnitude—a number and a unit. Examples are speed (15 mi/h), distance (12 km), and volume (200 cm³).

Scalar quantities that are measured in the same units may be added or subtracted in the usual way. For example,

$$14 \text{ mm} + 13 \text{ mm} = 27 \text{ mm}$$
$$20 \text{ ft}^2 - 4 \text{ ft}^2 = 6 \text{ ft}^2$$

Some physical quantities, such as force and velocity, have direction as well as magnitude. In such cases, they are called **vector quantities.** The direction must be a part of any calculations involving such quantities.

> A *vector quantity* is specified completely by a magnitude and a direction. It consists of a number, a unit, and a direction. Examples are displacement (20 m, N) and velocity (40 mi/h, 30°N of W).

The direction of a vector may be given by reference to conventional north, east, west, and south directions. Consider, for example, the vectors 20 m, W and 40 m at 30° N of E, as shown in Fig. 3.7. The expression "north of east" indicates that the angle is formed by rotating a line northward from the easterly direction.

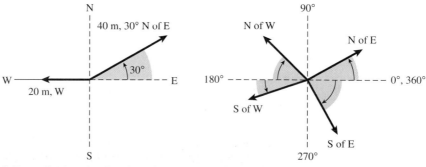

Figure 3.7 Indicating the direction of a vector by reference to north (N), south (S), east (E), and west (W).

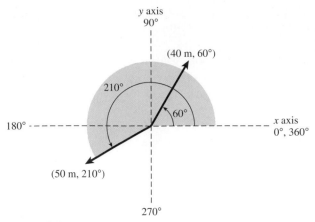

Figure 3.8 Indicating the direction of a vector as an angle measured from the positive *x* axis.

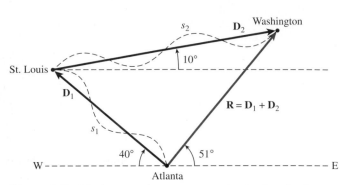

Figure 3.9 Displacement is a vector quantity; its direction is indicated by a solid arrow. Distance is a scalar quantity, indicated above by a dotted line.

Another method of specifying direction that will be particularly useful later on is to make reference to perpendicular lines called *axes*. These imaginary lines are usually chosen to be horizontal and vertical, but they may be oriented along other directions as long as the two lines remain perpendicular. An imaginary horizontal line is usually called the *x* axis, and an imaginary vertical line is called the *y* axis. See Fig. 3.8. Directions are given by angles measured counterclockwise from the positive *x* axis. The vectors 40 m at 60° and 50 m at 210° are shown in the figure.

Assume that a person travels by car from Atlanta to St. Louis. The *displacement* from Atlanta can be represented by a line segment drawn to scale from Atlanta to St. Louis (see Fig. 3.9). An arrowhead is drawn on the St. Louis end to denote the direction. It is important to note that the displacement, represented by the vector D_1, is completely independent of the actual path or the mode of transportation. The odometer would show that the car had actually traveled a scalar distance s_1 of 541 mi. The magnitude of the displacement is only 472 mi.

Another important difference between a vector displacement and a scalar displacement is that the vector component has a constant direction of 140° (or 40° N of W). The direction of the car at any instant on the trip is not important, however, when considering the scalar distance.

Now, let us suppose that our traveler continues the drive to Washington. This time the vector displacement D_2 is 716 mi at a constant direction of 10° N of E. The corresponding ground distance s_2 is 793 mi. The total distance traveled for the entire trip from Atlanta is the arithmetic sum of the scalar quantities s_1 and s_2.

$$s_1 + s_2 = 541 \text{ mi} + 793 \text{ mi} = 1334 \text{ mi}$$

The *vector sum* of the two displacements D_1 and D_2 must take note of direction, however, as well as magnitudes. The question now is not the distance traveled but the resultant displacement from Atlanta. This vector sum is represented in Fig. 3.9 by the symbol **R,** where

$$\mathbf{R} = \mathbf{D_1} + \mathbf{D_2}$$

Methods we will discuss in Section 3.8 will allow us to determine the magnitude and direction of **R.** Using a ruler and a device for measuring angles, we would see that

$$\mathbf{R} = 545 \text{ mi}, 51°$$

Remember that, in performing vector additions, both the magnitude and the direction of the displacements must be considered. The additions are geometric instead of algebraic.

It is possible for the magnitude of a vector sum to be less than the magnitude of either of the component displacements.

A vector is usually denoted in print by boldface type. For example, the symbol $\mathbf{D}_1$ denotes a displacement vector in Fig. 3.9. A vector can be denoted conveniently in handwriting by underscoring the letter or by putting an arrow over it. In print, the magnitude of a vector is usually indicated by italics; thus, D denotes the magnitude of the vector $\mathbf{D}$. A vector is often specified by a pair of numbers (R, θ). The first number and unit give the magnitude, and the second number gives the angle measured counterclockwise from the positive x axis. For example,

$$\mathbf{R} = (R, \theta) = (200 \text{ km}, 114°)$$

Note that the magnitude R of a vector is always positive. A negative sign before the symbol of a vector merely reverses its direction; in other words, it interchanges the arrow tip without affecting the length. If $\mathbf{A} = (10 \text{ m}, \text{E})$, then $-\mathbf{A}$ would be $(10 \text{ m}, \text{W})$.

3.8 Addition of Vectors by Graphical Methods

In this section, we discuss two common graphical methods for finding the geometric sum of vectors. The *polygon method* is the more useful one, since it can be readily applied to more than two vectors. The *parallelogram method* is useful for the addition of two vectors at a time. In each case, the magnitude of a vector is indicated to scale by the length of a line segment. The direction is denoted by an arrow tip at the end of the line segment.

Example 3.3

A ship travels 100 km due north on the first day of a voyage, 60 km northeast on the second day, and 120 km due east on the third day. Find the resultant displacement by the polygon method.

Plan: Set the starting point at the origin of the voyage and decide on an appropriate scale. Use a protractor to measure the angles and a ruler to draw the length of each vector in proportion to its magnitude. The resultant displacement will be a vector drawn from the origin to the tip of the last vector.

Solution: Suppose we let 1 cm correspond to a distance of 20 km, as shown in Fig. 3.10. Using this scale, we find that

$$100 \text{ km} = 100 \text{ km} \times \frac{1 \text{ cm}}{20 \text{ km}} = 5 \text{ cm}$$

$$60 \text{ km} = 60 \text{ km} \times \frac{1 \text{ cm}}{20 \text{ km}} = 3 \text{ cm}$$

$$120 \text{ km} = 120 \text{ km} \times \frac{1 \text{ cm}}{20 \text{ km}} = 6 \text{ cm}$$

By measuring with a ruler, we find from the scale diagram that the arrow for the resultant is 10.8 cm long. Therefore, the magnitude is

$$10.8 \text{ cm} = 10.8 \text{ cm} \times \frac{20 \text{ km}}{1 \text{ cm}} = 216 \text{ km}$$

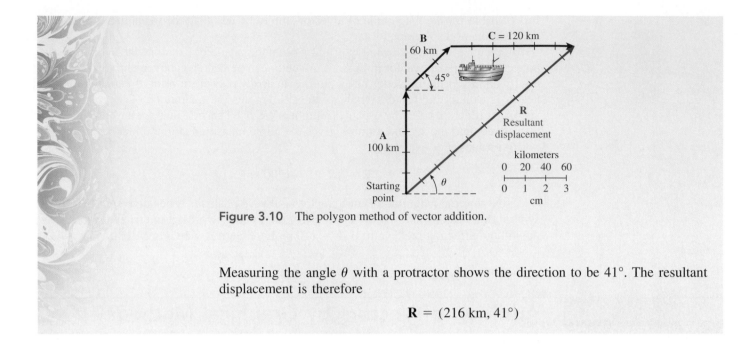

Figure 3.10 The polygon method of vector addition.

Measuring the angle θ with a protractor shows the direction to be $41°$. The resultant displacement is therefore

$$\mathbf{R} = (216 \text{ km}, 41°)$$

Note that the order in which the vectors are added does not change the resultant in any way. We could have begun with any of the three distances traveled by the ship in the previous example.

Graphical methods can be used to find the resultant of all kinds of vectors. They are not restricted to measuring displacement, and they are particularly useful in finding the resultant of a number of *forces*. For now, consider a force to be defined as a push or pull that tends to produce motion. The force vector is also specified by a number, unit, and angle, as with displacements, and they can be added in the same manner as for displacement vectors.

Problem-Solving Strategy

The Polygon Method of Vector Addition

1. Choose a scale and determine the length of the arrows that correspond to each vector.

2. Draw to scale an arrow representing the magnitude and direction of the first vector.

3. Draw the arrow of the second vector so that its tail is joined to the tip of the first vector.

4. Continue the process of joining tail to tip until the magnitude and direction of all vectors have been represented.

5. Draw the resultant vector with its tail at the origin (starting point) and its tip joined to the tip of the last vector.

6. Measure with ruler and protractor to determine the magnitude and direction of the resultant vector.

In Example 3.4, we determine the resultant force on a donkey being pulled in different directions with two ropes (see Fig. 3.11). This time we will apply the *parallelogram method,* which is useful only for two vectors at a time. Each vector is drawn to scale, and both tails are at a common origin. The two arrows then form two adjoining sides of a parallelogram. The other two sides are constructed by drawing parallel lines of equal length. The resultant is represented by the diagonal of the parallelogram included between the two vector arrows.

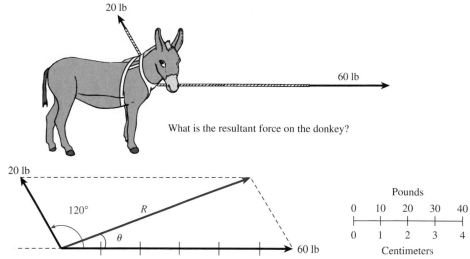

Figure 3.11 The parallelogram method of vector addition.

Example 3.4

Find the resultant force on the donkey in Fig. 3.11 if the angle between the two ropes is 120°. One end is pulled with a force of 60 lb, and the other with a force of 20 lb. Use the parallelogram method of vector addition.

Plan: Construct a parallelogram with two of the sides formed by vectors drawn proportional to the magnitudes of the forces. The resultant force can then be found by measuring the diagonal of the parallelogram.

Solution: Using a scale of 1 cm corresponds to 10 lb gives

$$60 \text{ lb} \times \frac{1 \text{ cm}}{10 \text{ lb}} = 6 \text{ cm} \qquad 20 \text{ lb} \times \frac{1 \text{ cm}}{10 \text{ lb}} = 2 \text{ cm}$$

A parallelogram is constructed in Fig. 3.11 by drawing the two forces to scale from a common origin. Use a protractor to make sure that the angle between the vectors is 120°. Completing the parallelogram allows the resultant to be drawn as a diagonal from the origin. Measurement of R and θ with a ruler and protractor and converting as before gives 52.9 lb for the magnitude and 19.1° for the direction.

$$\mathbf{R} = (52.9 \text{ lb}, 19.1°)$$

A second look at the parallelogram will show that the same answer would result from applying the polygon method and attaching the 20-lb vector to the tip of the 60-lb vector.

3.9 Force and Vectors

As you saw in Section 3.8, *force* vectors can be added graphically, just as was done earlier for displacements. Because forces are such an important part of the study of mechanics, it is wise to develop vector skills by using force applications in addition to displacement applications. A stretched spring exerts forces on the objects to which its ends are attached, compressed air exerts forces on the walls of its container, and a tractor exerts a force on the trailer it is pulling. Probably the most familiar force is that of gravitational attraction exerted on every body by the Earth. This force is called the *weight* of the body. A definite force exists even though there is no contact between the

Earth and the bodies it attracts. Weight as a vector quantity is directed toward the center of the Earth.

The SI unit of force is the newton (N), which will be properly defined in a later chapter. For the moment, consider its relationship to the pound:

$$1 \text{ N} = 0.225 \text{ lb} \qquad 1 \text{ lb} = 4.45 \text{ N}$$

A 120-lb woman has a weight of 534 N. If the weight of a pipe wrench were 20 N, it would weigh about 4.5 lb in USCS units. Until all industries have converted completely to SI units, the pound will still be with us, and frequent conversions are necessary. We will use both units of force in this chapter as we learn to work with vector quantities.

Two of the measurable effects of forces are (1) changing the dimensions or shape of a body and (2) changing a body's motion. If in the first case, there is no resultant displacement of the body, the push or pull causing the change in shape is called a ***static force.*** If a force changes the motion of a body, it is called a ***dynamic force.*** Both types of forces are conveniently represented by vectors, as in Example 3.4.

The effectiveness of any force depends on the direction in which it acts. For example, it is easier to pull a sled along the ground with an inclined rope, as shown in Fig. 3.12, than to push it. In each case, the applied force is producing more than a single effect. That is, the pull on the cord is both lifting the sled and moving it forward. Similarly, pushing the sled would have the effect of adding to the weight of the sled. We are thus led to the idea of ***components*** of a force: the effective values of a force in directions other than that of the force itself. In Fig. 3.12, the force **F** can be replaced by its horizontal and vertical components, $\mathbf{F}_x$ and $\mathbf{F}_y$.

If a force is represented graphically by its magnitude and an angle (R, θ), its components along the x and y directions can be determined. A force **F** acting at an angle θ above the horizontal is drawn in Fig. 3.13. The meaning of the x and y components, $\mathbf{F}_x$ and $\mathbf{F}_y$, can be seen in this diagram. The segment from O to the perpendicular dropped from A to the x axis is called the x *component* of **F** and is labeled $\mathbf{F}_x$. The segment from O to the perpendicular line from A to the y axis is called the y *component* of **F** and is labeled $\mathbf{F}_y$. By drawing the vectors to scale, we can determine the magnitudes of the components graphically. These two components acting together would have the same effect as the original force **F**.

PHYSICS TODAY

Rock climbers use a combination of forces to scale steep surfaces. By pushing against protruding rocks, climbers use the rocks' horizontal and vertical forces to move themselves upward.

(Photo © Vol. 20/Corbis.)

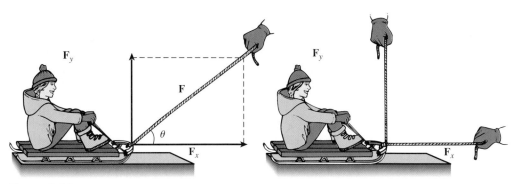

Figure 3.12

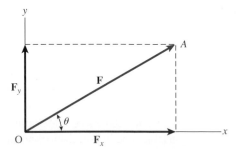

Figure 3.13   Graphical representation of the x and y components of **F.**

Example 3.5

A lawn mower is pushed downward along the handle with a force of 160 N at an angle of 30° with the horizontal. What is the magnitude of the horizontal component of this force?

Plan: From the picture in Fig. 3.14a, we recognize that the force exerted along the handle acts on the body of the mower. We will use a ruler and a protractor to draw the forces and angles to scale, as shown in Fig. 3.14b. Finally, we will measure the components and convert them to newtons to obtain the two components.

Solution: A suitable scale might be 1 cm = 40 N, meaning that the vector **F** would be 4 cm long and at an angle of 30° with the horizontal. The x component of the force is drawn and labeled F_x. Measurement of this line reveals that

$$F_x \text{ corresponds to 3.46 cm}$$

Now, since 1 cm = 40 N, we obtain

$$F_x = 3.46 \text{ cm}\left(\frac{40 \text{ N}}{1 \text{ cm}}\right) = 138 \text{ N}$$

Notice that the effective force is quite a bit less than the applied force. As an additional exercise, you should show that the magnitude of the downward component of the 160-N force is $F_y = 80.0$ N.

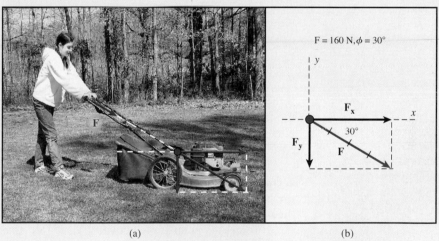

(a)   (b)

Figure 3.14   Finding the components of a force by the graphical method. (*Photo by Paul E. Tippens.*)

3.10   # The Resultant Force

When two or more forces act at the same point on an object, they are said to be ***concurrent forces.*** The combined effect of such forces is called the ***resultant force.***

> The *resultant force* is the single force that will produce the same effect in both magnitude and direction as two or more concurrent forces.

Resultant forces may be calculated graphically by representing each concurrent force as a vector. The polygon or parallelogram method of vector addition will then give the resultant force.

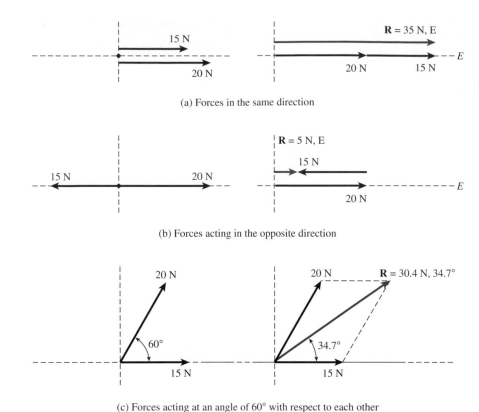

(a) Forces in the same direction

(b) Forces acting in the opposite direction

(c) Forces acting at an angle of 60° with respect to each other

Figure 3.15   The effect of direction on the resultant of two forces.

Often forces act in the same line, either together or in opposition to each other. If two forces act on a single object in the same direction, the resultant force is equal to the sum of the magnitudes of the forces. The direction of the resultant would be the same as that of either force. Consider, for example, a 15-N force and a 20-N force acting in the same easterly direction. Their resultant is 35 N, E, as demonstrated in Fig. 3.15a.

If the same two forces act in opposite directions, the magnitude of the resultant force is equal to the *difference* of the magnitudes of the two forces, and it acts in the direction of the larger force. Suppose the 15-N force in our example were changed so that it pulled to the west. As seen in Fig. 3.15b, the resultant would be 5 N, E.

If forces act at an angle between 0 and 180° to each other, their resultant is the vector sum. The polygon method or the parallelogram method of vector addition may be used to find the resultant force. In Fig. 3.15c, our two forces of 15 and 20 N act at an angle of 60° with each other. The resultant force, calculated by the parallelogram method, is found to be 30.4 N at 34.7°.

3.11   Trigonometry and Vectors

Graphical treatments of vectors are effective for visualizing forces, but they are usually not very accurate. A much more useful approach is to take advantage of simple right-triangle trigonometry, and today's calculators have simplified the process considerably. Familiarity with the *Pythagorean theorem* and some experience with the *sine, cosine,* and *tangent* functions is all you will need for this unit of study.

Trigonometric methods can improve your accuracy and speed in determining the resultant vector or in finding the components of a vector. In most cases, it is helpful to use imaginary *x* and *y* axes when working with vectors in an analytical way. Any vector can then be drawn with its tail at the center of these imaginary lines. Components of the vector might be seen as effects along the x and y axis.

Example 3.6

What are the x and y components of a force of 200 N at an angle of 60°?

Plan: Draw the vector diagram and use trigonometry to find the components.

Solution: Place the tail of the 200-N vector at the origin and indicate the components as shown in Fig. 3.16.

We first compute the x component, F_x, by noting that it is the side adjacent. The 200-N vector is the hypotenuse. Using the cosine function, we obtain

$$\cos 60° = \frac{F_x}{200 \text{ N}}$$

from which

$$F_x = (200 \text{ N}) \cos 60° = 100 \text{ N}$$

For purposes of calculation, we recognize that the side opposite to 60° is equal in length to F_y. Thus, we may write

$$\sin 60° = \frac{F_y}{200 \text{ N}}$$

or

$$F_y = (200 \text{ N}) \sin 60° = 173 \text{ N}$$

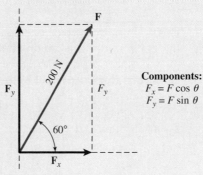

Figure 3.16   Using trigonometry to find the x and y components of a vector.

In general, we may write the x and y components of a vector in terms of its magnitude F and direction θ.

$$F_x = F \cos \theta$$
$$F_y = F \sin \theta$$

Components of a Vector   **(3.1)**

where θ is the angle between the vector and the positive x axis, measured in a counter-clockwise direction.

The sign of a given component can be determined from a vector diagram. The four possibilities are shown in Fig. 3.17. In addition to the *polar angle* θ, we show the *reference angle* ϕ for each quadrant. When the polar angle is greater than 90°, it is easier to view the directions of the components when you work with the reference angle ϕ. Applications of trigonometry using the polar angle θ will also give the correct signs, but a visual check on the direction of the components is helpful.

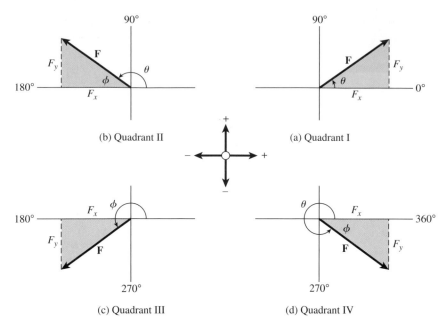

Figure 3.17   (a) In the first quadrant, angle θ is between $0°$ and $90°$; both F_x and F_y are positive. (b) In the second quadrant, angle θ is between $90°$ and $180°$; F_x is negative and F_y is positive. (c) In the third quadrant, angle θ is between $180°$ and $270°$; F_x and F_y are negative. (d) In the fourth quadrant, angle θ is between $270°$ and $360°$; F_x is positive and F_y is negative.

Example 3.7   Find the x and y components of a 400-N force when the polar angle θ is $220°$.

Plan: Draw the vector and its components, indicating both the reference angle and the polar angle. Use trigonometry to find the components.

Solution: Refer to Fig. 3.17, where we can find the reference angle ϕ as follows:

$$\phi = 220° - 180° = 40°$$

From the figure, it is seen that both components are negative.

$$F_x = -|F \cos \phi| = -(400 \text{ N}) \cos 40°$$
$$= -306 \text{ N}$$
$$F_y = -|F \sin \phi| = -(400 \text{ N}) \sin 40°$$
$$= -257 \text{ N}$$

Note that the signs were determined from Fig. 3.17. With electronic calculators, both the magnitude and the sign of F_x and F_y can be found directly from Eq. (3.1) using the polar angle $\theta = 220°$. You should verify this fact.

Trigonometry is also useful in calculating the resultant force. In the special case for two forces $\mathbf{F}_x$ and $\mathbf{F}_y$ at right angles to each other, as in Fig. 3.18, the resultant (R, θ) may be found from

$$R = \sqrt{F_x^2 + F_y^2} \qquad \tan \theta = \frac{F_y}{F_x} \qquad \text{(3.2)}$$

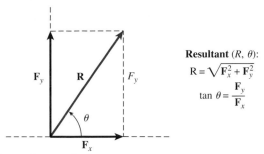

Figure 3.18 The resultant of perpendicular vectors.

If either F_x or F_y is negative, it is usually better to determine the reference angle ϕ, as described in Fig. 3.17. The sign (or direction) of the forces F_x and F_y determines which of the four quadrants is used. Then, Eq. (3.2) becomes

$$\tan \phi = \left| \frac{F_y}{F_x} \right|$$

Only the absolute values of F_x and F_y are needed. If desired, the angle θ from the positive x axis may be determined. In any case, the direction must be clearly identified.

Example 3.8

What is the resultant of a 5-N force directed horizontally to the right and a 12-N force directed downward?

Plan: Noting that the forces are right and downward, we draw a fourth-quadrant vector diagram like the one shown in Fig. 3.17d. Apply Eq. (3.2) to find the resultant.

Solution: Treat the two force vectors as components $F_x = 65$ N and $F_y = -12$ N of the resultant force **R.** Then the magnitude of **R** becomes

$$R = \sqrt{F_x^2 + F_y^2} = \sqrt{(5 \text{ N})^2 + (-12 \text{ N})^2}$$
$$= \sqrt{169 \text{ N}^2} = 13.0 \text{ N}$$

To find the direction of R, we first find the reference angle ϕ.

$$\tan \phi = \left| \frac{-12 \text{ N}}{5 \text{ N}} \right| = 2.40$$

$$\phi = 67.4° \text{ S of E}$$

The polar angle θ measured counterclockwise from the positive x-axis is

$$\theta = 360° - 67.4° = 292.6°$$

The resultant force is 13.0 N at 292.6°. Angles should be reported to the nearest tenth of a degree even if it requires four significant figures to show the required precision. Other answers can be reported with only three significant figures.

3.12 | The Component Method of Vector Addition

Often it is necessary to add a series of displacements or to find the resultant of a number of forces using mathematical methods. In such cases, one should begin with a rough graphical sketch using the polygon method of vector addition. However, since trigonometry will be used to insure that final results are accurate, we need only estimate the lengths of each

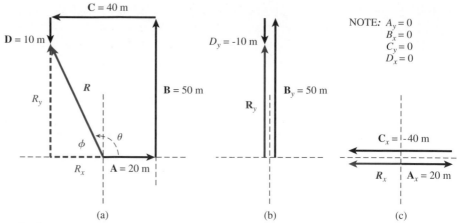

Figure 3.19 The x component of the resultant vector is equal to the sum of the x components of each vector. The y component of the resultant is equal to the sum of the y components.

vector. For example, a 60-m displacement or a 60-N force should be drawn as a vector approximately three times as long as the vector for a 20-m displacement or a 20-N force. The given angles should also be estimated. Vectors of 30°, 160°, 240°, and 324° should be drawn in appropriate quadrants and as close as possible to the actual directions. Learn to draw these rough diagrams quickly, so that you will have an idea of the direction of the resultant before calculations are made.

It is useful to recognize that the x component of the resultant, or the sum of a number of vectors is given by the sum of the x components of each vector. Similarly, the y component of the resultant is the sum of the y components. Suppose we wish to add a number of vectors **A, B, C,** . . . to find their resultant **R.** We might write

$$R_x = A_x + B_x + C_x + \cdots \tag{3.3}$$
$$R_y = A_y + B_y + C_y + \cdots \tag{3.4}$$

The magnitude of the resultant R and its direction θ can then be found from Eq. (3.2).

We will consider an example to illustrate the component method of vector addition. Suppose a surveyor walks 20 m, E; 50 m, N; 40 m, W; and finally 10 m, S. It is our goal to find the resultant displacement.

First, we draw each vector approximately to scale using the polygon approach. In that manner, we see from Fig. 3.19 that the resultant **R** should be in the second quadrant.

Finding the components of each vector is simple in this problem, because each vector lies completely along a given axis so that one component is zero in each case. Note that components are positive or negative, whereas the vector magnitudes are always positive. It is sometimes useful to construct a table of components, such as Table 3.5. For each vector, we list its magnitude, its reference angle, and the x and y components.

Table 3.5

A Table of Components

Vector	Angle θ	x component	y component
$A = 20$ m	0°	$A_x = +20$ m	$A_y = 0$
$B = 50$ m	90°	$B_x = 0$	$B_y = +50$ m
$C = 40$ m	180°	$C_x = -40$ m	$C_y = 0$
$D = 10$ m	270°	$D_x = 0$	$D_y = -10$ m
R	θ	$R_x = \Sigma F_x = -20$ m	$R_y = \Sigma F_y = +40$ m

Note carefully the representation of each of these components in Fig. 3.19. It is easy to see the meaning of the net x component and of the net y component.

The resultant can now be found from the components R_x and R_y of the resultant vector.

$$R = \sqrt{R_x^2 + R_y^2} = \sqrt{(-20 \text{ m})^2 + (40 \text{ m})^2}$$
$$R = \sqrt{400 \text{ m}^2 + 1600 \text{ m}^2} = \sqrt{2000 \text{ m}^2}; \; R = 44.7 \text{ m}$$

Next, the direction can be found from the tangent function.

$$\tan \phi = \left| \frac{R_y}{R_x} \right| = \left| \frac{40 \text{ m}}{-20 \text{ m}} \right| = 2.00$$
$$\phi = 63.4° \text{ N of W (or } 116.6°)$$

The procedure followed in the previous example can also be used to solve the more general problems that may involve vectors not along perpendicular axes. Remember that the components are found using the sine and cosine functions, and these components must be given appropriate algebraic signs before addition occurs. Also, you should recall that in this text, we assume that *each given magnitude is accurate to three significant digits and that each angle is accurate to the nearest tenth of a degree.*

Problem-Solving Strategy

The Component Method for Vector Addition

(These steps are illustrated in Example 3.9)

1. Construct a rough vector polygon, drawing each vector at proportional lengths and angles. Indicate the resultant as a line drawn from the tail of the first vector to the tip of the last vector.

2. Find the x and y components of each vector, using trigonometry if necessary. Verify that the algebraic signs are correct before proceeding.

$$A_x = A \cos \theta; \quad A_y = A \sin \theta$$

3. Make a table of x and y components and add algebraically to find the magnitude and sign of the resultant components:

$$R_x = A_x + B_x + C_x + \cdots$$
$$R_y = A_y + B_y + C_y + \cdots$$

4. Find the magnitude and direction of the resultant from its perpendicular components R_x and R_y.

$$R = \sqrt{R_x^2 + R_y^2}; \quad \tan \phi = \left| \frac{R_y}{R_x} \right|$$

Example 3.9

Three ropes are tied to a stake, and the following forces are exerted. $A = 20$ N, E; $B = 30$ N, 30° N of W; and $C = 40$ N, 52° S of W. Determine the resultant force using the component method.

Plan: We will draw a rough sketch of the problem as shown in Fig. 3.20. The forces are represented as a proportional vectors, and their directions are indicated by angles with respect to the x axis. We will then solve for the resultant force utilizing the problem-solving strategy.

Solution: The details of the procedure are outlined in the following steps:

1. Construct a proportional vector polygon, adding the forces as in Fig. 3.20b. The resultant is estimated to be in the third quadrant.

2. Construct a table of the x and y components of each vector. Note from Fig. 3.21 that the reference angles ϕ are determined from the x axes for trig purposes. Care must be taken to list the correct sign for each component. For example, B_x, C_x, and C_y are each negative. The results are shown in Table 3.6.

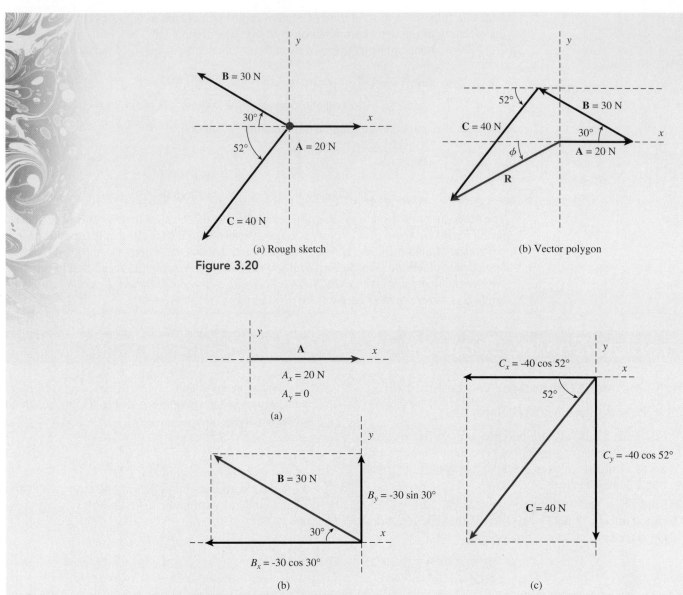

(a) Rough sketch

(b) Vector polygon

Figure 3.20

(a)

$A_x = 20 \text{ N}$

$A_y = 0$

(b)

$B_y = \text{-}30 \sin 30°$

$B_x = \text{-}30 \cos 30°$

$C_x = \text{-}40 \cos 52°$

$C_y = \text{-}40 \cos 52°$

(c)

Figure 3.21 Finding components of vectors.

Table 3.6

A Table of Components

Vector	Angle ϕ_x	x component	y component
$A = 20$ N	0°	$A_x = +20$ N	$A_y = 0$
$B = 30$ N	30°	$B_x = -(30 \text{ N})(\cos 30°)$	$B_y = (30 \text{ N})(\sin 30°)$
		$= -26.0$ N	$= 15.0$ N
$C = 40$ N	52°	$C_x = -(40 \text{ N})(\cos 52°)$	$C_y = -(40 \text{ N})(\sin 52°)$
		$= -24.6$ N	$= -31.5$ N
R	θ	$R_x = \Sigma F_x = -30.6$ N	$R_y = \Sigma F_y = -16.5$ N

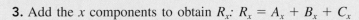

3. Add the x components to obtain R_x: $R_x = A_x + B_x + C_x$

$$R_x = 20.0\,\text{N} - 26.0\,\text{N} - 24.6\,\text{N}; \qquad R_x = -30.6\,\text{N}$$

4. Add the y components to obtain R_y: $R_y = A_y + B_y + C_y$

$$R_y = 0\,\text{N} + 15.0\,\text{N} - 31.5\,\text{N}; \qquad R_y = -16.5\,\text{N}$$

5. Now we find R and θ from R_x and R_y.

A separate figure (see Fig. 3.22) is often helpful to aid in calculating the magnitude and direction of the resultant force.

$$R = \sqrt{R_x^2 + R_y^2} = \sqrt{(-30.6\,\text{N})^2 + (-16.5\,\text{N})^2}; \qquad R = 34.8\,\text{N}$$

Next, the direction can be found from the tangent function.

$$\tan \phi = \left| \frac{R_y}{R_x} \right| = \left| \frac{-16.5\,\text{N}}{-30.6\,\text{N}} \right| = 0.539$$

$$\phi = 28.3°\ \text{S of W} \qquad \text{or} \qquad 180° - 28.3° = 208.3°$$

Thus, the resultant force is 34.8 N at 208.3°.

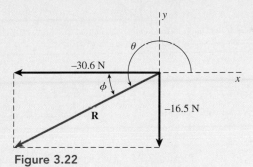

Figure 3.22

3.13 Unit Vector Notation (Optional)

A useful tool for many vector applications is to specify the direction by using what is called a ***unit vector.*** Such an approach clearly separates the magnitude of a vector from its direction.

> **Unit Vector:** A dimensionless vector whose magnitude is exactly 1 and whose direction is fixed by definition.

The symbols **i, j, k** can be used to describe unit vectors in the positive x, y, and z directions as indicated by Fig. 3.23. For example, a displacement of 40 m, E could be written simply as $+40\,\mathbf{i}$, and a displacement of 40 m, W could be given as $-40\,\mathbf{i}$. For convenience, units are usually omitted when using the **i, j** notation. Study each example given in Fig. 3.23 until you understand the meaning and use of unit vectors.

Consider a vector **A** in Fig. 3.24 lying in the xy plane and having the components A_x and A_y. We can represent the x and y components of vector **A** by using products of their magnitudes and the appropriate unit vector. Then we could write the vector **A** in what is called unit vector notation:

$$\mathbf{A} = A_x\mathbf{i} + A_y\mathbf{j}$$

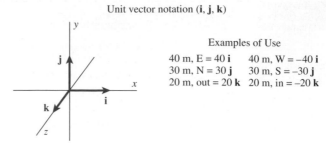

Figure 3.23   Unit vectors are useful when working with components of vectors.

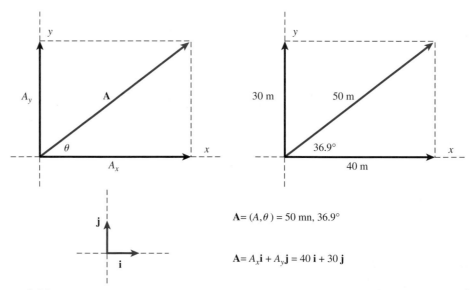

Figure 3.24   Two ways of representing a vector.

Thus, a vector $(\mathbf{A}, \theta)$ can now be completely described using products of its components and appropriate unit vectors.

In Fig. 3.24, if the magnitude of vector $\mathbf{A}$ is equal to 50 m and the angle is 36.9°, the components are $A_x = +40$ m and $A_y = +30$ m. The vector can now be written in two acceptable forms:

$$\mathbf{A} = (50 \text{ m}, 36.9°) \qquad \text{or} \qquad \mathbf{A} = 40\,\mathbf{i} + 30\,\mathbf{j}$$

The unit-vector approach is convenient when applying the component method of vector addition because the components of the resultant can be found by adding polynomials.

Consider Table 3.6, which was compiled for Example 3.9. The resultant could be found by adding the unit vector polynomials as follows:

$$
\begin{array}{ll}
A = A_x\mathbf{i} + A_y\mathbf{j} & A = +20.0\,\mathbf{i} + 0 \\
B = B_x\mathbf{i} + B_y\mathbf{j} & B = -26.0\,\mathbf{i} + 15.0\,\mathbf{j} \\
\underline{C = C_x\mathbf{i} + C_y\mathbf{j}} & \underline{C = -24.6\,\mathbf{i} - 31.5\,\mathbf{j}} \\
R = R_x\mathbf{i} + R_y\mathbf{j} & R = -30.6\,\mathbf{i} - 16.5\,\mathbf{j}
\end{array}
$$

The magnitude and direction in polar coordinates could then be calculated as before from Eq. (3.2). Unit vectors help to organize the data without the need for making a table.

3.14   Vector Difference

When we study relative velocity, acceleration, and certain other quantities, we must be able to find the difference of two vector quantities. The difference between two vectors is obtained by adding one vector to the negative of the other. The negative of a vector is found by constructing a vector equal in magnitude but opposite in direction. For example, if **A** is a vector whose magnitude is 40 m and whose direction is east, then the vector $-$**A** is a displacement of 40 m directed to the west. Just as we have in algebra that

$$a - b = a + (-b)$$

we have in vector subtraction that

$$\mathbf{A} - \mathbf{B} = \mathbf{A} + (-\mathbf{B})$$

The process of subtracting vectors is illustrated in Fig. 3.25. The given vectors are shown in Fig. 3.25a; Fig. 3.25b shows the vectors **A** and $-$**B.** The vector sum by the polygon method is pictured in Fig. 3.25c.

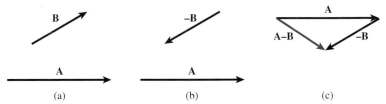

(a)   (b)   (c)

Figure 3.25   Finding the difference of two vectors.

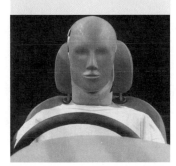

Summary and Review

Summary

Technical measurement is essential for the application of physics. You have learned that there are seven fundamental quantities and that each has a single approved SI unit. In mechanics, the three fundamental quantities of length, mass, and time are the only ones required for most applications. Some of these applications involve *vectors* and some involve *scalars*. Since vector quantities have direction, they must be added and subtracted by special methods. The following points summarize this unit of study:

- The SI prefixes used to express multiples and subdivisions of the base units are given below:

    giga (G) = 10^9 milli (m) = 10^{-3}
    mega (M) = 10^6 micro (μ) = 10^{-6}
    kilo (k) = 10^3 nano (n) = 10^{-9}
    centi (c) = 10^{-2} pico (p) = 10^{-12}

- To convert one unit to another,

    a. Write down the quantity to be converted (number and unit).

    b. Recall the necessary definitions.

    c. Form two conversion factors for each definition.

    d. Multiply the quantity to be converted by those conversion factors that cancel all but the desired units.

- The *polygon method* of vector addition: The *resultant vector* is found by drawing each vector to scale, placing the tail of one vector to the tip of another until all vectors are drawn. The resultant is the straight line drawn from the starting point to the tip of the last vector (Fig. 3.26).

- The *parallelogram method* of vector addition: The resultant of two vectors is the diagonal of a parallelogram formed by the two vectors as adjacent sides. The direction is away from the common origin of the two vectors (Fig. 3.27).

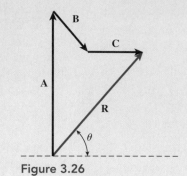

Figure 3.26

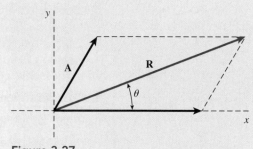

Figure 3.27

- The x and y *components* of a vector (R, θ):

$$R_x = R \cos \theta \qquad R_y = R \sin \theta$$

- The *resultant* of two perpendicular vectors R_x and R_y:

$$R = \sqrt{R_x^2 + R_y^2} \qquad \tan \phi = \left| \frac{R_y}{R_x} \right|$$

- The **component method** of vector addition:

$$R_x = A_x + B_x + C_x + \cdots$$
$$R_y = A_y + B_y + C_y + \cdots$$
$$R = \sqrt{R_x^2 + R_y^2}$$
$$\tan \phi = \left| \frac{R_y}{R_x} \right|$$

Key Terms

Review Questions

3.1. Express the following measurements in proper SI form using the appropriate prefixes. The symbol for the base unit is given in parentheses:

   a. 298,000 meters (m)

   b. 7600 volts (V)

   c. 0.000067 amperes (A)

   d. 0.0645 newtons (N)

   e. 43,000,000 grams (g)

   f. 0.00000065 farads (F)

3.2. What three fundamental quantities appear in the definition of most laws of mechanics? Name the three fundamental units associated with each quantity in the SI and USCS units.

3.3. A unit of specific heat capacity is cal/g · C°. How many definitions are needed to convert these units to their corresponding units in the USCS, where the units are Btu/lb · F°? Show by a series of products how you would perform such a conversion.

3.4. Given that the units of s, v, a, and t are meters (m), meters per second (m/s), meters per second squared (m/s^2), and seconds (s), respectively, what are the dimensions of each quantity? Accept or reject the following equations on the basis of dimensional analysis:

   a. $s = vt + \frac{1}{2}at^2$ c. $v_f = v_0 + at^2$

   b. $2as = v_f^2 - v_0^2$ d. $s = vt + 4at^2$

3.5. Distinguish between vector and scalar quantities, and give examples of each. Explain the difference between adding vectors and adding scalars. Is it possible for the sum of two vectors to have a magnitude less than either of the original vectors?

3.6. What are the minimum and maximum resultants of two forces of 10 N and 7 N if they act on the same object?

3.7. Look up a section on rectangular and polar coordinates in a math book. What are the similarities between components of a vector and the rectangular and polar coordinates of a point?

3.8. If a vector has a direction of 230° from the positive x axis, what are the signs of its x and y components? If the ratio R_y/R_x is negative, what are the possible angles for R as measured from the positive x axis?

Problems

Note: In this chapter and others, the assumption is that all numbers are accurate to three significant digits unless otherwise indicated. Answers are given to odd-numbered problems and to selected Critical Thinking Questions.

Section 3.6 Unit Conversions

3.1. What is the height in centimeters of a woman who is 5 feet and 6 inches tall?   Ans. 168 cm

3.2. A single floor tile measures 8 in. on each side. If the tiles are laid side by side, what distance in meters can be covered by a single row of 20 tiles?

3.3. A soccer field is 100 m long and 60 m across. What are the length and width of the field in feet?   Ans. 328 ft, 197 ft

3.4. A wrench has a handle 8 in. long. What is the length of the handle in centimeters?

3.5. A 19-in. computer monitor has a viewable area that measures 18 in. diagonally. Express this distance in meters.   Ans. 0.457 m

3.6. The length of a notebook is 234.5 mm and the width is 158.4 mm. Express the surface area in square meters.

3.7. A cube has 5 in. on a side. What is the volume of the cube in SI units and in USCS units?
        Ans. 0.00205 m^3, 0.0723 ft^3

3.8. The speed limit on an interstate highway is posted at 75 mi/h. (a) What is this speed in kilometers per hour? (b) In feet per second?

3.9. A Nissan engine has a piston displacement (volume) of 1600 cm^3 and a bore diameter of 84 mm. Express these measurements in cubic inches and inches.   Ans. 97.6 in.^{3}, 3.31 in.

3.10. An electrician must install an underground cable from the highway to a home located 1.20 mi into the woods. How many feet of cable will be needed?

3.11. One U.S. gallon is a volume equivalent to 231 in.^{3}. How many gallons are needed to fill a tank that is 18 in. long, 16 in. wide, and 12 in. high?
        Ans. 15.0 gal

3.12. The density of brass is 8.89 g/cm^3. What is the density in kilograms per cubic meter?

Section 3.8 Addition of Vectors by Graphical Methods

3.13. A woman walks 4 km east and then 8 km north. (a) Use the polygon method to find her resultant displacement. (b) Verify the result by the parallelo-gram method. Ans. 8.94 km, 63.4° N of E

3.14. A land rover on the surface of Mars moves a distance of 38 m at an angle of 180°. It then turns and moves a distance of 66 m at an angle of 270°. What is the displacement from the starting position?

3.15. A surveyor starts at the southeast corner of a lot and charts the following displacements: $A = 600$ m, N; $B = 400$ m, W; $C = 200$ m, S; and $D = 100$ m, E. What is the net displacement from the starting point? Ans. 500 m, 126.9°

3.16. A downward force of 200 N acts simultaneously with a 500-N force directed to the left. Use the polygon method to find the resultant force.

3.17. The following three forces act simultaneously on the same object. $A = 300$ N, 30° N of E; $B = 600$ N, 270°; and $C = 100$ N due east. Find the resultant force using the polygon method.
  Ans. 576 N, 51.4° S of E

3.18. A boat travels west a distance of 200 m, then north for 400 m, and finally 100 m at 30° S of E. What is the net displacement?

3.19. Two ropes A and B are attached to a mooring hook so that an angle of 60° exists between the two ropes. The tension in rope A is 80 N, and the tension in rope B is 120 N. Use the parallelogram method to find the resultant force on the hook.
  Ans. 174 N

3.20. Two forces A and B act on the same object and produce a resultant force of 50 N at 36.9° N of W. The force $A = 40$ N due west. Find the magnitude and direction of force B.

Section 3.11 Trigonometry and Vectors

3.21. Find the x and y components of (a) a displacement of 200 km at 34°, (b) a velocity of 40 km/h at 120°, and (c) a force of 50 N at 330°.
  Ans. 166 km, 112 km; −20 km/h, 34.6 km/h; 43.3 N, −25 N

3.22. A sled is pulled with a force of 540 N at an angle of 40° with the horizontal. What are the horizontal and vertical components of this force?

3.23. The hammer in Fig. 3.28 applies a force of 260 N at an angle of 15° with the vertical. What is the upward component of the force on the nail?
  Ans. 251 N

3.24. A boy attempts to lift his sister from the pavement (Fig. 3.29). If the vertical component of his pull **F** has a magnitude of 110 N and the horizontal component has a magnitude of 214 N, what are the magnitude and direction of the force **F**?

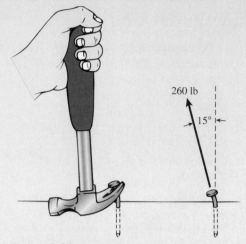

260 lb

15°

Figure 3.28

F

θ

Figure 3.29
(*Photo by Paul E. Tippens.*)

3.25. A river flows south with a velocity of 20 km/h. A boat has a maximum speed of 50 km/h in still water. In the river, at maximum throttle, the boat heads due west. What are the resultant speed and direction of the boat?
  Ans. 53.9 km/h, 21.8° S of W

3.26. A rope that makes an angle of 30° with the horizontal drags a crate along the floor. What must be the tension in the rope if a horizontal force of 40 N is needed to drag the crate?

3.27. A long pole is used to lift a window. If a vertical lift of 80 N is needed to raise the window, what force must be exerted along the pole? Assume the pole makes an angle of 34° with the vertical wall.

Ans. 96.5 N

3.28. The resultant of two forces **A** and **B** is 40 N at 210°. If force **A** is 200 N at 270°, what are the magnitude and direction of force **B?**

Section 3.12 The Component Method of Vector Addition

3.29. Find the resultant of the following perpendicular forces: (a) 400 N, 0°; (b) 820 N, 270°; and (c) 500 N, 90°.

Ans. 512 N, 321.3°

3.30. Four ropes, all at right angles to each other, pull on a ring. The forces are 40 N, east; 80 N, north; 70 N, west; and 20 N, south. Find the magnitude and direction of the resultant force acting on the ring.

3.31. Two forces act on the car, as shown in Fig. 3.30. Force **A** is equal to 120 N, west, and force **B** is equal to 200 N at 60° north of west. What are the magnitude and direction of the resultant force on the car?

Ans. 280 N, 38.2° N of W

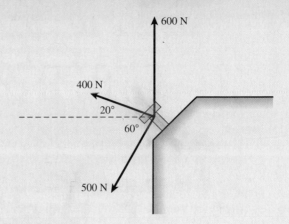

Figure 3.31

3.35. Three boats exert forces on a mooring hook, as shown in Fig. 3.32. Find the resultant of these three forces.

Ans. 853 N, 101.7°

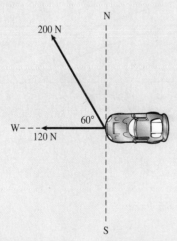

Figure 3.30

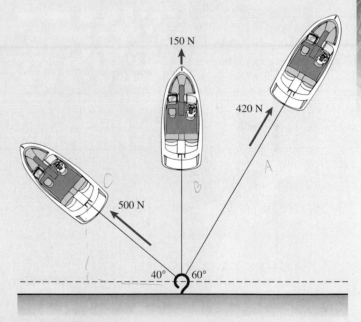

Figure 3.32

3.32. Suppose that the direction of force **B** in Prob. 3.31 is reversed (+180°) and other parameters remain the same. What is the new resultant? (This is the vector difference **A** − **B.**)

3.33. Determine the resultant force on the bolt in Fig. 3.31.

Ans. 69.6 N, 154.1°

3.34. Determine the resultant of the following forces by the component method of vector addition: **A** = (200 N, 30°), **B** = (300 N, 330°), and **C** = (400 N, 250°).

Section 3.14 Vector Difference

3.36. Two displacements are **A** = 9 m, N and **B** = 12 m, S. Find the magnitude and direction of (**A** + **B**) and (**A** − **B**).

3.37. Given that **A** = 24 m, E; **B** = 50 m, S, find the magnitude and direction of (a) **A** + **B** and (b) **B** − **A**.

Ans. (a) 55.5 m, 64.4° N of W, (b) 55.5 m, 64.4° N of E

3.38. Velocity has a magnitude and a direction that can be represented by a vector. Consider a boat moving initially with a velocity of 30 m/s directly west. At some later instant, the boat is found to have a velocity of 12 m/s at 30° S of W. What is the change in velocity?

3.39. Consider four vectors: **A** = 450 N, W; **B** = 160 N, 44° N of W; **C** = 800 N, E; and **D** = 100 m, 34° N of E. Determine the magnitude and direction of **A** − **B** + **C** − **D**. Draw the vector polygon. Ans. 417 N, 17.0° S of E

Additional Problems

3.40. Find the horizontal and vertical components of the following vectors: **A** = (400 N, 37°), **B** = (90 m, 320°), and **C** = (70 km/h, 150°).

3.41. A cable is attached to the end of a beam. What pull at an angle of 40° with the horizontal is needed to produce an effective horizontal force of 200 N? Ans. 261 N

3.42. A fishing dock runs north and south. What must be the speed of a boat heading at an angle of 40° E of N if its velocity component along the dock is to be 30 km/h?

3.43. Find the resultant **R** = **A** + **B** for the following pairs of vectors: (a) **A** = (520 N, south), **B** = 269 N, west, (b) **A** = 18 m/s, north, **B** = 15 m/s, west. Ans. 585 N, 246.6°; 23.4 m/s, 129.8°

3.44. Determine the vector difference (**A** − **B**) for the pairs of forces in Prob. 3.43.

3.45. A traffic light is attached to the midpoint of a rope so that each segment makes an angle of 10° with the horizontal. The tension in each rope segment is 200 N. If the resultant force at the midpoint is zero, what must be the weight of the traffic light? Ans. 69.5 N

3.46. Determine the resultant of the forces shown in Fig. 3.33.

3.47. Find the resultant force acting on the ring for Fig. 3.34. Ans. 315°, 26.9° N of W

3.48. A 200-N block rests on a 30° inclined plane. If the weight of the block acts vertically downward, what are the components of the weight down the plane and perpendicular to the plane?

3.49. Find the resultant of the following three displacements: **A** = 220 m, 60°; **B** = 125 m, 210°; and **C** = 175 m, 340°. Ans. 180 m, 22.3°

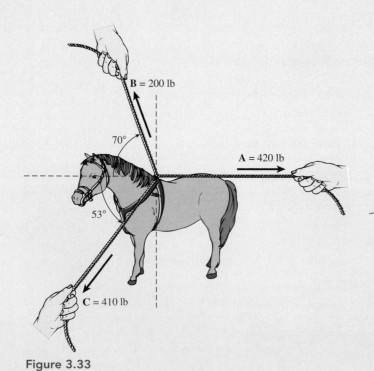

Figure 3.33

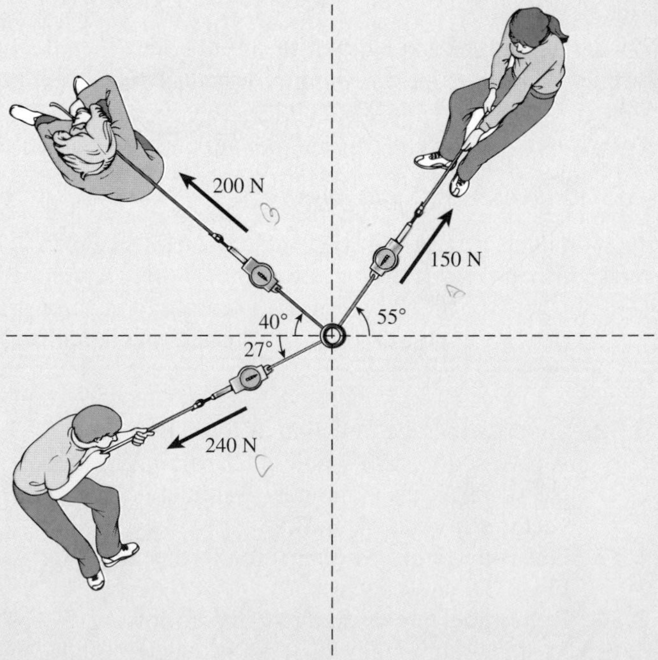

Figure 3.34

Critical Thinking Questions

*3.50. Consider three vectors: **A** = 100 m, 0°; **B** = 400 m, 270°; and **C** = 200 m, 30°. Choose an appropriate scale and show graphically that the order in which these vectors is added does not matter; that is, **A** + **B** + **C** = **C** + **B** + **A**. Is this also true for subtracting vectors? Show graphically how **A** − **C** differs from **C** − **A**.

3.51. Two forces A = 30 N and B = 90 N can act on an object in any direction desired. What is the maximum resultant force? What is the minimum resultant force? Can the resultant force be zero?

> Ans. 120 N, 60 N, no

3.52. Consider two forces A = 40 N and B = 80 N. What must be the angle between these two forces in order to produce a resultant force of 60 N?

*3.53. What third force **F** must be added to the following two forces so that the resultant force is zero? **A** = 120 N, 110° and **B** = 60 N, 200°?

> Ans. 134 N, 316.6°

*3.54. An airplane needs a resultant heading of due west. The speed of the plane is 600 km/h in still air. If the wind has a speed of 40 km/h and blows in a direction of 30° S of W, in what direction should the aircraft be pointed, and what will be its speed relative to the ground?

*3.55. What are the magnitude F and direction θ of the force needed to pull the car of Fig. 3.35 directly east with a resultant force of 400 lb?

> Ans. 223 lb, 17.9°

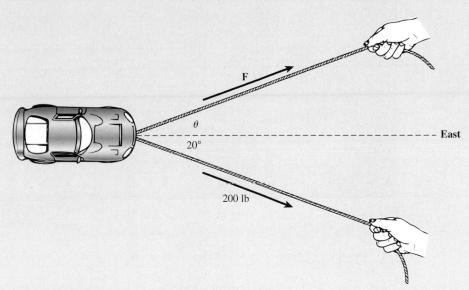

Figure 3.35

4 Translational Equilibrium and Friction

A mountain climber exerts action forces on crevices and ledges that produce reaction forces on the climber, allowing him to scale the cliffs. (*Photo © vol. 1 PhotoDisc/Getty.*)

Objectives

After completing this chapter, you should be able to

1. Demonstrate by example or experiment your understanding of Newton's first and third laws of motion.
2. State the first condition for equilibrium, give a physical example, and demonstrate graphically that the first condition is satisfied.
3. Construct a free-body diagram representing all forces acting on an object that is in translational equilibrium.
4. Solve for unknown forces by applying the first condition for equilibrium.
5. Apply your understanding of kinetic and static friction to the solution of equilibrium problems.

Forces may act in such a manner as to cause motion or to prevent motion. Large bridges must be designed so that the overall effect of forces is to prevent motion. Every truss, girder, beam, and cable must be in *equilibrium.* In other words, the resultant forces acting at any point on the entire structure must be balanced. Shelves, chain hoists, hooks, lifting cables, and even large buildings must be constructed so that the effects of forces are controlled and understood. In this chapter, we will continue our study of forces by studying objects at rest. The friction force that is so essential for equilibrium in many applications will also be introduced in this chapter as a natural extension of our work with all forces.

4.1 Newton's First Law

We know from experience that a stationary object remains at rest unless acted on by some outside force. A can of oil will stay on a workbench until someone tips it over. A suspended weight will hang until it is released. We know that forces are necessary to cause anything to move if it is originally at rest.

Less obvious is the fact that an object in motion will continue in motion until an outside force changes the motion. For example, a steel bar that slides on the shop floor soon comes to rest because of its interaction with the floor. The same bar would slide much farther on ice before stopping. This is because the horizontal interaction, called *friction,* between the floor and the bar is much greater than the friction between the ice and the bar. This leads to the idea that a sliding bar on a perfectly frictionless horizontal plane would stay in motion forever. These ideas are a part of *Newton's first law* of motion.

> **Newton's first law:** A body at rest remains at rest and a body in motion remains in uniform motion in a straight line unless acted on by an external unbalanced force.

Because of the existence of friction, no actual body is ever completely free from external forces. But there are situations in which it is possible to make the resultant force zero or approximately zero. In such cases, the body will behave in accordance with the first law of motion. Since we recognize that friction can never be eliminated completely, we also recognize that Newton's first law is an expression of an *ideal* situation. A flywheel rotating on lubricated ball bearings tends to keep on spinning, but even the slightest friction will eventually bring it to rest.

Newton called the property of a particle that allows it to maintain a constant state of motion or rest *inertia.* His first law is sometimes called the *law of inertia.* When an automobile is accelerated, the passengers obey this law by tending to remain at rest until the external force of the seat compels them to move. Similarly, when the automobile stops, the passengers continue in motion with constant speed until they are restrained by their seat belts or through their own efforts. All matter has inertia. The concept of *mass* is introduced later as a measure of a body's inertia.

4.2 Newton's Second Law

Since an object at rest or in constant motion will not alter its state without the action of an unbalanced force, we should next consider what happens if there is a resultant force. Experience tells us that larger and larger resultant forces on the same object will result in greater and greater changes in the velocity of the object. (See Fig. 4.1.) Additionally, if we keep the resultant force constant and apply it to greater and greater masses, the change in velocity decreases. The change in velocity per unit of time is defined as its *acceleration a.*

Newton was able to demonstrate a direct relationship between the applied force and the resulting acceleration. Further, he showed that the acceleration decreased proportionally with the inertia or mass m of the object. This principle is stated in *Newton's second law.*

 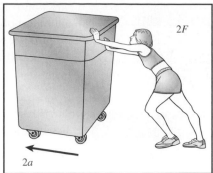

Figure 4.1 If we ignore friction forces, pushing the cart with twice the force produces twice the acceleration. Three times the force triples the acceleration.

(Photo © SS34 PhotoDisc/Getty.)

Newton's second law: The acceleration *a* of an object in the direction of a resultant force **F**, is directly proportional to the magnitude of the force and inversely proportional to the mass *m*.

$$\mathbf{a} = \frac{\mathbf{F}}{m} \quad \text{or} \quad \mathbf{F} = m\mathbf{a}$$

Note that when the velocity does not change, $a = 0$, and Newton's first law becomes a special case of the second law. Without the unbalanced force, the motion of the object will not change. The important word here is *change*. It helps to keep in mind that there is no resultant force either on objects at rest or on those moving at constant speed.

A mathematical treatment of Newton's second law of motion will be given later, along with more complete definitions of *force* and *mass*. We will first consider in detail the treatment of objects at rest, or more specifically, objects with no acceleration. Then, after you are able to deal effectively with a full vector treatment of forces, we will consider the implications of changing motion.

4.3   Newton's Third Law

There can be no force unless two bodies are involved. When a hammer strikes a nail, it exerts an "action" force on the nail. But the nail must also "react" by pushing back against the hammer. In all cases, there must be an *acting* force and a *reacting* force. Whenever two bodies interact, the force exerted by the second body on the first (the **reaction force**) is equal in magnitude and opposite in direction to the force exerted by the first body on the second (the action force). This principle is stated in **Newton's third law.**

Newton's third law: For every action force there must be an equal and opposite reaction force.

Therefore, there can never be a single isolated force. Consider the examples of action and reaction forces in Fig. 4.2.

Note that the acting and reacting forces do not cancel each other. They are equal in magnitude and opposite in direction, but they act on *different* objects. For two forces to cancel, they must act on the same object. It might be said that the action forces create the reaction forces.

For example, someone starting to climb a ladder begins by putting one foot on the rung and pushing on it. The rung must exert an equal and opposite force on the foot to prevent collapse. The greater the force exerted by the foot on the rung, the greater the reaction

Figure 4.2 Examples of action and reaction forces. (*Photos by Hemera, Inc.*)

against the foot must be. Of course the rung cannot create a reaction force until the force of the foot is applied. The action force acts on the object, and the reacting force acts on the agent that applies the force.

4.4 Equilibrium

The *resultant force* was defined as a single force whose effect is the same as a given system of forces. If the tendency of a number of forces is to cause motion, the resultant will also produce this tendency. A condition of equilibrium exists where the resultant of all external forces is zero. This is the same as saying that each external force is balanced by the sum of all the other external forces when equilibrium exists. Therefore, according to Newton's first law, a body in equilibrium must be either at rest or in motion with constant velocity since there is no unbalanced force.

Consider the system of forces in Fig. 4.3a. The vector polygon solution shows that, regardless of the sequence in which the vectors are added, their resultant is always zero. The tip of the last vector lands on the tail of the first vector (see Sec. 3.7).

A system of forces not in equilibrium can be put in equilibrium by replacing their resultant force with an equal but opposite force called the ***equilibrant.*** For instance, the two forces **A** and **B** in Fig. 4.4a have a resultant **R** in a direction 30° above the horizontal. If we add **E,** which is equal in magnitude to **R** but which has an angle 180° greater, the system will be in equilibrium, as shown in Fig. 4.4b.

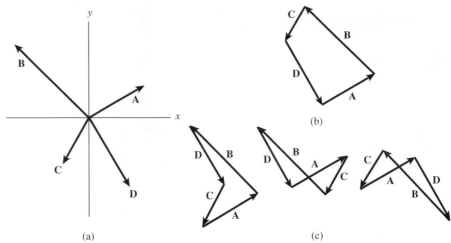

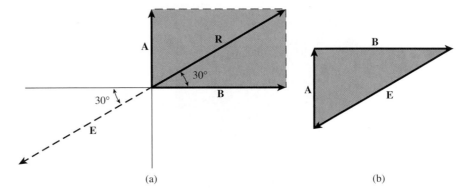

Figure 4.3   Forces in equilibrium.

Figure 4.4   The equilibrant.

PHYSICS TODAY

NASA is developing space propulsion alternatives. Solar electric propulsion uses solar cells to generate electricity to ionize krypton or xenon atoms. Once electrically charged, these ions generate thrust by being accelerated through an electromagnetic field and then expelled.

We have shown in Chapter 3 that the magnitudes of the x and y components of any resultant **R** are given by

$$R_x = \sum F_x = A_x + B_x + C_x + \ldots$$
$$R_y = \sum F_y = A_y + B_y + C_y + \ldots$$

When a body is in equilibrium, the resultant of all forces acting on it is zero. In this case, both R_x and R_y must be zero; hence, for a body in equilibrium,

$$\sum F_x = 0 \qquad \sum F_y = 0 \qquad \textbf{(4.1)}$$

These two equations represent a mathematical statement of the *first condition for equilibrium,* which can be stated as follows:

A body is in translational equilibrium if and only if the vector sum of the forces acting on it is zero.

The term ***translational equilibrium*** is used to distinguish the first condition from the second condition for equilibrium, which involves rotational motion, discussed in Chapter 5.

4.5   # Free-Body Diagrams

Before applying the first condition for equilibrium to the solution of physical problems, you must be able to construct vector diagrams. Consider, for example, the 400 N weight suspended by ropes shown in Fig. 4.5a. There are three forces acting on the knot—those

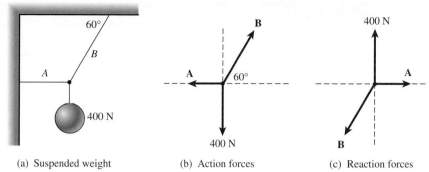

(a) Suspended weight   (b) Action forces   (c) Reaction forces

Figure 4.5   Free-body diagrams showing action and reaction forces.

exerted by the ceiling, the wall, and the Earth (weight). If each of these forces is labeled and represented as a vector, we can draw a vector diagram such as the one in Fig. 4.5b. Such a diagram is called a ***free-body diagram.***

A free-body diagram is a vector diagram that describes all forces acting on a particular body or object. Note that in the case of concurrent forces, all vectors point away from the center of the *x* and *y* axes, which cross at a common origin.

In drawing free-body diagrams, it is important to distinguish between action and reaction forces. In our example, there are forces on the knot, but there are also three equal and opposite reaction forces exerted *by* the knot. From Newton's third law, the reaction forces exerted *by* the knot *on* the ceiling, wall, and Earth are shown in Fig. 4.5c. To avoid confusion, it is important to pick a point at which all forces are acting and draw those forces that act *on* the body at that point.

Problem-Solving Strategy

How to Construct a Free-Body Diagram

1. Draw a sketch and label the conditions of the problem. Be sure to indicate all known and unknown forces and angles.

2. Isolate each body of the system to be studied. Do this either mentally or by drawing a light circle around the point where all forces are applied.

3. Construct a force diagram for each body to be studied. The forces are represented as vectors with their tails placed at the center of a rectangular coordinate system. (See examples in Figs. 4.6 and 4.8.)

4. Represent the *x* and *y* axes with dotted lines. These axes need not necessarily be drawn horizontally and vertically, as we shall see.

5. Dot in rectangles corresponding to the *x* and *y* components of each vector, and determine known angles from given conditions.

6. Label all known and unknown components opposite and adjacent to known angles.

Although this process may appear laborious, it is helpful and sometimes necessary for a clear understanding of a problem. As you gain practice drawing free-body diagrams, their use will become routine.

The two types of forces that act on a body are *contact forces* and *field forces*. Both must be considered in the construction of a force diagram. For example, the gravitational attraction on a body by the Earth, called its ***weight,*** does not have a point of contact with the body. Nevertheless, it exerts a real force and must be considered an important factor in any force problem. The direction of the weight vector should always be assumed to be downward.

Example 4.1   A block of weight W hangs from a cord that is knotted to two other cords, A and B, fastened to the ceiling. If cord B makes an angle of 60° with the ceiling and cord A forms a 30° angle, draw the free-body diagram of the knot.

Plan: We will follow the step-by-step procedure for drawing a free-body diagram.

Solution: A sketch is drawn and labeled as shown in Fig. 4.6a, and we draw a light circle around the knot where each force is acting. The completed free-body diagram is then drawn as in Fig. 4.6b. Note that all components are clearly identified opposite and adjacent to the given angles.

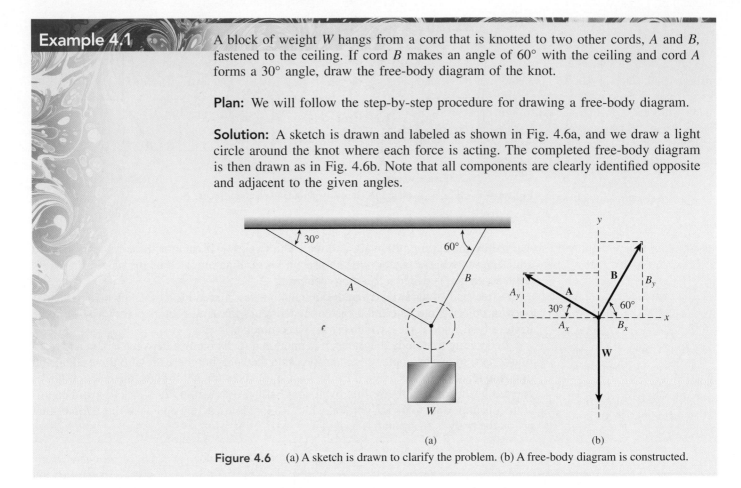

(a)   (b)

Figure 4.6   (a) A sketch is drawn to clarify the problem. (b) A free-body diagram is constructed.

The free-body diagram drawn for Example 4.1 is valid and workable, but choosing the x and y axes along the vectors **B** and **A,** instead of horizontally and vertically, can make the solution easier. By rotating the perpendicular axes as shown in Fig. 4.7, we find that we need only to resolve the weight vector **W** into its components. The vectors **A** and **B** are now entirely along a particular axis. As a general rule, we should choose the x and y axes so as to maximize the number of unknown forces that lie along an axis.

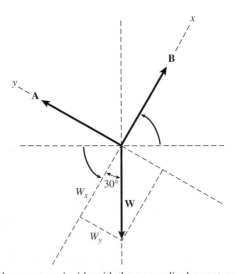

Figure 4.7   Rotation of x and y axes to coincide with the perpendicular vectors **A** and **B.**

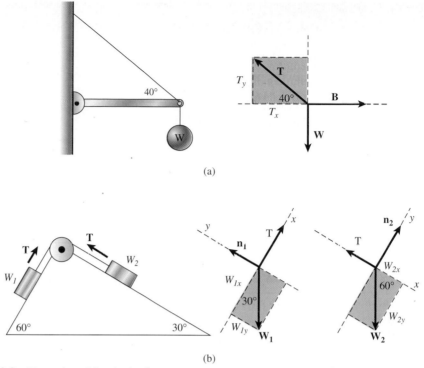

(a)

(b)

Figure 4.8   Examples of free-body diagrams. Note that the components of vectors are labeled opposite and adjacent to known angles.

Probably the most difficult part of constructing vector diagrams is the visualization of forces. In drawing free-body diagrams, it is helpful to imagine that the forces are acting on *you*. Become the knot in a rope or the block on a table and try to see the forces you would experience. Two additional examples are shown in Fig. 4.8. Note that the force exerted by the light boom in Fig. 4.8a is outward and not toward the wall. This is because we are interested in forces exerted *on* the end of the boom and not those exerted *by* the end of the boom. We pick a point at the end of the boom where the two ropes are attached. The 60-N weight and the tension **T** are action forces exerted by the ropes at this point. If the end of the boom is not to move, these forces must be balanced by a third force—the force exerted by the wall (through the boom). This third force **B**, acting on the end of the boom, must not be confused with the inward *reaction force* that acts *on* the wall.

The second example (Fig. 4.8b) also shows action forces acting on two weights connected by a light cord. Friction forces, which are discussed later, are not included in these diagrams. The tension in the cord on either side is shown as **T**, and the normal forces n_1, and n_2 are perpendicular forces exerted by the plane on the blocks. If these forces were absent, the blocks would swing together. (Note the choice of axes in each diagram.)

4.6   Solution of Equilibrium Problems

In Chapter 3, we discussed a procedure for finding the resultant of a number of forces by rectangular resolution. A similar procedure can be used to add forces that are in equilibrium. In this case, the first condition for equilibrium tells us that the resultant is zero, or

$$R_x = \sum F_x = 0 \qquad R_y = \sum F_y = 0 \tag{4.2}$$

Thus, we have two equations that can be used to find unknown forces.

Problem-Solving Strategy

Translational Equilibrium

1. Draw a sketch and label the conditions of the problem.

2. Draw a free-body diagram. (See Sec. 4.5.)

3. Resolve all forces into their x and y components, even though they may contain unknown factors, such as

$A \cos 60°$ or $B \sin 60°$. (You may wish to construct a force table like Table 4.1.)

4. Use the first condition for equilibrium [Eq. (4.1)] to set up two equations in terms of the unknown forces.

5. Solve algebraically for the unknown factors.

Table 4.1

Force	θ_x	x component	y component
A	60°	$A_x = -A \cos 60°$	$A_y = A \sin 60°$
B	0°	$B_x = B$	$B_y = 0$
W	−90°	$W_x = 0$	$W_y = -100 \text{ N}$
		$\Sigma F_x = B - A \cos 60°$	$\Sigma F_y = A \sin 60° - 100 \text{ N}$

Example 4.2

A 100-N ball suspended by a rope A is pulled aside by a horizontal rope B and held so that rope A forms an angle of 30° with the vertical wall. (See Fig. 4.9.) Find the tensions in ropes A and B.

Plan: Follow the problem-solving strategy.

Solution

1. Draw a sketch (Fig. 4.9a).

2. Draw a free-body diagram (Fig. 4.9b).

3. Resolve all forces into their components (Table 4.1). Note from the figure that A_x and W_y are negative.

4. We now apply the first condition for equilibrium. Summing the forces along the x axis yields

$$\sum F_x = B - A \cos 60° = 0$$

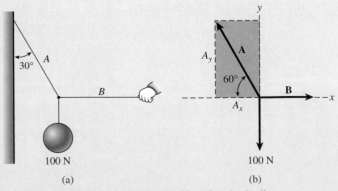

(a)　　　　　(b)

Figure 4.9 Forces acting on the knot are represented in a free-body diagram.

from which we obtain

$$B = A \cos 60° = 0.5A \tag{4.3}$$

since $\cos 60° = 0.5$. A second equation results from summing the y components.

$$\sum F_y = A \sin 60° - 100 \text{ N} = 0$$

from which

$$A \sin 60° = 100 \text{ N} \tag{4.4}$$

5. Finally, we solve for the unknown forces. Since $\sin 60° = 0.866$, we have from Eq. (4.4)

$$0.866A = 100 \text{ N}$$

or

$$A = \frac{100 \text{ N}}{0.866} = 115 \text{ N}$$

Now that the value of A is known, Eq. (4.3) can be solved for B as follows:

$$B = 0.5A = (0.5)(115 \text{ N})$$
$$= 57.5 \text{ N}$$

Example 4.3

A 200-N ball hangs from a cord knotted to two other cords, as shown in Fig. 4.10. Find the tensions in ropes A, B, and C.

Plan: We will construct a free-body diagram and then use the first condition for equilibrium to solve for the unknown rope tensions.

Solution: The free-body diagram is constructed from the given sketch (Fig. 4.10b). The x and y components are calculated from the figure and are given in Table 4.2.
Summing the forces along the x axis, we obtain

$$\sum F_x = -A \cos 60° + B \cos 45° = 0$$

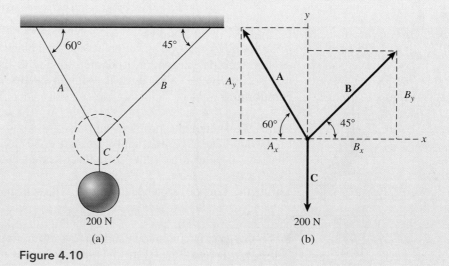

(a)   (b)

Figure 4.10

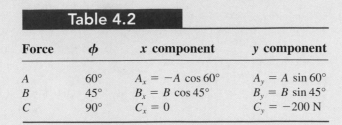

Force	ϕ	x component	y component
A	60°	$A_x = -A \cos 60°$	$A_y = A \sin 60°$
B	45°	$B_x = B \cos 45°$	$B_y = B \sin 45°$
C	90°	$C_x = 0$	$C_y = -200$ N

which can be simplified by substituting known trigonometric functions. Hence,

$$-0.5A + 0.707B = 0 \tag{4.5}$$

More information is needed to solve this equation. We obtain a second equation by summing the forces along the y axis, giving

$$0.866A + 0.707B = 200 \text{ N} \tag{4.6}$$

Eqs. (4.5) and (4.6) are now solved simultaneously for A and B by the process of substitution. Solving for A in Eq. (4.5) gives

$$A = \frac{0.707B}{0.5} \quad \text{or} \quad A = 1.414B \tag{4.7}$$

Now we substitute this equality into Eq. (4.6), obtaining

$$0.866(1.414B) + 0.707B = 200 \text{ N}$$

which can be solved for B as follows:

$$1.225B + 0.707B = 200 \text{ N}$$
$$1.93B = 200 \text{ N}$$
$$B = \frac{200 \text{ N}}{1.93} = 104 \text{ N}$$

The tension A can now be found by substituting $B = 104$ N into Eq. (4.7):

$$A = 1.414B = 1.414(104 \text{ N}) \quad \text{or} \quad A = 146 \text{ N}$$

The tension in cord C is, of course, 200 N because it must be equal to the weight.

Example 4.4

A 200-N block rests on a frictionless inclined plane of slope angle 30°. A cord attached to the block passes over a frictionless pulley at the top of the plane and is attached to a second block. What must the weight of the second block be if the system is in equilibrium?

Plan: We sketch the problem and draw a free-body diagram for each block (see Fig. 4.11). Then we will apply the first condition for equilibrium to each diagram in order to determine value of the suspended weight W_2.

Solution: For the suspended weight, $\Sigma F_y = 0$ gives

$$T - W_2 = 0 \quad \text{or} \quad T = W_2$$

Since the rope is continuous and the system is frictionless, the tension applied to the 200-N block (see Fig. 4.11b) must also be equal to W_2.

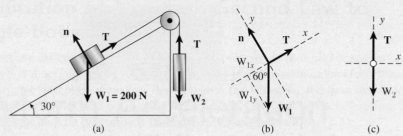

Figure 4.11 A free-body diagram is drawn for each block in the problem.

Table 4.3

Force	ϕ	x component	y component
T	0°	$T_x = T = W_2$	$T_y = 0$
n	90°	$n_x = 0$	$n_y = n$
W₁	60°	$W_{1x} = -(200 \text{ N}) \cos 60°$	$W_{1y} = -(200 \text{ N}) \sin 60°$

Considering the diagram for the block on the incline, we determine the components of each force acting on it as shown in Table 4.3.

Applying the first condition for equilibrium yields

$$\sum F_x = 0: \qquad T - (200 \text{ N}) \cos 60° = 0 \qquad \textbf{(4.8)}$$

$$\sum F_y = 0: \qquad n - (200 \text{ N}) \sin 60° = 0 \qquad \textbf{(4.9)}$$

From Eq. (4.8), we obtain

$$T = (200 \text{ N}) \cos 60° = 100 \text{ N}$$

and since tension T in the rope is equal to the weight W_2, we say that a weight of 100 N is needed to maintain equilibrium.

The normal force exerted by the plane on the 200-N block can be found from Eq. (4.9), although this calculation was not necessary to determine the weight W_2.

$$n = (200 \text{ N}) \sin 60°$$
$$= 173 \text{ N}$$

4.7 Friction

Whenever a body moves while it is in contact with another object, **friction forces** oppose the relative motion. These forces are caused by the adhesion of one surface to the other and by the interlocking of irregularities in the rubbing surfaces. It is friction that holds a nail in a board, allows us to walk, and makes automobile brakes work. In all these cases, friction has a desirable effect.

In many other instances, however, friction must be minimized. For example, it increases the work necessary to operate machinery, it causes wear, and it generates heat, which often causes additional damage. Automobiles and airplanes are streamlined to decrease air friction, which is large at high speeds.

Whenever one surface moves past another, the frictional force exerted by each body on the other is parallel or tangent to the two surfaces and acts in such a manner as to oppose relative motion. It is important to note that these forces exist not only when there is relative motion but even when one object only *tends* to slide past another.

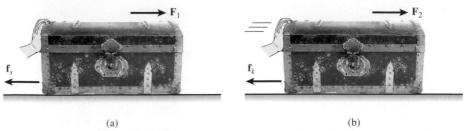

(a)   (b)

Figure 4.12   (a) In static friction, motion is impending. (b) In kinetic friction, the two surfaces are in relative motion. (*Photo by Hemera, Inc.*)

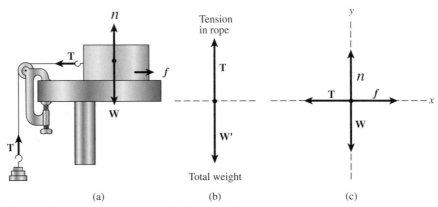

(a)   (b)   (c)

Figure 4.13   Experiment to determine the force of friction.

Suppose a force is exerted on a trunk, as shown in Fig. 4.12. At first the block will not be moved because of the action of a force called the *force of **static friction** f_s*. But as the applied force is increased, motion eventually occurs, and the friction force exerted by the horizontal surface while the truck is moving is called the *force of **kinetic friction** f_k*.

The laws governing friction forces can be determined experimentally in the laboratory by using an apparatus similar to the one shown in Fig. 4.13a. A box of weight W is placed on a horizontal table, and a string attached to the box is passed over a light, frictionless pulley and attached to a weight hanger. All forces acting on the box and hanger are shown in their corresponding free-body diagrams (Fig. 14.13b and c).

Let us consider that the system is in equilibrium, which requires the box to be stationary or moving with a constant velocity. In either case, we may apply the first condition for equilibrium. Consider the force diagram of the box as shown in Fig. 4.13c.

$$\sum F_x = 0: \qquad f - T = 0 \qquad \text{or} \qquad f = T$$
$$\sum F_y = 0: \qquad n - W = 0 \qquad \text{or} \qquad n = W$$

Thus, the force of friction is equal in magnitude to the tension in the string, and the normal force exerted by the table on the box is equal to the weight of the box. Note that the tension in the string is determined by the weight of the hanger plus the weight on the hanger.

Suppose we begin by slowly adding weights to the hanger, thus gradually increasing the tension in the string. As the tension is increased, the equal but oppositely directed force of static friction is also increased. If **T** is increased sufficiently, the box will start to move, indicating that **T** has overcome the *maximum* force of static friction $f_{s,\text{max}}$. Although the force of static friction f_s will vary according to the values of tension in the string, there exists a single maximum value $f_{s,\text{max}}$.

Now, we will continue the experiment by adding weights to the box, thereby increasing the normal force n between the box and the table. Our normal force becomes

$$n = W + \text{added weights}$$

Repeating the above procedure will show that a proportionately larger value for **T** will be necessary to overcome the maximum force of static friction. In other words, if we double the normal force between the two surfaces, the maximum force of static friction that must be overcome is also doubled. If n is tripled, f_s is tripled, and so it will be for other factors. Therefore, it can be said that the maximum force of static friction is directly proportional to the normal force between the two surfaces. We can write this proportionality as

$$f_{s,\text{max}} \propto n$$

The force of static friction is always less than or equal to the maximum force:

$$f_s \le \mu_s n \qquad (4.10)$$

Unless otherwise specified, we will write Eq. (4.10) as an equality and assume that the reference is to the *maximum* value of static friction. The symbol μ_s is a proportionality constant called the **coefficient of static friction.** Since μ_s is the constant ratio of two forces, it is a dimensionless quantity.

In the previous experiment, it will be noticed that after the maximum value of static friction has been overcome, the box will increase its speed, or accelerate, until it is stopped by the pulley. This indicates that a lesser value of **T** would be necessary to keep the box moving with a constant speed. Thus, the force of kinetic friction is smaller than the maximum value of f_s for the two surfaces. In other words, it requires more force to just start a block moving than it does to keep it moving with constant speed. In the latter case, the first condition for equilibrium is still satisfied. Hence, the same reasoning that led to Eq. (4.10) for static friction will yield the following equation for kinetic friction.

$$f_k = \mu_k n \qquad (4.11)$$

where μ_k is a proportionality constant called the **coefficient of kinetic friction.**

The proportionality coefficients μ_s and μ_k can be shown to depend on the roughness of the surfaces but not on the area of contact between the two surfaces. It can be seen from the equations above that μ depends only on the frictional force f and the normal force n between the surfaces. Of course, it must be realized that Eqs. (4.10) and (4.11) are not fundamentally rigorous, like other physical equations. Many variables interfere with the general application of these formulas. No one who has experience in automobile racing, for instance, will believe that the friction force is *completely* independent of the contact area. Nevertheless, the equations are useful tools for estimating resistive forces in specific cases.

Table 4.4 shows some representative values for the coefficients of static and kinetic friction between different types of surfaces. These values are approximate and depend on the condition of the surfaces. For our purposes, however, we will assume these coefficients have three significant figures.

Table 4.4

Approximate Coefficients of Friction

Material	μ_s	μ_k
Wood on wood	0.7	0.4
Steel on steel	0.15	0.09
Metal on leather	0.6	0.5
Wood on leather	0.5	0.4
Rubber on dry concrete	0.9	0.7
Rubber on wet concrete	0.7	0.57

Problem-Solving Strategy

Considerations for Problems Involving Friction

1. Friction forces are parallel to the surfaces and directly oppose motion or impending motion.

2. The maximum force of static friction is usually larger than the force of kinetic friction for the same materials.

3. In drawing free-body diagrams, it is usually better to choose the x axis along the direction of motion and the y axis normal to the direction of motion or impending motion.

4. The first condition for equilibrium can be applied to set up two equations representing forces along and perpendicular to the plane of motion.

5. The relations $f_s = \mu_s n$ and $f_k = \mu_k n$ can be applied to solve for the desired quantity.

6. The **normal force** should never be assumed to be equal to the weight. Determine its magnitude by summing forces along the normal axis.

Example 4.5

A 50-N sled rests on a horizontal surface. A horizontal pull of 10 N is required just to start the sled moving. After motion is started, only a 5-N force is needed to move the sled with a constant velocity. Find the coefficients of static and kinetic friction.

Plan: We recognize the key phrases *just to start moving* and *move with constant speed*. The first implies *static friction;* the last implies *kinetic friction*. In each case, a condition of equilibrium exists, and we can find the values for the normal force and for the friction force, which are needed to find the coefficients.

Solution: We superimpose free-body diagrams over the sketches for each situation as shown in Fig. 4.14a and Fig. 4.14b. Applying the first condition for equilibrium to Fig. 4.14a gives

$$\sum F_x = 0: \qquad 10\ \text{N} - f_s = 0 \qquad \text{or} \qquad f_s = 10\ \text{N}$$

$$\sum F_y = 0: \qquad n - 50\ \text{N} = 0 \qquad \text{or} \qquad n = 50\ \text{N}$$

We can find the coefficient of static friction from Eq. (4.10).

$$\mu_s = \frac{f_s}{n} = \frac{10\ \text{N}}{50\ \text{N}}; \qquad \mu_s = 0.20$$

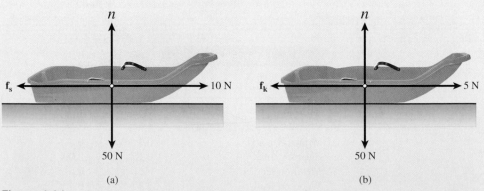

(a)   (b)

Figure 4.14   (a) A force of 10 N is needed to overcome the maximum force of static friction. (b) A force of only 5 N is required to move the sled with constant speed. (*Photo by Hemera, Inc.*)

The force that overcomes kinetic friction is only 5 N. Hence, summing the forces along the x axis yields

$$5\,N - f_k = 0 \qquad \text{or} \qquad f_k = 5\,N$$

The normal force is still 50 N, so that

$$\mu_k = \frac{f_k}{n} = \frac{5\,N}{50\,N}; \qquad \mu_k = 0.10$$

Example 4.6

What force **T** at an angle of 30° above the horizontal is required to drag a 40-lb chest to the right at constant speed if $\mu_k = 0.2$?

Plan: We shall first sketch the problem and then construct a free-body diagram, such as the one shown in Fig. 4.15. Then we will apply the first condition for equilibrium to find the force **T**.

Solution: Motion is at constant speed, so $\Sigma F_x = \Sigma F_y = 0$

$$\sum F_x = 0 \qquad\qquad T_x - f_k = 0 \tag{4.12}$$
$$\sum F_y = 0 \qquad n + T_y - 40\,lb = 0$$

The latter equation shows the normal force to be

$$n = 40\,lb - T_y \tag{4.13}$$

It should be noted that the normal force is decreased by the y component of **T**. Substituting $f_k = \mu_k n$ into Eq. (4.12) gives

$$T_x - \mu_k n = 0$$

But $n = 40\,lb - T_y$ from Eq. (4.13), so

$$T_x - \mu_k(40\,lb - T_y) = 0 \tag{4.14}$$

From the free-body diagram, it is noted that

$$T_x = T \cos 30° = 0.866T$$

and

$$T_y = T \sin 30° = 0.5T$$

Figure 4.15 The force **T** at an angle *above* the horizontal reduces the normal force needed for equilibrium, causing the friction force to be smaller. (*Photos by Hemera, Inc.*)

thus, recalling that $\mu_k = 0.2$, we can write Eq. (4.14) as

$$0.866T - (0.2)(40 \text{ lb} - 0.5T) = 0$$

which can be solved for T as follows:

$$0.866T - 8 \text{ lb} + 0.1T = 0$$
$$0.966T - 8 \text{ lb} = 0$$
$$0.966T = 8 \text{ lb}$$
$$T = \frac{8 \text{ lb}}{0.966} = 8.3 \text{ lb}$$

Therefore, a force of 8.3 lb is required to pull the chest with constant speed if the rope makes an angle of 30° above the horizontal.

Example 4.7

A 120-N concrete block rests on a 30° inclined plane. If $\mu_k = 0.5$, what push **P** parallel to the plane and directed up the plane will cause the block to move (a) up the plane with constant speed and (b) down the plane with constant speed?

Plan: We will first sketch the problem (Fig. 4.16a) and then construct a free-body diagram for each of the two situations. For motion up the plane, we might draw Fig. 4.16b, and for motion down the plane, we would draw Fig. 4.16c. Note that the friction force opposes motion in each case and that we have chosen the x axis along the plane. The required force **P** is found by assuming equilibrium (constant motion) in each case. To be consistent with signs, we consider forces directed *up* the plane to be positive.

Solution (a): Applying the first condition for equilibrium to Fig. 4.16b, we obtain

$$\sum F_x = 0 \qquad P - f_k - W_x = 0 \tag{4.15}$$
$$\sum F_y = 0 \qquad n - W_y = 0 \tag{4.16}$$

From the figure, the x and y components of the weight are

$$W_x = (120 \text{ N}) \cos 60° = 60.0 \text{ N}$$
$$W_y = (120 \text{ N}) \sin 60° = 104 \text{ N}$$

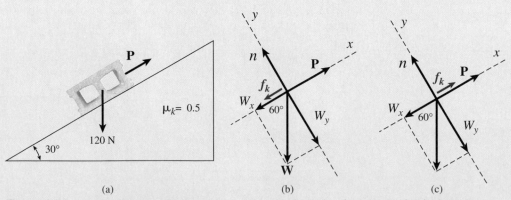

Figure 4.16   (a) Friction on an inclined plane; (b) motion is *up* the plane; (c) motion is *down* the plane. (*Photo by Hemera, Inc.*)

Substituting $W_y = 104$ N into Eq. (4.16) allows us to solve for the normal force n.

$$n - W_y = n - 104 \text{ N} = 0; \quad \text{or} \quad n = 104 \text{ N}$$

From Eq. (4.15), we next solve for the push P, giving

$$P = f_k + W_x$$

But $f_k = \mu_k n$, so that

$$P = \mu_k n + W_x$$

Now, P can be found by substituting $\mu_k = 0.5$, $n = 104$ N, and $W_x = 60.0$ N:

$$P = (0.4)(104 \text{ N}) + 60.0 \text{ N}$$
$$P = 52.0 \text{ N} + 60.0 \text{ N} \quad \text{or} \quad P = 112 \text{ N}$$

Note that the push P *up* the plane in this case must overcome both the 52-N friction force and the 60-N component of the weight *down* the plane.

Solution (b): In the second case, the push P is necessary to retard the natural downward motion of the block until its speed remains constant. The friction force is now directed *up* the incline in the same direction as the push P. The normal force and the weight components will not change. Therefore, summing the forces along the x axis gives

$$\sum F_x = 0; \quad P + f_k - W_x = 0$$

We now may solve for P and substitute the values for f_k and W_x.

$$P = W_x - f_k = 60 \text{ N} - 52 \text{ N}$$
$$P = 8.00 \text{ N}$$

The 8.00 N force directed up the plane and the 52.0 N friction force directed up the plane exactly balance the 60-N component of the weight directed down the plane.

Example 4.8

What is the maximum slope angle θ for an inclined plane such that a block of weight W will not slide down the plane?

Plan: The maximum slope angle would be the one for which the component of the weight directed down the incline would be sufficient to overcome the maximum force of static friction. As usual, our first approach will be to draw a sketch and a free-body diagram (Fig. 4.17). Then, by applying the conditions for equilibrium, we can use trigonometry to solve for the inclination angle.

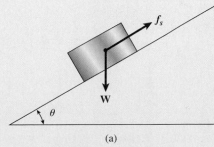
(a)

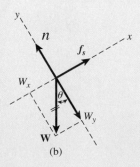
(b)

Figure 4.17   The limiting angle of repose.

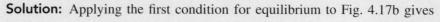

Solution: Applying the first condition for equilibrium to Fig. 4.17b gives

$$\sum F_x = 0: \qquad f_s - W_x = 0 \qquad \text{or} \qquad f_s = W_x$$

$$\sum F_y = 0: \qquad n - W_y = 0 \qquad \text{or} \qquad n_s = W_y$$

From Fig. 4.17b, it is noted that the slope angle θ is the angle adjacent to the negative y axis, making W_x the side opposite and W_y the side adjacent. In this case,

$$\tan\theta = \frac{W_x}{W_y}$$

But we have seen that $W_x = f_k$ and $W_y = n,$ so that

$$\tan\theta = \frac{W_x}{W_y} = \frac{f_s}{n}$$

Finally, we recall that the ratio of f_s to n defines the coefficient of static friction. Hence,

$$\tan\theta = \mu_s$$

Therefore, a block, regardless of its weight, will remain at rest on an inclined plane unless $\tan\theta$ equals or exceeds μ_s. The angle θ in this case is called the *limiting angle,* or the **angle of repose.**

Summary and Review

Summary

In this chapter, we have defined objects that are at rest or in motion with constant speed to be in equilibrium. Through the use of vector diagrams and Newton's laws, we have found it possible to determine unknown forces for systems that are known to be in equilibrium. The following items will summarize the more important concepts to be remembered:

- *Newton's first law of motion* states that an object at rest and an object in motion with constant speed will maintain the state of rest or constant motion unless acted on by a resultant force.
- *Newton's second law of motion* states that the acceleration **a** of an object in the direction of a resultant force **F** is directly proportional to the magnitude of the force and inversely proportional to the mass *m*.
- *Newton's third law of motion* states that every action must produce an equal and opposite reaction. The action and reaction forces do not act on the same body.
- Free-body diagrams: From the conditions of the problem, a neat sketch is drawn and all known quantities are labeled. Then a force diagram indicating all forces and their components is constructed. All information such as that given in Fig. 4.18 should be part of the diagram.
- Translational equilibrium: A body in translational equilibrium has no resultant force acting on it. In such cases, the sum of all the *x* components is zero, and the sum of all the *y* components is zero. This is known as the first condition for equilibrium and is written

$$R_x = \sum F_x = 0 \qquad R_y = \sum F_y = 0$$

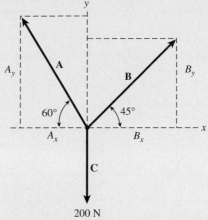

Figure 4.18

- Applying these conditions to Fig. 4.18, for example, we obtain two equations in two unknowns:

$$B \cos 45° - A \cos 60° = 0$$
$$B \sin 45° + A \sin 60° - 200 \text{ N} = 0$$

These equations can be solved to find *A* and *B*.
- *Static friction* exists between two surfaces when motion is impending; *kinetic friction* occurs when the two surfaces are in relative motion. The force of static friction is less than or equal to the *maximum* force of static friction, which is proportional to the normal force. The force of kinetic friction is also proportional to the normal force.

$$f_s \le \mu_s n \qquad f_k = \mu_k n$$

- Friction forces are often considered in equilibrium problems, but they are difficult to quantify, and, in practice, there are many outside factors that may interfere with their strict application.

Key Terms

Review Questions

4.1. A popular stunt consists of placing a coin on a card and the card on the top of a glass. The edge of the card is flipped briskly with the forefinger, causing the card to fly off the top of the glass as the coin drops into the glass. Explain. What law does this illustrate?

4.2. When the head of a hammer becomes loose, you can reseat it by holding the hammer vertically and tapping the base of the handle against the floor. Explain. What law does this illustrate?

4.3. Explain the part played by Newton's third law of motion in the following activities: (a) walking, (b) rowing, (c) rocket launching, and (d) parachuting.

4.4. Can a moving body be in equilibrium? Give several examples.

4.5. According to Newton's third law of motion, every force has an equal and oppositely directed reaction force. Therefore, the concept of a resultant, unbalanced force must be an illusion that does not hold up under close examination. Do you agree with this statement? Give the reasons for your answer.

4.6. A brick is suspended from the ceiling by a light string. A second identical string is attached to the bottom of the brick and hangs within the reach of a student. When the student pulls the lower string slowly, the upper string breaks, but when the lower string is jerked, it breaks. Explain why the string breaks in each case.

4.7. A long steel cable is stretched between two buildings. Show by diagrams and discussion why it is not possible to pull the cable so taut that it will be perfectly horizontal with no sag in the middle.

4.8. We have seen that it is often advantageous to choose the x and y axes so that as many forces as possible are completely specified along an axis. Suppose that no two forces are perpendicular to each other. Will there still be an advantage to rotating axes to align an unknown force with an axis, as opposed to aligning a known force? Test this approach by applying it to one of the text examples.

4.9. Discuss a few beneficial uses of the force of friction.

4.10. Why do we speak of a *maximum* force of static friction? Why do we not discuss a maximum force of kinetic friction?

4.11. Why is it easier to pull a sled at an angle than it is to push a sled at the same angle? Draw free-body diagrams to show what the normal force would be in each case.

4.12. Is the normal force acting on a body always equal to its weight?

4.13. When walking across a frozen pond, should you take short steps or long ones? Why? If the ice were completely frictionless, would it be possible for you to get off the pond? Explain.

Problems

Note: For all of the problems at the end of this chapter, the rigid booms or struts are considered to be of negligible weight. All forces are considered to be concurrent forces.

Section 4.5 Free-Body Diagrams

4.1. Draw a free-body diagram for the arrangements shown in Fig. 4.19a and b. Isolate a point where the important forces are acting, and represent each force as a vector. Determine the reference angle and label the components.

4.2. Study each force acting at the end of the light strut in Fig. 4.20. Draw the appropriate free-body diagram.

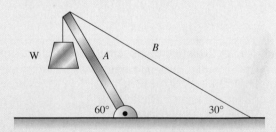

Figure 4.20

Section 4.6 Solution of Equilibrium Problems

4.3. Three identical bricks are strung together with cords and hung from a scale that reads a total of 24 N. What is the tension in the cord that supports the lowest brick? What is the tension in the cord between the middle brick and the top brick?

Ans. 8 N, 16 N

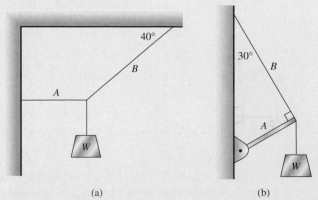

(a)   (b)

Figure 4.19

4.4. A single chain supports a pulley whose weight is 40 N. Two identical 80-N weights are then connected with a cord that passes over the pulley. What is the tension in the supporting chain? What is the tension in each cord?

4.5. If the weight of the block in Fig. 4.19a is 80 N, what are the tensions in ropes *A* and *B?*
Ans. *A* = 95.3 N, *B* = 124 N

4.6. If rope *B* in Fig. 4.19a will break for tensions greater than 200 lb, what is the maximum weight *W* that can be supported?

4.7. If *W* = 600 N in Fig. 4.19b, what is the force exerted by the rope on the end of the boom *A?* What is the tension in rope *B?*
Ans. *A* = 300 N, *B* = 520 N

4.8. If rope *B* in Fig. 4.19a will break if its tension exceeds 400 N, what is the maximum weight *W?*

4.9. What is the maximum weight *W* for Fig. 4.19b if the rope can sustain a maximum tension of only 800 N? Ans. 924 N

4.10. A 70-N block rests on a 35° inclined plane. Determine the normal force and find the friction force that keeps the block from sliding.

4.11. A wire is stretched between two poles 10 m apart. A sign is attached to the midpoint of the line causing it to sag vertically a distance of 50 cm. If the tension in each line segment is 2000 N, what is the weight of the sign? Ans. 398 N

4.12. An 80-N traffic light is supported at the midpoint of a 30-m length of cable between two poles. Find the tension in each cable segment if the cable sags a vertical distance of 1 m.

*4.13. The ends of three 8-ft studs are nailed together, forming a tripod with an apex that is 6 ft above the ground. What is the compression in each of these studs if a 100-lb weight is hung from the apex?
Ans. 44.4 lb

4.14. A 20-N picture is hung from a nail as in Fig. 4.21, so that the supporting cords make an angle of 60°. What is the tension of each cord segment?

Section 4.7 Friction

4.15. A horizontal force of 40 N will just start an empty 600-N sled moving across packed snow. After motion is begun, only 10 N is needed to keep motion at constant speed. Find the coefficients of static and kinetic friction. Ans. 0.0667, 0.0167

4.16. Suppose 200-N of supplies are added to the sled in Prob. 4.15. What new force is needed to drag the sled at constant speed?

4.17. Assume surfaces where μ_s = 0.7 and μ_k = 0.4. What horizontal force is needed to just start a 50-N

Figure 4.21

block moving along a wooden floor? What force will move it at constant speed?
Ans. 35 N, 20 N

4.18. A dockworker finds that a horizontal force of 60 lb is needed to drag a 150-lb crate across the deck at constant speed. What is the coefficient of kinetic friction?

4.19. The dockworker in Prob. 4.18 finds that a smaller crate of similar material can be dragged at constant speed with a horizontal force of only 40 lb. What is the weight of this crate? Ans. 100 lb

4.20. A steel block weighing 240 N rests on a level steel beam. What horizontal force will move the block at constant speed if the coefficient of kinetic friction is 0.12?

4.21. A 60-N toolbox is dragged horizontally at constant speed by a rope making an angle of 35° with the floor. The tension in the rope is 40 N. Determine the magnitude of the friction force and the normal force. Ans. 32.8 N, 37.1 N

4.22. What is the coefficient of kinetic friction for the example in Prob. 4.21?

*4.23. The coefficient of static friction for wood on wood is 0.7. What is the maximum angle for an inclined wooden plane if a wooden block is to remain at rest on the plane? Ans. 35°

*4.24. A roof is sloped at an angle of 40°. What is the maximum coefficient of static friction between the sole of a shoe and the roof to prevent slipping?

*4.25. A 200-N sled is pushed along a horizontal surface at constant speed with a 50-N force that makes an angle of 28° below the horizontal. What is the coefficient of kinetic friction?   Ans. 0.198

*4.26. What is the normal force on the block in Fig. 4.22? What is the component of the weight acting down the plane?

*4.27. What push **P** directed up the plane will cause the block in Fig. 4.22 to move up the plane with constant speed?   Ans. 54.1 N

*4.28. If the block in Fig. 4.22 is released, it will overcome static friction and slide rapidly down the plane. What push **P** directed up the incline will retard the downward motion until the block moves at constant speed?

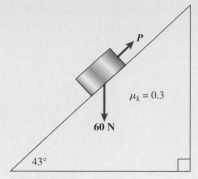

Figure 4.22

Additional Problems

4.29. Determine the tension in rope *A* and the force *B* exerted by the strut on the rope in Fig. 4.23.
   Ans. *A* = 231 N, *B* = 462 N

4.30. If the breaking strength of cable *A* in Fig. 4.24 is 200 N, what is the maximum weight that can be supported by this apparatus?

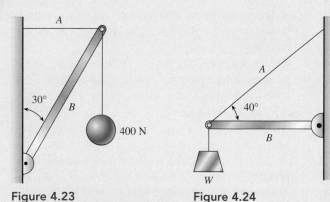

Figure 4.23   **Figure 4.24**

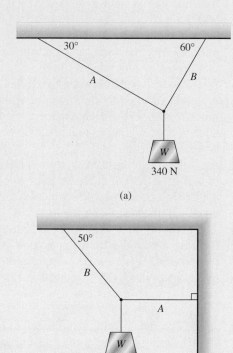

(a)

(b)

Figure 4.25

4.31. What is the minimum push **P** parallel to a 37° inclined plane if a 90-N wagon is to be rolled up the plane at constant speed? Ignore friction.
   Ans. 54.2 N

4.32. A horizontal force of only 8 lb moves a cake of ice with constant speed across a floor ($\mu_k = 0.1$). What is the weight of the ice?

4.33. Find the tension in ropes *A* and *B* for the arrangement shown in Fig. 4.25a.   Ans. 170 N, 294 N

4.34. Find the tension in ropes *A* and *B* in Fig. 4.25b.

4.35. A cable is stretched horizontally across the top of two vertical poles 20 m apart. A 250-N sign suspended from the midpoint causes the rope to sag a vertical distance of 1.2 m. What is the tension in each cable segment?   Ans. 1049 N

4.36. Assume that the cable in Prob. 4.35 has a breaking strength of 1200 N. What is the maximum weight that can be supported at the midpoint?

4.37. Find the tension in the cable and the compression in the light boom for Fig. 4.26a.

Ans. A = 43.2 lb, B = 34.5 lb

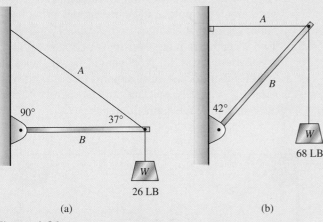

Figure 4.26

4.38. Find the tension in the cable and the compression in the light boom for Fig. 4.26b.

4.39. Determine the tension in ropes A and B for Fig. 4.27a.   Ans. A = 1410 N, B = 1150 N

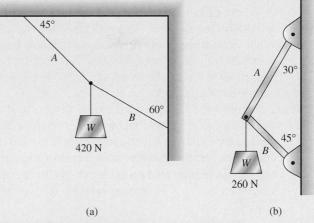

Figure 4.27

***4.40.** Find the forces in the light boards of Fig. 4.27b and state whether the boards are under tension or compression.

Critical Thinking Questions

4.41. Study the structure drawn in Fig. 4.28 and analyze the forces acting at the point at which the rope is attached to the light poles. What is the direction of the forces acting *on* the ends of the poles? What is the direction of the forces exerted *by* the poles at that point? Draw the appropriate free-body diagram.

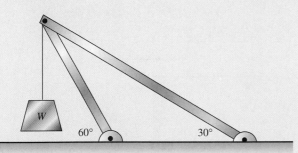

Figure 4.28

***4.42.** Determine the forces acting *on* the ends of the poles in Fig. 4.28 if W = 500 N.

***4.43.** A 2-N eraser is pressed against a vertical chalkboard with a horizontal push of 12 N. If μ_s = 0.25, find the horizontal force required to start motion parallel to the floor. What if you want to start its motion up or down? Find the vertical forces required to just start motion up the board and then down the board.

Ans. 3.00 N, up = 5 N, down = 1 N

***4.44.** It is determined experimentally that a 20-lb horizontal force will move a 60-lb lawn mower at a constant speed. The handle of the mower makes an angle of 40° with the ground. What push along the handle will move the mower at a constant speed? Is the normal force equal to the weight of the mower? What is the normal force?

***4.45.** Suppose the lawn mower of Prob. 4.44 is to be moved backward. What pull along the handle is required to move with constant speed? What is the normal force in this case? Discuss the differences between this example and the one in the previous problem.   Ans. 20.4 N, 46.9 N

***4.46.** A truck is pulled out of mud by a rope attached to the truck and to a tree. When the angles are as shown in Fig. 4.29, a force of 40 lb is exerted at the midpoint of the rope. What force is exerted on the truck?

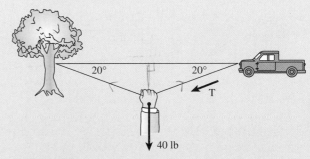

Figure 4.29

*4.47. Suppose a force of 900 N is required to move the truck in Fig. 4.29. What force is required at the midpoint of the line for the angles shown?

Ans. 616 N

4.48. A 70-N block of steel is at rest on a 40° incline. What is the magnitude of the static friction force directed up the plane? Is this necessarily the maximum force of static friction? What is the normal force at this angle?

*4.49. Determine the compression in the center strut *B* and the tension in the rope *A* for the situation described by Fig. 4.30. Distinguish clearly the difference between the compression force in the strut and the force indicated on your free-body diagram.

Ans. A = 643 N, B = 940 N

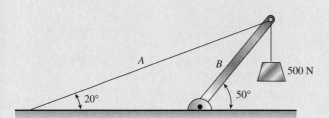

Figure 4.30

*4.50. What *horizontal* push **P** is required to just prevent a 200-N block from slipping down a 60° inclined plane where $\mu_s = 0.4$? Why does it take a lesser force if **P** acts *parallel* to the plane? Is the friction force greater, less, or the same for these two cases?

*4.51. Find the tension in each cord of Fig. 4.31 if the suspended weight is 476 N.

Ans. A = 476 N, B = 275 N, C = 275 N

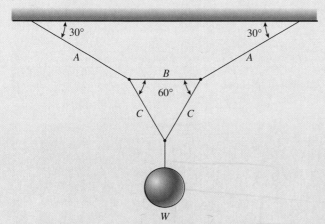

Figure 4.31

*4.52. Find the force required to pull a 40-N sled horizontally at a constant speed by exerting a pull along a pole that makes a 30° angle with the ground ($\mu_k = 0.4$). Now find the force required if you push along the pole at the same angle. What is the major factor that changes in these cases?

*4.53. Two weights are hung over two frictionless pulleys as shown in Fig. 4.32. What weight *W* will cause the 300-lb block to just start moving to the right? Assume $\mu_s = 0.3$. *Note:* The pulleys merely change the direction of the applied forces.

Ans. 108 lb

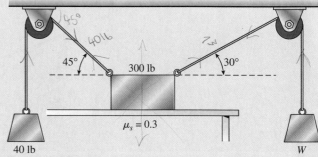

Figure 4.32

*4.54. Find the maximum weight that can be hung at point *O* in Fig. 4.33 without upsetting the equilibrium. Assume that $\mu_s = 0.3$ between the block and table.

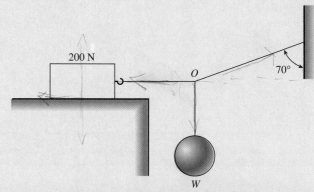

Figure 4.33

5 Torque and Rotational Equilibrium

Golden Gate Bridge: Mechanical engineers must ensure that all forces and torques are balanced in the design and construction of bridges.

(*Photo © vol. 44 PhotoDisc/Getty.*)

Objectives

After completing this chapter, you should be able to

1. Illustrate by example and definition your understanding of the terms *moment arm* and *torque.*
2. Calculate the resultant torque about any axis when given the magnitude and position of forces on an extended object.
3. Solve for unknown forces or distances by applying the first and second conditions for equilibrium.
4. Define and illustrate by example what is meant by the center of gravity.

In previous chapters, we have discussed forces that act at a single point. Translational equilibrium exists when the vector sum of forces is zero. There are many cases, however, in which the forces acting on an object do not have a common point of application. Such forces are said to be *nonconcurrent.* For example, a mechanic exerts a force on the handle of a wrench to tighten a bolt. A carpenter uses a long lever to pry the lid from a wooden box. The engineer considers twisting forces that tend to snap a beam attached to a wall. The steering wheel of an automobile is turned by forces that do not have a common point of application. In such cases, one or more forces may cause a *tendency to rotate* that we will define as *torque.* If we learn to measure or predict the torques produced by certain forces, we can obtain desired rotational effects. If no rotation is desired, there must be no resultant torque. This leads naturally to a condition for **rotational equilibrium** that is important for industrial and engineering applications.

5.1   Conditions for Equilibrium

When a body is in equilibrium, it is either at rest or in uniform motion. According to Newton's first law, only the application of a resultant force can change this condition. We have seen that, if all forces acting on such a body intersect at a single point and their vector sum is zero, the system must be in equilibrium. When a body is acted on by forces that do not have a common **line of action,** it may be in translational equilibrium but not in rotational equilibrium. In other words, it may not move to the right or left or up or down, but it may still rotate. In studying equilibrium, we must consider the point of application of each force as well as its magnitude.

Consider the forces exerted on the lug wrench in Fig. 5.1a. Two equal opposing forces **F** are applied to the right and to the left. The first condition for equilibrium tells us that the vertical and horizontal forces are balanced. Hence, the system is said to be in equilibrium. If the same two forces are applied as shown in Fig. 5.1b, however, the wrench has a definite tendency to rotate. This is true even though the vector sum of the forces is still zero. Clearly, we need a second condition for equilibrium to cover rotational motion. A formal statement of this condition will be given later. First, we need to define some terms.

In Fig. 5.1b, the forces **F** do not have the same *line of action.*

The line of action of a force is an imaginary line extended indefinitely along the vector in both directions.

When the lines of action of forces do not intersect at a common point, rotation may occur about a point called the **axis of rotation.** In our example, the axis of rotation is an imaginary line passing through the bolt perpendicular to the page.

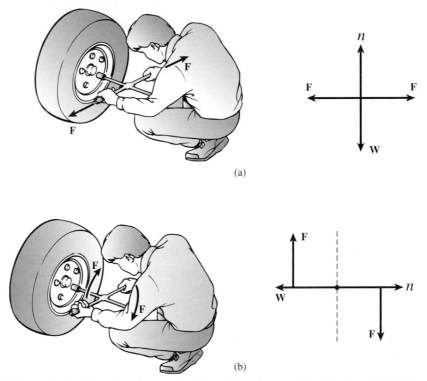

(a)

(b)

Figure 5.1   (a) Equilibrium exists because the forces have the same line of action. (b) Equilibrium does not exist because opposing forces do not have the same line of action.

5.2   The Moment Arm

The perpendicular distance from the axis of rotation to the line of action of a force is called the ***moment arm*** of that force. It is the moment arm that determines the effectiveness of a given force in causing rotational motion. For example, if we exert a force **F** at increasing distances from the center of a large wheel, it becomes easier and easier to rotate the wheel about its center. (See Fig. 5.2.)

> The moment arm of a force is the perpendicular distance from the line of action of the force to the axis of rotation.

If the line of action of a force passes through the axis of rotation (point *A* of Fig. 5.2), the moment arm is zero. No rotational effect is observed, regardless of the magnitude of the

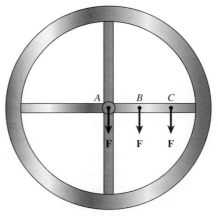

Figure 5.2   The unbalanced force **F** has no rotational effect at point *A* but becomes increasingly effective as the moment arm lengthens.

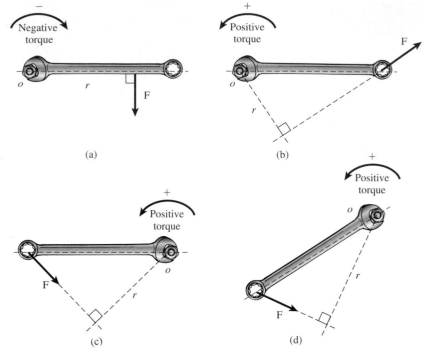

Figure 5.3 Examples of moment arms *r*.

force. In this simple example, the moment arms at points *B* and *C* are simply the distance from the axis of rotation to the point of application of the force. Note, however, that the line of action of a force is a mere geometrical construction. The moment arm is drawn perpendicular to this line. It may be equal to the distance from the axis to the point of application of a force, but this is true only when the applied force is directed perpendicular to this distance. In the examples of Fig. 5.3, *r* represents the moment arm and *O* represents the axis of rotation. Study each example, observing how the moment arms are drawn and reasoning whether the rotation is clockwise or counterclockwise about *O*.

5.3   Torque

PHYSICS TODAY

The International Space Station was put together using a high-tech version of a cordless drill. The battery-powered pistol grip tool, or PGT, can count the number of turns and limit the amount of torque applied to a bolt. NASA requires designers to use only a single type of bolt, one that can easily be grasped by astronauts in EVA (extravehicular activity) suits.

Force has been defined as a push or pull that tends to cause motion. *Torque* τ can be defined as the tendency to produce a change in rotational motion. It is also called the *moment of force* in some textbooks. As we have seen, rotational motion is affected by both the magnitude of a force *F* and its moment arm *r*. Thus, we will define torque as the product of a force and its moment arm.

$$Torque = force \times moment\ arm$$

$$\tau = Fr \tag{5.1}$$

It must be understood that *r* in Eq. (5.1) is measured perpendicular to the line of action of the force **F**. The units of torque are the units of force times distance, for example, *newton-meters* (N · m) and *pound-feet* (lb · ft).

Earlier, we established a sign convention to indicate the direction of forces. The direction of torque depends on whether it tends to produce clockwise (cw) or counterclockwise (ccw) rotation. We shall follow the same convention we used for measuring angles. If the force **F** tends to produce counterclockwise rotation about an axis, the torque will be considered positive. Clockwise torques will be considered negative. In Fig. 5.3, all the torques are positive (ccw) except for that in Fig. 5.3a.

Example 5.1 A force of 250 N is exerted on a cable wrapped around a drum that has a diameter of 120 mm. What is the torque produced about the center of the drum?

Plan: Make a drawing, as in Fig. 5.4, and extend the line of action of the force. Determine the moment arm r and then find the torque from Eq. (5.1).

Solution: Notice that the line of action of the 250-N force is perpendicular to the diameter of the drum. The moment arm is therefore equal to the radius of the drum.

$$r = \frac{D}{2} = \frac{120 \text{ mm}}{2} \qquad \text{or} \qquad r = 60 \text{ mm} = 0.06 \text{ m}$$

The magnitude of the torque is found from Eq. (5.1).

$$\tau = Fr = (250 \text{ N})(0.06 \text{ m}) = 15.0 \text{ N} \cdot \text{m}$$

Finally, we determine that the *sign* of the torque is negative because it tends to cause *clockwise* motion about the center of the drum. Thus, the answer should be written as

$$\tau = -15.0 \text{ N} \cdot \text{m}$$

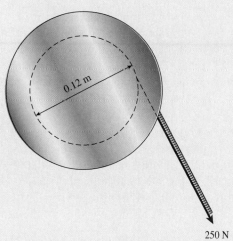

250 N

Figure 5.4 The tangential force exerted by a cable wrapped around a drum.

Example 5.2 A mechanic exerts a 20-lb force at the end of a 10-in. wrench, as shown in Fig. 5.5. If this pull makes an angle of 60° with the handle, what is the torque produced on the nut?

Plan: From the sketch, we will determine the moment arm, multiply it by the magnitude of the force, and then affix the appropriate sign by convention.

Solution: First, we draw a neat sketch, extend the line of action of the 20-lb force, and draw in the moment arm as shown. Note that the moment arm r is perpendicular both to the line of action of the force and to the axis of rotation. You must remember that the moment arm is a geometrical construction and may or may not

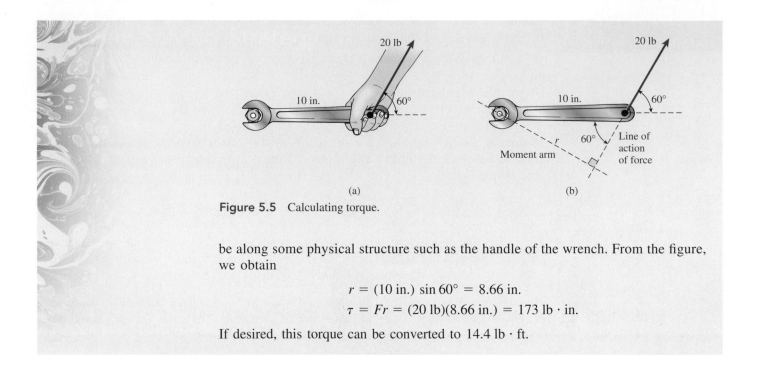

(a)   (b)

Figure 5.5 Calculating torque.

be along some physical structure such as the handle of the wrench. From the figure, we obtain

$$r = (10 \text{ in.}) \sin 60° = 8.66 \text{ in.}$$

$$\tau = Fr = (20 \text{ lb})(8.66 \text{ in.}) = 173 \text{ lb} \cdot \text{in.}$$

If desired, this torque can be converted to 14.4 lb · ft.

In some applications, it is more useful to work with the *components* of a force to obtain the resultant torque. In Example 5.2, for instance, we could have resolved the 20-lb vector into its horizontal and vertical components. Instead of finding the torque of a single force, we would then need to find the torque of two component forces. As shown in Fig. 5.6, the 20-lb vector has components F_x and F_y, which are found from trigonometry:

$$F_x = (20 \text{ lb})(\cos 60°) = 10 \text{ lb}$$

$$F_y = (20 \text{ lb})(\sin 60°) = 17.3 \text{ lb}$$

Notice from Fig. 5.6b that the line of action of the 10-lb force passes through the axis of rotation. It does not produce any torque because its moment arm is zero. The entire torque is, therefore, due to the 17.3-lb component perpendicular to the handle. The moment arm of this force is the length of the wrench, and the torque is

$$\tau = Fr = (17.3 \text{ lb})(10 \text{ in.}) = 173 \text{ lb} \cdot \text{in.}$$

Note that the same result is obtained using this method. No more calculations are required because the horizontal component has a zero moment arm. If we choose the components of a force along and perpendicular to a known distance, we need concern ourselves only with the torque of this perpendicular component.

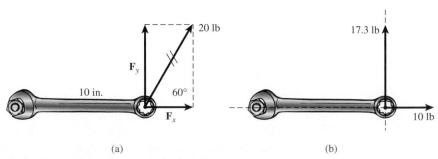

(a)   (b)

Figure 5.6 The component method of calculating torque.

5.4 Resultant Torque

In Chapter 3, we demonstrated that the resultant of a number of forces could be obtained by adding the x and y components of each force to find the components of the resultant.

$$R_x = A_x + B_x + C_x + \cdots \qquad R_y = A_y + B_y + C_y + \cdots$$

This procedure applies to forces that have a common point of intersection. Forces that do not have a common line of action may produce a resultant torque in addition to a resultant translational force. When the applied forces act in the same plane, the resultant torque is the algebraic sum of the positive and negative torques due to each force.

$$\tau_R = \sum \tau = \tau_1 + \tau_2 + \tau_3 + \cdots \qquad (5.2)$$

Remember that counterclockwise torques are positive and clockwise torques are negative.

One essential element of effective problem-solving techniques is organization. The following procedure is useful for calculating resultant torque.

Problem-Solving Strategy

Calculating Resultant Torque

1. Read the problem and then draw and label a figure.

2. Construct a free-body diagram showing all forces, distances, and the axis of rotation.

3. Extend the lines of action of each force using dotted lines.

4. Draw and label the moment arms for each force.

5. Calculate the moment arms, if necessary.

6. Calculate the torques due to each force independent of other forces; make sure to affix the proper sign (ccw = + and cw = −).

7. The resultant torque is the algebraic sum of the torques due to each force. See Eq. (5.2).

Example 5.3

A piece of angle iron is hinged at point A, as shown in Fig. 5.7. Determine the resultant torque at A due to the 60- and 80-N forces acting together.

Plan: Extend the lines of actions for the two forces and determine their moment arms using trigonometry and the given angles. For each force, note whether the tendency

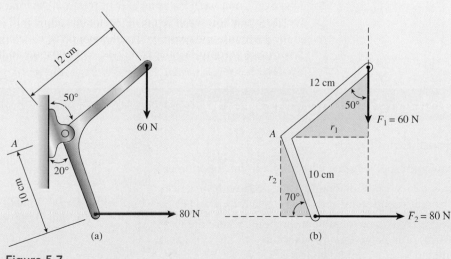

(a)   (b)

Figure 5.7

to rotate about point A will be positive or negative by convention. The resultant torque is the algebraic sum of the individual torques.

Solution: The moment arms r_1 and r_2 are labeled, as shown in Fig. 5.7b. The lengths are

$$r_1 = (12 \text{ cm}) \sin 50° = 9.19 \text{ cm}$$
$$r_2 = (10 \text{ cm}) \sin 70° = 9.40 \text{ cm}$$

Considering A as the axis of rotation, the torque due to $\mathbf{F}_1$ is negative (cw), and the torque due to $\mathbf{F}_2$ is positive (ccw). The resultant torque is found as follows:

$$
\begin{aligned}
\tau_R = \tau_1 + \tau_2 &= F_1 r_1 + F_2 r_2 \\
&= -(60 \text{ N})(9.19 \text{ cm}) + (80 \text{ N})(9.40 \text{ cm}) \\
&= -552 \text{ N} \cdot \text{cm} + 752 \text{ N} \cdot \text{cm} \\
&= 200 \text{ N} \cdot \text{cm}
\end{aligned}
$$

The resultant torque is $200 \text{ N} \cdot \text{cm}$, counterclockwise. This answer is best expressed as $2.00 \text{ N} \cdot \text{m}$ in SI units.

5.5   Equilibrium

We are now ready to discuss the necessary condition for rotational equilibrium. The condition for translational equilibrium was stated in equation form as

$$\sum F_x = 0 \qquad \sum F_y = 0 \tag{5.3}$$

If we are to ensure that the rotational effects are also balanced, we must stipulate that there is no resultant torque. Hence, the second condition for equilibrium is

The algebraic sum of all the torques about any axis must be zero.

$$\sum \tau = \tau_1 + \tau_2 + \tau_3 + \cdots = 0 \tag{5.4}$$

The second condition for equilibrium simply tells us that the clockwise torques are exactly balanced by the counterclockwise torques. Whether or not rotation is occurring, the rotational speed will not change. Moreover, since the rotation is not changing, we may choose whatever point we wish as an axis of rotation. As long as the moment arms are measured to the same point for each force, the resultant torque will be zero. Choosing the axis of rotation at the point of application of an unknown force is a way to simplify problems. If a particular force has a zero moment arm, it does not contribute to torque, regardless of its magnitude.

Problem-Solving Strategy

Rotational Equilibrium

1. Draw and label a sketch with all given information.

2. Draw a free-body diagram (if needed) with distances given between forces.

3. Choose an axis of rotation at the point where the least information is given, for example, at the point of application of an unknown force.

4. Sum the torques due to each force about the chosen axis of rotation and set the result equal to zero.

$$\tau_R = \tau_1 = \tau_2 + \tau_3 + \cdots = 0$$

5. Apply the first condition for equilibrium to obtain two additional equations.

$$\sum F_x = 0 \qquad \sum F_y = 0$$

6. Solve for the unknown quantities.

Example 5.4

A 300-N girl and a 400-N boy stand on a 16-m platform supported by posts *A* and *B*, as described in Fig. 5.8. The platform itself weighs 200 N. What are the forces exerted by the supports on the platform?

Plan: A free-body diagram is drawn (see Fig. 5.8b), clearly showing all forces and distances between forces. If the weight of the board is uniformly distributed, we can consider that the entire weight of the board acts at its geometric center. See *center of gravity* in Section 5.6. The unknown forces can be determined by applying the two conditions for equilibrium.

Solution: Applying the first condition for equilibrium to vertical forces, we obtain

$$\sum F_y = 0; \quad A + B - 300 \text{ N} - 200 \text{ N} - 400 \text{ N} = 0$$

Simplifying this equation gives

$$A + B = 900 \text{ N}$$

Since this equation has two unknowns, we need more information. We will next apply the second condition for equilibrium.

Since rotation is not occurring about any point, we can choose an axis of rotation anywhere we wish. A logical choice would be at a point where one of the unknown forces acts because it would then have a zero moment arm. Let us choose to sum the torques about the *B* support. The second condition for equilibrium gives

$$\sum \tau_B = 0; \quad -A(12 \text{ m}) + (300 \text{ N})(10 \text{ m}) + (200 \text{ N})(4 \text{ m}) - (400 \text{ N})(4 \text{ m}) = 0$$

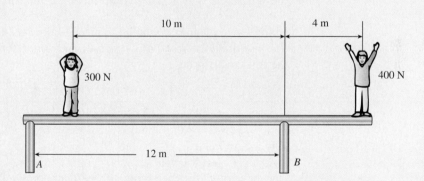

(a) Sketch of the system

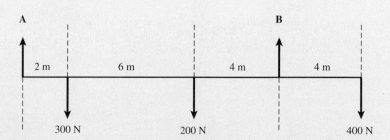

(b) Free-body diagram

Figure 5.8 Draw a free-body diagram indicating all forces and the distances between them. Assume that all the weight of the board acts at its geometric center for purposes of torque calculations.

Note that the 400-N force and the force **A** tend to produce clockwise rotation about *B*. (Their torques were negative.) Simplifying gives

$$-(12 \text{ m})A + 3000 \text{ N} \cdot \text{m} - 1600 \text{ N} \cdot \text{m} + 800 \text{ N} \cdot \text{m} = 0$$

Adding (12 m)*A* to both sides and simplifying, we obtain

$$2200 \text{ N} \cdot \text{m} = (12 \text{ m})A$$

Dividing both sides by 12 m, we find that

$$A = 183 \text{ N}$$

Now, to find the force exerted by support *B,* we can return to the equation obtained from the first condition for equilibrium.

$$A + B = 900 \text{ N}$$

Solving for *B,* we obtain

$$B = 900 \text{ N} - A = 900 \text{ N} - 183 \text{ N}$$
$$= 717 \text{ N}$$

As a check on this solution, we could choose the axis of rotation at *A* and then apply the second condition for equilibrium to find *B*.

Example 5.5

A uniform 500-N boom, 3 m long is supported by a cable, as shown in Fig. 5.9. The horizontal boom is hinged at the wall, and the cable makes a 30° angle with the boom. If a load of 900 N is hung from the right end, what is the tension *T* in the cable? Find the horizontal and vertical components of the force exerted by the hinge.

Plan: Once again we assume that the entire weight of the boom acts at its midpoint. We will draw the free-body diagram and apply both conditions for equilibrium to find the unknown forces.

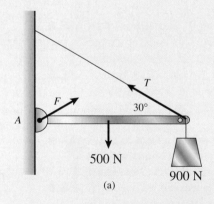

(a)

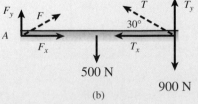

(b)

Figure 5.9   Forces on a horizontal boom.

Solution: When working with forces that are at an angle with the boom, it is sometimes helpful to draw a free-body diagram depicting the components of those forces along or perpendicular to the boom (see Fig. 5.9b). Note that we know neither the magnitude nor the direction of the force **F** exerted by the wall on the left end of the boom. (Do not make the mistake of assuming that the force is exerted entirely along the hinge as it was in Chapter 4 when we neglected the weight of the boom.) The left end is a logical choice for our axis of rotation because, regardless of the angle, that force still has a zero moment arm. Its torque about point A will also be zero.

We will first calculate the cable tension by summing torques about the left end and setting the result equal to zero.

$$F(0) - (500 \text{ N})(1.5 \text{ m}) - (900 \text{ N})(3 \text{ m}) + T_x(0) + T_y(3 \text{ m}) = 0$$

$$0 - 750 \text{ N} \cdot \text{m} - 2700 \text{ N} \cdot \text{m} + 0 + T_y(3 \text{ m}) = 0$$

Simplifying, we obtain an expression for T_y:

$$3T_y = 3450 \text{ N} \qquad \text{or} \qquad T_y = \frac{3450 \text{ N}}{3} = 1150 \text{ N}$$

Now, from Fig. 5.9b, we see that

$$T_y = T \sin 30° \qquad \text{or} \qquad T_y = 0.5T$$

And since $T_y = 1150 \text{ N}$, we write

$$(0.5)T = 1150 \text{ N} \qquad \text{or} \qquad T - \frac{3450 \text{ N} \cdot \text{m}}{0.5(3 \text{ m})} = 2300 \text{ N}$$

Next, we apply the first condition for equilibrium, using the horizontal and vertical components of F and T along with the given forces. The horizontal component F_x of the force exerted by the wall on the boom is found by summing forces along the x axis.

$$\sum F_x = 0 \qquad \text{or} \qquad F_x - T_x = 0$$

from which

$$F_x = T_x = T \cos 30°$$
$$= (2300 \text{ N}) \cos 30° = 1992 \text{ N}$$

The vertical component of the force F_y is found by summing forces along the y axis.

$$\sum F_y = 0 \qquad \text{or} \qquad F_y + T_y - 500 \text{ N} - 900 \text{ N} = 0$$

Solving for F_y, we obtain

$$F_y = 1400 \text{ N} - T_y$$

Using trigonometry, we find T_y from Fig. 5.9b:

$$T_y = (2300 \text{ N}) \sin 30° = 1150 \text{ N}$$

which can be substituted to find F_y.

$$F_y = 1400 \text{ N} - 1150 \text{ N} = 250 \text{ N}$$

It is left as an exercise for you to show that the magnitude and direction of the force **F** from its components is 2010 N at an angle of 7.2° above the horizontal.

Before leaving this section, it would be good to recall the assumptions made about significant figures in this text. In all calculations, we are assuming three significant figures, so you must keep at least four significant figures in all calculations before rounding the final answer. All angles are to be reported to the nearest tenth of one degree. Thus, the force exerted by the wall on the boom is written as 2010 N at 7.2°.

5.6 Center of Gravity

Every particle on the Earth has *weight*. In the case of a body made up of many particles, these forces are essentially parallel and directed toward the center of the Earth. Regardless of the shape and size of the body, there exists a point at which the entire weight of the body may be considered to be concentrated. This point is called the ***center of gravity*** of the body. Of course, all of the weight does not in fact act at this point. But we would calculate the same torque about a given axis if we considered the entire weight to act at that point.

The center of gravity of a regular body, such as a uniform sphere, cube, rod, or beam, is located at its geometric center. This fact was used in the examples of Section 5.5, where we considered the weight of an entire beam as acting at its center. Although the center of gravity is a fixed point, it does not necessarily lie within the body. For example, a hollow sphere, a circular hoop, and a rubber tire all have centers of gravity outside the material of the body.

From the definition of the center of gravity, it is recognized that any body that is suspended at this point will be in equilibrium. This is true because the weight vector, which represents the sum of forces acting on each portion of the body, will have a zero moment arm. Thus, we can compute the center of gravity of a body by determining the point at which an upward force will produce rotational equilibrium.

Example 5.6

Compute the center of gravity of the barbell system drawn in Fig. 5.10. Assume that the weight of the 36-in. connecting rod is negligible.

Plan: The center of gravity is the point where a single upward force **F** will balance the entire system. Superimposing a free-body diagram on the barbells, we draw the upward force **F** at a point located an unknown distance x from a chosen reference point. In this case, the reference point is chosen at the center of the left masses. Finally, we apply the conditions of equilibrium to solve for that distance.

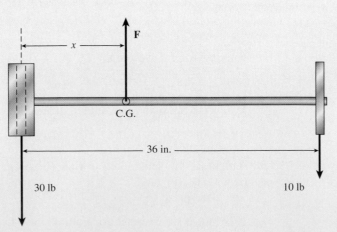

Figure 5.10   Computing the center of gravity.

Solution: Since the resultant force is zero, the upward force **F** must equal the sum of the downward forces, and we may write

$$F = 30 \text{ lb} + 10 \text{ lb} = 40 \text{ lb}$$

The sum of the torques about the geometric center of the left masses must also be equal to zero, so

$$\sum \tau = (40 \text{ lb})x + (30 \text{ lb})(0) - (10 \text{ lb})(36 \text{ in.}) = 0$$

$$(40 \text{ lb})x = 360 \text{ lb} \cdot \text{in.}$$

$$x = 9.00 \text{ in.}$$

If the barbells were suspended from the ceiling at a point 9 in. from the center of the left mass, the system would be in equilibrium. This point is the center of gravity. You should show that the same conclusions would be drawn if you were to choose the axis at the right end or at any other location.

Summary and Review

Summary

When forces acting on a body do not have the same line of action or do not intersect at a common point, rotation may occur. In this chapter, we introduced the concept of torque as a measure of the tendency to rotate. The major concepts are summarized below:

- The *moment arm* of a force is the perpendicular distance from the line of action of the force to the axis of rotation.
- The *torque* about a given axis is defined as the product of the magnitude of a force and its moment arm:

$$Torque = force \times moment\ arm$$

It is positive if it tends to produce counterclockwise motion and negative if the motion produced is clockwise.

$(+)$ ccw

$(-)$ cw

$$\tau = Fr$$

- The *resultant torque* τ_R about a particular axis A is the algebraic sum of the torques produced by each force. The signs are determined by convention.

$$\tau_R = \sum \tau_A = F_1 r_1 + F_2 r_2 + F_3 r_3 + \cdots$$

- *Rotational equilibrium:* A body in rotational equilibrium has no resultant torque acting on it. In such cases, the sum of all the torques about *any* axis must equal zero. The axis may be chosen anywhere because the system is not tending to rotate about any point. This is called the second condition for equilibrium and may be written

The sum of all torques about any point is zero.

$$\sum \tau = 0$$

- *Total equilibrium* exists when the first and second conditions are satisfied. In such cases, three independent equations can be written:

(a) $\sum F_x = 0$ (b) $\sum F_y = 0$ (c) $\sum \tau = 0$

By writing these equations for a given situation, unknown forces, distances, or torques can be found.

- The *center of gravity* of a body is the point through which the resultant weight acts, regardless of how the body is oriented. For applications involving torque, the entire weight of the object may be considered as acting at this point.

Key Terms

axis of rotation   94
center of gravity   104
force   96
line of action   94
moment arm   95
rotational equilibrium   94
torque   96

Review Questions

5.1. You lift a heavy suitcase with your right hand. Describe and explain the position of your body.

5.2. A parlor trick consists of asking you to stand against a wall with your feet together so that the side of your right foot rests against the wall. You are then asked to raise your left foot off the floor. Why can't you do this without falling?

5.3. Why is a minivan more likely to roll over than a Corvette or some other sports car?

5.4. If you know the weight of a brick to be 25 N, describe how you could use a meterstick and a pivot to measure the weight of a baseball.

5.5. Describe and explain the arm and leg motions used by a tightrope walker to maintain balance.

5.6. Discuss the following items and their use of the principle of torque: (a) screwdriver, (b) wrench, (c) pliers, (d) wheelbarrow, (e) nutcracker, and (f) crowbar.

Problems

Section 5.2 The Moment Arm

5.1. Draw and label the moment arm of the force F about an axis at point A in Fig. 5.11a. What is the magnitude of the moment arm?   Ans. 0.845 ft

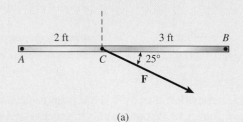

(a)

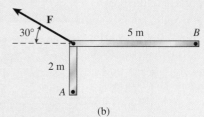

(b)

Figure 5.11

5.2. Find the moment arm about axis B in Fig. 5.11a.

5.3. Draw and label the moment arm if the axis of rotation is at point A in Fig. 5.11b. What is the magnitude of the moment arm?   Ans. 1.73 m

5.4. Find the moment arm about axis B in Fig. 5.11b.

Section 5.3 Torque

5.5. If the force F in Fig. 5.11a is equal to 80 lb, what is the resultant torque about axis A (neglecting the weight of the rod)? What is the resultant torque about axis B?   Ans. −67.5 lb · ft, 101 lb · ft

5.6. The force F in Fig. 5.11b is 400 N, and the angle iron is of negligible weight. What is the resultant torque about axis A and about axis B?

5.7. A leather belt is wrapped around a pulley 20 cm in diameter. A force of 60 N is applied to the belt. What is the torque at the center of the shaft?   Ans. 6 N · m

5.8. The light rod in Fig. 5.12 is 60 cm long and pivoted about point A. Find the magnitude and sign of the torque due to the 200-N force if angle θ is (a) 90°, (b) 60°, (c) 30°, and (d) 0°.

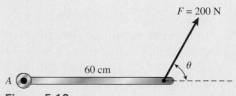

Figure 5.12

5.9. A person who weighs 650 N rides a bicycle. The pedals move in a circle of radius 40 cm. If the entire weight acts on each downward-moving pedal, what is the maximum torque?   Ans. 260 N · m

5.10. A single belt is wrapped around two pulleys. The drive pulley has a diameter of 10 cm, and the output pulley has a diameter of 20 cm. If the top belt tension is essentially 50 N at the edge of each pulley, what are the input and output torques?

Section 5.4 Resultant Torque

5.11. What is the resultant torque about point A in Fig. 5.13? Neglect the weight of the bar.   Ans. 90 N · m

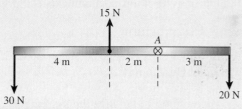

Figure 5.13

5.12. Find the resultant torque in Fig. 5.13 if the axis is moved to the left end of the bar.

5.13. What horizontal force must be exerted at point A in Fig. 5.11b to make the resultant torque about point B equal to zero when the force $F = 80$ N?   Ans. 100 N

5.14. Two wheels of diameters 60 cm and 20 cm are fastened together and turn on the same axis as in Fig. 5.14. What is the resultant torque about a central axis for the shown weights?

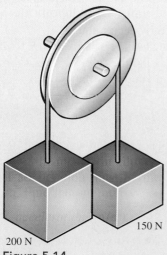

Figure 5.14

5.15. Suppose you remove the 150-N weight from the small wheel in Fig. 5.14. What new weight can you hang to produce zero resultant torque?

Ans. 400 N

5.16. Determine the resultant torque about the corner A for Fig. 5.15.

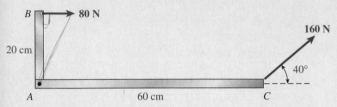

Figure 5.15

5.17. Find the resultant torque about point C in Fig. 5.15.

Ans. -16.0 N $\cdot$ m

*5.18. Find the resultant torque about axis B in Fig. 5.15.

Section 5.5 Equilibrium

5.19. A uniform meterstick is balanced at its midpoint with a single support. A 60-N weight is suspended at the 30-cm mark. At what point must a 40-N weight be hung to balance the system?

Ans. At the 80-cm mark

5.20. Weights of 10 N, 20 N, and 30 N are placed on a meterstick at the 20-cm, 40-cm, and 60-cm marks, respectively. The meterstick is balanced by a single support at its midpoint. At what point may a 5-N weight be attached to produce equilibrium?

5.21. An 8-m board of negligible weight is supported at a point 2 m from the right end, where a 50-N weight is attached. What downward force must be exerted at the left end to produce equilibrium?

Ans. 16.7 N

5.22. A 4-m pole is supported at each end by hunters carrying an 800-N deer, which is hung at a point 1.5 m from the left end. What are the upward forces required by each hunter?

5.23. Assume that the bar in Fig. 5.16 is of negligible weight. Find the forces F and A, provided the system is in equilibrium.

Ans. $A = 26.7$ N, $F = 107$ N

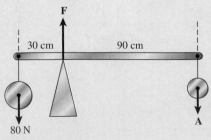

Figure 5.16

5.24. For equilibrium, what are the forces F_1 and F_2 in Fig. 5.17? Neglect the weight of the bar.

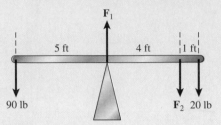

Figure 5.17

5.25. Consider the light bar supported as shown in Fig. 5.18. What are the forces exerted by the supports A and B?

Ans. $A = 50.9$ N, $B = 49.1$ N

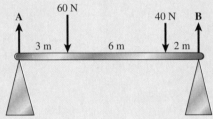

Figure 5.18

5.26. A V-belt is wrapped around a pulley 16 in. in diameter. If a resultant torque of 4 lb ft is required, what force must be applied along the belt?

5.27. A bridge whose total weight is 4500 N is 20 m long and supported at each end. Find the forces exerted at each end when a 1600-N tractor is located 8 m from the left end.　　　Ans. 2890 N, 3210 N

5.28. A 10-ft platform weighing 40 lb is supported at each end by stepladders. A 180-lb painter is located 4 ft from the right end. Find the forces exerted by the supports.

*5.29. A horizontal, 6-m boom weighing 400 N is hinged at the wall, as shown in Fig. 5.19. A cable is attached at a point 4.5 m away from the wall, and a 1200-N weight is attached to the right end. What is the tension in the cable?　　　Ans. 2340 N

*5.30. What are the horizontal and vertical components of the force exerted by the wall on the boom? What are the magnitude and direction of this force?

Section 5.6 Center of Gravity

5.31. A uniform 6-m bar has a length of 6 m and weighs 30 N. A 50-N weight is hung from the left end, and a 20-N force is hung at the right end. How far from the left end will a single upward force produce equilibrium?　　　Ans. 2.10 m

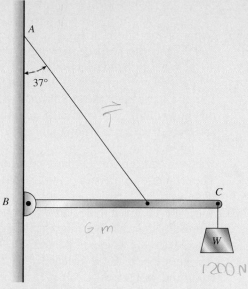

Figure 5.19

5.32. A 40-N sphere and a 12-N sphere are connected by a light rod 200 mm in length. How far from the middle of the 40-N sphere is the center of gravity?

5.33. Weights of 2, 5, 8, and 10 N are hung from a 10-m rod at distances of 2, 4, 6, and 8 m from the left end. Neglecting the weight of the rod, how far from the left end is the center of gravity?

Ans. 6.08 m

5.34. Compute the center of gravity of a sledgehammer if the metal head weighs 12 lb and the 32-in. supporting handle weighs 2 lb. Assume that the handle is of uniform construction and weight.

Additional Problems

5.35. What is the resultant torque about the hinge in Fig. 5.20? Neglect the weight of the curved bar.

Ans. −3.42 N · m

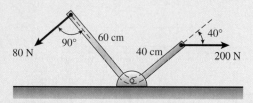

Figure 5.20

5.36. What horizontal force applied to the left end of the curved bar in Fig. 5.20 will produce rotational equilibrium? The bar forms a right angle.

5.37. Weights of 100, 200, and 500 N are placed on a light board resting on two supports, as shown in Fig. 5.21. What are the forces exerted by the supports?

Ans. 375 N, 425 N

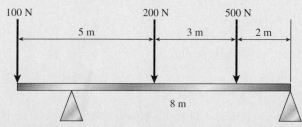

Figure 5.21

5.38. An 8-m steel metal beam weighs 2400 N and is supported 3 m from the right end. If a 9000-N

weight is placed on the right end, what force must be exerted at the left end to balance the system?

*5.39. Find the resultant torque about point A in Fig. 5.22.

Ans. −3.87 N · m

*5.40. Find the resultant torque about point B in Fig. 5.22.

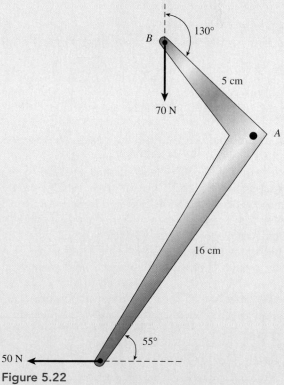

Figure 5.22

Critical Thinking Questions

*5.41. A 30-lb box and a 50-lb box are on opposite ends of a 16-ft board that is supported only at its midpoint. How far from the left end should a 40-lb box be placed to produce equilibrium? Would the result be different if the board weighed 90 lb? Why, or why not?   Ans. 4.00 ft, no

5.42. On a lab bench, you have a small rock, a 4-N meterstick, and a single knife-edge support. Explain how you can use these three items to find the weight of the small rock.

5.43. Find the forces F_1, F_2, and F_3 such that the system drawn in Fig. 5.23 is in equilibrium.
     Ans. F_1 = 213 lb, F_2 = 254 lb, F_3 = 83.9 lb

5.44. (a) What weight W will produce a tension of 400 N in the rope attached to the boom in Fig. 5.24? (b) What would be the tension in the rope if W = 400 N? Neglect the weight of the boom in each case.

5.45. Suppose the boom in Fig. 5.24 has a weight of 100 N and the suspended weight W is equal to 40 N. What is the tension in the cord?
     Ans. T = 234 N

5.46. For the conditions set in Prob. 5.45, what are the horizontal and vertical components of the force exerted by the floor hinge on the base of the boom?

*5.47. What is the tension in the cable for Fig. 5.25? The weight of the boom is 300 N, but its length is unknown.   Ans. 360 N

*5.48. What are the magnitude and direction of the force exerted by the wall on the boom in Fig. 5.25? Again, assume that the weight of the board is 300 N.

*5.49. An automobile has a distance of 3.4 m between front and rear axles. If 60 percent of the weight rests on the front wheels, how far is the center of gravity located from the front axle?
     Ans. 1.36 m

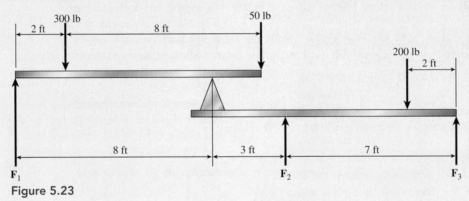

Figure 5.23

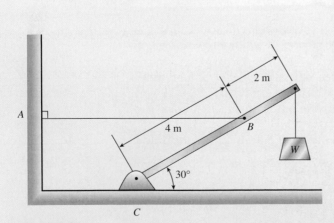

Figure 5.24

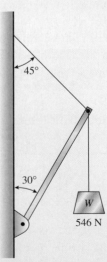

Figure 5.25

6

Uniform Acceleration

The cheetah: A cat that is built for speed. Its strength and agility allow it to attain a top speed of over 100 km/h. Such speeds can be sustained for about only 10 seconds. (*Photo © vol. 44 PhotoDisc/Getty.*)

Objectives

After completing this chapter, you should be able to

1. Define and apply definitions for average velocity and average acceleration.
2. Solve problems involving time, displacement, average velocity, and average acceleration.
3. Apply one of the five general equations for uniform acceleration to solve for one of the five parameters: initial velocity, final velocity, acceleration, time, and displacement.
4. Solve general problems involving free-falling bodies in a gravitational field.
5. Explain with equations and diagrams the horizontal and vertical motion of a projectile launched at various angles.
6. Determine the position and velocity of a projectile when its initial velocity and position are given.
7. Determine the range, the maximum height, and the time of flight for projectiles when the initial velocity and angle of projection are given.

6.1   Speed and Velocity

The simplest kind of motion an object can experience is uniform motion in a straight line. If the object covers the same distances in each successive unit of time, it is said to move with constant speed. For example, if a train covers 8 m of track every second that it moves, we say that it has a **constant speed** of 8 m/s. Whether the speed is constant or not, the average speed of a moving object is defined by

$$Average\ speed = \frac{distance\ traveled}{time\ elapsed}$$

$$\bar{v} = \frac{x}{t} \tag{6.1}$$

The bar over the symbol v means that the speed represents an average value for the time interval t.

Remember that the dimension of speed is the ratio of a length to a time interval. Hence, the units of miles per hour, feet per second, meters per second, and centimeters per second are all typical units of speed.

Example 6.1

A golfer sinks a putt 3 s after the ball leaves the club face. If the ball traveled with an average speed of 0.8 m/s, how long was the putt?

Solution: Solving Eq. (6.1) for x, we have

$$x = \bar{v}t = (0.8\ \text{m/s})(3\ \text{s})$$

Therefore, the distance of the putt is

$$x = 2.4\ \text{m}$$

It is important to recognize that speed is a scalar quantity that is completely independent of direction. In Example 6.1, it was not necessary for us to know either the speed of the golf ball at any instant or the nature of its path. Similarly, the **average speed** of a car traveling from Atlanta to Chicago is a function only of the distance registered on its odometer and the time required to make the trip. It makes no difference as far as computation is concerned whether the driver of the car took the direct or scenic route or even if he or she stopped for meals.

We must make a clear distinction between the scalar quantity *speed* and its directional counterpart *velocity.* This is best done by recalling the difference between *distance* and *displacement,* as discussed in Chapter 3. Suppose, in Fig. 6.1, an object moves along the

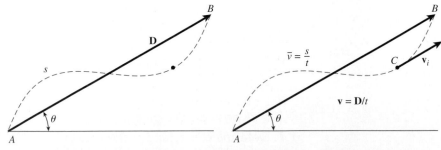

Figure 6.1   Displacement and velocity are vector quantities, whereas distance and speed are independent of direction: s, distance; $\mathbf{D}$, displacement; $\mathbf{v}$, velocity; t, time.

broken path from A to B. The actual distance traveled is denoted by s, whereas the displacement is represented by the polar coordinates

$$\mathbf{D} = (D, \theta)$$

As an example, suppose the distance s in Fig. 6.1 is 500 km, and the displacement is 350 km at 45°. If the actual traveling time is 8 h, the average speed is

$$\bar{v} = \frac{s}{t} = \frac{500 \text{ km}}{8 \text{ h}} = 62.5 \text{ km/h}$$

In this text, we will follow the practice of using the symbol s for curved paths and the symbols x or y for straight-line distances.

The average *velocity*, however, must consider the displacement magnitude and direction. The average velocity is given by

$$\bar{\mathbf{v}} = \frac{\mathbf{D}}{t} = \frac{350 \text{ km}, 45°}{8 \text{ h}}$$

$$\bar{\mathbf{v}} = 43.8 \text{ km/h}, 45°$$

Therefore, if the path of the moving object is curved, the difference between speed and velocity is one of magnitude as well as direction.

Automobiles cannot always travel at constant speeds for long periods of time. In traveling from point A to B, we may be required to slow down or speed up because of road conditions. For this reason, it is sometimes useful to talk of **instantaneous speed** or **instantaneous velocity.**

> The instantaneous speed is a scalar quantity representing the speed at the instant the car is at an arbitrary point C. It is, therefore, the time rate of change in distance.

> The instantaneous velocity is a vector quantity representing the velocity v_i at any point C. It is the time rate of change in displacement.

In this chapter, we will be concerned with motion along a straight line so that the magnitudes of speed and velocity are the same at any instant. If the direction does not change, the instantaneous speed is the scalar part of the the instantaneous velocity. It is good practice, however, to reserve the term *velocity* for the more complete description of motion. As we shall see in the coming sections, a change in velocity may result in a change of direction as well. In such cases, the terms *velocity* and *displacement* are more appropriate than the terms *speed* and *distance*.

PHYSICS TODAY

In front-end collisions, airbags have proven useful in preventing injuries to the head and chest. In crashes with a sudden decrease in velocity of 10 to 15 mi/h, a sensor device located at the front of the vehicle triggers an igniter that causes sodium azide pellets to decompose. This produces nitrogen gas that inflates the nylon airbags and forces them out of their storage compartments. The time between impact and airbag inflation is less than 40 ms. When the occupant and the airbag make contact, the gas is forced out, deflating the bag in 2 s.

6.2 Acceleration

In most cases, the velocity of a moving object changes as motion continues. Motion in which the magnitude or direction changes with time is called *acceleration*. For example, suppose we observe the motion of a sprinter for a period of time t. The initial velocity $\mathbf{v}_0$ of an object is defined as the velocity at the beginning of the time interval (usually $t = 0$). The final velocity is defined as the velocity at the end of the time interval (when $t = t$). Thus, if we are able to measure initial and final velocities of a moving object, we can say that its acceleration is given by

$$Acceleration = \frac{change \text{ } in \text{ } velocity}{time \text{ } interval}$$

$$\mathbf{a} = \frac{\mathbf{v}_f - \mathbf{v}_0}{t} \qquad \qquad (6.2)$$

Acceleration written in the manner of Eq. (6.2) is a vector quantity and thus depends on changes in direction as well as changes in magnitude. If the direction does not change and the motion is in a straight line, only the *speed* of the object changes. However, if the path is curved, there must be acceleration, even if the speed is not changed. For motion in a perfect circle and constant speed, the acceleration will be at right angles to the velocity. Such uniform circular motion will be treated later.

6.3   Uniform Acceleration

The simplest kind of acceleration is motion in a straight line in which the speed changes at a constant rate. This special kind of motion is generally called **uniform acceleration** or **constant acceleration.** Since the direction does not change, the vector difference in Eq. (6.2) becomes simply the difference between the signed values of final and initial velocities. However, it must be remembered that velocity is still a *vector* quantity and that signs affixed to velocity denote *direction* rather than *magnitude.* For constant acceleration, we can write

$$a = \frac{v_f - v_0}{t} \tag{6.3}$$

As an example, consider the car in Fig. 6.2, which moves with constant acceleration from an initial speed of 12 m/s to a final speed of 22 m/s. Considering the right direction as positive, the car's velocity at A is $+12$ m/s and its final velocity at B is $+22$ m/s. If the increase in velocity requires 5 s, the acceleration can be found from Eq. (6.2).

$$a = \frac{v_f - v_0}{t} = \frac{22 \text{ m/s} - 12 \text{ m/s}}{5 \text{ s}}$$

$$= \frac{10 \text{ m/s}}{5 \text{ s}} = 2 \text{ m/s}^2$$

The answer is read as *two meters per second per second* or as *two meters per second squared.* The meaning is that each second, the car increases its speed by 2 m/s. Since the car began at 12 m/s, its speed after times of 1, 2, and 3 s, would be 14, 16, and 18 m/s, respectively.

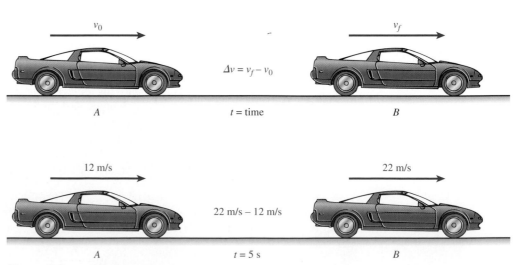

Figure 6.2   Uniformly accelerated motion.

Example 6.2

A train reduces its velocity from 60 km/h to 20 km/h in a time of 8 s. Find the acceleration in SI units.

Plan: The velocities must be converted to SI units (m/s). Then we recognize that acceleration is the change in velocity per unit of time.

Solution: The initial velocity is

$$60 \, \frac{km}{h} \times \frac{1000 \, m}{1 \, km} \times \frac{1 \, h}{3600 \, s} = 16.7 \, m/s$$

Similarly, we find that 20 km/h is equal to 5.56 m/s. Since the velocities are in the same direction and acceleration is assumed to be constant, Eq. (6.3) gives

$$a = \frac{v_f - v_0}{t} = \frac{5.56 \, m/s - 16.7 \, m/s}{8 \, s}$$

$$a = -1.39 \, m/s^2$$

Since the original direction of the train in Example 6.2 was considered positive, the negative sign for acceleration means that the train reduced its speed by 1.39 m/s every second. Such motion is sometimes called *deceleration,* but this term is problematic because $a = -1.39 \, m/s^2$ really means that the velocity *becomes more negative* by that amount every second. If the speed increases in a negative direction, the acceleration is also negative. Acceleration refers to the *change* in velocity, which can mean either an increase or a decrease in speed.

Many times the same equation is used to solve for different quantities. You should therefore solve each equation literally for each symbol in the equation. A convenient form for Eq. (6.3) arises when it is solved explicitly for the final velocity. Thus,

Final velocity = initial velocity + change in velocity

$$v_f = v_0 + at \qquad\qquad (6.4)$$

Example 6.3

An automobile maintains a constant acceleration of 8 m/s². If its initial velocity was 20 m/s due north, what will its velocity be after 6 s?

Plan: The initial velocity will be increased by 8 m/s every second the automobile travels. To get the final velocity, we need only add this change to the initial velocity.

Solution: The final velocity is obtained from Eq. (6.4).

$$v_f = v_0 + at = 20 \, m/s + (8 \, m/s^2)(6 \, s)$$
$$= 20 \, m/s + 48 \, m/s = 68 \, m/s$$

Hence, the final velocity is 68 m/s, also due north.

Now that the concept of initial and final velocities in understood, let us return to the equation for the *average* velocity and express it in terms of initial and final values. As long as the acceleration is constant, the average velocity of an object can be found just like the arithmetic mean of two numbers. Given an initial velocity and a final velocity, the average velocity is simply

$$\overline{v} = \frac{v_f + v_0}{2} \qquad\qquad (6.5)$$

You will recall that the distance x is the product of average velocity and time. Therefore, we can substitute from Eq. (6.5) to obtain a more useful expression for computing distance when the acceleration is uniform.

$$x = \left(\frac{v_f + v_0}{2} \right) t \tag{6.6}$$

Example 6.4

A moving object increases its velocity uniformly from 20 to 40 m/s in 2 min. What is its average velocity, and how far did it travel in the 2 min?

Plan: First, we convert the time of 2 min to 120 s for consistency of units. Then, we recognize that the average velocity is midway between the initial and final values for constant acceleration. Finally, the distance traveled will be the product of average velocity and time.

Solution: The average velocity is found from Eq. (6.5).

$$\bar{v} = \frac{v_f + v_0}{2} = \frac{40 \text{ m/s} + 20 \text{ m/s}}{2}$$

$$\bar{v} = 30 \text{ m/s}$$

Equation (6.6) is applied to find the distance traveled for the 120 s.

$$x = (30 \text{ m/s})(120 \text{ s}) = 3600 \text{ m}$$

6.4   Other Useful Relations

Thus far, we have presented two fundamental relations. One arises from the definition of velocity and the other from the definition of acceleration. They are

$$x = \bar{v}t = \left(\frac{v_f + v_0}{2} \right) t \tag{6.6}$$

and

$$v_f = v_0 + at \tag{6.4}$$

Although these are the only formulas necessary to attack the many problems presented in this chapter, three other useful relationships can be obtained from them. The first is derived by eliminating the final velocity from Eqs. (6.6) and (6.4). Substituting the latter into the former yields

$$x = \left[\frac{(v_0 + at) + v_0}{2} \right] t$$

Simplifying gives

$$x = v_0 t + \frac{1}{2} a t^2 \tag{6.7}$$

A similar equation is obtained by eliminating v_0 from the same two equations:

$$x = v_f t - \frac{1}{2} a t^2 \tag{6.8}$$

The third additional equation comes from the elimination of time t from the basic equations. A little algebra yields

$$2ax = v_f^2 - v_0^2 \qquad\qquad (6.9)$$

Although these equations add no new information, they are useful for solving problems in which three of the parameters are given and one of the other two is to be found.

6.5 Solutions to Acceleration Problems

Although the solution to problems involving constant acceleration depends primarily on choosing the correct formula and substituting known values, there are several suggestions to help the beginning student. Problems frequently occur with motion that either started from rest or was brought to a stop from some initial velocity. In either case, the formulas discussed can be simplified by the substitution of either $v_0 = 0$ or $v_f = 0$, as the case may be. Table 6.1 summarizes the general formulas.

Table 6.1

Summary of Acceleration Formulas

(1) $x = \left(\dfrac{v_0 + v_f}{2}\right)t$ (4) $x = v_f t - \dfrac{1}{2}at^2$

(2) $v_f = v_0 + at$ (5) $2ax = v_f^2 - v_0^2$

(3) $x = v_0 t + \dfrac{1}{2}at^2$

A close look at five general equations will reveal a total of five parameters: x, v_0, v_f, a, and t. Given any three of these quantities, the remaining two can be found from the general equations. Therefore, a starting point in solving any problem is to read it thoroughly with a view to establishing the three quantities required for solution. It is also important to choose a direction to call positive and apply it consistently to velocity, displacement, and acceleration when inserting the values into equations.

If you have difficulty in deciding which equation should be used, it may help to recall the conditions such an equation must satisfy. First, it must contain the unknown parameter. Second, all other parameters that appear in the equation must be known. For example, if a problem gives you values for v_f, v_0, and t, you may solve for a in Eq. (2) of Table 6.1.

Problem-Solving Strategy

Constant Acceleration Problems

1. Read the problem and then draw and label a rough sketch.

2. Indicate the consistent positive direction.

3. Establish three given parameters and two that are unknown. Make sure the signs and units are consistent.

Given: _____   Find: _____

_____   _____

_____

4. Select the equation that contains one of the unknown parameters but not the other one.

$$x = \left(\dfrac{v_0 + v_f}{2}\right)t \qquad 2ax = v_f^2 - v_0^2$$

$$v_f = v_0 + at \qquad x = v_f t - \dfrac{1}{2}at^2$$

$$x = v_0 t + \dfrac{1}{2}at^2$$

5. Substitute known quantities and solve the equation.

The examples that follow are abbreviated and do not include figures, but they do illustrate the approach.

Example 6.5

A motorboat starting from rest attains a velocity of 15 m/s in a time of 6 s. What was the acceleration, and how far did the boat travel?

Plan: We will draw and label a sketch indicating the positive direction as consistent with the initial velocity. Then we will organize the given data, choose the appropriate equations, and solve for the acceleration and the distance traveled.

Solution: All given parameters are positive in this instance.

$$\text{Given: } v_0 = 0 \qquad \text{Find: } a = ?$$
$$v_f = 15 \text{ m/s} \qquad x = ?$$
$$t = 6 \text{ s}$$

In solving for acceleration, we must choose an equation that contains a but not x. Equation (2) in Table 6.1 can be used, where $v_0 = 0$. Hence,

$$v_f = 0 + at \qquad \text{or} \qquad v_f = at$$

Solving for the acceleration a, we have

$$a = \frac{v_f}{t} = \frac{15 \text{ m/s}}{6 \text{ s}}$$
$$= 2.50 \text{ m/s}^2$$

The displacement can be found from an equation that contains x and not a. Equation (1) in Table 6.1 yields

$$x = \left(\frac{v_f + v_0}{2}\right)t = \frac{(15 \text{ m/s} + 0)(6 \text{ s})}{2}$$
$$x = 45.0 \text{ m}$$

Note that because we now know the acceleration a, we could also have solved for x by using Eqs. (3), (4), or (5). However, that would require using the *calculated* value for a, which might be incorrect. It's better to stay with the original information.

Example 6.6

An airplane lands on a carrier deck with an initial velocity of 90 m/s and is brought to a stop in a distance of 100 m. Find the acceleration and the stopping time.

Plan: The procedure is the same as for previous problems. We will take care in selecting the equation that contains only the original information.

Solution:

$$\text{Given: } v_0 = 90 \text{ m/s} \qquad \text{Find: } a = ?$$
$$v_f = 0 \text{ m/s} \qquad t = ?$$
$$x = 100 \text{ m}$$

Looking at Table 6.1, we select Eq. (5) as the one that contains a and not t.

$$2ax = v_f^2 - v_0^2$$
$$a = \frac{v_f - v_0}{t} = \frac{(0)^2 - (90 \text{ m/s})^2}{2(100 \text{ m})}$$
$$a = -40.5 \text{ m/s}^2$$

The negative acceleration is due to the fact that the stopping force is directed opposite to the initial velocity. A person encountering such acceleration would experience a stopping force equal to a little more than four times their weight.

Next, we find the stopping time by selecting the equation that contains t and not a. Once again, Eq. (1) is the appropriate choice.

$$x = \left(\frac{v_f + v_0}{2}\right)t \quad \text{or} \quad t = \frac{2x}{v_f + v_0}$$

$$t = \frac{2(100 \text{ m})}{0 + 90 \text{ m/s}} = 2.22 \text{ s}$$

The airplane experiences an acceleration of -40.5 m/s^2 and stops in a time of 2.22 s.

Example 6.7

A train traveling initially at 16 m/s undergoes a constant acceleration of 2 m/s^2 in the same direction. How far will it travel in 20 s, and what will be its final velocity?

Plan: Organize the data and solve for the unknowns.

Solution:

$$\text{Given: } v_0 = 16 \text{ m/s} \qquad \text{Find: } x = ?$$
$$a = 2 \text{ m/s}^2 \qquad\qquad v_f = ?$$
$$t = 20 \text{ s}$$

Selecting Eq. (3) from Table 6.1 as the one containing x and not v_f, we have

$$x = v_0 t + \frac{1}{2}at^2$$

$$= (16 \text{ m/s})(20 \text{ s}) + \frac{1}{2}(2 \text{ m/s}^2)(20 \text{ s})^2$$

$$= 320 \text{ m} + 400 \text{ m} = 720 \text{ m}$$

The final velocity is found from Eq. (2):

$$v_f = v_0 + at$$
$$= 16 \text{ m/s} + (2 \text{ m/s}^2)(20 \text{ s}) = 56.0 \text{ m/s}$$

The train travels for a distance of 720 m and attains a velocity of 56 m/s.

6.6 Sign Convention in Acceleration Problems

The signs of acceleration a, displacement x, and velocity v are mutually independent and are each determined by different criteria. Probably no other point is more confusing for beginning students. Whenever the direction of motion changes, as with an object tossed into the air or an object attached to an oscillating spring, the signs of displacement and acceleration are particularly difficult to visualize. It is useful to remember that *only* the sign of velocity is dependent on the direction of motion. The sign of displacement depends on the location or position of the object, and the sign of acceleration is determined by the force that causes the velocity to change.

Consider, for example, a baseball tossed upward, as shown in Fig. 6.3. The ball moves upward along a straight line until it stops and returns in a downward path along the same line. We will take the point of release as zero for displacement ($y = 0$). Now, the sign of displacement will be *positive* at any point *above* the release point and *negative* at any point *below* the release point. It doesn't matter whether the ball is moving *upward* or *downward;* it is only the *location* (the y coordinate of its position) that determines the sign of displacement. The value of y may be $+1$ m on the way up and $+1$ m on the way back down. Its displacement will become negative only if the ball is physically located *below* its release point.

$y = +, v = 0, a = -$

$y = +, v = +, a = -$ $y = +, v = -, a = -$

UP = +

Release point $y = 0, v = -, a = -$

$y = -, v = -, a = -$

Figure 6.3
(*Photo by Paul E. Tippens.*)

Observe the signs of the velocity during the flight of the ball. Assuming the upward direction to be positive, the velocity of the ball is positive any time its motion is directed upward and negative any time its motion is directed downward. It doesn't matter that the velocity is changing with time, nor does its location in space matter.

Finally, consider the acceleration of the ball during its flight. The only force acting on the ball during its flight is its weight, which is always directed downward. Thus, the sign of the acceleration is negative (downward) during the entire motion. Note that the acceleration is negative when the ball is moving in an upward direction and also as it moves in a downward direction. In essence, the velocity is always becoming more negative. Even when its velocity passes through zero at the top, the acceleration is still constant and in a downward direction. To determine whether the acceleration of an object is positive or negative, we must not look at its location or its direction of motion; we must look, instead, at the direction of the force that is causing the velocity to change. In this example, that force is its weight.

Once a choice is made for a positive direction, the following conventions will determine the signs of *velocity, displacement,* and *acceleration:*

Displacement is positive or negative depending on the location or coordinate of position of the object relative to its zero position.

Velocity is positive or negative depending on whether the direction of motion is with or against the chosen positive direction.

Acceleration is positive or negative depending on whether the resultant force is with or against the positive direction.

6.7   Gravity and Free-Falling Bodies

Much of our knowledge about the physics of falling bodies originated with the Italian scientist Galileo Galilei (1564–1642). He was the first to deduce that in the absence of friction, all bodies, large or small, heavy or light, fall to the earth with the same acceleration. This was a revolutionary idea, for it contradicted what a person might expect. Until the time of Galileo, people followed the teachings of Aristotle—that heavy objects fall proportionally faster than lighter objects. The classic explanation for the paradox rests with the fact that heavier bodies are proportionately more difficult to accelerate. This resistance to a change in motion is a property of a body called its *inertia.* Thus, in a vacuum, a feather will fall at the same rate as a steel ball because the larger inertial effect of the steel ball compensates exactly for its larger weight. (See Fig. 6.4.)

In the treatment of falling bodies given in this chapter, the effects of air friction are neglected entirely. Under these circumstances, *gravitational acceleration* is uniformly accelerated motion. At sea level and 45° latitude, this acceleration has been measured to be 32.17 ft/s^2, or 9.806 m/s^2, and is denoted by g. For our purposes, the following values will be sufficiently accurate:

Figure 6.4   All bodies fall with the same acceleration in a vacuum.

$$g = \pm 9.80 \text{ m/s}^2$$
$$g = \pm 32.0 \text{ ft/s}^2$$

(6.10)

Since gravitational acceleration is constant acceleration, the same general equations of motion apply. One of the parameters is known in advance, however, and need not be stated in the problem. If the constant g is inserted into the general equations (Table 6.1), the following modified forms will result:

(1a) $y = \dfrac{v_f + v_0}{2}t \qquad y = \bar{v}t$

(2a) $v_f = v_0 + gt$

(3a) $y = v_0 t + \dfrac{1}{2}gt^2$

(4a) $y = v_f t - \dfrac{1}{2}gt^2$

(5a) $2gy = v_f^2 - v_0^2$

Before these equations are used, a few general comments are in order. In problems dealing with free-falling bodies, it is extremely important to choose a direction to call positive and to follow through consistently in the substitution of known values. The sign of the answer is necessary to determine the location of a point or the direction of the velocity at specific times. For example, the distance y in the previous equations represents the displacement above or below the origin. If the upward direction is chosen as positive, a positive value for y indicates a displacement above the starting point; if y is negative, it represents a displacement below the starting point. Similarly, the signs of v_0, v_f, and g indicate their directions.

Example 6.8

A rubber ball is dropped from rest, as shown in Fig. 6.5. Find its velocity and position after 1, 2, 3, and 4 s.

Plan: Since all parameters will be measured in a *downward* direction, it is more convenient to choose downward as *positive*. In that way, all parameters will be positive.

Solution: Organizing the data, we have

$$\text{Given: } v_0 = 0 \qquad\qquad \text{Find: } v_f = ?$$
$$g = +9.80 \text{ m/s}^2 \qquad\qquad y = ?$$
$$t = 1, 2, 3, 4 \text{ s}$$

The downward velocity as a function of time is given by Eq. (2a), where $v_0 = 0$.

$$v_f = v_0 + gt = 0 + gt$$
$$= (9.80 \text{ m/s}^2)t$$

After 1 s, we have

$$v_f = (9.80 \text{ m/s}^2)(1 \text{ s}) = 9.80 \text{ m/s} \qquad \text{(downward)}$$

Substitutions of $t = 2, 3,$ and 4 s give final velocities of 19.6 m/s, 29.4 m/s, and 39.2 m/s, respectively. Each velocity is positive because the downward direction was chosen as positive.

The positive y as a function of time is calculated from Eq. (3a). Since the initial velocity is zero, we write

$$y = v_0 t + \frac{1}{2}gt^2 = \frac{1}{2}gt^2$$

After a time of 1 s, the downward displacement will be

$$y = \frac{1}{2}(9.80 \text{ m/s}^2)(1 \text{ s})^2 = 4.90 \text{ m}$$

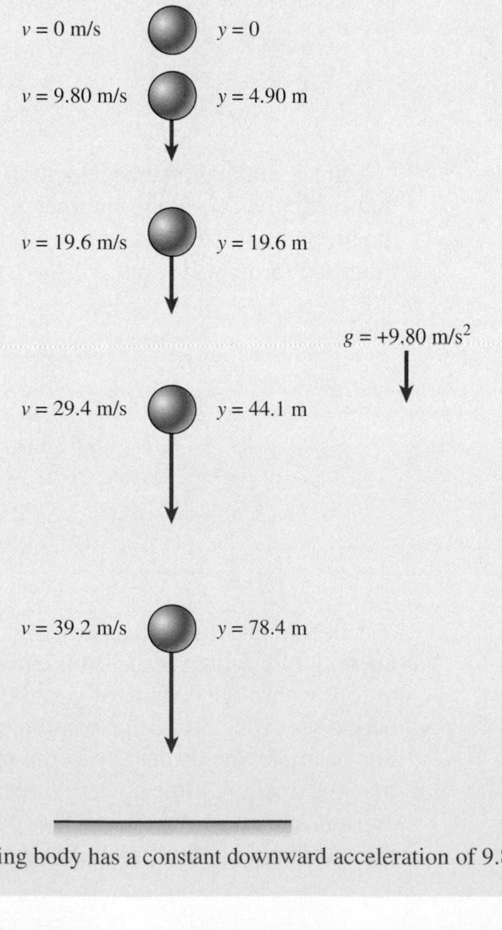

$v = 0$ m/s $y = 0$

$v = 9.80$ m/s $y = 4.90$ m

$v = 19.6$ m/s $y = 19.6$ m

$g = +9.80$ m/s^2

$v = 29.4$ m/s $y = 44.1$ m

$v = 39.2$ m/s $y = 78.4$ m

Figure 6.5   A free-falling body has a constant downward acceleration of 9.80 m/s^2.

Similar calculations for $t = 2, 3$, and 4 s will yield displacements of 19.6, 44.1, and 78.4 m, respectively. Note that each displacement is positive (in the downward direction). The results are summarized in Table 6.2.

Table 6.2

Velocities and Displacements of a Ball Dropped from Rest

Time t, s	Velocity at End of Time t, m/s	Displacement at End of Time t, m
0	0	0
1	9.80	4.90
2	19.6	19.6
3	29.4	44.1
4	39.2	78.4

Example 6.9

Assuming a ball is projected upward with an initial velocity of 96 ft/s, explain without using equations how its upward motion is just the reverse of its downward motion.

Solution: We shall assume the upward direction to be positive, making the acceleration due to gravity equal to -32 ft/s^2. The negative sign indicates that an object projected vertically will have its velocity reduced by 32 ft/s every second it rises. (Refer to Fig. 6.6).

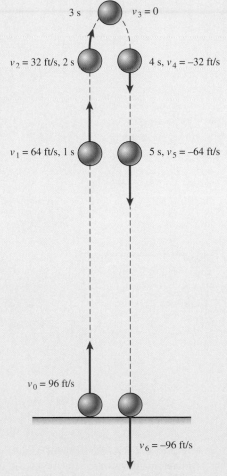

Figure 6.6 A ball thrown vertically upward returns to the ground with the same speed.

If its initial velocity is 96 ft/s, its velocity after 1 s will be reduced to 64 ft/s. After 2 s, its velocity will be 32 ft/s, and after 3 s, its velocity will be reduced to zero. When the velocity becomes zero, the ball has reached its maximum height and begins to fall freely from rest. Now, however, the velocity of the ball will be increasing by 32 ft/s every second since both the direction of motion and the acceleration due to gravity are in the negative direction. Its velocity after 4, 5, and 6 s will be -32, -64, and -96 ft/s^2, respectively. Except for the sign, which indicates the direction of motion, the velocities are the same at equal heights above the ground.

Example 6.10

A baseball thrown vertically upward from the roof of a tall building has an initial velocity of 20 m/s. (a) Calculate the time required to reach its maximum height. (b) Find the maximum height. (c) Determine its position and velocity after 1.5 s. (d) What are its position and velocity after 5 s? (See Fig. 6.7.)

Plan: We will choose the upward direction as positive since the initial velocity is directed upward. That means that the acceleration will be -9.8 m/s^2 for each part. For each part of the problem, we follow the same strategy as that applied to general acceleration problems.

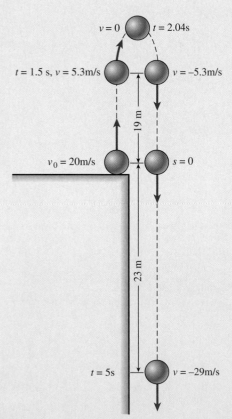

Figure 6.7   A ball projected vertically upward rises until its velocity is zero; then it falls with increasing downward velocity.

Solution (a): The time to reach the maximum height is found by recognizing that the velocity of the ball will be zero at that point. The data are organized as follows:

$$\text{Given: } v_0 = 20 \text{ m/s} \qquad \text{Find: } t = ?$$
$$v_f = 0 \qquad\qquad y = ?$$
$$g = -9.8 \text{ m/s}^2$$

The time required to reach the maximum height can be determined from Eq. (2a):

$$t = \frac{v_f - v_0}{g} = -\frac{v_0}{g}$$

$$= \frac{-20 \text{ m/s}}{-9.8 \text{ m/s}^2} = 2.04 \text{ s}$$

Solution (b): The maximum height is found by setting $v_f = 0$ in Eq. (1a).

$$y = \left(\frac{v_f + v_0}{2}\right)t = \frac{v_0}{2}t$$

$$= \frac{20 \text{ m/s}}{2}(2.04 \text{ s}) = 20.4 \text{ m}$$

Solution (c): To find the position and velocity after 1.5 s, we must establish new conditions.

$$\text{Given: } v_0 = 20 \text{ m/s} \qquad \text{Find: } y = ?$$
$$g = -9.8 \text{ m/s}^2 \qquad v_f = ?$$
$$t = 1.5 \text{ s}$$

We can now calculate the position as follows:

$$y = v_0 t + \frac{1}{2}gt^2$$

$$= (20 \text{ m/s})(1.5 \text{ s}) + \frac{1}{2}(-9.8 \text{ m/s}^2)(1.5 \text{ s})^2$$

$$= 30 \text{ m} - 11 \text{ m} = 19 \text{ m}$$

The velocity after 1.5 s is given by

$$v_f = v_0 + gt$$
$$= 20 \text{ m/s} + (-9.8 \text{ m/s}^2)(1.5 \text{ s})$$
$$= 20 \text{ m/s} - 14.7 \text{ m/s} = 5.3 \text{ m/s}$$

Solution (d): The same equations apply to find the position and velocity after 5 s. Thus,

$$y = v_0 t + \frac{1}{2}gt^2$$

$$= (20 \text{ m/s})(5 \text{ s}) + \frac{1}{2}(-9.8 \text{ m/s}^2)(5 \text{ s})^2$$

$$= 100 \text{ m} - 123 \text{ m} = -23 \text{ m}$$

The negative sign indicates that the ball is located 23 m below the point of release.

The velocity after 5 s is given by

$$v_f = v_0 + gt$$
$$= 20 \text{ m/s} + (-9.8 \text{ m/s}^2)(5 \text{ s})$$
$$= 20 \text{ m/s} - 49 \text{ m/s} = -29 \text{ m/s}$$

In this case, the negative sign indicates that the ball is traveling downward.

6.8   Projectile Motion

We have seen that objects projected vertically upward or downward or dropped from rest are accelerated uniformly in the Earth's gravitational field. Now we consider the more general problem of an object projected freely into a gravitational field in a nonvertical direction, as in Fig. 6.8 in which a football is kicked into space. In the absence of friction, such motion is another example of uniform or constant acceleration.

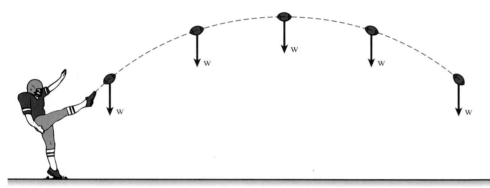

Figure 6.8   A kicked football is a projectile launched freely into space under the influence of gravity alone. If we neglect air resistance, the only force acting on such an object is its weight.

An object launched into space without motive power of its own is called a ***projectile.*** Neglecting air resistance, the only force acting on a projectile is its weight **W,** which causes its path to deviate from a straight line. It receives constant downward acceleration because of the gravitational force acting toward the center of the Earth. But it differs from the motion previously studied in that the acceleration is usually not in line with the direction of initial velocity. Since no force acts horizontally to change the velocity, the horizontal acceleration is zero; this results in constant horizontal velocity. The downward force of gravity, on the other hand, causes the vertical velocity to change uniformly. Thus, the motion of a projectile is normally in two dimensions, and it must be studied that way.

6.9   Horizontal Projection

If an object is projected horizontally, its motion can best be described by considering its horizontal and vertical motion separately. For example, in Fig. 6.9, an electronic device is rigged to simultaneously project one ball horizontally while releasing another from rest at the same height. The horizontal velocity of the projected ball is unchanged, as indicated by the arrows of the same length throughout its trajectory. The vertical velocity, on the other hand, is initially zero and increases uniformly, in accordance with equations derived earlier for one-dimensional motion. The balls will strike the floor at the same instant, even though one is also moving horizontally. Thus, problems are simplified greatly by finding separate solutions for horizontal and vertical components.

The same general equations listed in Table 6.1 for uniform acceleration also apply for projectile motion. However, we know in advance that if the particle is projected near the

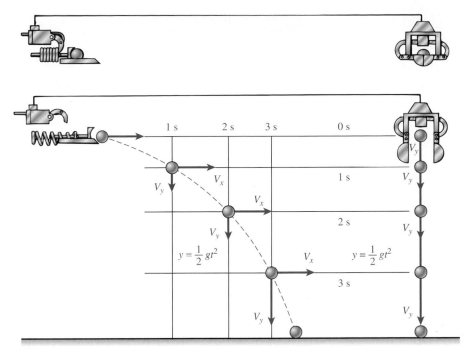

Figure 6.9

Earth, the vertical acceleration is equal to 9.8 m/s² or 32 ft/s² and that it is always directed downward. Thus, if the upward direction is chosen as positive, the acceleration of a projectile will be negative and equal to the gravitational acceleration. We can indicate the components of velocity with appropriate subscripts. For example, if we wish to express the vertical motion as a function of time, we might write

$$y = v_{0y}t + \frac{1}{2}gt^2$$

Note that y represents the vertical displacement, v_{0y} the initial vertical velocity, and g the acceleration due to gravity (normally -9.8 m/s²).

Let's first consider the motion of a projectile that is launched horizontally in a gravitational field. In this case, we note that

$$v_{0x} = v_x \qquad v_{0y} = 0 \qquad a_x = 0 \qquad a_y = g = -9.8 \text{ m/s}^2$$

since the horizontal velocity is constant and the vertical velocity is zero initially. Thus, the horizontal and vertical displacements of such a projectile can be found from

$$x = v_{0x}t \qquad \textit{Horizontal Displacement}$$
$$y = \frac{1}{2}gt^2 \qquad \textit{Vertical Displacement} \qquad \textbf{(6.11)}$$

Since $v_f = v_0 + gt$, we can find the horizontal and vertical components of the final velocity from

$$v_x = v_{0x} \qquad \textit{Horizontal Velocity}$$
$$v_y = v_{0y} + gt \qquad \textit{Vertical Velocity} \qquad \textbf{(6.12)}$$

The final displacement and velocity of a horizontally projected particle can be found from their components. In all of the previous equations, the symbol g may be taken as the acceleration due to gravity, which is always directed downward. Thus, g will be negative when up is chosen as the positive direction.

Table 6.3 provides a summary of how the general formulas for uniform acceleration may be modified for applications to projectiles.

Table 6.3

Uniform Acceleration and Projectiles

General Acceleration Formulas	Modified for Projectile Motion	
(1) $x = \left(\dfrac{v_f + v_0}{2}\right)t$	$x = v_{0x}t$	$y = \left(\dfrac{v_y + v_{0y}}{2}\right)t$
(2) $v_f = v_0 + at$	$v_x = v_{0x}$	$v_y = v_{0y} + gt$
(3) $x = v_0t + \dfrac{1}{2}at^2$	$x = v_{0x}t$	$y = v_{0y}t + \dfrac{1}{2}gt^2$
(4) $x = v_ft - \dfrac{1}{2}at^2$	$x = v_{0x}t$	$y = v_yt - \dfrac{1}{2}gt^2$
(5) $2ax = v_f^2 - v_0^2$	$a_x = 0$	$2gy = v_y^2 - v_{0y}^2$

Example 6.11

A skier leaves the jump horizontally with an initial velocity of 25 m/s, as shown in Fig. 6.10. The initial height at the end of the ramp is 80 m above the landing position. (a) How long is the skier in the air? (b) How far does the skier travel horizontally? (c) What are the horizontal and vertical components of the final velocity?

Plan: For horizontal projection, we note that the initial vertical velocity is zero and that the horizontal velocity does not change. Also, if we choose the downward direction as positive, the acceleration will be $+9.8$ m/s², and other parameters will also be positive. Each part of the problem can be solved by substituting given data into the appropriate equations from Table 6.3.

Solution (a): The time spent in the air is a function of only vertical parameters.

Given: $v_{0y} = 0$, $a = +9.8$ m/s², $y = +80$ m   Find: $t = ?$

We need an equation that contains t and not v_f.

$$y = v_{0y}t + \frac{1}{2}gt^2$$

Setting $v_{0y} = 0$ and solving for t, we obtain

$$y = \frac{1}{2}gt^2 \qquad \text{and} \qquad t = \sqrt{\frac{2y}{g}}$$

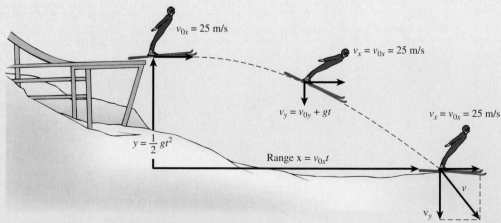

Figure 6.10   A skier leaves the jump with an initial horizontal velocity of 25 m/s. His position and velocity can be found as a function of time.

Substituting given values for y and g, we can find t.

$$t = \sqrt{\frac{2(80 \text{ m})}{9.8 \text{ m/s}^2}} \quad \text{or} \quad t = 4.04 \text{ s}$$

It takes a time of 4.04 s for the skier to reach the landing point.

Solution (b): Since the horizontal velocity is constant, the *range* is determined only by the time of flight.

$$x = v_{0x}t = (25 \text{ m/s})(4.04 \text{ s}) = 101 \text{ m}$$

Solution (c): The horizontal component of the velocity is unchanged and, therefore, equal to 25 m/s at the landing point. The final vertical component is given by

$$v_y = gt = (9.8 \text{ m/s}^2)(4.04 \text{ s}) = 39.6 \text{ m/s}$$

The horizontal component is 25 m/s to the right, and the vertical component is 39.6 m/s directed *downward*. Remember that the downward direction was chosen as positive for convenience. It is left as an exercise for you to show that the final velocity is 46.8 m/s at an angle of 57.7° below the horizontal. This will be the velocity the instant before striking the ground.

6.10 The More General Problem of Trajectories

The more general case of projectile motion occurs when the projectile is thrown at an angle. This problem is illustrated in Fig. 6.11, where the motion of a projectile thrown at an angle θ with an initial velocity v_0 is compared with the motion of an object projected vertically upward. Once again, it is easy to see the advantage of treating the horizontal and vertical motions separately. In this case, the equations listed in Table 6.3 may be used, and we should consider the upward direction as positive. Thus, if the vertical position y is above the origin, it will be positive; if it is below the origin, it will be negative. Similarly, upward velocities will be positive. Since the acceleration is always directed downward, we must substitute a negative value for g.

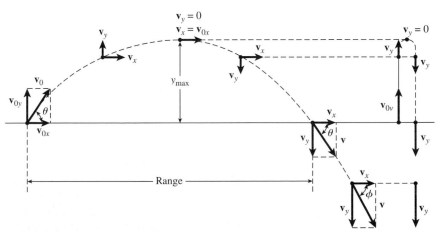

Figure 6.11 The motion of a projectile thrown at an angle is compared with the motion of an object thrown vertically upward.

Problem-Solving Strategy

Projectile Motion

1. Resolve the initial velocity v_0 into its x and y components:

$$v_{0x} = v_0 \cos\theta \qquad v_{0y} = v_0 \sin\theta$$

2. The horizontal and vertical components of the displacement at any instant are given by

$$x = v_{0x}t$$

$$y = v_{0y}t + \frac{1}{2}gt^2$$

3. The horizontal and vertical components of the velocity at any instant are given by

$$v_x = v_{0x}$$

$$v_y = v_{0y} + gt$$

4. The final position and velocity can then be obtained from their components.

5. Make sure that you use the appropriate signs and consistent units. Remember that gravity g may be positive or negative depending on your initial choice.

Example 6.12

A projectile is fired with an initial velocity of 80 m/s at an angle of 30° above the horizontal. Find (a) its position and velocity after 6 s, (b) the time required to reach the maximum height, and (c) the horizontal range R, as indicated in Fig. 6.11.

Plan: This time we choose the upward direction as positive, making $g = -9.8$ m/s^2. Since the projection is at an angle, we will work with initial and final components of velocity. By treating the vertical motion separately from the horizontal motion, we can solve for each of the unknowns in the example.

Solution (a): The horizontal and vertical components of the initial velocity are

$$v_{0x} = v_0 \cos\theta = (80 \text{ m/s}) \cos 30° = 69.3 \text{ m/s}$$

$$v_{0y} = v_0 \sin\theta = (80 \text{ m/s}) \sin 30° = 40.0 \text{ m/s}$$

The x component of its position after 6 s is given by

$$x = v_{0x}t = (69.3 \text{ m/s})(6 \text{ s}) = 416 \text{ m}$$

The y component of its position at this time is

$$y = v_{0y}t + \frac{1}{2}gt^2$$

from which

$$y = (40 \text{ m/s})(6 \text{ s}) + \frac{1}{2}(-9.8 \text{ m/s}^2)(6 \text{ s})^2$$

$$y = 240 \text{ m} - 176 \text{ m} = 64.0 \text{ m}$$

The position after 6 s is 416 m downrange and 64.0 m above its initial position.

In computing its velocity at this point, we first recognize that the x component of the velocity does not change. Thus,

$$v_x = v_{0x} = 69.3 \text{ m/s}$$

The y component of velocity must be found from

$$v_y = v_{0y} + gt$$

so that

$$v_y = 40 \text{ m/s} + (-9.8 \text{ m/s}^2)(6 \text{ s})$$
$$v_y = 40 \text{ m/s} - 58.8 \text{ m/s}$$
$$v_y = -18.8 \text{ m/s}$$

The negative sign indicates that the projectile has passed its highest point and is now on its way down. Finally, we can compute the resultant velocity after 6 s from its components, as shown in Fig. 6.12.

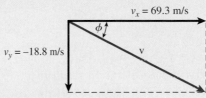

Figure 6.12

The magnitude of the velocity is

$$v = \sqrt{v_x^2 + v_y^2} = \sqrt{(69.3 \text{ m/s})^2 + (-18.8 \text{ m/s})^2} = 71.8 \text{ m/s}$$

The angle is found from the tangent function

$$\tan \phi = \frac{v_y}{v_x} = \left| \frac{-18.8 \text{ m/s}}{69.3 \text{ m/s}} \right| = 0.271$$

$$\phi = 15.2° \text{ S of E}$$

Solution (b): At the maximum point in the path, the y component of the velocity is zero. Hence, the time to reach the maximum height can be found from

$$v_y = v_{0y} + gt \qquad \text{where} \qquad v_y = 0$$
$$0 = (40 \text{ m/s}) + (-9.8 \text{ m/s}^2)t$$

Solving for t, we obtain

$$t = \frac{40 \text{ m/s}}{9.8 \text{ m/s}^2} = 4.08 \text{ s}$$

As an exercise, you should use this result to show that the maximum height in the trajectory is 81.6 m.

Solution (c): The range of the projectile can be found by recognizing that the total time t' of the entire flight is equal to twice the time to reach the highest point. Therefore,

$$t' = 2(4.08 \text{ s}) = 8.16 \text{ s}$$

The range is

$$R = v_{0x}t' = (69.3 \text{ m/s})(8.16 \text{ s})$$
$$= 565 \text{ m}$$

In this example, we see that the projectile rises to a maximum height of 81.6 m in a time of 4.08 s. After 6 s, it reaches a point 416 m downrange and 64.0 m above its starting position. Its velocity at that point is 71.8 m/s in a direction of 15° below the horizontal. The total range is 565 m.

Summary and Review

Summary

A convenient way of describing objects in motion is to discuss their *velocity* or their *acceleration*. In this chapter, a number of applications have been presented that involve these physical quantities.

- *Average velocity* is the distance traveled per unit of time, and *average acceleration* is the change in velocity per unit of time.

$$\bar{v} = \frac{x}{t} \qquad a = \frac{v_f - v_0}{t}$$

- The definitions of velocity and acceleration result in the establishment of five basic equations involving uniformly accelerated motion:

$$x = \left(\frac{v_0 + v_f}{x}\right)t$$

$$v_f = v_0 + at$$

$$x = v_0 t + \frac{1}{2}at^2$$

$$x = v_f t - \frac{1}{2}at^2$$

$$2ax = v_f^2 - v_0^2$$

Given any three of the five parameters (v_0, v_f, a, x, and t), the other two can be determined from one of these equations.

- To solve acceleration problems, read the problem with a view to establishing the three given parameters and the two that are unknown. You might set up columns like this:

Given: $a = 4$ m/s^2 Find: $v_f = ?$
$x = 500$ m $v_0 = ?$
$t = 20$ s

This procedure helps you choose the appropriate equation. Remember to choose a direction as positive and to apply it consistently throughout the problem.

- *Gravitational acceleration:* Problems involving gravitational acceleration can be solved like other acceleration problems. In this case, one of the parameters is known in advance to be

$$a = g = 9.80 \text{ m/s}^2 \text{ or } 32.0 \text{ ft/s}^2$$

The sign of gravitational acceleration is $+$ or $-$ depending on whether you choose up or down as the positive direction.

- *Projectile motion:* The key to problems involving projectile motion is to treat the horizontal motion and the vertical motion separately. Most projectile problems are solved using the following approach:
  - Resolve the initial velocity v_0 into its x and y components:

$$v_{0x} = v_0 \cos \theta \qquad v_{0y} = v_0 \sin \theta$$

- The horizontal and vertical components of its position at any instant are given by

$$x = v_{0x}t \qquad y = v_{0y}t + \frac{1}{2}gt^2$$

- The horizontal and vertical components of its velocity at any instant are given by

$$v_x = v_{0x} \qquad v_y = v_{0y} + gt$$

- The final position and velocity can then be obtained from their components.

The important point to remember in applying these equations is to be consistent throughout with units and sign conversion.

Key Terms

Review Questions

6.1. Distinguish clearly between the terms *speed* and *velocity*. A stock car racer drives 500 laps around a 1-mi track in a time of 5 h. What was the average speed? What was the average velocity?

6.2. A bus driver travels a distance of 300 km in 4 h. At the same time, a tourist travels the same 300 km in a car but stops for two 30-min breaks. Still, the tourist arrives at the destination at the same instant as the bus driver. Compare the average speeds of each driver.

6.3. Give some examples of motion in which the speed is constant but the velocity is not.

6.4. A bowling ball rolls down a 30° incline for a short distance and then levels out for a few seconds before starting up another 30° incline. Discuss the acceleration of the ball as it rolls along each segment of track. Is the acceleration the same going up the incline as it is going down the incline? Suppose the ball is given an initial push at the top of the first incline and then released to travel at greater speeds. Will the acceleration be any different on any of the segments?

6.5. A rock is thrown vertically upward. It moves to the maximum height, returns to its release point, and continues downward. Discuss the signs of its displacement, velocity, and acceleration at each point in its path. What is its acceleration when the velocity reaches zero at the highest point?

6.6. In the absence of air resistance, the same time is required for a projectile to reach its highest point as is required for it to return. Will this be true if air resistance is not neglected? Draw free-body diagrams for each situation.

6.7. Is the motion of a projectile fired at an angle an example of uniform acceleration? What is it if it is fired vertically upward or vertically downward? Explain.

6.8. At what angle should a baseball be thrown to achieve its maximum range? At what angle must it be thrown to attain the maximum height?

6.9. A hunter shoots an arrow directly at a squirrel on a tree limb, and the squirrel drops at the instant the arrow leaves the bow. Will the arrow strike the squirrel? Draw the trajectories that might be expected. Under what conditions will the arrow not strike the squirrel?

6.10. A child drops a ball from the window of a car traveling at a constant speed of 60 km/h. What is the initial velocity of the ball relative to the ground? Describe the motion.

6.11. A toy car is pulled across the floor with uniform speed. A spring attached to the car projects a marble vertically upward. Describe the motion of the marble relative to the floor and relative to the car.

6.12. Explain the adjustment of sights on a rifle for increasing target distances.

6.13. Explain the reasoning behind the use of high or low trajectories for punting in a football game.

Problems

Section 6.1 Speed and Velocity

6.1. A car travels a distance of 86 km at an average speed of 8 m/s. How many hours were required for the trip? *Ans. 2.99 h*

6.2. Sound travels at an average speed of 340 m/s. Lightning from a distant thundercloud is seen almost immediately. If the sound of thunder reaches the ear 3 s later, how far away is the storm?

6.3. A small rocket leaves its pad and travels a distance of 40 m vertically upward before returning to the Earth 5 s after it was launched. What was the average velocity for the trip? *Ans. 16 m/s*

6.4. A car travels along a U-shaped curve for a distance of 400 m in 30 s. Its final location, however, is only 40 m from the starting position. What is the average speed, and what is the magnitude of the average velocity?

6.5. A woman walks for 4 min directly north with an average velocity of 6 km/h; then she walks eastward at 4 km/h for 10 min. What is her average speed for the trip? *Ans. 4.57 km/h*

6.6. What is the average velocity for the entire trip described in Prob. 6.5?

6.7. A car travels at an average speed of 60 mi/h for 3 h and 20 in. What was the distance traveled? *Ans. 200 mi*

6.8. How long will it take to travel 400 km if the average speed is 90 km/h?

6.9. A marble rolls up an inclined ramp a distance of 5 m and then stops and returns to a point 5 m below its starting point. Assume that $x = 0$ when $t = 0$. The entire trip took only 2 s. What was the average speed, and what was the average velocity? *Ans. 7.5 m/s, -2.5 m/s*

Section 6.3 Uniform Acceleration

6.10. The tip of a robot arm is moving to the right at 8 m/s. Four seconds later, it is moving to the left at 2 m/s. What is the change in velocity, and what is the acceleration?

6.11. An arrow accelerates from zero to 40 m/s in the 0.5 s it is in contact with the bow string. What is the average acceleration? Ans. 80 m/s^2

6.12. A car traveling initially at 50 km/h accelerates at a rate of 4 m/s^2 for 3 s. What is the final speed?

6.13. A truck traveling at 60 mi/h brakes to a stop in 180 ft. What were the average acceleration and stopping time? Ans. -21.5 ft/s^2, 4.09 s

6.14. An arresting device on a carrier deck stops an airplane in 1.5 s. The average acceleration was 49 m/s^2. What was the stopping distance? What was the initial speed?

6.15. In a braking test, a car traveling at 60 km/h is stopped in a time of 3 s. What were the acceleration and stopping distance?
Ans. -5.56 m/s^2, 25.0 m

6.16. A bullet leaves a 28-in. rifle barrel at 2700 ft/s. What were its acceleration and time in the barrel?

6.17. The ball in Fig. 6.13 is given an initial velocity of 16 m/s at the bottom of an inclined plane. Two seconds later it is still moving up the plane, but with a velocity of only 4 m/s. What is the acceleration? Ans. -6.00 m/s^2

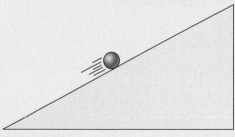

Figure 6.13

6.18. For Prob. 6.17, what is the maximum displacement from the bottom, and what is the velocity 4 s after leaving the bottom?

6.19. A monorail train traveling at 22 m/s must be stopped in a distance of 120 m. What average acceleration is required, and what is the stopping time? Ans. -2.02 m/s^2, 10.9 s

Section 6.7 Gravity and Free-Falling Bodies

6.20. A ball is dropped from rest and falls for 5 s. What are its position and velocity at that instant?

6.21. A rock is dropped from rest. When will its displacement be 18 m below the point of release? What is its velocity at that time?
Ans. 1.92 s, -18.8 m/s

6.22. A woman drops a weight from the top of a bridge while a friend below measures the time it takes for the object to strike the water below. What is the height of the bridge if the time is 3 s?

6.23. A brick is given an initial downward velocity of 6 m/s. What is its final velocity after falling a distance of 40 m? Ans. 28.6 m/s, downward

6.24. A projectile is thrown vertically upward and returns to its starting position in 5 s. What was its initial velocity, and how high did it rise?

6.25. An arrow is shot vertically upward with an initial velocity of 80 ft/s. What is its maximum height?
Ans. 100 ft

6.26. In Prob. 6.25, what are the position and velocity of the arrow after 2 s and after 6 s?

6.27. A hammer is thrown vertically upward to the top of a roof 16 m high. What minimum initial velocity was required? Ans. 17.7 m/s

Section 6.9 Horizontal Projection

6.28. A baseball leaves a bat with a horizontal velocity of 20 m/s. In a time of 0.25 s, how far will it have traveled horizontally, and how far will it have fallen vertically?

6.29. An airplane traveling at 70 m/s drops a box of supplies. What horizontal distance will the box travel before striking the ground 340 m below?
Ans. 583 m

6.30. At a lumber mill, logs are discharged horizontally at 15 m/s from a greased chute that is 20 m above a mill pond. How far do the logs travel horizontally?

6.31. A steel ball rolls off the edge of a tabletop 4 ft above the floor. If it strikes the floor 5 ft from the base of the table, what was its initial horizontal speed? Ans. $v_{0x} = 10.0 \text{ ft/s}$

6.32. A bullet leaves the barrel of a weapon with an initial horizontal velocity of 400 m/s. Find the horizontal and vertical displacements after 3 s.

6.33. A projectile has an initial horizontal velocity of 40 m/s at the edge of a rooftop. Find the horizontal and vertical components of its velocity after 3 s.
Ans. 40 m/s, -29.4 m/s

Section 6.10 The More General Problem of Trajectories

6.34. A stone is given an initial velocity of 20 m/s at an angle of 58°. What are its horizontal and vertical displacements after 3 s?

6.35. A baseball leaves the bat with a velocity of 30 m/s at an angle of 30°. What are the horizontal and vertical components of its velocity after 3 s?

Ans. 26.0 m/s, −14.4 m/s

6.36. For the baseball in Prob. 6.35, what is the maximum height, and what is the range?

6.37. An arrow leaves the bow with an initial velocity of 120 ft/s at an angle of 37° with the horizontal. What are the horizontal and vertical components of its displacement two seconds later?

Ans. $x = 208$ ft, $y = 56.0$ ft

*6.38. In Prob. 6.37, what are the magnitude and direction of the arrow's velocity after 2 s?

*6.39. A golf ball in Fig. 6.14 leaves the tee with a velocity of 40 m/s at 65°. If it lands on a green located 10 m higher than the tee, what was the time of flight, and what was the horizontal distance to the tee?

Ans. 7.11 s, 120 m

*6.40. A projectile leaves the ground with a velocity of 35 m/s at an angle of 32°. What is the maximum height attained?

*6.41. The projectile in Prob. 6.40 rises and falls, striking a billboard at a point 8 m above the ground. What was the time of flight, and how far did it travel horizontally?

Ans. 3.29 s, 97.7 m

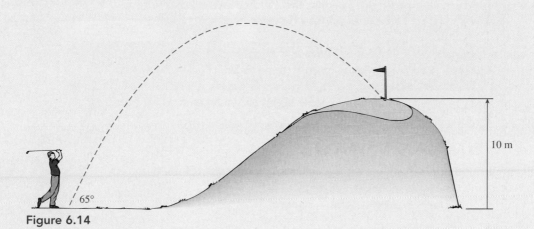

65°

10 m

Figure 6.14

Additional Problems

6.42. A rocket travels in space at 60 m/s before it is given a sudden acceleration. If its velocity increases to 140 m/s in 8 s, what was its average acceleration, and how far did it travel in this time?

6.43. A railroad car starts from rest and coasts freely down an incline. With an average acceleration of 4 ft/s², what will be the velocity after 5 s? What distance does it travel?

Ans. 20 ft/s, 50 ft

*6.44. An object is projected horizontally at 20 m/s. At the same time, another object located 12 m downrange is dropped from rest. When will they collide, and how far are they located below the release point?

6.45. A truck moving at an initial velocity of 30 m/s is brought to a stop in 10 s. What was the acceleration of the car, and what was the stopping distance?

Ans. −3.00 m/s², 150 m

6.46. A ball is thrown vertically upward with an initial velocity of 23 m/s. What are its position and velocity after 2 s, after 4 s, and after 8 s?

6.47. A stone is thrown vertically downward from the top of a bridge. Four seconds later it strikes the water below. If the final velocity was 60 m/s, what was the initial velocity of the stone, and how high was the bridge?

Ans. $v_0 = 20.8$ m/s, downward; $y = 162$ m

6.48. A ball is thrown vertically upward with an initial velocity of 80 ft/s. What are its position and velocity after (a) 1 s; (b) 3 s; and (c) 6 s?

6.49. An aircraft flying horizontally at 500 mi/h releases a package. Four seconds later, the package strikes the ground below. What was the altitude of the plane?

Ans. −256 ft

6.50. In Prob. 6.49, what was the horizontal range of the package, and what are the components of its final velocity?

6.51. A putting green is located 240 ft horizontally and 64 ft vertically from the tee. What must be the magnitude and direction of the initial velocity if a ball is to strike the green at this location after a time of 4 s?

Ans. 100 ft/s, 53.1°

Critical Thinking Questions

6.52. A long strip of pavement is marked off in 10-m intervals. Students use stopwatches to record the times at which a car passes each mark. The following data is listed:

Distance, m	0	10	20	30	40	50
Time, s	0	2.1	4.2	6.3	8.4	10.5

Plot a graph with distance along the y axis and time along the x axis. What is the significance of the slope of this curve? What is the average speed of the car? At what instant in time is the distance equal to 34 m? What is the acceleration of the car?
Ans. Slope is v, 4.76 m/s, 7.14 s, 0

6.53. An astronaut tests gravity on the Moon by dropping a tool from a height of 5 m. The following data are recorded electronically.

Height, m	5.00	4.00	3.00	2.00	1.00	0
Time, s	0	1.11	1.56	1.92	2.21	2.47

Plot the graph of this data. Is it a straight line? What is the average speed for the entire fall? What is the acceleration? How would you compare this with gravity on Earth?

6.54. A car is traveling initially north at 20 m/s. After traveling a distance of 6 m, the car passes point A, where its velocity is still northward but is reduced to 5 m/s. (a) What are the magnitude and direction of the acceleration of the car? (b) What time was required? (c) If the acceleration is held constant, what will be the velocity of the car when it returns to point A?
Ans. (a) 31.2 m/s², south; (b) 0.48 s; (c) −5 m/s

*6.55. A ball moving up an incline is initially located 6 m from the bottom of an incline and has a velocity of 4 m/s. Five seconds later, it is located 3 m from the bottom. Assuming constant acceleration, what was the average velocity? What is the meaning of a negative average velocity? What are the average acceleration and final velocity?

*6.56. The acceleration due to gravity on a distant planet is determined to be one-fourth its value on Earth. Does this mean that a ball dropped from a height of 4 m above this planet will strike the ground in one-fourth the time? What are the times required on the planet and on Earth?
Ans. $t_p = 1.81$ s, $t_e = 0.904$ s

*6.57. Consider the two balls A and B shown in Fig. 6.15. Ball A has a constant acceleration of 4 m/s² directed to the right, and ball B has a constant acceleration of 2 m/s² directed to the left. Ball A is initially traveling to the left at 2 m/s, and ball B is traveling to the left initially at 5 m/s. Find the time t at which the balls collide. Also, assuming $x = 0$ at the initial position of ball A, what is their common displacement when they collide?
Ans. $t = 2.0$ s, $x = +4$ m

Figure 6.15

*6.58. Initially, a truck with a velocity of 40 ft/s is located a distance of 500 ft ahead of a car. If the car begins at rest and accelerates at 10 ft/s², when will it overtake the truck? How far is the point from the initial position of the car?

*6.59. A ball is dropped from rest at the top of a 100-m-tall building. At the same instant, a second ball is thrown upward from the base of the building with an initial velocity of 50 m/s. When will the two balls collide, and at what distance above the street?
Ans. 2.00 s, 80.4 m

*6.60. A balloonist rising vertically with a velocity of 4 m/s releases a sandbag at the instant when the balloon is 16 m above the ground. Compute the position and velocity of the sandbag relative to the ground after 0.3 s and 2 s. How many seconds after its release will it strike the ground?

*6.61. An arrow is shot vertically upward with a velocity of 40 m/s. Three seconds later, another arrow is shot upward with a velocity of 60 m/s. At what time and position will they meet?
Ans. 4.54 s, 80.6 m

*6.62. Someone wishes to strike a target that has a horizontal range of 12 km. What must be the velocity of an object projected at an angle of 35° if it is to strike the target? What is the time of flight?

*6.63. A wild boar charges directly at a hunter with a constant speed of 60 ft/s. At the instant the boar is 100 yd away, the hunter fires an arrow at 30° with the ground. What must be the velocity of the arrow if it is to strike its target?
Ans. 76.2 ft/s

Newton's Second Law

The space shuttle *Endeavor* lifts off for an 11-day mission in space. All of Newton's laws of motion—the law of inertia, action-reaction, and the acceleration produced by a resultant force—are exhibited during this liftoff.
(*Photo by NASA Marshall Space Flight Center.*)

Objectives

After completing this chapter, you should be able to

1. Describe the relationships among force, mass, and acceleration and give the consistent units for each in the metric and U.S. customary systems of units.

2. Define the units *newton* and *slug* and explain why they are derived units rather than fundamental units.

3. Demonstrate by definition and example your understanding of the distinction between mass and weight.

4. Determine mass from weight and weight from mass at a point at which the acceleration due to gravity is known.

5. Draw a free-body diagram for objects in motion with constant acceleration, set the resultant force equal to the total mass times the acceleration, and solve for unknown parameters.

According to Newton's first law of motion, an object will undergo a change in its state of rest or motion *only* when acted on by a resultant, unbalanced force. We now know that a change in motion, for example, a change in speed, results in *acceleration*. In many industrial applications, we need to be able to predict the acceleration that will be produced by a given force. For example, the forward force required to accelerate a car from rest to a speed of 60 km/h in 8 s is of interest to the automotive industry. In this chapter, you will study the relationships among force, mass, and acceleration.

7.1 Newton's Second Law of Motion

Before studying the relationship between a resultant force and acceleration in a formal way, let us first consider a simple experiment. A linear air track is an apparatus for studying the motion of objects under conditions that approximate zero friction. Hundreds of tiny air jets create an upward force that balances the weight of the glider in Fig. 7.1. A string is attached to the front of the glider, and a spring scale of negligible weight is used to measure the applied horizontal force as shown. The acceleration the glider receives can be measured by determining the change in speed for a known time interval. The first applied force $\mathbf{F}_1$ in Fig. 7.1a causes an acceleration $\mathbf{a}_1$. Twice the force $2\mathbf{F}_1$ will produce twice the acceleration $2\mathbf{a}_1$, and three times the force $3\mathbf{F}_1$ gives three times the acceleration $3\mathbf{a}_1$.

Such observations show that the acceleration of a given body is directly proportional to the applied force. This means that the ratio of force to acceleration is always constant:

$$\frac{\mathbf{F}_1}{\mathbf{a}_1} = \frac{\mathbf{F}_2}{\mathbf{a}_2} = \frac{\mathbf{F}_3}{\mathbf{a}_3} = \text{constant}$$

The constant is a measure of how effective a given force is in producing acceleration. We shall see that this ratio is a property of the body called its mass *m*, where

$$m = \frac{F}{a}$$

A mass of one kilogram (1 kg) was defined in Chapter 3 by comparison with a standard artifact. Keeping this definition, we can now define a new force unit such that it will give a unit mass a unit acceleration.

A force of one newton (1 N) is that resultant force that will give a 1-kg mass an acceleration of 1 m/s^2.

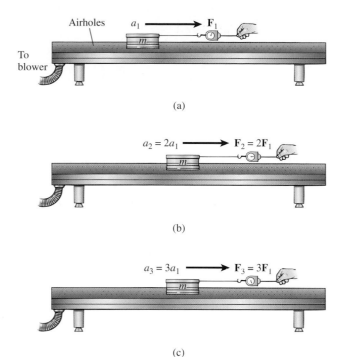

Figure 7.1   Variation of acceleration with force.

The ***newton*** is declared the SI unit of force. A resultant force of 2 N will produce an acceleration of 2 m/s², and a force of 3 N will give an acceleration of 3 m/s² to a 1-kg mass.

Now let's return to our air track experiment to see how acceleration is affected by increasing the mass. This time we shall keep the applied force **F** constant. The mass will be changed by adding in succession more gliders of equal size and weight. Note from Fig. 7.2 that if the force is unchanged, an increase in mass will result in a proportionate *decrease* in the acceleration. Applying a constant force of 12 N in succession to 1-, 2-, and 3-kg masses will produce accelerations of 12 m/s², 6 m/s², and 4 m/s², respectively. These three cases are shown in Fig. 7.2a, b, and c.

From the above observations, we are now prepared to state ***Newton's second law*** of motion.

Newton's Second Law of Motion: Whenever an unbalanced force acts on a body, it produces in the direction of the force an acceleration that is directly proportional to the force and inversely proportional to the mass of the body.

By using our newly defined unit, the newton, we can write this law as the following equation:

$$Resultant\ force = mass \times acceleration$$

$$F = ma \qquad Newton's\ Second\ Law \qquad \textbf{(7.1)}$$

Since this relationship depends on the definition of a new unit, we may substitute only units consistent with that definition. For example, if the mass is in kilograms (kg), the unit of force must be newtons (N), and the unit of acceleration must be meters per second squared (m/s²).

$$Force\ (N) = mass\ (kg) \times acceleration\ (m/s^2)$$

In USCS, the mass unit is defined from chosen units of *pound* (lb) for force and *feet per second squared* (ft/s²) for acceleration. This unit of mass is called the ***slug*** (from the sluggish or inertial property of mass).

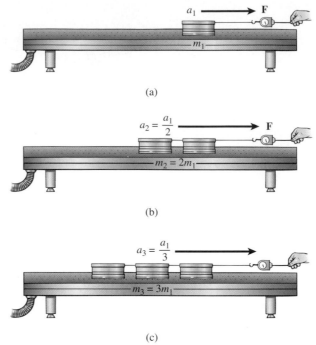

Figure 7.2 Variation of acceleration with mass.

A mass of one slug is that mass to which a resultant force of 1 lb will give an acceleration of 1 ft/s^2.

Force (lb) = *mass* (slug) $\times$ *acceleration* (ft/s^2)

The SI unit of force is less than the USCS unit, and a mass of 1 slug is much greater than a mass of 1 kilogram. The following conversion factors might be helpful:

1 lb = 4.448 N   1 slug = 14.59 kg

A 1-lb bag of apples might contain four or five apples—each weighing about a newton. A person weighing 160 lb on earth would have a mass of 5 slugs or 73 kg.

It is important to recognize that the *F* in Newton's second law represents a *resultant,* or unbalanced, *force.* If more than one force acts on an object, it will be necessary to determine the resultant force *along the direction of motion.* The resultant force will always be along the direction of motion, since it is the *cause* of the acceleration. All components of forces that are perpendicular to the acceleration will be balanced. If the *x* axis is chosen along the direction of motion, we can determine the *x* component of each force and write

$$\sum F_x = ma_x \qquad\qquad \textbf{(7.2)}$$

A similar equation could be written for the *y* components if the *y* axis were chosen along the direction of motion.

7.2   The Relationship Between Weight and Mass

Before we consider examples of Newton's second law, we must have a clear understanding of the difference between the **weight** of a body and its **mass.** Perhaps no other point is more confusing to the beginning student. The *pound* (lb), which is a unit of force, is often used as a mass unit, the pound-mass (lb$_m$). The kilogram, which is a unit of mass, is often used in industry as a unit of force, the kilogram-force (kgf). These seemingly inconsistent units result from the fact that there are many different systems of units in use. In this text, there should be less cause for confusion because we use only the SI and U.S. customary (British gravitational) units. Therefore, in this book the *pound* (lb) will always refer to *weight,* which is a force, and the unit *kilogram* (kg) will always refer to the *mass* of a body.

The weight of any body is the force with which the body is pulled vertically downward by gravity. When a body falls freely to the Earth, the only force acting on it is its weight **W.** This net force produces an acceleration **g,** which is the same for all falling bodies. Thus, from Newton's second law, we can write the relationship between a body's weight and its mass:

$$W = mg \qquad \text{or} \qquad m = \frac{W}{g} \tag{7.3}$$

In either system of units, (1) the mass of a particle is equal to its weight divided by the acceleration of gravity, (2) weight has the same units as the unit of force, and (3) the acceleration of gravity has the same units as acceleration.

Therefore, we can summarize as follows:

$$\text{SI: } W \text{ (N)} = m \text{ (kg)} \times g \text{ (9.8 m/s}^2)$$
$$\text{USCS: } W \text{ (lb)} = m \text{ (slug)} \times g \text{ (32 ft/s}^2)$$

The values for *g,* and hence the weights, in the above relationships apply only at points on Earth near sea level, where *g* has these values.

Two things must be remembered to understand fully the difference between mass and weight:

Mass is a universal constant equal to the ratio of a body's weight to the gravitational acceleration due to its weight.

Weight is the force of gravitational attraction and varies depending on the acceleration of gravity.

Therefore, the mass of a body is a measure only of its inertia and is not in any way dependent on gravity. In outer space, a hammer has negligible weight, but it serves to drive nails just the same since its mass is unchanged.

To reinforce the distinction between weight and mass, consider the examples shown in Fig. 7.3, where a 10-kg ball is placed at three very different locations. If we take the 10-kg ball from a point near the Earth's surface ($g = 9.8$ m/s^2) and move it to a point where gravity is reduced by one-half to 4.9 m/s^2, we notice that its weight is also reduced by one-half. The illustration in Fig. 7.3 is not drawn to scale because an object would have to be very far from the Earth's surface before a significant change in gravity would occur. Still, it does help in understanding the distinction between *weight,* which is gravity-dependent, and *mass,* which is a constant ratio of *W* to *g.* Even on the surface of the Moon, where gravity is only one-sixth of its value on Earth, the mass of the ball is still 10 kg. However, its weight is reduced to 16 N.

In the SI system of units, objects are generally described in terms of their mass in kilograms, which is constant. In USCS units, however, a body is often described by stating its weight at a point where gravity is equal to 32 ft/s^2. This often causes confusion if the object is transported to a location where gravity is significantly more or less than 32 ft/s^2 and the *actual weight* changes. Such confusion is merely one of the many reasons for discarding these older units. They are included here only to provide the degree of familiarity necessary to work with them as they are sometimes encountered in U.S. commerce and industry.

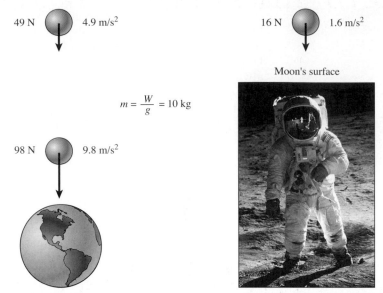

49 N   4.9 m/s²

16 N   1.6 m/s²

Moon's surface

$$m = \frac{W}{g} = 10 \text{ kg}$$

98 N   9.8 m/s²

Figure 7.3 The weight of an object varies with location. However, the mass is the constant ratio of weight to the acceleration due to gravity. (*Photo by NASA.*)

Example 7.1

Find the mass of a body whose weight on Earth is 100 N. If this mass were taken to a distant planet where $g = 2.0 \text{ m/s}^2$, what would be its weight on that planet?

Plan: We first find the mass on Earth, where $g = 9.8 \text{ m/s}^2$. Since the mass is constant, we can use that same value to find the weight on the distant planet where $g = 2.0 \text{ m/s}^2$.

Solution:

$$m = \frac{W}{g} = \frac{100 \text{ N}}{9.8 \text{ m/s}^2} = 10.2 \text{ kg}$$

The weight on the planet is

$$W = mg = (10.2 \text{ kg})(2 \text{ m/s}^2); \qquad W = 20.4 \text{ N}$$

Example 7.2

A 150-lb astronaut finds that his weight is reduced to 60 lb at a distant location. What is the acceleration due to gravity at that location?

Plan: In the USCS units, the astronaut is assumed to weigh 150 lb only where $g = 32 \text{ ft/s}^2$. Therefore, we can find his mass. Then we find gravity by recognizing that mass is the same at the new location.

Solution: On Earth, the mass is

$$m = \frac{W}{g} = \frac{150 \text{ lb}}{32 \text{ ft/s}^2} = 4.69 \text{ slugs}$$

Now, since $W = mg$, we find that gravity at the new location is

$$g = \frac{W}{m} = \frac{60 \text{ lb}}{4.69 \text{ slugs}}; \qquad g = 12.8 \text{ ft/s}^2$$

7.3 Application of Newton's Second Law to Single-Body Problems

The primary difference between the problems discussed in this chapter and earlier problems is that a net, unbalanced force is acting to produce an acceleration. Thus, after constructing the free-body diagrams that describe the situation, the first step is to determine the unbalanced force and set it equal to the product of mass and acceleration. The desired quantity can then be determined from the relation given as Eq. (7.1):

$$Resultant\ force = mass \times acceleration$$

$$F\ (resultant) = ma$$

The following examples will serve to demonstrate the relationships among force, mass, and acceleration.

Example 7.3

A resultant force of 29 N acts in an easterly direction on a 7.5-kg mass. What is the resulting acceleration?

Plan: The resultant force is given by $F = ma$, and the acceleration is in the same direction as the resultant force.

Solution: Solving for a, we obtain.

$$a = \frac{F}{m} = \frac{29\ N}{7.5\ kg}; \qquad a = 3.87\ m/s^2$$

Thus, the resulting acceleration is 3.87 m/s^2 directed toward the east.

Example 7.4

What resultant force will give a 24-lb sled an acceleration of 5 ft/s^2?

Plan: We first find the *mass* of an object whose weight on Earth is 24 lb. Then we use the mass to find the resultant force from $F = ma$.

Solution:

$$m = \frac{W}{g} = \frac{24\ lb}{32\ ft/s^2} = 0.75\ slug$$

$$F = ma = (0.75\ slug)(5\ ft/s^2) \qquad F = 3.75\ lb$$

Example 7.5

In an experiment aboard a space shuttle, an astronaut observes that a resultant force of only 12 N will give a steel box an acceleration of 4 m/s^2. What is the mass of the box?

Solution:

$$F = ma \qquad or \qquad m = \frac{F}{a}$$

$$m = \frac{12\ N}{4\ m/s^2}; \qquad m = 3\ kg$$

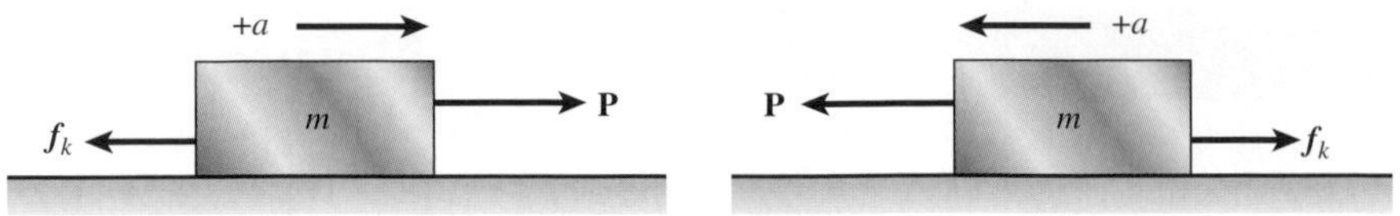

(a) Net force $\mathbf{P} - f_k$ (right)   (b) Net force $\mathbf{P} - f_k$ (left)

Figure 7.4 The direction of acceleration should be chosen as positive.

In Examples 7.3–7.5, the unbalanced forces were easily determined. As the number of forces acting on a body increases, however, the problem of determining the resultant force is less obvious. In these cases, perhaps it is wise to point out a few considerations.

According to Newton's second law, a resultant force always produces an acceleration *in the direction of the resultant force*. This means that the net force and the acceleration it causes are of the same algebraic sign, and they each have the same line of action. Therefore, if the direction of motion (acceleration) is considered positive, fewer negative factors will be introduced into the equation $F = ma$. For example, in Fig. 7.4b, it is preferable to choose the direction of motion (left) as positive since the equation

$$P - f_k = ma$$

is preferable to the equation

$$f_k - P = -ma$$

which would result if we chose the right direction as positive.

Another consideration that results from the previous discussion is that the forces acting normal to the line of motion will be in equilibrium if the resultant force is constant. Thus, in problems involving friction, normal forces can be determined from the first condition for equilibrium.

In summary, we apply the following equations to acceleration problems:

$$\sum F_x = ma_x \qquad \sum F_y = ma_y \qquad \qquad (7.4)$$

One of these equations is chosen along the line of motion, and the other will be perpendicular to the line of motion. This simplifies the problem by ensuring that forces perpendicular to the motion are balanced.

Example 7.6

A horizontal force of 200 N drags a 12-kg chest across a level floor where $\mu_k = 0.4$. Find the resulting acceleration.

Plan: Since the acceleration is produced by a *resultant* force, we will draw a free-body diagram (see Fig. 7.5b) and choose the positive x axis along the direction of motion. Following procedures learned in previous chapters, we will calculate the resultant force and set it equal to the product of the mass and acceleration.

Solution: Applying Newton's second law to the x axis, we have

$$Resultant\ force = mass \times acceleration$$
$$200\ N - f_k = ma$$

Next, we substitute $f_k = \mu_k n$ to obtain

$$200\ N - \mu_k n = ma$$

Since the vertical forces are balanced, we see from Fig. 7.5b that $\Sigma F_y = ma_y = 0$.

$$n - mg = 0 \qquad or \qquad n = mg$$

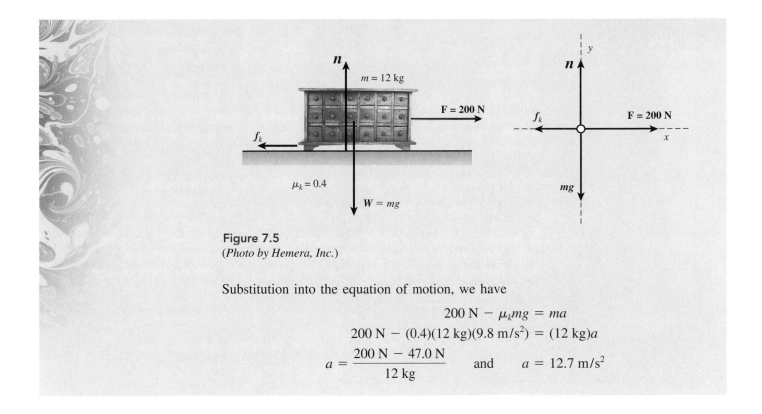

Figure 7.5
(*Photo by Hemera, Inc.*)

Substitution into the equation of motion, we have

$$200 \text{ N} - \mu_k mg = ma$$
$$200 \text{ N} - (0.4)(12 \text{ kg})(9.8 \text{ m/s}^2) = (12 \text{ kg})a$$
$$a = \frac{200 \text{ N} - 47.0 \text{ N}}{12 \text{ kg}} \quad \text{and} \quad a = 12.7 \text{ m/s}^2$$

7.4   Problem-Solving Techniques

Solving all physics problems requires an ability to organize the given data and to apply formulas in a consistent manner. Often a procedure is helpful to the beginning student. This is particularly true for problems in this chapter. A logical sequence of operations for problems involving Newton's second law is outlined in the procedure below.

Problem-Solving Strategy

Newton's Second Law of Motion

1. Read the problem carefully and then draw and label a rough sketch.

2. Indicate all given information and state what is to be found.

3. Construct a free-body diagram for each object undergoing acceleration and choose an x or y axis along the continuous line of motion.

4. Indicate a consistent choice for the positive direction along the continuous line of motion.

5. Be careful not to confuse the weight of an object with its mass:

$$W = mg \qquad m = \frac{W}{g}$$

6. From the free-body diagram(s), determine the resultant force along the assumed positive line of motion (ΣF).

7. Determine the total mass of the system $(m_1 + m_2 + m_3 + \cdots)$

8. Set the resultant force (ΣF) equal to the total mass times the acceleration a.

$$\Sigma F = (m_1 + m_2 + m_3 + \cdots)a$$

9. Substitute all given quantities and solve for the unknown.

Example 7.7

A freight elevator shown in Fig. 7.6 is lifted upward with an acceleration of 2.5 m/s². If the tension in the supporting cable is 9600 N, what is the mass of the elevator and its contents?

Plan: We will follow the problem-solving strategy by drawing an appropriate free-body diagram (see Fig. 7.6) and writing Newton's second law for the line of acceleration. The mass of the elevator can be then be found as the only unknown in that equation.

Solution: We organize the given data and state what is to be found.

$$\text{Given: } T = 9600 \text{ N}; g = 9.8 \text{ m/s}^2; a = 2.5 \text{ m/s}^2 \qquad \text{Find: } m = ?$$

In solving for mass, it is useful to write the weight as the product of mass and gravity (mg). Note that the positive direction of acceleration (upward) is indicated on the free-body diagram. The acceleration along the x axis is zero, and the entire acceleration a is along the y axis.

$$a_y = a \qquad \text{and} \qquad a_x = 0$$

The resultant force is, therefore, the sum of the forces along the y axis.

$$\sum F_y = T - mg$$

From Newton's second law, we now write

$$Resultant\ force = total\ mass \times acceleration$$
$$T - mg = ma$$

Finally, we can solve for the mass as in the following steps:

$$T = mg + ma = m(g + a)$$
$$m = \frac{T}{g + a} = \frac{9600 \text{ N}}{9.8 \text{ m/s}^2 + 2.5 \text{ m/s}^2}$$
$$m = \frac{9600 \text{ N}}{12.3 \text{ m/s}^2} = 780 \text{ kg}$$

Since the resultant force must be in the same direction as the acceleration, it will often be to our advantage to select a single axis along the motion as we did in this example. We should also choose the direction of motion as positive.

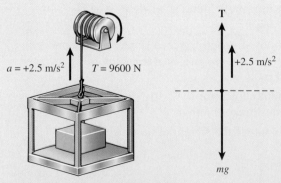

$$a = +2.5 \text{ m/s}^2 \qquad T = 9600 \text{ N}$$

T

$+2.5$ m/s²

mg

Figure 7.6   Upward acceleration in a gravitational field.

Example 7.8

A 100-kg ball is lowered by means of a cable with a downward acceleration of 5 m/s². What is the tension in the cable?

Plan: Follow the strategy as before. However, be careful of the signs given for each of the vectors as we apply Newton's second law.

Solution: As in Example 7.7, we construct a rough sketch and a free-body diagram, as in Fig. 7.7. Organizing the information, we write

$$\text{Given: } m = 100 \text{ kg}; g = 9.8 \text{ m/s}^2; a = 5 \text{ m/s}^2 \qquad \text{Find: } T = ?$$

The *downward* direction of motion is chosen as the positive direction. This means that downward vectors will be positive and upward vectors negative. The resultant force is, therefore, $mg - T$ rather than $T - mg$. Again it is useful to write the weight of the elevator as mg. Now, from Newton's second law, we write

$$Resultant\ downward\ force = total\ mass \times downward\ acceleration$$
$$mg - T = ma$$
$$T = mg - ma = m(g - a)$$
$$T = (100 \text{ kg})(9.8 \text{ m/s}^2 - 5 \text{ m/s}^2)$$
$$T = 480 \text{ N}$$

If you prefer, this problem can be worked by substituting the known values at the start of the problem rather than first finding the algebraic solution. However, it is usually better to substitute at the very end.

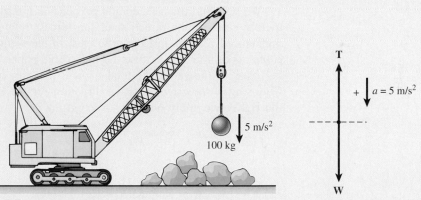

Figure 7.7   Downward acceleration.

Example 7.9

An *Atwood machine* consists of a single pulley with masses suspended at the endpoints of a connecting cable. It is a simplified version of many industrial systems in which counterweights are used for balance. Assume that the mass on the right side is 10 kg and that the mass on the left side is 2 kg. (a) What is the acceleration of the system? (b) What is the tension in the cable?

Plan: In this problem, there are two masses—one moving upward and the other moving downward. In such cases, it is often better to apply Newton's second law first to the entire moving system. The line of motion is then upward on the left (+) and downward on the right (+). Thus, the mass of the system becomes everything that is moving, and the resultant force is merely the difference in the weights. Once we find the

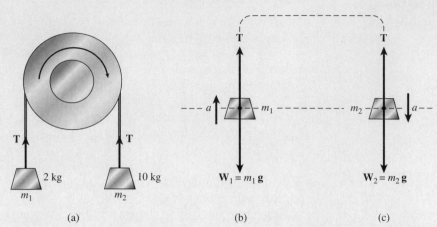

(a)   (b)   (c)

Figure 7.8   Two masses suspended from a single pulley. Free-body diagrams are drawn; the positive direction of acceleration is chosen to be upward on the left and downward on the right.

acceleration of the entire system, we can then apply Newton's second law separately to either of the two masses to find the tension in the rope.

Solution (a): First, we draw a sketch and free-body diagrams for each of the attached masses, as shown in Fig. 7.8. The weights of the masses are written as m_1g and m_2g, respectively. Organizing the data, we write

Given: $m_1 = 2$ kg; $m_2 = 10$ kg; $g = 9.8$ m/s^2;   Find: a and T.

We will find the acceleration by considering the *entire* system. Note that the pulley merely changes the direction of the forces. If we neglect the mass of the cable, the unbalanced force is just the difference in the weights $(m_2g - m_1g)$. The tension in the cable is not a factor because the light cable is part of the moving system. If we neglect the mass of the cable, the total mass is the sum of all the moving masses $(m_1 + m_2)$. Choosing a *consistent* positive direction for the motion (*up* on the left and *down* on the right), we apply Newton's second law:

$$m_2g - m_1g = (m_1 + m_2)a$$

$$a = \frac{m_2g - m_1g}{m_1 + m_2} = \frac{(m_2 - m_1)g}{m_1 + m_2}$$

$$a = \frac{(10\text{ kg} - 2\text{ kg})(9.8\text{ m/s}^2)}{10\text{ kg} + 2\text{ kg}}$$

$$a = 6.53\text{ m/s}^2$$

Solution (b): To solve for the tension T in the cable, we must consider *either* of the masses by itself because looking at the system as a whole does not involve the tension. Suppose we consider just those forces acting on the left mass m_1.

Resultant force on m_1 = mass m_1 × acceleration of m_1

The acceleration of m_1 is, of course, the same as for the total system (6.53 m/s^2). So,

$$T - m_1g = m_1a \quad \text{or} \quad T = m_1a + m_1g = m_1(a + g)$$

$$T = (2\text{ kg})(6.53\text{ m/s}^2 + 9.8\text{ m/s}^2)$$

$$T = 32.7\text{ N}$$

You should demonstrate that the same answer would result if we applied Newton's second law to the second mass.

Example 7.10

A 5-kg block m_1 rests on a frictionless tabletop. A cord attached to it passes over a light frictionless pulley and is attached to a mass m_2, as shown in Fig. 7.9. (a) What must be the mass m_2 such that the system accelerates at 2 m/s^2? (b) What is the tension in the cord for this arrangement?

Plan: Again we find it convenient to calculate the acceleration of the *entire* system by summing forces along the entire line of motion. After finding the acceleration, we will then apply Newton's second law to only one of the masses to find the tension in the connecting cord.

Solution (a): The sketch and free-body diagrams are shown for each body in Fig. 7.9. The consistent positive direction is to the right on the tabletop and downward for the suspended mass. The given information is

Given: $m_1 = 5$ kg; $a = 2$ m/s^2; $g = 9.8$ m/s^2;   Find: m_2 and T.

The normal force n balances the weight m_1g of the table mass, and the cord is part of the system, so the resultant force on the system is just the weight of the suspended mass m_2g. Applying Newton's second law to the entire system gives

Resultant force on system = total mass × acceleration

$$m_2g = (m_1 + m_2)a$$

To find the mass m_2, we must isolate that variable as follows:

$$m_2g = m_1a + m_2a$$
$$m_2g - m_2a = m_1a$$
$$m_2(g - a) = m_1a$$
$$m_2 = \frac{m_1a}{g - a}$$

Now, substitution of known values gives

$$m_2 = \frac{(5 \text{ kg})(2 \text{ m/s}^2)}{9.8 \text{ m/s}^2 - 2 \text{ m/s}^2} = 1.28 \text{ kg}$$

Solution (b): Now that we know the mass m_2, we can find the tension in the cord by considering independently either mass m_1 or m_2. The simplest choice

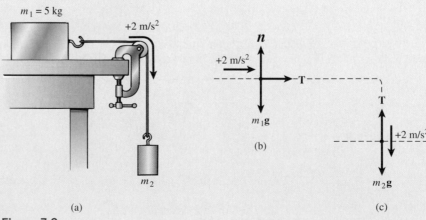

Figure 7.9

would be the table mass m_1, since the resultant force on that mass *is* the tension in the cord.

$$T = m_1a = (5 \text{ kg})(2 \text{ m/s}^2)$$
$$T = 10.0 \text{ N}$$

The same answer results if we apply Newton's second law just for the suspended mass. Choosing the downward direction positive, we have

$$m_2g - T = m_2a \quad \text{or} \quad T = m_2(g - a)$$

Substitution will again show that the tension must be 10.0 N.

Example 7.11

Consider the masses $m_1 = 20$ kg and $m_2 = 18$ kg in the system represented by Fig. 7.10. Assume that the coefficient of kinetic friction is 0.1 and the inclination angle θ is 30°. (a) Find the acceleration of the system. (b) What is the tension in the cord?

Plan: This problem is similar to Example 7.10, except that one of the masses moves up the incline against friction. We will take care to choose a consistent line of motion for the entire system. Newton's law will be applied first to the entire system and then for a single mass.

Solution (a): Draw the free-body diagrams for each object and then list the given information.

Given: $m_1 = 20$ kg; $m_2 = 18$ kg; $g = 9.8$ m/s^2; Find: a and T.

Note the positive line of motion shown in Fig. 7.10. We will need to work with components of vectors that are along or perpendicular to this line. The slope angle is 30°, which means that the reference angle for the weight m_1g is 60° or the complement of the slope angle. Thus, the resultant force on the system is the difference between the suspended weight m_2g and the opposing forces of friction f_k and the weight component m_1g down the incline. Applying Newton's law gives

Resultant force on entire system = total mass × acceleration of system

$$m_2g - f_k - m_1g \cos 60° = (m_1 + m_2)a \tag{7.5}$$

Looking at the free-body diagram and recalling the definition of the friction force, we see that

$$f_k = \mu_k n \quad \text{and} \quad n = m_1g \sin 60°$$

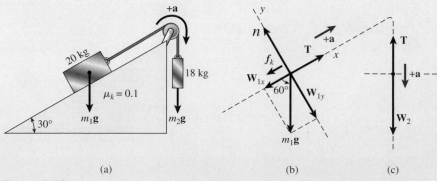

(a)   (b)   (c)

Figure 7.10

Substitution of these quantities into Eq. (7.5) gives

$$m_2g - \mu_k(m_1g \sin 60°) - m_1g \cos 60° = (m_1 + m_2)a$$

Solving for a, we have

$$a = \frac{m_2g - \mu_k m_1g \sin 60° - m_1g \cos 60°}{m_1 + m_2}$$

Finally, we substitute all the given information to find a:

$$a = \frac{(18 \text{ kg})(9.8 \text{ m/s}^2) - 0.1(20 \text{ kg})(9.8 \text{ m/s}^2) \sin 60° - (20 \text{ kg})(9.8 \text{ m/s}^2) \cos 60°}{20 \text{ kg} + 18 \text{ kg}}$$

$$a = 1.62 \text{ m/s}^2$$

You may substitute given information along the way if you prefer, but the danger is that one early mistake will be compounded in all the work that follows.

Solution (b): To find the tension in the cord, we apply Newton's law only for the 18-kg mass. From Fig. 7.10c, we obtain

$$Resultant\ force\ on\ m_2 = mass\ m_2 \times acceleration\ of\ m_2$$

$$m_2g - T = m_2a$$

$$T = m_2g - m_2a = m_2(g - a)$$

$$T = (18 \text{ kg})(9.8 \text{ m/s}^2 - 1.62 \text{ m/s}^2)$$

$$T = 147 \text{ N}$$

Verify this result by applying Newton's law to the mass on the incline.

Summary and Review

Summary

In this chapter, we have considered the fact that a resultant force will always produce an acceleration in the direction of the force. The magnitude of the acceleration is directly proportional to the force and inversely proportional to the mass, according to Newton's second law of motion. The following concepts are essential to applications of this fundamental law:

- The mathematical formula that expresses Newton's second law of motion may be written as follows:

$$Force = mass \times acceleration$$

$$F = ma \qquad m = \frac{F}{a} \qquad a = \frac{F}{m}$$

In SI units: $1 \text{ N} = (1 \text{ kg})(1 \text{ m/s}^2)$

In USCS units: $1 \text{ lb} = (1 \text{ slug})(1 \text{ ft/s}^2)$

- Weight is the force due to a particular acceleration g. Thus, weight W is related to mass m by Newton's second law:

$$W = mg \qquad m = \frac{F}{g} \qquad g = 9.8 \text{ m/s}^2 \text{ or } 32 \text{ ft/s}^2$$

For example, a mass of 1 kg has a weight of 9.8 N. A weight of 1 lb has a mass of $\frac{1}{32}$ slug. In a given problem, you must determine whether weight or mass is given. Then you must determine what is needed in an equation. Conversions of mass to weight and weight to mass are common.

- Application of Newton's second law:

  1. Construct a free-body diagram for each body undergoing an acceleration. Indicate on this diagram the direction of positive acceleration.
  2. Determine an expression for the net resultant force on a body or a system of bodies.
  3. Set the resultant force equal to the total mass of the system multiplied by the acceleration of the system.
  4. Solve the resulting equation for the unknown quantity.

Key Terms

mass   140

newton   139

Newton's second law   139

slug   139

weight   140

Review Questions

7.1. Distinguish clearly between the mass of an object and its weight and give the appropriate units for each in the SI and USCS systems of units.

7.2. What exactly do we mean when we describe an athlete as a 160-lb person? What would be the mass of this person on the Moon if $g_m = 6 \text{ ft/s}^2$?

7.3. A round piece of brass found in the laboratory is labeled 500 g. Is this its weight or its mass? How can you be sure?

7.4. A state of equilibrium is maintained on a force table by hanging masses from pulleys mounted at various locations on the circular edge. In calculating the equilibrant, we sometimes use grams instead of newtons. Are we justified in doing this?

7.5. When drawing free-body diagrams, why is it usually to our advantage to choose either the x or y axis along the direction of motion, even if it means rotated axes? Use the example of motion along an inclined plane as an illustration.

7.6. In the example of an Atwood machine (Example 7.9), we neglected the mass of the cord that connects the two masses. Discuss how this problem is altered if the mass of the cord is large enough to affect the motion.

7.7. In industry, we often hear of a kilogram-force (kg_f) unit, which is defined as a force equivalent to the weight of a 1-kg mass near the Earth's surface. In the United States, we also speak often of the pound-mass (lb_m) unit, which is the mass of an object that has a weight of 1 lb near the surface of the Earth. Find the value of these quantities in appropriate SI units and discuss the problems caused by their use.

Problems

Section 7.1 Newton's Second Law

7.1. A 4-kg mass is acted on by a resultant force of (a) 4 N, (b) 8 N, and (c) 12 N. What are the resulting accelerations?

Ans. (a) 1 m/s^2, (b) 2 m/s^2, (c) 3 m/s^2

7.2. A constant force of 20 N acts on a mass of (a) 2 kg, (b) 4 kg, and (c) 6 kg. What are the resulting accelerations?

7.3. A constant force of 60 lb acts on each of three objects, producing accelerations of 4, 8, and 12 ft/s^2. What are the masses?

Ans. 15, 7.5, and 5 slugs

7.4. What resultant force is necessary to give a 4-kg hammer an acceleration of 6 m/s^2?

7.5. It is determined that a resultant force of 60 N will give a wagon an acceleration of 10 m/s^2. What force is required to give the wagon an acceleration of only 2 m/s^2? Ans. 12 N

7.6. A 1000-kg car moving north at 100 km/h brakes to a stop in 50 m. What are the magnitude and direction of the force? Ans. 7720 N, south

Section 7.2 The Relationship Between Weight and Mass

7.7. What is the weight of a 4.8-kg mailbox? What is the mass of a 40-N tank? Ans. 47.0 N, 4.08 kg

7.8. What is the mass of a 60-lb child? What is the weight of a 7-slug man?

7.9. A woman weighs 800 N on Earth. When she walks on the Moon, she weighs only 133 N. What is the acceleration due to gravity on the Moon, and what is her mass on the Moon? On the Earth?

Ans. 1.63 m/s^2, 81.6 kg both places

7.10. What is the weight of a 70-kg astronaut on the surface of the Earth? Compare the *resultant* force required to give him or her an acceleration of 4 m/s^2 on the Earth with the *resultant* force required to give the same acceleration in space where gravity is negligible.

7.11. Find the mass and the weight of a body if a resultant force of 16 N will give it an acceleration of 5 m/s^2. Ans. 3.20 kg, 31.4 N

7.12. Find the mass and weight of a body if a resultant force of 850 N causes its speed to increase from 6 m/s to 15 m/s in a time of 5 s near the surface of the Earth.

7.13. Find the mass and weight of a body if a resultant force of 400 N causes it to decrease its velocity by 4 m/s in 3 s. Ans. 300 kg, 2940 N

Section 7.3 Application of Newton's Second Law to Single-Body Problems

7.14. What horizontal pull is required to drag a 6-kg sled with an acceleration of 4 m/s^2 if a friction force of 20 N opposes the motion?

7.15. A 1200-kg automobile is speeding at 25 m/s. What resultant force is required to stop the car in 70 m on a level road? What must be the coefficient of kinetic friction? Ans. 5360 N, 0.456

7.16. A 10-kg mass is lifted upward by a light cable. What is the tension in the cable if the acceleration is (a) zero, (b) 6 m/s^2 upward, and (c) 6 m/s^2 downward?

7.17. A 20-kg mass hangs at the end of a rope. Find the acceleration of the mass if the tension in the cable is (a) 196 N, (b) 120 N, and (c) 260 N.

Ans. 0, −3.8 m/s^2, +3.2 m/s^2

7.18. An 800-kg elevator is lifted vertically by a strong rope. Find the acceleration of the elevator if the rope tension is (a) 9000 N, (b) 7840 N, and (c) 2000 N.

7.19. A horizontal force of 100 N pulls an 8-kg cabinet across a level floor. Find the acceleration of the cabinet if $\mu_k = 0.2$. Ans. 10.5 m/s^2

7.20. In Fig. 7.11, an unknown mass slides down the 30° inclined plane. What is the acceleration in the absence of friction?

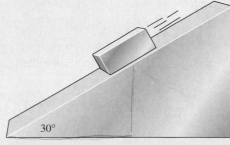

Figure 7.11

7.21. Assume that $\mu_k = 0.2$ in Fig. 7.11. What is the acceleration? Why did you not need to know the mass of the block?

Ans. 3.20 m/s^2, down the plane

*7.22. Assume that $m = 10$ kg and $\mu_k = 0.3$ in Fig. 7.11. What push **P** directed up and along the incline in Fig. 7.11 will produce an acceleration of 4 m/s^2 also up the incline?

*7.23. What force **P** down the incline in Fig. 7.11 is required to cause the acceleration *down* the plane to be 4 m/s^2? Assume that $m = 10$ kg and $\mu_k = 0.3$.

Ans. 16.5 N

Section 7.4 Applications of Newton's Second Law to Multibody Problems

7.24. Assume zero friction for the system shown in Fig. 7.12. What is the acceleration of the system? What is the tension T in the connecting cord?

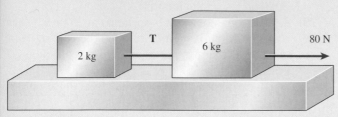

Figure 7.12

7.25. What force does block A exert on block B in Fig. 7.13?   Ans. 15.0 N

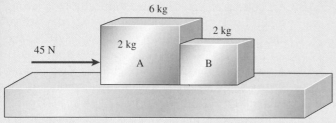

Figure 7.13

*7.26. What are the acceleration of the system and the tension in the connecting cord for the arrangement shown in Fig. 7.14? Surfaces are frictionless.

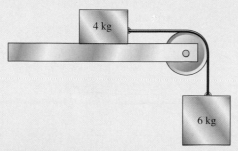

Figure 7.14

*7.27. If the coefficient of kinetic friction between the table and the 4-kg block is 0.2 in Fig. 7.14, what is the acceleration of the system? What is the tension in the cord?   Ans. 5.10 m/s^2, 28.2 N

*7.28. Assume that the masses $m_1 = 2$ kg and $m_2 = 8$ kg are connected by a cord that passes over a light frictionless pulley as in Fig. 7.15. What are the acceleration of the system and the tension in the cord?

Figure 7.15

*7.29. The system described in Fig. 7.16 starts from rest. What is the acceleration assuming zero friction?   Ans. 2.69 m/s^2 down the incline

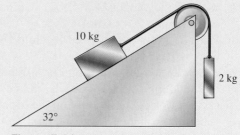

Figure 7.16

*7.30. What is the acceleration in Fig. 7.16 as the 10-kg block moves down the plane against friction? Assume $\mu_k = 0.2$.

*7.31. What is the tension in the cord for Prob. 7.30?   Ans. 22.2 N

Additional Problems

7.32. In Fig. 7.15, assume the mass m_2 is three times that of mass m_1. Find the acceleration of the system.

7.33. A 200-lb worker stands on weighing scales in an elevator where the upward acceleration is 6 ft/s^2. The elevator stops and then accelerates downward at 6 ft/s^2. What is the reading of the scales on the way up, and what is the reading on the way down?   Ans. 238 lb, 163 lb

7.34. A 8-kg load is accelerated upward with a cord whose breaking strength is 200 N. What is the maximum acceleration?

7.35. For rubber tires on a concrete road $\mu_k = 0.7$. What is the minimum horizontal stopping distance for a 1600-kg truck traveling at 20 m/s?   Ans. 29.2 m

*7.36. Suppose the 4- and 6-kg masses in Fig. 7.14 are switched so that the larger mass is on the table. What would be the acceleration and tension in the cord (neglecting friction)?

*7.37. Consider two masses A and B connected by a cord and hung over a single pulley. If mass A is twice that of mass B, what is the acceleration of the system?   Ans. 3.27 m/s^2

*7.38. A 5-kg mass rests on a 34° inclined plane where $\mu_k = 0.2$. What push up the incline will cause the block to accelerate at 4 m/s^2?

*7.39. A 96-lb block rests on a table where $\mu_k = 0.2$. A cord tied to this block passes over a light, frictionless pulley. What weight must be attached to the free end if the system is to accelerate at 4 ft/s^2?   Ans. 35.7 lb

Critical Thinking Questions

7.40. In a laboratory experiment, the acceleration of a small car is measured by the separation of spots burned at regular intervals in a paraffin-coated tape. Larger and larger weights are transferred from the car to a hanger at the end of a tape that passes over a light, frictionless pulley. In this manner, the mass of the entire system is kept constant. Since the car moves on a horizontal air track with negligible friction, the resultant force is equal to the weights at the end of the tape. The following data are recorded:

Weight, W	2	4	6	8	10	12
Acceleration, m/s^2	1.4	2.9	4.1	5.6	7.1	8.4

Plot a graph of weight versus acceleration. What is the significance of the slope of this curve? What is the mass of the system?

7.41. In the experiment described in Prob. 7.40, the student places a constant weight of 4 N at the free end of the tape. Several runs are made, increasing the mass of the car each time by adding weights. What happens to the acceleration as the mass of the system is increased? What should the value of the product of the mass of the system and the acceleration be for each run? Is it necessary to include the mass of the constant 4-N weight in these experiments?

7.42. An arrangement similar to that described by Fig. 7.14 is set up except that the masses are replaced. What is the acceleration of the system if the suspended mass is three times that of the mass on the table and $\mu_k = 0.3$?   Ans. 6.62 m/s^2

7.43. Three masses, 2 kg, 4 kg, and 6 kg, are connected (in order) by strings and hung from the ceiling with another string so that the largest mass is in the lowest position. What is the tension in each cord? If they are then detached from the ceiling, what must be the tension in the top string in order that the

system accelerate upward at 4 m/s^2? In the latter case, what are the tensions in the strings that connect masses?

7.44. An 80-kg astronaut on a space walk pushes against a 200-kg solar panel that has become dislodged from a spacecraft. The force causes the panel to accelerate at 2 m/s^2. What acceleration does the astronaut receive? Do they continue to accelerate after the push? Why or why not?
   Ans. -5.00 m/s^2, no

7.45. A 400-lb sled slides down a hill ($\mu_k = 0.2$) inclined at an angle of 60°. What is the normal force on the sled? What is the force of kinetic friction? What is the resultant force down the hill? What is the acceleration? Is it necessary to know the weight of the sled to determine its acceleration?

7.46. Three masses, $m_1 = 10$ kg, $m_2 = 8$ kg, and $m_3 = 6$ kg, are connected as shown in Fig. 7.17. Neglecting friction, what is the acceleration of the system? What are the tensions in the cord on the left and in the cord on the right? Would the acceleration be the same if the middle mass m_2 were removed?

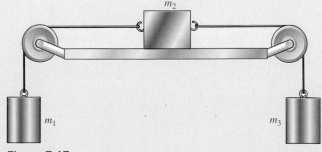

Figure 7.17

*7.47. Assume that $\mu_k = 0.3$ between the mass m_2 and the table in Fig. 7.17. The masses m_2 and m_3 are 8 and 6 kg, respectively. What mass m_1 is required to cause the system to accelerate to the left at 2 m/s^2?

*7.48. A block of unknown mass is given a push up a 40° inclined plane and then released. It continues to move up the plane (+) at an acceleration of −9 m/s². What is the coefficient of kinetic friction? Ans. 0.360

*7.49. Block A in Fig. 7.18 has a weight of 64 lb. What must be the weight of block B if block A moves up the plane with an acceleration of 6 ft/s²? Neglect friction.

*7.50. The mass of block B in Fig. 7.18 is 4 kg. What must be the mass of block A if it is to move down the plane at an acceleration 2 m/s²? Neglect friction. Ans. 7.28 kg

*7.51. Assume that the masses A and B in Fig. 7.18 are 4 kg and 10 kg, respectively. The coefficient of kinetic friction is 0.3. Find the acceleration if (a) the system is initially moving up the plane, and (b) if the system is initially moving down the plane.

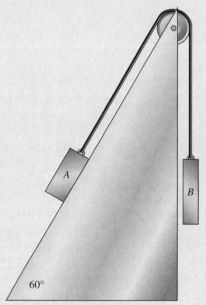

A

B

60°

Figure 7.18

8

Work, Energy, and Power

The Ninja, a roller coaster at Six Flags over Georgia, has a height of 122 ft and a speed of 52 mi/h. The gravitational potential energy due to its height changes into kinetic energy of motion, and the interchange between the two forms of energy continues until the end of the ride.
(*Photo by Paul E. Tippens.*)

Objectives

After completing this chapter, you should be able to

1. Define and write mathematical formulas for work, potential energy, kinetic energy, and power.
2. Apply the concepts of work, energy, and power to the solution of problems similar to those given as examples in the text.
3. Define and demonstrate by example your understanding of the following units: joule, foot-pound, watt, horsepower, and foot-pound per second.
4. Discuss and apply your knowledge of the relationship between the performance of work and the corresponding change in kinetic energy.
5. Discuss and apply your knowledge of the principle of conservation of mechanical energy.
6. Determine the power of a system and understand its relationship to time, force, distance, and velocity.

The principal reason for the application of a resultant force is to cause a displacement. For example, a large crane lifts a steel beam to the top of a building; the compressor in an air conditioner forces a fluid through its cooling cycle; and electromagnetic forces move electrons across a television screen. Whenever a force acts through a distance, we will learn, *work* is done in a way that can be measured or predicted. The capacity for doing work will be defined as ***energy,*** and the rate at which it is accomplished will be defined as ***power.*** The control and use of energy is probably the major concern of industry today, and a thorough understanding of the three concepts of work, energy, and power is essential.

8.1   Work

When we attempt to drag a wagon with a rope, as in Fig. 8.1a, nothing happens. We are exerting a force, but the wagon has not moved. On the other hand, if we continually increase our pull, eventually the wagon will be displaced. In this case, we have actually accomplished something in return for our efforts. This accomplishment is defined in physics as **work.** The term *work* has an explicit, quantitative, operational definition. For work to be done, three things are necessary:

1. There must be an applied force.

2. The force must act through a certain distance, called the ***displacement.***

3. The force must have a component along the displacement.

Assuming that we are given these conditions, a formal definition of work may be stated:

> Work is a scalar quantity equal to the product of the magnitudes of the displacement and the component of the force in the direction of the displacement.

$$Work = force\ component \times displacement$$

$$Work = F_x x \tag{8.1}$$

In this equation, F_x is the component of **F** along the displacement x. In Fig. 8.1, only F_x contributes to work. Its magnitude can be found from trigonometry, and work can be expressed in terms of the angle θ between **F** and x:

$$Work = (F \cos \theta)x \tag{8.2}$$

Quite often the force causing the work is directed entirely along the displacement. This happens when a weight is lifted vertically or when a horizontal force drags an object along the floor. In these simple cases, $F_x = F$, and the work is the simple product of force and displacement:

$$Work = Fx \tag{8.3}$$

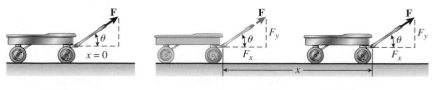

(a) Work = 0   (b) Work = $(F \cos \theta)x$

Figure 8.1   The work done by a force **F** undergoing a displacement x.

Another special case occurs when the applied force is perpendicular to the displacement. In this instance, the work will be zero, since $F_x = 0$. An example is motion parallel to the Earth's surface in which gravity acts vertically downward and is perpendicular to all horizontal displacements. Then the force of gravity does no work.

Example 8.1 What work is done by a 60-N force in pulling the wagon in Fig. 8.1 a distance of 50 m when the force transmitted by the handle makes an angle of 30° with the horizontal?

Plan: Only the component of the applied force **F** that lies along the displacement contributes to work. The work will be found as the product of this component $F \cos \theta$ and the linear displacement x.

Solution: Applying Eq. (8.1) gives

$$\text{Work} = (F \cos \theta)x = (60 \text{ N})(\cos 30°)(50 \text{ m})$$
$$\text{Work} = 2600 \text{ N} \cdot \text{m}$$

Note that the units of work are the units of force times distance. Thus, in SI units, work is measured in *newton-meters* ($\text{N} \cdot \text{m}$). By agreement, this combination unit is renamed the *joule,* which is denoted by the symbol J.

> One joule (1 J) is equal to the work done by a force of 1 newton in moving an object through a parallel distance of 1 meter.

In Example 8.1, the work done in pulling the wagon would be written as 2600 J.

In the United States, work is sometimes also given in USCS units. When the force is given in *pounds* (lb) and the displacement is given in *feet* (ft), the corresponding work unit is called the *foot-pound* ($\text{ft} \cdot \text{lb}$).

> One foot-pound (1 ft · lb) is equal to the work done by a force of 1 pound in moving an object through a parallel distance of 1 foot.

No special name is given to this unit.

The following conversion factors will be useful when comparing work units in the two systems:

$$1 \text{ J} = 0.7376 \text{ ft} \cdot \text{lb} \qquad 1 \text{ ft} \cdot \text{lb} = 1.356 \text{ J}$$

8.2 Resultant Work

When we consider the work of several forces acting on the same object, it is often useful to distinguish between positive and negative work. In this text, we will follow the convention that the work of a particular force is positive if the force component is in the same direction as the displacement. Negative work is done by a force component that opposes the actual displacement. Hence, work done by a crane in lifting a load is positive, but the gravitational force exerted by the Earth on the load is doing negative work. Similarly, when we stretch a spring, the work on the spring is positive; the work on the spring is negative when the spring contracts, pulling us back. Another important example of negative work is that performed by a frictional force that is opposite to the direction of displacement.

If several forces act on a body in motion, the ***resultant work*** (total work) is the algebraic sum of the works of the individual forces. This will also be equal to the work of the resultant force. The accomplishment of net work requires the existence of a resultant force. These ideas are clarified in Example 8.2.

Example 8.2

A push of 80 N moves a 5-kg block up a 30° inclined plane, as shown in Fig. 8.2. The coefficient of kinetic friction is 0.25, and the length of the plane is 20 m. (a) Compute the work done by each force acting on the block. (b) Show that the net work done by these forces is the same as the work of the resultant force.

Plan: Construct and label a free-body diagram (see Fig. 8.2b) showing each of the forces acting throughout the displacement x. It will be important to distinguish between the work of an individual force, such as $\mathbf{P}$, f_k, n, or $\mathbf{W}$ and the *resultant work*. In the first part of the problem, we will consider the work of each of these forces independent of the others. Then, recognizing that all the forces have a common displacement, we will show that the resultant work is the same as the sum of the individual works.

Solution (a): Note that the normal force does zero work because it is perpendicular to the displacement and $\cos 90° = 0$.

$$(\text{Work})_n = (n \cos 90°)x \qquad \text{or} \qquad (\text{Work})_n = 0$$

The push $\mathbf{P}$ is entirely along the displacement and in the same direction. Thus,

$$(\text{Work})_P = (P \cos 0°)_x = (80 \text{ N})(1)(20 \text{ m})$$
$$(\text{Work})_P = 1600 \text{ J}$$

To find the work of the friction force f_k and the work of the weight $\mathbf{W}$, we must first determine the components of the weight along and perpendicular to the plane.

$$W = mg = (5 \text{ kg})(9.8 \text{ m/s}^2); \quad W = 49.0 \text{ N}$$
$$W_x = (49.0 \text{ N}) \sin 30° = 24.5 \text{ N}$$
$$W_y = (49.0 \text{ N}) \cos 30° = 42.4 \text{ N}$$

Notice that the reference angle 30° is with respect to the y axis in this instance to avoid cluttering the diagram. That means that the opposite side is the x component and the adjacent side is the y component. Be careful to choose the correct trig functions.

The forces normal to the plane are balanced so that $n = W_y$ and

$$n = W_y = 42.4 \text{ N}$$

That means that the friction force f_k is

$$f_k = \mu_k n = (0.25)(42.4 \text{ N}) \qquad \text{or} \qquad f_k = -10.6 \text{ N}$$

The negative sign indicates that the friction force is directed down the plane. Therefore, the work done by the friction force is

$$(\text{Work})_f = f_k x = (-10.6 \text{ N})(20 \text{ m}); \qquad (\text{Work})_f = -212 \text{ J}$$

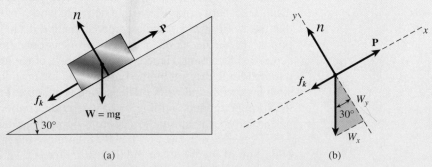

(a)   (b)

Figure 8.2 The work required to push a block up a 30° inclined plane.

The weight **W** of the block also does negative work since its component W_x is directed opposite to the displacement.

$$(\text{Work})_w = -(24.5 \text{ N})(20 \text{ m}) = -490 \text{ J}$$

Solution (b): The net work is equal to the sum of the works done by each force.

$$\text{Net work} = (\text{work})_n + (\text{work})_P + (\text{work})_f + (\text{work})_w$$
$$= 0 + 1600 \text{ J} - 212 \text{ J} - 490 \text{ J}$$
$$= 898 \text{ J}$$

To show that this is also the work of the resultant force, we first compute the resultant force, which is the sum of the forces along the incline.

$$F_R = P - f_k - W_x$$
$$= 80 \text{ N} - 10.6 \text{ N} - 24.5 \text{ N} = 44.9 \text{ N}$$

The work of F_R is therefore

$$\text{Net work} = F_R x = (44.9 \text{ N})(20 \text{ m}) = 898 \text{ J}$$

which compares with the value obtained by computing the work of each force separately.

It is important to distinguish between the *resultant* or *net* work and the work of an individual force. If we speak of the work required to move an object through a distance, the work done by the pulling force is not necessarily the resultant work. Work may be done by a friction force or by other forces. *Resultant work* is simply the work done by a resultant force. If the resultant force is zero, then the resultant work is zero, even though individual forces may be doing positive or negative work.

8.3 Energy

Energy may be thought of as *anything that can be converted to work.* When we say that an object has energy, we mean that it is capable of exerting a force on another object to do work on it. Conversely, if we do work on some object, we have added to it an amount of energy equal to the work done. The units of energy are the same as those for work: the *joule* and the *foot-pound.*

In mechanics, we are concerned with two kinds of energy:

Kinetic Energy K: Energy possessed by a body by virtue of its motion.

Potential Energy U: Energy possessed by a system by virtue of position or condition.

Any mass m that has velocity also is said to have kinetic energy. However, in order for potential energy to be present, there must be the *potential* of an applied force. Thus, an object cannot *have* potential energy. Rather the potential energy must belong to a *system.* A box held a certain distance above the surface of the Earth is an example of a system that has *potential energy.* If the box is released, the Earth can exert a force on the box. Without the Earth, there is no potential energy.

One can readily think of many examples for each kind of energy. A moving car, a moving bullet, and a rotating flywheel all have the ability to do work because of their motion. Similarly, a lifted object, a compressed spring, and a stretched rubber band offer the potential for work, provided a force is activated. Several examples of each kind of energy are provided in Fig. 8.3.

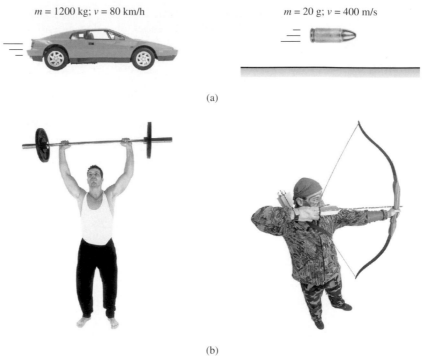

$m = 1200$ kg; $v = 80$ km/h

$m = 20$ g; $v = 400$ m/s

(a)

(b)

Figure 8.3 (a) The kinetic energy of a moving car or bullet. (b) The potential energy of a suspended weight or a cocked bow. (*Photos by Hemera, Inc.*)

| 8.4 | # Work and Kinetic Energy |

We have defined kinetic energy as the capability for performing work as a result of the motion of a body. To see the relationship between motion and work, consider a force **F** acting on the cart in Fig. 8.4. We will assume that this force is the *resultant force* on the cart, ignoring any friction forces. Consider that the cart and its load have a combined mass m and its initial and final velocities are v_0 and v_f, respectively. According to Newton's second law of motion, acceleration will occur at a rate given by

$$a = \frac{F}{m} \qquad (8.4)$$

Continuing with this example, we recall from Chapter 5 that

$$2ax = v_f^2 - v_0^2$$

which can be expressed in terms of a as follows:

$$a = \frac{v_f^2 - v_0^2}{2x}$$

If we substitute this expression into Eq. (8.4), we have

$$\frac{F}{m} = \frac{v_f^2 - v_0^2}{2x}$$

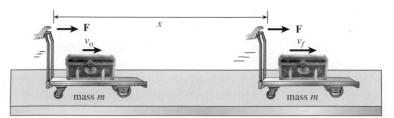

Figure 8.4 The work done by the resultant force **F** results in a change in the kinetic energy of the total mass m. (*Photos by Hemera, Inc.*)

Rearranging the factors and simplifying gives

$$Fx = \frac{1}{2}mv_f^2 - \frac{1}{2}mv_0^2$$

A close look at this result shows us that the left side of the equation represents the *resultant work* done by a constant force that acts through a displacement *x*. The terms on the right side appear to be final and initial values of some important quantity ($\frac{1}{2}mv^2$). We will call this quantity the kinetic energy *K* and write the formula

$$K = \frac{1}{2}mv^2 \qquad\qquad \textit{Kinetic Energy} \quad \textbf{(8.5)}$$

Given this definition, we can now state that *the resultant work done on a mass **m** by a constant force **F** acting through a distance **x** is equal to its change in kinetic energy Δ**K**.* This is a statement of what we will call the **work-energy theorem.**

> **The Work-Energy Theorem:** The work of a resultant external force on a body is equal to the change in kinetic energy of that body.

$$Fx = \frac{1}{2}mv_f^2 - \frac{1}{2}mv_0^2 \qquad\qquad \textbf{(8.6)}$$

In many applications, the force *F* in Eq. (8.6) is not constant, but varies significantly over time. In such cases, the work-energy theorem can be applied to find the *average force,* which might be thought of as that constant force that would accomplish the same amount of work.

A close look at the work-energy theorem will show that an increase in kinetic energy ($v_f > v_0$) will result from *positive work,* whereas a *decrease* in kinetic energy ($v_f < v_0$) will result from *negative* work. In the special case where zero work is done, the kinetic energy is constant and equal to the value given in Eq. (8.5). It should also be noted that the unit of kinetic energy must be the same as the unit of work. As an exercise, you should show that $1 \text{ kg} \cdot \text{m/s}^2 = 1 \text{ J}$.

Example 8.3

Find the kinetic energy of a 4-kg sledgehammer at the instant its velocity is 24 m/s.

Solution: Direct application of Eq. (8.5) yields

$$K = \frac{1}{2}mv^2 = \frac{1}{2}(4 \text{ kg})(24 \text{ m/s})^2$$

$$K = 1150 \text{ J}$$

Example 8.4

Compute the kinetic energy for a 3200-lb automobile traveling at 88 ft/s (60 mi/h).

Plan: Since the automobile is described by its weight in USCS units, we must divide by gravity to find its mass. The kinetic energy is then found as usual.

Solution:

$$K = \frac{1}{2}mv^2 = \frac{1}{2}\left(\frac{W}{g}\right)v^2$$

$$= \frac{1}{2}\left(\frac{3200 \text{ lb}}{32 \text{ ft/s}^2}\right)(88 \text{ ft/s})^2 = 3.87 \times 10^5 \text{ ft} \cdot \text{lb}$$

The use of the ft · lb as a unit of energy is outdated and strongly discouraged. However, it remains in limited use today, and conversions are sometimes necessary.

Example 8.5

What average force F is necessary to stop a 16-g bullet traveling at 260 m/s as it penetrates into a block of wood for a distance of 12 cm?

Plan: The force exerted by the block on the bullet is by no means constant, but we can assume an *average* stopping force. Then, the work required to stop the bullet will be equal to the change in kinetic energy. (See Fig. 8.5.)

Solution: Noting that the velocity of the bullet changes from an initial value of $v_0 = 260$ m/s to a final value of zero, direct application of Eq. (8.6) yields

$$Fx = \frac{1}{2}m(0)^2 - \frac{1}{2}mv_f^2 \qquad \text{or} \qquad Fx = -\frac{1}{2}mv_f^2$$

Solving explicitly for F, we obtain

$$F = \frac{-mv_f^2}{2x}$$

The given quantities in SI units are

$$m = 16 \text{ g} = 0.016 \text{ kg}; \quad x = 12 \text{ cm} = 0.12 \text{ m}; \quad v_0 = 260 \text{ m/s}$$

Substitution yields the average stopping force.

$$F = \frac{-mv_f^2}{2x} = \frac{-(0.016 \text{ kg})(260 \text{ m/s})^2}{2(0.12 \text{ m})}$$

$$F = -4510 \text{ N}$$

The negative sign indicates that the force was directed opposite to the displacement. It should be noted that this force is about 30,000 times the weight of the bullet.

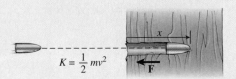

Figure 8.5   The work done in stopping the bullet is equal to the change in kinetic energy of the bullet.

8.5

Potential Energy

The energy that systems possess by virtue of their positions or conditions is called *potential energy*. Since energy expresses itself in the form of work, potential energy implies that there must be a potential for doing work. For example, suppose the pile driver in Fig. 8.6 is used to lift a body of weight **W** to a height h above the ground stake. We say that the body-Earth system has gravitational potential energy. When such a body is released, it will do work when it strikes the stake. If it is heavy enough and if it has fallen from a great enough height, the work done will result in driving the stake through a distance y.

The external force **F** required to lift the body must at least be equal to the weight **W**. Thus, the work done on the system is given by

$$\text{Work} = Wh = mgh$$

This amount of work can also be done *by* the body after it has dropped a distance h. Thus, the body has potential energy equal in magnitude to the external work required to lift it. This energy does not come from the Earth-body system, but results from work done on the system by an external agent. Only *external* forces, such as **F** in Fig. 8.6 or friction, can add energy to or remove energy from the system made up of the body and the Earth.

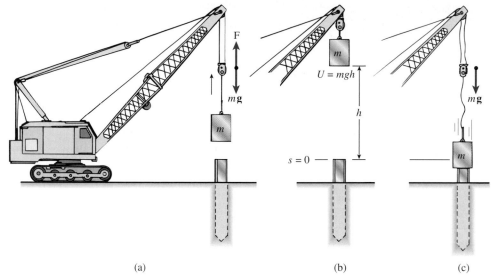

Figure 8.6 (a) Lifting a mass m to a height h requires the work mgh. (b) The potential energy, therefore, is mgh. (c) When the mass is released, it has the capacity for doing the work mgh on the stake.

PHYSICS TODAY

Water above a waterwheel is potential energy. As the water falls, it becomes kinetic energy, which is used to turn the wheel. The potential energy decreases as the kinetic energy increases.

Note from the preceding discussion that potential energy U can be found from

$$U = Wh = mgh \qquad \textit{Potential Energy} \quad (8.7)$$

where W and m are the weight and the mass of an object located a distance h above some reference point.

The potential energy depends on the choice of a particular reference level. The gravitational potential energy for an airplane is quite different when measured to a mountain peak, a skyscraper, or the ocean. The capacity for doing work is much greater if the aircraft falls to sea level. Potential energy has physical significance only in the event that a reference is established.

Example 8.6

A 1.2-kg toolbox is held 2 m above the top of a table that is 80 cm from the floor. Find the potential energy relative to the top of the table and relative to the floor.

Plan: The height above the table and the height above the floor are the two reference points for potential energy. The product of weight and height will give the potential energy relative to those positions.

Solution (a): The potential energy relative to the tabletop is

$$U = mgh = (1.2 \text{ kg})(9.8 \text{ m/s}^2)(2 \text{ m})$$
$$= 23.5 \text{ J}$$

Notice that kilograms, meters, and seconds are the only units of mass, length, and time that are consistent with the definition of a joule.

Solution (b): The total height in the second case is the sum of the height of the tabletop from the floor and the height of the toolbox above the tabletop.

$$U = mgh = mg(2 \text{ m} + 0.80 \text{ m})$$
$$= (1.2 \text{ kg})(9.8 \text{ m/s}^2)(2.8 \text{ m})$$
$$= 32.9 \text{ J}$$

Example 8.7

A 300-kg commercial air conditioner is lifted by a chain hoist until its potential energy is 26 kJ relative to the shop floor. What is its height above the floor?

Plan: We will solve Eq. (8.7) for the height h and substitute the known values.

Solution: We are given that $U = 26$ kJ or 26,000 J and $m = 300$ kg.

$$U = mgh; \qquad h = \frac{U}{mg}$$

$$h = \frac{26,000 \text{ J}}{(300 \text{ kg})(9.8 \text{ m/s}^2)} = 8.84 \text{ m}$$

PHYSICS TODAY

The stones in an Egyptian pyramid built 2500 years ago have the same potential energy as when they were first built.

We have stated that the potential for doing work is a function only of the weight mg and the height h above some reference level. The potential energy at a particular position above a reference point is not dependent on the path taken to reach that position because the same work must be done against gravity regardless of the path. In Example 8.7, 26 kJ of work was required to lift the air conditioner to a vertical height of 8.84 m. If we choose to exert a lesser force by moving it up an incline, a greater distance would be required. In either case, the work done against gravity is 26 kJ because the end result is the placement of a 300-kg mass at a height of 8.84 m.

8.6   Conservation of Energy

Quite often, at relatively low speeds, an interchange takes place between kinetic and potential energies. For example, consider a mass m lifted to a height h and dropped, as shown in Fig. 8.7. An external force has increased the energy of the system, giving it a potential energy $U = mgh$ at the highest point. This is the total energy available to the system, and it cannot change unless an external resistive force is encountered. As the mass falls, its potential energy decreases because its height above the ground is reduced. The lost

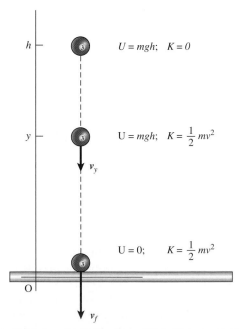

Figure 8.7 In the absence of friction, the total energy $(U + K)$ is constant. It is the same at the top, at the bottom, or at any other point in the path.

potential energy reappears in the form of kinetic energy of motion. In the absence of air resistance, the total energy ($U + K$) remains the same. Potential energy continues to be converted into kinetic energy until the mass reaches the ground ($h = 0$).

At this final position, the kinetic energy is equal to the total energy, and the potential energy is zero. The important point to be made is that the sum of U and K is the same at any point during the fall (see Fig. 8.7). If we represent the total energy of a system by the symbol E, we can write

$$Total\ energy = kinetic\ energy + potential\ energy = constant$$
$$E = K + U = \text{constant}$$

In the example of the falling ball, we say that mechanical energy is *conserved*. At the top, the total energy is mgh, and at the bottom, the total energy is $\frac{1}{2}mv_f^2$, if we neglect air resistance. We are now prepared to state the principle of ***conservation of mechanical energy.***

Conservation of Mechanical Energy: In the absence of air resistance or other dissipative forces, the sum of the potential and kinetic energies is a constant, provided that no energy is added to the system.

In applying this principle, it is useful to think of a beginning point and an ending point for any process. At either point, if there is velocity v, there is kinetic energy K; if there is height h, there is potential energy U. If we assign the subscripts 0 to the beginning point and f to the final point, we might write

$$Total\ energy\ at\ beginning\ point = total\ energy\ at\ ending\ point$$
$$U_0 + K_0 = U_f + K_f$$

Or, based on the appropriate formulas

$$mgh_0 + \frac{1}{2}mv_0^2 = mgh_f + \frac{1}{2}mv_f^2 \qquad \textbf{(8.8)}$$

Of course, this equation strictly applies only to those cases where no friction forces are involved and no energy is added to the system.

In the example of an object falling from rest at an initial height h_0, the total initial energy is mgh_0 ($v_0 = 0$), and the total final energy is $\frac{1}{2}mv_f^2$ ($h_f = 0$). Thus,

$$mgh_0 = \frac{1}{2}mv_f^2$$

Solving this relationship for v_f gives a useful equation for determining the final velocity from energy considerations for an object falling from rest with no friction.

$$v_f = \sqrt{2gh_0}$$

Note that mass is not a factor in determining the final velocity because it appears in each of the formulas for energy. A great advantage of this method is that the final velocity is determined from the initial and final energy states. In the absence of friction, the actual path taken does not matter. For example, the same final velocity would result if the object followed a curved path from the same initial height.

Example 8.8

In Fig. 8.8, a 40-kg wrecking ball is pulled to one side until it is 1.6 m above its lowest point. Neglecting all friction, what will be its velocity as it passes through its lowest point?

Plan: Conservation of total energy requires that the sum of U and K be the same at the beginning and ending points. We can find the velocity by recognizing that the final kinetic energy must be equal to the initial potential energy if energy is conserved.

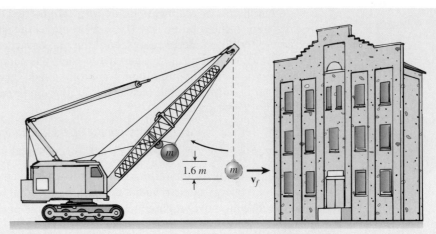

Figure 8.8 The velocity of a suspended mass as it passes through its lowest point can be found from energy considerations.

Solution: Applying Eq. (8.8), we have

$$mgh_0 + 0 = 0 + \frac{1}{2}mv_f^2 \quad \text{or} \quad mgh_0 = \frac{1}{2}mv_f^2$$

Solving for the final velocity and substituting known values, we obtain

$$v = \sqrt{2gh_0} = \sqrt{2(9.8 \text{ m/s}^2)(1.6 \text{ m})}$$
$$v_f = 5.60 \text{ m/s}$$

As an added example, you should show that the total energy E at the beginning and end of this process is 627 J.

8.7 Energy and Friction Forces

It is helpful to consider the conservation of mechanical energy as an accounting process in which one keeps track of what happens to the energy of a system from beginning to end. For example, suppose you withdraw $1000 from a bank and then pay $400 for an airline ticket to New York. You would have $600 left to spend on entertainment. The $400 was spent and cannot be refunded, but you still must account for it. Now consider a sled at the top of a hill and assume a total energy of 1000 J. If 400 J of energy is lost to friction forces as the sled moves downhill, the sled would arrive at the bottom with 600 J to be used for velocity. The 400 J lost to work done against friction forces cannot be regained, so the final total energy E_f is less than the initial total energy E_0. The heat and other dissipative losses still must be accounted for in the process. We might write the following statement:

Initial total energy = final total energy + losses due to friction

$$U_0 + K_0 = U_f + K_f + |\text{work against friction}| \tag{8.9}$$

The work done by friction forces is always negative, so we have used the absolute value brackets to indicate that we are accounting for the positive value of the loss in energy.

By accounting for friction, we can now write a more general statement of the ***conservation of energy:***

Conservation of Energy: The total energy of a system is always constant, although energy changes from one form to another may occur within the system.

In real-world applications, it is not possible to remove external forces from consideration. A more general statement of the principle of conservation can be obtained by rewriting Eq. (8.9) in terms of the initial and final values of height and velocity.

$$mgh_0 + \frac{1}{2}mv_0^2 = mgh_f + \frac{1}{2}mv_f^2 + |f_k x| \tag{8.10}$$

The energy loss term has been replaced with the absolute value of the work done by a kinetic friction force acting through a distance x.

Of course, if an object begins at rest ($v_0 = 0$) from a height h_0 above its final position ($h_f = 0$), Eq. (8.10) simplifies to

$$mgh_0 = \frac{1}{2}mv_f^2 + |f_k x| \tag{8.11}$$

In working problems, it is useful to establish the sum of potential and kinetic energies at some beginning point. Then determine the total energy at the ending point and add the absolute value of any energy losses. Conservation of energy demands that these two expressions be equal. From such a statement, you may then solve for the unknown parameter.

Example 8.9

A 20-kg sled rests at the top of a 30° slope 80 m in length, as shown in Fig. 8.9. If $\mu_k = 0.2$, what is the velocity at the bottom of the incline?

Plan: The total energy E at the beginning is the potential energy $U = mgh_0$. Some of this energy is lost doing work against friction $f_k x$, leaving the remainder for kinetic energy $K - \frac{1}{2}mv^2$. We will draw a free-body diagram, as shown in Fig. 8.9, and use it to calculate the magnitude of the friction force. Finally, applying the law of conservation of energy, we can solve for the velocity at the bottom.

Solution: Before making any calculations, let's write the conservation equation in general terms. The total energy at the top must equal the total energy at the bottom plus the *loss* doing work against friction.

$$mgh_0 + \frac{1}{2}mv_0^2 = mgh_f + \frac{1}{2}mv_f^2 + |f_k x|$$

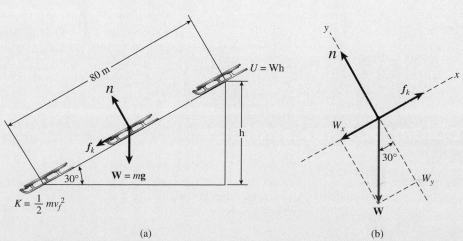

(a)   (b)

Figure 8.9   Some of the initial potential energy at the top of the incline is lost because of doing work against friction as the sled slides down.

Recognizing that $v_0 = 0$ and $h_f = 0$, we can simplify to

$$mgh_0 = \frac{1}{2}mv_f^2 + |f_k x|$$

Now we can visualize what is needed to solve for the final velocity. We will need to calculate the initial height and the friction force. From the triangle drawn in Fig. 8.9, we can find the height h_0 as follows:

$$h_0 = (80 \text{ m}) \sin 30° = 40 \text{ m}$$

The friction force depends on the value for the normal force. A study of the free-body diagram reveals that forces are balanced perpendicular to the incline, so the normal force is equal to the y component of the weight. Thus,

$$n = W_y = mg \cos 30°$$
$$= (20 \text{ kg})(9.8 \text{ m/s}^2) \cos 30° = 170 \text{ N}$$

The force of friction is the product of μ_k and n, so that

$$f_k = \mu_k n = (0.2)(170 \text{ N}) = 34.0 \text{ N}$$

Using this information, we return to the conservation equation

$$mgh_0 = \frac{1}{2}mv_f^2 + |f_k x|$$

$$(20 \text{ kg})(9.8 \text{ m/s}^2)(40 \text{ m}) = \frac{1}{2}(20 \text{ kg})v_f^2 + |(34 \text{ N})(80 \text{ m})|$$

$$7840 \text{ J} = \frac{1}{2}(20 \text{ kg})v_f^2 + 2720 \text{ J}$$

Of the original 7840 J available to the system, 2720 J was lost doing work against friction, and the remaining energy was kinetic energy. We can now solve for the final velocity.

$$\frac{1}{2}(20 \text{ kg})v_f^2 = 7840 \text{ J} - 2720 \text{ J}$$

$$(10 \text{ kg})v_f^2 = 5120 \text{ J}$$

$$v_f = \sqrt{\frac{5120 \text{ J}}{10 \text{ kg}}} = 22.6 \text{ m/s}$$

As an additional exercise, you should show that the final velocity would have been 28.0 m/s if there had been no friction forces.

Problem Solving Strategy

Conservation of Energy

1. Read the problem and then draw and label a rough sketch, identifying each object whose height or velocity changes.

2. Determine a reference point for measuring gravitational potential energy, for example, the bottom on an incline, the floor of a room, or the lowest point in the path of a particle.

3. For each object, make a note of the beginning and ending heights and velocities: h_0, v_0, h_f, and v_f. The heights are each measured from the chosen reference position and only the magnitudes are needed for velocities.

4. The total energy of the system at any instant is the sum of the kinetic and potential energies. Thus, the initial total energy E_0 and the final total energy E_f are

$$E_0 = mgh_0 + \frac{1}{2}mv_0^2 \qquad E_f = mgh_f + \frac{1}{2}mv_f^2$$

5. Determine whether or not friction forces are present. If friction or air resistance is present, then the energy losses must be given or determined. Often, the energy loss in doing work against friction is the simple product of the friction force f and the displacement x. Remember that $f = \mu_k n$.

6. Write the conservation of energy equation and solve for the unknown.

$$mgh_0 + \frac{1}{2}mv_0^2 = mgh_f + \frac{1}{2}mv_f^2 + |\text{energy losses}|$$

7. Remember to use the absolute value of the energy lost when applying the relationship in step 6. The actual work against friction is always negative, but you are accounting for it here as a *loss*.

8.8 Power

In our definition of work, *time* is not involved in any way. The same amount of work is done whether the task takes an hour or a year. Given enough time, even the weakest motor can lift an enormous load. If we wish to perform a task efficiently, however, the *rate* at which work is done becomes an important engineering quantity.

Power is the rate at which work is accomplished.

$$P = \frac{\text{work}}{t} \qquad\qquad (8.12)$$

The SI unit for power is the *joule per second,* which is renamed the **watt** (W). Thus, an 80-W lightbulb burns energy at the rate of 80 J/s.

$$1 \text{ W} = 1 \text{ J/s}$$

In USCS units, we use the *foot-pound per second* (ft · lb/s). No special name is given to this unit of power.

The watt and the foot-pound per second are inconveniently small units for most industrial purposes. Therefore, the **kilowatt** (kW) and the **horsepower** (hp) are defined:

$$1 \text{ kW} = 1000 \text{ W}$$
$$1 \text{ hp} = 550 \text{ ft} \cdot \text{lb/s}$$

In the United States, the watt and kilowatt are used almost exclusively in connection with electric power; horsepower is reserved for mechanical power. This practice is purely a convention and by no means necessary. It is perfectly proper to speak of an 0.08-hp bulb or to brag about a 238-kW engine. The conversion factors are

$$1 \text{ hp} = 746 \text{ W} = 0.746 \text{ kW}$$
$$1 \text{ kW} = 1.34 \text{ hp}$$

Since work is frequently done in a continuous fashion, an expression for power that involves velocity is useful. Thus,

$$P = \frac{\text{work}}{t} = \frac{Fx}{t} \qquad\qquad (8.13)$$

from which

$$P = F\frac{x}{t} = Fv \qquad\qquad (8.14)$$

where v is the velocity of the body on which the parallel force F is applied.

Example 8.10   A loaded elevator has a total mass of 2800 kg and is lifted to a height of 200 m in a time of 45 s. Express the average power in SI units and in USCS units.

Solution: This is a direct application of Eq. (8.13) where the distance x becomes the height h above the ground.

$$P = \frac{Fx}{t} = \frac{mgh}{t}$$

$$P = \frac{(2800 \text{ kg})(9.8 \text{ m/s}^2)(200 \text{ m})}{45 \text{ s}} = 1.22 \times 10^5 \text{ W}$$

$$P = 122 \text{ kW}$$

Since 1 hp = 746 W, the horsepower developed is

$$P = (1.22 \times 10^5 \text{ W}) \left(\frac{1 \text{ hp}}{746 \text{ W}} \right) = 164 \text{ hp}$$

Example 8.11   A 280-kg piano is being lifted at a steady speed to an apartment 10 m above the ground. The crane lifting the piano expends an average power of 600 W. How much time is required?

Plan: We write the power equation and then solve explicitly for time.

Solution: Given $h = 10$ m, $m = 280$ kg, and $P = 400$ W, we have

$$P = \frac{mgh}{t} \qquad \text{or} \qquad t = \frac{mgh}{P}$$

$$t = \frac{(280 \text{ kg})(9.8 \text{ m/s}^2)(10 \text{ m})}{600 \text{ W}}$$

$$t = 45.7 \text{ s}$$

Power companies charge customers for *kilowatt-hours,* and it is often called a *power bill.* As an exercise, you should use unit analysis to show that the product of a power unit and a time unit is really a unit of work or energy.

Summary and Review

Summary

The concepts of work, energy, and power have been discussed in this chapter. The major points to remember are summarized as follows:

- The *work* done by a force $\mathbf{F}$ acting through a distance x is found from the following equations (refer to Fig. 8.1):

$$\boxed{\text{Work} = F_x x \qquad \text{Work} = (F \cos \theta)x}$$

SI unit: joule (J)   USCS unit: foot-pound (ft · lb)

- *Kinetic energy K* is the capacity for doing work as a result of motion. It has the same units as work and is found from

$$\boxed{K = \frac{1}{2}mv^2 \qquad K = \frac{1}{2}\left(\frac{W}{g}\right)v^2}$$

- Gravitational *potential energy* is the energy that results from the position of an object relative to the Earth. Potential energy U has the same units as work and is found from

$$\boxed{U = Wh \qquad U = mgh}$$

where W or mg is the weight of the object and h is the height above some reference position.

- Net work is equal to the change in kinetic energy.

$$Fx = \frac{1}{2}mv_f^2 - \frac{1}{2}mv_0^2$$

- Conservation of mechanical energy with no friction:

$$U_0 + K_0 = U_f + K_f$$

$$mgh_0 + \frac{1}{2}mv_0^2 = mgh_f + \frac{1}{2}mv_f^2$$

- Conservation of energy including friction:

$$E_0 = E_f + |\text{energy losses}|$$

$$mgh_0 + \frac{1}{2}mv_0^2 = mgh_f + \frac{1}{2}mv_f^2 + f_k x$$

- Power is the rate at which work is done:

$$\boxed{P = \frac{\text{work}}{t} \qquad P = \frac{Fx}{t} \qquad P = Fv}$$

SI unit: watt (W)   USCS unit: ft · lb/s
Other units: $1 \text{ kW} = 10^3 \text{ W}$ $1 \text{ hp} = 550 \text{ ft · lb/s}$

Key Terms

conservation of energy   168
conservation of mechanical
  energy   167
displacement   158
energy   158
foot-pound   159

horsepower   171
joule   159
kilowatt   171
kinetic energy   161
potential energy   161
power   158

resultant work   159
watt   171
work   158
work-energy theorem   163

Review Questions

8.1. Distinguish clearly between the physicist's concept of work and the general concept of work.

8.2. Two teams are engaged in a tug of war. Is work done? When?

8.3. Whenever net work is done on a body, will the body necessarily undergo acceleration? Discuss.

8.4. A diver stands on a board 10 ft above the water. What kind of energy results from this position? What happens to this energy as she dives into the water? Is work done? If so, what is doing the work, and on what is the work done?

8.5. Compare the potential energies for two bodies A and B if (a) A is twice as high as B but of the same

mass; (b) B is twice as heavy as A but at the same height; and (c) A is twice as heavy as B, but B is twice as high as A.

8.6. Compare the kinetic energies of two bodies A and B if (a) A has twice the speed of B, (b) A has half the mass of B, and (c) A has twice the mass and half the speed of B.

8.7. In stacking 8-ft boards, you lift an entire board at its center and lay it on the pile. Your helper lifts one end, rests it on the pile, and then lifts the other end. Compare the work done.

8.8. In the light of what you have learned about work and energy, describe the most efficient procedure for ringing the bell with a sledgehammer at the fair. What precautions should you take?

8.9. A roller coaster at the fair boasts "a maximum height of 100 ft with a maximum speed of 60 mi/h." Do you believe the advertisement? Explain.

8.10. A man mows his yard for years using a 4-hp mower. He then buys a 6-hp mower. After using the new mower for a while, it seems to him as if he has twice the power as before. Why do you think he is convinced of the increase in power?

Problems

Section 8.1 Work

8.1. What is the work done by a force of 20 N acting through a parallel distance of 8 m? What force will do the same work through a distance of 4 m?
Ans. 160 J, 40 N

8.2. A worker lifts a 40-lb weight through a height of 10 ft. How many meters can a 10-kg block be lifted by the same amount of work?

8.3. A tugboat exerts a constant force of 4000 N on a ship, moving it a distance of 15 m. What work is done? Ans. 60 kJ

8.4. A 5-kg hammer is lifted to a height of 3 m. What is the minimum required work?

8.5. A push of 120 N is applied along the handle of a lawn mower, producing a horizontal displacement of 14 m. If the handle makes an angle of 30° with the ground, what work was done by the 120-N force? Ans. 1460 J

8.6. The trunk in Fig. 8.10 is dragged a horizontal distance of 24 m by a rope that makes an angle θ with the floor. If the rope tension is 80 N, what works are done for the following angles: 0°, 30°, 60°, 90°?

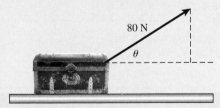

80 N

θ

Figure 8.10

8.7. A horizontal force pushes a 10-kg sled along a driveway for a distance of 40 m. If the coefficient of sliding friction is 0.2, what work is done by the friction force? Ans. −784 J

8.8. A sled is dragged a distance of 12.0 m by a rope under constant tension of 140 N. The task requires 1200 J of work. What angle does the rope make with the ground?

Section 8.2 Resultant Work

8.9. An average force of 40 N compresses a coiled spring a distance of 6 cm. What is the work done by the 40-N force? What work is done by the spring? What is the resultant work?
Ans. 2.40 J, −2.40 J, 0

8.10. A horizontal force of 20 N drags a small sled 42 m across the ice at constant speed. Find the work done by the pulling force and by the friction force. What is the resultant force?

8.11. A 10-kg block is dragged 20 m by a parallel force of 26 N. If $\mu_k = 0.2$, what is the resultant work, and what acceleration results?
Ans. 128 J, 0.640 m/s^2

8.12. A rope making an angle of 35° with the horizontal drags a 10-kg toolbox a horizontal distance of 20 m. The tension in the rope is 60 N, and the constant friction force is 30 N. What work is done by the rope and by friction? What is the resultant work?

8.13. For the example described in Prob. 8.12, what is the coefficient of friction between the toolbox and the floor? Ans. 0.472

*8.14. A 40-kg sled is pulled horizontally for 500 m ($\mu_k = 0.2$). If the resultant work is 50 kJ, what was the parallel pulling force?

*8.15. Assume that $m = 8$ kg in Fig. 8.11 and $\mu_k = 0$. What minimum work is required by the force P to reach the top of the inclined plane? What work is required to lift the 8-kg block vertically to the same height? Ans. 941 J, 941 J

*8.16. What is the minimum work by the force P to move the 8-kg block to the top of the incline if $\mu_k = 0.4$?

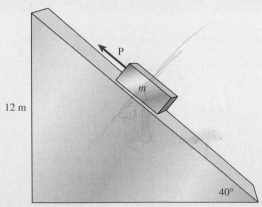

P

m

12 m

40°

Figure 8.11

Compare this with the work needed to lift it vertically to the same height.

*8.17. What is the resultant work when the 8-kg block slides from the top to the bottom of the incline in Fig. 8.11? Assume that $\mu_k = 0.4$ Ans. 492 J

Section 8.4 Work and Kinetic Energy

8.18. What is the kinetic energy of a 6-g bullet at the instant its speed is 190 m/s? What is the kinetic energy of a 1200-kg car traveling at 80 km/h?

8.19. What is the kinetic energy of a 2400-lb automobile when its speed is 55 mi/h? What is the kinetic energy of a 9-lb ball when its speed is 40 ft/s?
Ans. 244,000 ft · lb; 225 ft · lb

8.20. What is the change in kinetic energy when a 50-g ball hits the pavement with a velocity of 16 m/s and rebounds with a velocity of 10 m/s?

8.21. A runaway 400-kg wagon enters a cornfield with a velocity of 12 m/s and eventually comes to rest. What work was done on the wagon?
Ans. −28.8 kJ

8.22. A 2400-lb car increases its speed from 30 mi/h to 60 mi/h. What resultant work was required? What is the equivalent work in joules?

8.23. A 0.6-kg hammer head is moving at 30 m/s just before striking the head of a spike. Find the initial kinetic energy. What work can be done by the hammer head?   Ans. 270 J, 270 J

8.24. A 12-lb hammer moving at 80 ft/s strikes the head of a nail moving it into the wall a distance of $\frac{1}{4}$ in. What was the average stopping force?

8.25. What average force is needed to increase the velocity of a 2-kg object from 5 m/s to 12 m/s over a distance of 8 m?   Ans. 14.9 N

*8.26. Verify the answer to Prob. 8.25 by applying Newton's second law of motion.

*8.27. A 20-g projectile strikes a mud bank in Fig. 8.12, penetrating a distance of 6 cm before stopping.

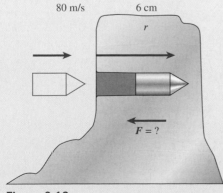

80 m/s   6 cm

r

$F = ?$

Figure 8.12

Find the stopping force F if the entrance velocity is 80 m/s.   Ans. 1070 N

*8.28. A 1500-kg car is moving along a level road at 60 km/h. What work is required to stop the car? If $\mu_k = 0.7$, what is the stopping distance?

Section 8.5 Potential Energy

8.29. A 2-kg block rests on top of a table 80 cm from the floor. Find the potential energy of the block relative to (a) the floor, (b) the seat of a chair 40 cm from the floor, and (c) relative to the ceiling 3 m from the floor.   Ans. 15.7 J, 7.84 J, −43.1 J

8.30. A 1.2-kg brick is held a distance of 2 m above a manhole and then dropped into it. The bottom of the manhole is 3 m below the street. Relative to the street, what is the potential energy at each location? What is the change in potential energy?

8.31. At a particular instant, a mortar shell has a velocity of 60 m/s. If its potential energy at that point is one-half of its kinetic energy, what is its height above the Earth?   Ans. 91.8 m

*8.32. A 20-kg sled is pushed up a 34° slope to a vertical height of 140 m. A constant friction force of 50 N acts for the entire distance. What external work was required? What was the change in potential energy?

*8.33. An average force of 600 N is required to compress a coiled spring a distance of 4 cm. What work is done *by* the spring? What is the change in potential energy of the compressed spring?
Ans. −24 J, +24 J

Section 8.6 Conservation of Energy

8.34. An 18-kg weight is lifted to a height of 12 m and then released to fall freely. What is the potential energy, the kinetic energy, and the total energy at (a) the highest point, (b) 3 m above the ground, and (c) at the ground?

8.35. A 4-kg hammer is lifted to a height of 10 m and dropped. What are the potential and kinetic energies of the hammer when it has fallen to a point 4 m from the Earth?   Ans. 157 J, 235 J

8.36. What will be the velocity of the hammer in Prob. 8.35 just before it strikes the ground? What is the velocity at the 4-m location?

8.37. What initial velocity must be given to a 5-kg mass if it is to rise to a height of 10 m? What is the total energy at any point in its path?
Ans. 14 m/s, 490 J

8.38. A simple pendulum 1 m long has an 8-kg bob. How much work is needed to move the pendulum from its lowest point to a horizontal position?

From energy considerations, find the velocity of the bob as it swings through the lowest point.

8.39. A ballistic pendulum is illustrated in Fig. 8.13. A 40-g ball is caught by a 500-g suspended mass. After impact, the two masses rise a vertical distance of 45 mm. Find the velocity of the combined masses just after impact. Ans. 93.9 cm/s

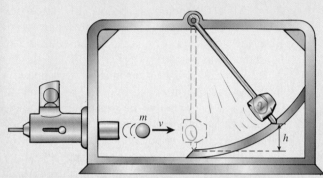

Figure 8.13

*8.40. A 100-lb sled slides from rest at the top of a 37° inclined plane. The original height is 80 ft. In the absence of friction, what is the velocity of the sled when it reaches the bottom of the incline?

*8.41. An 8-kg wagon in Fig. 8.14 has an initial downward velocity of 7 m/s. Neglecting friction, find the velocity when it reaches point B. Ans. 21.0 m/s.

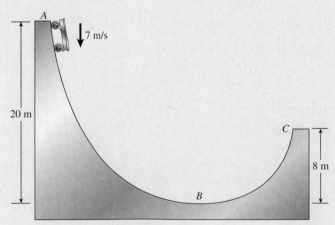

Figure 8.14

*8.42. What is the velocity of the 8-kg block at point C in Prob. 8.41?

*8.43. An 80-lb girl sits in a swing of negligible weight. If she is given an initial velocity of 20 ft/s, to what height will she rise? Ans. 6.25 ft

Section 8.7 Energy and Friction Forces

8.44. A 60-kg sled slides from rest to the bottom of a 25° slope of length 30 m. A 100-N friction force acts for the entire distance. What is the total energy at the top of the slope and at the bottom? What is the velocity of the sled at the bottom?

8.45. A 500-g block is released from the top of a 30° incline and slides 160 cm to the bottom. A constant friction force of 0.9 N acts the entire distance. What is the total energy at the top? What work is done by friction? What is the velocity at the bottom? Ans. 3.92 J, −1.44 J, 3.15 m/s

8.46. What initial velocity must be given to the 500-g block in Prob. 8.45 if it is to just reach the top of the same slope?

8.47. A 64-lb cart starts up a 37° incline with an initial velocity of 60 ft/s. If it comes to rest after moving a distance of 70 ft, how much energy was lost to friction? Ans. 906 ft · lb

*8.48. A 0.4-kg ball drops a vertical distance of 40 m and rebounds to a height of 16 m. How much energy was lost in collision with the floor?

*8.49. A 4-kg sled is given an initial velocity of 10 m/s at the top of a 34° slope. If $\mu_k = 0.2$, how far must the sled travel until its velocity reaches 30 m/s? Ans. 104 m

*8.50. Assume in Fig. 8.14 that the mass of the wagon is 6 kg and that 300 J of energy is lost doing work against friction. What is the velocity when the mass reaches point C?

*8.51. A bus brakes to avoid an accident. The tread marks of the tires are 80 ft long. If $\mu_k = 0.7$, what was the speed before applying brakes? Ans. 59.9 ft/s

Section 8.8 Power

8.52. A power station conveyor belt lifts 500 tons of ore to a height of 90 ft in 1 hr. What average horsepower is required?

8.53. A 40-kg mass is lifted through a distance of 20 m in a time of 3 s. What average power is employed? Ans. 2.61 kW

8.54. A 300-kg elevator is lifted a vertical distance of 100 m in 2 min. What is the output power?

8.55. A 90-kW engine is used to lift a 1200-kg load. What is the average velocity of the lift? Ans. 7.65 m/s

8.56. To what height can a 400-W engine lift a 100-kg mass in 3 s?

8.57. An 800-N student runs up a flight of stairs rising 6 m in 8 s. What is the average power expended? Ans. 600 W

*8.58. A speedboat must develop 120 hp in order to move at a constant speed of 15 ft/s through the water. What is the average resistive force due to the water?

Additional Problems

8.59. A worker lifts a 20-kg bucket from a well at constant speed and does 8 kJ of work. How deep is the well? **Ans. 40.8 m**

8.60. A horizontal force of 200 N pushes an 800-N crate horizontally for a distance of 6 m at constant speed. What work is done by the 200-N force? What is the resultant work?

*8.61. A 10-kg mass is lifted to a height of 20 m and released. What is the total energy of the system? What is the velocity of the mass when it is located 5 m from the floor? **Ans. 1960 J, 17.1 m/s**

8.62. A crate is lifted at a constant speed of 5 m/s by an engine whose output power is 4 kW. What is the mass of the crate?

8.63. A roller coaster boasts a maximum height of 100 ft. What is the maximum speed in miles per hour when it reaches its lowest point? **Ans. 54.4 mi/h**

8.64. A 20-N force drags an 8-kg block a horizontal distance of 40 m by a rope at an angle of 37° with the horizontal. Assume $\mu_k = 0.2$ and that the time required is 1 min. What resultant work is done?

8.65. What is the velocity of the block in Prob. 8.64 at the end of the trip? What *resultant* power was expended? **Ans. 5.20 m/s, 1.80 W**

8.66. A 70-kg skier slides from rest down a 30-m slope that makes an angle of 28° with the horizontal. Assume that $\mu_k = 0.2$. What is the velocity of the skier at the bottom of the slope?

*8.67. A 0.3-mg flea can jump to a height of about 3 cm. What must be the takeoff speed? Do you really need to know the mass of the flea?
Ans. 76.7 cm/s, no

*8.68. A roller coaster goes through a low point and barely makes the next hill 15 m higher. What is the minimum speed at the bottom of the loop?

*8.69. The hammer of a pile driver weighs 800 lb and falls a distance of 16 ft before striking the pile. The impact drives the pile 6 in. deeper into the ground. What was the average force driving the pile?
Ans. 25,600 lb

*8.70. Suppose that water from the top of the waterfall shown in Fig. 8.15 is delivered to a turbine located at the base of the falls—a vertical distance of 94 m (308 ft). Assume that 20 percent of the available energy is lost to friction and other resistive forces. If 3000 kg of water enters the turbine every minute, what is the output power of the turbine?

Figure 8.15 Lower Falls at Yellowstone National Park. (*Photo by Paul E. Tippens.*)

Critical Thinking Questions

*8.71. An inclined board is used to unload boxes of nails from the back of a truck. The height of the truck bed is 60 cm, and the board is 1.2 m in length. Assume that $\mu_k = 0.4$ and the boxes are given an initial push to start sliding. What is their speed when they reach the ground below? What initial speed would they need at the bottom in order to slide back into the truck bed? In the absence of friction, would these two questions have the same answer?   Ans. 1.90 m/s, 4.46 m/s, yes

*8.72. A 96-lb safe is pushed with negligible friction up a 30° incline for a distance of 12 ft. What is the increase in potential energy? Would the same change in potential energy occur if a 10-lb friction force opposed the motion up the incline? Why? Would the same work be required?

*8.73. A 2-kg ball is suspended from a 3-m cable attached to a spike in the wall. The ball is pulled out, so the cable makes an angle of 70° with the wall and then is released. If 10 J of energy is lost during the collision with the wall, what is the maximum angle between the cable and the wall after the first rebound?   Ans. 59.2°

*8.74. A 3-kg ball dropped from a height of 12 m has a velocity of 10 m/s just before hitting the ground. What is the average retarding force due to the air? If the ball rebounds from the surface with a speed of 8 m/s, what energy was lost on impact? How high will it rebound if the average air resistance is the same as before?

*8.75. Consider a roller coaster where the first hill is 34 m high. If the coaster loses only 8 percent of its energy between the first two hills, what is the maximum height possible for the second hill?
         Ans. 31.3 m

*8.76. A 4-kg block is compressed against a spring at the bottom of the inclined plane in Fig. 8.16. A force of 4000 N was required to compress the spring a distance of 6 cm. If it is then released and the coefficient of friction is 0.4, how far up the incline will the block move?

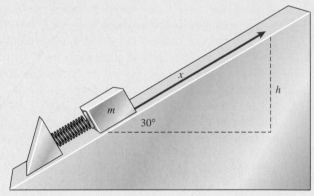

Figure 8.16

9

Impulse and Momentum

Astronaut Edward H. White II, pilot for the *Gemini-Titan 4* space flight, floats in zero gravity of space. White is attached to the spacecraft by a 25-ft umbilical line and a 23-ft tether line, both wrapped in gold tape to form one cord. In his right hand, White carries a hand-held self-maneuvering unit (HHSMU). Firing the gas-powered gun transferred momentum to the astronaut. (*Photo by NASA.*)

Objectives

After completing this chapter, you should be able to

1. Define and give examples of *impulse* and *momentum* as vector quantities.

2. Write and apply a relationship between impulse and the resulting *change in momentum*.

3. State the law of *conservation of momentum* and apply it to the solution of physical problems.

4. Define and be able to calculate the *coefficient of restitution* for two surfaces.

5. Distinguish by example and definition between elastic and inelastic collisions.

6. Predict the velocities of two colliding bodies after impact when the coefficient of restitution, masses, and initial velocities are given.

Energy and work are scalar quantities that say absolutely nothing about direction. The law of conservation of energy describes only the relationship between initial and final states; it says nothing about how the energies are distributed. For example, when two objects collide, we can say that the total energy before collision must equal the energy after collision if we neglect friction and other heat losses. But we need a new concept if we are to determine how the total energy is divided between the objects or even their relative directions after impact. The concepts of *impulse* and *momentum* presented in this chapter add a vector description to our discussion of energy and motion.

9.1   Impulse and Momentum

When a golf ball is driven from the ground, as in Fig. 9.1, a large average force **F** acts on the ball during a short interval of time Δt, causing the ball to accelerate from rest to a final velocity $\mathbf{v}_f$. It is extremely difficult to measure either the force or its duration, but their product $\mathbf{F} \Delta t$ can be determined from the resulting change in velocity of the golf ball. From Newton's second law, we have

$$\mathbf{F} = m\mathbf{a} = m\frac{\mathbf{v}_f - \mathbf{v}_0}{\Delta t}$$

Multiplying by Δt gives

$$\mathbf{F} \Delta t = m(\mathbf{v}_f - \mathbf{v}_0)$$

or

$$\mathbf{F} \Delta t = m\mathbf{v}_f - m\mathbf{v}_0 \tag{9.1}$$

This equation is so useful in solving problems involving impact that the terms are given special names.

The impulse **F** Δt is a vector quantity equal in magnitude to the product of the force and the time interval in which it acts. Its direction is the same as that of the force.

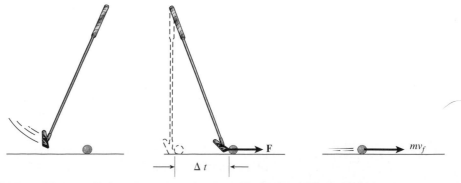

Figure 9.1   When a golf club strikes the ball, a force **F** acting through the time interval results in a change in its momentum.

The momentum **p** of a particle is a vector quantity equal in magnitude to the product of its mass m and its velocity **v.**

$$\mathbf{p} = m\mathbf{v}$$

Thus, Eq. (9.1) can be stated verbally:

Impulse $(\mathbf{F}\,\Delta t)$ = *change in momentum* $(m\mathbf{v}_f - m\mathbf{v}_0)$

The SI unit of impulse is the *newton-second* (N · s). The unit of momentum is the *kilogram-meter per second* (kg · m/s). It is convenient to distinguish these units, even though they are actually the same:

$$\text{N} \cdot \text{s} = \frac{\text{kg} \cdot \text{m}}{\text{s}^2} \times \text{s} = \text{kg} \cdot \text{m/s}$$

The corresponding units in the USCS are the *pound-second* (lb · s) and *slug-foot per second* (slug · ft/s).

Example 9.1

The head of a 3-kg sledgehammer is moving at a velocity of 14 m/s as it strikes a steel spike. It is brought to a stop in 0.02 s. Determine the average force on the spike.

Plan: First, we will find the impulse $\mathbf{F}\,\Delta t$, which is equal to the change in momentum $m\mathbf{v}$ for the hammer. Then, we will solve for the time by dividing the average force into the impulse. Since both momentum and impulse are vector quantities, we must be careful of signs.

Solution: Consider that the upward direction is positive and that the head is initially moving downward. This means that $v_0 = -14$ m/s, $v_f = 0$, $m = 3$ kg, and $\Delta t = 0.02$ s.

$$F\,\Delta t = mv_f - mv_0 = 0 - (3\text{ kg})(-14\text{ m/s})$$
$$= +42.0\text{ N} \cdot \text{m}$$

Dividing this impulse by 0.02 s gives

$$F\,\Delta t = 42\text{ N} \cdot \text{m} \qquad \text{or} \qquad F = \frac{42\text{ N} \cdot \text{m}}{0.02\text{ s}} = 2100\text{ N}$$

The average stopping force acting *on the spike* is 2100 N directed upward (+). The reaction force exerted *on the hammer* is equal in magnitude but opposite in direction. It must be emphasized that these forces are *average* forces. Near the beginning of contact with the spike, the stopping force will be much greater than 2100 N.

Example 9.2

A 0.15-kg baseball moving toward the batter with a velocity of 30 m/s is struck with a bat that causes it to reverse its direction with a velocity of 42 m/s. (See Fig. 9.2.) Find the impulse and the average force exerted on the ball if the bat is in contact with the ball for 0.002 s.

Plan: Draw a sketch like the one shown in Fig. 9.2. Notice that the directions and signs of velocity are indicated. We recognize that the impulse imparted to the ball must be equal to the change in momentum of the ball, and we must be consistent with the signs given for velocity before and after the bat strikes the ball.

Solution: Choosing right as positive and organizing the data, we have

Given: $m = 0.15$ kg; $v_0 = -30$ m/s;   Find: $F\,\Delta t$ and Δt
$v_f = +42$ m/s; $\Delta t = 0.002$ s

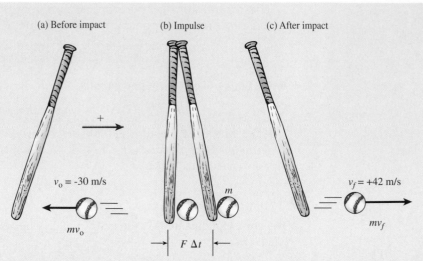

(a) Before impact (b) Impulse (c) After impact

$v_o = -30$ m/s m $v_f = +42$ m/s

mv_o $F \Delta t$ mv_f

Figure 9.2 The impulse $F \Delta t$ is equal to the change in momentum.

Substituting into Eq. (9.1), we first calculate the impulse.

$$F \Delta t = mv_f - mv_0$$
$$= (0.15 \text{ kg})(42 \text{ m/s}) - (0.15 \text{ kg})(-30 \text{ m/s})$$
$$= 6.30 \text{ kg} \cdot \text{m/s} + 4.50 \text{ kg m/s} = 10.8 \text{ kg} \cdot \text{m/s}$$

The velocity changes from -30 m/s to $+42$ m/s, a total change of $+72$ m/s. It is easy to see how the incorrect use of signs can lead to a significant error.

Next, we are asked to find the average force exerted by the bat while it is in contact with the ball for 0.002 s. Solving for F, we obtain

$$F \Delta t = 6.75 \text{ kg} \cdot \text{m/s}$$

$$F = \frac{10.8 \text{ kg} \cdot \text{m/s}}{0.002 \text{ s}} = 5400 \text{ N}$$

Once again, we must recognize that this is the *average* force on the ball.

9.2 The Law of Conservation of Momentum

Let us consider the *head-on* collision of the masses m_1 and m_2, as shown in Fig. 9.3. Consider the surfaces to be frictionless. We denote their velocities before impact as $\mathbf{u}_1$ and $\mathbf{u}_2$ and after impact as $\mathbf{v}_1$ and $\mathbf{v}_2$. The impulse of the force $\mathbf{F}_1$ acting on the right mass is

$$F_1 \Delta t = m_1v_1 - m_1u_1$$

Similarly, the impulse of the force $\mathbf{F}_2$ on the left mass is

$$F_2 \Delta t = m_2v_2 - m_2u_2$$

During the time Δt, $\mathbf{F}_1 = -\mathbf{F}_2$, so that

$$F_1 \Delta t = -F_2 \Delta t$$

or

$$m_1v_1 - m_1u_1 = -(m_2v_2 - m_2u_2)$$

and finally, after rearranging,

$$m_1u_1 + m_2u_2 = m_1v_1 + m_2v_2 \tag{9.2}$$

Total momentum before impact = total momentum after impact

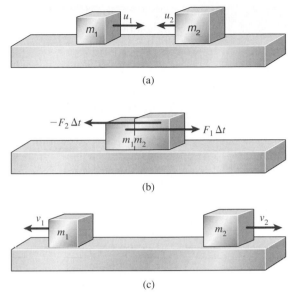

Figure 9.3   (a) Before impact: $m_1u_1 + m_2u_2$; (b) during impact: $F_1 \Delta t = -F_2 \Delta t$; (c) after impact: $m_1v_1 + m_2v_2$.

Thus, we have derived a statement of the law of ***conservation of momentum:***

> The total momentum of colliding bodies before impact is equal to their total momentum after impact.

Example 9.3

Assume that an 8-kg mass m_1 moving to the right at 4 m/s collides with a 6-kg mass m_2 moving to the left at 5 m/s. What is the total momentum before and after the collision?

Plan: We will draw and label a sketch for this problem similar to that shown in Fig. 9.3. Then, choosing the rightward direction as positive, we will organize the data and sum the momentums of the two masses before the collision. Finally, assuming that momentum is conserved, we will give the same value for the final momentum.

Solution: Taking positive motion to the right, we organize the data

$$\text{Given:} \quad m_1 = 8 \text{ kg}, u_1 = +4 \text{ m/s} \qquad \text{Find:} \quad p_0 = ?$$
$$m_2 = 6 \text{ kg}, u_2 = -5 \text{ m/s} \qquad\qquad p_f = ?$$

Now, the total momentum *before* the collision is

$$p_0 = m_1u_1 + m_2u_2$$
$$= (8 \text{ kg})(4 \text{ m/s}) + (6 \text{ kg})(-5 \text{ m/s})$$
$$= 32 \text{ kg} \cdot \text{m/s} - 30 \text{ kg} \cdot \text{m/s} = +2 \text{ kg} \cdot \text{m/s}$$

Finally, conservation of momentum means that the same value applies for the total momentum *after* the collision.

If the velocity of either mass after the collision can be determined, the other can also be found from the principle of conservation of momentum.

Example 9.4

A 1400-kg cannon mounted on wheels fires a 60-kg ball in a horizontal direction with a velocity of 50 m/s, as shown in Fig. 9.4. Assuming that the cannon can move freely, what will be its recoil velocity?

Plan: Draw and label a sketch as in Fig. 9.4, labeling the right direction as positive. We will then organize the data and substitute into the conservation equation to solve for the recoil velocity of the cannon. It is helpful to choose a subscript for each mass that identifies it, such as m_c for the cannon and m_b for the cannon ball. For example, we can represent the momentum for the cannon *before* the collision as $m_c u_c$. Since the recoil velocity should be to the *left* and the mass of the cannon is *much larger* than that of the projectile, we will make sure that our answer is consistent.

Solution: Remember that right is positive and u applies before and v applies after collision.

$$\text{Given:}\quad m_c = 1400 \text{ kg}, u_c = 0 \text{ m/s} \qquad \text{Find:}\quad v_c = ?$$
$$m_b = 60 \text{ kg}, u_b = 0 \text{ m/s}$$
$$v_b = 50 \text{ m/s}$$

Substitution into the conservation equation gives

$$m_c u_c + m_b u_b = m_c v_c + m_b v_b$$
$$0 + 0 = m_c v_c + m_b v_b$$

Solving for the velocity of the cannon after the collision, we have

$$m_c v_c = -m_b v_b \qquad \text{or} \qquad v_c = \frac{-m_b v_b}{m_c}$$

$$v_c = \frac{-(60 \text{ kg})(50 \text{ m/s})}{1400 \text{ kg}} = -2.14 \text{ m/s}$$

The sign and magnitude of the recoil velocity is reasonable for the given information.

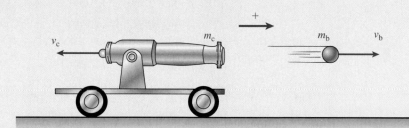

Figure 9.4   Calculating the recoil velocity of a cannon.

An interesting experiment that demonstrates the conservation of momentum can be performed with eight small marbles and a grooved track, as shown in Fig. 9.5. If one marble is released from the left, it will be stopped upon colliding with the others, and one on the right end will roll out with the same velocity. Similarly, when two, three, four, or five marbles are released from the left, the same number will roll out to the right with the same velocity, the others remaining at rest in the center.

You might reasonably ask why two marbles roll off in Fig. 9.5 instead of one with twice the velocity, since this condition would also conserve momentum. For example, if each marble has a mass of 50 g, and if two marbles approach from the left at a velocity of 20 cm/s, the total momentum before impact is 2000 g · cm/s. This same momentum can be

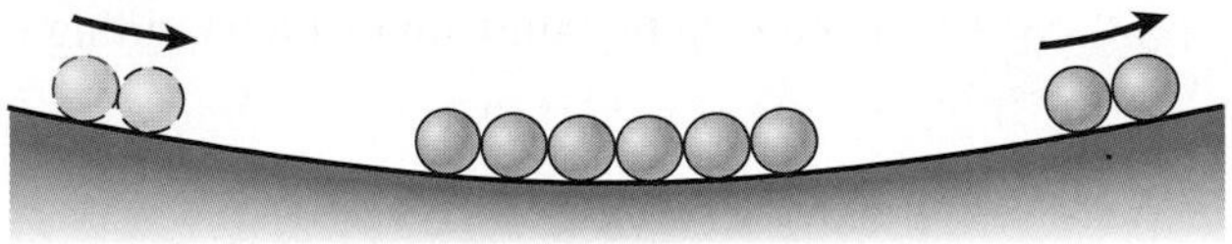

Figure 9.5 Conservation of momentum.

achieved after impact if only one marble is left, assuming that it had a velocity of 40 cm/s. The answer lies in the fact that energy must also be conserved. If one marble came off with twice the velocity, its kinetic energy would be much greater than that available from the other two. The kinetic energy put into the system would be

$$E_0 = \frac{1}{2}mv^2 = \frac{1}{2}(0.1\text{ kg})(0.2\text{ m/s})^2$$
$$= 2 \times 10^{-3}\text{ J}$$

The kinetic energy of one marble traveling at 40 cm/s is exactly twice this value.

$$E_f = \frac{1}{2}mv^2 = \frac{1}{2}(0.05\text{ kg})(0.4\text{ m/s})^2$$
$$= 4 \times 10^{-3}\text{ J}$$

Therefore, energy as well as momentum is important in describing impact phenomena.

9.3 Elastic and Inelastic Impacts

PHYSICS TODAY

Because of its momentum, a fully loaded supertanker cruising at 16 knots will take 20 minutes to stop. If an immobile object were to appear 3 nautical miles away, there would be a collision.

From the experiment in Section 9.2, you might assume that kinetic energy as well as the momentum is unchanged by a collision. Although this assumption is approximately true for hard bodies like marbles and billiard balls, it is not true for soft bodies, which rebound much more slowly on impact. During impact, all bodies become slightly deformed, and small amounts of heat are liberated. The vigor with which a body restores itself to its original shape after deformation is a measure of its ***elasticity,*** or restitution.

If the kinetic energy remains constant in a collision (an ideal case), the collision is said to be ***completely elastic.*** In this instance, no energy is lost through heat or deformation in a collision. A hardened steel ball dropped on a marble slab would approximate a completely elastic collision.

If the colliding bodies stick together and move off as a unit afterward, the collision is said to be ***completely inelastic.*** A bullet that becomes embedded in a wooden block is an example of this type of impact. The majority of collisions fall between these two extremes.

In a completely elastic collision between two masses m_1 and m_2, we can say that both energy and momentum will be conserved. Hence, we can apply two equations:

$$\text{Energy:}\quad \frac{1}{2}m_1u_1^2 + \frac{1}{2}m_2u_2^2 = \frac{1}{2}m_1v_1^2 + \frac{1}{2}m_2v_2^2$$

$$\text{Momentum:}\quad m_1u_1 + m_2u_2 = m_1v_1 + m_2v_2$$

which can be simplified to give

$$m_1(u_1^2 - v_1^2) = m_2(v_2^2 - u_2^2)$$
$$m_1(u_1 - v_1) = m_2(v_2 - u_2)$$

Dividing the first equation by the second gives

$$\frac{u_1^2 - v_1^2}{u_1 - v_1} = \frac{v_2^2 - u_2^2}{v_2 - u_2}$$

Factoring the numerators and dividing out, we obtain

$$u_1 + v_1 = u_2 + v_2$$

or

$$v_1 - v_2 = u_2 - u_1 = -(u_1 - u_2) \tag{9.3}$$

Thus, in the ideal case of a completely elastic collision, the relative velocity after collision, $v_1 - v_2$, is equal to the negative of the relative velocity before collision. The closer these quantities are to being equal, the more elastic the collision. The negative ratio of the relative velocity after collision to the relative velocity before collision is, called the *coefficient of restitution.*

> The coefficient of restitution e is the negative ratio of the relative velocity after impact to the relative velocity before impact.

$$e = -\frac{v_1 - v_2}{u_1 - u_2}$$

Incorporating the minus sign into the numerator of this equation yields

$$e = \frac{v_2 - v_1}{u_1 - u_2} \tag{9.4}$$

If the collision is completely elastic, $e = 1$. If the collision is completely inelastic, $e = 0$. In the inelastic case, the two bodies move off with the same velocity, so that $v_2 = v_1$. In general, the coefficient of restitution has some value between 0 and 1.

A simple method of determining the coefficient of restitution is shown in Fig. 9.6. A sphere of the material being measured is dropped onto a fixed plate from a height h_1. Its rebound is measured to a height h_2. In this case, the mass of the plate is so large that v_2 is approximately 0. Therefore,

$$e = \frac{v_2 - v_1}{u_1 - u_2} = -\frac{v_1}{u_1}$$

The velocity u_1 is simply the velocity acquired in falling from a height h_1, which is found from

$$u_1^2 - u_0^2 = 2gh_1$$

But the initial velocity $u_0 = 0$, so

$$u_1^2 = 2gh_1$$

or

$$u_1 = \sqrt{2gh_1}$$

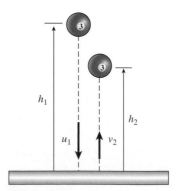

Figure 9.6

We have considered the downward direction as positive. If the ball rebounds to a height h_2, its rebound velocity v_1 must be $-\sqrt{2gh_2}$. (The minus sign indicates the change in direction.) Hence, the coefficient of restitution is given by

$$e = -\frac{v_1}{u_1} = -\frac{-\sqrt{2gh_2}}{\sqrt{2gh_1}}$$

or

$$e = \sqrt{\frac{h_2}{h_1}} \tag{9.5}$$

The resulting coefficient is a joint property of the ball and the rebounding surface.

For an extremely elastic surface, e has a value of 0.95 or greater (steel or glass), whereas for less resilient substances, e may be extremely small. It is interesting to note that the rebound height is a function of the vigor with which the impact deformation is restored. Contrary to popular conception, a steel ball or glass marble will bounce to a greater height than most rubber balls.

Problem-Solving Strategy

Conservation of Momentum: Collisions

1. Read the problem and then draw and label a rough sketch. Indicate the direction of motion for each mass by drawing vectors on the diagram.

2. Choose the x axis along the line of collision and indicate the positive direction. Velocities must be reckoned as positive or negative based on this choice.

3. Make a list of the given masses and velocities, taking care to use the appropriate sign and units for each velocity. The use of appropriate subscripts and letters will help you keep track of the different masses and velocities before and after a collision.

4. Write the conservation of momentum equation:

$$m_1 u_1 + m_2 u_2 = m_1 v_1 + m_2 v_2$$

5. Substitute all given quantities and simplify the resulting expression. When you substitute velocities, it is essential to include the appropriate sign for each velocity.

6. If the collision is completely *inelastic,* you proceed by solving the momentum equation for the unknown quantity.

7. If the collision is elastic, conservation of energy will yield a second independent equation:

$$v_2 - v_1 = e(u_1 - u_2)$$

where e is the *coefficient of restitution.* (For perfectly elastic collisions, $e = 1$.) Then proceed by solving this equation simultaneously with the momentum equation. Take care not to confuse the signs of substitution with the signs of operation.

$m_1 = 2\ kg \quad M_1 = -24\ m/s$
$m_2 = 4\ kg \quad M_2 = 16\ m/s$

Example 9.5

A 2-kg ball traveling to the left with a speed of 24 m/s collides head on with a 4-kg ball traveling to the right at 16 m/s. (a) Find the resulting velocity if the two balls stick together on impact. (b) Find their final velocities if the coefficient of restitution is 0.80.

Plan: We will draw and label a sketch indicating the rightward direction as positive. Then, after listing the given information, we will apply the conservation equation for a completely inelastic collision in which the combined velocity after the collision can be found directly from conservation. Second, we will use the definition of the coefficient of restitution to establish another relationship between final velocities. That will solve two simultaneous equations to find the final velocities of each mass.

Solution (a): Organizing the data, we write

Given: $m_1 = 2$ kg, $u_1 = -24$ m/s, $e = 0.8$ Find: $v_1 = ?$
          $m_2 = 4$ kg, $u_2 = +16$ m/s $v_2 = ?$

For the inelastic case, $e = 0$, and the combined velocity after collision is

$$v_c = v_1 = v_2$$

Thus, we may write Eq. (9.2) as follows:

$$m_1u_1 + m_2u_2 = m_1v_1 + m_2v_2 = (m_1 + m_2)v_c$$

Given that the rightward direction is positive, substitution yields

$$(2 \text{ kg})(-24 \text{ m/s}) + (4 \text{ kg})(16 \text{ m/s}) = (2 \text{ kg} + 4 \text{ kg})v_c$$
$$-48 \text{ kg} \cdot \text{m/s} + 64 \text{ kg} \cdot \text{m/s} = (6 \text{ kg})v_c$$
$$16 \text{ kg} \cdot \text{m/s} = (6 \text{ kg})v_c$$

from which

$$v_c = 2.67 \text{ m/s}$$

The fact that this velocity is also positive indicates that both bodies move together to the right after collision.

Solution (b): In this case, e is not zero, and the balls rebound after collision with different velocities. Therefore, more information is needed than that derived from the momentum equation alone. We resort to the given value of $e = 0.80$ and Eq. (9.4) to give us more information.

$$e = 0.80 = \frac{v_2 - v_1}{u_1 - u_2}$$

or

$$v_2 - v_1 = (0.80)(u_1 - u_2)$$

Substituting the known values for u_1 and u_2, we obtain

$$v_2 - v_1 = (0.80)(-24 \text{ m/s} - 16 \text{ m/s})$$
$$= (0.80)(-40 \text{ m/s})$$

or, finally,

$$v_2 - v_1 = -32 \text{ m/s}$$

We can now use the momentum equation to arrive at another relation between v_2 and v_1, allowing us to solve the two equations simultaneously.

$$m_1u_1 + m_2u_2 = m_1v_1 + m_2v_2$$

The left side of this equation has already been found in Part (a) to equal 16 kg · m/s. Therefore, we have on substitution for m_1 and m_2 on the right side

$$16 \text{ kg} \cdot \text{m/s} = (2 \text{ kg})v_1 + (4 \text{ kg})v_2$$

from which

$$2v_1 + 4v_2 = 16 \text{ m/s}$$

or

$$v_1 + 2v_2 = 8 \text{ m/s}$$

Hence, we have two equations:

$$v_2 - v_1 = -32 \text{ m/s} \qquad v_1 + 2v_2 = 8 \text{ m/s}$$

Solving simultaneously, we obtain

$$v_1 = 24 \text{ m/s} \qquad v_2 = -8 \text{ m/s}$$

Thus, after colliding, the balls reverse their directions, m_1 moving to the right with a velocity of 24 m/s and m_2 moving to the left with a velocity of 8 m/s.

Example 9.6

A 12-g bullet is fired into a 2-kg block of wood suspended from a cord, as shown in Fig. 9.7. The impact of the bullet causes the block to swing 10 cm above its original level. Compute the velocity of the bullet as it strikes the block.

Plan: The problem needs to be separated into two parts: the conservation of momentum during impact and the conservation of energy during the rise of block and bullet. The beginning velocity for the rise is the same as the ending velocity for the impact process. Therefore, we will find the velocity v_c required to reach the maximum height and use conservation of momentum to find the entrance velocity of the bullet required to impart that velocity to the combined masses.

Solution: We will use the symbol m_b for the mass of the bullet and m_w for the mass of the wood block. The kinetic energy of the combined masses must be equal to the potential energy at the highest point. Thus,

$$\frac{1}{2}(m_b + m_w)v_c^2 - (m_b + m_w)gh$$

Dividing by the combined mass $(m_b + m_w)$ and simplifying, we obtain

$$v_c^2 = 2gh \qquad \text{or} \qquad v_c = \sqrt{2gh}$$

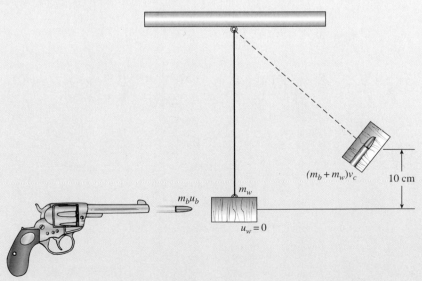

Figure 9.7 Computing the entrance velocity of a bullet fired into a suspended block.

from which

$$v_c = \sqrt{2(9.8 \text{ m/s}^2)(0.10 \text{ m})} = 1.40 \text{ m/s}$$

Now, we can use this as the final combined velocity after impact. Conservation of momentum requires that

$$m_b u_b + m_w u_w = (m_b + m_w)v_c$$

We now know that $m_b = 0.012$ kg, $m_w = 2$ kg, $u_w = 0$ and $v_c = 1.40$ m/s. Substitution of these values yields

$$(0.012 \text{ kg})u_b + 0 = (0.012 \text{ kg} + 2 \text{ kg})(1.40 \text{ m/s})$$
$$(0.012 \text{ kg})u_b + 0 = 0.0168 \text{ kg} \cdot \text{m/s} + 2.8 \text{ kg} \cdot \text{m/s}$$
$$(0.012 \text{ kg})u_b = 2.82 \text{ kg} \cdot \text{m/s}$$

which gives the entrance velocity for the bullet as

$$u_b = 235 \text{ m/s}$$

Summary and Review

Summary

In this chapter, you learned the relationship between impulse and momentum. Physical problems were then introduced to deal with elastic and inelastic collisions. The major concepts are summarized as follows:

- The *impulse* is the product of the average force $\mathbf{F}$ and the time interval Δt through which it acts.

$$\text{Impulse} = F\,\Delta t$$
$$\text{SI units: N} \cdot \text{s}$$
$$\text{USCS units: lb} \cdot \text{s}$$

- The *momentum* of a particle is its mass times its velocity.

$$\text{Momentum } p = mv$$
$$\text{SI units: kg} \cdot \text{m/s}$$
$$\text{USCS units: slug} \cdot \text{ft/s}$$

- The impulse is equal to the change in momentum:

$$F\,\Delta t = mv_f - mv_0$$

Note: $\text{N} \cdot \text{s} = \text{kg} \cdot \text{m/s}$ (equivalent units)

- *Conservation of momentum:* The total momentum before impact is equal to the total momentum after impact (see Fig. 9.3).

$$m_1 u_1 + m_2 u_2 = m_1 v_1 + m_2 v_2$$

- The *coefficient of restitution* is found from relative velocities before and after collision or from the rebound height:

$$e = \frac{v_2 - v_1}{u_1 - u_2} \qquad e = \sqrt{\frac{h_2}{h_1}}$$

- For a completely elastic collision, $e = 1$. For a completely inelastic collision, $e = 0$.

Key Terms

coefficient of restitution   186
completely elastic collision   185
completely inelastic collision   185
conservation of momentum   183
clasticity   185
impulse   180
momentum   180

Review Questions

9.1. Show the equivalence of the units of impulse with the units of momentum in USCS units.

9.2. Discuss the vector nature of impulse and momentum.

9.3. How does the magnitude of the impulse 1 lb · s compare with the magnitude of the impulse 1 N · s?

9.4. Discuss the conservation of energy and momentum for (a) an elastic collision and (b) an inelastic collision.

9.5. If you hold a weapon loosely when firing, it appears to give a greater kick than when you hold it tight against your shoulder. Explain. What effect does the weight of the weapon have?

9.6. A mortar shell explodes in midair. How is momentum conserved? How is energy conserved?

9.7. Suppose you hit a tennis ball into the air with a racket. The ball first strikes the concrete court and then bounds over the fence, landing in the grass. How many impulses were involved, and which impulse was the greatest?

9.8. A father and his daughter stand facing each other on a frozen pond. If the girl pushes her father backward, describe their relative motion and velocities. Would they differ if the father pushed the daughter?

9.9. Two small cars of masses m_1 and m_2 are tied together with a compressed spring between them. The cord is burned with a match, releasing the spring and imparting an equal impulse to each car. Compare the ratio of their displacements to the ratio of their masses at some later instant.

Problems

Section 9.1 Impulse and Momentum

9.1. A 0.5-kg wrench is dropped from a height of 10 m. What is its momentum just before it strikes the floor?   Ans. 7 kg · m/s, down

9.2. Compute the momentum and kinetic energy of a 2400-lb car moving north at 55 mi/h.

9.3. A 2500-kg truck traveling at 40 km/h strikes a brick wall and comes to a stop in 0.2 s. (a) What is the change in momentum? (b) What is the impulse? (c) What is the average force on the wall during the crash?
Ans. (a) -2.78×10^4 kg · m/s,
(b) -2.78×10^4 N · s, (c) 1.39×10^5 N

9.4. What is the momentum of a 3-g bullet moving at 600 m/s in a direction 30° above the horizontal? What are the horizontal and vertical components of this momentum?

9.5. A 0.2-kg baseball traveling to the left at 20 m/s is driven in the opposite direction at 35 m/s when it is hit by a bat. The average force on the ball is 6400 N. How long was it in contact with the bat?
Ans. 1.72 ms

9.6. A bat exerts an average force of 248 lb on a 0.6-lb ball for 0.01 s. The incoming velocity of the ball was 44 ft/s. If it leaves in the opposite direction, what is its velocity?

*9.7. A 500-g ball travels from left to right at 20 m/s. A bat drives the ball in the opposite direction with a velocity of 36 m/s. The time of contact was 0.003 s. What was the average force on the ball?
Ans. 9333 N

*9.8. A 400-g rubber ball is dropped a vertical distance of 12 m onto the pavement. It is in contact with the pavement for 0.01 s and rebounds to a height of 10 m. What is the total change in momentum? What average force is exerted by the pavement on the ball?

*9.9. A cue stick strikes an eight ball with an average force of 80 N over a time of 12 ms. If the mass of the ball is 200 g, what will be its velocity?
Ans. 4.80 m/s

9.10. A golfer hits a 46-g golfball with an initial velocity of 50 m/s at 30°. What are the horizontal and vertical components of the momentum imparted to the ball?

*9.11. The face of the club in Prob. 9.10 is in contact with the ball for 1.5 ms. What are the horizontal and vertical components of the average force on the ball? (*Hint:* Work horizontal and vertical impulse and momentum independently.)
Ans. $F_x = 1330$ N, $F_y = 767$ N

Section 9.2 Conservation of Momentum

9.12. A 20-kg girl and a boy on roller skates stand at rest facing each other. They push each other as hard as they can, and the boy moves to the left with a velocity of 2 m/s, while the girl moves to the right with a velocity of 3 m/s. What is the mass of the boy?

9.13. The mass of the toy truck in Fig. 9.8 is three times that of the car, and they are tied together against a compressed spring. When the string breaks, the car moves to the left at 6 m/s. What was the velocity imparted to the truck?   Ans. 2 m/s

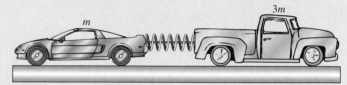

Figure 9.8   A toy car and truck are tied together with a string after being compressed against a spring.

9.14. A 70-kg person standing on a frictionless platform of ice throws a football forward with a velocity of 12 m/s. If the person moves backward with a velocity of 34 cm/s, what is the mass of the football?

9.15. A 20-kg child is at rest in a wagon. The child jumps forward at 2 m/s, sending the wagon backward at 12 m/s. What is the mass of the wagon?   Ans. 3.33 kg

9.16. Two children, one weighing 80 lb and the other 50 lb, are at rest on roller skates. The larger child pushes the smaller one, who moves away at 6 mi/h. What is the velocity of the larger child?

9.17. A 60-g firecracker explodes, sending a 45-g piece to the left and another to the right with a velocity of 40 m/s. What is the velocity of the 45-g piece?
Ans. −13.3 m/s

*9.18. A 24-g bullet is fired with a muzzle velocity of 900 m/s from a 5-kg rifle. Find the recoil velocity of the rifle. What is the ratio of the kinetic energy of the bullet to that of the rifle?

*9.19. A 6-kg bowling ball collides head on with 1.8-kg pin. The pin moves forward at 3 m/s and the ball slows to 1.6 m/s. What was the initial velocity of the bowling ball?   Ans. 2.50 m/s

*9.20. A 60-kg man on a lake of ice catches a 2-kg ball. The ball and man each move at 8 cm/s after the ball is caught. What was the velocity of the ball before it was caught? What energy was lost in the process?

*9.21. A 200-g rock traveling south at 10 m/s strikes a 3-kg block initially at rest. (a) If the two stick together on collision, what will be their common velocity? (b) What energy was lost in the collision?

Ans. 62.5 cm/s, 9.38 J

Section 9.3 Elastic and Inelastic Impacts

9.22. A car traveling at 8 m/s crashes into a car of identical mass stopped at a traffic light. What is the velocity of the wreckage immediately after the crash, assuming that the cars stick together?

9.23. A 2000-kg truck traveling at 10 m/s crashes into a 1200-kg car initially at rest. What is the common velocity after the collision if they stick together? What is the loss in kinetic energy?

Ans. 6.25 m/s, 37,500 J

9.24. A 30-kg child stands on a frictionless surface. The father throws a 0.8-kg football with a velocity of 15 m/s. What velocity will the child have after catching the football?

9.25. A 20-g object traveling to the left at 8 m/s collides head on with a 10-g object traveling to the right at 5 m/s. What is their combined velocity if they stick together after impact?

Ans. 3.67 m/s to the left

*9.26. Two metal balls A and B are suspended, as shown in Fig. 9.9, so that each touches the other. The masses are given in the figure. Ball A is pulled to the side until it is 12 cm above its initial position, and then it is released. If it strikes ball B in a completely elastic collision, find the height h reached by ball B assuming zero friction.

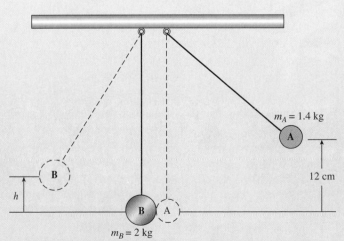

Figure 9.9 The completely elastic collision of two suspended metal balls.

*9.27. A 2-kg block of clay is suspended from the ceiling by a long cord, as shown in Fig. 9.10. A 500-g steel ball, fired horizontally, imbeds itself into the clay, causing both masses to rise to a height of 20 cm. Find the entrance velocity of the ball.

Ans. 9.90 m/s

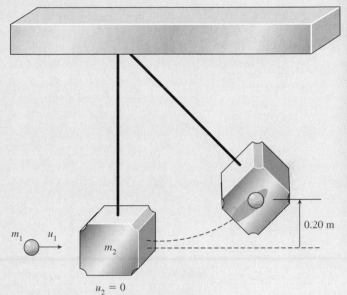

Figure 9.10

*9.28. In Prob. 9.27, suppose the 500-g ball passes entirely through the clay and emerges with a velocity of 10 m/s. What must be the new entrance velocity if the block is to rise to the same height of 20 cm?

*9.29. A 9-g bullet is embedded into a 2.0-kg ballistic pendulum similar to the one shown in Fig. 9.7. What was the initial velocity of the bullet if the combined masses rise to a height of 9 cm?

Ans. 297 m/s

*9.30. A billiard ball moving to the left at 30 cm/s collides head on with another ball moving to the right at 20 cm/s. The masses of the balls are identical. If the collision is perfectly elastic, what is the velocity of each ball after impact?

9.31. The coefficient of restitution for steel is 0.90. If a steel ball is dropped from a height of 7 m, how high will it rebound?

Ans. 5.67 m

*9.32. What is the time between the first contact with the surface and the second contact for Prob. 9.31?

*9.33. A ball dropped from rest onto a fixed horizontal plate rebounds to a height that is 81 percent of its original height. What is the coefficient of restitution? What is the required velocity on the first impact to cause the ball to rebound to a height of 8 m?

Ans. 0.9, 13.9 m/s

*9.34. A 300-g block moving north at 50 cm/s collides with a 200-g block moving south at 100 cm/s. If the collision is completely inelastic, what is the common velocity of the blocks after they stick together? What is the loss in energy?

*9.35. Suppose the collision in Prob. 9.34 is perfectly elastic. What are the velocities after impact?

Ans. −80 cm/s, +70 cm/s

*9.36. A 5-kg and a 12-kg object approach each other with equal velocities of 25 m/s. What will be their velocities after impact if the collision is (a) completely inelastic or (b) perfectly elastic?

Additional Problems

9.37. An average force of 4000 N acts on a 400-g piece of metal, causing it to move from rest to a velocity of 30 m/s. What was the time of contact for this force? Ans. 3.00 ms

9.38. A 600-g object whose velocity is initially 12 m/s collides with a wall and rebounds with half of its original kinetic energy. What impulse was applied by the wall?

9.39. A 10-kg block at rest on a horizontal surface is struck by a 20-g projectile bullet moving at 200 m/s. The bullet passes entirely through the block and exits with a velocity of 10 m/s. What is the velocity of the block? Ans. 38.0 cm/s

9.40. In Prob. 9.39, how much kinetic energy was lost?

9.41. A 60-g body having an initial velocity of 100 cm/s to the right collides with a 150-g body moving to the left at 30 cm/s. The coefficient of restitution is 0.8. What are the velocities after impact? What percent of the energy is lost in collision?

Ans. −67.1 cm/s, +36.9 cm/s; 35.5 percent

9.42. The block in Fig. 9.10 weighs 1.5 kg. How high will it rise if a 40 g projectile with an initial velocity of 80 m/s embeds itself into the block?

Ans. 22.0 cm

9.43. A single railroad car moving north at 10 m/s strikes two identical, coupled cars initially moving south at 2 m/s. If all three cars couple after colliding, what is their common velocity?

Ans. 2.00 m/s, north

*9.44. An atomic particle of mass 2.00×10^{-27} kg moves with a velocity of 4.00×10^6 and collides head on with a particle of mass 1.20×10^{-27} kg initially at rest. Assuming a perfectly elastic collision, what is the velocity of the incident particle after the collision?

*9.45. A bat strikes a 400-g ball moving horizontally to the left at 20 m/s. It leaves the bat with a velocity of 60 m/s at an angle of 30° with the horizontal. What are the horizontal and vertical components of the impulse imparted to the ball?

Ans. 28.8 N · s, 12 N · s

*9.46. If the bat in Prob. 9.45 was in contact with the ball for 5 ms, what was the magnitude of the average force on the ball?

*9.47. Cart A has a mass of 300 g and moves on a frictionless air track at 1.4 m/s when it hits Cart B at rest. The collision is perfectly elastic, and the 300-g cart's velocity is reduced to 0.620 m/s after the collision. What is the mass of the other cart, and what was its velocity after the collision?

Ans. 116 g, 2.02 m/s

*9.48. In Fig. 9.11, a 1-kg mass slides with a velocity of 15 m/s toward a 2-kg mass initially at rest. All surfaces are frictionless. What is their common velocity if they stick together on colliding? What is the ratio of the final kinetic energy to the initial kinetic energy for the system?

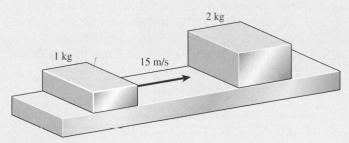

Figure 9.11

*9.49. Assume that the collision in Prob. 9.48 is completely elastic. Find the velocity of each mass after the collision. Ans. $v_1 = -5$ m/s, $v_2 = 10$ m/s

*9.50. A 2-kg mass moves to the right at 2 m/s and collides with a 6-kg mass moving to the left at 4 m/s. If the collision is completely inelastic, what is the common velocity of the two masses after they collide, and how much energy is lost in the collision?

*9.51. In Prob. 9.50, assume that the collision is perfectly elastic. What are the velocities after the collision?

Ans. −1 m/s, −7 m/s

Critical Thinking Questions

*9.52. An astronaut in orbit outside a capsule uses a revolver to control motion. The astronaut with gear weighs 200 lb on earth. If the revolver fires 0.05-lb bullets at 2700 ft/s and ten shots are fired, what is the final velocity of the astronaut? Compare the final kinetic energy of the ten bullets with that of the astronaut. Why is the difference so great?

Ans. −6.75 ft/s, bullets = 57,000 ft lb,
astronaut = 142 ft · lb

*9.53. In applying conservation of momentum for colliding objects to find final velocities, could we use the weight of the objects instead of the mass? Why, or why not? Verify your answer by applying it to one of the examples in the text.

*9.54. A 20-g bullet, moving at 200 m/s, strikes a 10-kg wooden block and passes entirely through it, emerging with a velocity of 10 m/s. What is the velocity of the block after impact? How much energy was lost? Ans. 0.380 m/s, 399 J

*9.55. A 0.30-kg baseball moving horizontally at 40 m/s is struck by a bat. If the ball is in contact with the bat for a time of 5 ms and it leaves with a velocity of 60 m/s at an angle of 30°, what are the horizontal and vertical components of the average force acting on the bat?

*9.56. When two masses collide, they produce equal but opposite impulses. The masses do not change in the collision, so the change in momentum of one should be the negative of the change for the other. Is this true whether the collision is elastic or inelastic? Verify your answer by using the data in Probs. 9.50 and 9.51.

*9.57. Two toy cars of masses m and $3m$ approach each other, each traveling at 5 m/s. If they couple together, what is their common speed afterward? What are the velocities of each car if the collision is perfectly elastic?

*9.58. An 8-g bullet is fired horizontally at two blocks resting on a frictionless surface. The first block has a mass of 1 kg, and the second has a mass of 2 kg. The bullet passes completely through the first block and lodges into the second. After their collisions, the 1-kg block moves with a velocity of 1 m/s, and the 2-kg block moves with 2 m/s. What is the velocity of the bullet before and after emerging from the first block? Ans. 627 m/s, 502 m/s

*9.59. A 1-kg mass A is attached to a support by a cord of length 80 cm and is held horizontally as in Fig. 9.12. After release, it swings downward, striking the 2-kg mass B, which is at rest on a frictionless tabletop. Assuming that the collision is perfectly elastic, what are the velocities of each mass immediately after impact?

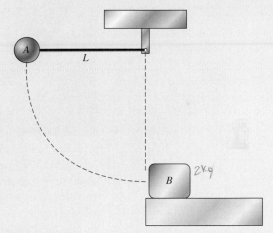

Figure 9.12

10

Uniform Circular Motion

Centrifuges are used to remove solid particles from a liquid. The more massive particles have greater inertia and hence move to the outside of the tubes until the necessary centripetal force causes them to move in a circle. (There is no outward force on these particles.) In biology and biochemistry, centrifuges are used for isolating and separating biocompounds on the basis of molecular weight. The same principle can be applied on a much larger scale. Giant centrifuges are used to test the reactions of pilots and astronauts to forces above those experienced in the Earth's gravity. (*Photo © vol. 29 PhotoDisc/Getty.*)

Objectives

After completing this chapter, you should be able to

1. Demonstrate by definition and examples your understanding of the concepts of *centripetal acceleration* and *centripetal force.*

2. Apply your knowledge of centripetal force and centripetal acceleration to the solution of problems similar to those in this text.

3. Define and apply the concepts of *frequency* and *period of rotation* and relate them to the linear speed of an object in uniform circular motion.

4. Apply your knowledge of centripetal force to problems involving *banking angles*, the *conical pendulum*, and motion in a *vertical circle.*

5. State and apply the universal law of gravitation.

In previous chapters, we considered primarily motion in a straight line. This approach is sufficient to describe and apply most technical concepts. Bodies in the natural world, however, generally move along curved paths. Artillery shells travel along parabolic paths under the influence of the Earth's gravitational field. Planets revolve around the Sun in paths that are nearly circular. On the atomic level, electrons circle about the nucleus of an atom. In fact, it is difficult to imagine any phenomenon in physics that does not involve motion in at least two dimensions.

10.1   Motion in a Circular Path

Newton's first law tells us that all bodies moving in a straight line with constant speed will maintain their velocity unaltered unless acted on by an external force. The velocity of a body is a vector quantity consisting of both its speed and its direction. Just as a resultant force is required to change its speed, a resultant force must be applied to change its direction. Whenever this force acts in a direction other than the original direction of motion, the path of a moving particle is changed.

The simplest kind of two-dimensional motion occurs when a constant external force always acts at right angles to the path of a moving particle. In this case, the resultant force will produce an acceleration that alters only the direction of motion, leaving the speed constant. This simple type of motion is referred to as ***uniform circular motion.***

Uniform circular motion is motion in which there is no change in speed, only a change in direction.

An example of uniform circular motion is afforded by swinging a rock in a circular path with a string, as shown in Fig. 10.1. As the rock revolves with constant speed, the inward force of the tension in the string constantly changes the direction of the rock, causing it to move in a circular path. If the string broke, the rock would fly off at a tangent perpendicular to the radius of its circular path.

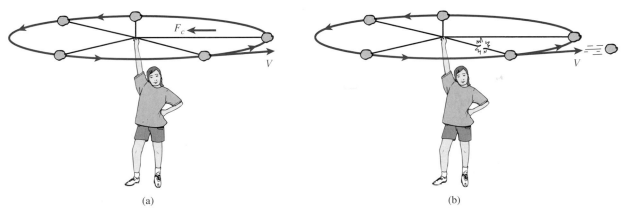

Figure 10.1   (a) The inward pull of the string on the rock causes it to move in a circular path. (b) If the string breaks, the rock will fly off at a tangent to the circle.

10.2   Centripetal Acceleration

Newton's second law of motion states that a resultant force must produce an acceleration in the direction of the force. In uniform circular motion, the acceleration changes the velocity of a moving particle by altering its direction.

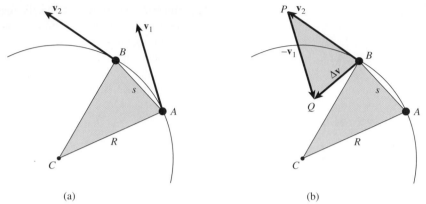

(a)   (b)

Figure 10.2 (a) *A* and *B* are the positions at two instants separated by a time interval Δt. (b) The change in velocity v is represented graphically. The vector will point directly toward the center if Δt is made small enough for the chord s to equal the arc joining points *A* and *B*.

The position and velocity of a particle moving in a circular path of radius *R* are shown at two instants in Fig. 10.2. When the particle is at point *A*, its velocity is represented by the vector $\mathbf{v}_1$. After a time interval Δt, its velocity is represented by the vector $\mathbf{v}_2$. The acceleration, by definition, is the change in velocity per unit time. Thus,

$$\mathbf{a} = \frac{\Delta \mathbf{v}}{\Delta t} = \frac{\mathbf{v}_2 - \mathbf{v}_1}{\Delta t} \tag{10.1}$$

The change in velocity $\Delta \mathbf{v}$ is represented graphically in Fig. 10.2b. The difference between the two vectors $\mathbf{v}_2$ and $\mathbf{v}_1$ is constructed according to the methods introduced in Chapter 2. Since the velocities $\mathbf{v}_2$ and $\mathbf{v}_1$ have the same magnitude, they form the legs of an isosceles triangle *BPQ* whose base is $\Delta \mathbf{v}$. If we construct a similar triangle *ABC*, it can be noted that the magnitude of $\Delta \mathbf{v}$ has the same relationship to the magnitude of either velocity as the chord s has to the radius *R*. This proportionality is stated symbolically as

$$\frac{\Delta v}{v} = \frac{s}{R} \tag{10.2}$$

where v represents the absolute magnitude of either $\mathbf{v}_1$ or $\mathbf{v}_2$.

The distance the particle actually covers in traveling from point *B* to point *B* is not the distance s but the length of the arc from *A* to *B*. The shorter the time interval Δt, the closer these points are until in the limit the chord length becomes equal to the arc length. In this case, the length s is given by

$$s = v \, \Delta t$$

which, when substituted into Eq. (10.2), yields

$$\frac{\Delta v}{v} = \frac{v \, \Delta t}{R}$$

Since the acceleration from Eq. (10.1) is $\Delta v / \Delta t$, we can rearrange terms and obtain

$$\frac{\Delta v}{\Delta t} = \frac{v^2}{R}$$

Therefore, the rate of change in velocity, or the ***centripetal acceleration,*** is given by

$$a_c = \frac{v^2}{R} \tag{10.3}$$

where v is the ***linear speed*** of a particle moving in a circular path of radius *R*.

The term *centripetal* means that the acceleration is always directed toward the center. Notice in Fig. 10.2b that the vector $\Delta\mathbf{v}$ does not point toward the center. This is because we have considered a long time interval between the measurements at A and B. If we restrict the separation of these points to an infinitesimal distance, the vector $\Delta\mathbf{v}$ would point toward the center.

The units of centripetal acceleration are the same as those for linear acceleration. For example, in the SI, v^2/R would have the units

$$\frac{(\text{m/s})^2}{\text{m}} = \frac{\text{m}^2/\text{s}^2}{\text{m}} = \text{m/s}^2$$

Example 10.1

A 2-kg body is tied to the end of a cord and whirled in a horizontal circle of radius 1.5 m. If the body makes three complete revolutions every second, determine its linear speed and the centripetal acceleration.

Plan: The distance traveled by the body in one revolution is equal to the circumference of a circle ($C = 2\pi R$), and since three revolutions are made each second, the time for one revolution must be one-third of 1 second, or 0.333 s. This information allows us to find the linear speed of the body, and we can find the acceleration from Eq. (10.3).

Solution: First, we find the circumference of the circular path.

$$C = 2\pi R = 2\pi(1.5 \text{ m}) \qquad \text{or} \qquad C = 9.43 \text{ m}$$

Dividing the distance by the 0.333 s needed for one revolution, we obtain

$$v = \frac{9.43 \text{ m}}{0.333 \text{ s}} = 28.3 \text{ m/s}$$

Next, we obtain the acceleration from Eq. (10.3).

$$a_c = \frac{v^2}{R} = \frac{(28.3 \text{ m/s})^2}{1.5 \text{ m}} \qquad \text{or} \qquad a_c = 534 \text{ m/s}^2$$

The procedure used to compute linear speed in Example 10.1 is so useful that it is worth remembering. If we define the *period* as the time for one complete revolution and designate it by the letter T, the linear speed can be computed by dividing the period into the circumference. Thus,

$$v = \frac{2\pi R}{T} \tag{10.4}$$

Another useful parameter in engineering problems is the rotational speed, expressed in *revolutions per minute* (rpm) or *revolutions per second* (rev/s). This quantity is called the *frequency of rotation* and is given by the reciprocal of the period.

$$f = \frac{1}{T} \tag{10.5}$$

The validity of this relation is demonstrated by noting that the reciprocal of seconds per revolution (s/rev) is revolutions per second (rev/s). Substitution of this definition into Eq. (10.4) yields an alternative equation for determining the linear speed.

$$v = 2\pi f R \tag{10.6}$$

For example, if the frequency is 1 rev/s and the radius 1 m, the linear speed would be 2π m/s.

Centripetal Force

The inward force necessary to maintain uniform circular motion is defined as the *centripetal force.* From Newton's second law of motion, the magnitude of this force must equal the product of mass and centripetal acceleration. Thus,

$$F_c = ma_c = \frac{mv^2}{R} \qquad (10.7)$$

where m is the mass of an object moving with a velocity v in a circular path of radius R. The units chosen for the quantities F_c, m, v, and R must be consistent for the system chosen. For example, the SI units for mv^2/R are

$$\frac{kg \cdot m^2/s^2}{m} = kg \cdot m/s^2 = N$$

An inspection of Eq. (10.7) reveals that the inward force $\mathbf{F}_c$ is directly proportional to the square of the velocity of the moving object. This means that increasing the linear speed to twice its original value will require four times the original force. Similar reasoning will show that doubling the mass or halving the radius will require twice the original centripetal force.

For problems in which the rotational speed is expressed in terms of the *frequency,* the centripetal force can be determined from

$$F_c = \frac{mv^2}{R} = 4\pi^2 f^2 mR \qquad (10.8)$$

This relation results from substitution of Eq. (10.6), which expresses the linear speed in terms of the frequency of revolution.

Example 10.2   A 4-kg ball is swung in a horizontal circle by a cord 2 m long. Find the tension in the cord if the period is 0.5 s.

Plan: The tension in the cord will be equal to the centripetal force necessary to maintain circular motion. The linear speed will be found by dividing the circumference of the path by the period or time for one revolution.

Solution: The velocity around the path is

$$v = \frac{2\pi R}{T} = \frac{2\pi(2\ m)}{0.5\ s} = 25.1\ m/s$$

from which the centripetal force is

$$F_c = \frac{mv^2}{R} = \frac{(4\ kg)(25.1\ m/s)^2}{2\ m}$$

$$F_c = 1260\ N$$

Example 10.3

Two 500-g masses rotate about a central axis at 12 rev/s, as shown in Fig. 10.3. (a) What is the constant force acting on each mass? (b) What is the tension in the supporting rod?

Plan: The total downward force of the weights and rod is balanced by the upward force of the center support. Therefore, the resultant force acting on each revolving mass is the pull toward the center equal to the centripetal force. We will find the velocity from the radius and frequency of revolution. Then we can calculate the centripetal force on each mass.

Solution (a): The velocity of each mass is

$$v = 2\pi f R = 2\pi(12 \text{ rev/s})(0.30 \text{ m})$$
$$= 22.6 \text{ m/s}$$

Now, the centripetal force is found from Eq. (10.7)

$$F_c = \frac{mv^2}{R} = \frac{(0.500 \text{ kg})(22.6 \text{ m/s})^2}{0.300 \text{ m}}$$

$$F_c = 853 \text{ N}, \textit{toward center}$$

The same calculation holds for either mass.

Solution (b): The resultant force on each mass is equal to 853 N directed *toward* the center. It is exerted *by* the rod *on* the mass. The outward force we often *think* acts on the mass is actually the *reaction* force exerted *by* the mass *on* the rod. The tension in the rod is due to this outward force, and its is, of course, equal in magnitude to the centripetal force of 853 N.

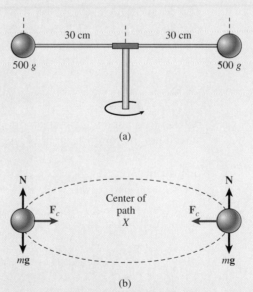

(a)

(b)

Figure 10.3   Objects traveling in a circular path. The resultant force of the rod on the objects provides the necessary centripetal force. According to Newton's third law, the objects exert an equal and opposite reaction force called the *centrifugal force*. These forces do not cancel each other because they act on different objects.

10.4   Banking Curves

When an automobile is driven around a sharp turn on a perfectly level road, friction between the tires and the road provides a centripetal force (see Fig. 10.4). If this central force becomes large enough, the car may slide off the road. The maximum force of static friction determines the maximum speed with which a car can negotiate a turn of a given radius.

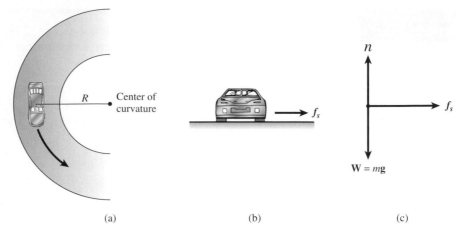

(a)   (b)   (c)

Figure 10.4 The centripetal force of friction. Note that there is no outward force on the car.

Example 10.4

What is the maximum speed at which an automobile can negotiate a curve of radius 100 m without sliding if the coefficient of static friction is 0.7?

Plan: The centripetal force necessary to maintain circular motion in this example is provided by static friction. As the car increases its speed, the centripetal (friction) force eventually becomes large enough to just overcome the maximum force of static friction. At that instant, the centripetal force equals the maximum force of static friction. Thus, there are two formulas that may be used to calculate the same force:

$$f_{s,\text{max}} = \mu_s n \qquad \text{and} \qquad F_c = \frac{mv^2}{R}$$

and since $F_c = f_{s,\text{max}}$, we may write

$$\frac{mv^2}{R} = \mu_s n \qquad\qquad (10.9)$$

Next, we can apply the first condition for equilibrium to determine the normal force and substitute given information to determine the velocity at the instant that sliding occurs.

Solution: Since vertical forces are balanced, we know that

$$n = W = mg \qquad \text{and} \qquad f_{s,\text{max}} = \mu_0 mg$$

so Eq. (10.9) becomes

$$\frac{mv^2}{R} = \mu_s mg \qquad \text{or} \qquad v^2 = \mu_s gR$$

from which

$$v = \sqrt{\mu_s gR} \qquad\qquad (10.10)$$

Finally, we substitute the known values for g, R, and μ_s to find the maximum speed.

$$v = \sqrt{(0.7)(9.8 \text{ m/s}^2)(100 \text{ m})} = 26.2 \text{ m/s}$$

or approximately 94.3 km/h (58.6 mi/h).

To some, it is quite surprising that the weight of the automobile was not considered in determining the maximum speed. Our experience tends to contradict the independence of weight. However, we must not confuse the stability of an automobile with the conditions for slipping. The force exerted by the road on the tires acts at the bottom of the tire, considerably

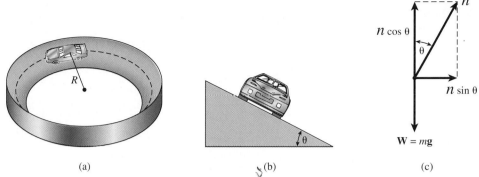

Figure 10.5 Effects of banking a turn. The horizontal component of the normal force n sin θ provides the necessary centripetal acceleration.

below the center of gravity for the car. Thus, a bus is much more likely to turn over than a corvette. You must also remember that there are many factors affecting friction that are not controlled when applying Eq. (10.10) to any given situation. Such things as tread patterns, temperature, and variation in surfaces might also influence the strict application of such equations. Still, they can frequently be used to give engineers reliable estimates.

Now let us consider the effects of banking turns to reduce or even eliminate the need for friction as the provider of a centripetal force. Consider the circular path followed by the car in Fig. 10.5a. When the car is at rest or traveling at a slow speed, the friction force is directed up the incline. As the speed of the car increases, the static friction force diminishes until it eventually reverses direction and acts *down* the incline. The optimal speed is that speed for which the friction force is zero, and the entire centripetal force is provided by the central component of the normal force exerted by the road on the car. The components of the normal force are

$$n_x = n \sin \theta \quad \text{and} \quad n_y = n \cos \theta$$

Note that the angle θ of inclination for the roadway is the same as the angle with respect to the y axis in the free-body diagram (Fig. 10.5c). It is the horizontal component, $n \sin \theta$, that provides the centripetal force. If we represent the linear velocity by v and the radius of the turn by R, we can write

$$n \sin \theta = \frac{mv^2}{R}$$

And since the vertical forces are balanced,

$$n \cos \theta = mg$$

Here, the angle θ represents the angle for which the friction force is zero. We call this the *optimum banking angle.* Dividing the first of these equations by the second and recalling that tan $\theta = (\sin \theta / \cos \theta)$, we can obtain the following expression:

$$\tan \theta = \frac{v^2}{gR} \quad \textit{Optimum Banking Angle} \quad \textbf{(10.11)}$$

Example 10.5 The speed limit on a given highway is 80 km/h. Find the optimum banking angle for a curve of radius 300 m.

Plan: We convert the velocity of 80 km/h to consistent SI units and then apply Eq. (10.11) to find the optimum banking angle.

Solution: Recalling that 1 km = 1000 m and 1 h = 3600 s, we write

$$v = 80\frac{km}{h}\left(\frac{1000\ m}{1\ km}\right)\left(\frac{1\ h}{3600\ s}\right) = 22.2\ m/s$$

Substitution into Eq. (10.11) gives

$$\tan\theta = \frac{v^2}{gR} = \frac{(22.2\ m/s)^2}{(9.8\ m/s^2)(300\ m)} = 0.168$$

The optimal banking angle is, therefore,

$$\theta = 9.5°$$

Actual roadways are not typically sloped for *optimum* banking angles, because at turns of small radiuses, the angles would be quite large. If the radius of the turn in Example 10.5 was changed from 300 m to 30 m, the optimum banking angle would be a colossal 59°! However, you do find examples where such angles are approximated. For example, consider the motorcycle inside a circular bowl at the county fair, or look at the end zones for major NASCAR speedways.

10.5 The Conical Pendulum

A ***conical pendulum*** consists of a mass m revolving in a horizontal circle with constant speed v at the end of a cord of length L. A comparison of Fig. 10.6 with Fig. 10.5 shows that the formula derived for the banking angle also applies for the angle the cord makes with the vertical in a conical pendulum. In the latter problem, the necessary centripetal force is provided by the horizontal component of the tension in the cord. The vertical component is equal to the weight of the revolving mass. Hence,

$$T\sin\theta = \frac{mv^2}{R} \qquad T\cos\theta = mg$$

from which

$$\tan\theta = \frac{v^2}{Rg}$$

is obtained as in Section 10.4.

A careful study of Eq. (10.11) will show that, as the linear speed increases, the angle that the cord makes with the vertical also increases. Therefore, the vertical position of the mass (indicated in Fig. 10.6) is raised, causing a reduction in the distance h below the point of support. If we wish to express Eq. (10.11) in terms of the vertical position h, we should note that

$$\tan\theta = \frac{R}{h}$$

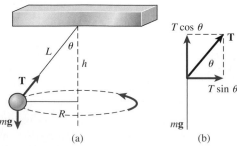

Figure 10.6

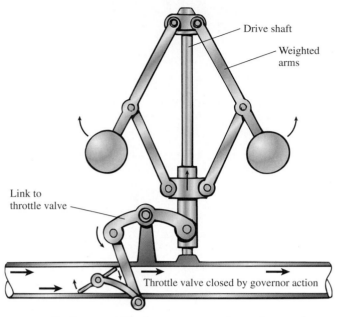

Figure 10.7   In some applications, centrifugal governors may be used to regulate speeds by opening or closing valves.

from which we obtain

$$\frac{R}{h} = \frac{v^2}{gR}$$

Thus, the distance of the weight below the support is a function of the linear speed and is given by

$$h = \frac{gR^2}{v^2}$$

A more useful form for this equation can be obtained by expressing the linear speed in terms of the rotational frequency. Since $v = 2\pi f R$, we can write

$$h = \frac{gR^2}{4\pi^2 f^2 R^2} = \frac{g}{4\pi^2 f^2}$$

Solving for f, we obtain

$$f = \frac{1}{2\pi} \sqrt{\frac{g}{h}} \qquad\qquad (10.12)$$

Early applications of the relationship between frequency of rotation and height were exemplified by the centrifugal governor, as shown in Fig 10.7. The position of the weights is used to open or close fuel valves. Such devices remain in use today for many engineering and scientific applications. However, they are rarely used in modern automobiles where speed is now controlled by electronic control units (ECU's).

10.6   Motion in a Vertical Circle

Motion in a vertical circle is different from the circular motion discussed in earlier sections. Since gravity always acts downward, the direction of the weight is the same at the top of the path as it is at the bottom. The forces that maintain the circular motion must always be directed, however, toward the center of the path. When more than one force acts on an object, it is the *resultant force* that produces the centripetal force.

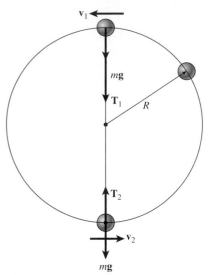

Figure 10.8

Consider a mass m tied to the end of a rope and swung in a circle of radius R, as shown by Fig. 10.8. We assign v_1 as the velocity at the top of the circular path and v_2 as the velocity at the bottom. Let us first consider the resultant force on the object as it passes through its highest point. The weight mg and the tension T_1 in the rope are each directed downward. The resultant of these forces is the centripetal force. Hence,

$$T_1 + mg = \frac{mv_1^2}{R} \tag{10.13}$$

On the other hand, when the mass passes through its lowest point, the weight mg is still directed downward, but the tension T_2 is directed upward. The resultant is still the necessary centripetal force, so we have

$$T_2 - mg = \frac{mv_2^2}{R} \tag{10.14}$$

It should be apparent from these equations that the tension in the cord at the bottom is greater than it is at the top. In one case, the weight adds to the tension, whereas in the other, the weight subtracts from the tension. The centripetal force (resultant) is a function of velocity, mass, and radius at any location.

Problem-Solving Strategy

Uniform Circular Motion

1. Read the problem and then draw and label a rough sketch.

2. Choose an axis perpendicular to the circular motion at the point where the centripetal force acts on a given mass.

3. Consider the direction of the centripetal force (toward the center) as positive.

4. The resultant force toward the center is the necessary centripetal force. If more than one force acts on the mass, the *net* force toward the center will be equal to

mv^2/R. This is actually a statement of Newton's second law for circulation motion.

$$\sum F = \frac{mv^2}{R} \qquad a_c = \frac{v^2}{R}$$

5. In calculating the resultant central force ($\sum F$), consider forces toward the center as positive and forces away from the center as negative. The right side of the equation, mv^2/R, will always be positive.

6. Substitute known quantities and solve the unknown factor. Be careful to distinguish between the weight and the mass of an object.

Example 10.6

In Fig. 10.8, assume that a 2-kg ball has a velocity of 5 m/s when it rounds the top of a circle whose radius is 80 cm. (a) What is the tension in the cord at that instant? (b) What is the minimum speed at the top necessary to maintain circular motion?

Plan: We will superimpose a free-body diagram for the forces acting on the ball at the top of the loop (see Fig. 10.8), indicating the direction of acceleration (downward) as positive. Then we will solve for the unknown tension T_1 by setting the resultant force equal to the mass times the centripetal acceleration. The same procedure will be followed to find the minimum speed, recognizing that the tension in the cord will be zero at that instant.

Solution (a): At the top of the path, both the weight mg and the tension in the cord are directed downward and toward the center of the circle. Thus, Newton's second law gives

$$T_1 + mg = \frac{mv_1^2}{R} \qquad \text{or} \qquad T_1 = \frac{mv_1^2}{R} - mg$$

Substituting $R = 0.80$ m, $g = 9.8$ m/s^2, $v = 5$ m/s, and $m = 2$ kg, we have

$$T_1 = \frac{(2 \text{ kg})(5 \text{ m/s})^2}{(0.80 \text{ m})} - (2 \text{ kg})(9.8 \text{ m/s}^2)$$

$$= 62.5 \text{ N} - 19.6 \text{ N} = 42.9 \text{ N}$$

Note that the centripetal force of 62.5 N is the *resultant* force at all points in the circular path where the speed of the mass is 5 m/s. At the top, the weight of 19.6 N provides part of this needed force, and the rope tension (42.9 N) provides the rest.

Solution (b): The **critical velocity** v_c occurs when the tension in the rope reduces to zero ($T_1 = 0$), and the entire centripetal force is provided by the weight mg. In this case,

$$T_1 + mg = \frac{mv_c^2}{R} \qquad \text{becomes} \qquad mg = \frac{mv_c^2}{R}$$

Dividing out the mass m and solving for v_c, this equation reduces to

$$v_c = \sqrt{gR} \qquad\qquad \textit{Critical Velocity} \quad \textbf{(10.15)}$$

Substituting the given information, we can now find v_c.

$$v_c = \sqrt{(9.8 \text{ m/s}^2)(0.80 \text{ m})} = 2.80 \text{ m/s}$$

If the velocity at the top drops below 2.80 m/s, the force required for circular motion no longer exists, and the ball will become a free-falling body.

10.7 Gravitation

The Earth and planets follow approximately circular orbits around the Sun. Newton suggested that the inward force that maintains planetary motion is only one example of a universal force called **gravitation,** which acts between all masses in the universe. He stated his thesis in the following **universal law of gravitation:**

> Each particle in the universe attracts each other particle with a force that is directly proportional to the product of their masses and inversely proportional to the square of the distance between them.

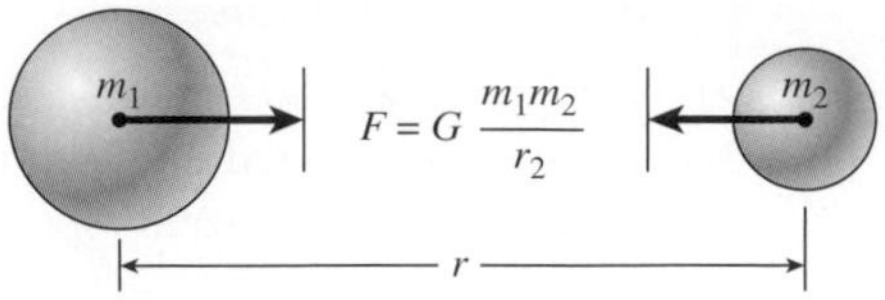

Figure 10.9   The universal law of gravitation.

The proportionality is usually stated in the form of an equation:

$$F = G\frac{m_1 m_2}{r^2} \tag{10.16}$$

where m_1 and m_2 are the masses of any two particles separated by a distance r, as shown in Fig. 10.9.

The proportionality constant G is a universal constant equal to

$$G = 6.67 \times 10^{-11}\ \text{N} \cdot \text{m}^2/\text{kg}^2$$
$$= 3.44 \times 10^{-8}\ \text{lb} \cdot \text{ft}^2/\text{slug}^2$$

Example 10.7

A 4-kg ball and a 2-kg ball are positioned so that their centers are 40 cm apart. With what force do they attract each other?

Solution: The force of attraction is found from Eq. (10.16):

$$F = G\frac{m_1 m_2}{r^2} = \frac{(6.67 \times 10^{-11}\ \text{N} \cdot \text{m}^2/\text{kg}^2)(4\ \text{kg})(2\ \text{kg})}{(0.40\ \text{m})^2}$$

$$F = 3.34 \times 10^{-9}\ \text{N}$$

The gravitational force is actually a small force. Because of the relatively large mass of the Earth in comparison with objects on its surface, we often assume that gravitational forces are strong. If we consider two marbles closely separated on a horizontal surface, however, our experience certainly substantiates that the gravitational attraction is weak.

Example 10.8

At the surface of the Earth, the acceleration of gravity is 9.8 m/s². If the radius of the Earth is 6.38×10^6 m, compute the mass of the Earth.

Plan: To find the radius of the Earth, we will consider a small test mass m' near the surface of the Earth whose radius is R_e. For our purposes, we will consider that the entire mass of the earth m_e is at its geometric center. This assumption is reasonable, since it would represent the average distance of the mass m' from each of the bits and pieces that make up the entire Earth. The gravitational force on our test mass can be found from $W = m'g$, but it can also be found from Eq. (10.16). By forming an equality of these two expressions, the test mass divides out, and the only unknown will be the mass of the Earth.

Solution: Assume the mass of the Earth is given by m_e and its radius is R_e. For a given test mass m' near the surface of the Earth,

$$W = m'g = \frac{Gm'm_e}{R_e^2}$$

Dividing out the test mass m' and solving for m_e, we have

$$m_e = \frac{gR_e^2}{G}$$

from which the mass of the Earth can be found by substitution.

$$m_e = \frac{(9.8 \text{ m/s}^2)(6.38 \times 10^6 \text{ m})^2}{6.67 \times 10^{-11} \text{ N} \cdot \text{m}^2/\text{kg}} = 5.98 \times 10^{24} \text{ kg}$$

10.8 The Gravitational Field and Weight

In earlier chapters, we defined *weight* as the attraction that the Earth has on masses located near the surface of the Earth. Perhaps it would be wise now to revisit this concept in light of Newton's law of gravitation. The attraction that any large spherical mass (such as the Earth) has on another mass located outside the sphere can be calculated by assuming that the entire mass of the large sphere is concentrated at its center. Assume, as in Example 10.8, that a mass m is located on the surface of the Earth whose mass is m_e. Setting the weight mg equal to the gravitational force, we obtain

$$mg = \frac{Gmm_e}{R_e^2}$$

The radius of the Earth is denoted by the symbol R_e. Now, if we divide out the mass m, we have the following value for the acceleration due to gravity.

$$g = \frac{Gm_e}{R_e^2} \qquad \textbf{(10.17)}$$

In Section 10.7, we used Eq. (10.16) to find the mass of the Earth from the given radius. We can also see from this equation that gravity and, therefore, the weight of an object depend on its location above the Earth's surface.

Example 10.9

How far above the surface of the Earth will a person's weight be reduced to one-half its value at the surface?

Plan: The weight mg at the surface will be reduced by one-half when the acceleration due to gravity g becomes $\frac{1}{2}(9.8 \text{ m/s}^2)$, or 4.9 m/s^2. We apply Eq. (10.17), this time for the general distance $r = R_e + h$, where h is the height above the Earth.

$$g = \frac{Gm_e}{r^2} = 4.9 \text{ m/s}^2 \qquad r = R_e + h$$

By solving for r, we can then subtract the radius of the Earth to determine h.

Solution:

$$r^2 = \frac{Gm_e}{4.9 \text{ m/s}^2} \qquad \text{or} \qquad r = \sqrt{\frac{Gm_e}{4.9 \text{ m/s}^2}}$$

From previous examples, we know that $m_e = 5.98 \times 10^{24}$ kg, so

$$r = \sqrt{\frac{(6.67 \times 10^{-11} \text{ N} \cdot \text{m}^2/\text{kg}^2)(5.98 \times 10^{24} \text{ kg})}{4.9 \text{ m/s}^2}} = 9.02 \times 10^6 \text{ m}$$

Finally, we subtract $R_e = 6.38 \times 10^6$ m to find the height h above the Earth:

$$h = 9.02 \times 10^6 \text{ m} - 6.38 \times 10^6 \text{ m} = 2.64 \times 10^6 \text{ m}$$

At a point a distance of 2640 km above the Earth, the weight of an object will be one-half of its weight at the surface.

If we know the acceleration due to gravity at any location above the Earth's surface, we can determine the gravitational force (weight) acting on that object. The direction of this force will be toward the center of the Earth. Refer to Fig. 10.10. We will find it useful to define a ***gravitational field*** as the *force per unit of mass* at a given location. The magnitude of this field is simply the acceleration due to gravity:

$$g = \frac{F_g}{m} = \frac{Gm_e}{r^2} \tag{10.18}$$

where r is the distance from the center of the Earth to the point where gravity is to be determined. It should be noted that the gravitational field is a *property of space* and it exists at a certain point above the Earth whether or not a mass is placed at that point. By knowing the gravitational field or the acceleration due to gravity at a point, we can immediately find the weight of a given mass placed at that point.

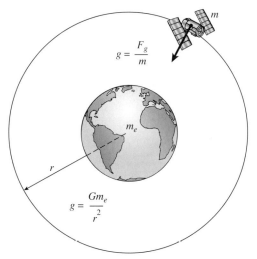

Figure 10.10   The gravitational field above the Earth can be represented by the acceleration g that a small mass m would experience if it were placed at that point. The magnitude of the field can be found from the mass m_e of the Earth and the distance r from the center of the Earth.

10.9   Satellites in Circular Orbits

An Earth satellite is just a projectile that "falls" around the Earth. In a thought experiment demonstrated by Fig. 10.11, imagine that you stand on the Earth and throw baseballs at increasing speeds. The more velocity you give to the baseball, the longer is the curved path to the ground. Since the Earth's surface is curved, one can at least imagine that if the velocity were great enough, the falling ball would simply follow the curved surface around the Earth. Of course, there are at least two serious problems with this example. First, the surface of the Earth is not uniform, and there will definitely be obstructions. Second, because of the large acceleration near the Earth's surface, the velocity would have to be

Figure 10.11 A baseball thrown with increasing horizontal velocity would eventually become a satellite as it "fell" around the Earth.

exceptionally large. Calculations will show that velocities of the order of 29,000 km/h or 18,000 mi/h are needed. The baseball would be burned to a cinder quickly at such speeds due to atmospheric friction. There are many satellites orbiting the Earth today, however, at altitudes where resistance and excessive speeds are not a problem. Some are moving in orbits that are nearly circular as they "fall" around the Earth. If a space station is in a circular orbit around the Earth, neither the space vehicle nor the passengers are "weightless." On the contrary, the gravitational force (weight) provides the centripetal force necessary for circular motion.

Consider for a moment the satellite of mass m that moves around the Earth in a circular orbit of radius r, as shown in Fig. 10.12. The centripetal force mv^2/r is determined from Newton's law of gravitation:

$$\frac{mv^2}{r} = \frac{Gmm_e}{r^2}$$

Simplifying and solving for speed v gives

$$v = \sqrt{\frac{Gm_e}{r}} \qquad\qquad \textbf{(10.19)}$$

Notice that there is only one speed v that a satellite can have if it is to remain in orbit of fixed radius r. If the speed changes, then so does the radius of the orbit.

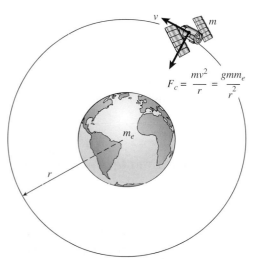

Figure 10.12 The centripetal force necessary for circular motion is provided by the gravitational force of attraction. Thus, there is only one speed v that a satellite may have if it is to remain in an orbit of fixed radius.

Example 10.10

A 100-kg astronaut rides in a space station that is moving in a circular orbit 900 km above the surface of the Earth. (a) What is the speed of the space station? (b) What is the weight of the astronaut?

Plan: First, we will determine the radius r of the orbit, which is equal to the sum of the height h and the radius R_e of the Earth. Then, we can find the speed from Eq. (10.19) and the weight of the astronaut from Newton's law of gravitation. The mass of the Earth, from Example 10.8 is 5.98×10^{24} kg.

Solution (a): Given that $R_e = 6.38 \times 10^6$ m and that $h = 900$ km, we calculate r as follows:

$$r = R_e + h = 6.38 \times 10^6 \text{ m} + 0.900 \times 10^6 \text{ m}; \qquad r = 7.28 \times 10^6 \text{ m}$$

Now, the speed is found by substitution into Eq. (10.19).

$$v = \sqrt{\frac{Gm_e}{r}} = \sqrt{\frac{(6.67 \times 10^{-11} \text{ N} \cdot \text{m}^3 \text{kg}^2)(5.98 \times 10^{24} \text{ kg})}{7.28 \times 10^6 \text{ m}}}$$

$$= 7400 \text{ m/s } (16{,}600 \text{ mi/h})$$

Solution (b): The weight of the 100-kg astronaut in orbit can be found from Newton's law of gravitation.

$$W = \frac{Gmm_e}{r^2} = \sqrt{\frac{(6.67 \times 10^{-11} \text{ N} \cdot \text{m}^3 \text{kg}^2)(5.98 \times 10^{24} \text{ kg})}{7.28 \times 10^6 \text{ m}}}$$

$$= 753 \text{ N}$$

As an additional exercise, you should verify the same result from mv^2/R. Notice that the astronaut is anything but "weightless." The astronaut is merely in a free-fall situation, which gives the appearance of weightlessness because there is no normal force acting to balance the weight.

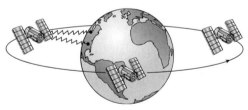

Figure 10.13 Geosynchronous satellites are positioned so that they move around the Earth in equatorial orbits with a period that is equal to that of the Earth (one day).

For many satellites, the period T, or the time required for one orbital revolution of the satellite, is of significant interest. For example, communication satellites often need to circle the Earth with a period equal to that of the Earth, in other words, one day. Such orbits are said to be *geosynchronous*, and the satellites are called *synchronous satellites*. As illustrated in Fig. 10.13, such satellites will remain accessible at an essentially constant latitude, permitting easy, direct-line communication between two points on the Earth. Three such satellites are needed to provide direct-line communication between all points on the entire Earth.

The derivation of a relationship between the period T of a satellite (or planet) and the radius r of its orbit can be accomplished by applying concepts previously discussed. Assuming a circular orbit, the velocity of the satellite is

$$v = \frac{2\pi r}{T}$$

Setting this equal to v, as given by Eq. (10.19), we have

$$\sqrt{\frac{Gm_e}{r}} = \frac{2\pi r}{t}$$

Solving for T, we obtain the following equation:

$$T^2 = \left(\frac{4\pi^2}{Gm_e}\right)r^3 \tag{10.20}$$

The square of the period of revolution is proportional to the cube of the orbital radius.

Example 10.11

What must be the altitude of all synchronous satellites that are to be placed in orbit about the Earth?

Plan: The period of a synchronous satellite is equal to one day, or 8.64×10^4 s. Using this time, we will use Eq. (10.20) to find the distance r from the center of the Earth. Then we will subtract the radius of the Earth to obtain the height h above the Earth's surface.

Solution: The distance r from the center of the Earth to the satellite is found from

$$T^2 = \left(\frac{4\pi^2}{Gm_e}\right)r^3 \quad \text{or} \quad r^3 = \left(\frac{Gm_e T^2}{4\pi^2}\right)$$

$$r^3 = \frac{(6.67 \times 10^{-11} \text{ N} \cdot \text{m}^2/\text{kg}^2)(5.98 \times 10^{24} \text{ kg})(8.64 \times 10^4 \text{ s})^2}{4\pi^2}$$

$$= 7.54 \times 10^{22} \text{ m}^3$$

Taking the cube root of both sides, we obtain

$$r = 4.23 \times 10^7 \text{ m}$$

Finally, after subtracting the radius of the Earth, we find that

$$h = 42.3 \times 10^6 \text{ m} - 6.38 \times 10^6 \text{ m} = 35.8 \times 10^6 \text{ m}$$

The geocentric orbit must be 35,800 km or more than 22,000 miles above the surface of the Earth.

10.10 Kepler's Laws

The movements of the planets and the stars have been studied for thousands of years. As early as the second century A.D., the Greek astronomer Claudius Ptolemy established the theory that the Earth was at the center of the universe. Many centuries later, Nicolaus Copernicus (1473–1543) was able to show that the Earth and other planets actually moved in circular orbits about the Sun.

The Danish astronomer Tycho Brahe (1546–1601) conducted extensive measurements of the motion of the planets over a period of 20 years. He was able to give fairly precise measurements of the planets and more than 700 stars visible to the naked eye. Since the telescope had not yet been invented, Brahe made his measurements using a large sextant and compass. The currently accepted model of the solar system has evolved from these early observations.

The German astronomer Johannes Kepler, who was Brahe's student, took the immense data compiled by Brahe and worked for many years trying to develop a mathematical model to agree with the observed data. Early in his investigation, it became obvious to Kepler that the orbits of the planets could not be circular. In fact, his studies demonstrated that the orbit of the planet Mars was really an ellipse, with the Sun at one focus. This conclusion was later generalized to apply to all planets orbiting the Sun, and Kepler was able to establish several mathematical statements relative to the solar system. Today these statements are known as ***Kepler's laws of planetary motion.***

Kepler's First Law: All planets move in elliptical orbits with the Sun at one focus. This law is sometimes called the *law of orbits.*

Figure 10.14 shows a planet of mass m_p moving in an elliptical orbit about the Sun whose mass is M_s. The semimajor axis is labeled a, and the semiminor axis is b. The smallest

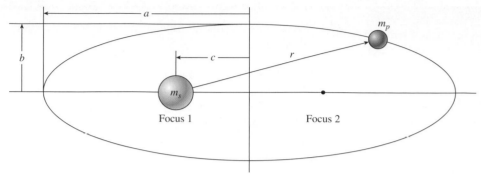

Figure 10.14 Kepler's first law states that all planets move in elliptical orbits with the Sun at one focus. The semimajor axis a and the semiminor axis b are indicated.

value of the distance r of the planet from the Sun is called the *perihelion,* and the largest value is called the *aphelion.* The distance c from the Sun to the center of the ellipse must obey the equation $a^2 = b^2 + c^2$. The ratio c/a is defined as the *eccentricity* of the orbit. With the exception of Mars, Mercury, and Pluto, most of the planetary orbits are nearly circular. They have an eccentricity that is approximately equal to 1, since c is about the same as a.

Kepler's Second Law: A line that connects a planet to the Sun sweeps out equal areas in equal times. This law can be called the *law of areas.*

The second law is illustrated by Fig. 10.15. It means that the planet must move more slowly when it is at a great distance from the Sun and more quickly when it is closer to the Sun. Newton was able to show later that this observation as well as those from the other two laws were a consequence of his law of universal gravitation.

Kepler's Third Law: The square of the period of any planet is proportional to the cube of the average distance from the planet to the Sun. This law is also known as the *law of periods.*

Kepler's third law is indicated clearly by Eq. (10.20), which was derived for a satellite in a circular orbit. It will also be true for ellipses if we replace R (the average distance from the planet to the Sun) with a, the semimajor axis of the ellipse. Thus, a more general form for Eq. (10.20) might be written as

$$T^2 = \frac{4\pi^2 a^3}{Gm_s} \qquad \textbf{(10.21)}$$

Note that when the path of the planet is circular, $a = R$, and Eq. (10.21) becomes the same as Eq. (10.20).

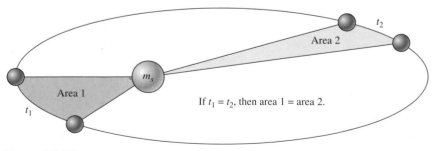

Figure 10.15

Summary and Review

Summary

We have defined uniform circular motion as motion in a circle where the speed is constant and only the direction changes. The change in direction produced by a central force is referred to as *centripetal acceleration.* The major concepts presented in this chapter are as follows:

- The linear speed v of an object in uniform circular motion can be calculated from the period T or frequency f:

$$v = \frac{2\pi R}{T} \qquad v = 2\pi f R$$

- The centripetal acceleration a_c is found from the linear speed, the period, or the frequency as follows:

$$a_c = \frac{v^2}{R} \qquad a_c = \frac{4\pi^2 R}{T^2} \qquad a_c = 4\pi^2 f^2 R$$

- The centripetal force F_c is equal to the product of the mass m and the centripetal acceleration a_c. It is given by

$$F_c = \frac{mv^2}{R} \qquad F_c = 4\pi^2 f^2 mR$$

- Other useful formulas are as follows:

$$v = \sqrt{\mu_s g R} \qquad \begin{array}{l} \textit{Maximum Speed} \\ \textit{Without Slipping} \end{array}$$

$$\tan \theta = \frac{v^2}{gR} \qquad \begin{array}{l} \textit{Banking Angle or} \\ \textit{Conical Pendulum} \end{array}$$

$$f = \frac{1}{2\pi} \sqrt{\frac{g}{h}} \qquad \begin{array}{l} \textit{Frequency of a} \\ \textit{Conical Pendulum} \end{array}$$

- **Newton's law of gravitation:** Each particle in the universe attracts each other particle with a force that is directly proportional to the product of their masses and inversely proportional to the square of the distance between them.

$$F = \frac{Gm_1 m_2}{r^2}; \qquad G = 6.67 \times 10^{-11}\ \text{N} \cdot \text{m}^2/\text{kg}^2$$

- **Kepler's first law:** All planets move in elliptical orbits with the Sun at one focus.
- **Kepler's second law:** A line that connects a planet to the Sun sweeps out equal areas in equal times.
- **Kepler's third law:** The square of the period T of any planet is proportional to the cube of the average distance from the planet to the Sun.

$$T^2 = \frac{4\pi^2 r^3}{Gm_s}$$

Key Terms

centripetal   199
centripetal acceleration   198
centripetal force   200
conical pendulum   204
critical velocity   207
frequency   200

geosynchronus   212
gravitation   207
gravitational field   210
Kepler's laws of planetary
  motion   213
linear speed   198

period   199
uniform circular motion   197
universal law of gravitation   207
weight   209

Review Questions

10.1. Explain with diagrams why the acceleration of a body moving in a circle at constant speed is directed toward the center.

10.2. A bicyclist leans to the side when negotiating a turn. Why? Describe with a free-body diagram the forces acting on the rider.

10.3. In negotiating a circular turn, a car hits a patch of ice and skids off the road. According to Newton's first law, the car will move forward in a direction tangent to the curve, not outward at right angles to it. Why?

10.4. If the force causing circular motion is directed toward the center of rotation, why is water thrown off clothes during the spin cycle of a washing machine?

10.5. When a ball tied at the end of a string is revolved in a circle at constant speed, the inward centripetal force is equal in magnitude to the outward centrifugal force. Does this represent a condition of equilibrium? Explain.

10.6. What factors contribute to the most desirable banking angles on roadways?

10.7. Does the centripetal force do work in uniform circular motion?

10.8. A motorcyclist negotiates a circular track at constant speed. What exerts the centripetal force, and on what does the force act? What exerts the centrifugal reaction force, and on what does it act?

10.9. A rock at the end of a string moves in a vertical circle. Under what conditions can its linear speed be constant? On what does the centripetal force act? On what does the centrifugal force act?

10.10. What is the value of the gravitational constant G on the Moon?

10.11. Given the mass of the Earth, its distance from the Sun, and its orbital speed, explain how you would determine the mass of the Sun.

10.12. What provides the force that keeps a satellite in orbit about the Earth? Is the satellite actually weightless in orbit? What happens to the orbit if the satellite increases or decreases its velocity?

10.13. What happens to the gravitational force between two masses when the distance between them is doubled? What happens if each mass is doubled?

10.14. The Earth moves more slowly in its orbit during the summer than it does during the winter. According to Kepler's laws, would you say that the Earth is closer or farther away from the Sun in the winter months?

Problems

Section 10.2 Centripetal Acceleration

10.1. A ball is attached to the end of a 1.5-m string, and it swings in a circle with a constant speed of 8 m/s. What is the centripetal acceleration?
> Ans. 42.7 m/s^2

10.2. What are the period and frequency of rotation for the ball in Prob. 10.1?

10.3. A drive pulley 6 cm in diameter is set to rotate at 9 rev/s. What is the centripetal acceleration of a point on the edge of the pulley? What would be the linear speed of a belt around the pulley?
> Ans. 95.9 m/s^2, 1.70 m/s

10.4. An object revolves in a circle of diameter 3 m at a frequency of 6 rev/s. What is the period of revolution, the linear speed, and the centripetal acceleration?

10.5. A car moves around a curve 50 m in radius and undergoes a centripetal acceleration of 2 m/s^2. What is its constant speed?
> Ans. 10.0 m/s

10.6. A 1500-kg car travels at a constant speed of 22 m/s along a circular track. If the centripetal acceleration is 6 m/s^2, what is the radius of the track?

10.7. An airplane dives along a curved path of radius R and velocity v. The centripetal acceleration is 20 m/s^2. If both the velocity and the radius are doubled, what will be the new acceleration?
> Ans. 40 m/s^2

Section 10.3 Centripetal Force

10.8. A 20-kg child riding a loop-the-loop at a fair moves at 16 m/s through a track of radius 16 m. What is the resultant force on the child?

10.9. A 3-kg rock, attached to a 2-m cord, swings in a horizontal circle so that it makes one revolution in 0.3 s. What is the centripetal force on the rock? Is there an outward force on the rock?
> Ans. 2630 N, no

10.10. A 5 kg object swings in a horizontal circle with a speed of 30 m/s. What is the radius of the path if the centripetal force is 2000 N?

10.11. Two 8-kg masses are attached to the end of a thin rod 400 mm long. The rod is supported in the middle and whirled in a circle. The rod can support a maximum tension of only 800 N. What is the maximum frequency of revolution?
> Ans. 3.56 rev/s

10.12. A 500-g damp shirt rotates against the wall of a washing machine at 300 rpm. The diameter of the rotating drum is 70 cm. What are the magnitude and direction of the resultant force on the shirt?

10.13. A 70-kg runner rounds a track of radius 25 m at a speed of 8.8 m/s. What is the central force causing the runner to turn, and what exerts the force?
> Ans. 217 N, friction

10.14. In Olympic bobsled competition, a team takes a turn of radius 24 ft at a speed of 60 mi/h. What is the acceleration? How many g's do passengers experience?

Section 10.4 Banking Curves

10.15. On a rainy day, the coefficient of static friction between tires and the roadway is only 0.4. What is the maximum speed at which a car can negotiate a turn of radius 80 m?   Ans. 63.8 km/h

10.16. A bus negotiates a turn of radius 120 m while traveling at a speed of 96 km/h. If slipping just begins at this speed, what is the coefficient of static friction between the tires and the road?

10.17. Find the coefficient of static friction necessary to sustain motion at 20 m/s around a turn of radius 84 m.   Ans. 0.486

10.18. A 20-kg child sits 3 m from the center of a rotating platform. If $\mu_s = 0.4$, what is the maximum number of revolutions per minute that can be achieved before the child slips?

10.19. A platform rotates freely at 100 rev/m. If the coefficient of static friction is 0.5, how far from the center of the platform can a bolt be placed without slipping?   Ans. 4.45 cm

10.20. Find the optimum banking angle for the car to negotiate the curve described in Prob. 10.15.

10.21. Find the optimum banking angle for the bus in Prob. 10.16.   Ans. 31.2°

10.22. The optimum banking angle for a curve of radius 20 m is found to be 28°. For what speed was this angle designed?

10.23. A curve in a road 9 m wide has a radius of 96 m. How much higher than the inside edge should the outside edge be so that an automobile can travel at the optimum speed of 40 km/h?   Ans. 1.17 m

Section 10.5 The Conical Pendulum

10.24. A conical pendulum swings in a horizontal circle of radius 30 cm. What angle does the supporting cord make with the vertical when the linear speed of the mass is 12 m/s?

10.25. What is the linear speed of the flyweights in Fig. 10.16 if $L = 20$ cm and $\theta = 60°$? What is the frequency of revolution?
          Ans. 1.71 m/s, 1.58 rev/s

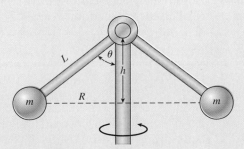

Figure 10.16

10.26. If the length of L in Fig. 10.16 is 60 cm, what velocity is required to cause the flyweights to move to an angle of 30° with the vertical?

10.27. Each of the flyweights in Fig. 10.16 has a mass of 2 kg. Length $L = 40$ cm, and the shaft rotates at 80 rev/min. What is the tension in each arm? What is the angle θ? What is the height h?
          Ans. 56.1 N, 69.6°, 14 cm

10.28. In Fig. 10.16, assume that $L = 6$ in., each flyweight is 1.5 lb, and the shaft is rotating at 100 rev/min. What is the tension in each arm? What is the angle θ? What is the distance h?

10.29. Consider the rotating swings in Fig. 10.17. Length $L = 10$ m, and the distance $a = 3$ m. What must be the linear velocity of the seat if the rope is to make an angle of 30° with the vertical?
          Ans. 6.73 m/s

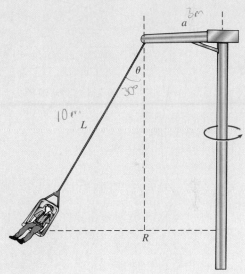

Figure 10.17

10.30. What must be the frequency of revolution for the swing in Fig. 10.17 if the angle θ is to be equal to 25°?

Section 10.6 Motion in a Vertical Circle

10.31. A rock rests on the bottom of a bucket moving in a vertical circle of radius 70 cm. What is the least speed the bucket must have as it rounds the top of a circle if the rock is to remain in the bucket?   Ans. 2.62 m/s

10.32. A 1.2-kg rock is tied to the end of a 90-cm length of string. The rock is then whirled in a vertical circle at a constant speed. What is the critical velocity at the top of the path if the string is not to become slack?

10.33. Assume that the rock of Prob. 10.32 moves in a vertical circle at a constant speed of 8 m/s. What are the tensions in the rope at the top and bottom of the circle?   Ans. 73.6 N, 97.1 N

*10.34. A test pilot in Fig. 10.18 goes into a dive at 620 ft/s and pulls out in a curve of radius 2800 ft. If the pilot weighs 160 lb, what acceleration will be experienced at the lowest point? What is the force exerted by the seat on the pilot?

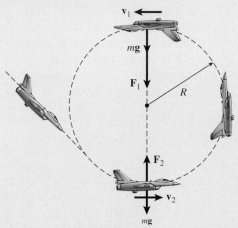

Figure 10.18 Forces exerted on an airplane at the upper and lower limits of a vertical loop.

10.35. If the pilot in Prob. 10.34 is not to experience an acceleration greater than seven times gravity (7 g), what is the maximum velocity for pulling out of a dive of radius 1 km?   Ans. 943 km/h

10.36. A 3-kg ball swings in a vertical circle at the end of an 8-m cord. When it reaches the top of its path, its velocity is 16 m/s. What is the tension in the cord? What is the critical speed at the top?

10.37. A 36-kg girl rides on the seat of a swing attached to two chains that are each 20 m long. If she is released from a position 8 m below the top of the swing, what force does the swing exert on the girl as she passes the lowest point?   Ans. 776 N

Section 10.7 Gravitation

10.38. How far apart should a 2-ton weight be from a 3-ton weight if their mutual force of attraction is equal to 0.0004 lb?

10.39. A 4-kg mass is separated from a 2-kg mass by a distance of 8 cm. Compute the gravitational force of attraction between the two masses.
                    Ans. 8.34×10^{-8} N

10.40. A 3-kg mass is located 10 cm away from a 6-kg mass. What is the resultant gravitational force on a 2-kg mass located at the midpoint of a line joining the first two masses?

10.41. On a distant planet, the acceleration due to gravity is 5.00 m/s^2, and the radius of the planet is roughly 4560 km. Use the law of gravitation to estimate the mass of this planet.
                    Ans. 1.56×10^{24} kg

*10.42. The mass of the Earth is about 81 times the mass of the Moon. If the radius of the Earth is four times that of the Moon, what is the acceleration due to gravity on the Moon?

*10.43. A 60-kg mass and a 20-kg mass are separated by 10 m. At what point on a line joining these charges will another mass experience zero resultant force?
                    Ans. 6.34 m from the 60-kg mass

Section 10.10 Kepler's Laws

10.44. What speed must a satellite have if it is to move in a circular orbit of 800 km above the surface of the Earth?

10.45. The mass of Jupiter is 1.90×10^{27} kg, and its radius is 7.15×10^7 m. What speed must a spacecraft have in order to circle Jupiter at a height of 6.00×10^7 m above the surface of Jupiter?
                    Ans. 31,000 m/s (about 69,800 mi/h)

10.46. What is the orbital speed of a satellite that moves in an orbit 1200 km above the Earth's surface?

10.47. The radius of the Moon is 1.74×10^6 m, and the acceleration due to its gravity is 1.63 m/s^2. Apply the law of universal gravitation to find the mass of the Moon.   Ans. 7.40×10^{22} kg

10.48. A satellite is located at a distance of 900 km above the Earth's surface. What is the period of the satellite's motion?

10.49. How far above the Earth's surface must a satellite be located if it is to circle the Earth in a time of 28 h?   Ans. 4.04×10^7 m

Additional Problems

10.50. At what frequency should a 6-lb ball be revolved in a radius of 3 ft to produce a centripetal acceleration of 12 ft/s^2? What is the tension in the cord?

10.51. What centripetal acceleration is required to move a 2.6-kg mass in a horizontal circle of radius 300 mm if its linear speed is 15 m/s? What is the centripetal force?   Ans. 750 m/s^2, 1950 N

10.52. What must be the speed of a satellite located 1000 mi above the Earth's surface if it is to travel in a circular path?

10.53. A 2-kg ball swings in a vertical circle at the end of a cord 2 m in length. What must the critical velocity be at the top if the orbit is to remain circular?
Ans. 4.43 m/s

10.54. A 4-kg rock swings at a constant speed of 10 m/s in a vertical circle at the end of a 1.4-m cord. What are the tensions in the cord at the top and bottom of the circular path?

10.55. What frequency of revolution is required to raise the flyweights in Fig. 10.16 a vertical distance of 25 mm above their lowest position? Assume that $L = 150$ mm.
Ans. 84.6 rev/min

10.56. The combined mass of a motorcycle and driver is 210 kg. If the driver is to negotiate a loop-the-loop of radius 6 m, what is the critical speed at the top?

10.57. If the speed at the top of the loop in Prob. 10.56 is 12 m/s, what is the normal force at the top of the loop?
Ans. 2980 N

10.58. The speed limit at a certain turn of radius 200 ft is 45 mi/h. What is the optimum banking angle for this situation? Are roads actually constructed at the optimum angles?

10.59. For the conical pendulum shown in Fig. 10.17, assume that $a = 2$ m and $L = 4$ m. What linear speed is required to cause the swing to move out to an angle of 20°?
Ans. 3.47 m/s

Critical Thinking Questions

10.60. A coin rests on a rotating platform distance of 12 cm from the center of rotation. If the coefficient of static friction is 0.6, what is the maximum frequency of rotation so that the coin does not slip? Suppose the frequency is cut in half. Now, how far from the center can the coin be placed?
Ans. 1.11 rev/s, 48 cm

*10.61. The laboratory apparatus shown in Fig. 10.19 allows a rotating mass to stretch a spring so that the supporting cord is vertical at a particular frequency of rotation. Assume that the mass of the bob is 400 g and the radius of revolution is 14 cm. With a stopwatch, the time for 50 revolutions is found to be 35 s. What are the magnitude and direction of the force acting on the bob?

10.62. In Prob. 10.59, assume that a 100-g mass is added to the 400-g bob. The force required to stretch the spring should be the same as before, but the rotating mass has increased. What changes when the experiment is performed again so that the centripetal force is the same as before? On what does the centripetal force act in this experiment?

*10.63. A 10-in. diameter platform turns at 78 rev/min. A bug rests on the platform 1 in. from the outside edge. If the bug weighs 0.02 lb, what force acts on it? What exerts this force? Where should the bug crawl in order to reduce this force by one-half?

10.64. The diameter of Jupiter is 11 times that of the Earth, and its mass is about 320 times that of the Earth. What is the acceleration due to gravity near the surface of Jupiter?
Ans. 25.9 m/s²

10.65. Assume that $L = 50$ cm and $m = 2$ kg in Fig. 10.16. How many revolutions per second are needed to make the angle $\theta = 30°$? What is the tension in the supporting rod at that point?
Ans. 0.757 rev/s, 22.6 N

*10.66. A 9-kg block rests on the bed of a truck as it turns a curve of radius 86 m. Assume that $\mu_k = 0.3$ and that $\mu_s = 0.4$. Does the friction force on the block act toward the center of the turn or away from it? What is the maximum speed with which the truck can make the turn without slipping? If the truck makes the turn at a much greater speed, what would be the resultant force on the block?

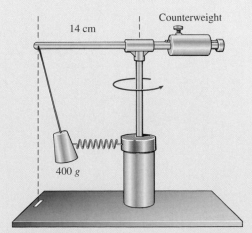

14 cm
Counterweight
400 g

Figure 10.19

Rotation of Rigid Bodies

Wind turbines such as these can generate significant energy in a way that is environmentally friendly and renewable. Such power sources account for almost 20 percent of the energy needs in Denmark. The concepts of rotational acceleration, angular velocity, angular displacement, rotational inertia, and other topics discussed in this chapter are useful in describing the operation of wind turbines. (*Photo © vol. 29 PhotoDisc/Getty.*)

Objectives

After completing this chapter, you should be able to

1. Define *angular displacement, angular velocity,* and *angular acceleration* and apply these concepts to the solution of physical problems.

2. Draw analogies relating rotational-motion parameters (θ, ω, α) to linear-motion parameters, and solve angular acceleration problems in a manner similar to that learned in Chapter 6 for linear acceleration problems (refer to Table 11.1 on p. 224).

3. Write and apply the relationships between linear speed or acceleration and angular speed or acceleration.

4. Define the *moment of inertia* of a body and describe how this quantity and the angular speed can be used to calculate *rotational kinetic energy.*

5. Apply the concepts of *Newton's second law, rotational work, rotational power,* and *angular momentum* to the solution of physical problems.

We have been considering only **translational motion,** in which an object's position is changing along a straight line. But it is possible for an object to move in a curved path or to undergo rotational motion. For example, wheels, shafts, pulleys, gyroscopes, and many other mechanical devices rotate about their axes without translational motion. The generation and transmission of power are nearly always dependent on rotational motion of some kind. It is essential for you to be able to predict and control such motion. The concepts and formulas presented in this chapter are designed to provide you with these essential skills.

11.1 Angular Displacement

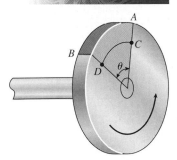

Figure 11.1 Angular displacement θ is indicated by the shaded portion of the disk. The angular displacement is the same from C to D as it is from A to B for a rigid body.

The **angular displacement** of a body describes the amount of rotation. If point A on the rotating disk in Fig. 11.1 rotates on its axis to point B, the angular displacement is denoted by the angle θ. There are several ways of measuring this angle. We are already familiar with the units of degrees and revolutions, which are related according to the definition

$$1 \text{ rev} = 360°$$

Neither of these units is useful in describing rotation of rigid bodies. A more applicable measure of angular displacement is the **radian** (rad). An angle of 1 rad is a central angle whose arc s is equal in length to the radius R. (See Fig. 11.2.) More generally, the radian is defined by the following equation:

$$\theta = \frac{s}{R} \tag{11.1}$$

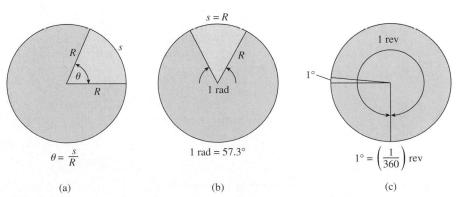

(a)   (b)   (c)

Figure 11.2 The measure of angular displacement and a comparison of units.

where s is the arc of a circle described by the angle θ. Since the ratio of s to R is the ratio of two distances, the radian is a unitless quantity.

The conversion factor that relates radians to degrees is found by considering an arc length s equal to the circumference of a circle $2\pi R$. Such an angle in radians is given from Eq. (11.1) by

$$\theta = \frac{2\pi \cancel{R}}{\cancel{R}} = 2\pi \text{ rad}$$

Hence,

$$1 \text{ rev} = 360° = 2\pi \text{ rad}$$

from which we note that

$$1 \text{ rad} = \frac{360°}{2\pi} = 57.3°$$

Example 11.1

One end of a rope is attached to a bucket of water, and the other end is wrapped many times around a circular drum of radius 12 cm. How many revolutions of the drum are required to lift the bucket a vertical distance of 5 m?

Plan: The vertical distance lifted must be equal to the length of rope wrapped around the drum so that the arc length $s = 5$ m. We will first calculate the rotation in *radians* needed for an arc length of 5 m. Remember to set the angle mode on your calculator to radians (it is normally set for degrees). Then a conversion of this angle to revolutions will give the required answer.

Solution: From Eq. (11.1), we have

$$\theta = \frac{s}{R} = \frac{5 \text{ m}}{0.12 \text{ m}} = 41.7 \text{ rad}$$

Recalling that 1 rev $= 2\pi$ rad, we convert to find the angle in revolutions.

$$\theta = 41.7 \text{ rad}\left(\frac{1 \text{ rev}}{2\pi \text{ rad}}\right) = 6.63 \text{ rev}$$

Thus, approximately six and two-thirds revolutions will lift the bucket 5 m.

Example 11.2

A seat on the perimeter of a Ferris wheel at the county fair experiences an angular displacement of 37°. If the radius of the wheel is 20 m, what arc length is described by the seat?

Plan: Since angular displacement was defined in terms of radians, we must first convert degrees to radians. The arc length can then be found by solving Eq. (11.1) for s.

$$\theta = (37°)\left(\frac{2\pi \text{ rad}}{360°}\right) = 0.646 \text{ rad}$$

The arc length is given by

$$s = R\theta = (20 \text{ m})(0.646 \text{ rad})$$
$$s = 12.9 \text{ m}$$

The radian is dropped as a unit since it is a ratio of length to length (m/m $= 1$).

11.2 | Angular Velocity

The time rate of change in angular displacement is called the **angular velocity.** Thus, if an object rotates through an angle θ in a time t, its average angular velocity is given by

$$\overline{\omega} = \frac{\theta}{t} \qquad\qquad \textit{Angular Velocity} \quad \textbf{(11.2)}$$

The symbol ω, the Greek letter *omega,* is used to denote angular velocity. When a bar appears over the symbol, it indicates that the angular velocity is an average value. Although angular velocity may be expressed in *revolutions per minute (rpm)* or *revolutions per second,* in most physics applications it is necessary to use *radians per second* to conform to the basic choice of angular displacement θ in *radians.* You should also keep in mind that angular velocity can be clockwise or counterclockwise; that is, it has direction. We must choose a positive direction for rotation and substitute signs that are consistent with that choice.

Since the rate of rotation for many technical applications is expressed in revolutions per minute or revolutions per second, it is useful to find an expression for converting to radians per second. If the frequency of revolution in rev/s is denoted by the symbol f, the angular speed in rad/s is given by

$$\omega = 2\pi f \qquad\qquad \textbf{(11.3)}$$

If the frequency is rpm instead of rev/s, the conversion factor is $(2\pi/60)$.

Example 11.3

A bicycle wheel with a radius of 33 cm rotates at 40 rpm. What distance will the bicycle travel in 30 s?

Plan: First, we will convert the angular velocity of the wheel to radians per second. Then, we can use the definition of average velocity to calculate the arc distance s described by a point on the edge of the wheel. This distance will be the same as that traveled by the bicycle along a horizontal path.

Solution: First convert the frequency from rpm to rev/s.

$$f = \left(\frac{40 \text{ rev}}{1 \text{ min}}\right)\left(\frac{1 \text{ min}}{60 \text{ s}}\right) = 0.667 \text{ rev/s}$$

Substituting this frequency into Eq. (11.3) gives the angular velocity.

$$\omega = 2\pi f = (2\pi \text{ rad})(0.667 \text{ rev/s}) = 4.19 \text{ rad/s}$$

Now, we rewrite Eq. (11.1) and Eq. (11.2), giving

$$s = \theta R \qquad \text{and} \qquad \theta = \overline{\omega} t$$

This means that the linear distance s is

$$s = (\omega t)R = (4.19 \text{ rad/s})(30 \text{ s})(0.33 \text{ m})$$
$$s = 41.5 \text{ m}$$

It is important to remember that the angular velocity described by Eq. (11.2) is an *average* value (or a constant value). The same distinction must be made between *average* and *instantaneous* angular velocity as that discussed in Chapter 6 for average and instantaneous linear velocity.

11.3   Angular Acceleration

Like linear motion, rotational motion may be uniform or accelerated. The rate of rotation may increase or decrease under the influence of a resultant torque. For example, if the angular velocity changes from an initial value ω_0 to a final value ω_f in a time t, the angular acceleration is given by

$$\alpha = \frac{\omega_f - \omega_0}{t}$$

The Greek letter α (*alpha*) denotes angular acceleration. A more useful form for this equation is

$$\omega_f = \omega_0 + \alpha t \qquad \qquad \textbf{(11.4)}$$

A comparison of Eq. (11.4) with Eq. (6.4) for linear acceleration will show that their forms are identical if we draw analogies between angular and linear parameters.

Now that the concept of initial and final angular velocities has been introduced, we can express the average angular velocity in terms of its initial and final values:

$$\overline{\omega} = \frac{\omega_f + \omega_0}{2}$$

Substituting this equality for $\overline{\omega}$ in Eq. (11.2) yields a more useful expression for the angular displacement:

$$\theta = \overline{\omega}t = \left(\frac{\omega_f + \omega_0}{2}\right)t \qquad \qquad \textbf{(11.5)}$$

This equation is also similar to an equation derived for linear motion. In fact, the equations for angular acceleration have the same basic form as those derived in Chapter 6 for linear acceleration if we draw the following analogies:

$$s \text{ (m)} \leftrightarrow \theta \text{ (rad)}$$
$$v \text{ (m/s)} \leftrightarrow \omega \text{ (rad/s)}$$
$$a \text{ (m/s}^2) \leftrightarrow \alpha \text{ (rad/s}^2)$$

Time, of course, is the same for both types of motion and is measured in seconds. Table 11.1 illustrates the similarities between rotational and linear motion.

Table 11.1

Comparison of Linear Acceleration and Angular Acceleration Formulas

Constant Linear Acceleration	Constant Angular Acceleration
(1) $s = \left(\dfrac{v_0 + v_f}{2}\right)t$	$\theta = \left(\dfrac{\omega_f + \omega_0}{2}\right)t$
(2) $v_f = v_0 + at$	$\omega_f = \omega_0 + \alpha t$
(3) $s = v_0 t + \dfrac{1}{2}at^2$	$\theta = \omega_0 t + \dfrac{1}{2}\alpha t^2$
(4) $s = v_f t - \dfrac{1}{2}at^2$	$\theta = \omega_f t - \dfrac{1}{2}\alpha t^2$
(5) $2as = v_f^2 - v_0^2$	$2\alpha\theta = \omega_f^2 - \omega_0^2$

In applying these formulas, we must be careful to choose the appropriate units for each quantity. It is also important to choose a direction (clockwise or counterclockwise) as positive and to follow through consistently in affixing the appropriate sign to each quantity.

Example 11.4

A flywheel increases its rate of rotation from 6 to 12 rev/s in 8 s. Find the angular acceleration in radians per second squared.

Plan: When applying the equations for uniform angular acceleration, the only acceptable angular units are radians. We must first change the units for final and initial angular velocities. Then we will organize the given data, choose an appropriate equation, and solve for the angular acceleration.

Solution: The angular velocities are

$$\omega_0 = 2\pi f = \left(\frac{2\pi \text{ rad}}{1 \text{ rev}}\right)\left(\frac{6 \text{ rev}}{\text{s}}\right) = 37.7 \text{ rad/s}$$

$$\omega_f = 2\pi f = \left(\frac{2\pi \text{ rad}}{1 \text{ rev}}\right)\left(\frac{12 \text{ rev}}{\text{s}}\right) = 75.4 \text{ rad/s}$$

Now, we may solve for α using the definition of angular acceleration.

Given: $\omega_0 = 37.7$ rad/s; $\omega_f = 75.4$ rad/s; $t = 8$ s   Find: $\alpha = ?$

We select Eq. (2) from Table 11.1 as the one that contains α and not θ. Solving for α, we obtain

$$\alpha = \frac{\omega_f - \omega_0}{t} = \frac{75.4 \text{ rad/s} - 37.7 \text{ rad/s}}{8 \text{ s}}$$

$$\alpha = 4.71 \text{ rad/s}^2$$

Example 11.5

A grinding disk rotating initially at 6 rad/s receives a constant acceleration of 2 rad/s² for a time of 3 s. Find its angular displacement and its final angular velocity.

Plan: Organize the given data, select the appropriate equation, and solve for the unknown.

Solution:

Given: $\omega_0 = 6$ rad/s; $\alpha = 2$ rad/s; $t = 3$ s   Find: $\theta = ?$

Equation (3) contains α and not ω_f. The angular displacement is

$$\theta = \omega_0 t + \frac{1}{2}\alpha t^2$$

$$\theta = (6 \text{ rad/s})(3 \text{ s}) + \frac{1}{2}(2 \text{ rad/s}^2)(3 \text{ s})^2 = 27.0 \text{ rad}$$

The final angular velocity ω_f is found from Eq. (2)

$$\omega_f = \omega_0 + \alpha t$$
$$= 6 \text{ rad/s} + (2 \text{ rad/s}^2)(3 \text{ s})$$
$$= 12.0 \text{ rad/s}$$

11.4 Relationship Between Rotational and Linear Motion

The *axis of rotation* of a rigid rotating body can be defined as that line of particles that remains stationary during rotation. This may be a line through the body, as with a spinning top, or it may be a line through space, as with a rolling hoop. In any case, our experience tells us that the farther a particle is from the axis of rotation, the greater its linear velocity. This fact was expressed in Chapter 10 by the formula

$$v = 2\pi f R$$

where f is the frequency of rotation. We now derive a similar relation in terms of angular velocity. The rotating particle in Fig. 11.3 turns through an arc s, which is given by

$$s = \theta R$$

from Eq. (11.1). If this distance is traversed in a time t, the linear velocity of the particle is given by

$$v = \frac{s}{t} = \frac{\theta R}{t}$$

Since $\theta/t = \omega$, the linear velocity can be expressed as a function of the angular velocity.

$$v = \omega R \qquad (11.6)$$

This result also follows from Eq. (11.3), in which the angular velocity is expressed as a function of the frequency of revolution.

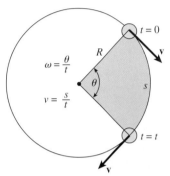

Figure 11.3 The relationship between angular velocity and linear velocity.

In the figure: $t = 0$, $\omega = \dfrac{\theta}{t}$, $v = \dfrac{s}{t}$, R, θ, s, v, $t = t$

Example 11.6

A drive shaft has an angular velocity of 60 rad/s. At what distance should flyweights be positioned from the axis if they are to have a linear velocity of 12 m/s?

Solution: Solving for R in Eq. (11.6), we obtain

$$R = \frac{v}{\omega} = \frac{12 \text{ m/s}}{60 \text{ rad/s}} = 0.200 \text{ m}$$

Let us now return to a particle moving in a circle of radius R and assume that the linear velocity changes from some initial value v_0 to a final value v_f in a time t. The *tangential acceleration* a_T of such a particle is given by

$$a_T = \frac{v_f - v_0}{t}$$

Because of the close relationship between linear velocity and angular velocity, as represented by Eq. (11.6), we can also express the tangential acceleration in terms of a change in angular velocity.

$$a_T = \frac{\omega_f R - \omega_0 R}{t} = \frac{\omega_f - \omega_0}{t} R$$

or

$$a_T = \alpha R \qquad (11.7)$$

where α represents the **angular acceleration.**

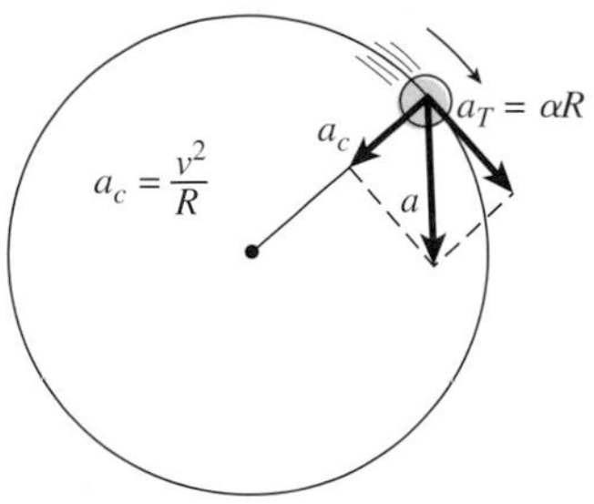

Figure 11.4 The relationship between tangential and centripetal acceleration.

We must be careful to distinguish between the tangential acceleration, as defined in Eq. (11.7), and the centripetal acceleration defined by

$$a_c = \frac{v^2}{R} \qquad (11.8)$$

The *tangential acceleration* represents a change in linear velocity, whereas the centripetal acceleration represents only a change in the direction of motion. The distinction is shown graphically in Fig. 11.4. The resultant acceleration can be found by computing the vector sum of the tangential and centripetal accelerations.

Example 11.7

Find the magnitude of the resultant acceleration of a particle moving in a circle of radius 0.5 m at the instant when its angular velocity is 3 rad/s and its angular acceleration is 4 rad/s^2.

Plan: We will draw a figure similar to the one shown in Fig. 11.4. Next, we will determine the linear velocity v as the product ωR. The centripetal acceleration a_c will then be found from Eq. (11.8). The tangential acceleration a_T is given by Eq. (11.7). The resultant of these perpendicular vectors will give us the net angular acceleration.

Solution: Given that $R = 0.5$ m and $\omega = 3$ rad/s, we find that

$$v = \omega R = (3 \text{ rad/s})(0.5 \text{ m}) = 1.50 \text{ m/s}$$

The centripetal acceleration, from Eq. (11.8), is, therefore,

$$a_c = \frac{v^2}{R} = \frac{(1.50 \text{ m/s})^2}{(0.5 \text{ m})} = 4.50 \text{ m/s}^2$$

Now, from Eq. (11.7), the tangential acceleration is

$$a_T = \alpha R = (4 \text{ rad/s}^2)(0.5 \text{ m}); \qquad a_T = 2.00 \text{ m/s}^2$$

Finally, the magnitude of the resultant acceleration is obtained from the Pythagorean theorem.

$$a = \sqrt{a_T^2 + a_c^2} = \sqrt{(2.00 \text{ m/s}^2)^2 + (4.50 \text{ m/s}^2)^2}$$
$$a = 4.92 \text{ m/s}^2$$

The direction of the acceleration, if desired, can be found from its components in the usual manner.

11.5 Rotational Kinetic Energy: Moment of Inertia

We have seen that a particle moving in a circle of radius R has a linear speed given by

$$v = \omega R$$

If the particle has a mass m, it will have a kinetic energy given by

$$K = \frac{1}{2}mv^2 = \frac{1}{2}m\omega^2 R^2$$

A rigid body like that in Fig. 11.5 can be considered as consisting of many particles of varying masses located at different distances from the axis of rotation O. The total kinetic

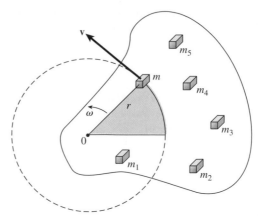

Figure 11.5 Rotation of an extended body. The body can be thought of as many individual masses all rotating with the same angular velocity.

energy of such an extended body will be the sum of the kinetic energies of each particle making up the body. Hence,

$$K = \sum \frac{1}{2} m \omega^2 r^2$$

Since the constant $\frac{1}{2}$ and the angular velocity ω are the same for all particles, we can rearrange to obtain

$$K = \frac{1}{2} \left(\sum mr^2 \right) \omega^2$$

The quantity in parentheses, $\sum mr^2$, has the same value for a given body regardless of its state of motion. We define this quantity as the ***moment of inertia*** and represent it by *I*:

$$I = m_1 r_1^2 + m_2 r_2^2 + m_3 r_3^2 + \cdots$$

or

$$I = \sum mr^2 \qquad (11.9)$$

The SI unit for *I* is the *kilogram-square meter,* and the USCS unit is the *slug-square foot.*

Using this definition, we can express the ***rotational kinetic energy*** of a body in terms of its moment of inertia and its angular velocity:

$$K = \frac{1}{2} I \omega^2 \qquad (11.10)$$

Note the similarity between the terms *m* for linear motion and *I* for rotational motion.

Example 11.8

Find the moment of inertia for the system illustrated in Fig. 11.6. The rods joining the masses are of negligible weight, and the system rotates with an angular velocity of 6 rad/s. What is the rotational kinetic energy? (Consider the masses to be point masses.)

Plan: The moment of inertia for the system is equal to the sum of the moments of inertia due to each mass about the center of rotation. The rotational kinetic energy is given by Eq. (11.10) using the calculated value for *I*.

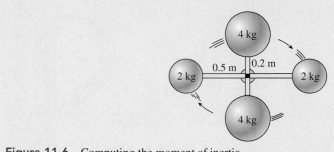

Figure 11.6 Computing the moment of inertia.

Solution: From Eq. (11.9), we obtain

$$I = \sum mr^2 = m_1r_1^2 + m_2r_2^2 + m_3r_3^2 + m_4r_4^2$$

$$I = (2 \text{ kg})(0.5 \text{ m})^2 + (4 \text{ kg})(0.2 \text{ m})^2 + (2 \text{ kg})(0.5 \text{ m})^2 + (4 \text{ kg})(0.2 \text{ m})^2$$

$$I = 1.32 \text{ kg} \cdot \text{m}^2$$

Using this result and the fact that $\omega = 6$ rad/s, the rotational kinetic energy is

$$K = \frac{1}{2}I\omega^2 = \frac{1}{2}(1.32 \text{ kg} \cdot \text{m}^2)(6 \text{ rad/s})^2 \quad \text{or} \quad K = 23.8 \text{ J}$$

For bodies that are not composed of separate masses but are actually continuous distributions of matter, the calculation of moments of inertia is more difficult and usually involves calculus. A few simple cases are given in Fig. 11.7, along with the formulas for computing their moments of inertia.

Sometimes it is desirable to express the rotational inertia of a body in terms of its *radius of gyration k.* This quantity is defined as the radial distance from the center of rotation to a circumference at which the total mass of the body might be concentrated without changing its moment of inertia. According to this definition, the moment of inertia can be found from the formula

$$I = mk^2 \tag{11.11}$$

where m represents the total mass of the rotating body and k is its radius of gyration.

11.6 The Second Law of Motion and Rotation

Suppose we analyze the motion of the rotating rigid body in Fig. 11.8. Consider a force **F** acting on the small mass m, indicated by a shaded portion of the object, at a distance r from the axis of rotation.

The force **F** applied perpendicular to r causes the body to rotate with a tangential acceleration:

$$a_T = \alpha r$$

where α is the angular acceleration. From Newton's second law of motion,

$$F = ma_T = m\alpha r$$

Multiplying both sides of this relation by r yields

$$Fr = (mr^2)\alpha$$

The quantity Fr is recognized as the torque produced by the force **F** about the axis. Thus, for the mass m, we write

$$\tau = (mr^2)\alpha$$

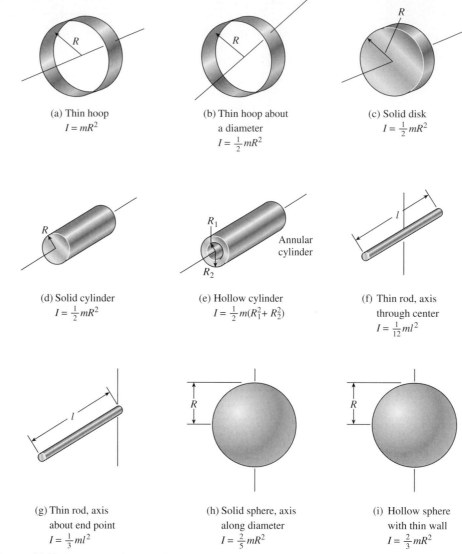

(a) Thin hoop
$I = mR^2$

(b) Thin hoop about
a diameter
$I = \frac{1}{2}mR^2$

(c) Solid disk
$I = \frac{1}{2}mR^2$

(d) Solid cylinder
$I = \frac{1}{2}mR^2$

(e) Hollow cylinder
$I = \frac{1}{2}m(R_1^2 + R_2^2)$

(f) Thin rod, axis
through center
$I = \frac{1}{12}ml^2$

(g) Thin rod, axis
about end point
$I = \frac{1}{3}ml^2$

(h) Solid sphere, axis
along diameter
$I = \frac{2}{5}mR^2$

(i) Hollow sphere
with thin wall
$I = \frac{2}{3}mR^2$

Figure 11.7 Moments of inertia for bodies about their indicated axes.

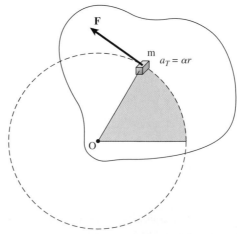

Figure 11.8 Newton's second law for rotational motion states the relationship between torque Fr and angular acceleration α.

A similar equation may be derived for all other portions of the rotating object. The angular acceleration will be constant for every portion, however, regardless of its mass or distance from the axis. Therefore, the resultant torque on the whole body is

$$\tau = \left(\sum mr^2 \right) \alpha$$

or

$$\tau = I\alpha \qquad\qquad (11.12)$$

Torque = moment of inertia × angular acceleration

Note the similarity of Eq. (11.12) with the second law for linear motion, $F = ma$. Newton's *law for rotational motion* is as follows:

A resultant torque applied to a rigid body will always result in an angular acceleration that is directly proportional to the applied torque and inversely proportional to the body's moment of inertia.

In applying Eq. (11.12), it is important to recall that the torque produced by a force is equal to the product of its distance from the axis and the perpendicular component of the force. It must also be remembered that the angular acceleration is expressed in radians per second per second.

Example 11.9

A circular grinding disk of radius 0.6 m and mass 90 kg is rotating at 460 rev/min. What force, applied tangent to the edge, will cause the disk to stop in 20 s?

Plan: The rotational inertia I can be found from the formula for a disk given in Fig. 11.7. Then, the angular acceleration α can be calculated from the change in angular velocity per unit of time. To find the force F at the edge, we will recognize that the torque (FR) must be equal to the product $I\alpha$, according to Newton's second law.

Solution: The rotational inertia of a disk is

$$I = \frac{1}{2}mR^2 = \frac{1}{2}(90 \text{ kg})(0.60 \text{ m})^2 = 16.2 \text{ kg} \cdot \text{m}^2$$

Converting 460 rpm to units of rad/s, we write the initial angular velocity as

$$\omega_0 = \left(460\frac{\text{rev}}{\text{min}} \right)\left(\frac{2\pi \text{ rad}}{\text{rev}} \right)\left(\frac{1 \text{ min}}{60 \text{ s}} \right); \qquad \omega_0 = 48.2 \text{ rad/s}$$

Noting that $\omega_f = 0$ and $t = 20$ s, we can find the angular acceleration α.

$$\alpha = \frac{\omega_f - \omega_0}{t} = \frac{0 - (48.2 \text{ rad/s})}{20 \text{ s}}$$

$$= -2.41 \text{ rad/s}^2$$

From Newton's second law, we recall that the resultant torque ($\tau = FR$), must be equal to the product of the rotational inertia and the angular acceleration ($\tau = I\alpha$). Thus,

$$FR = I\alpha \qquad \text{or} \qquad F = \frac{I\alpha}{R}$$

$$F = \frac{(16.2 \text{ kg} \cdot \text{m}^2)(-2.41 \text{ rad/s}^2)}{0.60 \text{ m}} = -65.0 \text{ N}$$

The negative sign appears because the force must be directed opposite to the direction of rotation of the disk.

11.7 Rotational Work and Power

Work was defined in Chapter 8 as the product of a displacement and the component of the force in the direction of the displacement. We now consider the work done in rotational displacement under the influence of a resultant torque.

Consider a force **F** acting at the edge of a pulley of radius r, as shown in Fig. 11.9. The effect of such a force is to rotate the pulley through an angle θ while the point at which the force is applied moves through a distance s. The arc distance s is related to θ by

$$s = r\theta$$

Hence, the work of the force **F** is by definition

$$\text{Work} = Fs = Fr\theta$$

but Fr is the torque due to the force, so we obtain

$$\text{Work} = \tau\theta \tag{11.13}$$

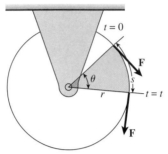

Figure 11.9 Work and power in rotation.

The angle θ must be expressed in radians for either system of units so that the work can be expressed in foot-pounds or joules.

Mechanical energy is usually transmitted in the form of **rotational work.** When we speak of the power output of engines, we are concerned with the rate at which rotational work is done. Thus, rotational power can be determined by dividing both sides of Eq. (11.13) by the time t required for the torque τ to effect a displacement θ:

$$\text{Power} = \frac{\text{work}}{t} = \frac{\tau\theta}{t} \tag{11.14}$$

Since θ/t represents the average angular velocity $\overline{\omega}$, we write

$$\text{Power} = \tau\overline{\omega} \tag{11.15}$$

Notice the similarity of this relation with its analog, $P = Fv$, derived earlier for linear motion. Both are measures of *average* power.

Example 11.10

A wheel of radius 60 cm has a moment of inertia of $5 \text{ kg} \cdot \text{m}^2$. A constant force of 60 N is applied tangent to the rim. Assuming it starts from rest, how much work is done in 4 s, and how much power is developed?

Plan: The work is a product of torque and angular displacement. We will first calculate the torque by multiplying the rim force by the radius of the wheel. Then, we will find the angular acceleration from Newton's second law. Once we know the acceleration, we can find the linear displacement and determine the work and power expended.

Solution: The given information is organized as follows:

Given: $R = 0.60 \text{ m}, F = 60 \text{ N}, I = 5 \text{ kg} \cdot \text{m}^2, t = 4 \text{ s}$ Find: *work* and *power*

The torque applied to the edge of the wheel is

$$\tau = FR = (60 \text{ N})(0.60 \text{ m}) = 36.0 \text{ N} \cdot \text{m}$$

Next, we find α from Newton's second law ($\tau = I\alpha$).

$$\alpha = \frac{\tau}{I} = \frac{36 \text{ N} \cdot \text{m}}{5 \text{ kg} \cdot \text{m}^2}; \qquad \alpha = 7.20 \text{ rad/s}^2$$

The angular displacement θ is

$$\theta = \omega_0 t + \frac{1}{2}\alpha t^2$$

$$= 0 + \frac{1}{2}(7.20 \text{ rad/s}^2)(4 \text{ s})^2 = 57.6 \text{ rad}$$

The work is, therefore,

$$\text{Work} = \tau\theta = (36 \text{ N} \cdot \text{m})(57.6 \text{ rad}) = 2070 \text{ J}$$

Finally, the average power is the work per unit time, or

$$P = \frac{\text{Work}}{t} = \frac{2070 \text{ J}}{4 \text{ s}}; \qquad P = 518 \text{ W}$$

The same result could be found by calculating the average angular velocity ω and using Eq. (11.15). As an additional example, you might show that the work done is equal to the change in rotational kinetic energy.

11.8 Combined Rotation and Translation

To understand the relationship between linear and angular motion of a rotating object, let us first consider that a circular disk of radius R slides along a horizontal surface without rotation and without friction. As shown by Fig. 11.10a, any piece of this disk will be traveling at a velocity equal to that of the center of mass.

Now, assume the same disk rotates freely but without slipping along the same surface, as in Fig. 11.10b. Greater energy is required to maintain the same horizontal speed because now there is rotation as well as translation. Since there is no slipping, the center of mass of the disk is rotating relative to the point of contact P with the same angular velocity as that of the rotating disk. Thus, we can write a familiar relationship between the linear velocity v of the center of mass of the disk and its rotational speed ω.

$$v = \omega R \qquad \text{or} \qquad \omega = \frac{v}{R}$$

To check your understanding of this equation, consider a bicycle wheel of radius 50 cm that rotates at 20 rad/s. Verify that the horizontal speed of the bike is 10 m/s.

In working with problems that involve both rotation and translation, we must remember to add the rotational kinetic energy K_R to the translational kinetic energy K_T. For example, in applying the principle of conservation of total energy, we know that the total of all kinds of energy before an event must equal the total after the event plus any losses due to friction or other dissipative forces.

$$(U_0 + K_{T0} + K_{R0}) = (U_f + K_{Tf} + K_{Rf}) + |\text{Losses}| \qquad \textbf{(11.16)}$$

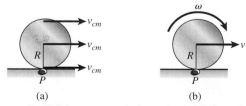

(a)   (b)

Figure 11.10   (a) All parts of a disk in pure translation move with the velocity v_{cm} of the center of mass. (b) A rolling object is a combination of translation and rotation such that the linear horizontal velocity is given by $v = \omega R$.

The subscripts 0 and f refer to initial and final values of potential energy U, rotational kinetic energy K_R, and translational kinetic energy K_T. The "losses" term can be set to zero if we assume rolling without friction.

Example 11.11

A circular hoop and a circular disk are designed so that each has a mass of 2 kg and a radius of 10 cm. They are then allowed to roll from rest at a height of 20 m to the bottom of an incline, as shown in Fig. 11.11. Compare their final speeds.

Plan: Since we are interested in finding the speed v at the bottom of the incline, we will convert rotational parameters into their corresponding linear parameters. For example, the rotational inertia I of a hoop is mR^2 and the rotational inertia I of a disk is $\frac{1}{2}mR^2$. Also, the rotational velocity ω is the ratio v/R. Conservation of energy demands that the sum of potential, kinetic, and rotational energy at the top of the incline must equal the sum of these energies at the bottom. Thus, we can apply Eq. (11.16) first for the hoop and then for the disk, assuming negligible friction losses for each case.

Solution: In each case, $U = mgh$; $K_R = \frac{1}{2}mv^2$, and $K_T = \frac{1}{2}I\omega^2$. Conservation of energy with no friction losses gives

$$(U_0 + K_{T0} + K_{R0}) = (U_f + K_{Tf} + K_{Rf})$$

$$mgh_0 + 0 + 0 = \frac{1}{2}mv_f^2 + \frac{1}{2}I\omega_f^2$$

For hoop: $I = mR^2$, so substitution yields

$$mgh_0 = \frac{1}{2}mv^2 + \frac{1}{2}(mR^2)\left(\frac{v^2}{R^2}\right)$$

$$mgh_0 = \frac{1}{2}mv^2 + \frac{1}{2}mv^2$$

Simplifying and solving for v, we obtain

$$v = \sqrt{gh_0} = \sqrt{(9.8 \text{ m/s}^2)(20 \text{ m})} \qquad \text{or} \qquad v = 14.0 \text{ m/s}$$

For disk: $I = \frac{1}{2}mR^2$, and

$$mgh_0 = \frac{1}{2}mv^2 + \frac{1}{2}\left(\frac{1}{2}mR^2\right)\left(\frac{v^2}{R^2}\right)$$

This can be solved for v to give

$$v = \sqrt{\frac{4}{3}gh_0} = \sqrt{\frac{4}{3}(9.8 \text{ m/s}^2)(20 \text{ m})} \qquad \text{or} \qquad v = 16.2 \text{ m/s}$$

Notice that even though the masses and radii were the same, the disk has a lower rotational inertia, resulting in a greater final speed. It will reach the bottom before the hoop.

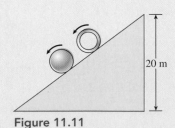

Figure 11.11

Angular Momentum

Consider a particle of mass m moving in a circle of radius r, as shown in Fig. 11.12a. If its linear velocity is v, it will have a linear momentum $p = mv$. With reference to the fixed axis of rotation, we define the **angular momentum L** of the particle as the product of its linear momentum mv and the perpendicular distance r from the axis to the revolving particle.

$$L = mvr \qquad (11.17)$$

Now, let us consider the definition of angular momentum as it applies to an extended, rigid body. Figure 11.12b describes such a body, which is rotating about its axis O. Each particle in the body has an angular momentum, given by Eq. (11.17). Substituting $v = \omega r$, each particle has an angular momentum given by

$$mvr = m(\omega r)r = (mr^2)\omega$$

Since the body is rigid, all particles within the body have the same angular velocity, and the total angular momentum of the body is

$$L = \left(\sum mr^2 \right)\omega$$

Thus, the total angular momentum is equal to the product of a body's angular velocity and its moment of inertia:

$$L = I\omega \qquad (11.18)$$

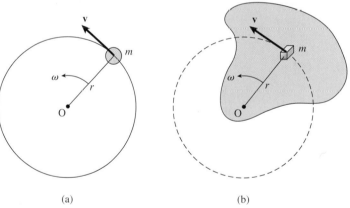

(a)   (b)

Figure 11.12 Defining angular momentum.

Example 11.12

A thin uniform rod is 1 m long and has a mass of 6 kg. If the rod is pivoted at its center and set into rotation with an angular velocity of 16 rad/s, compute its angular momentum.

Solution: The moment of inertia of a thin rod is, from Fig. 11.7,

$$I = \frac{ml^2}{12} = \frac{(6 \text{ kg})(1 \text{ m})^2}{12} = 0.5 \text{ kg} \cdot \text{m}^2$$

Hence, its angular momentum is

$$L = I\omega = (0.5 \text{ kg} \cdot \text{m}^2)(16 \text{ rad/s})$$
$$= 8 \text{ kg} \cdot \text{m}^2/\text{s}$$

Notice that the SI unit of angular momentum is kg $\cdot$ m²/s. The USCS unit is slug $\cdot$ ft²/s.

11.10   Conservation of Angular Momentum

We can understand the definition of angular momentum better if we return to the fundamental equation for angular motion, $\tau = I\alpha$. Recalling the defining equation for angular acceleration,

$$\alpha = \frac{\omega_f - \omega_0}{t}$$

we can write Newton's second law as

$$\tau = I\left(\frac{\omega_f - \omega_0}{t}\right)$$

Multiplying by t, we obtain

$$\tau t = I\omega_f - I\omega_0 \qquad\qquad (11.19)$$

Angular impulse = change in angular momentum

The product τt is defined as the *angular impulse.* Note the similarity between this equation and the one derived in Chapter 9 for linear impulse.

If no external torque is applied to a rotating body, we can set $\tau = 0$ in Eq. (11.19), yielding

$$0 = I\omega_f - I\omega_0$$
$$I\omega_f = I\omega_0 \qquad\qquad (11.20)$$

Final angular momentum = initial angular momentum

Thus, we are led to a statement of the ***conservation of angular momentum:***

> If the sum of the external torques acting on a body or system of bodies is zero, the angular momentum remains unchanged.

This statement holds true even if the rotating body is not rigid but is altered so that its moment of inertia changes. In this case, the angular speed also changes so that the product $I\omega$ is always constant. Skaters, divers, and acrobats all control the rate at which their bodies rotate by extending or retracting their limbs to decrease or increase their angular speed.

An interesting experiment illustrating the conservation of angular momentum is shown in Fig. 11.13. A woman stands on a rotating platform with heavy weights in each hand. She

Low angular velocity   High angular velocity

Figure 11.13   Experiment to demonstrate the conservation of angular momentum. The woman controls her rate of rotation by moving the weights inward to increase rotational speed or outward to decrease rotational speed.

is first set into rotation with her arms fully extended. By drawing her hands closer to her body, she decreases her moment of inertia. Since her angular momentum cannot change, she will notice a considerable increase in her angular speed. By extending her arms, she will be able to decrease her angular speed.

Example 11.13

Assume that the woman holding the extended weights in Fig. 11.13 has a rotational inertia of 6 kg · m² and that the rotational inertia is decreased to 2 kg · m² when she brings them close to her body. With the weights in their extended position, she is set into rotation at 1.4 rev/s. What will be her rate of rotation when she brings the weights close to her body?

Plan: In the absence of external torque, the rotational equilibrium of the system cannot change. This means that the angular momentum with the weights outstretched must be the same as with them near her side. The change in angular speed must compensate for the reduction in rotational inertia. Moreover, if we are happy with the final rate in rev/s, there will be no need to change to rad/s.

Solution: Conservation of angular momentum demands that

$$I_f \omega_f = I_0 \omega_0 \quad \text{or} \quad \omega_f = \frac{I_0 \omega_0}{I_f}$$

$$\omega_f = \frac{(6 \text{ kg} \cdot \text{m}^2)(1.4 \text{ rev/s})}{(2 \text{ kg} \cdot \text{m}^2)} = 4.20 \text{ rev/s}$$

Basically, we observe that decreasing the rotational inertia to one-third causes the angular speed to triple in order to conserve angular momentum.

Problem-Solving Strategy

Rotation of Rigid Bodies

1. It is helpful to realize that problems involving the rotation of a rigid body are similar to those you have worked before for constant linear acceleration. Review the analogies given in the summary at the end of this chapter.

2. When converting from linear to angular motion, or vice versa, you should recall the following relationships:

$$s = \theta R \qquad v = \omega R \qquad a = \alpha R$$

When these relations are applied, the angular measures must be in radians (rad).

3. Uniform angular acceleration problems are approached in the same way as they were for linear acceleration in Chapter 6. (Refer to Table 11.1.) You need find only three of the given quantities and select the appropriate equation that contains one unknown factor and not the other. Be careful to use consistent units for displacement, velocity, and acceleration.

4. Applications of Newton's second law, work, energy, power, and momentum are also solved in the same manner as for previous chapters. Just recall that for rotation, we use the rotational inertia instead of linear mass and we use the angular measures for displacement, velocity, and acceleration.

Summary and Review

Summary

In this chapter, we extended the concept of circular motion to include the rotation of a rigid body composed of many particles. We found that many problems can be solved by methods discussed earlier for linear motion. The essential concepts are now summarized.

- The similarities between rotational and linear motion:

Rotational	θ	ω	α	I	$I\omega$	τ	$I\alpha$	$\tau\theta$	$\frac{1}{2}I\omega^2$	$\tau\omega$
Linear	s	v	a	m	mv	F	ma	Fs	$\frac{1}{2}mv^2$	Fv

- The angle in radians is the ratio of the arc distance s to the radius R of the arc. Symbolically we write:

$$\theta = \frac{s}{R} \qquad s = \theta R$$

The radian is a unitless ratio of two lengths.

- Angular velocity, which is the rate of angular displacement, can be calculated from θ or from the frequency of rotation:

$$\overline{\omega} = \frac{\theta}{t} \qquad \overline{\omega} = 2\pi f \qquad \begin{array}{c}\textit{Average}\\ \textit{Angular Velocity}\end{array}$$

- Angular acceleration is the time rate of change in angular speed:

$$\alpha = \frac{\omega_f - \omega_0}{t} \qquad \textit{Angular Acceleration}$$

- By comparing θ to s, ω to v, and α to a, the following equations can be utilized for angular acceleration problems:

$$\theta = \left(\frac{\omega_f + \omega_0}{2}\right)t$$
$$\omega_f = \omega_0 + \alpha t$$
$$\theta = \omega_0 t + \frac{1}{2}\alpha t^2$$
$$2\alpha\theta = \omega_f^2 - \omega_0^2$$

When any three of the five parameters θ, α, t, ω_f, and ω_0 are given, the other two can be found from one of these equations. Choose a direction of rotation as being positive throughout your calculations.

- The following equations are useful when comparing linear motion with rotational motion:

$$v = \omega R \qquad a_T = \alpha R$$

- Other useful relationships:

$$I = \sum mR^2 \qquad \textit{Moment of Inertia}$$

$$I = mk^2 \qquad \textit{Radius of Gyration}$$

$$\text{Work} = \tau\theta \qquad \textit{Work}$$

$$L = I\omega \qquad \textit{Angular Momentum}$$

$$K = \frac{1}{2}I\omega^2 \qquad \textit{Rotational Kinetic Energy}$$

$$\tau = I\alpha \qquad \textit{Newton's Law}$$

$$P = \tau\omega \qquad \textit{Power}$$

$$I_f\omega_f = I_0\omega_0 \qquad \begin{array}{l}\textit{Conservation of Angular}\\ \textit{Momentum}\end{array}$$

Key Terms

angular acceleration   226
angular displacement   221
angular momentum L 235
angular velocity   223
axis of rotation   226

conservation of angular
  momentum   236
moment of inertia   228
radian   221
radius of gyration k 229

rotational kinetic energy   228
rotational work   232
tangential acceleration   227
translational motion   221

Review Questions

11.1. Make a list of the SI and USCS units for angular velocity, angular acceleration, moment of inertia, torque, and rotational kinetic energy.

11.2. State the angular analogies for the following translational equations:

    a. $r_f = v_0 + at$

    b. $s = v_0 t + \dfrac{1}{2}at^2$

    c. $F = ma$

    d. $K = \dfrac{1}{2}mv^2$

    e. Work $= Fs$

    f. Power $=$ work$/t = Fv$

11.3. A sphere, a cylinder, a disk and a hollow ring all have identical masses and are rotating with constant angular velocity about a common axis. Compare their individual rotational kinetic energies, assuming their outside diameters are equal.

11.4. Explain how a diver controls her motion so that she can strike the water feet first or head first.

11.5. A cat held feet up and dropped from any elevation above a few feet will always land feet first. How does it accomplish this?

11.6. When energy is supplied to a body so that both translation and rotation result, its total kinetic energy is given by

$$K = \frac{1}{2}mv^2 + \frac{1}{2}I\omega^2$$

How the energy is divided between rotational and translational effects is determined by the distribution of mass (moment of inertia). From these statements, which of the following objects will roll to the bottom of an inclined plane first?

    a. A solid disk of mass M

    b. A circular hoop of mass M

11.7. Refer to Question 11.6. If a solid sphere, a solid disk, a solid cylinder, and a hollow cylinder, all of the same radius, are released simultaneously from the top of an inclined plane, in what order will they reach the bottom?

11.8. A disk whose moment of inertia is I_1 and whose angular velocity is ω_1 is meshed with a disk whose moment of inertia is I_2 and whose angular velocity is ω_2. Write the conservation equation by denoting their combined angular velocity by ω.

11.9. Refer to Question 11.8. Suppose $\omega_1 = \omega_2$ and $I_1 = 2I_2$. How would their combined velocities compare with their initial velocity? Suppose $\omega_1 = 3\omega_2$ and $I_1 = I_2$?

Problems

Section 11.3 Angular Acceleration and
Section 11.4 Relationship Between Rotational and Linear Motion

11.1. A cable is wrapped around a drum 80 cm in diameter. How many revolutions of this drum will cause an object attached to the cable to move a linear distance of 2 m? What is the angular displacement?   Ans. 0.796 rev, 5 rad

11.2. A bicycle wheel is 26 in. in diameter. If the wheel makes 60 revolutions, what linear distance will it travel?

11.3. A point on the edge of a large wheel of radius 3 m moves through an angle of 37°. Find the length of the arc described by the point.   Ans. 1.94 m

11.4. A person sitting on the edge of a 6-ft-diameter platform moves a linear distance of 2 ft. Express the angular displacement in radians, degrees, and revolutions.

11.5. An electric motor turns at 600 rpm. What is the angular velocity? What is the angular displacement after 6 s?   Ans. 62.8 rad/s, 377 rad

11.6. A rotating pulley completes 12 revolutions in 4 s. Determine the average angular velocity in revolutions per second, revolutions per minute, and radians per second.

11.7. A bucket is hung from a rope that is wrapped many times around a circular drum of radius 60 cm. The bucket starts from rest and is lifted to a height of 20 m in 5 s. (a) How many revolutions were made by the drum? (b) What was the average angular speed of the rotating drum?   Ans. (a) 5.31 rev; (b) 6.67 rad/s

11.8. A wheel of radius 15.0 cm starts from rest and makes 2.00 revolutions in 3.00 s. (a) What is its *average* angular velocity in radians per second? (b) What is the final *linear* velocity of a point on the rim of the wheel?

11.9. A cylindrical piece of stock 6 in. in diameter rotates in a lathe at 800 rev/min. What is the linear velocity at the surface of the cylinder?
Ans. 20.9 ft/s

11.10. The proper tangential velocity for machining steel stock is about 70 cm/s. At what revolutions per minute should a steel cylinder 8 cm in diameter be turned in a lathe?

11.11. For the wheel in Prob. 11.8, what is the angular acceleration? What is the *linear* acceleration of a point on the edge of the wheel?
Ans. 2.79 rad/s^2, 0.419 m/s^2

11.12. A circular drum of radius 40 cm is initially rotating at 400 rev/min. It is then brought to a stop after making 50 revolutions. What were the angular acceleration and the stopping time?

11.13. A belt is wrapped around the edge of a pulley that is 40 cm in diameter. The pulley rotates with a constant angular acceleration of 3.50 rad/s^2. At $t = 0$, the rotational speed is 2 rad/s. What are the angular displacement and angular velocity of the pulley 2 s later?
Ans. 11.0 rad, 9.00 rad/s

11.14. In Prob. 11.13, what are the final linear speed and the final linear acceleration of the pulley belt as it moves around the edge of the pulley?

11.15. A wheel initially rotates at 6 rev/s and then undergoes a constant angular acceleration of 4 rad/s^2. What is the angular velocity after 5 s? How many revolutions will the wheel make?
Ans. 57.7 rad/s, 38.0 rev

11.16. A grinding disk is brought to a stop in 40 revolutions. If the braking acceleration was -6 rad/s^2, what was the initial frequency or revolution in revolutions per second?

11.17. A pulley 320 mm in diameter and rotating initially at 4 rev/s receives a constant angular acceleration of 2 rad/s^2. What is the linear velocity of a belt around the pulley after 8 s? What is the tangential acceleration of the belt?
Ans. 6.58 m/s, 0.320 m/s^2

*11.18. A person initially at rest 4 m from the center of a rotating platform covers a distance of 100 m in 20 s. What is the angular acceleration of the platform? What is the angular velocity after 4 s?

Section 11.5 Rotational Kinetic Energy: Moment of Inertia

11.19. A 2-kg mass and a 6-kg mass are connected by a light 30-cm bar. The system is then rotated horizontally at 300 rpm about an axis 10 cm from the 6-kg mass. What is the moment of inertia about this axis? What is the rotational kinetic energy?
Ans. 0.140 kg m^2, 69.1 J

11.20. A 1.2-kg bicycle wheel has a radius of 70 cm with spokes of negligible weight. If it starts from rest and receives an angular acceleration of 3 rad/s^2, what will be its rotational kinetic energy after 4 s?

11.21. A 16-lb grinding disk is rotating at 400 rev/min. What is the radius of the disk if its kinetic energy is 54.8 ft · lb? What is the moment of inertia?
Ans. 6.00 in., 0.0625 slug ft^2

11.22. What must be the radius of a 4-kg circular disk if it is to have the same moment of inertia as a 1-kg rod 1 m long and pivoted at its midpoint?

*11.23. A wagon wheel 60 cm in diameter is mounted on a central axle where it spins at 200 rev/min. The wheel can be thought of as a circular hoop of mass 2 kg, and each of 12 wooden 500-*g* spokes can be thought of as thin rods rotating about their ends. Calculate the moment of inertia for the entire wheel. What is the rotational kinetic energy?
Ans. 0.360 kg m^2, 78.9 J

11.24. Compare the rotational kinetic energies of three objects of equal radius and mass: a circular hoop, a circular disk, and a solid sphere.

Section 11.6 The Second Law of Motion and Rotation

11.25. A rope wrapped around a 5-kg circular drum pulls with a tension of 400 N. If the radius of the drum is 20 cm and it is free to rotate about its central axis, what is the angular acceleration?
Ans. 800 rad/s^2

11.26. The flywheel of an engine has a moment inertia of 24 slug ft^2. What torque is required to accelerate the wheel from rest to an angular velocity of 400 rpm in 10 s?

11.27. A 3-kg thin rod is 40 cm long and pivoted about its midpoint. What torque is required to cause it to make 20 revolutions while its rotational speed increases from 200 to 600 rev/min?
Ans. 0.558 N · m

11.28. A large 120-kg turbine wheel has a radius of gyration of 1 m. A frictional torque of 80 N · m opposes the rotation of the shaft. What torque must be applied to accelerate the wheel from rest to 300 rev/min in 10 s?

11.29. A 2-kg mass swings in a circle of radius 50-cm at the end of a light rod. What resultant torque is required to give an angular acceleration of 2.5 rad/s²?   Ans. 1.25 N · m

11.30. A rope is wrapped several times around a cylinder of radius 0.2 m and mass 30 kg. What is the angular acceleration of the cylinder if the tension in the rope is 40 N and it turns without friction?

11.31. An 8-kg grinding disk has a diameter of 60 cm and is rotating at 600 rev/min. What braking force must be applied tangentially to the disk if it is to stop rotating in 5 s?   Ans. 15.1 N

11.32. An unbalanced torque of 150 N · m imparts an angular acceleration of 12 rad/s² to the rotor of a generator. What is the moment of inertia?

Section 11.7 Rotational Work and Power

11.33. A rope wrapped around a 3-kg disk 20 cm in diameter is pulled for a linear distance of 5 m with a force of 40 N. What is the linear work done by the 40-N force? What is the rotational work done on the disk?   Ans. 200 J, 200 J

11.34. Use the work-energy theorem to calculate the final angular velocity of the disk in Prob. 11.33 if it starts from rest.

11.35. A 1.2-kW motor acts for 8 s on a wheel having a moment of inertia of 2 kg · m². Assuming the wheel was initially at rest, what is the final angular speed?   Ans. 98.0 rad/s

11.36. A cord is wrapped around the rim of a cylinder that has a mass of 10 kg and a radius of 30 cm. If the rope is pulled with a force of 60 N, what is the angular acceleration of the cylinder? What is the linear acceleration of the rope?

11.37. A 600-W motor drives a pulley with an average angular velocity of 20 rad/s. What torque is developed?   Ans. 30 N · m

11.38. The crankshaft on an automobile develops 350 lb · ft of torque at 1800 rpm. What is the output horsepower?

Section 11.8 Combined Rotation and Translation

11.39. A 2-kg cylinder has a radius of 20 cm. It rolls without slipping along a horizontal surface at a velocity of 12 m/s. (a) What is its translational

kinetic energy? (b) What is its rotational kinetic energy? (c) What is the total kinetic energy?
Ans. (a) 144 J; (b) 72 J; (c) 216 J

11.40. A circular ring is designed so that it has the same mass and radius as the cylinder of Prob. 11.39. What is the *total* kinetic energy if it rolls with the same horizontal velocity?

11.41. Consider an inclined plane of height 16 m. Using different materials, four objects are designed to have the same mass of 3 kg: a circular hoop, a disk, a sphere, and a box. Assume friction is negligible for the box, but for the rolling objects, assume there is enough friction that they roll without slipping. By calculating the final velocities in each case, determine the order in which they arrive at the bottom.   Ans. v_b = 17.7 m/s; v_s = 12.5 m/s; v_d = 14.5 m/s; v_h = 15.0 m/s.

*11.42. What is the height of an inclined plane in order that a circular disk will roll from rest to the bottom with a final velocity of 20 m/s?

Section 11.9 Angular Momentum and Section 11.10 Conservation of Angular Momentum

11.43. A 500-g steel rod 30 cm in length is pivoted about its center and rotated at 300 rev/min. What is the angular momentum?   Ans. 0.118 kg m/s²

11.44. In Prob. 11.43, what average braking torque must be applied to stop the rotation in 2 s?

11.45. A sudden torque of 400 N · m is applied to the edge of a disk initially at rest. If the rotational inertia of the disk is 4 kg · m² and the torque acts for 0.02 s, what is the change in angular momentum? What is the final angular speed?
Ans. 8.00 kg · m²/s, 2.00 rad/s

11.46. In Fig. 11.14, a 6-kg disk A rotating clockwise at 400 rev/min engages with a 3-kg disk B

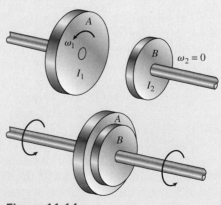

Figure 11.14

initially at rest. The radius of disk A is 0.4 m, and the radius of disk B is 0.2 m. What is the combined angular speed after the two disks are meshed?

11.47. Assume that disk B in Prob. 11.46 is initially rotating clockwise at 200 rev/min in the same direction as disk A. What is the common angular speed as they mesh?   Ans. 378 rev/min

11.48. Assume the same conditions as in Prob. 11.46 except that the disk B is rotating counterclockwise and A is rotating clockwise. What is the combined angular velocity after the disks are meshed?

11.49. The rod connecting the two weights in Fig. 11.15 has negligible weight but is configured to allow the weights to slip outward. At the instant when the angular speed is 600 rev/min, the 2-kg masses are 10 cm apart. What is the rotational speed later when the masses are 34 cm apart?
          Ans. 51.9 rpm

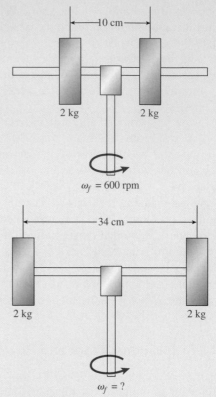

Figure 11.15

Additional Problems

11.50. A 6-kg circular grinding disk is rotating initially at 500 rev/min. The radius of the disk is 40 cm. What is the angular acceleration of the disk if the axe exerts a tangential force of 120 N at the edge? How many revolutions will the disk make before stopping? What work is done, and what power is lost in the process?

11.51. A 3-kg wheel with spokes of negligible mass is free to rotate about its center without friction. The edge of the wheel of radius 40 cm is struck suddenly with an average tangential force of 600 N lasting for 0.002 s. (a) What angular impulse is imparted to the wheel? (b) If the wheel was initially at rest, what was its angular speed at the end of the 0.002-s interval?
          Ans. (a) 0.48 N · m · s; (b) 1 rad/s

11.52. Disk A has three times the rotational inertia of disk B. Disk A is rotating initially clockwise at 200 rev/min, and disk B is rotating in the opposite direction at 800 rev/min. If the two are meshed together, what will be the common rate of rotation of the combined disks?

11.53. If the disks in Prob. 11.52 are initially rotating in the same direction, what would be the common angular speed after meshing?
          Ans. 350 rev/min clockwise

11.54. The radius of gyration of an 8-kg wheel is 50 cm. Find its moment of inertia and its kinetic energy if it is rotating at 400 rev/min.

11.55. How much work is required to slow the wheel in Prob. 11.54 to 100 rev/min?   Ans. 1644 J

11.56. A wheel of radius 2 ft has a moment of inertia of 8.2 slug ft². A constant force of 12 lb acts tangentially at the edge of the wheel, which is initially at rest. What is the angular acceleration?

11.57. In Prob. 11.56, the wheel was brought to rest in 5 s. How much work was done? What horsepower was developed?   Ans. 878 ft · lb, 0.319 hp

11.58. An engine operating at 1800 rev/min develops 200 hp. What is the torque developed?

11.59. A constant force of 200 N acts at the edge of a wheel 36 cm in diameter causing it to make 20 revolutions in 5 s. What power is developed?
          Ans. 905 W

*11.60. A 2-kg circular hoop rolls down an inclined plane from an initial height of 20 m. The kinetic energy that develops is shared between rotation and translation. What will be the speed when it reaches the bottom of the incline?

11.61. Suppose a circular disk rolls down the same incline as in Prob. 11.60. What is its speed when it reaches the bottom? Ans. 16.2 m/s

Critical Thinking Questions

11.62. A circular hoop of mass 2 kg and radius 60 cm spins freely about its center connected by light central spokes. A force of 50 N acts tangent to the edge of the wheel for a time of 0.02 s. (a) What is the angular impulse? (b) What is the change in angular momentum? (c) If the hoop was initially at rest, what was the final angular speed? (d) Use the work-energy theorem to calculate the angular displacement.
Ans. (a) 0.60 N · ms; (b) 0.60 kg · m²/s, (c) 0.833 rad/s, (d) 0.00833 rad

11.63. The spin cycle on a washing machine slows from 900 to 300 rev/min in 4 s. Determine the angular acceleration. Does a force act to throw off the water from the clothes or does the lack of a force do this? When the cycle operates at 900 rev/min, the output power is 4 kW. What is the torque developed? If the radius of the tub is 30 cm, what is the linear speed of the clothes near the inside edge?

11.64. A block is attached to a cord passing over a pulley through a hole in a horizontal tabletop as shown in Fig. 11.16. Initially the block is revolving at 4 rad/s at a distance r from the center of the hole. If the cord is pulled from below until its radius is $r/4$, what is the new angular velocity?
Ans. 64 rad/s

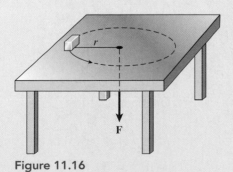

Figure 11.16

11.65. Suppose in Fig. 11.16 the block has a mass of 2 kg and is rotating at 3 rad/s when $r = 1$ m. At what distance r will the tension in the cord be 25 N?

*11.66. Consider Fig. 11.17, in which $m = 2$ kg, $M = 8$ kg, $R = 60$ cm, and $h = 6$ m. Write Newton's second law for the disk in terms of the tension in the rope, the moment of inertia of the disk, and the angular acceleration. Next write Newton's second law for falling mass in terms of the tension in the rope, the mass, and the linear acceleration. Eliminate T from these two equations. Find the linear acceleration of the 2-kg mass by recalling that $v = \omega R$, $a = \alpha R$, and $I = \frac{1}{2}mR^2$.
Ans. 3.27 m/s²

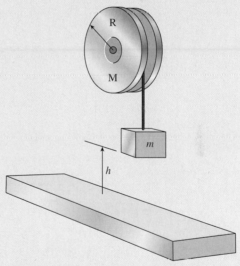

Figure 11.17

11.67. Use conservation of energy to find the velocity of the 2-kg mass in Fig. 11.17 just before it strikes the floor 6 m below. Use data given in Prob. 11.66.

11.68. A student stands on a platform with arms out-stretched, holding weights in each hand so that the rotational inertial is 6.0 kg · m². The platform is then set into constant, frictionless rotation at 90 rev/min. Now the student is able to reduce the rotational inertia to 2 kg · m² by pulling the weights toward the body. (a) What will be the new rate of rotation in the absence of external torque? (b) What is the ratio of the final kinetic energy to the initial kinetic energy? (c) Explain the increase in energy.

Ans. (a) 270 rev/min; (b) 3; (c) work done on masses

*11.69. Consider the apparatus shown in Fig. 11.18. Consider the large pulley as a 6-kg disk of radius 50 cm. The right mass is 4 kg, and the left mass is 2 kg. Consider both rotational and translational energies and find the velocity just before the 4-kg mass strikes the floor.

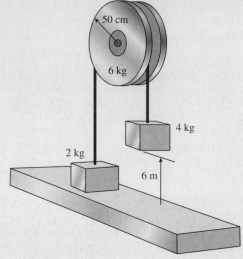

Figure 11.18

12 Simple Machines

Simple machines are used to perform a variety of tasks with considerable efficiency. In this example, a system of gears, pulleys, and levers function to produce accurate time measurements.

(*Photo © vol. 1 PhotoDisc/Getty.*)

Objectives

After completing this chapter, you should be able to

1. Describe a simple machine and its operation in general terms to the extent that *efficiency* and *conservation of energy* are explained.

2. Write and apply formulas for computing the efficiency of a simple machine in terms of work or power.

3. Distinguish by definition and example between *ideal* mechanical advantage and *actual* mechanical advantage.

4. Draw a diagram of each of the following simple machines and beside each diagram, write a formula for computing the ideal mechanical advantage: (a) lever, (b) inclined plane, (c) wedge, (d) gears, (e) pulley systems, (f) wheel and axle, (g) screw jack, (h) belt drive.

5. Compute the mechanical advantage and the efficiency of each of the simple machines listed in the previous objective.

A *simple machine* is any device that transmits the application of a force into useful work. With a chain hoist, we can transmit a small downward force into a large upward force for

245

lifting. In industry, delicate samples of radioactive material are handled by machines that allow an applied force to be reduced significantly. Single pulleys may be used to change the direction of an applied force without affecting its magnitude. A study of machines and their efficiency is essential for the productive use of energy. In this chapter, you will become familiar with levers, gears, pulley systems, inclined planes, and other machines used routinely for many industrial applications.

12.1   Simple Machines and Efficiency

In a simple machine, input work is done by the application of a single force, and the machine performs output work by means of a single force. During any such operation (Fig. 12.1), three processes occur:

1. Work is supplied to the machine.

2. Work is done against friction.

3. Output work is done by the machine.

According to the principle of conservation of energy, these processes are related as follows:

Input work = work against friction + output work

The amount of useful work performed by a machine can never be greater than the work supplied to it. There will always be some loss due to friction or other dissipative forces. For example, when pumping up a bicycle tire with a small hand pump, we exert a downward force on the plunger, causing air to be forced into the tire. That some of our input work is lost to friction can be verified easily by feeling how warm the wall of the hand pump becomes. The smaller we can make the friction loss in a machine, the greater the return for our effort. In other words, the effectiveness of a given machine can be measured by comparing its output work with the work supplied to it.

The efficiency e of a machine is defined as the ratio of the work output to the work input.

$$e = \frac{\text{work output}}{\text{work input}} \qquad \textbf{(12.1)}$$

The ***efficiency*** as defined in Eq. (12.1) will always be a number between 0 and 1. Common practice is to express this decimal as a percentage by multiplying by 100. For example, a machine that does 40 J of work when 80 J of work is supplied to it has an efficiency of 50 percent.

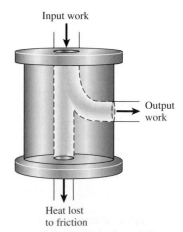

Input work

Output work

Heat lost to friction

Figure 12.1   Three processes occur in the operation of a machine: (1) the input of a certain amount of work, (2) the loss of energy in doing work against friction, and (3) the output of useful work.

Another useful expression for efficiency can be noted from the definition of power as work per unit time. We can write

$$P = \frac{\text{work}}{t} \quad \text{or} \quad \text{Work} = Pt$$

The efficiency in terms of power input P_i and power output P_o is given by

$$e = \frac{\text{work output}}{\text{work input}} = \frac{P_o t}{P_i t}$$

or

$$e = \frac{\text{power output}}{\text{power input}} = \frac{P_o}{P_i} \tag{12.2}$$

Example 12.1

A 45-kW motor winds a cable around a drum while lifting a 2700-kg mass to a height of 6 m in a time of 3 s. Determine the efficiency of the motor and how much work is done against friction forces.

Plan: The input power (45 kW) is given, and the output power is the rate at which work is done in lifting the mass. The efficiency of the motor is found by taking the ratio of output power to input power. Finally, we calculate the power lost by subtracting the output power from the input power.

Solution: The output power lifts the mass to a height $h = 6$ m in 3 s. Therefore,

$$P_o = \frac{Fh}{t} = \frac{mgh}{t}$$

$$= \frac{(2000 \text{ kg})(9.8 \text{ m/s}^2)(6 \text{ m})}{3 \text{ s}} = 39200 \text{ W}$$

$$= 39.2 \text{ kW}$$

The efficiency can now be found from Eq. (12.2).

$$e = \frac{P_o}{P_i} = \frac{39.2 \text{ kW}}{45 \text{ kW}} = 0.871$$

Thus, the efficiency of the motor is 87.1 percent.
  The power lost to friction is the difference between input and output powers.

$$P_i - P_o = 45 \text{ kW} - 39.2 \text{ kW} = 5.80 \text{ kW}$$

$$\textit{Power lost to friction} = 5.80 \text{ kW}$$

12.2 Mechanical Advantage

Simple machines such as the lever, block and tackle, chain hoist, gears, inclined plane, and screw jack all play important roles in modern industry. We can illustrate the operation of any machine by the general diagram in Fig. 12.2. An input force $\mathbf{F}_i$ acts through a distance s_i performing useful work $F_i s_i$.

The actual mechanical advantage M_A of a machine is defined as the ratio of the output force F_o to the input force F_i.

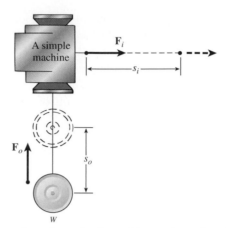

Figure 12.2 During the operation of any simple machine, an input force $\mathbf{F}_i$ acts through a distance s_i, while an output force $\mathbf{F}_o$ acts through a distance s_o.

$$M_A = \frac{\text{output force}}{\text{input force}} = \frac{F_o}{F_i} \tag{12.3}$$

An ***actual mechanical advantage*** greater than 1 indicates that the output force is greater than the input force. Although most machines have values of M_A greater than 1, this is not always the case. With small, fragile objects, it is sometimes desirable to make the output force smaller than the input force.

In Section 12.1, we noted that the efficiency of a machine increases as frictional effects become small. Applying the conservation-of-energy principle to the simple machine in Fig. 12.2 yields

$$Work\ input = work\ against\ friction + work\ output$$
$$F_i s_i = (\text{work})f + F_o s_o$$

The most efficient engine possible would realize no losses due to friction. We can represent this ideal case by setting $(\text{work})f = 0$ in the above equation. Thus,

$$F_o s_o = F_i s_i$$

Since this equation represents an ideal case, we define the ***ideal mechanical advantage*** M_I as

$$M_I = \frac{F_o}{F_i} = \frac{s_i}{s_o} \tag{12.4}$$

The ideal mechanical advantage of a simple machine is equal to the ratio of the distance the input force moves to the distance the output force moves.

The efficiency of a simple machine is the ratio of output work to input work. Therefore, for the general machine of Fig. 12.2, we have

$$e = \frac{F_o s_o}{F_i s_i} = \frac{F_o / F_i}{s_i / s_o}$$

Finally, utilizing Eqs. (12.3) and (12.4), we obtain

$$e = \frac{M_A}{M_I} \tag{12.5}$$

All the above concepts have been treated as they apply to a general machine. In the following sections, we shall apply them to specific machines.

12.3 | The Lever

Possibly the oldest and most generally useful machine is the simple lever. A **lever** consists of any rigid bar pivoted at a certain point called the **fulcrum.** Fig. 12.3 illustrates the use of a long rod to lift a weight **W.** We can calculate the ideal mechanical advantage of such a device in two ways. The first method involves the principle of equilibrium, and the second uses the principle of work, as discussed in Section 12.2. Since the equilibrium method is easier for the lever, we shall apply it first.

Because no translational motion is involved during the application of a lever, the condition for equilibrium is that the input torque equal the output torque:

$$F_i r_i = F_o r_o$$

The ideal mechanical advantage can be found from

$$M_I = \frac{F_o}{F_i} = \frac{r_i}{r_o} \tag{12.6}$$

The ratio F_o/F_i is considered the *ideal* case because no friction forces are considered.

The same result is obtained from work considerations. Note from Fig. 12.3b that the force $\mathbf{F}_i$ moves through the arc distance s_i, while the force $\mathbf{F}_o$ moves through the arc distance s_o. The two arcs are subtended by the same angle θ, however, and so we can write the proportion

$$\frac{s_i}{s_o} = \frac{r_i}{r_o}$$

Substitution into Eq. (12.4) will verify the result obtained from equilibrium considerations; that is, $M_I = r_i/r_o$.

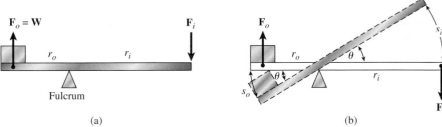

(a)   (b)

Figure 12.3   The lever.

Example 12.2

An iron bar 1.2 m long is used to lift a 60-kg crate. The bar is used as a lever, like the one shown in Fig. 12.3. The fulcrum is placed 30 cm from the crate. What is the ideal mechanical advantage, and what input force is required?

Plan: The input and output distances will determine the *ideal* mechanical advantage, which will also be the *actual* mechanical advantage in this case. By setting the ratio of output force to input force equal to the ideal mechanical advantage, we can find the necessary input force.

Solution: The output distance is $r_o = 0.30$ m, and the input distance $r_i = 1.2$ m $-$ 0.30 m or 0.90 m. Thus, the ideal mechanical advantage is

$$M_I = \frac{r_i}{r_o} = \frac{0.90 \text{ m}}{0.30 \text{ m}} = 3$$

Substitution into Eq. (12.6) gives

$$\frac{F_o}{F_i} = 3 \qquad \text{or} \qquad F_i = \frac{F_o}{3}$$

The output force is equal to weight mg. Thus,

$$F_i = \frac{mg}{3} = \frac{(60 \text{ kg})(9.8 \text{ m/s}^2)}{3} = 1960 \text{ N}$$

Before leaving the subject of the lever, it should be noted that little of the input work is lost to friction forces. For all practical purposes, the *actual* mechanical advantage of a simple lever is equal to the *ideal* mechanical advantage. Other examples of the lever are illustrated in Fig. 12.4.

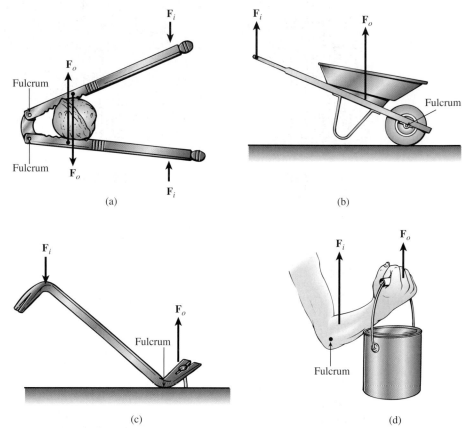

Figure 12.4   The lever forms the operating principle of many simple machines.

12.4   Applications of the Lever Principle

A serious limitation of the elementary lever is the fact that it operates through a small angle. There are many ways of overcoming this restriction by allowing for continuous rotation of the lever arm. For example, the **wheel and axle** (Fig. 12.5) allows for the continued action of the input force. Applying the reasoning described in Section 12.2 for a general machine, it can be shown that

$$M_I = \frac{F_o}{F_i} = \frac{R}{r} \qquad\qquad (12.7)$$

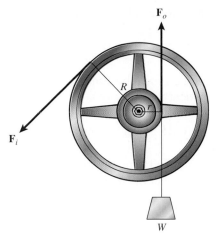

Figure 12.5 The wheel and axle.

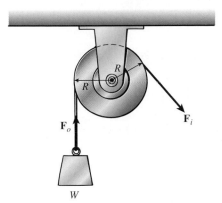

Figure 12.6 A single fixed pulley serves only to change the direction of the input force.

Thus, the ideal mechanical advantage of a wheel and axle is the ratio of the radius of the wheel to the radius of the axle.

Another application of the lever concept is through the use of pulleys. A single *pulley,* as shown in Fig. 12.6, is simply a lever whose input moment arm is equal to its output moment arm. From the principle of equilibrium, the input forces will equal the output force, and the ideal mechanical advantage will be

$$M_I = \frac{F_o}{F_i} = 1 \qquad \qquad \textbf{(12.8)}$$

The only advantage of such a device lies in its ability to change the direction of an input force.

A single movable pulley (Fig. 12.7), on the other hand, has an ideal mechanical advantage of 2. Note that the two supporting ropes must each be shortened by 1 ft to lift the load through a distance of 1 ft. Therefore, the input force moves through a distance of 2 ft, while the output force moves a distance of only 1 ft. Applying the principle of work, we have

$$F_i(2 \text{ ft}) = F_o(1 \text{ ft})$$

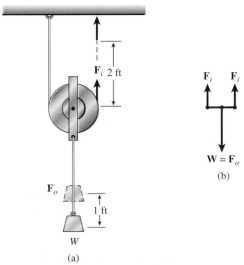

Figure 12.7 A single movable pulley. (a) The input force moves through twice the distance that the output force travels. (b) The free-body diagram shows that $2F_i = F_o$.

from which the ideal mechanical advantage is

$$M_I = \frac{F_o}{F_i} = 2 \qquad \qquad \textbf{(12.9)}$$

The same result can be shown by constructing a free-body diagram, as in Fig. 12.7b. From the figure, it is evident that

$$2F_i = F_o \qquad \text{and} \qquad M_I = \frac{F_o}{F_i} = 2$$

This approach is usually applied to problems involving movable pulleys since it allows one to associate M_I with the number of strands supporting the movable pulley.

Example 12.3

Calculate the ideal mechanical advantage of the block-and-tackle arrangement shown in Fig. 12.8a.

Plan: Draw a sketch and construct a free-body diagram (see Fig. 12.8b). We will find the actual mechanical advantage as a ratio of output to input forces.

Solution: From the diagram, we see that

$$4F_i = F_o$$

So the mechanical advantage is

$$M_I = M_A = \frac{F_o}{F_i} = \frac{4F_i}{F_i}$$

$$M_I = 4$$

Note that the uppermost pulley serves only to change the direction of the input force. The same mechanical advantage would result if $\mathbf{F}_i$ were applied upward at point a.

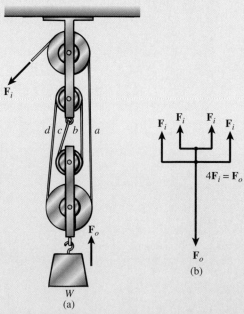

Figure 12.8   The block and tackle. This arrangement has an ideal mechanical advantage of 4, since four strands support the movable block.

12.5 | The Transmission of Torque

The simple machines discussed so far are used to transmit and apply forces to move loads. In most mechanical applications, work is done by transmitting torque from one drive to another. For example, the belt drive (Fig. 12.9) transmits the torque from a driving pulley to an output pulley. The mechanical advantage of such a system is the ratio of the torques between the output pulley and the driving pulley:

$$M_I = \frac{\text{output torque}}{\text{input torque}} = \frac{\tau_o}{\tau_i}$$

From the definition of torque, we can write this expression in terms of the radii of the pulleys:

$$M_I - \frac{\tau_o}{\tau_i} = \frac{F_o r_o}{F_i r_i}$$

If there is no slippage between the belt and pulleys, it is safe to say that the tangential input force $\mathbf{F}_i$ is equal to the tangential output force $\mathbf{F}_o$. Thus,

$$M_I = \frac{F_o r_o}{F_i r_i} = \frac{r_o}{r_i}$$

Since the diameters of pulleys are usually specified instead of their radii, a more convenient expression is

$$M_I = \frac{D_o}{D_i} \qquad (12.10)$$

where D_i is the diameter of the driving pulley and D_o is the diameter of the output pulley.

Suppose we now apply the principle of work to the ***belt drive.*** Remember that work is defined in rotary motion as the product of torque τ and angular displacement θ. For the belt drive, assuming ideal conditions, the input work equals the output work. Thus,

$$\tau_i \theta_i = \tau_o \theta_o$$

The power input must also be equal to the power output. Dividing the above equation by the time t required to rotate through the angles θ_i and θ_o, we obtain

$$\tau_i \frac{\theta_i}{t} = \tau_o \frac{\theta_o}{t} \qquad \text{or} \qquad \tau_i \omega_i = \tau_o \omega_o$$

where ω_i and ω_o are the angular speeds of the input and output pulleys. Note that the ratio τ_o/τ_i represents the ideal mechanical advantage. Therefore, we can add another expression to Eq. (12.10) to obtain

$$M_I = \frac{D_o}{D_i} = \frac{\omega_i}{\omega_o} \qquad (12.11)$$

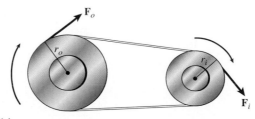

Figure 12.9 The belt drive.

This important result shows that the mechanical advantage is achieved at the expense of rotary motion. In other words, if the mechanical advantage is 2, the input shaft must rotate with twice the angular speed of the output shaft. The ratio ω_i/ω_o is sometimes referred to as the *speed ratio*.

If the speed ratio is greater than 1, the machine produces an output torque that is greater than the input torque. As we have seen, this feat is accomplished at the expense of rotation. On the other hand, many machines are designed to increase the rotational output speed. In these cases, the speed ratio is less than 1, and the increased rotational speed is accomplished with reduced torque output.

Example 12.4

Consider the belt drive illustrated in Fig. 12.9, where the diameter of the small driving pulley is 6 in. and the diameter of the driven pulley is 18 in. A 6-hp motor drives the input pulley at 600 rpm. Calculate the rpm and torque delivered to the driven wheel if the system is 75 percent efficient.

Plan: First, we will calculate the ideal mechanical advantage (100 percent efficiency) from the ratio of pulley diameters. By multiplying this value by the given efficiency, we will obtain the *actual* mechanical advantage, which is the ratio of output torque to input torque. That will allow us to solve for the output torque. Finally, the output rate of rotation can be solved from a ratio of the pulley diameters.

Solution: The ideal mechanical advantage in given from Eq. (12.11)

$$M_I = \frac{D_o}{D_i} = \frac{18 \text{ in.}}{6 \text{ in.}} = 3$$

Since the efficiency is 75 percent, the actual mechanical advantage is given from Eq. (12.5).

$$M_A = eM_I = (0.75)(3) = 2.25$$

Now the actual mechanical advantage is the simple ratio of output torque τ_o to input torque τ_i. Recalling that the power in rotational motion is equal to the product of torque and angular velocity, we can solve for τ_i as follows:

$$\tau_i = \frac{P_i}{\omega_i} = \frac{(6 \text{ hp})[(550 \text{ ft} \cdot \text{lb/s})/\text{hp}]}{(600 \text{ rev/min})(2\pi \text{ rad/rev})(1 \text{ min/60 s})}$$

$$= \frac{(6)(550 \text{ ft} \cdot \text{lb/s})}{20\pi \text{ rad/s}} = 52.5 \text{ ft} \cdot \text{lb}$$

Since $M_A = \tau_o/\tau_i$, the output torque is given by

$$\tau_o = M_A\tau_i = (2.25)(52.5 \text{ ft} \cdot \text{lb})$$
$$= 118 \text{ ft} \cdot \text{lb}$$

Assuming the belt does not slip, it will move with the same linear velocity v around each pulley. Since $v = \omega r$, we can write the equality

$$\omega_i r_i = \omega_o r_o \qquad \text{or} \qquad \omega_i D_i = \omega_o D_o$$

from which

$$\omega_o = \frac{\omega_i D_i}{D_o} = \frac{(600 \text{ rpm})(6 \text{ in.})}{18 \text{ in.}} = 200 \text{ rpm}$$

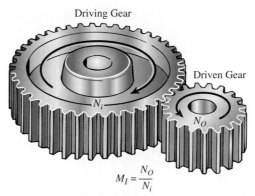

$$M_I = \frac{N_O}{N_i}$$

Figure 12.10 Spur gears. The ideal mechanical advantage is the ratio of the number of teeth on the output gear to the number of teeth on the input gear.

Note that the ratio of ω_i to ω_o yields the ideal mechanical advantage and not the actual mechanical advantage. The difference between M_I and M_A is due to friction, both in the belt and in the shaft bearings. Since greater tension on the belt will result in greater friction forces, maximum efficiency is obtained by reducing the belt tension until it just prevents the belt from slipping on the pulleys.

Before leaving our discussion of the transmission of torque, we must consider the application of gears. A **_gear_** is simply a notched wheel that can transmit torque by meshing with another notched wheel, as shown in Fig. 12.10. A pair of meshing gears differs from a belt drive only in the sense that the gears rotate in opposite directions. The same relationships derived for the belt drive hold for gears:

$$M_I = \frac{D_o}{D_i} = \frac{\omega_i}{\omega_o} \tag{12.12}$$

A more useful expression makes use of the fact that the number of teeth N on the rim of a gear is proportional to its diameter D. Because of this dependence, the ratio of the number of teeth on the driven gear N_o to the number of teeth on the driving gear N_i is the same as the ratio of their diameters. Hence, we can write

$$M_I = \frac{N_o}{N_i} = \frac{D_o}{D_i} \tag{12.13}$$

The use of gears avoids the problem of slippage, which is common with belt drives. It also conserves space and allows for a greater torque to be transmitted.

In addition to the *spur* gears illustrated in Fig. 12.10, there are several other types of gears. Four common types are worm, helical, bevel, and planetary gears. Examples of each are shown in Fig. 12.11. The same general relationships apply for all these gears.

12.6 The Inclined Plane

The only machines we have discussed so far involve application of the lever principle. A second fundamental machine is the **_inclined plane._** Suppose you have to move a heavy load from the ground to a truck bed without hoisting equipment. You should probably select a few long boards and form a ramp from the ground to the bed of the truck. Experience has taught you that it takes less effort to push a load up a small elevation than it does to lift the load directly. Since a smaller input force results in the same output force, a mechanical advantage is realized. The smaller input force is accomplished, however, at the cost of greater distance.

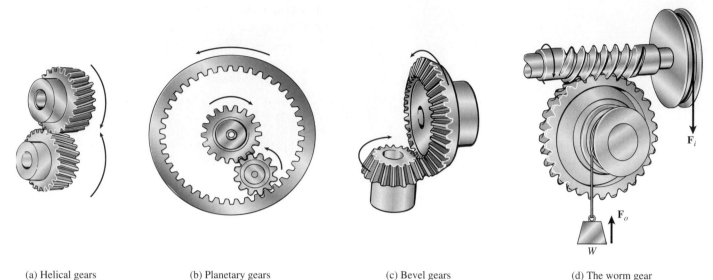

(a) Helical gears   (b) Planetary gears   (c) Bevel gears   (d) The worm gear

Figure 12.11   Four common types of gears: (a) helical, (b) planetary, (c) bevel, (d) worm. (The spur gear, which is the most common type, is shown in Fig. 12.10.)

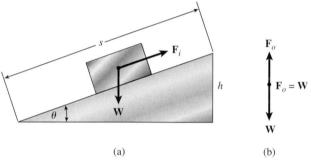

(a)   (b)

Figure 12.12   The inclined plane. The input force represents the effort required to push the block up the plane; the output force is equal to the weight of the block.

Consider the movement of a weight **W** up the inclined plane in Fig. 12.12. The slope angle θ is such that the weight must be moved through a distance s to reach a height h at the top of the incline. If we neglect friction, the work required to push the weight up the plane is the same as the work required to lift it up vertically. We can express this equality as

$$Work\ input\ =\ work\ output$$
$$F_i s\ =\ Wh$$

where $\mathbf{F}_i$ is the input force and **W** is the output force. Since friction was neglected, the ideal mechanical advantage will be the same as the actual mechanical advantage.

$$M_I = \left(\frac{W}{F_i}\right)_{ideal} = \frac{s}{h}$$

Thus, in the absence of friction, the ideal mechanical advantage of an inclined plane is simply the ratio of the output distance (up the incline) to the input distance (the height).

$$M_I = \frac{s}{h} \qquad \begin{array}{l} \textit{Ideal Mechanical Advantage} \\ \textit{Inclined Plane} \end{array} \qquad \textbf{(12.14)}$$

In most applications, there will be significant friction forces that must be overcome, making the required input force F_i greater and the actual mechanical advantage significantly less than the ratio of length to height. Example 12.5 illustrates this point.

Example 12.5

An 88-kg crate of beer bottles is to be raised to a loading platform that is 2 m above the ground. The length of the ramp is 4 m, and the coefficient of kinetic friction is 0.3. What are the *ideal* and *actual* mechanical advantages?

Plan: We will draw a sketch and free-body diagram similar to those shown in Fig. 12.13. The ideal mechanical advantage M_I is found by direct substitution into Eq. (12.14). However, the *actual* mechanical advantage is less due to fact that the input force must overcome the friction force and not just the component of the weight down the incline. Assuming constant motion up the plane, we will apply the first condition for equilibrium to determine the input force, which can then be used to find M_A.

Solution: The ideal mechanical advantage is

$$M_I = \frac{s}{h} = \frac{4 \text{ m}}{2 \text{ m}}; \qquad M_I = 2$$

The inclination angle is found from Fig. 12.13.

$$\sin \theta = \frac{2 \text{ m}}{4 \text{ m}} = 0.5; \qquad \theta = 30°$$

Using this angle, we find the components of the weight perpendicular to the incline.

$$W_y = mg \cos \theta = (88 \text{ kg})(9.8 \text{ m/s}^2) \cos 30°; \qquad W_y = 747 \text{ N}$$
$$W_x = mg \sin \theta = (88 \text{ kg})(9.8 \text{ m/s}^2) \sin 30°; \qquad W_x = 431 \text{ N}$$

To obtain the actual mechanical advantage, we wish to find the minimum force up the incline, which means we consider the forces on the crate to be in equilibrium, and

$$\sum F_x = 0; \qquad P - W_x - f_k = 0 \qquad \text{or} \qquad P = 431 \text{ N} + f_k$$
$$\sum F_y = 0; \qquad n - W_y = 0 \qquad \text{or} \qquad n = W_y = 747 \text{ N}$$

The last of these equations allows us to find the friction force f_k.

$$f_k = \mu_k n = (0.3)(747 \text{ N}); \qquad f_k = 224 \text{ N}$$

The first equilibrium equation gives us the input force P.

$$P = 431 \text{ N} + f_k = 431 \text{ N} + 224 \text{ N}; \qquad P = 755 \text{ N}$$

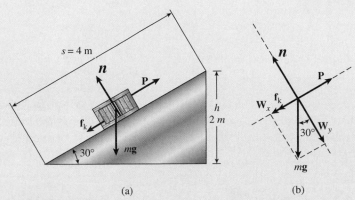

(a)   (b)

Figure 12.13

Finally, the actual mechanical advantage is the ratio of the output force W to the input force P. We find

$$M_A = \frac{W}{P} = \frac{mg}{P} = \frac{(88\ \text{kg})(9.8\ \text{m/s}^2)}{755\ \text{N}}$$

$$M_I = 1.14$$

It is left as an exercise for you to show that the efficiency of this ramp is only 57 percent.

Many machines apply the principle of the inclined plane. The simplest is the **wedge** (Fig. 12.14), which is actually a double inclined plane. In the ideal case, the mechanical advantage of a wedge of length L and thickness t is given by

$$M_I = \frac{L}{t} \tag{12.15}$$

This equation is a direct consequence of the general relation expressed by Eq. (12.14). The ideal mechanical advantage is always much greater than the actual mechanical advantage because of the large friction forces between the surfaces in contact. The wedge finds its application in axes, knives, chisels, planers, and all other cutting tools. A *cam* is a kind of rotary wedge that is used to lift valves in internal combustion engines.

One of the most useful applications of the inclined plane is the **screw.** This principle can be explained by examining a common tool known as the *screw jack* (Fig. 12.15). The threads are essentially an inclined plane wrapped continuously around a cylindrical shaft. When the input force $\mathbf{F}_i$ turns through a complete revolution ($2\pi R$), the output force $\mathbf{F}_o$ will advance through the distance p. This distance p is actually the distance between two adjacent threads, and it is called the **pitch** of the screw. The ideal mechanical advantage is the ratio of the input distance to the output distance.

$$M_I = \frac{s_i}{s_o} = \frac{2\pi R}{p} \tag{12.16}$$

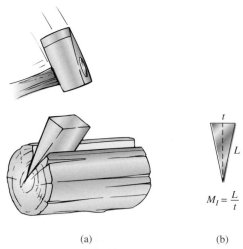

(a)   (b)

Figure 12.14   The wedge is actually a double inclined plane.

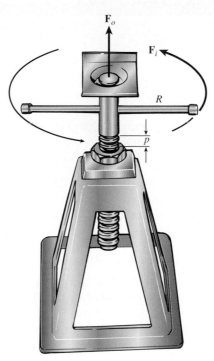

Figure 12.15 The screw jack.

The screw is an example of an inefficient machine, but in this case, it is usually an advantage since friction forces are needed to hold in place while the input force is not being applied.

Summary and Review

Summary

A simple machine has been defined as a device that converts a single input force F_i into a single output force F_o. In general, the input force moves through a distance s_i, and the output force moves through a distance s_o. The purpose is to accomplish useful work in a manner suited to a particular application. The major concepts are given as follows.

- A simple machine is a device that converts a single input force F_i into a single output force F_o. The input force moves through a distance s_i, and the output force moves a distance s_o. There are two mechanical advantages:

$$M_A = \frac{F_o}{F_i} \quad \text{Actual Mechanical Advantage} \\ \text{(friction considered)}$$

$$M_I = \frac{s_i}{s_o} \quad \text{Ideal Mechanical Advantage} \\ \text{(assumes no friction)}$$

- The efficiency of a machine is a ratio of output to input work. It is normally expressed as a percentage and can be calculated from any of the following relations:

$$e = \frac{\text{work output}}{\text{work input}} \qquad e = \frac{\text{power output}}{\text{power input}}$$

$$e = \frac{M_A}{M_I}$$

- The ideal mechanical advantages for a number of simple machines are given as follows.

$$M_I = \left(\frac{F_o}{F_i}\right)_{\text{ideal}} = \frac{r_i}{r_o} \qquad \textit{Lever}$$

$$M_I = \left(\frac{F_o}{F_i}\right)_{\text{ideal}} = \frac{R}{r} \qquad \textit{Wheel and Axle}$$

$$M_I = \frac{D_o}{D_i} = \frac{\omega_i}{\omega_o} \qquad \textit{Belt Drive}$$

$$M_I = \frac{W}{F_i} = \frac{s}{h} \qquad \textit{Inclined Plane}$$

$$M_I = \frac{L}{t} \qquad \textit{Wedge}$$

$$M_I = \frac{N_o}{N_i} = \frac{D_o}{D_i} \qquad \textit{Gears}$$

$$M_I = \frac{s_i}{s_o} = \frac{2\pi R}{p} \qquad \textit{Screw Jack}$$

Key Terms

actual mechanical advantage   248
belt drive   253
efficiency   246
fulcrum   249
gear   255
ideal mechanical advantage   248
inclined plane   255
lever   249
pitch   258
pulley   251
screw   258
simple machine   245
wedge   258
wheel and axle   250

Review Questions

12.1. What is meant by *useful work* or *output work?* What is meant by *input work?* Write the general relationship between input work and output work.

12.2. Two jacks are operated simultaneously to lift the front end of a car. Immediately afterward, the left jack feels warmer than the right one. Which jack is more efficient? Explain.

12.3. A machine may alter the magnitude and/or the direction of an input force. (a) Name several examples in which both changes occur. (b) Give examples in which only the magnitude of the input force is altered. (c) Give some examples in which only the direction is altered.

12.4. A machine lifts a load through a vertical distance of 4 ft, while the input force moves through a distance of 2 ft. Would this machine be helpful in lifting large weights? Explain.

12.5. A bicycle can be operated in three gear ranges. In *low range,* the pedals describe two complete revolutions, while the rear wheel turns through one revolution. In *medium range,* the pedals and the wheels turn at the same rate. In *high range,* the rear wheel of the bicycle completes two revolutions for every complete pedal revolution. Discuss the advantages and disadvantages of each range.

12.6. What happens to the ideal mechanical advantage if a simple machine is operated in reverse? What happens to its efficiency?

12.7. Give several examples of machines that have an actual mechanical advantage less than 1.

12.8. Why do buses and trucks often use larger steering wheels than those found on automobiles? What principle is used?

12.9. Draw diagrams of pulley systems that have ideal mechanical advantages of 2, 3, and 5.

12.10. Usually the road to the top of a mountain winds around the mountain instead of going straight up the side. Why? If we neglect friction, is more work required to reach the top along the spiral road? Is more power required? If we consider friction, would it require less work to drive straight up the side of the mountain? Explain.

Problems

Section 12.1 Simple Machines and Efficiency and Section 12.2 Mechanical Advantage

12.1. A 25 percent efficient machine performs external work of 200 J. What input work is required?
Ans. 800 J

12.2. What is the input work of a 30 percent efficient gasoline engine if during each cycle it performs 400 J of useful work?

12.3. A 60-W motor lifts a 2-kg mass to a height of 4 m in 3 s. Compute the output power. Ans. 26.1 W

12.4. What is the efficiency of the motor in Prob. 12.3? What is the rate at which work is done against friction?

12.5. A 60 percent efficient machine lifts a 10-kg mass at a constant speed of 3 m/s. What is the required input power? Ans. 490 W

12.6. During the operation of a 300-hp engine, energy is lost to friction at the rate of 200 hp. What is the useful output power, and what is the efficiency of the engine?

12.7. A frictionless machine lifts a 200-lb load through a vertical distance of 10 ft. The input force moves through a distance of 300 ft. What is the ideal mechanical advantage of the machine? What is the magnitude of the input force? Ans. 30, 6.67 lb

Section 12.4 Applications of the Lever Principle

12.8. One edge of a 50-kg safe is lifted with a 1.2-m steel rod. What input force is required at the end of the rod if a fulcrum is placed 12 cm from the safe? (*Hint:* To lift one edge, a force equal to one-half the weight of the safe is required.)

12.9. For the nutcracker in Fig. 12.4a, the nut is located 2 cm from the fulcrum, and an input force of 20 N is applied at the handles, which

are 10 cm from the fulcrum. What is the force applied to crack the nut? Ans. 100 N

12.10. For the wheelbarrow in Fig. 12.4b, the center of gravity of a net load of 40 kg is located 50 cm from the wheel. What upward lift must be applied at a point on the handles that is 1.4 m from the wheel?

12.11. What is the ideal mechanical advantage of the wheelbarrow in Prob. 12.10? Ans. 2.80

12.12. Find the ideal mechanical advantage of the crowbar in Fig. 12.4c if the input force is applied 30 cm from the nail and the fulcrum is located 2 cm from the nail.

12.13. The input force exerted by a muscle in the forearm (see Fig. 12.4d) is 120 N and acts at a distance of 4 cm from the elbow. The total length of the forearm is 25 cm. What weight is being lifted? Ans. 19.2 N

12.14. A wheel 20 cm in diameter is attached to an axle that has a diameter of 6 cm. If a weight of 400 N is attached to the axle, what force must be applied to the rim of the wheel to lift the weight at constant speed? Neglect friction.

12.15. A 20-kg mass is to be lifted with a rod 2 m long. If you can exert a downward force of 40 N on one end of the rod, where should you place a block of wood to act as a fulcrum? Ans. 33.9 cm from mass

12.16. Determine the force **F** required to lift a 200-N load *W* with the pulley shown in Fig. 12.16a.

12.17. What input force is needed to lift the 200-N load with the arrangement drawn in Fig. 12.16b?
Ans. 50.0 N

12.18. What are the input forces needed to lift the 200-N load for the arrangements in Fig. 12.16c and d?

12.19. What is the mechanical advantage of a screwdriver used as a wheel and axle if its blade is 0.3 in. wide and the diameter of its handle is 0.8 in.? Ans. 2.67

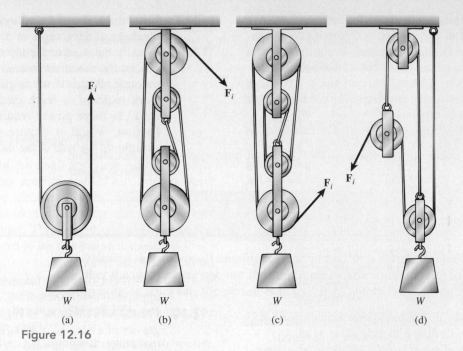

Figure 12.16

12.20. The chain hoist in Fig. 12.17 is a combination of the wheel and axle and the block and tackle. Show that the ideal mechanical advantage of such a device is given by

$$M_I = \frac{2R}{R - r}$$

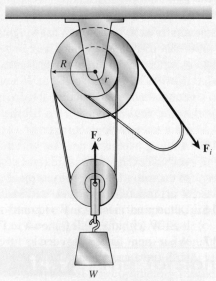

Figure 12.17

12.21. Assume that the larger radius in Fig. 12.17 is three times the smaller radius. What input force is required to lift a 10-kg load with no friction?

Ans. 32.7 N

Section 12.5 The Transmission of Torque

12.22. A 1500-rev/min motor has a drive pulley 3 in. in diameter, and the driven pulley is 9 in. in diameter. What is the ideal mechanical advantage, and what are the revolutions per minute for the output pulley?

12.23. A 30-cm diameter input pulley turns at 200 rev/min on a belt drive connected to an output pulley 60 cm in diameter. What is the ratio of the output torque to the input torque? What are the output revolutions per minute?

Ans. 100 rev/min

12.24. A V-belt pulley system has output and input drives of diameters 6 in. and 4 in., respectively. A torque of 200 lb · in. is applied to the input drive. What is the output torque?

12.25. The ratio of output speed to input speed for a gear drive is 2:1. What is the mechanical advantage?

Ans. $\frac{1}{2}$

12.26. A set of two spur gears has 40 teeth and 10 teeth. What are the possible ideal mechanical advantages?

12.27. For the spur gears in Prob. 12.26, what is the rotational speed of the smaller gear if the speed of the larger gear is 200 rev/min? Ans. 800 rev/min

Section 12.7 Applications of the Inclined Plane

12.28. What must be the thickness of the base if a wedge is 20 cm long and it is desired that the input force be one-tenth of the output force?

12.29. What should be the apex angle of a wedge if it is to have a mechanical advantage of 10?

Ans. 5.71°

12.30. A 10-kg crate is moved from the ground to a loading platform by means of a ramp 6 m long and 2 m high. Assume that $\mu_k = 0.25$. What are the ideal and actual mechanical advantages of the ramp?

12.31. For the ramp in Prob. 12.30, what is the efficiency of the ramp?

Ans. 58.6 percent

*12.32. An input force of 20 lb is applied to the 6-in. handle of a wrench used to tighten a $\frac{1}{4}$-in.-diameter bolt. An actual output force of 600 lb is produced. If the bolt has 10 threads per inch, what is the ideal mechanical advantage and what is the efficiency?

12.33. The lever of a screw jack is 24 in. long. If the screw has six threads per inch, what is the ideal mechanical advantage?

Ans. 904

12.34. If the screw jack in Prob. 12.33 is 15 percent efficient, what force is needed to lift 2000 lb?

Additional Problems

*12.35. An inclined plane is 6 m long and 1 m high. The coefficient of kinetic friction is 0.2. What force is required to pull a weight of 2400 N up the incline at constant speed? What is the efficiency of the inclined plane?

Ans. 873 N, 45.8 percent

*12.36. A wheel and axle are used to raise a mass of 700 kg. The radius of the wheel is 0.50 m, and the radius of the axle is 0.04 m. If the actual efficiency is 60 percent, what input force must be applied to the wheel?

*12.37. A shaft rotating at 800 rev/min delivers a torque of 240 N · m to an output shaft that is rotating at 200 rev/min. If the efficiency of the machine is 70 percent, compute the output torque. What is the output power?

Ans. 672 N · m, 14.1 kW

*12.38. A screw jack has a screw with a pitch of 0.25 in. Its handle is 16 in. long, and a load of 1.9 tons is being lifted. Neglecting friction, what force is required at the end of the handle? What is the mechanical advantage?

*12.39. A certain refrigeration compressor comes equipped with a 250-mm-diameter pulley and is designed to operate at 600 rev/min. What should be the diameter of the motor pulley if the motor speed is 2000 rev/min?

Ans. 75.0 mm

*12.40. In a fan belt, the driving wheel is 20 cm in diameter and the driven wheel is 50 cm in diameter. The power input comes from a 4-kW motor that causes the driving wheel to rotate at 300 rev/min. If the efficiency is 80 percent, calculate the revolutions per minute and the torque delivered to the driven wheel.

12.41. A log-splitting wedge has a side length of 16 cm, and the angle at the apex is 10°. What is the ideal mechanical advantage?

Ans. 5.76

12.42. A machine has an efficiency of 72 percent. An input force of 500 N moves through a parallel distance of 40 cm. How much energy is lost in the process?

*12.43. A motor with an efficiency of 80 percent operates a winch with an efficiency of 50 percent. If the power supplied to the motor is 6 kW, how far will the winch lift a 400-kg mass in a time of 4 s?

Ans. 2.45 m

Critical Thinking Questions

12.44. A 60-N weight is lifted in the three different ways shown in Fig. 12.18. Compute the ideal mechanical advantage and the required input force for each application.

Ans. 2, 30 N; 3, 20 N; 0.33, 180 N

12.45. A worm drive similar to that shown in Fig. 12.11 has n teeth in the gear wheel. (If $n = 80$, one complete turn of the worm will advance the wheel one-eightieth of a revolution.) Derive an expression for the ideal mechanical advantage of the worm gear in terms of the radius of the input pulley R, the radius of the output shaft r, and the number of teeth n in the gear wheel.

12.46. The worm drive of Prob. 12.45 has 80 teeth in the gear wheel. If the radius of the input wheel is 30 cm and the radius of the output shaft is 5 cm, what input force is required to lift a 1200-kg load? Assume an efficiency of 80 percent.

Ans. 30.6 N

12.47. The oarlock on a 3.5 m oar is 1 m from the handle end. A person in a rowboat applies a force of 50 N to the handle end. What is the ideal mechanical advantage and output force? Does moving the oarlock closer to the handle end increase or decrease the mechanical advantage? How far from the handle end should the oarlock be placed

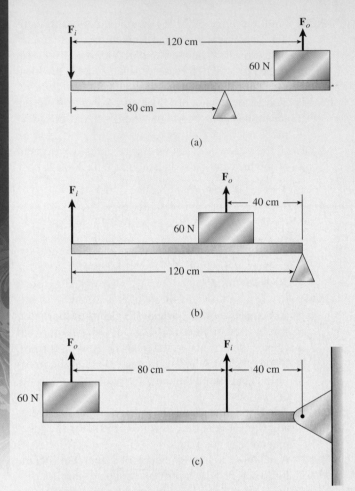

(a)

(b)

F_o F_i

|← 80 cm →|← 40 cm →|

60 N

(c)

Figure 12.18

to produce a 20 percent increase in the output force?

12.48. Draw sketches showing possible dimensions that will produce a mechanical advantage of 5 for a lever, a wheel and axle, and an inclined plane.

12.49. A 60-W motor drives the input pulley of a belt drive at 150 rev/min. The diameters of the input and output pulleys are 60 cm and 20 cm, respectively. Assume an actual mechanical advantage of 0.25. (a) What is the output torque? (b) What is the output power? (c) What is the efficiency?

*12.50. A pair of step pulleys (Fig. 12.19) makes it possible to change output speeds merely by shifting the belt. If an electric motor turns the input pulley at 2000 rev/min, find the possible angular speeds of the output shaft. The pulley diameters are 4, 6, and 8 cm.

Ans. (a) Small input pulley: 2000, 1330, 1000 rev/min; (b) middle input pulley: 3000, 2000, 1500 rev/min; (c) large input pulley: 4000, 2670, 2000 rev/min.

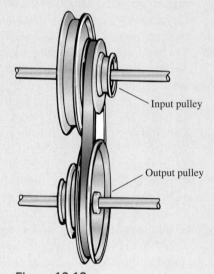

Figure 12.19

13 Elasticity

Bungee jumping utilizes a long elastic strap that stretches until it reaches a maximum length that is proportional to the weight of the jumper. The elasticity of the strap determines the amplitude of the resulting vibrations. If the *elastic limit* for the strap is exceeded, the rope will break. (*Photo © vol. 10 PhotoDisc/Getty.*)

Objectives

After completing this chapter, you should be able to

1. Demonstrate by example and discussion your understanding of *elasticity*, *elastic limit*, *stress*, *strain*, and *ultimate strength*.

2. Write and apply formulas for calculating Young's modulus, shear modulus, and bulk modulus.

3. Define and discuss the meanings of *hardness*, *malleability*, and *ductility* as applied to metals.

Until now, we discussed objects in motion or at rest. The objects have been assumed to be rigid and absolutely solid. But we know that wire can be stretched, that rubber tires will compress, and that bolts will sometimes break. A more complete understanding of nature requires a study of the mechanical properties of matter. The concepts of *elasticity, tension,* and *compression* are analyzed in this chapter. As the kinds of alloys increase and the demands on them become greater, our knowledge of such concepts becomes more important. For example, the stress placed on spaceships or on cables in modern bridges is of a magnitude unheard of a few years ago.

13.1 Elastic Properties of Matter

We define an ***elastic body*** as one that returns to its original size and shape when a deforming force is removed. Rubber bands, golf balls, trampolines, diving boards, footballs, and springs are common examples of elastic bodies. Putty, dough, and clay are examples of inelastic bodies. For all elastic bodies, we shall find it convenient to establish a cause-and-effect relationship between a deformation and the deforming forces.

Consider the coiled spring of length *l* shown in Fig. 13.1. We can study its ***elasticity*** by adding successive weights and observing the increase in length. A 20 N weight lengthens the spring by 1 cm, a 40-N weight lengthens the spring by 2 cm, and a 60 N weight lengthens the spring by 3 cm. Evidently, there is a direct relationship between the elongation of a spring and the applied force.

Robert Hooke first stated this relationship in connection with the invention of a balance spring for a clock. In general, he found that a force **F** acting on a spring (Fig. 13.2) produces an elongation *s* that is directly proportional to the magnitude of the force. ***Hooke's law*** can be written

$$F = ks \qquad (13.1)$$

The proportionality constant *k* varies extremely with the type of material and is called the ***spring constant.*** For the example illustrated in Fig. 13.1, the spring constant is

$$k = \frac{F}{s} = 20 \text{ N/cm}$$

Hooke's law is by no means restricted to coiled springs; it will apply to the deformation of all elastic bodies. To make the law more generally applicable, it will be convenient to define the terms *stress* and *strain.* ***Stress*** refers to the *cause* of an elastic deformation, whereas ***strain*** refers to the *effect,* in other words, the deformation itself.

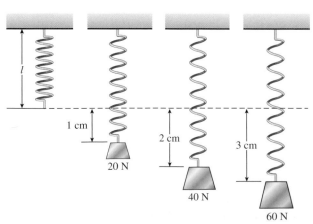

Figure 13.1 The uniform elongation of a spring.

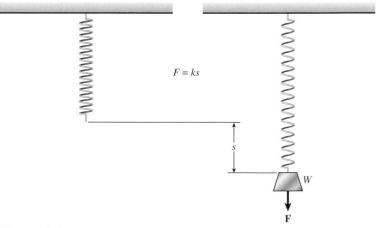

Figure 13.2 The relation between a stretching force **F** and the elongation it produces.

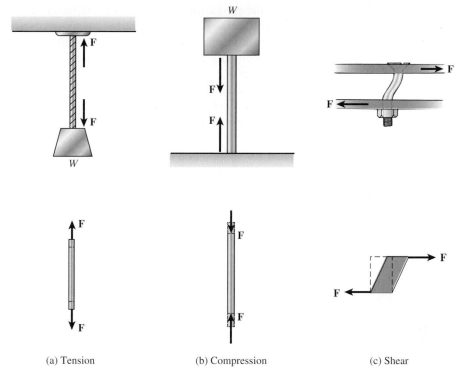

(a) Tension	(b) Compression	(c) Shear

Figure 13.3 Three common stresses, shown with their corresponding deformations: (a) tension, (b) compression, and (c) shear.

Three common types of stresses and their corresponding deformations are shown in Fig. 13.3. A ***tensile stress*** occurs when equal and opposite forces are directed away from each other. A ***compressive stress*** occurs when equal and opposite forces are directed toward each other. A ***shearing stress*** occurs when equal and opposite forces do not have the same line of action.

The effectiveness of any force producing a stress is highly dependent on the area over which the force is distributed. For this reason, a more complete definition of *stress* can be stated as follows:

> Stress is the ratio of an applied force to the area over which it acts, for example, newtons per square meter or pounds per square foot.

As mentioned earlier, the term *strain* must represent the effect of a given stress. The general definition of strain might be as follows:

> Strain is the relative change in the dimensions or shape of a body as the result of an applied stress.

In the case of a tensile or compressive stress, strain may be considered a change in length per unit length. A shearing stress, on the other hand, may alter only the shape of a body without changing its dimensions. Shearing strain is usually measured in terms of an angular displacement.

The ***elastic limit*** is the maximum stress a body can experience without becoming permanently deformed. For example, an aluminum rod whose cross-sectional area is 1 in.^{2} will become permanently deformed by the application of a tensile force greater than 19,000 lb. This does not mean that the aluminum rod will break at this point; it means only that the rod will not return to its original size. In fact, the tension can be increased to about 21,000 lb before the rod breaks. It is this property of metals that allows them to be drawn out into wires of smaller cross sections. The greatest stress a wire can withstand without breaking is known as its ***ultimate strength.***

If the elastic limit of a material is not exceeded, we can apply Hooke's law to any elastic deformation. Within the limits of a given material, it has been experimentally verified that the ratio of a given stress to the strain it produces is a constant. In other words, the stress is directly proportional to the strain. *Hooke's law* states

Provided that the elastic limit is not exceeded, an elastic deformation (strain) is directly proportional to the magnitude of the applied force per unit area (stress).

If we call the proportionality constant the **modulus of elasticity,** we can write Hooke's law in its most general form:

$$\text{Modulus of elasticity} = \frac{stress}{strain} \qquad \textbf{(13.2)}$$

In the following sections, we will discuss the specific applications of this fundamental relation.

13.2 Young's Modulus

In this section, we consider longitudinal stresses and strains as they apply to wires, rods, or bars. For example, in Fig. 13.4, a force **F** is applied to the end of a wire of cross-sectional area A. The longitudinal stress is given by

$$\text{Longitudinal stress} = \frac{F}{A}$$

The metric unit for stress is the *newton per square meter,* which is identical to the *pascal* (Pa).

$$1 \text{ Pa} = 1 \text{ N/m}^2$$

The USCS unit for stress is the *pound per square inch* (lb/in.^{2}). Since the pound per square inch remains in common use, it will be helpful to compare it with the SI unit:

$$1 \text{ lb/in.}^2 = 6895 \text{ Pa} = 6.895 \text{ kPa}$$

The effect of such a stress is to stretch the wire, in other words, to increase its length. Hence, the longitudinal strain can be represented by the change in length per unit length. We can write

$$\text{Longitudinal strain} = \frac{\Delta l}{l}$$

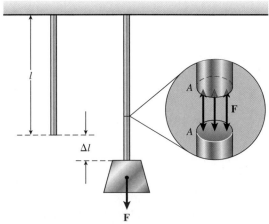

Figure 13.4 Computing Young's modulus for a wire of cross section A. The elongation Δl is exaggerated for clarity.

where l is the original length and Δl is the elongation. Experimentation has shown that a comparable decrease in length occurs for a compressive stress. The same equations will apply whether we are discussing an object under tension or an object under compression.

If we define the longitudinal modulus of elasticity of *Young's modulus Y*, we can write Eq. (13.2) as

$$Young's\ modulus = \frac{longitudinal\ stress}{longitudinal\ strain}$$

$$Y = \frac{F/A}{\Delta l/l} = \frac{Fl}{A\ \Delta l} \tag{13.3}$$

The units of Young's modulus are the same as the units of stress: pounds per square inch or pascals. This makes sense because the longitudinal strain is a unitless quantity. Representative values for some of the most common materials are listed in Tables 13.1 and 13.2.

Table 13.1

Elastic Constants for Various Materials in SI Units

Material	Young's Modulus Y, MPa*	Shear Modulus S, MPa	Bulk Modulus B, MPa	Elastic Limit, MPa	Ultimate Strength, MPa
Aluminum	68,900	23,700	68,900	131	145
Brass	89,600	35,300	58,600	379	455
Copper	117,000	42,300	117,000	159	338
Iron	89,600	68,900	96,500	165	324
Steel	207,000	82,700	159,000	248	489

*(1 MPa = 10^6 Pa)

Table 13.2

Elastic Constants for Various Materials in USCS Units

Material	Young's Modulus Y, lb/in.^{2}	Shear Modulus S, lb/in.^{2}	Bulk Modulus B, lb/in.^{2}	Elastic Limit, lb/in.^{2}	Ultimate Strength, lb/in.^{2}
Aluminum	10×10^6	3.44×10^6	10×10^6	19,000	21,000
Brass	13×10^6	5.12×10^6	8.5×10^6	55,000	66,000
Copper	17×10^6	6.14×10^6	17×10^6	23,000	49,000
Iron	13×10^6	10×10^6	14×10^6	24,000	47,000
Steel	30×10^6	12×10^6	23×10^6	36,000	71,000

Example 13.1

A vertical telephone wire 120 m long and 2.2 mm in diameter is stretched by a force of 380 N along the length of the wire. What is the longitudinal stress, and what is the longitudinal strain if the length after stretching is 120.10 m? Determine Young's modulus for the wire.

Plan: We will calculate the cross-sectional area of the wire and determine the stress as the force per unit of area. Then, the strain is recognized as the change in length per unit of initial length. Finally, Young's modulus is the ratio of stress to strain.

Solution: The cross section of a wire 2.2×10^{-3} m in diameter is

$$A = \frac{\pi D^2}{4} = \frac{\pi (2.2 \times 10^{-3} \text{ m})^2}{4}; \qquad A = 3.80 \times 10^{-6} \text{ m}^2$$

$$\text{Stress} = \frac{F}{A} = \frac{380 \text{ N}}{3.80 \times 10^{-6} \text{ m}^2}$$

$$\text{Stress} = 100 \times 10^6 \text{ N/m}^2 = 100 \text{ MPa}$$

The change in length is $(120.10 \text{ m} - 120.00 \text{ m})$ or 0.100 m. Therefore,

$$\text{Strain} = \frac{\Delta l}{l} = \frac{0.10 \text{ m}}{120 \text{ m}}; \qquad \text{Strain} = 8.3 \times 10^{-4}$$

Finally, Young's modulus is the ratio of stress to strain.

$$Y = \frac{\text{stress}}{\text{strain}} = \frac{100 \text{ MPa}}{8.3 \times 10^{-4}}; \qquad Y = 120,000 \text{ MPa}$$

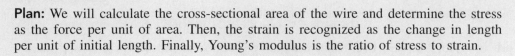

Example 13.2

What is the maximum mass that can be hung from a steel wire 2 m in length and 6 mm in diameter if its elastic limit is not to be exceeded? Determine the increase in length under this load.

Plan: Again, we will need to find the cross-sectional area of the wire. Then, from Table 13.2, we note that the elastic limit for steel is 248,000 MPa. The weight W of the suspended mass must not produce a stress greater than this limit. Therefore, we can solve for the mass in the equation for stress. The increase in length can be found directly from Eq. (13.3).

Solution: The area of the wire is

$$A = \frac{\pi D^2}{4} = \frac{\pi (6 \times 10^{-3} \text{ m})^2}{4}; \qquad A = 2.83 \times 10^{-5} \text{ m}^2$$

The limiting stress in this case is the weight per unit area, so

$$\textit{Elastic limit} = \frac{W}{A} \qquad \text{or} \qquad W = \textit{elastic limit} \times A$$

$$W = (2.48 \times 10^8 \text{ Pa})(2.83 \times 10^{-5} \text{ m}^2) = 7.01 \times 10^3 \text{ N}$$

The largest mass that can be supported is found from this weight:

$$m = \frac{W}{g} = \frac{7.01 \times 10^3 \text{ N}}{9.8 \text{ m/s}^2} = 716 \text{ kg}$$

The increase in length under such a load is found from Eq. (13.3), as follows:

$$\Delta L = \frac{l}{Y}\left(\frac{F}{A}\right) = \frac{2.00 \text{ m}}{2.07 \times 10^{11} \text{ Pa}} (2.48 \times 10^8 \text{ Pa})$$

$$\Delta L = 2.40 \times 10^{-3} \text{ m}$$

The length increases by 2.40 mm to a new length of 2.0024 m.

13.3 Shear Modulus

Compressive and tensile stresses produce a slight change in linear dimensions. As mentioned earlier, a shearing stress alters only the shape of a body, leaving its volume unchanged. For example, consider the parallel noncurrent forces acting on the cube in Fig. 13.5. The applied force causes each successive layer of atoms to slip sideways, much like the pages of a book under similar stress. The interatomic forces restore the block to its original shape when the stress is relieved.

The *shearing stress* is defined as the ratio of the tangential force F to the area A over which it is applied. The **shearing strain** is defined as the angle ϕ (in radians), which is called the *shearing angle* (refer to Fig. 13.5b). Applying Hooke's law, we can now define the **shear modulus** S as follows:

$$S = \frac{\text{shearing stress}}{\text{shearing strain}} = \frac{F/A}{\phi} \qquad (13.4)$$

The angle ϕ is usually so small that it is approximately equal to $\tan \phi$. Making use of this fact, we can rewrite Eq. (13.4) in the form

$$S = \frac{F/A}{\tan \phi} = \frac{F/A}{d/l} \qquad (13.5)$$

Since the value of S is an indication of the rigidity of a body, it is sometimes referred to as the **modulus of rigidity.**

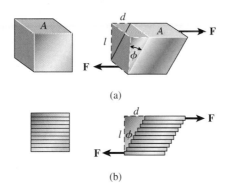

Figure 13.5   Shearing stress and shearing strain.

The steel bolt in Fig. 13.6 has a cross section of $1.8 \times 10^{-4} \text{ m}^2$ and projects 3.8 cm out from the wall. If the end of the bolt is subjected to a sheering force of 35 kN, what will be the downward deflection of the bolt?

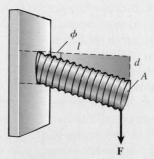

Figure 13.6

Plan: The shearing stress is the shearing force **F** per unit of area A, and the shearing strain is the deflection d per unit length l of the bolt. We can determine the shear modulus for steel from Table 13.1 and then calculate the deflection from Eq. (13.5). Be careful to use the SI base units for all substituted quantities.

Solution: We solve for the deflection d as follows:

$$S = \frac{F/A}{d/l} = \frac{Fl}{Ad}$$

$$d = \frac{Fl}{AS} = \frac{(35{,}000 \text{ N})(0.038 \text{ m})}{(1.8 \times 10^{-4} \text{ m}^2)(8.27 \times 10^{10} \text{ Pa})}$$

$$d = 8.94 \times 10^{-5} \text{ m}$$

The end of the bolt whose diameter is around 1.5 cm deflects about 0.09 mm. Shearing strains are generally very small.

13.4 Volume Elasticity; Bulk Modulus

So far we have considered stresses that cause a change in the shape of an object or result in primarily one-dimensional strains. In this section, we will be concerned with changes in volume. For example, consider the cube in Fig. 13.7 on which forces are applied uniformly over the surface. The initial volume of the cube is denoted by V, and the area of each face is represented by A. The resultant force **F** applied normally to each face causes a change in volume $-\Delta V$. The minus sign indicates that the change represents a volume reduction. The *volume stress* F/A is the normal force per unit area, whereas the *volume strain* $-\Delta V/V$ is the change in volume per unit volume. Applying Hooke's law, we define the modulus of volume elasticity, or ***bulk modulus,*** as follows:

$$B = \frac{\text{volume stress}}{\text{volume strain}} = -\frac{F/A}{\Delta V/V} \qquad \textbf{(13.6)}$$

This type of strain applies to liquids as well as solids. Table 13.3 lists the bulk moduli for a few of the more common liquids. When working with liquids, it is sometimes more convenient to represent the stress as pressure P, which is defined as the force per unit area F/A. With this definition, we can rewrite Eq. (13.6):

$$B = \frac{-P}{\Delta V/V} \qquad \textbf{(13.7)}$$

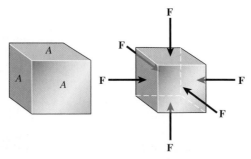

Figure 13.7 The bulk modulus. The original volume is reduced by the application of a uniform compressive force on each face.

Table 13.3

Bulk Moduli for Liquids

Liquid	Bulk Modulus B	
	MPa	lb/in.^{2}
Benzene	1,050	1.5×10^5
Ethyl alcohol	1,100	1.6×10^5
Mercury	27,000	40×10^5
Oil	1,700	2.5×10^5
Water	2,100	3.1×10^5

The reciprocal of the bulk modulus is called the *compressibility k*. Often it is more convenient to study the elasticity of materials by measuring their compressibilities. By definition,

$$k = \frac{1}{B} = -\left(\frac{1}{P}\right)\frac{\Delta V}{V_0} \tag{13.8}$$

Equation (13.8) indicates that the compressibility is the fractional change in volume per unit increase in pressure.

Example 13.4

A hydrostatic press contains 5 liters of water. Find the decrease in volume of the water when it is subjected to a pressure of 2000 kPa.

Solution: The decrease in volume is found by solving for ΔV from Eq. (13.7):

$$\Delta V = -\frac{PV}{B} = \frac{(2 \times 10^6 \text{ Pa})(5 \text{ L})}{2.1 \times 10^9 \text{ Pa}}$$
$$= -0.00476 \text{ L} = -4.76 \text{ mL}$$

13.5 Other Physical Properties of Metals

In addition to the elasticity, tensile strength, and shearing strength of materials, there are other important properties of metals. A solid consists of molecules arranged so closely together that they attract each other strongly. This attraction, called **cohesion,** gives a solid a definite shape and size. It also affects its usefulness to industry as a working material. Properties such as hardness, ductility, malleability, and conductivity must be understood before metals are chosen for specific applications. Three of these properties are illustrated in Fig. 13.8.

Hardness is an industrial term used to describe the ability of metals to resist forces that tend to penetrate them. Hard materials resist being scratched, worn away, penetrated, or otherwise damaged physically. Some metals such as sodium and potassium are soft, while iron and steel are two of the hardest materials. The hardness of metals is tested with machines that push a cone-shaped diamond point into test materials. The penetration is measured, and a hardness reading is taken directly from a dial.

Two other special properties of materials are **ductility** and **malleability.** The meaning of each of these terms is seen from Fig. 13.8. Ductility is defined as the ability of a metal

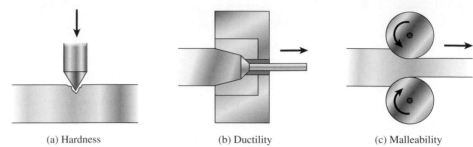

(a) Hardness   (b) Ductility   (c) Malleability

Figure 13.8   Illustrating the working properties of metals: (a) a hard metal resists penetration, (b) a ductile metal can be drawn into a wire, and (c) a malleable metal can be rolled into sheets.

to be drawn out into a wire. Tungsten and copper are extremely ductile. Malleability is the property that enables us to hammer or bend metals into any desired shape or to roll them into sheets. Most metals are malleable, gold being the most malleable.

Conductivity refers to the ability of metals to permit the flow of electricity. The best conductors are silver, copper, gold, and aluminum, in that order. More will be said about this property in later chapters.

Summary and Review

Summary

In industry, we must utilize materials effectively and for appropriate situations. Otherwise, metal failures will result in costly damage or serious injury to employees. In this chapter, we have discussed the elastic properties of matter and some of the formulas used to predict the effects of stress on certain solids. The following points will summarize this chapter:

- According to *Hooke's law,* an elastic body will deform or elongate an amount s under the application of a force **F.** The constant of proportionality k is the *spring constant:*

$$F = ks \qquad k = \frac{F}{s} \qquad \textit{Hooke's Law}$$

- *Stress* is the ratio of an applied force to the area over which it acts. *Strain* is the relative change in dimensions that results from the stress.

  For example,

$$\text{Longitudinal stress} = \frac{F}{A}$$

$$\text{Longitudinal strain} = \frac{\Delta l}{l}$$

- The *modulus of elasticity* is the constant ratio of stress to strain:

$$\textit{Modulus of elasticity} = \frac{stress}{strain}$$

- *Young's modulus Y* is for longitudinal deformations:

$$Y = \frac{F/A}{\Delta l/l}$$

or

$$Y = \frac{Fl}{A \cdot \Delta l} \qquad \textit{Young's Modulus}$$

- A shearing strain occurs when an angular deformation ϕ is produced:

$$S = \frac{F/A}{\tan \phi}$$

or

$$S = \frac{F/A}{d/l} \qquad \textit{Shear Modulus}$$

- Whenever an applied stress results in a change in volume ΔV, you will need the *bulk modulus B,* given by

$$B = -\frac{F/A}{\Delta V/V} \qquad \textit{Bulk Modulus}$$

The reciprocal of the bulk modulus is called the compressibility.

Key Terms

Review Questions

13.1. Explain clearly the relationship between stress and strain.

13.2. Two wires have the same length and cross-sectional area but are not of the same material. Each wire is hung from the ceiling with a 2000-lb weight attached. The wire on the left stretches twice as far as the one on the right. Which has the greater Young's modulus?

13.3. Does Young's modulus depend on the length and cross-sectional area? Explain.

13.4. Two wires, A and B, are made of the same material and are subjected to the same loads. Discuss their relative elongations when (a) wire A is twice as long as wire B and has twice the diameter of wire B and (b) wire A is twice as long as wire B and has one-half the diameter of wire B.

13.5. After studying the various elastic constants given in Tables 13.1 and 13.2, would you say that it was usually easier to stretch a material or to shear a material? Explain.

13.6. A 200-kg mass is supported evenly by three wires, one of copper, one of aluminum, and one of steel. If the wires have the same dimensions, which wire experiences the greatest stress, and which the least? Which has the greatest strain and the least strain?

13.7. Discuss the various stresses resulting when a machine screw is tightened.

13.8. Give several practical examples of longitudinal, shearing, and volume strains.

13.9. For a given metal, would you expect there to be any relationship between its modulus of elasticity and its coefficient of restitution? Discuss.

13.10. Which has the greater compressibility, steel or water?

Problems

Section 13.1 Elastic Properties of Matter

13.1. When a mass of 500 g is hung from a spring, the spring stretches 3 cm. What is the spring constant?
Ans. 163 N/m

13.2. What will be the increase in stretch for the spring of Prob. 13.1 if an additional 500-g mass is hung below the first?

13.3. The spring constant for a certain spring is found to be 3000 N/m. What force is required to compress the spring for a distance of 5 cm?
Ans. 150 N

13.4. A 6-in. spring has a 4-lb weight hung from one end, causing the new length to be 6.5 in. What is the spring constant? What is the strain?

13.5. A coil spring 12 cm long is used to support a 1.8-kg mass producing a strain of 0.10. How far did the spring stretch? What is the spring constant?
Ans. 1.2 cm, 1470 N/m

13.6. For the coil of Prob. 13.5, what total mass should be hung if an elongation of 4 cm is desired?

Section 13.2 Young's Modulus

13.7. A 60-kg mass is suspended by means of a cable that has a diameter of 9 mm. What is the stress?
Ans. 9.24 Mpa

13.8. A 50-cm length of wire is stretched to a new length of 50.01 cm. What is the strain?

13.9. A 12-m rod receives a compressional strain of −0.0004. What is the new length of the rod?
Ans. 11.995 m

13.10. Young's modulus for a certain rod is 4×10^{11} Pa. What strain will be produced by a tensile stress of 420 Mpa?

13.11. A 500-kg mass is hung from the end of a 2-m length of metal wire 1 mm in diameter. If the wire stretches by 1.40 cm, what are the stress and strain? What is Young's modulus for this metal?
Ans. 6.24×10^9 Pa, 7.00×10^{-3}, 8.91×10^{11} Pa

13.12. A 16-ft steel girder with a cross-sectional area of 10 in.^{2} supports a compressional load of 20 tons. What is the decrease in length of the girder?

13.13. How much will a 60-cm length of brass wire, 1.2 mm in diameter, elongate when a 3-kg mass is hung from an end? Ans. 0.174 mm

13.14. A wire of cross section 4 mm^2 is stretched 0.1 mm by a certain weight. How far will a wire of the same material and length stretch if its cross-sectional area is 8 mm^2 and the same weight is attached?

13.15. The compressive stress on the human thigh bone in Fig. 13.9 resembles that exerted on the cross section of a hollow cylinder. If the maximum

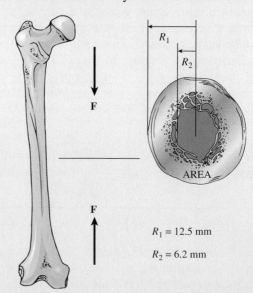

$R_1 = 12.5$ mm

$R_2 = 6.2$ mm

Figure 13.9 Compressive stress on the thigh bone (femur).
(*Photo by Hemera, Inc.*)

stress that can be sustained is 172 MPa, what is the force required to rupture the bone at the narrowest portion. Use the dimensions given in the figure.   Ans. 63.7 kN

Section 13.3 Shear Modulus

13.16. A shearing force of 40,000 N is applied to the top of a cube that is 30 cm on a side. What is the shearing stress?

13.17. If the cube in Prob. 13.16 is made of copper, what will be the lateral displacement of the upper surface of the cube?   Ans. 3.15 μm

13.18. A shearing force of 26,000 N is distributed uniformly over the cross section of a pin 1.3 cm in diameter. What is the shearing stress?

13.19. An aluminum rod 20 mm in diameter projects 4.0 cm from the wall. The end of the rod is subjected to a shearing force of 48,000 N. Compute the downward deflection.   Ans. 0.257 mm

13.20. A steel rod projects 1.0 in. above a floor and is 0.5 in. in diameter. The shearing force $\mathbf{F}$ is 6000 lb, and the shear modulus is 11.6×10^6 lb/in.^{2}. What is the shearing stress, and what is the horizontal deflection?

13.21. A 1500-kg load is supported at the end of a 5-m aluminum beam as shown in Fig. 13.10. The beam has a cross-sectional area of 26 cm^2, and the shear modulus is 23,700 MPa. What are the shearing stress and the downward deflection of the beam?   Ans. 5.65×10^6 Pa, 1.19 mm

13.22. A steel plate 0.5 in. thick has an ultimate shearing strength of 50,000 lb/in.^{2}. What force must be applied to punch a $\frac{1}{4}$-in. hole through the plate?

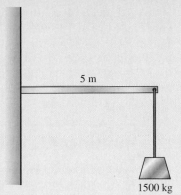

Figure 13.10

Section 13.4 Volume Elasticity; Bulk Modulus

13.23. A pressure of 3×10^8 Pa is applied to a block of volume 0.500 m^3. If the volume decreases by 0.004 m^3, what is the bulk modulus? What is the compressibility?
                Ans. 37,500 MPa, 2.67×10^{-11} Pa^{-1}

13.24. The bulk modulus for a certain grade of oil is 2.8×10^{10} Pa. What pressure is required to decrease its volume by a factor of 1.2 percent?

13.25. A solid brass sphere ($B = 35,000$ Mpa) of volume 0.8 m^3 is dropped into the ocean to a depth where the water pressure is 20 Mpa greater than it is at the surface. What is the change in volume of the sphere?   Ans. -4.57×10^{-4} m^3

13.26. A certain fluid compresses 0.40 percent under a pressure of 6 MPa. What is the compressibility of this fluid?

13.27. What is the fractional decrease in the volume of water when it is subjected to a pressure of 15 MPa?   Ans. -0.00714

Additional Problems

13.28. A 10-m steel wire, 2.5 mm in diameter, stretches a distance of 0.56 mm when a load is attached to its end. What was the mass of the load?

13.29. A shearing force of 3000 N is applied to the upper surface of a copper cube 40 mm on a side. Assume $S = 4.2 \times 10^{10}$ Pa. What is the shearing angle?   Ans. 4.46×10^{-5} rad

13.30. A solid cylindrical steel column is 6 m long and 8 cm in diameter. What is the decrease in length if the column supports a 90,000-kg load?

13.31. A piston, 8 cm in diameter, exerts a force of 2000 N on 1 liter of benzene. What is the decrease in volume of the benzene?   Ans. -3.79×10^{-7} m^3

13.32. How much will a 600-mm length of brass wire, 1.2 mm in diameter, stretch when a 4-kg mass is hung from its end?

13.33. A solid cylindrical steel column is 12 ft tall and 6 in. in diameter. What load does it support if its decrease in length is -0.0255 in.?
                           Ans. 1.50×10^5 lb

13.34. Compute the volume contraction of mercury if its original volume of 1600 cm^3 is subjected to a pressure of 400,000 Pa.

13.35. What is the minimum diameter of a brass rod if it is to undergo a 400-N tension without exceeding the elastic limit?   Ans. 1.16 mm

13.36. A cubical metal block 40 cm on a side is given a shearing force of 400,000 N at the top edge. What is the shear modulus for this metal if the upper edge deflects a distance of 0.0143 mm?

13.37. A steel piano wire has an ultimate strength of about 35,000 lb/in.². How large a load can a 0.5-in.-diameter steel wire hold without breaking?   Ans. 6870 lb

Critical Thinking Questions

13.38. A metal wire increases its length by 2 mm when subjected to tensile force. What elongation could be expected from this same force if the diameter of the wire was reduced to one-half of its initial value? Suppose the metal wire maintains its diameter, but doubles its length. What elongation would be expected for the same load?
                Ans. 0.500 mm, 4 mm

13.39. A cylinder 4 cm in diameter is filled with oil. What total force must be exerted on the oil to produce a 0.8 percent decrease in volume? Compare the forces necessary if the oil is replaced by water. By mercury.

*13.40. A 15-kg ball is connected to the end of a steel wire 6 m long and 1.0 mm in diameter. The other end of the wire is connected to a high ceiling, forming a pendulum. If we ignore the small change in length, what is the maximum speed that the ball can have as it passes through its lowest point without exceeding the elastic limit? How much will the length of the wire increase under the limiting stress? What effect will this change have on the maximum velocity?
                Ans. 4.38 m/s, 7.19 mm, slightly increases maximum velocity

13.41. A cylinder 10 in. in diameter is filled to a height of 6 in. with glycerin. A piston of the same diameter pushes downward on the liquid with a force of 800 lb. The compressibility of glycerin is 1.50×10^{-6} in.²/lb. What is the stress on the glycerin? How far down does the piston move?

*13.42. The twisting of a cylindrical shaft (Fig. 13.11) through an angle θ is an example of a shearing strain. An analysis of the situation shows that the angle of twist in radians is given by

$$\theta = \frac{2\tau l}{\pi S R^4}$$

where τ = applied torque
   l = length of cylinder
   R = radius of cylinder
   S = shear modulus

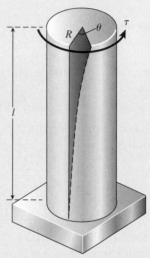

Figure 13.11 A torque τ applied at one end of a solid cylinder causes it to twist through an angle θ.

If a torque of 100 lb · ft is applied to the end of a cylindrical steel shaft 10 ft long and 2 in. in diameter, what will be the angle of twist in radians?
                Ans. 0.00764 rad

*13.43. An aluminum shaft 1 cm in diameter and 16 cm high is subjected to a torsional shearing stress as explained in Prob. 13.42. What applied torque will cause a twist of 1° as defined in Fig. 13.11?

*13.44. Two sheets of aluminum on an aircraft wing are to be held together by aluminum rivets of cross-sectional area 0.25 in.². The shearing stress on each rivet must not exceed one-tenth of the elastic limit for aluminum. How many rivets are needed if each rivet supports the same fraction of the total shearing force of 25,000 lb?
                Ans. 53 rivets

14

Simple Harmonic Motion

A trampoline exerts a restoring force on the jumper that is directly proportional to the average force required to displace the mat. Back and forth motion, due to restoring forces, provides the energy necessary for simple harmonic motion.
(*Photo by Mark Tippens.*)

Objectives

After completing this chapter, you should be able to

1. Define *simple harmonic motion* and describe and apply Hooke's law and Newton's second law to determine acceleration as a function of displacement.

2. Apply the principles of conservation of mechanical energy for a mass moving with simple harmonic motion.

3. Use the reference circle to describe the variation in magnitude and direction of displacement, velocity, and acceleration for simple harmonic motion.

4. Write and apply formulas for the determination of displacement x, velocity v, or acceleration a in terms of time, frequency, and amplitude.

5. Write and apply a relationship between the frequency of motion and the mass of a vibrating object when the spring constant is known.

279

6. Compute the frequency or period in simple harmonic motion when the position and acceleration are given.

7. Describe the motion of a simple pendulum and calculate the length required to produce a given frequency.

Until now, we have discussed the motion of objects under the influence of a constant, unchanging force. Such motion was described by calculating the position and velocity as functions of time. The real world often consists, however, of varying forces. Common examples are swinging pendulums, balance wheels of watches, tuning forks, a mass vibrating at the end of a coiled spring, and vibrating air columns in musical instruments. In these cases and many others, we need a more complete description of motion caused by a resultant force that varies in a predictable manner.

| **14.1** | **Periodic Motion** |

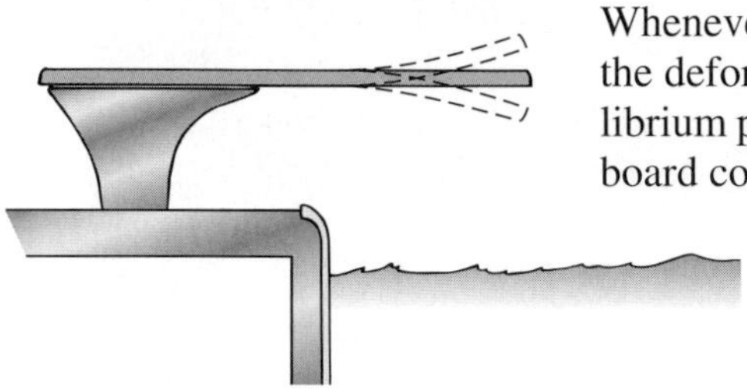

Whenever an object is deformed, an elastic restoring force appears that is proportional to the deformation. When released, such an object will vibrate back and forth about its equilibrium position. For example, after a swimmer springs from a diving board (Fig. 14.1), the board continues to vibrate about its normal position for a definite length of time.

This type of motion is said to be *periodic* because the position and velocity of the moving particles are repeated as a function of time. Since the restoring force is reduced after each vibration, the board eventually comes to rest.

Figure 14.1   The periodic vibration of a diving board.

Periodic motion is that motion in which a body moves back and forth over a fixed path, returning to each position and velocity after a definite interval of time.

An air table is a laboratory device on which objects may slide with very little friction. Suppose we attach one end of a light spring to the wall and the other end to a circular disk that is free to slide on an air table, such as shown in Fig. 14.2. We will represent its initial equilibrium position at $x = 0$, and then we will stretch it to the right a distance $x = A$. When released, we observe that the disk vibrates back and forth through the equilibrium position with negligible friction. According to Hooke's law (see Chapter 13), the restoring

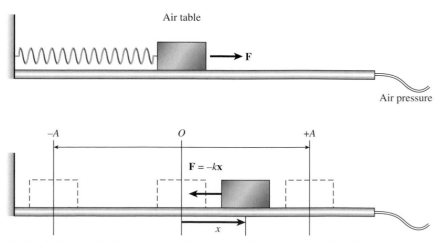

Figure 14.2   A disk attached to a spring glides back and forth on an air table, illustrating simple harmonic motion.

force **F** is directly proportional to the displacement from $x = 0$. This restoring force is always *apposed* to the displacement, so Hooke's law can be written as

$$F = -kx \qquad Hooke's \ Law \qquad (14.1)$$

The maximum displacement from its equilibrium position $x = \pm A$ is called the *amplitude.* At this position, the disk experiences its maximum force directed toward the center of oscillation. The force decreases as the disk approaches the center, becoming zero at that point. The momentum of the disk carries it past the midpoint, and the force then reverses its direction, slowing the motion until the disk reaches its amplitude in the other direction and the oscillation continues. In the absence of friction, such motion would continue indefinitely. This type of oscillatory motion in the absence of friction is known as *simple harmonic motion (SHM).*

> Simple harmonic motion (SHM) is periodic motion in the absence of friction and is produced by a restoring force that is directly proportional to the displacement and oppositely directed.

The *period T* is defined as the time for one complete oscillation for an object moving with SHM. Consider, for example, the mass attached to the end of a vertical spring, as shown in Fig. 14.3. If we pull the mass downward a distance $y = -A$ and then release it, the motion of the mass will approximate SHM. The time from its release at $y = -A$ until it returns to $y = -A$ represents the time for one complete oscillation (the period). Actually, we could select any position y during the vibration, and the period would be the time to return to that point *moving in the same direction.* It should be noted that the time required to move from the center of oscillation to the maximum displacement A in either direction is only one-fourth of the period.

For example, suppose the mass vibrating in Fig. 14.3 has a period equal to 4 s. At $t = 0$, it passes the center of oscillation in an upward direction. After 1 s, it will be located at $y = +A$. After 2 s, it will be passing through the center again. However, it will be moving in the *downward* direction, so the time is equal to only one-half of the period. The full period of 4 s is only realized when it returns to $y = 0$ *and* it is also moving in the same direction as it was when $t = 0$.

The *frequency f* is the number of complete oscillations per unit of time. Since the period is the number of seconds per oscillation, it follows that frequency will be the reciprocal of the period or the number of oscillations per second.

$$f = \frac{1}{T} \qquad (14.2)$$

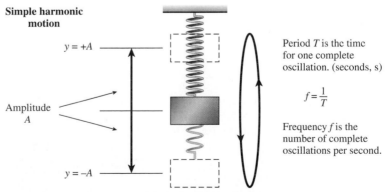

Figure 14.3 Simple harmonic motion (SHM) is periodic motion with a constant amplitude, frequency, and period.

The SI unit for frequency (oscillations/second) is the **hertz** (Hz)

$$\text{The Hertz} \qquad 1 \text{ Hz} = \frac{1}{s} = s^{-1}$$

Thus, a frequency of 20 oscillations per second would be written as 20 Hz.

Example 14.1

The suspended mass in Fig. 14.3 is pulled downward and then released to vibrate with SHM. A student determines that the time for 50 complete vibrations is 74.1 s. What are the period and frequency of the motion?

Plan: The given time is for 50 vibrations. We recognize the period is the time for *one* vibration and the frequency is the reciprocal of the period.

Solution: Dividing the total time by the total vibrations, we have

$$T = \frac{74.1 \text{ s}}{50 \text{ vib}} = 1.48 \text{ s}$$

Finally, from Eq. (13.2), we find the frequency.

$$f = \frac{1}{T} = \frac{1}{1.48 \text{ s}}; \qquad f = 0.675 \text{ Hz}$$

You should recognize that a *vibration* or an *oscillation* is a dimensionless unit such that vib/s becomes just s^{-1} or Hz.

Example 14.2

A light spring is attached to the ceiling, and its lowest position is marked on a meterstick. When a 3-kg mass is attached to bottom of the spring, it moves downward a vertical distance of 12 cm. Find the spring constant.

Plan: According to Hooke's law, the spring constant is the ratio of the change in force to the change in displacement. Note that the spring constant is an *absolute* quantity. The negative sign in Hooke's law indicates that the direction of the restoring force is opposite to the displacement.

Solution: The change in force is equal to the weight of the mass mg so that $\Delta F = mg$ and $\Delta x = 0.12$ m. From Hooke's law, we find that

$$k = \frac{\Delta F}{\Delta x} = \frac{mg}{x}$$

$$= \frac{(3 \text{ kg})(9.8 \text{ m/s}^2)}{0.12 \text{ m}} = 242 \text{ N/m}$$

The delta symbols in the previous calculation are important. It is the *change* in force and the *change* in displacement that determine the spring constant. In this example, if we add a second 3-kg mass to the bottom of the first, the spring will move downward another 12 cm.

Later, in Sec. 14.7, we will show that the period and frequency for a system vibrating with SHM can be determined from the mass and the spring constant.

14.2 Newton's Second Law and Hooke's Law

The restoring force for a system vibrating with SHM must obey Hooke's law, but any resultant force must also conform to Newton's second law of motion. Thus, the acceleration of a vibrating mass must be proportional to both the resultant force and to the displacement.

$$F = ma \quad \text{and} \quad F = -kx$$

Combining these two relations gives

$$ma = -kx$$

so the acceleration of a mass m moving with SHM is given by

$$a = -\frac{k}{m}x \tag{14.3}$$

The negative sign indicates that the acceleration (and the restoring force) is always directed opposite to the displacement. If the displacement is downward, the acceleration is upward; if the displacement is to the right, the acceleration is to the left.

Example 14.3

Suppose the circular disk in Fig. 14.4 has a mass of 1.5 kg and is pulled outward a distance of 12 cm and released to oscillate with SHM on the air table. The spring constant is 120 N/m. (a) What are the magnitude and direction of the acceleration? (b) What is the force on the mass when it is located at the following displacements: (a) +12 cm, (b) +8 cm, and (c) −4 cm?

Plan: The acceleration for each displacement is found from Eq. (14.3). The force can then be found from either Hooke's law or Newton's second law. However, we must be careful of signs because both the force and the acceleration must be opposite to the displacement. We will choose the rightward direction as positive.

Solution (a): We first calculate the acceleration and force for $x = +12$ cm $= 0.12$ m, which should represent the *maximum* acceleration and restoring force since mass is at the amplitude A.

$$a = -\frac{kx}{m} = -\frac{(120 \text{ N/m})(+0.12 \text{ m})}{1.5 \text{ kg}}; \quad a = -9.6 \text{ m/s}^2$$

$$F = -kx = -(120 \text{ N/m})(+0.12 \text{ m}); \quad F = -14.4 \text{ N}$$

Note that both force and acceleration are directed to the left at this displacement.

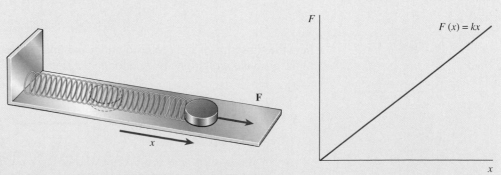

Figure 14.4 The work done by the stretching force **F** when it displaces the mass a distance x from its equilibrium position.

Solution (b): When $x = +8$ cm $= 0.08$ m, the acceleration and force are

$$a = -\frac{kx}{m} = -\frac{(120 \text{ N/m})(+0.08 \text{ m})}{1.5 \text{ kg}}; \qquad a = -6.4 \text{ m/s}^2$$

$$F = -kx = -(120 \text{ N/m}) + (0.08 \text{ m}); \qquad F = -9.6 \text{ N}$$

Solution (c): When $x = -4$ cm $= -0.04$ m, we obtain

$$a = -\frac{kx}{m} = -\frac{(120 \text{ N/m})(-0.04 \text{ m})}{1.5 \text{ kg}}; \qquad a = +3.2 \text{ m/s}^2$$

$$F = -kx = -(120 \text{ N/m})(-0.04 \text{ m}); \qquad F = +4.8 \text{ N}$$

Notice in the last example, the position of the disk is on the left side of the equilibrium position, meaning the spring is compressed, exerting a rightward restoring force. The signs of the answers give an indication of their directions.

14.3 Work and Energy in Simple Harmonic Motion

Suppose we consider the work done in stretching a spring, such as the one in Fig. 14.5. An external force **F** acts through a distance x in compressing the spring. This work is positive and equal to the product Fx. At the same time, the spring exerts an equal and opposite force (against the pusher) doing the same amount of *negative* work. If we plot a graph of the force **F** as a function of displacement x, it can be shown that the work done by this force is equal to $\frac{1}{2}kx^2$, which means that the potential energy U stored in the spring is given by

$$\text{Potential Energy} \qquad U = \frac{1}{2}kx^2 \qquad \qquad \textbf{(14.4)}$$

When a compressed spring is released, the potential energy is converted to kinetic energy ($\frac{1}{2}mv^2$) as the attached mass gains velocity. Assuming there is no friction, the final kinetic energy will be equal to the initial potential energy. Potential energy is stored in the spring only when it is either compressed or extended. Kinetic energy exists only if the mass has velocity.

Remember that the total energy ($U + K$) of a system does not change. Therefore, in the absence of friction, we write

$$\text{Conservation of Energy} \qquad U_0 + K_0 = U_f + K_f$$

$$\frac{1}{2}kx_0^2 + \frac{1}{2}mv_0^2 = \frac{1}{2}kx_f^2 + \frac{1}{2}mv_f^2 \qquad \qquad \textbf{(14.5)}$$

where the subscripts 0 and f refer to initial and final values. If friction is involved, we must add the absolute work done by friction to the right side of this equation.

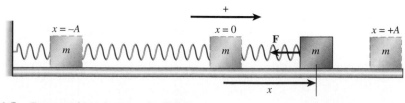

Figure 14.5 Conservation of energy for SHM.

We are now ready to consider the conservation of energy for a mass m oscillating with SHM, as illustrated in Fig. 14.5. Basically, at *any* point during the oscillation, the total energy ($E = U + K$) is

$$E = \frac{1}{2}kx^2 + \frac{1}{2}mv^2$$

Consider the total energy E for each of the following cases:

At $x = \pm A$: $E = \frac{1}{2}kA^2 + \frac{1}{2}m(0)^2$ or $E = \frac{1}{2}kA^2$

At $x = 0$: $E = \frac{1}{2}k(0)^2 + \frac{1}{2}mv_{max}^2$ or $E = \frac{1}{2}mv_{max}^2$

At $x = x$: $E = \frac{1}{2}kx^2 + \frac{1}{2}mv^2$

Next, we will derive an expression for finding the velocity v of a mass moving with SHM and no friction. Since the total energy at any point is the same as it is at the amplitude, we can write

$$\frac{1}{2}kx^2 + \frac{1}{2}mv^2 = \frac{1}{2}kA^2$$

Solving for the velocity v, we find that

$$v = \sqrt{\frac{k}{m}(A^2 - x^2)} \qquad \textbf{(14.6)}$$

Note, for the special case where $x = 0$, the velocity is a maximum and equal to

$$v_{max} = \sqrt{\frac{k}{m}}A \qquad \textbf{(14.7)}$$

Equations (14.6) and (14.7) are useful in repetitive calculations, but in most cases, it is better to just apply the conservation equation (14.5) because it is much easier to remember. Since energy is a scalar quantity, we do not learn the *direction* of the velocity from these equations. The square root of a number may be positive or negative.

Example 14.4

A 0.4-kg mass is connected to a light spring and oscillates with SHM along a frictionless surface, as in Fig. (14.5). The spring constant is 20 N/m, and the amplitude is 5 cm. (a) What is the maximum velocity of the mass? (b) What is the velocity when the mass is located a distance of +3 cm to the right of the equilibrium position?

Plan: The total energy is conserved, so each of these questions can be answered by applying Eq. (14.4) for the given locations. We recognize that the velocity is a maximum when $x = 0$ because the restoring force has been in a consistent direction for the longest period of time at that point. The velocity at the location $x = +3$ cm can be found by recognizing that the total energy at that position is the same as the total energy at either amplitude ($\frac{1}{2}kA^2$). It will be useful to organize the data before solving for v.

Given: $A = 0.05$ m, $x = 0$ and $+0.03$ m, $m = 0.4$ kg, $k = 20$ N/m

Solution (a): The maximum velocity occurs for $x = 0$, so conservation of energy requires that

$$\frac{1}{2}k(0)^2 + \frac{1}{2}mv_{max}^2 = \frac{1}{2}kA^2 \qquad \text{or} \qquad \frac{1}{2}mv_{max}^2 = \frac{1}{2}kA^2$$

Solving for v_{max}, yields

$$v_{max} = \sqrt{\frac{k}{m}}A = \sqrt{\frac{20\ N/m}{0.4\ kg}}(0.05\ m); \qquad v_{max} = \pm 0.354\ m/s$$

We do not know from this result whether the mass is moving to the right or to the left as it passes through $x = 0$.

Solution (b): The velocity at $x = +0.03$ m is found from the conservation equation.

$$\frac{1}{2}kx^2 + \frac{1}{2}mv_{max}^2 = \frac{1}{2}kA^2$$

$$\frac{1}{2}(20\ N/m)(0.03\ m)^2 + \frac{1}{2}(0.4\ kg)v^2 = \frac{1}{2}(20\ N/m)(0.05\ m)^2$$

Solving for v, we obtain

$$v = \pm 0.283\ m/s$$

Direct substitution into Eq. (14.6) gives the same answer. Once again, this tells us only the *speed* of the mass at this location. It may be moving to the right or to the left at the instant when it is located 3 cm to the right of center.

14.4 The Reference Circle and Simple Harmonic Motion

The laws for uniform acceleration do not apply for SHM because of the existence of a varying force. Simple harmonic motion results from a varying force that is proportional to the displacement. You will recall from Eq. (14.3) that

$$a = -\frac{k}{m}x$$

As long as the mass remains constant, the acceleration increases with the displacement and is always directed opposite to the displacement.

To determine new relationships that will allow us to predict the position, velocity, and displacement as a function of time, we must resort to calculus. Fortunately, these equations can also be deduced by comparing SHM to the periodic revolution of a mass about a given radius. Consider the apparatus shown in Fig. 14.6, where the shadow of a ball attached to a rotating disk moves back and forth with simple harmonic motion. The experiment suggests that our knowledge of uniform circular motion may be useful in describing SHM.

The *reference circle* in Fig. 14.7 is used to compare the motion of an object moving in a circle with its horizontal projection. Since it is the motion of the projection that we wish to study, we shall refer to the position P of the object moving in a circle as the *reference point*. The radius of the reference circle is equal to the amplitude of the horizontal oscillation. If the linear speed v_T and angular velocity ω of the reference point are constant, the projection Q will move back and forth with SHM. Time is assigned a zero value when the reference point is at B in Fig. 14.7. At any later time t, the reference point P will have moved through an angle θ. The displacement x of the projection Q is therefore

$$x = A \cos \theta$$

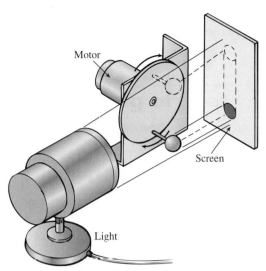

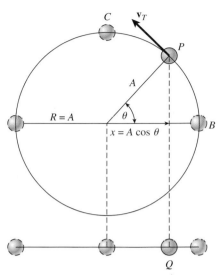

Figure 14.6 The projection or shadow of a ball attached to a rotating disk moves with simple harmonic motion.

Figure 14.7 Displacement in simple harmonic motion.

Recalling that the angle $\theta = \omega t$, we can now write the displacement as a function of the angular velocity of the reference point.

$$x = A \cos \theta = A \cos \omega t \qquad \textbf{(14.8)}$$

Although the angular velocity ω is useful for describing motion of the reference point P, it is not applied directly to the projection Q. We should recall, however, that the angular velocity is related to the frequency of revolution by

$$\omega = 2\pi f$$

where ω is expressed in radians per second and f is the number of revolutions per second. It should also be recognized that the projection Q will describe one complete oscillation, whereas the reference point describes one complete revolution. Thus, the frequency f is the same for each point. Substituting $\omega = 2\pi f$ into Eq. (14.8), we obtain

$$x = A \cos 2\pi f t \qquad \textbf{(14.9)}$$

This equation can be applied to compute the displacement of a body moving with SHM of amplitude A and frequency f. Remember that the displacement x is always measured from the center of oscillation.

14.5   Velocity in Simple Harmonic Motion

Let us consider a body moving back and forth with SHM under the influence of a restoring force. Since the direction of the vibrating body is reversed at the end points of its motion, its velocity must be zero when its displacement is a maximum. It is then accelerated toward the center by the restoring force until it reaches its maximum speed at the center of oscillation, when its displacement is zero.

In Fig. 14.8, the velocity of a vibrating body is compared at three instants with corresponding points on a reference circle. It will be noticed that the velocity $\mathbf{v}$ of the vibrating body at any instant is the horizontal component of the tangential velocity $\mathbf{v}_T$ of the reference point. At point B, the reference point is moving vertically upward and has no horizontal velocity. This point therefore corresponds to the zero velocity of the vibrating body when it reaches its amplitude A. At point C, the horizontal component of $\mathbf{v}_T$ is equal to its entire magnitude. This point corresponds to a position of maximum velocity for the vibrating

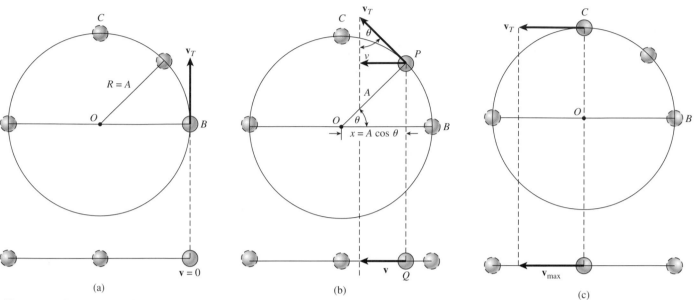

Figure 14.8 Velocity and the reference circle.

body, in other words, at its center of oscillation. In general, the velocity of the vibrating body at any point Q is determined from the reference circle to be

$$v = -v_T \sin\theta = -v_T \sin\omega t \qquad (14.10)$$

The minus sign appears since the direction of the velocity is toward the left. We can make this equation more useful by recalling the relationship between the tangential velocity v_T and the angular velocity:

$$v_T = \omega A = 2\pi f A$$

Substitution into Eq. (14.10) yields

$$v = -2\pi f A \sin 2\pi f t \qquad (14.11)$$

This equation will give the velocity of a vibrating body at any instant if it is remembered that $\sin\theta$ is negative when the reference point lies below the diameter of the reference circle.

Example 14.5

A mass m is attached to a spring, as in Fig. 14.4, is pulled to the right a distance of 6 cm and released. It returns to the point of release in a time of 2 s and continues to vibrate with SHM. (a) What is its maximum velocity? (b) What is its position and velocity 5.2 s after its release?

Plan: First, we recognize that the 2 s for the first complete oscillation is the *period* of the motion. The frequency is the reciprocal of the period, so $f = 0.5$ Hz. (If one oscillation takes 2 s, then there is one-half an oscillation every second.) We will organize the given information and decide which equations will involve the given quantities. The first maximum for velocity occurs when the displacement is zero, which corresponds to 90° on the reference circle. The position and velocity 5.2 s after release can be found from Eq. (14.9) and Eq. (14.11). Conservation of energy will not help here, because we do not know the spring constant or the position.

Solution (a): We are given that $f = 0.5$ Hz, $A = 0.06$ m, and $\theta = 90°$. The maximum velocity is found by substitution into Eq. (14.9). Recall that $\sin 90° = 1$.

$$v_{max} = -2\pi fA \sin 90° = -2\pi fA$$
$$= -2\pi(0.5 \text{ Hz})(0.06 \text{ m})$$
$$= -0.188 \text{ m/s}$$

The negative sign means that the first maximum velocity is -18.8 cm/s directed to the left. If we had substituted $270°$ for the angle θ, the maximum velocity would be $+18.8$ m/s, directed to the right.

Solution (b): Here, we are asked to find the position and velocity as at a particular time (5.2 s). When the reference angle θ is written as $2\pi ft$, it is important to remember that the angles must be in *radians—not* in degrees. Makes sure your calculator is set to read angles in radians. Also, a small error in the radian measure is significant, so be sure not to round your data until you reach the final answer. The displacement at $t = 5.2$ s is found from Eq. (14.9).

$$x = A\cos(2\pi ft) = (0.06 \text{ m}) \cos[2\pi(0.5 \text{ Hz})(5.2 \text{ s})]$$
$$= (0.06 \text{ m}) \cos(16.34 \text{ rad}) = (0.06 \text{ m})(-0.809)$$
$$= -0.0485 \text{ m} = -4.85 \text{ cm}$$

The velocity is found from Eq. (14.11) using the same angle in radians.

$$v = -2\pi fA \sin(16.34 \text{ rad})$$
$$= -2\pi(0.5 \text{ Hz})(0.06 \text{ m})(-0.588)$$
$$= +0.122 \text{ m} = +11.2 \text{ cm/s}$$

Note that the velocity after 5.2 s is positive, indicating that the mass is moving to the right at that instant.

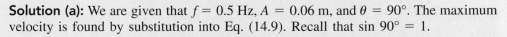

14.6 Acceleration in Simple Harmonic Motion

The velocity of a vibrating body is never constant. Therefore, acceleration plays an important role in the equations derived for position and velocity in the previous sections. We already have an expression for predicting the acceleration as a function of distance, and now we will derive relationship for time.

At the position of maximum displacement ($\pm A$), the velocity of a vibrating mass is zero. It is at this instant that the mass is acted on by the maximum restoring force. Thus, the acceleration of the mass is a maximum when its velocity is zero. As the mass approaches its equilibrium position, the restoring force (and therefore the acceleration) is reduced until it reaches zero at the center of oscillation. At this equilibrium position, the acceleration is zero, and the velocity is a maximum.

Let's look at the reference circle in Fig. 14.9, which is drawn for studying the acceleration a of a particle moving with SHM. Note that the centripetal acceleration a_c of a mass moving in a circle of radius $R = A$, is compared to the acceleration of its shadow. The acceleration a of the shadow represents of SHM, and it is equal to the horizontal component of the centripetal acceleration a_c of the mass. From the figure,

$$a = -a_c \cos\theta = -a_c \cos\omega t \qquad (14.12)$$

where $\omega = 2\pi f$. The negative sign shows that the acceleration is opposite to the displacement but consistent with the direction of velocity.

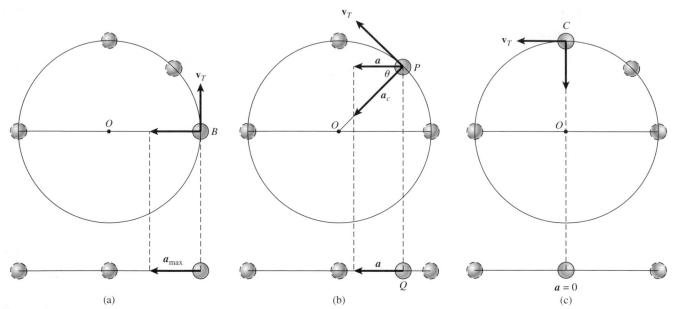

Figure 14.9 Acceleration and the reference circle.

From our study of rotation and circular motion, we recall that

$$v = \omega R \qquad \text{and} \qquad a_c = \frac{v^2}{R}$$

Combining these relations yields

$$a_c = \frac{(\omega R)^2}{R} \qquad \text{or} \qquad a_c = \omega^2 R$$

Since $a_c = \omega^2 R$ and $R = A$, we can rewrite Eq. (14.12) as

$$a = -\omega^2 A \cos \omega t$$

This relation expresses the acceleration of a body moving with SHM with amplitude A and angular frequency ω (in rad/s). The same equation expressed in terms of the frequency f (in Hz) can be found by substituting $\omega = 2\pi f$ to obtain

$$a = -4\pi^2 f^2 A \cos(2\pi ft) \tag{14.13}$$

We can simplify this expression even more by noting from Eq. (14.9) that

$$\cos\theta = \cos(2\pi ft) = \frac{x}{A}$$

Thus, Eq. (14.13) becomes

$$a = -4\pi^2 f^2 A \frac{x}{A}$$

or

$$a = -4\pi^2 f^2 x \tag{14.14}$$

We see that the acceleration is directly proportional to the displacement and opposite in direction, as it must be to conform with Hooke's law.

14.7 | The Period and Frequency

From the information now established concerning displacement, velocity, and acceleration of vibrating bodies, we can derive some useful formulas for computing the period or frequency of vibration. For example, if we solve Eq. (14.14) for the frequency f, we obtain

$$f = \frac{1}{2\pi}\sqrt{-\frac{a}{x}} \tag{14.15}$$

Since the ***displacement*** x and the acceleration are always opposite in signs, the term $-a/x$ is always positive.

The period T is the reciprocal of the frequency. Making use of this fact in Eq. (14.15), we define the period by

$$T = 2\pi\sqrt{-\frac{x}{a}} \tag{14.16}$$

Thus, if the acceleration is known at a particular displacement, the period of vibration can be computed.

When the motion of bodies under the influence of an elastic ***restoring force*** is considered, it is more convenient to express the period as a function of the spring constant and mass of the vibrating body. This can be done by comparing Eqs. (14.3) and (14.14):

$$a = -\frac{k}{m}x \qquad a = -4\pi^2 f^2 x$$

Combining these relations, we obtain

$$4\pi^2 f^2 = \frac{k}{m}$$

from which the frequency is

$$f = \frac{1}{2\pi}\sqrt{\frac{k}{m}} \tag{14.17}$$

Finally, the period T is given by the reciprocal of the frequency. Thus,

$$T = 2\pi\sqrt{\frac{m}{k}} \tag{14.18}$$

Note that neither the period nor the frequency depends on the amplitude (maximum displacement) of the vibrating body. They depend only on the spring constant and the mass of the vibrating body.

PHYSICS TODAY

Balconies and overhanging walkways must be carefully engineered with respect to their resonant frequency. People tapping their feet or marching can cause these structures to resonate and shake. The total height of the wave can then be greater than what would result from just the weight of the people added to the weight of the structure.

Example 14.6

A 0.2-kg steel ball is attached to the end of a flat strip of metal that is clamped at its base, as shown in Fig. 14.10. If a force of 5 N is required to displace the mass by 3 cm, what will its period of vibration be upon release? What is its maximum acceleration?

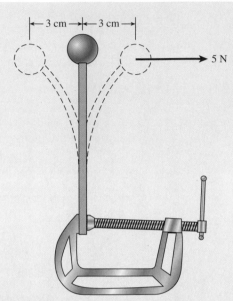

Figure 14.10

Plan: First, we will find the spring constant k from Hooke's law and the fact that a force of 5 N displaces the mass by 3 cm. The maximum acceleration occurs when the displacement is a maximum (at $x = 3$ cm).

Solution: From Hooke's law,

$$k = \frac{F}{x} = \frac{5 \text{ N}}{0.3 \text{ m}}; \qquad k = 167 \text{ N/m}$$

Now we substitute $k = 167$ N/m and $m = 0.2$ kg into Eq. (14.18) to find the period T.

$$T = 2\pi \sqrt{\frac{m}{k}} = 2\pi \sqrt{\frac{0.2 \text{ kg}}{167 \text{ N/m}}}$$

$$T = 0.218 \text{ s}$$

We recall that the frequency f is equal to $1/T$, and the maximum acceleration is found by substituting $x = \pm A = \pm 0.03$ m into Eq. (14.14).

$$a = -4\pi^2 f^2 x = -\frac{4\pi^2 A}{T^2}$$

$$a = -\frac{4\pi^2 (0.03 \text{ m})}{(0.218 \text{ s})^2}; \qquad a = -25.0 \text{ m/s}^2$$

Note that the negative sign results because we used the positive sign for the amplitude. When the ball reaches the left side, $x = -0.03$ m, and the acceleration will be $+25$ m/s^2.

14.8 | The Simple Pendulum

When a heavy pendulum bob is swinging at the end of a light cord or rod, as in Fig. 14.11, it approximates SHM. If we assume that all the mass is concentrated at the center of gravity of the bob and that the restoring force acts at a single point, we refer to this apparatus as a ***simple pendulum.*** Although this assumption is never strictly true, a close approximation is obtained by making the mass of the connecting rod or cord small in comparison to the pendulum bob.

Notice that the displacement x of the bob is not along a straight line but lies along the arc subtended by the angle θ. From methods discussed in Chapter 11, the length of the displacement is simply the product of the angle θ and the length of the cord. Thus,

$$x = L\theta$$

If the motion of the bob is SHM, the restoring force must be given by

$$F = -kx = -kL\theta \qquad \textbf{(14.19)}$$

which means that the restoring force should be proportional to θ since the length L is constant. Let us examine the restoring force to see if this is the case. In the back-and-forth movement of the pendulum bob, the necessary restoring force is provided by the tangential component of the weight. From Fig. 14.11, we can write

$$F = -mg \sin \theta \qquad \textbf{(14.20)}$$

Thus, the restoring force is proportional to $\sin \theta$ instead of to θ. The conclusion must be that the bob does not oscillate with SHM. If we stipulate that the angle θ is small, however, $\sin \theta$ will be approximately equal to the angle θ in radians. You can verify this approximation by considering several small angles:

$\sin \theta$	θ (rad)
$\sin 6° = 0.1045$	$6° = 0.1047$
$\sin 12° = 0.208$	$12° = 0.209$
$\sin 27° = 0.454$	$27° = 0.471$

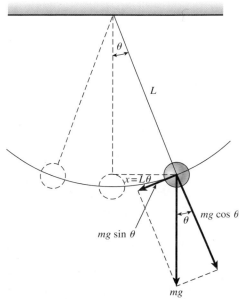

Figure 14.11

When the approximation $\sin \theta \approx \theta$ is used, Eq. (14.20) becomes

$$F = -mg \sin \theta = -mg\theta$$

Comparing this relation with Eq. (14.19), we obtain

$$F = -kL\theta = -mg\theta$$

from which

$$\frac{m}{k} = \frac{L}{g}$$

Substitution of this proportion into Eq. (14.18) yields an expression for the period of a *simple pendulum:*

$$T = 2\pi \sqrt{\frac{L}{g}} \qquad \text{(14.21)}$$

Notice that for small amplitudes, the period of a simple pendulum is a function of neither the mass of the bob nor the amplitude of vibration. In fact, since the acceleration of gravity is constant, the period is determined solely by the length of the connecting cord or rod.

Example 14.7

In a laboratory experiment, a student is given a stopwatch, a wooden bob, and a roll of string. To determine the acceleration due to gravity, simple pendulum is constructed with a length of 1 m. The wooden bob is attached to the end, and the pendulum is caused to vibrate with SHM. If the time for 20 complete oscillations is 40 s, what is the value obtained for *g?*

Plan: The period is the time for one vibration or 2 s in this case (40 s/20 vib = 2 s/vib). To find the acceleration due to gravity, we must solve Eq. (14.21) explicitly for *g* and then substitute the values for *T* and *L*.

Solution: Squaring both sides of Eq. (14.21) gives

$$T^2 = 4\pi^2 \frac{L}{g}$$

from which

$$g = \frac{4\pi^2 L}{T^2} = \frac{4\pi^2(1 \text{ m})}{(2 \text{ s})^2}$$

$$= 9.87 \text{ m/s}^2$$

14.9 The Torsion Pendulum

Another example of SHM is the torsion pendulum (Fig. 14.12), which consists of a solid disk or cylinder supported at the end of a thin rod. If the disk is twisted through an angle θ, the restoring torque τ is directly proportional to the angular displacement. Thus,

$$\tau = -k'\theta \qquad \text{(14.22)}$$

where k' is a constant that depends on the material from which the thin rod is made. (See Prob. 13.42.)

Figure 14.12

When the disk is released, the restoring torque produces an angular acceleration that is directly proportional to the angular displacement. The period of the simple angular harmonic motion thus produced is given by

$$T = 2\pi \sqrt{\frac{I}{k'}}$$ (14.23)

where I is the moment of inertia of the vibrating system and k' is the **torsion constant** defined in Eq. (14.22).

Example 14.8

A solid disk of mass 0.40 kg and radius 0.12 m is supported at its center by a thin, rigid rod, which is attached to the ceiling (see Fig. 14.12). The rod is twisted through an angle of 1 rad and released to vibrate. If the torsion constant is 0.025 N · m/rad, what is the maximum acceleration and the period of oscillation?

Plan: We will first calculate the moment of inertia for the disk ($\frac{1}{2}mR^2$). To find the angular acceleration as a function of angular displacement, we need to combine Newton's law and Hooke's law for rotation, similar to the method used for linear vibration. The period can be found by direct substitution into Eq. (14.23).

Solution: The moment of inertia for the disk is

$$I = \frac{1}{2}mR^2 = \frac{1}{2}(0.40 \text{ kg})(0.12 \text{ m})^2; \qquad I = 2.9 \times 10^{-3} \text{ kg} \cdot \text{m}^2$$

The torque is equal to $I\alpha$ from Newton's law and $-k'\theta$ from Hooke's law, so

$$I\alpha = -k'\theta \qquad \text{or} \qquad \alpha = \frac{-k'\theta}{I}$$

$$\alpha = \frac{-(0.025 \text{ N} \cdot \text{m/rad})(1 \text{ rad})}{2.9 \times 10^{-3} \text{ kg} \cdot \text{m}^2} = -8.68 \text{ rad/s}^2$$

Next, the period T is found by direct substitution.

$$T = 2\pi \sqrt{\frac{I}{k'}} = 2\pi \sqrt{\frac{2.9 \times 10^{-3} \text{ kg} \cdot \text{m}^2}{0.025 \text{ N} \cdot \text{m/rad}}}$$

$$T = 2.14 \text{ s}$$

Notice that the period is not a function of the angular displacement.

Summary and Review

Summary

Simple harmonic motion is periodic motion in which the restoring force is proportional to the displacement. Such vibratory motion without friction produces predictable variations in displacement and velocity. The major concepts discussed in this chapter are summarized as follows.

- Simple harmonic motion is produced by a *restoring force* **F** given by

$$F = -kx \qquad \textit{Restoring Force}$$

- Since $F = ma = -kx$, the acceleration produced by a restoring force is

$$a = -\frac{k}{m}x \qquad \textit{Acceleration}$$

- Energy is conserved during SHM without friction. For a mass m oscillating at the end of a spring, we find that the total energy E is constant.

$$E = \frac{1}{2}kx^2 + \frac{1}{2}mv^2 = \text{constant}$$

$$\frac{1}{2}kx^2 + \frac{1}{2}mv^2 = \frac{1}{2}mv_{\text{max}}^2 = \frac{1}{2}kA^2$$

In this relation, k is the spring constant, v is the velocity, x is the displacement, A is the amplitude, and m is the mass.

- A convenient way to study simple harmonic motion is to use the *reference circle*. The variations in displacement x, velocity v, and acceleration a can be seen by reference to Figs. 14.7, 14.8, and 14.9, respectively.

- For SHM, the displacement, velocity, and acceleration may be expressed in terms of the amplitude A, the time t, and the frequency of vibration f:

$$x = A\cos 2\pi ft \qquad \textit{Displacement}$$
$$v = -2\pi fA \sin 2\pi ft \qquad \textit{Velocity}$$
$$a = -4\pi^2 f^2 x \qquad \textit{Acceleration}$$

- The period T and the frequency f in simple harmonic motion are found from

$$f = \frac{1}{2\pi}\sqrt{-\frac{a}{x}}$$

or

$$f = \frac{1}{2\pi}\sqrt{\frac{k}{m}} \qquad \textit{Frequency}$$

$$T = 2\pi\sqrt{-\frac{x}{a}}$$

or

$$T = 2\pi\sqrt{\frac{m}{k}} \qquad \textit{Period}$$

SIGN CONVENTIONS IN SHM

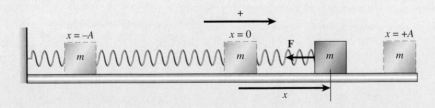

- *Displacement x* is positive when the mass is located to the right of $x = 0$ and negative when the displacement is to the left of zero. It is not determined by the direction of velocity or acceleration.
- *Velocity v* is positive when moving to the right and negative when moving to the left. The direction of acceleration or velocity is not a factor. The velocity is a maximum at the midpoint and zero at each end.
- *Acceleration a* and the restoring force **F** are positive when the displacement is negative and negative when the displacement is positive. Acceleration and force are a maximum at the endpoints and zero at the midpoint.

- For a simple pendulum of length L, the period is given by

$$T = 2\pi \sqrt{\frac{L}{g}} \quad \text{\textit{Period of Simple Pendulum}}$$

- A torsion pendulum consists of a solid disk or cylinder of moment of inertia I suspended at the end of a thin rod. If the torsion constant k' is known, the period is given by

$$T = 2\pi \sqrt{\frac{I}{k'}} \quad \text{\textit{Period of Torsion Pendulum}}$$

Key Terms

amplitude   281

displacement   291

frequency   281

hertz   282

period   281

periodic motion   280

restoring force   291

simple harmonic motion (SHM)   281

simple pendulum   293

torsion constant   295

Review Questions

14.1. Give several examples of motion that is SHM.

14.2. What effect will doubling the amplitude A of a body moving with SHM have on (a) the period, (b) the maximum velocity, and (c) the maximum acceleration?

14.3. A 2-kg mass m_1 moves in SHM with a frequency f_1. What mass m_2 will cause the system to vibrate with twice the frequency?

14.4. Explain, with the use of diagrams, why the velocity in SHM is greatest when the acceleration is the least.

14.5. A disk is attached to a spring of force constant k and set into vibration of amplitude A, as shown in Fig. 14.2. The spring is then replaced with one with a force constant of $4k$, and the disk is set into vibration at the same amplitude. Compare their periods and frequencies of oscillation.

14.6. A pendulum clock runs too slow and loses time. What adjustment should be made?

14.7. Given a spring of known force constant, a meterstick, and a stopwatch, how can you determine the value of an unknown mass?

14.8. How can the principle of the pendulum be used to compute (a) length, (b) mass, and (c) time?

14.9. Explain clearly why the motion of a pendulum is not simple harmonic when the amplitude is large. Is the period larger or smaller than it should be if the motion were strictly simple harmonic?

Problems

Section 14.1 Periodic Motion and Section 14.2 Newton's Second Law and Hooke's Law

14.1. A rock swings in a circle at constant speed on the end of a string, making 50 revolutions in 30 s. What are the frequency and the period for this motion?   Ans. 1.67 rev/s, 0.600 s

14.2. A child sits at the edge of a platform rotating at 30 rev/min. The platform is 10 m in diameter. What is the period of the motion, and what is the child's speed?

14.3. A rubber ball swings in a horizontal circle 2 m in diameter and makes 20 revolutions in 1 min. The shadow of the ball is projected on a wall by a distant light. What are the amplitude, frequency, and period for the motion of the shadow?
                Ans. 1.00 m, 0.333 Hz, 3.00 s

14.4. Assume that a ball makes 300 rev/min while moving in a circle of radius 12 cm. What are the amplitude, frequency, and period for the motion of its shadow projected on a wall?

14.5. A mass oscillates at a frequency of 3 Hz and an amplitude of 6 cm. What are its positions at time $t = 0$ and $t = 3.22$ s?   Ans. 6 cm, −3.22 cm

14.6. A 50-g mass oscillates with an SHM of frequency 0.25 Hz. Assume that $t = 0$ when the mass is at its maximum displacement. At what time will its displacement be zero? At what time will it be located at half of its amplitude?

14.7. When a mass of 200-g is hung from a spring, the spring is displaced downward a distance of 1.5 cm. What is the spring constant k?
                                           Ans. 131 N/m

14.8. An additional mass of 400 kg is added to the initial 200-g mass in Prob. 14.7. What will be the increase in downward displacement?

14.9. A 1.5-kg mass oscillates at the end of a spring in SHM. The amplitude of the vibration is 0.15 m, and the spring constant is 80 N/m. What are the magnitude and direction of the acceleration and force on the mass when it is located at the following displacements: (a) 0.15 m, (b) −0.09 m, and (c) +0.05 m?
Ans. (a) −8 m/s², −12 N; (b) +4.8 m/s², +7.2 N; (c) −2.67 m/s², −4 N

14.10. A light spring and a 0.65-kg block are placed on a horizontal frictionless surface. The spring is compressed a distance of 6 cm and released to vibrate with SHM. If the spring constant is 9 N/m, what is the initial acceleration of the block? What is the initial force on the block?

Section 14.3 Work and Energy in Simple Harmonic Motion

14.11. A spring is compressed a distance of 4 cm. If the spring constant is 200 N/m, how much work is done by the compressing force? What is the potential energy? Ans. 0.16 J, 0.16 J

14.12. A toy gun operates by pushing a 0.15-kg plastic ball against a spring, compressing it a distance of 8 cm. The spring constant is 400 N/m. When the ball is released, what will be the velocity as it leaves the end of the spring?

14.13. A 0.5-kg mass is attached to a light spring with a spring constant of 25 N/m. The mass is displaced a distance of 6 cm and released to vibrate with SHM on a horizontal frictionless surface. (a) What is the total energy of the system? (b) What is the maximum velocity? (c) What is the maximum acceleration?
Ans. (a) 45 mJ, (b) 0.424 m/s, (c) ±10 m/s.

14.14. Consider the same conditions as in Prob. 14.13. What is the velocity of the 0.5-kg mass when its position is $x = +5$ cm, and what is its velocity when $x = −3$ cm?

Section 14.5 Velocity in Simple Harmonic Motion

14.15. A body vibrates with a frequency of 1.4 Hz and an amplitude of 4 cm. What is the maximum velocity? What is its position when the velocity is zero? Ans. ±0.351 m/s, $x = 4$ cm

14.16. An object oscillates at a frequency of 5 Hz and an amplitude of 6 cm. What is the maximum velocity?

14.17. A smooth block on a frictionless surface is attached to a spring, pulled to the right a distance of 4 cm, and then released. Three seconds later, it returns to the point of release. What is the frequency, and what is the maximum speed?
Ans. 0.333 Hz, 8.38 cm/s

14.18. In Prob. 14.17, what are the position and velocity 2.55 s after release?

*14.19. A mass at the end of a spring vibrates up and down with a frequency of 0.600 Hz and amplitude of 5 cm. What is its displacement 2.56 s after it is released from $A = +5$ cm? Ans. −4.87 cm

*14.20. An object vibrates with SHM of amplitude 6 cm and frequency 0.490 Hz. At $t = 0$, the displacement is $x = +6$ cm. At what later time will its displacement first become $x = +2$ cm?

*14.21. Show that the velocity of an object in SHM can be written as a function of its amplitude and displacement:

$$v = \pm 2\pi f \sqrt{A^2 - x^2}$$

14.22. Use the relation derived in Prob. 14.21 to verify the answers obtained for position and velocity in Prob. 14.18.

14.23. A mass vibrating at a frequency of 0.5 Hz has a velocity of 5 cm/s as it passes the center of oscillation. What are the amplitude and the period of vibration? Ans. 1.59 cm, 2 s

*14.24. A body vibrates with a frequency of 8 Hz and an amplitude of 5 cm. At what time after it is released from $x = +5$ cm will its velocity be equal to +2.00 m/s?

Section 14.6 Acceleration in Simple Harmonic Motion

14.25. A 400-g mass is attached to a spring, causing it to stretch a vertical distance of 2 cm. The mass is then pulled downward a distance of 4 cm and released to vibrate with SHM as shown in Fig. 14.13. What is the spring constant? What are

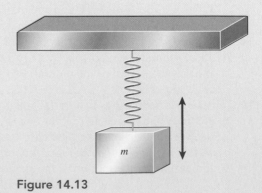

Figure 14.13

the magnitude and direction of the acceleration when the mass is located 2 cm below its equilibrium position?

Ans. 196 N/m, 9.8 m/s^2 upward

14.26. What is the maximum acceleration for the system described in Prob. 14.25, and what is its acceleration when it is located 3 cm above its equilibrium position?

14.27. A body makes one complete oscillation in 0.5 s. What is its acceleration when it is displaced a distance of $x = +2$ cm from its equilibrium position? Ans. −3.16 m/s^2

14.28. Find the maximum velocity and the maximum acceleration of an object moving in SHM of amplitude 16 cm and frequency 2 Hz.

*14.29. An object vibrating with a period of 2 s is displaced a distance of $x = +6$ cm and released. What are the velocity and acceleration 3.20 s after release? Ans. +11.1 cm/s, 0.479 m/s^2

*14.30. A body vibrates with SHM of period 1.5 s and amplitude of 6 in. What are its maximum velocity and acceleration?

*14.31. For the body in Prob. 14.30, what are its velocity and acceleration after a time of 7 s?

Ans. 1.81 ft/s, 4.26 ft/s^2

Section 14.7 The Period and Frequency

14.32. The prong of a tuning fork vibrates with a frequency of 330 Hz and an amplitude of 2 mm. What is the velocity when the displacement is 1.5 mm?

*14.33. A 400-g mass stretches a spring 20 cm. The 400-g mass is then removed and replaced with an unknown mass m. When the unknown mass is pulled down 5 cm and released, it vibrates with a period of 0.1 s. Compute the mass of the object. Ans. 4.96 g

*14.34. A long, thin piece of metal is clamped at its lower end and has a 2-kg ball fastened to its top end. When the ball is pulled to one side and released, it vibrates with a period of 1.5 s. What is the spring constant for this device?

*14.35. A car and its passengers have a total mass of 1600 kg. The frame of the car is supported by four springs, each having a force constant of

20,000 N/m. Find the frequency of vibration of the car when it drives over a bump in the road.

Ans. 1.13 Hz

Section 14.8 The Simple Pendulum

14.36. What are the period and frequency of a simple pendulum 2 m in length?

*14.37. A simple pendulum clock beats seconds every time the bob reaches its maximum amplitude on either side. What is the period of this motion? What should be the length of the pendulum at the point at which $g = 9.80$ m/s^2?

Ans. 2.00 s, 0.993 m

14.38. A 10-m length of cord is attached to a steel bob hung from the ceiling. What is the period of its natural oscillation?

*14.39. On the surface of the Moon, the acceleration due to gravity is only 1.67 m/s^2. A pendulum clock adjusted for the Earth is taken to the Moon. What fraction of its length on Earth must the new length be? Ans. 0.17

*14.40. A student constructs a pendulum of length 3 m and determines that it makes 50 complete vibrations in 2 min and 54 s. What is the acceleration due to gravity at this student's location?

Section 14.9 The Torsion Pendulum

*14.41. A torsion pendulum oscillates at a frequency of 0.55 Hz. What is the period of its vibration? What is the angular acceleration when its angular displacement is 60°? Ans. −12.5 rad/s^2

*14.42. The maximum angular acceleration of a torsional pendulum is 20 rad/s^2 when the angular displacement is 70°. What is the frequency of vibration?

*14.43. A disk 20 cm in diameter forms the base of a torsion pendulum. A force of 20 N applied to the rim causes it to twist an angle of 12°. If the period of the angular vibration after release is 0.5 s, what is the moment of inertia of the disk?

Ans. 0.0605 kg m^2

*14.44. An irregular object is suspended by a wire as a torsion pendulum. A torque of 40 lb · ft causes it to twist through an angle of 15°. When released, the body oscillates with a frequency of 3 Hz. What is the moment of inertia of the irregular body?

Additional Problems

14.45. The spring constant of a metal spring is 2000 N/m. What mass will cause this spring to stretch a distance of 4 cm? Ans. 8.16 kg

14.46. A 4-kg mass hangs from a spring whose constant k is 400 N/m. The mass is pulled downward a distance of 6 cm and released. What is the acceleration at the instant of release?

14.47. What is the natural frequency of vibration for the system described in Prob. 14.46? What is the maximum velocity?   Ans. 1.59 Hz, ± 59.9 cm/s

*14.48. A 50-g mass on the end of a spring ($k = 20$ N/m) is moving at a speed of 120 cm/s when located a distance of 10 cm from the equilibrium position. What is the amplitude of the vibration?

*14.49. A 40-g mass is attached to a spring ($k = 10$ N/m) and released with an amplitude of 20 cm. What is the velocity of the mass when it is halfway to the equilibrium position?   Ans. 2.74 m/s

14.50. What is the frequency of the motion for the mass in Prob. 14.49?

14.51. A 2-kg mass is hung from a light spring. When displaced and released, the mass makes 20 oscillations in 25 s. Find the period and the spring constant.   Ans. 1.25 s, 50.5 N/m

14.52. What length of pendulum is required if the period is to be 1.6 s at a point where $g = 9.80$ m/s^2?

*14.53. An object is moving with SHM of amplitude 20 cm and frequency 1.5 Hz. What are the maximum acceleration and velocity?
       Ans. ± 17.8 m/s^2, ± 188 cm/s

*14.54. For the object in Prob. 14.53, what are the position, velocity, and acceleration 1.4 s after it reaches its maximum displacement?

Critical Thinking Questions

*14.55. A mass m connected to the end of a spring oscillates with a frequency $f = 2$ Hz and an amplitude A. If the mass m is doubled, what is the new frequency for the same amplitude? If the mass is unchanged and the amplitude is doubled, what is the frequency?   Ans. 1.41 Hz, 2.00 Hz

*14.56. Consider a 2-kg mass connected to a spring whose constant is 400 N/m. What is the natural frequency of vibration? If the system is stretched by +8 cm and then released, at what points are its velocity and acceleration maximized? Will it reach half its maximum velocity at half the amplitude? Calculate the maximum velocity and the velocity at $x = 4$ cm to verify your answer.

*14.57. A 200-g mass is suspended from a long, spiral spring. When displaced downward 10 cm, the mass is found to vibrate with a period of 2 s. What is the spring constant? What are its velocity

and acceleration as it passes *upward* through the point +5 cm above its equilibrium position?
       Ans. 1.97 N/m, 27.2 cm/s, -49.3 cm/s^2

*14.58. A pendulum clock beats seconds every time the bob passes through its lowest point. What must be the length of the pendulum at a place where $g = 32.0$ ft/s? If the clock is moved to a point where $g = 31.0$ ft/s^2, how much time will it lose in 1 day?

*14.59. A 500-g mass is connected to a device having a spring constant of 6 N/m. The mass is displaced to the right a distance $x = +5$ cm from its equilibrium position and then released. What are its velocity and acceleration when $x = +3$ cm and when $x = -3$ cm?
       Ans. ± 0.139 m/s, -0.360 m/s^2;
       ± 0.139 m/s, $+0.360$ m/s^2

15

Fluids

Hot-air balloons use heated air, which is less dense than the surrounding air, to create an upward buoyant force. According to Archimedes' principle, the buoyant force is equal to the weight of the air displaced by the balloon. (*Photo by Paul E. Tippens.*)

Objectives

After completing this chapter, you should be able to

1. Define and apply the concepts of fluid pressure and buoyant force to the solution of physical problems similar to examples in the text.

2. Write and illustrate with drawings your understanding of the four basic principles of fluid pressure as summarized in Section 15.3.

3. Define *absolute pressure*, *gauge pressure*, and *atmospheric pressure* and demonstrate by examples your understanding of the relationships between these terms.

4. Write and apply formulas for calculating the mechanical advantage of a hydraulic press in terms of input and output forces or areas.

5. Define the rate of flow of a fluid and solve problems that relate the rate of flow to the velocity and cross-sectional area.

6. Write Bernoulli's equation in its general form and describe the equation as it would apply to (a) a fluid at rest, (b) fluid flow at constant pressure, and (c) flow through a horizontal pipe.

7. Apply Bernoulli's equation to the solution of problems involving absolute pressure P, density ρ, fluid elevation h, and fluid velocity v.

Liquids and gases are called fluids because they flow freely and fill their containers. In this chapter, you will learn that fluids may exert forces on the walls of their containers. Such forces acting on definite surface areas create a condition of pressure. A hydraulic press utilizes fluid pressure to lift heavy loads. The structure of water basins, dams, and large oil tanks is determined largely by pressure considerations. The design of boats, submarines, and weather balloons must take into account the pressure and density of the surrounding fluid. We will also study the fundamental aspects of fluid flow and Bernoulli's laws that govern such motion.

15.1 Density

Before discussing the statics and dynamics of fluids, it is important to understand the relation of a body's mass to its volume. We might say that a block of lead is *heavier* than a block of wood. What we really mean is that a block of lead is heavier than a wooden block *of similar size*. The terms *light* and *heavy* are comparative terms. As illustrated in Fig. 15.1, a 1-cm³ block of lead has a mass of 11.3 g, whereas a 1-cm³ block of oak has a mass of only 0.81 g. The volume of wood must be 14 times that of the lead if they are to have the same mass.

The *density* ρ of a body is defined as the ratio of its mass m to its volume V.

$$\rho = \frac{m}{V} \qquad m = \rho V \tag{15.1}$$

The SI unit for density is the *kilogram per cubic meter* (kg/m^3). Thus, if an object has a mass of 4 kg and a volume of 0.002 m³, it has a density of 2000 kg/m³. When dealing with small volumes, the density is often expressed in grams per cubic centimeter (g/cm^3).

Although the use of USCS units is to be discouraged, the older units are still very common in the United States, and it is wise to at least mention the concept of **weight density D,** which is often applied for the older units for weight (lb) and length (ft).

The weight density D of a body is defined as the ratio of its weight W to its volume V. *The common unit is the pound per cubic foot* (lb/ft³).

$$D = \frac{W}{V} \qquad W = DV \tag{15.2}$$

For example, the weight density of water is 62.4 lb/ft³.

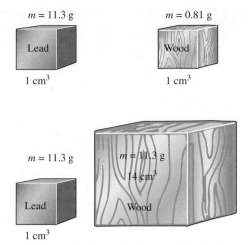

Figure 15.1 A comparison of mass and volume for blocks of lead and wood. The volume of wood must be 14 times that of the lead if they are to have the same mass.

The relationship between weight density and mass density is found by recalling that $W = mg$. Therefore,

$$D = \frac{mg}{V} = \rho g \qquad \textbf{(15.3)}$$

The densities for common solids, liquids, and gases are given in Table 15.1.

Table 15.1

Mass Density and Weight Density

Substance	ρ		D, lb/ft³
	kg/m³	g/cm³	
Solids:			
Aluminum	2,700	2.7	169
Brass	8,700	8.7	540
Copper	8,890	8.89	555
Glass	2,600	2.6	162
Gold	19,300	19.3	1,204
Ice	920	0.92	57
Iron	7,850	7.85	490
Lead	11,300	11.3	705
Oak	810	0.81	51
Silver	10,500	10.5	654
Steel	7,800	7.8	487
Liquids:			
Alcohol	790	0.79	49
Benzene	880	0.88	54.7
Gasoline	680	0.68	42
Mercury	13,600	13.6	850
Water	1,000	1.0	62.4
Gases (0°C):			
Air	1.29	0.00129	0.0807
Helium	0.178	0.000178	0.0110
Hydrogen	0.090	0.000090	0.0058
Nitrogen	1.25	0.00126	0.0782
Oxygen	1.43	0.00143	0.0892

Example 15.1 A cylindrical tank for gasoline is 3 m high and 1.2 m in diameter. How many kilograms of gasoline will the tank hold?

Plan: To find the mass, we must first determine the volume of the right circular cylinder ($V = \pi r^2 h$), where $r = \frac{1}{2}D = 0.60$ m. The mass is then found from Eq. (15.1).

Solution: The volume is

$$V = \pi r^2 h = \pi(0.6 \text{ m})^2(3 \text{ m}); \qquad V = 3.39 \text{ m}^3$$

Solving the density equation for m, we have

$$m = \rho V = (680 \text{ kg/m}^3)(3.39 \text{ m}^3); \qquad m = 2310 \text{ kg}$$

Another method of indicating the density of substances is by comparing their density to that of water. The ratio of the substance density to that of water then becomes the **specific gravity,** which is a dimensionless quantity. If an object is twice as dense as water, it has a specific gravity of 2; an object that is one-third as dense as water has a relative density of $\frac{1}{3}$.

The specific gravity of a substance is defined as the ratio of its density to the density of water at 4°C (1000 kg/m^3)

A better name for this quantity is *relative density,* but the term *specific gravity* is much more widely used.

15.2 Pressure

The effectiveness of a given force often depends on the area over which it acts. For example, a woman wearing narrow heels will do much more damage to floors than she would with flat heels. Even though she exerts the same downward force in each case, with the narrow heels, her weight is spread over a much smaller surface area. The *normal force per unit of area* is called **pressure.** Symbolically, the pressure P is given by

$$P = \frac{F}{A} \tag{15.4}$$

where A is the area over which the perpendicular force F is applied. The unit of pressure is the ratio of any force unit to a unit of area. Examples are *newtons per square meter* and *pounds per square inch.* In SI units, the N/m^2 is renamed the *pascal* (Pa).

$$1 \text{ pascal (Pa)} = 1 \text{ newton per square meter (N/m}^2)$$

When reporting pressure, the *kilopascal* (kPa) is a convenient size unit for most applications. However, only the Pa should be substituted into formulas.

$$1 \text{ kPa} = 1000 \text{ N/m}^2 = 0.145 \text{ lb/in.}^2$$

Example 15.2 A golf shoe has 10 cleats, each having an area of 6.5×10^{-6} m^2 in contact with the floor. Assume that, in walking, there is one instant when all 10 cleats support the entire weight of an 80-kg person. What is the pressure exerted by the cleats on the floor?

Plan: We will calculate the total force on the floor by finding the weight of an 80-kg mass. Then, we will divide that force by the area of 10 cleats to give the total pressure.

Solution: The total area is $10(6.5 \times 10^{-6} \text{ m}^2)$ or $65 \times 10^{-6} \text{ m}^2$. Thus, the pressure is

$$P = \frac{F}{A} = \frac{mg}{A}$$

$$= \frac{(80 \text{ kg})(9.8 \text{ m/s}^2)}{65.0 \times 10^{-6} \text{ m}^2} = 1.21 \times 10^7 \text{ N/m}^2$$

Recalling that a N/m^2 is a pascal (Pa), we can write the total pressure as

$$P = 1.21 \times 10^7 \text{ Pa} = 12.1 \text{ MPa}$$

When the area of a shoe in contact with the floor decreases (as with some high-heel shoes), the pressure becomes larger. It is easy to see why such factors must be considered in floor construction.

15.3   Fluid Pressure

There is a significant difference between the way a force acts on a fluid and on a solid. Since a solid is a rigid body, it can withstand the application of a force without a significant change in shape. A liquid, on the other hand, can sustain a force only at an enclosed surface or boundary. If a fluid is not restrained, it will flow under a shearing stress instead of being deformed elastically.

> The force exerted by a fluid on the walls of its container must always act perpendicular to the walls.

It is this characteristic property of fluids that makes the concept of pressure so useful. Holes bored in the bottom and sides of a barrel of water (Fig. 15.2) demonstrate that the force exerted by the water is everywhere perpendicular to the surface of the barrel.

A moment's reflection will show the student that a liquid also exerts an upward pressure. Anyone who has tried to keep a rubber float under the surface of water is immediately convinced of the existence of an upward pressure. In fact, we find that

> Fluids exert pressure in all directions.

Figure 15.3 shows a liquid under pressure. The forces acting on the face of the piston, the walls of the enclosure, and the surfaces of a suspended object are shown in the figure.

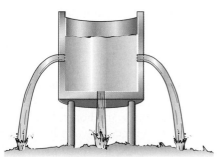

Figure 15.2 The forces exerted by a fluid on the walls of its container are perpendicular at every point.

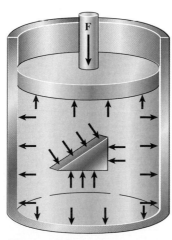

Figure 15.3 Fluids exert pressure in all directions.

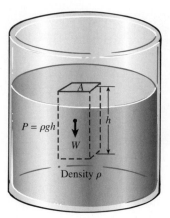

Figure 15.4 The relationship of pressure, density, and depth.

PHYSICS TODAY

Toothpaste coming out when the tube is squeezed, the Heimlich maneuver (which exerts a sharp upward pressure on a person's abdomen to dislodge an item stuck in the throat), and a hydraulic lift are all examples of Pascal's law.

Just as larger volumes of solid objects exert greater forces against their supports, fluids exert greater pressure at increasing depths. The fluid at the bottom of a container is always under a greater pressure than that near the surface. This is due to the weight of the overlying liquid. We must point out, however, a distinct difference between the pressure exerted by solids and that exerted by liquids. A solid object can exert only a *downward* force due to its weight. At any particular depth in a fluid, the pressure is the same in all directions. If this were not true, the fluid would flow under the influence of a resultant pressure until a new condition of equilibrium was reached.

Since the weight of the overlying fluid is proportional to its density, the pressure at any depth is also proportional to the density of the fluid. This can be seen by considering a rectangular column of water extending from the surface to a depth h, as shown in Fig. 15.4. The weight of the entire column acts on the surface area A at the bottom of the column.

From Eq. (15.1), we can write the weight of the column as

$$W = DV = DAh$$

where D is the weight density of the fluid. The pressure (weight per unit area) at the depth h is given by

$$P = \frac{W}{A} = Dh$$

or, in terms of mass density,

$$P = Dh = \rho gh \qquad (15.5)$$

The fluid pressure at any point is directly proportional to the density of the fluid and to the depth below the surface of the fluid.

Example 15.3

The water pressure in a certain house is 160 lb/in.^{2}. How high must the water level be above the point of release in the house?

Plan: We find from tables that the weight density D of water is 62.4 lb/ft^3. The given pressure in the house is 160 lb/in.^{2}, so we must convert to lb/ft^3 to obtain consistent units. Then, we will apply Eq. (15.5) to solve for the height h.

Solution: Converting units, we have

$$P = \left(160\frac{\text{lb}}{\text{in.}^2}\right)\left(\frac{144 \text{ in.}^2}{1 \text{ ft}^2}\right) = 23{,}040 \text{ lb/ft}^2$$

Now solving for h in Eq. (15.5) gives

$$h = \frac{P}{D} = \frac{23{,}040 \text{ lb/ft}^2}{62.4 \text{ lb/ft}^3}; \qquad h = 369 \text{ ft}$$

In Example 15.3, no mention was made of the size or shape of the reservoir containing the supply of water. Also, no information was given about the path of the water or size of the pipes connecting the reservoir to the home. Can we assume that our answer is correct when it is based only on the difference in water levels? Doesn't the shape or area of a container have any effect on liquid pressure? To answer these questions, we must recall some of the characteristics of fluids already discussed.

Consider a series of vessels of different areas and shapes interconnected, as shown in Fig. 15.5. It would seem at first glance that the greater volume of water in vessel A should

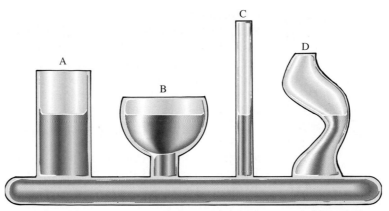

Figure 15.5 Water seeks its own level, indicating that the pressure is independent of the area or shape of the container.

develop a greater pressure at the bottom than vessel *D*. The effect of such a difference in pressure would then force the liquid to rise higher in vessel *D*. Filling the vessels with liquid shows the levels to be the same, however, for each vessel.

Part of the problem in understanding this paradox results from confusing the terms *pressure* and *total force*. Since pressure is measured in terms of a unit area, we do not consider the *total area* when solving problems involving pressure. For example, in vessel *A*, the area of the liquid at the bottom of the vessel is much greater than the area at the bottom of vessel *D*. This means the liquid in vessel *A* will exert a greater ***total force*** on the bottom than the liquid in vessel *D*. But the greater force is applied over a larger area, so the pressure is the same in both vessels.

If the bottoms of vessels *B, C,* and *D* have the same area, we can say that the total forces are also equal at the bottoms of these containers. (Of course, the pressures are equal at any particular depth.) You may wonder how the total forces can be equal when vessels *B* and *C* contain a greater volume of water. The additional water in each case is supported by vertical components of forces exerted by the walls of the container on the fluid. (See Fig. 15.6.) When the walls of a container are vertical, the forces acting on the sides have no upward components. The total force at the bottom of a container is therefore equal to the weight of a straight column of water above the base area.

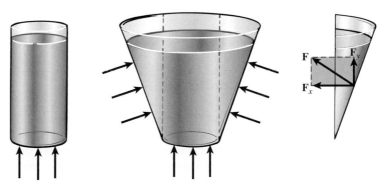

Figure 15.6 The pressure at the bottom of each vessel is a function only of the depth of the liquid and is the same in all directions. Since the area of the bottom is the same for both vessels, the total force exerted on the bottom of each container is also the same.

Example 15.4

Assume the vessels in Fig. 15.5 are filled with gasoline until the fluid level is 20 cm above the base of each vessel. The areas at the bases of vessels *A* and *B* are 20 cm^2 and 10 cm^2, respectively. Compare the pressure and the total force at the base of each container.

Plan: The density of gasoline is given in Table 15.1. The pressure is the same at the base of either container and is given by ρgh. However, the total force is not the same since the weight of the water above the base is different. The total force is found as the product of pressure and area.

Solution: The pressure at the base of either vessel is

$$P = \rho gh = (680 \text{ kg/m}^3)(9.8 \text{ m/s}^2)(0.20 \text{ m}); \qquad P = 1330 \text{ Pa}$$

We need to convert the areas from cm^2 to m^2, recalling that $1 \text{ cm}^2 = 1 \times 10^{-4} \text{ m}^2$. Then, the pressure is found by solving for force in Eq. (15.4).

$$F = PA = (1330 \text{ Pa})(20 \times 10^{-4} \text{ m}^2) = 2.67 \text{ N}$$
$$F = PA = (1330 \text{ Pa})(10 \times 10^{-4} \text{ m}^2) = 1.33 \text{ N}$$

Problem-Solving Strategy

Before considering other applications of fluid pressure, let us summarize the principles discussed in this section for fluids at rest.

1. The forces exerted by a fluid on the walls of its container are always perpendicular to the walls.

2. The fluid pressure is directly proportional to the depth of the fluid and to its density.

3. At any particular depth, the fluid pressure is the same in all directions.

4. Fluid pressure is independent of the shape or area of its container.

15.4 Measuring Pressure

The pressure discussed in Section 15.3 is due only to the fluid itself and can be calculated from Eq. (15.5). Unfortunately, this is usually not the case. Any liquid in an open container, for example, is subjected to atmospheric pressure in addition to the pressure of its own weight. Since the liquid is relatively incompressible, the external pressure of the atmosphere is transmitted equally throughout the volume of the liquid. This fact, first stated by the French mathematician Blaise Pascal (1623–1662), is called *Pascal's law.* Generally, it can be stated as follows:

> An external pressure applied to an enclosed fluid is transmitted uniformly throughout the volume of the liquid.

Most devices that measure pressure directly actually measure the difference between the *absolute pressure* and *atmospheric pressure.* The result is called the *gauge pressure.*

Absolute pressure = gauge pressure + atmospheric pressure

Atmospheric pressure at sea level is 101.3 kPa, or 14.7 lb/in.². Because atmospheric pressure enters into so many calculations, we often use a pressure unit of 1 *atmosphere* (atm), defined as the average pressure exerted by the atmosphere at sea level, that is, 101.3 kPa.

A common device for measuring gauge pressure is the open-tube *manometer* (muh-nom'-uh-ter), shown in Fig. 15.7. The manometer consists of a U-shaped tube containing a liquid, usually mercury. When both ends of the tube are open, the mercury seeks its own level because 1 atm of pressure is exerted at each of the open ends. When one end

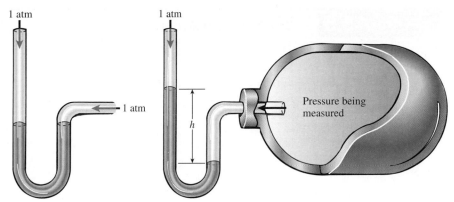

Figure 15.7 The open-tube manometer. Pressure is measured by the height *h* of the mercury column.

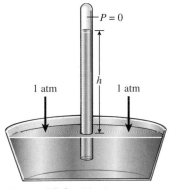

Figure 15.8 The barometer.

is connected to a pressurized chamber, the mercury will rise in the open tube until the pressures are equalized. The difference between the two levels of mercury is a measure of the gauge pressure: the difference between the absolute pressure in the chamber and the atmospheric pressure at the open end. The manometer is used so often in laboratory situations that atmospheric pressures and other pressures are often expressed in *centimeters of mercury* or *inches of mercury.*

Atmospheric pressure is usually measured in the laboratory with a mercury barometer. The principle of its operation is shown in Fig. 15.8. A glass tube, closed at one end, is filled with mercury. The open end is covered, and the tube is inverted in a bowl of mercury. When the open end is uncovered, the mercury flows out of the tube until the pressure exerted by the column of mercury exactly balances atmospheric pressure acting on the mercury in the bowl. Since the pressure in the tube above the column of mercury is zero, the height of the column above the level of mercury in the bowl indicates the atmospheric pressure. At sea level, an atmospheric pressure of 14.7 lb/in.² will cause the level of the mercury in the tube to stabilize at a height of 76 cm, or 30 in.

In summary, we can write the following equivalent measures of atmospheric pressure:

$$1 \text{ atm} = 101.3 \text{ kPa} = 14.7 \text{ lb/in.}^2 = 76 \text{ cm of mercury}$$
$$= 30 \text{ in. of mercury} = 2116 \text{ lb/ft}^2$$

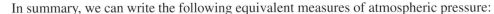

Example 15.5 The mercury manometer is used to measure the pressure of a gas inside a tank. (See Fig. 15.7.) If the difference between the two mercury levels is 36 cm, what is the absolute pressure inside the tank?

Plan: Remember the absolute pressure is the sum of gauge pressure and atmospheric pressure. The gauge reads 36 cm, which registers the *difference* between the pressure outside (1 atm) and the pressure inside the tank. One atmosphere of pressure is equivalent to a 76-cm column of mercury. The *absolute* pressure in the tank is, therefore, the sum of 36 cm and 76 cm, or 112 cm of mercury. The absolute pressure in the tank is the pressure due to a column of mercury 112 cm high.

Solution: The density of mercury is $1.36 \times 10^4 \text{ kg/m}^3$, and 112 cm is 1.12 m, so the absolute pressure inside the tank is found from Eq. (15.5).

$$P = \rho g h = (1.36 \times 10^4 \text{ kg/m}^3)(9.8 \text{ m/s}^2)(1.12 \text{ m})$$
$$P = 1.49 \times 10^5 \text{ Pa} \quad \text{or} \quad P = 149 \text{ kPa}$$

You should verify that this absolute pressure could also be expressed as 1.47 atm.

15.5 The Hydraulic Press

The most universal application of Pascal's law is found with the hydraulic press, shown in Fig. 15.9. According to Pascal's principle, a pressure applied to the liquid in the left column will be transmitted undiminished to the liquid in the column on the right. Thus, if an input force $\mathbf{F}_i$ acts upon a piston of area A_i, it will cause an output force $\mathbf{F}_o$ to act on a piston of area A_o, so

Input pressure = output pressure

$$\frac{F_i}{A_i} = \frac{F_o}{A_o} \tag{15.6}$$

The ideal mechanical advantage of such a device is equal to the ratio of the output force to the input force. Symbolically, we write

$$M_I = \frac{F_o}{F_i} = \frac{A_o}{A_i} \tag{15.7}$$

A small input force can be multiplied to yield a much larger output force simply by having the output piston much larger in area than the input piston. The output force is given by

$$F_o = F_i \frac{A_o}{A_i} \tag{15.8}$$

According to the methods developed in Chapter 12 for simple machines, the input work must equal the output work if we neglect friction. If the input force $\mathbf{F}_i$ travels through a distance s_i, while the output force $\mathbf{F}_o$ travels through a distance s_o, we can write

Input work = output work

$$F_i s_i = F_o s_o$$

This relation leads to another useful expression for the ideal mechanical advantage of a hydraulic press.

$$M_I = \frac{F_o}{F_i} = \frac{s_i}{s_o} \tag{15.9}$$

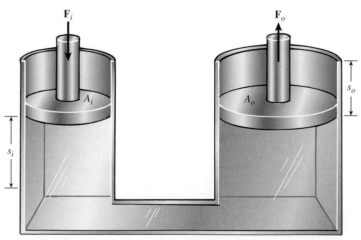

Figure 15.9 The hydraulic press.

Notice that the mechanical advantage is gained at the expense of input distance. For this reason, most applications utilize a system of valves to permit the output piston to be raised by a series of short input strokes.

Example 15.6

A hydraulic press has an input piston 5 cm in diameter and an output piston that is 60 cm in diameter. What input force is required to deliver a total output force capable of lifting a 950-kg automobile?

Plan: To find the input force, we must first use the diameters of the pistons to determine the ideal mechanical advantage from Eq. (15.7). We will assume negligible friction and recall that the area of each piston is $\pi d^2/4$. The necessary input force can be found from the calculated value of M_I.

Solution: The ideal mechanical advantage is

$$M_I = \frac{A_o}{A_i} = \frac{\pi d_o^2/4}{\pi d_i^2/4} = \frac{d_o^2}{d_i^2}; \qquad M_I = \left(\frac{d_o}{d_i}\right)^2$$

$$M_I = \left(\frac{60 \text{ cm}}{5 \text{ cm}}\right)^2 = 144$$

The necessary output force is $F_o = W = mg$, so Solving Eq. (15.7) for F_i gives

$$M_i = \frac{F_o}{F_i} = \frac{mg}{F_i} \qquad \text{or} \qquad F_i = \frac{mg}{M_I}$$

$$F_i = \frac{(950 \text{ kg})(9.8 \text{ m/s}^2)(1 \text{ m})}{144} = 64.7 \text{ N}$$

The principle of the hydraulic press is found in many engineering and mechanical applications. Backhoes, shock absorbers, and automobile braking systems are a few common examples.

15.6 Archimedes' Principle

Anyone familiar with swimming and other water sports has observed that objects seem to lose weight when submerged in water. In fact, the object may even float on the surface because of the upward pressure exerted by the water. An ancient Greek mathematician, Archimedes (287–212 B.C.), first studied the buoyant force exerted by fluids. *Archimedes' principle* can be stated as follows:

> An object that is completely or partly submerged in a fluid experiences an upward force equal to the weight of the fluid displaced.

Archimedes' principle can be demonstrated by studying the forces a fluid exerts on a suspended object. Consider a disk of area A and height H that is completely submerged in a fluid, as shown in Fig. 15.10. Recall that the pressure at any depth h in a fluid is given by

$$P = \rho g h$$

PHYSICS TODAY

You can test Archimedes' principle by submerging an object in a fluid such as water. If the object is not as dense as the fluid, the object will sink only to the point at which enough water has been displaced to equal the object's weight. The volume of the displaced water and the object's weight will prove to be equal.

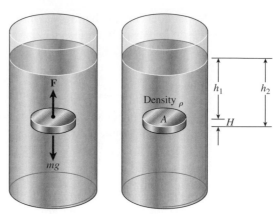

Figure 15.10   The buoyant force exerted on the disk is equal to the weight of the fluid it displaces.

where ρ is the mass density of the fluid and g is the acceleration of gravity. Of course, if we wish to represent the absolute pressure within the fluid, we must add the external pressure exerted by the atmosphere. The total downward pressure P_1 on top of the disk in Fig. 15.10 is, therefore,

$$P_1 = P_a + \rho g h_1 \qquad \text{(downward)}$$

where P_a is atmospheric pressure and h_1 is the depth at the top of the disk. Similarly, the upward pressure P_2 on the bottom of the disk is

$$P_2 = P_a + \rho g h_2 \qquad \text{(upward)}$$

where h_2 is the depth at the bottom of the disk. Since h_2 is greater than h_1, the pressure on the bottom of the disk will exceed the pressure at the top, resulting in a net upward force. If we represent the downward force by F_1 and the upward force by F_2, we can write

$$F_1 = P_1 A \qquad F_2 = P_2 A$$

The net upward force exerted *by* the fluid *on* the disk is called the ***buoyant force*** and is given by

$$
\begin{aligned}
F_B &= F_2 - F_1 = A(P_2 - P_1) \\
&= A(P_a + \rho g h_2 - P_a - \rho g h_1) \\
&= A\rho g(h_2 - h_1) = A\rho g H
\end{aligned}
$$

where $H = h_2 - h_1$ is the height of the disk. Finally, if we recall that the volume of the disk is $V = AH$, we obtain this important result:

$$F_B = \rho g V = mg \qquad\qquad \textbf{(15.10)}$$

Buoyant force = weight of displaced fluid

which is Archimedes' principle.

  In applying this result, it must be recalled that Eq. (15.10) allows us to compute only the *buoyant force* due to the difference in pressures. It does not represent the resultant force. A submerged body will sink if the weight of the fluid it displaces (the buoyant force) is less than the weight of the body. If the weight of the displaced fluid is exactly equal to the weight of the submerged body, it will neither sink nor rise. In this instance, the body will be in equilibrium. If the weight of displaced fluid exceeds the weight of a submerged body, the body will rise to the surface and float. When the floating body comes to equilibrium at the surface, it will displace its own weight of liquid. Figure 15.11 demonstrates this point with the use of an overflow can and a beaker to catch the fluid displaced by a wooden block.

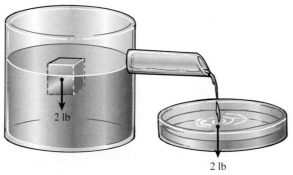

Figure 15.11   A floating body displaces its own weight of fluid.

Example 15.7   A cork float has a volume of 4 cm³ and a density of 207 kg/m³. (a) What volume of the cork is beneath the surface when the cork floats in water? (b) What downward force is needed to submerge the cork completely?

Plan: The floating cork will displace a volume of water equal to its own weight. We will use the density and volume of the cork to find its weight. Then, we will apply Archimedes' principle to find the volume of water required to provide a buoyant force equal to the weight of the cork. That volume of water is also equal to the volume of cork under the surface. In Part (a), the buoyant force must equal the sum of the weight of the block *and* the downward force to submerge the block at the surface. Therefore, we need to find the buoyant force on the completely submerged cork and then subtract the weight of the cork to find the extra force needed to keep it submerged.

Solution (a): The density of cork is 207 kg/m³, and its volume is 4 cm³. Recalling that 1 cm³ = 1 × 10⁻⁶ m³, we will find the weight of 4 × 10⁻⁶ m³ of cork.

$$\rho = \frac{m}{V}; \qquad m = \frac{W}{g} \qquad \text{so that} \qquad \rho = \frac{W}{gV}$$

$$W = \rho g V = (207 \text{ kg/m}^3)(9.8 \text{ m/s}^2)(4 \times 10^{-6} \text{ m}^3)$$
$$= 8.11 \times 10^{-3} \text{ N}$$

Now, since the same weight of water is displaced, $W_w = \rho_w g V$, we see that

$$V_w = \frac{W}{\rho g} = \frac{8.11 \times 10^{-3} \text{ N}}{(1000 \text{ kg/m}^3)(9.8 \text{ m/s}^2)}$$
$$= 8.28 \times 10^{-7} \text{ m}^3 \qquad \text{or} \qquad 0.828 \text{ cm}^3$$

The volume of cork under water must also be equal to 0.828 cm³.

If the area of the floating surface were known, you could calculate how deep the cork would sink into the water. Note that approximately 21 percent of the cork is under water. As an exercise, you should show that the fraction of volume submerged is equal to the specific gravity of an object.

Solution (b): When the cork is submerged, equilibrium demands that forces be balanced.

The sum of the downward forces must equal the buoyant force F_B. Thus,

$$F + W = F_B$$

The required downward force F is then equal to the difference between the buoyant force and the weight of the cork.

$$F = F_B - W$$

Archimedes principle gives the buoyant force as the weight of 4 cm³ of water.

$$F_B = \rho g V = (1000 \text{ kg/m}^3)(9.8 \text{ m/s}^2)(4 \times 10^{-6} \text{ m}^3)$$
$$= 39.2 \times 10^{-3} \text{ N}$$

The force F needed to submerge the cork, is, therefore

$$F = 39.2 \times 10^{-3} \text{ N} - 8.11 \times 10^{-3} \text{ N}$$
$$= 31.1 \times 10^{-3} \text{ N}$$

Example 15.8

A weather balloon is to operate at an altitude at which the density of air is 0.90 kg/m³. At this altitude, the balloon has a volume of 20 m³ and is filled with helium ($\rho_{He} = 0.178$ kg/m³). If the balloon skin weighs 88 N, what load can be supported at this level?

Plan: The balloon will be in equilibrium and become stable when the upward buoyant force on the balloon is equal to the downward forces due to the weights of the load, the balloon skin, and the helium inside the balloon. First, we will find the buoyant force due to the displaced air. Then, we will calculate the weight of helium inside the balloon.

The load that can be supported is determined by the weight needed to produce equilibrium.

Solution: The buoyant force is equal to the weight of the displaced air.

$$F_B = \rho_{air} g V = (0.9 \text{ kg/m}^3)(9.8 \text{ m/s}^2)(20 \text{ m}^3)$$
$$= 176 \text{ N}$$

The weight of the contained helium is

$$W_{He} = \rho_{He} g V_{He} = (0.178 \text{ kg/m}^3)(9.8 \text{ m/s}^2)(20 \text{ m}^3)$$
$$= 34.9 \text{ N}$$

Vertical forces are balanced, so

$$F_B = W_L + W_{He} + W_{balloon}$$

Solving for W_L gives

$$W_L = F_B - W_{He} - W_{balloon}$$
$$= 176 \text{ N} - 34.9 \text{ N} - 88 \text{ N}$$
$$= 53.1 \text{ N}$$

Large balloons can maintain a condition of equilibrium at any altitude by adjustment of their weight or buoyant force. The weight can be lightened by releasing the ballast provided for that purpose. The buoyant force can be decreased by releasing the gas from the balloon or increased by pumping more gas into the flexible balloon. Hot-air balloons use the lower density of heated air to provide their buoyancy.

Problem-Solving Strategy

Fluids at Rest

1. Draw a figure and label it with those quantities that are given and those that are to be found. Use consistent units for area, volume, density, and pressure.

2. Do not confuse *absolute* pressure with *gauge* pressure or *mass* density with *weight* density. You must use *absolute* pressure unless the problem involves a *difference* of pressure. Be careful with units if you attempt to use weight density, which is a *force* per unit volume.

3. The difference in pressure between two points is proportional to the density of the fluid and to the depth in the fluid:

$$P_2 - P_1 = \rho g h \qquad \rho = \frac{m}{V} \qquad P = \frac{F}{A}$$

4. Archimedes' principle states that an object completely or partially submerged in a fluid experiences an upward buoyant force equal to the weight of the displaced fluid:

$$F_B = mg = \rho g V \qquad \textit{(buoyant force)}$$

5. Remember that the buoyant force depends on the density of the *displaced fluid* and the volume of the *displaced fluid*. It has nothing to do with the mass or density of the object submerged in the fluid. If the object happens to be *totally* submerged, it is true that the volume of the object and of the displaced fluid are equal. This fact can be used to find the buoyant force in those cases.

6. For an object that is *floating* in a fluid, the upward buoyant force must equal the weight of the object. This means that the weight of the object must be equal to the weight of the *fluid displaced.* Therefore, we may write

$$m_x g = m_f g \qquad \text{or} \qquad \rho_x V_x = \rho_f V_f$$

The subscript *x* refers to the floating object, and the subscript *f* refers to the displaced fluid. For example, if an object with a volume of 3 m³ is floating with two-thirds of its volume submerged, then $V_x = 3 \text{ m}^3$ and $V_f = 2 \text{ m}^3$.

15.7 Fluid Flow

Until now, our study of fluids has been restricted to conditions of rest, which are considerably simpler than dealing with fluids in motion. The mathematical difficulties encountered when treating the motion of a fluid are formidable. We will find it easier to make a few assumptions. First of all, we will consider that all fluids in motion exhibit *streamline flow.*

> Streamline flow is the motion of a fluid in which every particle in the fluid follows the same path (past a particular point) as that followed by previous particles.

Figure 15.12 illustrates the *streamlines* of air flowing past two stationary obstacles. Note that the streamlines break down as air passes over the second obstacle, setting up whirls and eddies. These little whirlpools represent *turbulent flow,* and they absorb much of the fluid energy, increasing the frictional drag through the fluid.

We will further consider that fluids are incompressible and have essentially no internal friction. Under these conditions, certain predictions can be made about the *rate of fluid flow* through a pipe or other container.

> The rate of flow is defined as the volume of fluid that passes a certain cross section per unit of time.

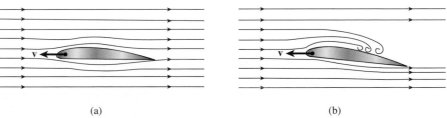

(a)   (b)

Figure 15.12 Streamline and turbulent flow of a fluid in its path.

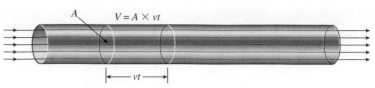

Figure 15.13   Computing the rate of flow of a fluid through a pipe.

To express this rate quantitatively, we will consider a liquid flowing through the pipe of Fig. 15.13 with an average velocity v. During a time interval t, each particle in the stream moves through a distance vt. The volume V flowing through a cross-section A is given by

$$V = Avt$$

Thus, the **rate of flow** (volume per unit time) can be calculated from

$$R = \frac{Avt}{t} = vA \qquad (15.11)$$

Rate of flow = velocity × cross section

The units of R express the ratio of a volume unit to a time unit. Common examples are cubic feet per second, cubic meters per second, liters per second, and gallons per minute.

If the fluid is incompressible and we ignore the effect of internal friction, the rate of flow R will remain constant. This means that a variation in the pipe cross section, as illustrated in Fig. 15.14, will result in a change in speed of the liquid so that the product vA remains constant. Symbolically, we write

$$R = v_1A_1 = v_2A_2 \qquad (15.12)$$

A liquid will flow faster through a narrow section of pipe and more slowly through the broad sections. This principle causes water to flow more rapidly when the banks of a small stream suddenly come closer together.

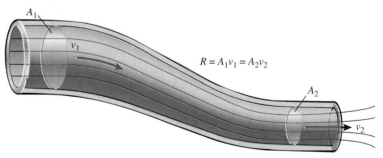

$$R = A_1v_1 = A_2v_2$$

Figure 15.14   In streamline flow, the product of the fluid velocity and the cross-sectional area of the pipe is constant at any point.

Example 15.9   Water flows through a rubber hose 2 cm in diameter at a velocity of 4 m/s. (a) What must be the diameter of the nozzle if water is to emerge at 20 m/s? (b) What is the rate of flow in cubic meters per minute?

Plan: The rate of flow must be the same in the hose and through the nozzle, so $A_1v_1 = A_2v_2$. From this, we will find the velocity through the nozzle. The, after determining the area of either opening, we can multiply by velocity to find the rate of flow.

Solution (a): Since the area A is proportional to the square of the diameter, we can write

$$d_1^2 v_1 = d_2^2 v_2 \qquad \text{or} \qquad d_2^2 = \frac{v_1 d_1^2}{v_2}$$

from which

$$d_2 = \sqrt{\frac{v_1 d_1^2}{v_2}} = \sqrt{\frac{(4 \text{ m/s})(2 \text{ cm})^2}{(20 \text{ m/s})}}$$

$$= \sqrt{0.80 \text{ cm}^2} = 0.894 \text{ cm}$$

Solution (b): To find the rate of flow, we will first find the area of the 2-cm hose.

$$A_1 = \frac{\pi d_1^2}{4} = \frac{\pi (2 \text{ cm})^2}{4} = 3.14 \text{ cm}^2$$

$$= 3.14 \text{ cm}^2 \left(\frac{1 \times 10^{-4} \text{ m}^2}{1 \text{ cm}^2} \right) = 3.14 \times 10^{-4} \text{ m}^2$$

The rate of flow is $R = A_1 v_1$, so

$$R = (3.14 \times 10^{-4} \text{ m}^2)(4 \text{ m/s}) = 1.26 \times 10^{-3} \text{ m}^3/\text{s}$$

$$= (1.26 \times 10^{-3} \text{ m}^3/\text{s})(60 \text{ s/min}) = 0.0754 \text{ m}^3/\text{min}$$

The same rate would be found by considering the product $A_2 v_2$.

Problem-Solving Strategy

Rate of Flow Problems

1. Read the problem carefully and, after drawing a rough sketch, list the given information.

2. Remember the rate of flow R represents the volume of fluid passing a given cross section in a unit of time.

3. When a volume of fluid passes from one cross section A_1 into another A_2, the rate of flow is unchanged.

$$R = v_1 A_1 = v_2 A_2$$

Be sure to use consistent units for volume and area.

4. Since the area A of a pipe is proportional to the square of its diameter d, a more useful form of the above equation might be

$$v_1 d_1^2 = v_2 d_2^2$$

5. The units chosen for velocity or diameter in one section of the pipe must be the same as those in the second section of pipe.

15.8   Pressure and Velocity

We have noted that a fluid's velocity increases when it flows through a constriction. An increase in velocity can result only through the presence of an accelerating force. To accelerate the liquid as it enters the constriction, the pushing force from the large cross section must be greater than the resisting force from the constriction. In other words, the pressure at points A and C in Fig. 15.15 must be greater than the pressure at B. The tubes inserted into the pipe above these points indicate clearly the difference in pressure. The fluid level in the tube above the restriction is lower than the level in the adjacent areas. If h is the difference in height, the pressure differential is given by

$$P_A - P_B = \rho g h \tag{15.13}$$

This assumes that the pipe is horizontal and that no pressure changes are introduced because of a change in potential energy.

The example given in Fig. 15.15 illustrates the principle of the *venturi meter.* From a determination of the difference in pressure, this device makes it possible to calculate the velocity of water in a horizontal pipe.

The ***venturi effect*** has many other applications for both liquids and gases. The carburetor in an automobile utilizes the venturi principle to mix gasoline vapor and air. Air

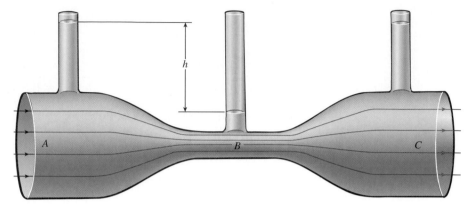

Figure 15.15   The increased velocity of a fluid flowing through a constriction causes a drop in pressure.

passing through a constriction on its way to the cylinders creates a low-pressure area as its velocity increases. The decrease in pressure is used to draw fuel into the air column, where it is readily vaporized.

Figure 15.16 shows two methods you can use to demonstrate the decrease in pressure due to an increase in velocity. The simplest example consists of blowing air past the top surface of a sheet of paper, as shown in Fig. 15.16a. The pressure in the airstream above the paper will be reduced. This allows the excess pressure on the bottom to force the paper upward.

A second demonstration requires a hollow spool, a cardboard disk, and a pin (Fig. 15.16b). The pin is driven through the cardboard disk and placed in one end of the hollow spool, as shown in the figure. If you blow through the open end, you will find that the disk becomes more tightly pressed to the other end. One would expect the cardboard disk to fly off immediately. The explanation is that air blown into the spool must escape through the narrow space between the disk and the end of the spool. This action creates a low-pressure area, allowing the external atmospheric pressure to push the disk tight against the spool.

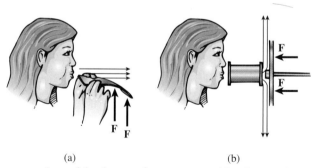

(a)   (b)

Figure 15.16   Demonstrations of the decrease in pressure resulting from an increase in air speeds.

15.9   Bernoulli's Equation

In our discussion of fluids, we have emphasized four quantities: the pressure P, the density ρ, the velocity v, and the height h above some reference level. The relationship between these quantities and their ability to describe fluids in motion was first established by Daniel Bernoulli (1700–1782), a Swiss mathematician. Steps leading to the development of this fundamental relationship can be understood by considering Fig. 15.17.

Since a fluid has mass, it must obey the same conservation laws established earlier for solids. Consequently, the work required to move a certain volume of fluid through a pipe must equal the total change in kinetic and potential energy. Let us consider the work required to move the fluid from point a to point b in Fig. 15.17a. The net work must be the sum of the work done by the input force $\mathbf{F}_1$ and the negative work done by the resisting force $\mathbf{F}_2$.

$$\text{Net work} = F_1 s_1 - F_2 s_2$$

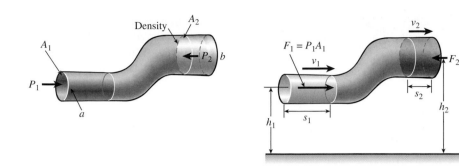

(a)   (b)

Figure 15.17   Derivation of Bernouilli's equation.

But $F_1 = P_1A_1$ and $F_2 = P_2A_2$, so

$$\text{Net work} = P_1A_1s_1 - P_2A_2s_2$$

The product of area and distance represents the volume V of the fluid moved through the pipe. Since this volume is the same at the bottom and at the top of the pipe, we can substitute

$$V = A_1s_1 = A_2s_2$$

obtaining

$$\text{Net work} = P_1V - P_2V = (P_1 - P_2)V$$

The kinetic energy K of a fluid is defined as $\frac{1}{2}mv^2$, where m is the mass of the fluid and v is its velocity. Since the mass remains constant, a change in kinetic energy ΔK results only from the difference in fluid velocity. In our example, the change in kinetic energy is

$$\Delta K = \frac{1}{2}mv_2^2 - \frac{1}{2}mv_1^2$$

The potential energy of a fluid at a height h above some reference point is defined as mgh, where mg represents the weight of the fluid. The volume of fluid moved through the pipe is constant. Therefore, the change in potential energy ΔU results from the increases in height of the fluid from h_1 to h_2:

$$\Delta U = mgh_2 - mgh_1$$

We are now prepared to apply the principle of conservation of energy. The net work done on the system must equal the sum of the increases in kinetic and potential energy. Thus,

$$\text{Net work} = \Delta K + \Delta U$$

$$(P_1 - P_2)V = \left(\frac{1}{2}mv_2^2 - \frac{1}{2}mv_1^2\right) + (mgh_2 - mgh_1)$$

If the density of the fluid is ρ, we can substitute $V = m/\rho$, giving

$$(P_1 - P_2)\frac{m}{\rho} = \frac{1}{2}mv_2^2 - \frac{1}{2}mv_1^2 + mgh_2 - mgh_1$$

Multiplying through by ρ/m and rearranging, we obtain *Bernoulli's equation:*

$$P_1 + \rho gh_1 + \frac{1}{2}\rho v_1^2 = P_2 + \rho gh_2 + \frac{1}{2}\rho v_2^2 \qquad \textbf{(15.14)}$$

Since the subscripts 1 and 2 refer to any two points, **Bernoulli's equation** can be stated more simply as

$$P + \rho gh + \frac{1}{2}\rho v^2 = \text{constant} \qquad \textit{Bernoulli's Equation} \quad \textbf{(15.15)}$$

Bernoulli's equation finds application in almost every aspect of fluid flow. The pressure P must be recognized as the *absolute* pressure and not the *gauge* pressure. Remember that ρ is the mass density and not the weight density of the fluid. Notice that the units of each term in Bernoulli's equation are units of pressure.

15.10

PHYSICS TODAY

Electric Dolphins
Dolphins achieve remarkable propulsive efficiency as they swim. Scientists who study fluid dynamics have long believed that dolphins control turbulence by moving their skins. Applying this theory to airplanes, the U.S. Air Force is using micromachining to make ICs and microsensors that will turn plane wings into sensitive electronic skin that could reduce turbulence drag.

Applications of Bernoulli's Equation

In many physical situations, the velocity, height, or pressure of a fluid is constant. In such cases, Bernoulli's equation holds in simpler form. For example, when a liquid is stationary, both v_1 and v_2 are zero. Bernoulli's equation will then show that the difference in pressure is

$$P_2 - P_1 = \rho g(h_1 - h_2) \tag{15.16}$$

This equation is identical to the relationship discussed for fluids at rest.

Another important result occurs when there is no change in pressure ($P_1 = P_2$). In Fig. 15.18 a liquid emerges from a hole, or orifice, near the bottom of an open tank. Its velocity as it emerges from the orifice can be determined from Bernoulli's equation. We will assume the liquid level in the tank falls slowly in comparison with the emergent velocity, so the velocity v_2 at the top is assumed to be zero. In addition, it is noted that the liquid pressure both at the top and at the orifice is equal to atmospheric pressure. Thus, $P_1 = P_2$ and $v_2 = 0$, reducing Bernoulli's equation to

$$\rho g h_1 + \frac{1}{2}\rho v_1^2 = \rho g h_2$$

or

$$v_1^2 = 2g(h_2 - h_1) = 2gh$$

This relationship is known as *Torricelli's* (toh-ree-chel'-ees) *theorem:*

$$v = \sqrt{2gh} \tag{15.17}$$

Note that the emergent velocity of a liquid at a depth h is the same as that of an object dropped from rest at a height h.

The rate at which a liquid flows from an orifice is given by vA from Eq. (15.11). Torricelli's relation allows us to express the rate of flow in terms of the height of a liquid above the orifice. Hence,

$$R = vA = A\sqrt{2gh} \tag{15.18}$$

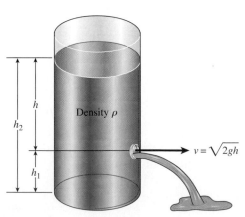

Figure 15.18   Torricelli's theorem.

Example 15.10

A crack in a water tank has a cross-sectional area of 1 cm². At what rate is water lost from the tank if the water level in the tank is 4 m above the opening?

Solution: The area $A = 1 \text{ cm}^2 = 10^{-4} \text{ m}^2$ and the height $h = 4$ m. Direct substitution into Eq. (15.18) gives

$$R = A\sqrt{2gh} = (10^{-4} \text{ m}^2)\sqrt{(2)(9.8 \text{ m/s}^2)(4 \text{ m})}$$
$$= (10^{-4} \text{ m}^2)(8.85 \text{ m/s}) = 8.85 \times 10^{-4} \text{ m}^3/\text{s}$$

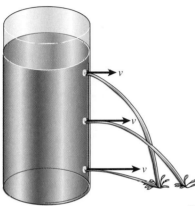

Figure 15.19 The discharge velocity increases with depth below the surface, but the range is a maximum at the midpoint.

An interesting example demonstrating Torricelli's principle is shown in Fig. 15.19. The discharge velocity increases with depth. Note that the maximum range occurs when the opening is at the middle of the water column. Although the discharge velocity increases below the midpoint, the water strikes the floor closer. This is because it strikes the floor sooner. Holes equidistant above and below the midpoint will have the same horizontal range.

As a final application, let us return to the venturi effect, which describes the motion of a fluid through a constriction. If the pipe in Fig. 15.20 is horizontal, we can set $h_1 = h_2$ in Bernoulli's equation, giving

$$P_1 + \frac{1}{2}\rho v_1^2 = P_2 + \frac{1}{2}\rho v_2^2 \qquad \textbf{(15.19)}$$

Since v_1 is greater than v_2, it follows that the pressure P_1 must be less than the pressure P_2 for Eq. (15.19) to hold. This relationship between velocity and pressure has already been discussed.

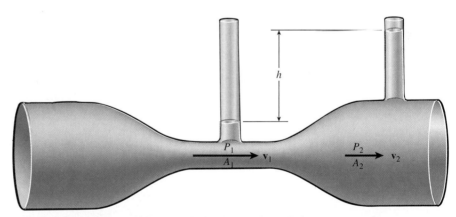

Figure 15.20 Fluid flow through a constriction in a horizontal pipe.

Problem-Solving Strategy

Applications of Bernoulli's Equation

1. Read the problem carefully and, after drawing a rough sketch, list the given information. Make sure that consistent units are used for pressure, height, and density.

2. The height h of a fluid is measured from a common reference point to the center of mass of the fluid. For example, a constriction in a horizontal pipe as in Fig. 15.20 does *not* represent a change in height ($h_1 = h_2$).

3. In Bernoulli's equation, the density ρ is *mass* density, and the appropriate units are kg/m³ and slug/ft³.

4. Write Bernoulli's equation for the problem and simplify by eliminating those factors that do not change:

$$P_1 + \rho g h_1 + \frac{1}{2}\rho v_1^2 = P_2 + \rho g h_2 + \frac{1}{2}\rho v_2^2$$

5. For a stationary fluid, $v_1 = v_2$, and the third term on each side is eliminated; the middle terms drop out for a horizontal pipe ($h_1 = h_2$); and, if there is no change in pressure ($P_1 = P_2$), the first terms do not appear, and the result is Torricelli's theorem (Eq. 15.17). Refer to the boxed equations in the summary.

6. Substitute the given quantities and solve for the unknown.

Example 15.11

As shown in Fig. 15.20, the water flowing through the constriction of a venturi tube has a velocity of $v_1 = 4$ m/s. If $h = 8$ cm, what will be the exit velocity v_2 when it flows into the larger tube?

Plan: We will first find the pressure difference between the wide and narrow regions based on the difference in the liquid heights h. Then, we will apply Bernoulli's equation for horizontal fluid flow to find another expression for the difference in pressure. Using the two equations, we can eliminate the need to know the pressure, and we can solve for the exit velocity.

Solution: The difference in pressure from Eq. (15.13) is

$$P_2 - P_1 = \rho g h$$

Using Bernoulli's equation where the center of the fluid flow does not change, we have

$$P_2 - P_1 = \frac{1}{2}\rho v_1^2 - \frac{1}{2}\rho v_2^2$$

Combining these two equations, we obtain

$$\rho g h = \frac{1}{2}\rho v_1^2 - \frac{1}{2}\rho v_2^2$$

Multiplying by 2 and dividing out the density ρ, we can simplify this expression to give

$$2gh = v_1^2 - v_2^2$$

Note that this relation is similar to that for a free-falling body. We can now solve this equation for the exit velocity v_2.

$$v_2^2 = v_1^2 - 2gh \qquad \text{or} \qquad v_2 = \sqrt{v_1^2 - 2gh}$$

$$v_2 = \sqrt{(4 \text{ m/s})^2 - 2(9.8 \text{ m/s}^2)(0.08 \text{ m})} = \sqrt{14.4 \text{ m}^2/\text{s}^2}$$

$$v_2 = 3.80 \text{ m/s}$$

The velocity is lower in the pipe of larger cross section.

In Example 15.11, the density ρ of the fluid did not enter into our calculations because the density of the fluid in the constriction was the same as that in the larger cross section. In such applications, it must be remembered that the density ρ in Bernoulli's equation is the *mass density* and not the weight density.

Summary and Review

Summary

We have presented the concepts of fluids at rest and in motion. Density, buoyant forces, and other quantities were defined and applied to many physical examples. The rate of flow of fluids was related to fluid velocity and to cross-sectional areas of pipes, and Bernoulli's equation was introduced to treat the more complete description of the dynamics of fluids. The essential concepts are summarized as follows.

- An important physical property of matter is its *density*. The weight density D and the mass density ρ are defined as follows:

$$Weight\ density = \frac{weight}{volume} \qquad \boxed{D = \frac{w}{V}}$$

$$N/m^3\ or\ lb/ft^3$$

$$Mass\ density = \frac{mass}{volume} \qquad \boxed{\rho = \frac{m}{V}}$$

$$kg/m^3\ or\ slug/ft^3$$

- Since $W = mg$, the relationship between D and ρ is

$$\boxed{D = \rho g}$$

$$Weight\ density = mass\ density \times gravity$$

- Important points to remember about fluid pressure:

  a. The forces exerted by a fluid on the walls of its container are always perpendicular to the walls.
  b. The fluid pressure is directly proportional to the depth of the fluid and to its density.

$$\boxed{P = \frac{F}{A} \qquad P = Dh \qquad P = \rho g h}$$

  c. At any particular depth, the fluid pressure is the same in all directions.
  d. Fluid pressure is independent of the shape or area of the container.

- Pascal's law states that *an external pressure applied to an enclosed fluid is transmitted uniformly throughout the volume of the liquid.*
- When measuring fluid pressure, be sure to distinguish between *absolute* pressure and *gauge* pressure:

Absolute pressure
$$= gauge\ pressure + atmospheric\ pressure$$
Atmospheric pressure
$$= 1\ atm = 1.013 \times 10^5\ N/m^2$$
$$= 1.013 \times 10^5\ Pa = 14.7\ lb/in.^2$$
$$= 76\ cm\ of\ mercury$$

- Applying Pascal's law to the hydraulic press gives the following for the ideal mechanical advantage:

$$\boxed{M_I = \frac{F_o}{F_i} = \frac{s_i}{s_o}} \qquad \begin{array}{l} Ideal\ Mechanical\ Advantage \\ for\ Hydraulic\ Press \end{array}$$

- Archimedes' principle: *An object that is completely or partly submerged in a fluid experiences an upward force equal to the weight of the fluid displaced.*

$$\boxed{F_B = mg} \quad or \quad \boxed{F_B = V\rho g} \quad Buoyant\ Force$$

- The *rate of flow* is defined as the volume of fluid that passes a certain cross section A per unit of time t. In terms of fluid velocity v, we write

$$\boxed{R = \frac{V}{t} = vA} \qquad \begin{array}{l} Rate\ of\ flow = velocity \\ \times\ cross\text{-}section \end{array}$$

- For an incompressible fluid flowing through pipes in which the cross sections vary, the rate of flow is constant:

$$\boxed{v_1 A_1 = v_2 A_2} \qquad \boxed{d_1^2 v_1 = d_2^2 v_2}$$

where v is the fluid velocity, A is the cross-sectional area of the pipe, and d is the diameter of the pipe.
- The net work done on a fluid is equal to the changes in kinetic and potential energy of the fluid. Bernoulli's equation expresses this fact in terms of pressure P, the density ρ, the height of the fluid h, and its velocity v.

$$\boxed{P + \rho g h + \frac{1}{2}\rho v^2 = constant} \qquad \begin{array}{l} Bernoulli's \\ Equation \end{array}$$

If a volume of fluid changes from a state 1 to a state 2, as shown in Fig. 15.17, we can write

$$\boxed{P_1 + \rho g h_1 + \frac{1}{2}\rho v_1^2 = P_2 + \rho g h_2 + \frac{1}{2}\rho v_2^2}$$

- Special applications of Bernoulli's equation occur when one of the parameters does not change:

For a stationary liquid,
$(v_1 = v_2)$

$$P_2 - P_1 = \rho g(h_1 - h_2)$$

If the pressure is constant,
$(P_1 = P_2)$

$$v = \sqrt{2gh}$$

For a horizontal pipe,
$(h_1 = h_2)$

$$P_1 + \frac{1}{2}\rho v_1^2 = P_2 + \frac{1}{2}\rho v_2^2$$

Key Terms

absolute pressure   308
Archimedes' principle   311
atmosphere   308
atmospheric pressure   308
Bernoulli's equation   319
buoyant force   312
density   302

gauge pressure   308
manometer   308
Pascal's law   308
pressure   304
rate of flow   316
specific gravity   304
streamline flow   315

Torricelli's theorem   320
total force   307
turbulent flow   315
venturi effect   317
weight density   302

Review Questions

15.1. Make a list of the units for weight density and the similar units for mass density.

15.2. Which is numerically larger: the weight density of an object or its mass density?

15.3. The density of water is given in Table 15.1 as 62.4 lb/ft³. In performing an experiment with water on the surface of the Moon, would you trust this value? Explain.

15.4. Which is heavier, 870 kg of brass or 3.5 ft³ of copper?

15.5. Why are dams so much thicker at the bottom than at the top? Does the pressure exerted on the dam depend on the length of the reservoir perpendicular to the dam?

15.6. A large block of ice floats in a bucket of water so the level of the water is at the top of the bucket. Will the water overflow when the ice melts? Explain.

15.7. A tub of water rests on weighing scales that indicate 40 lb total weight. Will the total weight increase when a 5-lb fish is floating on the surface of the water? Discuss.

15.8. Suppose an iron block supported by a string is submerged completely in the tub of Question 15.7. How will the reading on the scales be affected?

15.9. A boy just learning to swim finds that he can float on the surface more easily after inhaling air. He also observes that he can hasten his descent to the bottom of the pool by exhaling air on the way down. Explain his observations.

15.10. A toy sailboat filled with pennies floats in a small tub of water. If the pennies are thrown into the water, what happens to the water level in the tub?

15.11. Is it more difficult to hold a cork float barely under the surface than it is at a depth of 5 ft? Explain.

15.12. Is it possible to construct a barometer using water instead of mercury? How high will the column of water be if the external pressure is 1 atm?

15.13. Discuss the operation of a submarine and a weather balloon. Why will a balloon rise to a definite height and stop? Will a submarine sink to a particular depth and stop if no changes are made after submerging?

15.14. What assumptions and generalizations are made concerning the study of fluid dynamics?

15.15. Why does the flow of water from a faucet decrease when someone turns on another faucet in the same building?

15.16. Two rowboats moving parallel to each other in the same direction are drawn together. Explain.

15.17. Explain what would happen in a modern jet airliner at high speed if a hijacker fired a bullet through the window or broke open an escape hatch.

15.18. During high-velocity windstorms or hurricanes, the roofs of houses are sometimes blown off; yet the houses are otherwise undamaged. Explain with the use of diagrams.

15.19. A small child knocks a balloon over the heating duct in his home and is surprised to find that the balloon remains suspended above the duct, bobbing from one side to the other. Explain.

15.20. What conditions would determine the maximum lift capacity of a streamlined aircraft wing? Draw figures to justify your answer.

15.21. Explain with diagrams how a baseball pitcher throws a rising fast ball, an outside curve ball, and a sinking fast ball. Would a pitcher prefer to throw into the wind or with the wind when delivering the three pitches discussed above?

15.22. Two identical reservoirs are placed on the floor side by side. One is filled with mercury, and the other is filled with water. A hole is bored in each reservoir at the same depth below the surface. Compare the ranges of the emergent fluids.

Problems

Section 15.1 Density

15.1. What volume does 0.4 kg of alcohol occupy? What is the weight of this volume?
> Ans. 5.06×10^{-4} m³, 3.92 N

15.2. An unknown substance has a volume of 20 ft³ and weighs 3370 lb. What are the weight density and the mass density?

15.3. What volume of water has the same mass of 100 cm³ of lead? What is the weight density of lead?
> Ans. 1130 cm³, 1.11×10^5 N/m³

*15.4. A 200-mL flask (1 L = 1000 cm³) is filled with an unknown liquid. An electronic balance indicates that the added liquid has a mass of 176 g. What is the specific gravity of the liquid? Can you guess the identity of the liquid?

Section 15.3 Fluid Pressure

15.5. Find the pressure in kilopascals due to a column of mercury 60 cm high. What is this pressure in lb/in.² and in atmospheres?
> Ans. 80.0 kPa, 11.6 lb/in.², 0.79 atm

15.6. A pipe contains water under a gauge pressure of 400 kPa. If you patch a 4-mm-diameter hole in the pipe with a piece of tape, what force must the tape be able to withstand?

*15.7. A submarine dives to a depth of 120 ft and levels off. The interior of the submarine is maintained at atmospheric pressure. What are the pressure and the total force applied to a hatch 2 ft wide and 3 ft long? The weight density of sea water is around 64 lb/ft³.
> Ans. 53.3 lb/in.², 46,100 lb

15.8. If you constructed a barometer using water as the liquid instead of mercury, what height of water would indicate a pressure of 1 atmosphere?

15.9. A 20-kg piston rests on a sample of gas in a cylinder 8 cm in diameter. What is the gauge pressure on the gas? What is the absolute pressure?
> Ans. 39.0 kPa, 140.3 kPa

*15.10. An open U-shaped tube such as the one in Fig. 15.21 is 1 cm² in cross section. What volume of water must be poured into the right tube to cause the mercury in the left tube to rise 1 cm above its original position?

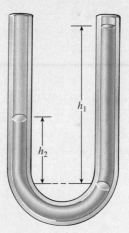

Figure 15.21

15.11. The gauge pressure in an automobile tire is 28 lb/in.². If the wheel supports 1000 lb, what area of the tire is in contact with the ground?
> Ans. 35.7 in.²

*15.12. Two liquids that do not react chemically are placed in a bent tube like the one in Fig. 15.21. Show that the heights of the liquids above their surface of separation are inversely proportional to their densities:

$$\frac{h_1}{h_2} = \frac{\rho_2}{\rho_1}$$

*15.13. Assume that the two liquids in the U-shaped tube in Fig. 15.21 are water and oil. Compute the density of the oil if the water stands 19 cm above the interface and the oil stands 24 cm above the interface. Refer to Prob. 15.12.   Ans. 792 kg/m^3

15.14. A water-pressure gauge indicates a pressure of 50 lb/in.^{2} at the foot of a building. What is the maximum height to which the water will rise in the building?

Section 15.5 The Hydraulic Press

15.15. The areas of the small and large pistons in a hydraulic press are 0.5 and 25 in.^{2}, respectively. What is the ideal mechanical advantage of the press? What force must be exerted to lift a 1-ton (2000-lb) load? Through what distance must the input force act to lift this load a distance of 1 in.?
Ans. 50, 40 lb, 50 in.

15.16. A force of 400 N is applied to the small piston of a hydraulic press whose diameter is 4 cm. What must be the diameter of the large piston if it is to lift a 200-kg load?

15.17. The inlet pipe that supplies air pressure to operate a hydraulic lift is 2 cm in diameter. The output piston is 32 cm in diameter. What air pressure (gauge pressure) must be used to lift an 1800-kg automobile?   Ans. 219 kPa

15.18. The area of a piston in a force pump is 10 in.^{2}. What force is required to raise water with the piston to a height of 100 ft?

Section 15.6 Archimedes' Principle

15.19. A 100-g cube 2 cm on each side is attached to a string and then totally submerged in water. What is the buoyant force, and what is the tension in the rope?   Ans. 0.0784 N, 0.902 N

*15.20. A solid object weighs 8 N in air. When this object is suspended from a spring scale and submerged in water, the apparent weight is only 6.5 N. What is the density of the object?

*15.21. A cube of wood 5.0 cm on each edge floats in water with three-fourths of its volume submerged. (a) What is the weight of the cube? (b) What is the mass of the cube? (c) What is the specific gravity of wood?   Ans. (a) 0.919 N, (b) 93.8 g, (c) 0.75

*15.22. A 20-g piece of metal has a density of 4000 kg/m^3. It is hung in a jar of oil (1500 kg/m^3) by a thin thread until it is completely submerged. What is the tension in the thread?

*15.23. The mass of a piece of rock is found to be 9.17 g in air. When the rock is submerged in a fluid of density 873 kg/m^3, its apparent mass is only 7.26 g. What is the density of this rock?
Ans. 4191 kg/m^3

*15.24. A balloon 40 m in diameter is filled with helium. The mass of the balloon and attached basket is 18 kg. What additional mass can be lifted by this balloon?

Section 15.7 Fluid Flow

15.25. Gasoline flows through a 1-in.-diameter hose at an average velocity of 5 ft/s. What is the rate of flow in gallons per minute (1 ft^3 = 7.48 gal)? How much time is required to fill a 20-gal tank?
Ans. 12.2 gal/min, 1.63 min

15.26. Water flows from a terminal 3 cm in diameter and has an average velocity of 2 m/s. What is the rate of flow in liters per minute (1 L = 0.001 m^3)? How much time is required to fill a 40-L container?

15.27. What must be the diameter of a hose if it is to deliver 8 L of oil in 1 min with an exit velocity of 3 m/s?   Ans. 7.52 mm

*15.28. Water flowing from a 2-in. pipe emerges horizontally at the rate of 8 gal/min. What is the emerging velocity? What is the horizontal range of the stream of water if the pipe is 4 ft from the ground?

15.29. Water flowing at 6 m/s through a 6-cm pipe is connected to a 3-cm pipe. What is the velocity in the small pipe? Is the rate of flow greater in the smaller pipe?   Ans. 24 m/s, no

Section 15.10 Applications of Bernoulli's Equation

15.30. Consider the situation described by Prob. 15.29. If the centers of each pipe are on the same horizontal line, what is the difference in pressure between the two connecting pipes?

15.31. What is the emergent velocity of water from a crack in its container 6 m below the surface? If the area of the crack is 1.3 cm^2, at what rate of flow does water leave the container?
Ans. 10.8 m/s, 1.41 × 10^{-3} m^3/s

15.32. A 2-cm-diameter hole is in the side of a water tank, and it is located 5 m below the water level in the tank. What is the emergent velocity of the water from the hole? What volume of water will escape from this hole in 1 min?

*15.33. Water flows through a horizontal pipe at the rate of 82 ft^3/min. A pressure gauge placed on a 6-in. cross section of this pipe reads 16 lb/in.^{2}. What is the gauge pressure in a section of pipe where the diameter is 3 in.?   Ans. 11.1 lb/in.^{2}

*15.34. Water flows at the rate of 6 gal/min through an opening in the bottom of a cylindrical tank. The water in the tank is 16 ft deep. What is the rate of escape if an added pressure of 9 lb/in.^{2} is applied to the source of the water?

*15.35. Water moves through a pipe at 4 m/s under an absolute pressure of 200 kPa. The pipe narrows to one-half of its original diameter. What is the absolute pressure in the narrow part of the pipe?

Ans. 80.0 kPa

*15.36. Water flows steadily through a horizontal pipe. At a point where the absolute pressure is 300 kPa, the velocity is 2 m/s. The pipe suddenly narrows, causing the absolute pressure to drop to 100 kPa. What will be the velocity of the water in this constriction?

Additional Problems

*15.37. Human blood of density 1050 kg/m^3 is held a distance of 60 cm above an arm of a patient to whom it is being administered. How much higher is the pressure at this position than it would be if it were held at the same level as the arm?

*15.38. A cylindrical tank 50 ft high and 20 ft in diameter is filled with water. (a) What is the water pressure on the bottom of the tank? (b) What is the total force on the bottom? (c) What is the pressure in a water pipe that is located 90 ft below the water level in the tank?

*15.39. A block of wood weighs 16 lb in air. A lead sinker, which has an apparent weight of 28 lb in water, is attached to the wood, and both are submerged in water. If their combined apparent weight in water is 18 lb, find the density of the wooden block. Ans. 38.4 lb/ft^3

*15.40. A 100-g block of wood has a volume of 120 cm^3. Will it float in water? Gasoline?

*15.41. A vertical test tube has 3 cm of oil (0.8 g/cm^3) floating on 9 cm of water. What is the pressure at the bottom of the tube? Ans. 1.12 kPa

15.42. What percentage of an iceberg will remain below the surface of seawater (1030 kg/m^3)?

*15.43. What is the smallest area of ice 30 cm thick that will support a 90-kg man? The ice is floating in freshwater. Ans. 3.75 m^2

15.44. A spring balance indicates a weight of 40 N when an object is hung in air. When the same object is submerged in water, the indicated weight is reduced to only 30 N. What is the density of the object?

15.45. A thin-walled metal cup has a mass of 100 g and a total volume of 250 cm^3. What is the maximum number of pennies that can be placed in the cup without sinking in water? The mass of a single penny is 3.11 g. Ans. 48

15.46. What is the absolute pressure at the bottom of a lake that is 30 m deep?

15.47. A fluid is forced out of a 6-mm-diameter tube so that 200 mL emerges in 32 s. What is the average velocity of the fluid in the tube? Ans. 0.221 m/s

*15.48. A pump of 2 kW output power discharges water from a cellar into a street 6 m above. How much higher is the pressure at this position than it would be if it were held at the same level as the arm? Ans. 6.17 kPa

At what rate in liters per second is the cellar emptied?

15.49. A horizontal pipe of diameter 120 mm has a constriction of diameter 40 mm. The velocity of water in the pipe is 60 cm/s and the pressure is 150 kPa. (a) What is the velocity in the constriction? (b) What is the pressure in the constriction?

Ans. (a) 540 cm/s, (b) 136 kPa

*15.50. The water column in the container shown in Fig. 15.20 stands at a height H above the base of the container. Show that the depth h required to give the horizontal range x is given by

$$h = \frac{H}{2} \pm \frac{\sqrt{H^2 - x^2}}{2}$$

How does this equation show that the holes equidistant above and below the midpoint will have the same horizontal range?

*15.51. A column of water stands 16 ft above the base of its container. What are two hole depths at which the emergent water will have a horizontal range of 8 ft? Ans. 1.07 ft, 14.9 ft

*15.52. Refer to Fig. 15.20 and Prob. 15.50. Show that the horizontal range is given by

$$x = 2 \sqrt{h(H - h)}$$

Use this relation to show that the maximum range is equal to the height H of the water column.

*15.53. Water flows through a horizontal pipe at the rate of 60 gal/min (1 ft^3 = 7.48 gal). What is the velocity in a narrow section of the pipe that is reduced from 6 to 1 in. in diameter?

Ans. 24.5 ft/s

15.54. What must be the gauge pressure in a fire hose if the nozzle is to force water to a height of 20 m?

*15.55. Water flows through the pipe shown in Fig. 15.22 at the rate of 30 liters per second. The absolute pressure at point A is 200 kPa, and point B is 8 m higher than point A. The lower section of pipe has a diameter of 16 cm, and the upper section narrows to a diameter of 10 cm. (a) Find the velocities of the stream at points A and B. (b) What is the absolute pressure at point B?

Ans. (a) 1.49 m/s, 3.82 m/s; (b) 115 kPa

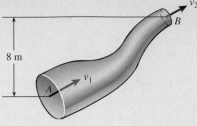

Figure 15.22

Critical Thinking Questions

15.56. A living-room floor has floor dimensions of 4.50 m and 3.20 m and a height of 2.40 m. The density of air is 1.29 kg/m^3. What does the air in the room weigh? What force does the atmosphere exert on the floor of the room?

Ans. 437 N, 1.46 × 10^6 N

15.57. A tin coffee can floating in water (1.00 g/cm^3) has an internal volume of 180 cm^3 and a mass of 112 g. How many grams of metal can be added to the can without causing it to sink in the water?

*15.58. A wooden block floats in water with two-thirds of its volume submerged. The same block floats in oil with nine-tenths of its volume submerged. What is the ratio of the density of the oil to the density of water (the specific gravity)?

Ans. 0.741

*15.59. An aircraft wing 25 ft long and 5 ft wide experiences a lifting force of 800 lb. What is the difference in pressure between the upper and lower surfaces of the wing?

*15.60. Assume that air (ρ = 1.29 kg/m^3) flows past the top surface of an aircraft wing at 36 m/s. The air moving past the lower surface of the wing has a

velocity of 27 m/s. If the wing has a weight of 2700 N and an area of 3.5 m^2, what is the buoyant force on the wing?

Ans. 1280 N

*15.61. Seawater has a weight density of 64 lb/ft^3. This water is pumped through a system of pipes (see Fig. 15.23) at the rate of 4 ft^3/min. Pipe diameters at the lower and upper ends are 4 in. and 2 in., respectively. The water is discharged into the atmosphere at the upper end, a distance of 6 ft higher than the lower section. What are the velocities of flow in the upper and lower pipes? What are the pressures in the lower and upper sections?

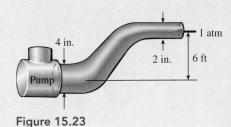

Figure 15.23

16

Temperature and Expansion

Temperature is a measure of the average kinetic energy per molecule. An infrared thermometer captures the invisible infrared energy naturally emitted from all objects. The infrared radiation coming from the air canal in the ear passes through the optical system of the thermometer and is converted to an electrical signal that is proportional to the energy radiated from that area. By calibrating this signal with known temperatures, a digital readout can be displayed. *(Photo by Blake Tippens.)*

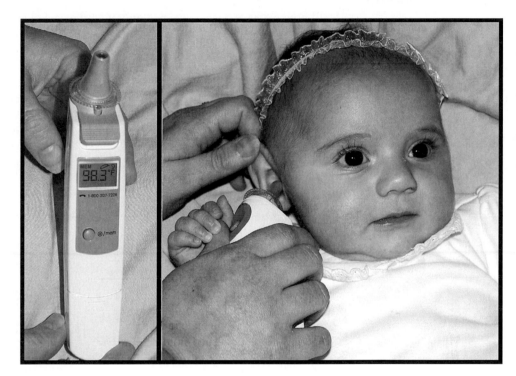

Objectives

After completing this chapter, you should be able to

1. Demonstrate your understanding of the Celsius, Fahrenheit, Kelvin, and Rankine temperature scales by converting from specific temperatures on one scale to corresponding temperatures on another scale.

2. Distinguish between specific temperatures and temperature intervals and convert an interval on one scale to the equivalent interval on another scale.

3. Write formulas for linear expansion, area expansion, and volume expansion and be able to apply them to the solution of problems similar to those given in this chapter.

We have discussed the behavior of systems at rest and in motion. The fundamental quantities of mass, length, and time were introduced to describe the state of a given mechanical system.

Consider, for example, a 10-kg block moving with a constant velocity of 20 m/s. The quantities of mass, length, and time are all present, and we find them sufficient to describe the motion. We can speak of the weight of the block, its kinetic energy, or its momentum, but a complete description of a system requires more than a simple statement of these quantities.

329

This becomes apparent when our 10-kg block encounters frictional forces. As the block slides to a stop, its energy seems to disappear, but the block and its supporting surface are slightly warmer. If energy is to be conserved, we must assume that the lost energy reappears in some form not yet considered. When energy disappears from the visible motion of objects and does not reappear in the form of visible potential energy, we frequently notice a rise in temperature. In this chapter, we introduce the concept of temperature as a fourth fundamental quantity.

16.1 Temperature and Thermal Energy

Until now we have been concerned only with the causes and the effects of *external* motion. A block resting on a table is in translational and rotational equilibrium insofar as its surroundings are concerned. A much closer study of the block reveals, however, that it is active internally. Figure 16.1 shows a simple model of a solid. Individual molecules are held together by elastic forces analogous to the springs in the figure. These molecules oscillate about their equilibrium positions with a particular frequency and amplitude A. Thus, both potential energy and kinetic energy are associated with the molecular motion. Since this internal energy is related to the hotness or coldness of a body, it is often referred to as ***thermal energy.***

Thermal energy represents the total internal energy of an object: the sum of its molecular kinetic and potential energies.

When two objects with different temperatures are placed in contact, energy is transferred from one to the other. For example, suppose hot coals are dropped into a container of water, as shown in Fig. 16.2. Thermal energy will be transferred from the coals to the

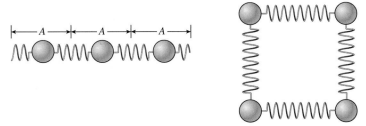

Figure 16.1   A simplified model of a solid in which the individual molecules are held together by elastic forces.

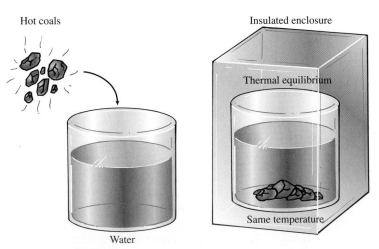

Figure 16.2   Thermal equilibrium.

water until the system reaches a stable condition, called ***thermal equilibrium.*** When touched, the coals and the water produce similar sensations, and there is no more transfer of thermal energy.

Such changes in thermal energy states cannot be explained satisfactorily in terms of classical mechanics alone. Therefore, all objects must have a new fundamental property that determines whether they will be in thermal equilibrium with other objects. This property is called ***temperature.*** In our example, the coals and the water are said to have the same temperature when the transfer of energy is zero.

> Two objects are said to be in thermal equilibrium if and only if they are at the same temperature.

Once we establish a means of measuring temperature, we have a necessary and sufficient condition for thermal equilibrium. The transfer of thermal energy that is due only to a difference in temperature is defined as ***heat.***

> Heat is defined as the transfer of thermal energy that is due to a difference of temperature.

Before discussing the measurement of temperature, we should distinguish clearly between temperature and thermal energy. It is possible for two objects to be in thermal equilibrium (same temperature) with different thermal energies. For example, consider a pitcher of water and a small cup of water, each having a temperature of 90°C. If they are mixed together, there will be no transfer of energy, but the thermal energy is much greater in the pitcher because it contains many more molecules. Remember that thermal energy represents the *sum* of the kinetic and potential energies of all the molecules. If we pour the water from each container onto separate blocks of ice, as shown in Fig. 16.3, more ice will be melted by the larger volume, indicating that it had more thermal energy.

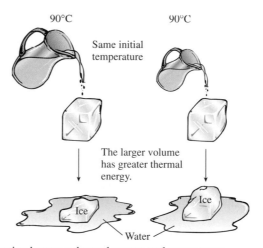

Figure 16.3   The distinction between thermal energy and temperature.

16.2   The Measurement of Temperature

Temperature is usually determined by measuring some mechanical, optical, or electrical quantity that varies with temperature. For example, most substances expand as their temperature increases. If a change in any dimension can be shown to have a one-to-one correspondence with changes in temperature, the variation can be calibrated to measure temperature. A device calibrated in this way is called a ***thermometer.*** The temperature of another object can then be measured by placing the thermometer in close contact with it and allowing the two to reach thermal equilibrium. The temperature indicated by a number on the graduated thermometer also corresponds to the temperature of the surrounding objects.

A thermometer is a device that, through marked scales, can give an indication of its own temperature.

Two things are necessary when constructing a thermometer. First, we must have confirmation that some thermometric property X varies with temperature t. If the variation is linear, we can write

$$t = kX$$

where k is the proportionality constant. The thermometric property should be one that is easily measured, for example, the expansion of a liquid, the pressure in a gas, or the resistance in an electric circuit. Other quantities that vary with temperature are radiated energy, the color of emitted light, vapor pressure, and magnetic susceptibility. Thermometers have been constructed for each of these thermometric properties. The choice is dictated by the range of temperatures over which the thermometer is linear and by the mechanics of its use.

The second requirement in constructing a thermometer is the establishment of standard temperatures. Early temperature scales were based on the choice of upper and lower *fixed points,* which are temperatures readily available for laboratory measurements. Two convenient temperatures reproduced easily are the ***lower fixed point*** and the ***upper fixed point.***

The lower fixed point (ice point) is the temperature at which water and ice coexist in thermal equilibrium under a pressure of 1 atm.

The upper fixed point (steam point) is the temperature at which water and steam coexist in equilibrium under a pressure of 1 atm.

A widely used temperature measurement for scientific work originated with a scale developed by Anders Celsius (1701–1744), a Swedish astronomer. The ***Celsius scale*** arbitrarily assigned the number 0 to the ice point and the number 100 to the steam point. Thus, at atmospheric pressure, there are 100 divisions between the freezing point and the boiling point of water. Each division or unit on the scale is called a ***degree*** (°). For example, room temperature is often taken as 20°C, which is read as *twenty degrees Celsius.*

Another scale for measuring temperature was developed in 1714 by Gabriel Daniel Fahrenheit. The development of this scale was based on the choice of different fixed points. Fahrenheit chose the temperature of a freezing solution of salt water as his lower fixed point and assigned it the number and unit of 0°F. The upper fixed point was chosen to be the temperature of the human body. For some unexplained reason, he assigned the number and unit of 96°F for body temperature. The fact that body temperature is really 98.6°F indicates an experimental error in establishing the scale. Relating the ***Fahrenheit scale*** to the universally accepted fixed points on the Celsius scale, it can be shown that 0 and 100°C correspond to 32 and 212°F.

We can compare the two scales by calibrating ordinary mercury-in-glass thermometers. This type of thermometer makes use of the fact that liquid mercury expands with increasing temperature. It consists of an evacuated glass capillary tube with a reservoir of mercury at the bottom and a closed top. Since mercury expands more than the glass tube, the mercury column rises in the tube until the mercury, glass, and its surroundings are in equilibrium.

Suppose we make two ungraduated thermometers and place them in a mixture of ice and water, as in Fig. 16.4. After allowing the mercury columns to stabilize, we mark 0°C on one thermometer and 32°F on the other. Next, we place the two thermometers directly above boiling water, allowing the mercury columns to stabilize at the steam point. Again

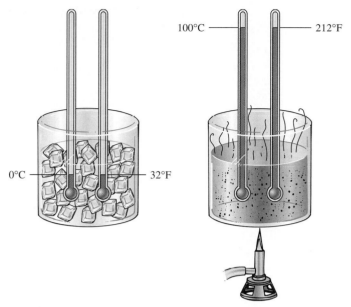

Figure 16.4   Calibration of Celsius and Fahrenheit thermometers.

we mark the two thermometers, inscribing 100°C and 212°F adjacent to the mercury level above the marks corresponding to the ice point. The level of mercury is the same in each thermometer. Thus, the only difference between the two thermometers will be in how they are graduated. There are 100 divisions, or Celsius degrees (C°), between the ice point and steam point on the Celsius thermometer, and there are 180 divisions, or Fahrenheit degrees (F°), on the Fahrenheit thermometer. Thus, 100 Celsius degrees represents the same temperature interval as 180 Fahrenheit degrees. Symbolically,

$$100 \ C° = 180 \ F° \qquad \text{or} \qquad 5 \ C° = 9 \ F° \qquad \textbf{(16.1)}$$

The degree mark (°) is placed after the C or F to emphasize that the numbers correspond to temperature intervals and not to specific temperatures. In other words, 20 F° is read "twenty Fahrenheit degrees" and corresponds to a *difference* between two temperatures on the Fahrenheit scale. The symbol 20°F, on the other hand, refers to a specific mark on the Fahrenheit thermometer. For example, suppose a pan of hot food cools from 98 to 76°F. These numbers correspond to specific temperatures, as indicated by the height of a mercury column. They represent a temperature interval, however, of

$$\Delta t = 98°F - 76°F = 22 \ F°$$

where Δt is used to denote a change in temperature.

The physics that treats the transfer of thermal energy is nearly always concerned with changes in temperature. Thus, it often becomes necessary to convert a temperature interval from one scale to the corresponding interval on the other scale. This can best be accomplished by recalling from Eq. (16.1) that an interval of 5 C° is equivalent to an interval of 9 F°. The appropriate conversion factors can be written as

$$\frac{5 \ C°}{9 \ F°} = 1 = \frac{9 \ F°}{5 \ C°} \qquad \textbf{(16.2)}$$

When F° is converted to C°, the factor on the left should be used; when C° is converted to F°, the factor on the right should be used.

It must be remembered that Eq. (16.2) applies for temperature intervals. It can be used only when working with *differences* in temperature. It is another matter entirely to find the temperature on the Fahrenheit scale that corresponds to the same temperature on the Celsius scale. Using ratio and proportion, we can develop an equation that will convert specific

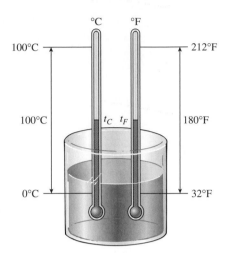

Figure 16.5   Comparison of the Celsius and Fahrenheit scales.

temperatures. Suppose, for example, we place two identical thermometers in a beaker of water, as shown in Fig. 16.5. One thermometer is graduated in Fahrenheit degrees and the other in Celsius degrees. The symbols t_C and t_F represent the same temperature (the temperature of the water), but they are on different scales. It should be apparent from the figure that the difference between t_C and 0°C corresponds to the same interval as the difference between t_F and 32°F. The ratio of the former to 100 divisions should be the same as the ratio of the latter to 180 divisions. Hence,

$$\frac{t_C - 0}{100} = \frac{t_F - 32}{180}$$

Simplifying and solving for t_C, we obtain

$$t_C = \frac{5}{9}(t_F - 32) \tag{16.3}$$

or, solving for t_F,

$$t_F = \frac{9}{5}t_C + 32 \tag{16.4}$$

Equations (16.3) and (16.4) are not true equalities because they result in a change of units. Rather than saying that 20°C is equal to 68°F, we should say that a temperature of 20°C *corresponds* to a temperature of 68°F.

Example 16.1   During a 24-h period, a steel rail varies in temperature from 20°F at night to 70°F in the middle of the day. Express this range of temperature in Celsius degrees.

Plan: We must recognize that the range of temperatures is an *interval* rather than a *specific temperature*. We determine the range in F° and then convert to C° by recognizing that an *interval* of 5 C° is the same *interval* as 9 F°.

Solution: The temperature interval in F° is

$$\Delta t = 70°F - 20°F = 50 \ F°$$

To find the corresponding interval in C°, we apply Eq. (16.2), choosing the conversion factor that cancels the Fahrenheit interval.

$$\Delta t = 50 \text{ F}° \left(\frac{5 \text{ C}°}{9 \text{ F}°} \right); \qquad \Delta t = 27.8 \text{ C}°$$

Example 16.2

The melting point of lead is 330°C. What is the corresponding temperature on the Fahrenheit scale?

Plan: Here we have a *specific temperature* on the Celsius scale that must be converted to a *corresponding* temperature on the Fahrenheit scale. We will first correct for the interval difference and then add 32°F to account for different zero points.

Solution: Substitution into Eq. (16.4) yields

$$t_F = \frac{9}{5} t_C + 32 = \frac{9}{5}(330) + 32$$

$$= 594 + 32 = 626°\text{F}$$

It is important to recognize that the t_F and t_C of Eqs. (16.3) and (16.4) represent corresponding temperatures. The numbers are different because the origin of each scale was at a different point and the degrees are of different size. What these equations tell us is the relationship between the *numbers* that are assigned to specific temperatures on two *different* scales.

16.3 The Gas Thermometer

Although the mercury-in-glass thermometer is the best known and most widely used, it is not as accurate as many other thermometers. In addition, mercury freezes at around −40°C, restricting the range over which it can be used. An accurate thermometer with an extensive measuring range can be constructed by utilizing the properties of a gas. All gases subjected to heating expand in nearly the same manner. If their expansion is prevented by maintaining a constant volume, the pressure will increase in proportion to temperature.

In general, there are two kinds of gas thermometers. One kind maintains a constant pressure and utilizes the increases in volume as an indicator. This type is called a ***constant-pressure thermometer.*** The other kind, called a ***constant-volume thermometer,*** measures the increase in pressure as a function of temperature. The constant-volume thermometer is illustrated in Fig. 16.6. The gas is contained in bulb *B,* and the pressure it exerts is measured by the mercury manometer. As the temperature of the gas increases, it expands, forcing the mercury down in the closed tube and up in the open tube. To maintain a constant volume of the gas, the open tube must be raised until the level of mercury in the closed tube is brought back to the reference mark *R*. The difference in the two mercury levels is then an indication of the gas pressure at constant volume. The instrument can be calibrated for temperature measurements through the use of fixed points, as mentioned in Section 16.2.

The same apparatus can be used as a constant-pressure thermometer, as illustrated in Fig. 16.7. In this instance, the volume of gas in bulb *B* is allowed to increase at constant pressure. The pressure exerted on the gas is maintained constant at 1 atm by lowering or raising the open tube until the mercury levels are the same in both tubes. The change in volume with temperature can then be indicated by the mercury level in the closed tube.

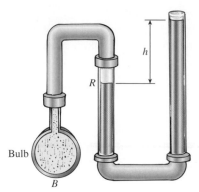

Figure 16.6 The constant-volume thermometer.

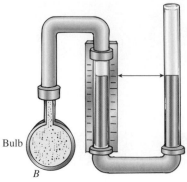

Figure 16.7 The constant-pressure thermometer.

Calibration consists of marking the mercury level at the ice point and again at the steam point.

Gas thermometers are useful because of their almost unlimited range. For this reason and because they are so accurate, they are commonly used in research laboratories and in bureaus of standards. They are also large and bulky, however, rendering them useless for many minute technical measurements.

16.4 The Absolute Temperature Scale

It has probably occurred to you that Celsius and Fahrenheit scales have a serious limitation. Neither 0°C nor 0°F represents a true zero of temperature. Consequently, for temperatures much lower than the ice point, a negative temperature results. Even more serious is the fact that a formula involving temperature as a variable will not work with the existing scales. For example, we have discussed the expansion of a gas with an increase in temperature. We can state this proportionality as

$$V = kt$$

where k is the proportionality constant and t is the temperature. Certainly, the volume of a gas is not zero at 0°C or negative at negative temperatures, and yet these conclusions might be drawn from the above relationship.

This example provides a clue for establishing an *absolute scale.* If we can determine the temperature at which the volume of a gas under constant pressure becomes zero, we can establish a true zero of temperature. Suppose we use a constant-pressure gas thermometer, like the one in Fig. 16.7. The volume of the gas in the bulb can be measured carefully, first at the ice point and then at the steam point. These two points can be plotted on a graph, as in Fig. 16.8, with the volume as the ordinate and the temperature as the abscissa. The points A and B correspond to the gas volume at temperatures of 0 and 100°C, respectively. A straight line through these two points, extended both to the left and to the right, provides a mathematical description of the change in volume as a function of temperature. Note that the line can be extended indefinitely to the right, indicating that there is no upper limit to temperature. We cannot extend the line indefinitely to the left, however, because it will eventually intercept the temperature axis. At this theoretical point, the gas would have zero volume. Further extension of the line would indicate a negative volume, which is meaningless. Therefore, the point at which the line intercepts the temperature axis is called the ***absolute zero*** of temperature. (Actually, any real gas would liquefy before reaching this point.)

If the above experiment is performed for several different gases, the slope of the curves will vary slightly. But the temperature intercept will always be the same and near −273°C. Ingenious theoretical and experimental procedures have established that the absolute zero of temperature is −273.15°C. In this text, we will assume that it is −273°C, without fear

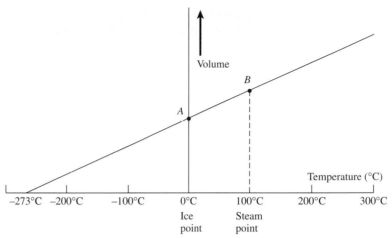

Figure 16.8   The variation of the volume of a gas as a function of temperature. Absolute zero can be defined by extrapolation to zero volume.

of significant error. Conversion to the Fahrenheit scale shows that absolute zero is $-460°F$ on that scale.

An absolute temperature scale has as its zero point the absolute zero of temperature. One such scale was devised by Lord Kelvin (1824–1907). The standard interval on this scale, the *kelvin,* has been adopted by the international (SI) metric system as the base unit for temperature measurement. The interval on the ***Kelvin scale*** represents the same change in temperature as the Celsius degree. Thus, an interval of 5 K (reads "five kelvins") is exactly the same as 5 C°.

The Kelvin scale is related to the Celsius scale by the formula

$$T_K = t_C + 273 \tag{16.5}$$

For example, 0°C would correspond to 273 K, and 100°C would correspond to 373 K. (See Fig. 16.9.) Hereafter, we will reserve the symbol T for absolute temperature and the symbol t for other temperatures.

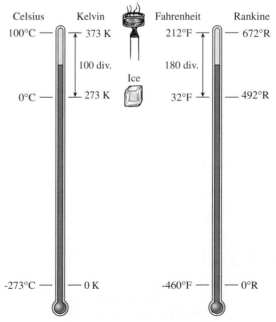

Figure 16.9   A comparison of the four common temperature scales.

Problems with the reproducibility of accurate measurements for the ice and steam points of water caused the International Committee on Weights and Measures to establish a new standard in 1954. This standard is based on the ***triple point of water,*** which is the single temperature and pressure at which water, water vapor, and ice can coexist in thermal equilibrium. This convenient event occurs at a temperature of around 0.01°C and at a pressure of 4.58 mm of mercury. To remain somewhat consistent with earlier measurements, the temperature for the triple point of water was set at exactly 273.16 K. Thus, the ***kelvin*** is now defined as the fraction 1/273.16 of the temperature of the triple point of water. The SI temperature is now fixed by this definition, and all other scales must be redefined based on this single standard temperature.

A second absolute scale, called the *Rankine scale,* remains in very limited use despite efforts by various organizations to eliminate its use entirely. The Rankine degree is included here only for historical purposes and will not be used in this text. Its absolute zero point is −460°F, and the degree interval is the same as the Fahrenheit interval. For example, 32°F corresponds to 492°R, and 212°F corresponds to 672°R.

$$T_R = t_F + 460 \qquad\qquad (16.6)$$

Remember that Eqs. (16.5) and (16.6) apply for specific temperatures. If we are concerned with a change in temperature or a difference in temperature, the absolute change or difference is the same in kelvins as it is in Celsius degrees. It is helpful to recall that

$$1\text{ K} = 1\text{ C}° \qquad 1\text{ R}° = 1\text{ F}° \qquad\qquad (16.7)$$

Example 16.3

A mercury-in-glass thermometer may not be used at temperatures below −40°C. This is because mercury freezes at this temperature. (a) What is the freezing point of mercury on the Kelvin scale? (b) What is the difference between this temperature and the freezing point of water? Express the answer in kelvins.

Solution (a): Direct substitution of −40°C into Eq. (16.5) yields

$$T_K = -40°\text{C} + 273 = 233\text{ K}$$

Solution (b): The difference in the freezing points is

$$\Delta t = 0°\text{C} - (-40°\text{C}) = 40\text{ C}°$$

Since the size of the kelvin is identical to that of the Celsius degree, the difference is also 40 kelvins.

At this point you may ask why we still retain the Celsius and Fahrenheit scales. When working with heat, one is nearly always concerned with differences in temperature. In fact, a difference in temperature is necessary for heat to be transferred. Otherwise, the system would be in thermal equilibrium. Since the Kelvin and Rankine scales are based on the same intervals as the Celsius and Fahrenheit scales, it makes no difference which scale is used for temperature intervals. On the other hand, if a formula calls for a specific temperature rather than a temperature difference, the absolute scale must be used.

16.5 Linear Expansion

The most common effect produced by temperature changes is a change in size. With a few exceptions, all substances increase in size with rising temperature. The atoms in a solid are held together in a regular pattern by electric forces. At any temperature, the atoms vibrate with a certain frequency and amplitude. As the temperature is increased, the amplitude

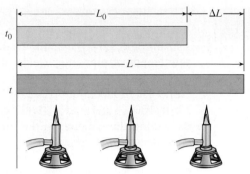

Figure 16.10 Linear expansion.

(maximum displacement) of the atomic vibrations increases. This results in an overall change in the dimensions of the solid.

A change in any *one* dimension of a solid is called ***linear expansion.*** It is found experimentally that an increase in a single dimension, for example, the length of a rod, is dependent on the original dimension and the change in temperature. Consider, for example, the rod in Fig. 16.10. The initial length is L_0, and the initial temperature is t_0. When the rod is heated to a temperature t, its new length is denoted by L. Thus, a change in temperature, $\Delta t = t - t_0$, has resulted in a change in length, $\Delta L = L - L_0$. The proportional change in length is given by

$$\Delta L = \alpha L_0 \, \Delta t \qquad \textbf{(16.8)}$$

where α is the proportionality constant called the ***coefficient of linear expansion.*** Since an increase in temperature does not produce the same increase in length for all materials, the coefficient α is a property of the material. Solving Eq. (16.8) for α, we obtain

$$\alpha = \frac{\Delta L}{L_0 \, \Delta t} \qquad \textbf{(16.9)}$$

The coefficient of linear expansion of a substance can be defined as the change in length per unit length per degree change in temperature. Since the ratio $\Delta L/L_0$ has no dimensions, the units of α are in inverse degrees, that is, $1/\text{C}^\circ$ or $1/\text{F}^\circ$. The expansion coefficients for many common materials are given in Table 16.1.

Table 16.1

Linear Expansion Coefficients

Substance	α	
	$10^{-5}/\text{C}^\circ$	$10^{-5}/\text{F}^\circ$
Aluminum	2.4	1.3
Brass	1.8	1.0
Concrete	0.7–1.2	0.4–0.7
Copper	1.7	0.94
Glass, Pyrex	0.3	0.17
Iron	1.2	0.66
Lead	3.0	1.7
Silver	2.0	1.1
Steel	1.2	0.66
Zinc	2.6	1.44

Example 16.4   An iron pipe is 60 m long at room temperature (20°C). If this pipe is to be used as a steam pipe, how much allowance must be made for expansion, and what will be the new length of pipe after steam has been flowing for a while?

Plan: The temperature of steam is 100°C, so the pipe's temperature will change from 20° to 100°C, an interval of 80 C°. The increase in length can be found from Eq. (16.7). By adding this amount to the initial length, we will find the new length of the pipe after steam flows through it.

Solution: From Table 16.1, we substitute $\alpha_{iron} = 1.2 \times 10^{-5}/C°$ to find the change in length.

$$\Delta L = \alpha_{iron} L_0 \, \Delta t = (1.2 \times 10^{-5}/C°)(60 \text{ m})(80 \text{ C}°); \qquad \Delta L = 0.0576 \text{ m}$$

The new length will be $L_0 + \Delta L$ or

$$L = 120 \text{ m} + 0.0576 \text{ m} = 120.0576 \text{ m}$$

An allowance of 5.76 cm is needed to accommodate the expansion.

We can see from Example 16.4 that the new length may be calculated by the following relation:

$$L = L_0 + \alpha L_0 \, \Delta t \qquad \qquad \textbf{(16.10)}$$

Remember, when calculating ΔL, that the units of α must be consistent with the units of Δt.

Linear expansion has both useful and destructive properties when applied to physical situations. The destructive effects require engineers to use expansion joints or rollers to make allowances for expansion and contraction. The predictable expansion of some materials, on the other hand, can be used to open or close switches at certain temperatures. Such devices are called *thermostats*.

Probably the most common application of the principle of linear expansion is the bimetallic strip. This device, shown in Fig. 16.11, consists of two flat strips of different metals welded or riveted together. The strips are fused so that they are the same length at a chosen temperature t_0. If we heat the strip, causing a rise in temperature, the material with the larger expansion coefficient will expand more. For example, a brass-iron strip will bend in an arc toward the iron side. When the source of heat is removed, the strip will gradually return to its original position. Cooling the strip below the initial temperature will cause the strip to bend in the other direction. The material with the higher coefficient of expansion also *decreases* in length at a faster rate. The bimetallic strip has many useful applications, from

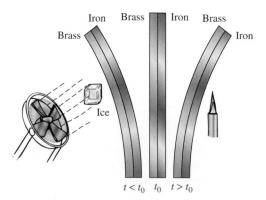

Figure 16.11   The bimetallic strip.

thermostatic control systems to blinking lights. Since the expansion is in direct proportion to an increase in temperature, the bimetallic strip can also be used as a thermometer.

16.6 Area Expansion

Linear expansion is by no means restricted to the length of a solid. Any line drawn through the solid will increase in length per unit length at the rate given by its expansion coefficient α. For example, in a solid cylinder, the length, diameter, and a diagonal drawn through the solid will all increase their dimensions in the same proportion. In fact, the expansion of a surface is exactly analogous to a photographic enlargement, as illustrated in Fig. 16.12. Notice also that if the material contains a hole, the area of the hole expands at the same rate it would if it were filled with material.

Let us consider the area expansion of the rectangular surface in Fig. 16.13. Both the length and the width of the material will expand at the rate given by Eq. (16.10). Thus, the new length and width are given, in factored form, by

$$L = L_0(1 + \alpha \, \Delta t)$$
$$W = W_0(1 + \alpha \, \Delta t)$$

We can now derive an expression for area expansion by finding the product of these two equations.

$$LW = L_0 W_0 (1 + \alpha \, \Delta t)^2$$
$$= L_0 W_0 (1 + 2\alpha \, \Delta t + \alpha^2 \, \Delta t^2)$$

Since the magnitude of α is of the order of 10^{-5}, we may certainly neglect the term containing α^2. Hence, we can write

$$LW = L_0 W_0 (1 + 2\alpha \, \Delta t)$$

or

$$A = A_0(1 + 2\alpha \, \Delta t)$$

where $A = LW$ represents the new area and $A_0 = L_0 W_0$ represents the original area. Rearranging terms, we obtain

$$A - A_0 = 2\alpha A_0 \, \Delta t$$

or

$$\Delta A = 2\alpha A_0 \, \Delta t \qquad \textbf{(16.11)}$$

The **coefficient of area expansion** γ (gamma) is approximately twice the coefficient of linear expansion. Symbolically,

$$\gamma = 2\alpha \qquad \textbf{(16.12)}$$

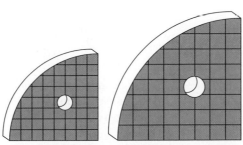

Figure 16.12 Thermal expansion is analogous to a photographic enlargement. Note that the hole increases in the same proportion as the material.

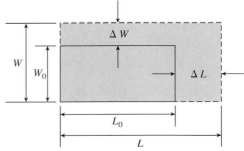

Figure 16.13 Area expansion.

where γ is the change in area per unit initial area per degree change in temperature. Using this definition, we may write the following formulas for area expansion.

$$\Delta A = \gamma A_0 \, \Delta t \tag{16.13}$$

$$A = A_0 + \gamma A_0 \, \Delta t \tag{16.14}$$

Example 16.5

A brass disk has a hole 80 mm in diameter punched in its center. Later, the disk at 23°C is placed into boiling water for a few minutes. What will be the new area of the hole?

Plan: We will first calculate the area of the hole at 23°C. Then we will calculate the increase in area due to the change in temperature. Remember that the area expansion coefficient is twice the linear value given in Table 16.1. We will express the new area also in mm^2, so there will be no need to change the area units.

Solution: The area at 23°C is

$$A_0 = \frac{\pi D^2}{4} = \frac{\pi (80 \text{ mm})^2}{4}; \qquad A_0 = 5027 \text{ mm}^2$$

The area expansion coefficient for brass is

$$\gamma = 2\alpha = 2(1.8 \times 10^{-5}/\text{C}°) = 3.6 \times 10^{-5} \text{ C}°$$

The increase in area is found from Eq. (16.13).

$$\Delta A = \gamma A_0 \, \Delta t = (3.6 \times 10^{-5}/\text{C}°)(5027 \text{ mm}^2)(100°\text{C} - 23°\text{C})$$
$$\Delta A = 13.9 \text{ mm}^2$$

The new area is found by adding the change to the original area.

$$A = A_0 + \Delta A = 5027 \text{ mm}^2 + 13.9 \text{ mm}^2; \qquad A = 5040.9 \text{ mm}^2$$

Another approach to Example 16.5, would be to use the linear expansion formula to find the increase in diameter and then calculate the new area based on the new diameter. That approach would actually be more accurate because $\gamma = 2\alpha$ is an *approximation* in the area expansion formula.

16.7 Volume Expansion

The expansion of heated material is the same in all directions. Therefore, the volume of a liquid, gas, or solid will have a predictable increase in volume with a rise in temperature. Reasoning similar to that of the previous sections will give us the following formulas for volume expansion.

$$\Delta V = \beta V_0 \, \Delta t \tag{16.15}$$

$$V = V_0 + \beta V_0 \, \Delta t \tag{16.16}$$

The symbol β (beta) is the **volume expansion coefficient.** It represents the change in volume per unit volume per degree change in temperature. For solid materials, it is approximately three times the linear expansion coefficient.

$$\beta = 3\alpha \tag{16.17}$$

When working with solids, we can compute β from the table of linear expansion coefficients (Table 16.1). For different liquids, the volume expansion coefficients are listed in

Table 16.2

Volume Expansion Coefficients

| | β | |
Liquid	$10^{-4}/C°$	$10^{-4}/F°$
Alcohol, ethyl	11	6.1
Benzene	12.4	6.9
Glycerin	5.1	2.8
Mercury	1.8	1.0
Water	2.1	1.2

Table 16.2. The molecular separation in gases is so great that they all expand at approximately the same rate. Volumetric expansion of gases will be discussed later in Chapter 19.

Example 16.6 A Pyrex glass bulb is filled with 50 cm³ of mercury at 20°C. What volume will overflow if the system is heated uniformly to a temperature of 60°? Refer to Fig. 16.14.

Plan: The inside volume of the glass bulb is the same as the volume of the contained fluid (50 cm³). The mercury has a higher coefficient of volume expansion, which means that the overflow will be equal to the difference between the expansion of the mercury ΔV_m and the expansion of the glass ΔV_g. Remember that $\beta_g = 3\alpha_g$.

Solution: We will first calculate the change in volume of the mercury.

$$\Delta V_m = \beta_m V_{0m} \, \Delta t = (1.8 \times 10^{-4}/C°)(50 \text{ cm}^3)(60°C - 20°C)$$
$$\Delta V_m = 0.360 \text{ cm}^3 \quad \text{(Increase in volume of mercury)}$$

Now, the change in volume of the inside of the glass bulb is

$$\Delta V_g = 3\alpha_g V_{0m} \, \Delta t = 3(0.3 \times 10^{-5}/C°)(50 \text{ cm}^3)(40 \text{ C}°)$$
$$\Delta V_g = 0.0180 \text{ cm}^3 \quad \text{(Increase in volume of glass)}$$

The volume overflow is the difference in the two expansions:

$$V_{overflow} = \Delta V_m - \Delta V_g = 0.360 \text{ cm}^3 - 0.018 \text{ cm}^3$$
$$V_{overflow} = 0.342 \text{ cm}^3$$

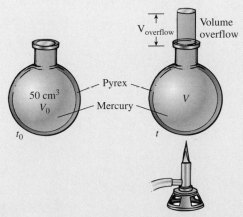

Figure 16.14 The volume overflow is found by subtracting the change in volume of the glass from the change in volume of the liquid.

Problem-Solving Strategy

Temperature and Expansion

1. Read the problem carefully and, after drawing a rough sketch, list the given information. Use zero subscripts to distinguish between initial and final values of length, area, volume, and temperature.

2. Do not confuse *specific* temperatures t with temperature *intervals* Δt. The practice of using the degree mark before and after the symbol is helpful; for example, $55°C - 22°C = 33\ C°$.

3. For expansion problems, be sure to include the temperature unit with the constant to avoid multiplying by the incorrect temperature interval. If the coefficient is $1/C°$, the interval Δt must be in Celsius degrees.

4. The area and volume expansion coefficients for solids can be found by multiplying two or three times, respectively, the linear values given in Table 16.1.

5. When you are asked to find an *initial* or *final* value of length, area, volume, or temperature, it is usually easier to calculate first the *change* in that parameter and then solve for the initial or final value. For example, you can find the final temperature t_f by first calculating Δt and then adding or subtracting to find t_f.

6. Simultaneous expansion of different materials must be adjusted to account for the differing rates of expansion. For a liquid in a solid container, the net increase or decrease in volume is equal to the difference in the changes experienced by each material. See Example 16.6.

16.8 The Unusual Expansion of Water

Suppose we fill the bulb of the tube in Fig. 16.15 with water at 0°C so that the narrow neck is partially filled. Expansion or contraction of the water can be measured easily by observing the water level in the narrow tube. As the temperature of the water increases, the water in the tube will gradually sink, indicating a contraction. The contraction continues until the temperatures of the bulb and the water are 4°C. As the temperature increases above 4°C, the water reverses its direction and rises continuously, indicating the normal expansion with an increase in temperature. This means that water has its minimum volume and its maximum density at 4°C.

The variation in the density of water with temperature is shown graphically in Fig. 16.16. If we study the graph from the high-temperature side, we note that the density gradually increases to a maximum of 1.0 g/cm³ at 4°C. The density then decreases gradually until the water reaches the ice point. Ice occupies a greater volume than water, and its formation sometimes results in cracked water pipes if proper precautions are not taken.

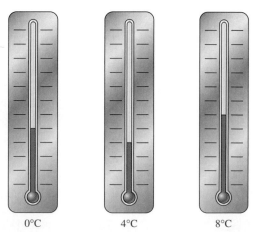

0°C   4°C   8°C

Figure 16.15 The irregular expansion of water. As the temperature of water is increased from 0°C to 8°C, it first contracts and then expands.

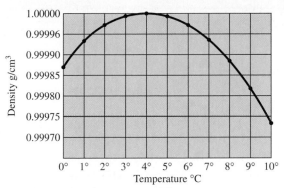

Figure 16.16   Variation in the density of water near 4°C.

The greater volume in the ice results from the way groups of its molecules are bonded in its crystalline structure. As the ice melts, the water formed still contains groups of molecules bonded in this open crystal structure. As these structures begin to collapse, the molecules move closer together, increasing the density. This is the dominant process until the water reaches a temperature of 4°C. From that point to higher temperatures, the increased amplitude of molecular vibrations takes over, and the water expands.

Once again, the beginning student may be tempted to marvel that science can be so exact. The very thought that the density of water at 4°C "turns out to be exactly 1.00 g/cm³" must certainly be a remarkable coincidence. Like the temperatures of the ice point and steam point, however, this result is also a consequence of a definition. The scientists who originally devised the metric system defined the *kilogram* as the mass of 1000 cm³ of water at 4°C. Later, the kilogram was redefined in terms of a platinum-iridium cylinder, which serves as a standard.

Summary and Review

Summary

We have seen that, because of the existence of four commonly used temperature scales, temperature conversions are important. You have also studied one important effect of changes in temperature of materials: a change in the physical dimensions. The major concepts are summarized as follows.

- There are four temperature scales with which you should be thoroughly familiar. These scales are compared in Fig. 16.9, giving values for the steam point, the ice point, and absolute zero on each scale. It is important for you to distinguish between a temperature interval Δt and a specific temperature t. For temperature intervals,

$$\frac{5\ C°}{9\ F°} = 1 = \frac{9\ F°}{5\ C°} \qquad 1\ K = 1\ C° \qquad 1\ R° = 1\ F°$$
Temperature Intervals

- For specific temperatures, you must correct for the interval difference, but you must also correct for the fact that different numbers are assigned for the same temperatures:

$$t_C = \frac{5}{9}(t_F - 32) \qquad t_F = \frac{9}{5}t_C + 32$$
Specific Temperatures

$$T_K = t_C + 273 \qquad T_R = t_F + 460$$
Absolute Temperatures

- The following relationships apply for thermal expansion of solids:

$$\Delta L = \alpha L_0\,\Delta t \qquad L = L_0 + \alpha L_0\,\Delta t$$
Linear Expansion

$$\Delta A = \gamma A_0\,\Delta t \qquad A = A_0 + \gamma A_0\,\Delta t \qquad \gamma = 2\alpha$$
Area Expansion

$$\Delta V = \beta V_0\,\Delta t \qquad V = V_0 + \beta V_0\,\Delta t \qquad \beta = 3\alpha$$
Volume Expansion

- The volume expansion of a liquid uses the same relation as for a solid except, of course, that there is no linear expansion coefficient α for a liquid. Only β is needed.

Key Terms

absolute zero   336
Celsius scale   332
coefficient of area expansion   341
coefficient of linear expansion   339
constant-pressure thermometer   335
constant-volume thermometer   335
degree   332
Fahrenheit scale   332

heat   331
kelvin   338
Kelvin scale   337
kilogram   345
linear expansion   339
lower fixed point (ice point)   332
temperature   331
thermal energy   330

thermal equilibrium   331
thermometer   331
triple point of water   338
upper fixed point (steam point)   332
volume expansion coefficient   342

Review Questions

16.1. Two lumps of hot iron ore are dropped into a container of water. The system is insulated and allowed to reach thermal equilibrium. Is it necessarily true that the iron ore and the water have the same thermal energy? Is it necessarily true that they have the same temperature? Discuss.

16.2. Distinguish clearly between thermal energy and temperature.

16.3. If a flame is placed underneath a mercury-in-glass thermometer, the mercury column first drops and then rises. Explain.

16.4. What factors must be considered in the design of a sensitive thermometer?

16.5. How good is our sense of touch as a means of judging temperature? Does the "hotter" object always have the higher temperature?

16.6. Given an unmarked thermometer, how would you proceed to graduate it in Celsius degrees?

16.7. A 6-in. ruler expands 0.0014 in. when the temperature is increased 1 C°. How much would a 6-cm ruler expand during the same temperature interval if it is made of the same material?

16.8. A brass rod connects the opposite sides of a brass ring. If the system is heated uniformly, will the ring remain circular?

16.9. A brass nut is used with a steel bolt. How is the closeness of fit affected when the bolt alone is heated? If the nut alone is heated? If they are both heated equally?

16.10. An aluminum cap is screwed tightly on the top of a pickle jar at room temperature. After the pickle jar has been stored in a refrigerator for a day or two, the cap cannot be removed easily. Explain. Suggest a way to remove the cap with little effort. How might this problem be solved by the manufacturer?

16.11. Describe the expansion of water near 4°C. Why does a lake freeze at the surface first? What temperature is likely to result at the bottom of the lake if its surface is frozen?

16.12. Follow reasoning similar to that for area expansion to derive Eqs. (16.15) and (16.17). In the text, it was stated that γ is only approximately equal to twice α. Why is it not exactly twice α? Is the error larger in Eq. (16.13) or in Eq. (16.15)?

Problems

Section 16.2 The Measurement of Temperature

16.1. Body temperature is normal at 98.6°F. What is the corresponding temperature on the Celsius scale? Ans. 37.0°C

16.2. The boiling point of sulfur is 444.5°C. What is the corresponding temperature on the Fahrenheit scale?

16.3. A steel rail cools from 70 to 30°C in 1 h. What is the change in temperature in Fahrenheit degrees for the same time period? Ans. 72 F°

*16.4. At what temperature will the Celsius scale and Fahrenheit scale have the same numerical reading?

16.5. A piece of charcoal initially at 180°F experiences a decrease in temperature of 120 F°. Express this change of temperature in Celsius degrees. What is the final temperature on the Celsius scale?
Ans. 66.7 C°, 15.6°C

16.6. Acetone boils at 56.5°C, and liquid nitrogen boils at −196°C. Express these specific temperatures on the Kelvin scale. What is the difference in these temperatures on the Celsius scale?

16.7. The boiling point of oxygen is −297.35°F. Express this temperature in kelvins and in degrees Celsius. Ans. 90.0 K, −183°C

16.8. If oxygen cools from 120 to 70°F, what is the change of temperature in kelvins?

16.9. A wall of firebrick has an inside temperature of 313°F and an outside temperature of 73°F. Express the difference of temperature in kelvins.
Ans. 133 K

16.10. Gold melts at 1336 K. What is the corresponding temperature in degrees Celsius and in degrees Fahrenheit?

16.11. A sample of gas cools from −120 to −180°C. Express the change of temperature in kelvins and in Fahrenheit degrees. Ans. −60 K, −108 F°

Section 16.5 Linear Expansion, Section 16.6 Area Expansion, and Section 16.7 Volume Expansion

16.12. A slab of concrete is 20 m long. What will be the increase in length if the temperature changes from 12°C to 30°C? Assume that $\alpha = 9 \times 10^{-6}/C°$.

16.13. A piece of copper tubing is 6 m long at 20°C. How much will it increase in length when heated to 80°C? Ans. 6.12 mm

16.14. A silver bar is 1 ft long at 70°F. How much will it increase in length when it is placed into boiling water (212°F)?

16.15. The diameter of a hole in a steel plate is 9 cm when the temperature is 20°C. What will be the diameter of the hole at 200°C?
Ans. 9.02 cm

*16.16. A brass rod is 2.00 m long at 15°C. To what temperature must the rod be heated so that its new length is 2.01 m?

16.17. A square copper plate 4 cm on a side at 20°C is heated to 120°C. What is the increase in the area of the copper plate? Ans. 0.0544 cm^2

*16.18. A circular hole in a steel plate has a diameter of 20.0 cm at 27°C. What must be the temperature of the plate such that the area of the hole will be 314 cm^2?

16.19. What is the increase in volume of 16.0 liters of ethyl alcohol when the temperature is increased by 30 C°?   Ans. 0.528 L

16.20. A Pyrex beaker has an inside volume of 600 mL at 20°C. At what temperature will the inside volume be 603 mL?

16.21. If 200 cm^3 of benzene exactly fills an aluminum cup at 40°C and the system cools to 18°C, how much benzene (at 18°C) can be added to the cup without overflowing?   Ans. 5.14 cm^3

16.22. A Pyrex glass beaker is filled to the top with 200 cm^3 of mercury at 20°C. How much mercury will overflow if the temperature of the system is increased to 68°C?

Additional Problems

*16.23. The diameter of a hole in a copper plate at 20°C is 3.00 mm. To what temperature must the copper be cooled if its diameter is to be 2.99 mm?
                                    Ans. −176°C

16.24. A rectangular sheet of aluminum measures 6 by 9 cm at 28°C. What is its area at 0°C?

*16.25. A steel tape measures the length of an aluminum rod as 60 cm when both are at 8°C. What will the tape read for the length of the rod if both are at 38°?   Ans. 60.022 cm

16.26. At 20°C, a copper cube measures 40 cm on a side. What is the volume of the cube when the temperature reaches 150°C?

16.27. A Pyrex beaker ($\alpha = 0.3 \times 10^{-5}$/C°) is filled to the top with 200 mL of glycerine ($\beta = 5.1 \times 10^{-4}$/C°). How much glycerine will overflow the top if the system is heated from 20°C to 100°C?
                                    Ans. 8.02 mL

16.28. An oven is set at 450°F. If the temperature drops by 50 kelvins, what is the new temperature in degrees Celsius?

*16.29. A 100-ft steel tape correctly measures distance when the temperature is 20°C. What is the true measurement if this tape indicates a distance of 94.62 ft on a day when the temperature is 36°C?   Ans. 94.64 ft

*16.30. The diameter of a steel rod is 3.000 mm when the temperature is 20°C. The diameter of a brass ring is 2.995 mm, also at 20°C. At what common temperature will the brass ring slip over the steel rod smoothly?

*16.31. The volume of a metal cube increases by 0.50 percent when the temperature of the cube increases by 100 C°. What is the linear expansion coefficient for this metal?   Ans. 1.67 × 10^{-5}/C°

16.32. By what percentage is the volume of a brass cube increased when heated from 20°C to 100°C?

16.33. A round brass plug has a diameter of 8.001 cm at 28°C. To what temperature must the plug be cooled if it is to fit snugly into an 8.000-cm hole?
                                    Ans. 21.1°C

*16.34. Five hundred cubic centimeters of ethyl alcohol completely fill a Pyrex beaker. If the temperature of the system increases by 70 C°, what volume of alcohol overflows?

Critical Thinking Questions

16.35. The laboratory apparatus for measuring the coefficient of linear expansion is illustrated in Fig. 16.17. The temperature of a metal rod is increased by passing steam through an enclosed jacket. The resulting increase in length is measured with a micrometer screw at one end. Since the original length and temperature are known, the expansion coefficient can be calculated from Eq. (16.8). The following data were recorded during an experiment with a rod of unknown metal:

$$L_0 = 600 \text{ mm} \qquad t_0 = 23°C$$
$$\Delta L = 1.04 \text{ mm} \qquad t_f = 98°C$$

What is the coefficient of linear expansion for this metal? Can you identify the metal?
                                    Ans. 2.3 × 10^{-5}/C°, Al

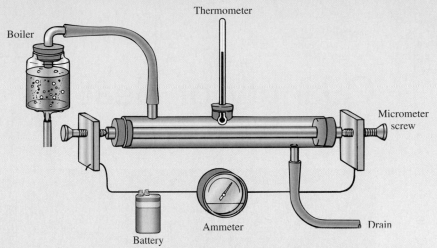

Boiler

Thermometer

Micrometer screw

Ammeter

Drain

Battery

Figure 16.17 Apparatus for measuring the coefficient of linear expansion.

*16.36. Assume that the end points of a rod are fixed rigidly between two walls to prevent expansion with increasing temperature. From the definitions of Young's modulus (Chapter 13) and your knowledge of linear expansion, show that the compressive force F exerted by the walls will be given by

$$F = \alpha A Y \, \Delta t$$

where A = cross section of the rod
Y = Young's modulus
Δt = increase in temperature of rod.

*16.37. Prove the density of a material changes with temperature so that the new density ρ is given by

$$\rho = \frac{\rho_0}{1 + \beta \Delta t}$$

where ρ_0 = original density
β = volume expansion coefficient
Δt = change in temperature

16.38. The density of mercury at $0°C$ is 13.6 g/cm^3. Use the relationship in Problem 16.37 to find the density of mercury at $60°C$?

16.39. A steel ring has an inside diameter of 4.000 cm at $20°C$. The ring is to be slipped over a copper shaft that has a diameter of 4.003 cm at $20°C$. To what temperature must the ring be heated? If the ring and the shaft are cooled uniformly, at what temperature will the ring just slip off the shaft?
Ans. $82.5°C$, $-130°C$

17

Quantity of Heat

Foundry: Workers observe the processing of molten steel in a foundry. It requires about 289 joules of heat to melt 1 gram of steel. In this chapter, we will define the quantity of heat to raise the temperature of a substance and to change the phase from a solid to a liquid or from a liquid to a gas. *(Photo © vol. 5 PhotoDisc/Getty.)*

Objectives

After completing this chapter, you should be able to

1. Define quantity of heat in terms of the *calorie*, the *kilocalorie*, the *joule*, and the *British thermal unit* (Btu).
2. Write a formula for the *specific heat capacity* of a material and apply it to the solution of problems involving the loss and gain of heat.
3. Write formulas for calculating the *latent heats of fusion* and *vaporization* and apply them to the solution of problems in which heat produces a change in phase of a substance.
4. Define the *heat of combustion* and apply it to problems involving the production of heat.

Thermal energy is the energy associated with molecular motion, but it is not possible to measure the position and velocity of every molecule in a substance to determine its thermal energy. We can measure *change* in thermal energy, however, by relating it to a change in temperature.

For example, when two systems at different temperatures are placed together, they eventually reach a common intermediate temperature. From this observation, it is safe to say that the system at the higher temperature has lost thermal energy to the system at the lower temperature. The thermal energy lost or gained by objects is called **heat.** This chapter is concerned with the quantitative measurement of heat.

17.1 The Meaning of Heat

It was originally believed that two systems reach thermal equilibrium through the transfer of a substance called *caloric.* It was postulated that all bodies contain an amount of caloric in proportion to their temperature. Thus, when two objects were placed in contact, the object of higher temperature transferred caloric to the object of lower temperature until they reached the same temperature. The idea that a substance is transferred carries with it the implication that there is a limit to the amount of heat energy that can be withdrawn from a body. It was this point that eventually led to the downfall of the caloric theory.

Count Rumford of Bavaria was the first to shed doubt on the caloric theory. He made his discovery in 1798 while supervising the boring of a cannon. The bore of the cannon was kept full of water to prevent overheating. As the water boiled away, it was replenished. According to the existing theory, caloric had to be supplied to boil the water. The apparent production of caloric was explained by supposing that when matter is finely divided, it loses some of its ability to retain caloric. Rumford devised an experiment to show that even a dull boring tool that did not cut the gun metal at all produced enough caloric to boil the water. In fact, it seemed that as long as mechanical work was supplied, the tool was an inexhaustible source of caloric. Rumford ruled out the caloric theory on the basis of his experiments and suggested that the explanation must be related to motion. Hence, the idea that mechanical work is responsible for the creation of heat was introduced. The equivalence of heat and work as two forms of energy was established later by Sir James Prescott Joule.

17.2 The Quantity of Heat

The idea of heat as a substance must be discarded. It is not something that an object *has* but something that it *gives up* or *absorbs.* Heat is simply another form of energy that can be measured only in terms of the effect it produces. The SI unit of energy, the *joule,* is also the preferred unit for heat since heat is a form of energy. There are three older units that remain in use today, however, and they will also be treated in this text. These earlier units were based on the thermal energy required to produce a standard change. They are the **calorie,** the **kilocalorie,** and the **British thermal unit.**

> One calorie (cal) is the quantity of heat required to change the temperature of 1 gram of water through 1 Celsius degree.

> One kilocalorie (kcal) is the quantity of heat required to change the temperature of 1 kilogram of water through 1 Celsius degree (kcal = 1000 cal).

> One British thermal unit (Btu) is the quantity of heat required to change the temperature of 1 standard pound (lb) through 1 Fahrenheit degree.

Aside from the fact that these older units imply that heat energy cannot be related to other forms of energy, there are other problems. The heat required to change the temperature of water from 92 to 93°C is not exactly the same as that needed to increase the temperature from 8 to 9°C. Thus, it is necessary to specify the interval of temperature in strict applications for the calorie and for the British thermal unit. The chosen intervals were 14.5 to 15.5°C and 63 to 64°F. Additionally, the pound unit that appears in the definition of the Btu must be recognized as the *mass of the standard pound.* This represents a departure from USCS units, in which the unit pound was reserved for weight. Therefore, in this chapter, when we refer to 1 lb of water, we shall be referring to a *mass* of water equivalent to about 1/32 slug. This distinction is necessary because the pound of water must represent a constant quantity of matter, independent of its location. By definition, the pound mass is related to the gram and kilogram as follows:

$$1 \text{ lb} = 454 \text{ g} = 0.454 \text{ kg}$$

The difference between these older units for heat results from the difference in masses and the difference in temperature scales. It is left as an exercise for you to show that

$$1 \text{ Btu} = 252 \text{ cal} = 0.252 \text{ kcal} \tag{17.1}$$

The first quantitative relationship between these older units and the traditional units for mechanical energy was established by Joule in 1843. Although Joule devised many different experiments to demonstrate the equivalence of heat units and energy units, the apparatus most often remembered is illustrated in Fig. 17.1. Falling weights supplied the mechanical energy, which rotated a set of paddles in a water can. The quantity of heat absorbed by the water was measured from the known mass and the measured increase in temperature of the water.

In modern times, the ***mechanical equivalent of heat*** has been accurately established by a variety of techniques. The accepted results are

$$1 \text{ cal} = 4.186 \text{ J}$$
$$1 \text{ kcal} = 4186 \text{ J}$$
$$1 \text{ Btu} = 778 \text{ ft} \cdot \text{lb}$$

Thus, 4.186 J of heat is needed to raise the temperature of 1 g of water from 14.5 to 15.5°C. Since each of the above units remains in common use, it will often become necessary to compare units or to convert from one unit to another.

Now that we have defined units for the quantitative measurement of heat, the distinction between quantity of heat and temperature should be clear. For example, suppose we pour 200 g of water into one beaker and 800 g of water into another beaker, as shown in Fig. 17.2. The initial temperature of the water in each beaker is measured to be 20°C. A

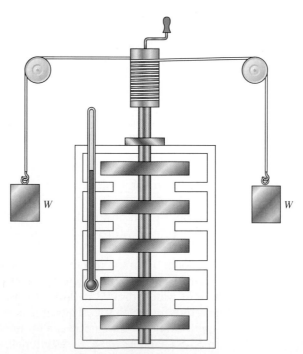

Figure 17.1   Joule's experiment for determining the mechanical equivalent of heat. The falling weights do work in stirring the water and raising its temperature.

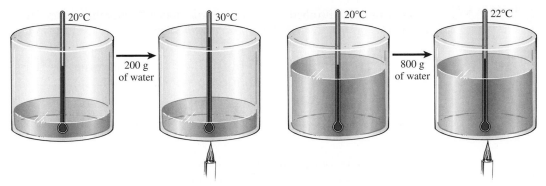

Figure 17.2 The same quantity of heat is applied to different masses of water. The larger mass experiences a smaller rise in temperature.

flame is placed under each beaker for the same length of time, delivering 8000 J of heat energy to the water in each beaker. The temperature of the 800-g quantity of water increases by a little more than 2 C°, but the temperature of the 200-g quantity increases by almost 10 C°. Yet the same quantity of heat was applied to the water in each beaker.

17.3 Specific Heat Capacity

We have defined a quantity of heat as the thermal energy required to raise the temperature of a given mass. The amount of thermal energy required to raise the temperature of a substance varies, however, for different materials. For example, suppose we apply heat to five balls, all of the same size but made of different materials, as shown in Fig. 17.3a. If we want to raise the temperature of each ball to 100°C, we will find that some of the balls must be heated longer than the others. For illustration, let us assume that each ball has a volume of 1 cm³ and an initial temperature of 0°C. Each ball is heated with a burner capable of delivering thermal energy at the rate of 1 cal/s. The approximate times required to obtain a temperature of 100°C for each ball are given in Fig. 17.3. Notice that the lead ball reaches the final temperature in only 37 s, whereas the iron ball requires 90 s of continuous heating. The glass, aluminum, and copper balls require intermediate times.

Since more heat was absorbed by the iron and copper balls, it might be expected that they would release more heat on cooling. This can be demonstrated by placing the five balls (at 100°C) simultaneously on a thin strip of paraffin, as in Fig. 17.3b. The iron and copper balls eventually melt through the paraffin and drop into the pan. The lead and glass balls never make it through. Clearly, there must be some property of a material that relates to the

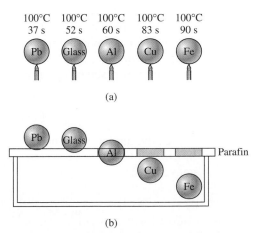

Figure 17.3 A comparison of heat capacities for five balls of different materials.

quantity of heat absorbed or released during a change in temperature. As a step in establishing this property, we first define **heat capacity.**

> The heat capacity of a body is the ratio of heat supplied to the corresponding rise in temperature of the body.

$$\text{Heat capacity} = \frac{Q}{\Delta t} \qquad \textbf{(17.2)}$$

The SI units for heat capacity are *joules per kelvin* (J/K), but since the Celsius interval is the same as the kelvin and more commonly used, the *joule per Celsius degree* (J/C°) is used in this text. Other units are *calories per Celsius degree* (cal/C°), *kilocalories per Celsius degree* (kcal/C°), and *Btu per Fahrenheit degree* (Btu/F°). In the preceding example, 89.4 cal of heat were required to raise the temperature of the iron ball by 100 C°. Thus, the heat capacity of this particular iron ball is 0.894 cal/C°.

The mass of an object is not included in the definition of heat capacity. Therefore, heat capacity is a property of the object. To make it a property of the material, the *heat capacity per unit mass* is defined. We call this property the **specific heat capacity,** denoted c.

> The specific heat capacity of a material is the quantity of heat required to raise the temperature of a unit mass through one degree.

$$c = \frac{Q}{m\,\Delta t} \qquad Q = mc\,\Delta t \qquad \textbf{(17.3)}$$

The SI unit of specific heat assigns the *joule* for heat, the *kilogram* for mass, and the *kelvin* for temperature. If we again replace the kelvin with the Celsius degree, the units of c are J/kg · C°. In industry, most temperature measurements are made in C° or F°, and the calorie and the Btu are still used more than the SI units. We will therefore continue to speak of specific heat in units of cal/g · C° and Btu/lb · F°, but we will also use the preferred SI units in some cases. In the example of the iron ball, the mass is found to be 7.85 g. The specific heat of iron is therefore

$$c = \frac{Q}{m\,\Delta t} = \frac{89.4\text{ cal}}{(7.85\text{ g})(100\text{ C}°)} = 0.114\text{ cal/g} \cdot \text{C}°$$

Notice that we speak of the heat capacity of the *ball* and the *specific* heat capacity of *iron.* The former relates to the object itself, whereas the latter relates to the material from which the object is made. In our experiment with the balls, we noted only the quantity of heat required to raise their temperature 100 C°. The density of the materials was not considered. If the sizes of the balls were adjusted so each had the same mass, we would observe different results. Since the specific heat of aluminum is highest, more heat would be required for the aluminum ball than for the others. Similarly, the aluminum ball would release more heat in cooling.

For most practical applications, the specific heat of water can be taken as

$$4186\text{ J/kg} \cdot \text{C}° \qquad 4.186\text{ J/g} \cdot \text{C}°$$
$$1\text{ cal/g} \cdot \text{C}° \qquad \text{or} \qquad 1\text{ Btu/lb} \cdot \text{F}°$$

Notice the numerical values are the same for the specific heat expressed in cal/g · C° and in Btu/lb · F°. This is a consequence of their definitions and can be demonstrated by unit conversion:

$$1\frac{\text{Btu}}{\text{lb} \cdot \text{F}°} \times \frac{9\text{ F}°}{5\text{ C}°} \times \frac{1\text{ lb}}{454\text{ g}} \times \frac{252\text{ cal}}{1\text{ Btu}} = 1\text{ cal/g} \cdot \text{C}°$$

The specific heats for many common substances are listed in Table 17.1.

| Table 17.1 | | |

Specific Heat Capacities

Substance	J/kg · C°	cal/g · C° or Btu/lb · F°
Aluminum	920	0.22
Brass	390	0.094
Copper	390	0.093
Ethyl alcohol	2500	0.60
Glass	840	0.20
Gold	130	0.03
Ice	2090	0.5
Iron	470	0.113
Lead	130	0.031
Mercury	140	0.033
Silver	230	0.056
Steam	2000	0.48
Steel	480	0.114
Turpentine	1800	0.42
Water	4186	1.00
Zinc	390	0.092

Once the specific heats of a large number of materials have been established, the thermal energy released or absorbed in many experiments can be determined. For example, the quantity of heat Q required to raise the temperature of a mass m through an interval t, from Eq. (17.3), is

$$Q = mc\,\Delta t \tag{17.4}$$

where c is the specific heat of the mass.

Example 17.1

How much heat is required to raise the temperature of 200 g of mercury from 20 to 100°C?

Solution: Substitution into Eq. (17.4) yields

$$Q = mc\,\Delta t = (0.2 \text{ kg})(140 \text{ J/kg} \cdot \text{C}°)(80 \text{ C}°)$$
$$= 2200 \text{ J}$$

17.4 The Measurement of Heat

We have often emphasized the distinction between thermal energy and temperature. The term *heat* has now been introduced as the thermal energy *absorbed* or *released* during a temperature change. The quantitative relationship between heat and temperature is best described by the concept of specific heat as it appears in Eq. (17.4). The physical relationships among all these terms are now beginning to fall into place.

The principle of thermal equilibrium tells us that whenever objects are placed together in an insulated enclosure, they will eventually reach the same temperature. This is the result of a transfer of thermal energy from the warmer bodies to the cooler bodies. If energy is to be conserved, we say that *the heat lost by the warm bodies must equal the heat gained by the cool bodies.* That is,

$$\text{Heat lost} = \text{heat gained} \qquad \text{(17.5)}$$

This equation expresses the net result of heat transfer within a system.

The heat lost or gained by an object is not related in a simple way to the molecular energies of the objects. Whenever thermal energy is supplied to an object, it can absorb the energy in many different ways. The concept of specific heat is needed to measure the abilities of different materials to utilize thermal energy to increase their temperatures. The same amount of applied thermal energy does not result in the same temperature increase for all materials. For this reason, we say that temperature is a *fundamental* quantity. Its measurement is necessary to determine the quantity of heat lost or gained in a given process.

In applying the general equation for the conservation of energy, Eq. (17.5), the quantity of heat *gained* or *lost* by each item is calculated from the equation

$$Q = mc\,\Delta t$$

When applied to *gains* and *losses,* the term Δt represents the *absolute* change of temperature. This means that we must think of *high temperature* minus *low temperature* rather than *final temperature* minus *initial temperature.* For example, suppose a heated bolt, initially at 80°C, is dropped into a cup of water whose initial temperature is 20°. Assume that the final equilibrium temperature is 30°. For purposes of determining the heat *lost* by the bolt, Δt is $+50°$, and for the calculation for the heat *gained* by the water, Δt is $+10°$.

Example 17.2 A handful of copper shot is heated to 98°C and then dropped into 160 g of water at 20°C. The final temperature of the mixture is 25°C. What was the mass of the shot?

Plan: To find the mass of the copper shot, we recognize that the heat lost by the shot must be equal to the heat gained by the water. Since no mention was made of the container, we will assume that negligible heat is exchanged elsewhere. We will set the heat lost equal to the heat gained and solve for the unknown mass.

Solution: Recalling that $Q = mc\,\Delta t$ for the shot and for the water, we write

$$\text{Heat lost by copper shot} = \text{heat gained by water}$$
$$m_s c_s\,\Delta t_s = m_w c_w\,\Delta t_w$$
$$m_s c_s (t_s - t_e) = m_w c_w (t_e - t_w)$$

From Table 17.1, we find that for copper, $c = 0.093$ cal/g · C°, and for water, $c = 1$ cal/g · C°. Substituting the other known quantities, we have

$$m_s(0.093 \text{ cal/g} \cdot \text{C°})(98° - 25°\text{C}) = (160 \text{ g})(1 \text{ cal/g} \cdot \text{C°})(25°\text{C} - 20°\text{C})$$
$$m_s(6.79 \text{ cal/g}) = 800 \text{ cal}$$
$$m_s = 118 \text{ g}$$

In this simple example, we have neglected two important facts: (1) the water must have a container that will also absorb heat from the shot; (2) the entire system must be insulated from external temperatures. Otherwise, the equilibrium temperature will always be room temperature. A laboratory device called a **calorimeter** (Fig. 17.4) is used to control these difficulties. The calorimeter consists of a thin metallic vessel K, generally aluminum, held centrally within an outer jacket A by a nonconducting rubber support H. Heat transfer is minimized in three ways: (1) the rubber gasket prevents transfer of heat by conduction,

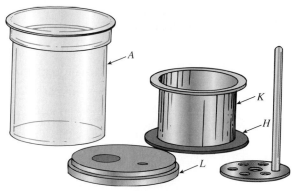

Figure 17.4   The laboratory calorimeter. *(Central Scientific Co.)*

(2) the dead air space between the container walls prevents heat transfer by air convection currents, and (3) highly polished metal vessels reduce heat transfer by radiation. These three methods of heat transfer will be discussed in Chapter 18. The wooden cover L has holes in the top for insertion of a thermometer and an aluminum stirrer.

Example 17.3   In a laboratory experiment, a calorimeter is used to find the specific heat of iron. Eighty grams of dry iron shot are placed in a cup and heated to a temperature of 95°. The mass of the inner aluminum cup and of the aluminum stirrer is 60 g. The calorimeter is partially filled with 150 g of water at 18°C. The hot shot is poured quickly into the cup, and the calorimeter is sealed, as shown in Fig. 17.5. After the system has reached thermal equilibrium, the final temperature is 22°C. Find the specific heat of iron.

Plan: The heat lost by the iron shot must equal the heat gained by the water plus the heat gained by the aluminum cup and stirrer. We will assume that the initial temperature of the cup is the same as that of the water and stirrer (18°C). Writing the conservation equation leaves the specific heat of iron as the only unknown.

Solution: Organizing the given data, we have

 Given: $m_s = 80$ g, $m_{Al} = 60$ g, $m_w = 150$ g, $t_w = t_{Al} = 18°C$, $t_s = 95°C$, $t_e = 22°C$

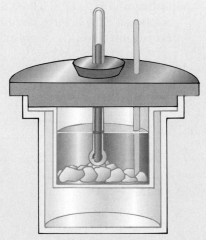

Figure 17.5   A calorimeter can be used to compute the specific heat of a substance.

We will calculate the heat gained by the water and by the aluminum separately:

$$Q_{water} = mc \, \Delta t = (150 \text{ g})(1 \text{ cal/g} \cdot \text{C}°)(22°\text{C} - 18°\text{C})$$
$$= (150 \text{ g})(1 \text{ cal/g} \cdot \text{C}°)(4 \text{ C}°) = 600 \text{ cal}$$
$$Q_{Al} = mc \, \Delta t = (60 \text{ g})(0.22 \text{ cal/g} \cdot \text{C}°)(22°\text{C} - 18°\text{C})$$
$$= (60 \text{ g})(0.22 \text{ cal/g} \cdot \text{C}°)(4 \text{ C}°) = 52.8 \text{ cal}$$

The total heat gained is the sum of these values.

$$\text{Heat gained} = 600 \text{ cal} + 52.8 \text{ cal} = 652.8 \text{ cal}$$

This amount must equal the heat lost by the iron shot:

$$\text{Heat lost} = Q_s = mc_s \, \Delta t = (80 \text{ g})c_s(95°\text{C} - 22°\text{C})$$

Setting the heat lost equal to the heat gained gives

$$(80 \text{ g})c_s(73 \text{ C}°) = 652.8 \text{ cal}$$

Solving for c_s, we obtain

$$c_s = \frac{652.8 \text{ cal}}{(80 \text{ g})(73 \text{ C}°)} = 0.11 \text{ cal/g} \cdot \text{C}°$$

In this experiment, the heat gained by the thermometer was neglected. In an actual experiment, the portion of the thermometer inside the calorimeter would absorb about the same amount of heat as an extra 0.5 g of water. This quantity, called the **water equivalent** of the thermometer, should be added to the mass of water in an accurate experiment.

17.5 Change of Phase

When a substance absorbs a given amount of heat, the speed of its molecules usually increases and its temperature rises. Depending on the specific heat of the substance, the rise in temperature is directly proportional to the quantity of heat supplied and inversely proportional to the mass of the substance. A curious thing happens, however, when a solid melts or when a liquid boils. In these cases, the temperature remains constant until all the solid melts or until all the liquid boils.

To understand what happens to the applied energy, let us consider a simple model, as illustrated in Fig. 17.6. Under the proper conditions of temperature and pressure, all substances can exist in three *phases*, solid, liquid, or gas. In the solid phase, the molecules are held together in a rigid, crystalline structure, so the substance has a definite shape and

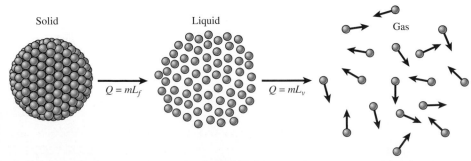

Figure 17.6 A simplified model showing the relative molecular separations in the solid, liquid, and gaseous phases. During a change of phase, the temperature remains constant.

volume. As heat is supplied, the energies of the particles in the solid gradually increase, and its temperature rises. Eventually, the kinetic energy becomes so great that some of the particles overcome the elastic forces that hold them in fixed positions. The increased separation gives them the freedom of motion that we associate with the liquid phase. At this point, the energy absorbed by the substance is used in separating the molecules more than in the solid phase. The temperature does not increase during such a change of phase. The change of phase from a solid to a liquid is called *fusion,* and the temperature at which this change occurs is called the ***melting point.***

The quantity of heat required to melt a unit mass of a substance at its melting point is called the ***latent heat of fusion*** for that substance.

> The latent heat of fusion L_f of a substance is the heat per unit mass required to change the substance from the solid to the liquid phase at its melting temperature.

$$L_f = \frac{Q}{m} \qquad Q = mL_f \tag{17.6}$$

The latent heat of fusion L_f is expressed in *joules per kilogram* (J/kg), *calories per gram* (cal/g), or *Btu per pound* (Btu/lb). At 0°C, 1 kg of ice will absorb about 334,000 J of heat in forming 1 kg of water at 0°C. Thus, the latent heat of fusion for water is 334,000 J/kg. The term *latent* arises from the fact that the temperature remains constant during the melting process. The heat of fusion for water is any of the following:

$$3.34 \times 10^5 \text{ J/kg} \qquad 334 \text{ J/g}$$
$$80 \text{ cal/g} \qquad \text{or} \qquad 144 \text{ Btu/lb}$$

After all of the solid melts, the kinetic energy of the particles in the resulting liquid increases in accordance with its specific heat, and the temperature rises again. Eventually the temperature will level off as the thermal energy is used to change the molecular structure, forming a gas or vapor. The change of phase from a liquid to a vapor is called *vaporization,* and the temperature associated with this change is called the ***boiling point*** of the substance.

The quantity of heat required to vaporize a unit mass is called the ***latent heat of vaporization.***

> The latent heat of vaporization L_v of a substance is the heat per unit mass required to change the substance from a liquid to a vapor at its boiling temperature.

$$L_v = \frac{Q}{m} \qquad Q = mL_v \tag{17.7}$$

The latent heat of vaporization L_v is expressed in units of *joules per kilogram, calories per gram,* or *Btu per pound*. It is found that 1 kg of water at 100°C absorbs 2,260,000 J of heat in forming 1 kg of steam at the same temperature. The heat of vaporization for water is

$$2.26 \times 10^6 \text{ J/kg} \qquad 2260 \text{ J/g}$$
$$540 \text{ cal/g} \qquad \text{or} \qquad 970 \text{ Btu/lb}$$

Values for the heat of fusion and the heat of vaporization of many substances are given in Table 17.2. They are given in SI units and in calories per gram. It should be noted that the Btu per pound (Btu/lb) equivalents can be obtained by multiplying the value in cal/g by (9/5). They differ only because of the difference in the temperature scales. The industrial use of the SI units of J/kg for both L_f and L_v is strongly encouraged, but few U.S. companies have made these conversions.

Table 17.2

Heats of Fusion and Heats of Vaporization for Various Substances

Substance	Melting Point, °C	Heat of Fusion		Boiling Point, °C	Heat of Vaporization	
		J/kg	cal/g		J/kg	cal/g
Alcohol, ethyl	−117.3	104×10^3	24.9	78.5	854×10^3	204
Ammonia	−75	452×10^3	108.1	−33.3	1370×10^3	327
Copper	1080	134×10^3	32	2870	4730×10^3	1130
Helium	−269.6	5.23×10^3	1.25	−268.9	20.9×10^3	5
Lead	327.3	24.5×10^3	5.86	1620	871×10^3	208
Mercury	−39	11.5×10^3	2.8	358	296×10^3	71
Oxygen	−218.8	13.9×10^3	3.3	−183	213×10^3	51
Silver	960.8	88.3×10^3	21	2193	2340×10^3	558
Water	0	334×10^3	80	100	2256×10^3	540
Zinc	420	100×10^3	24	918	1990×10^3	475

It is often helpful when studying the changes of phase of a substance to plot a graph showing how the temperature of the substance varies as thermal energy is applied at a steady rate. Such a graph is shown in Fig. 17.7 for water. If a quantity of ice is taken from a freezer at −20°C and heated, its temperature will increase gradually until the ice begins to melt at 0°C. For each degree rise in temperature, each gram of ice will absorb 0.5 cal of heat energy. During the melting process, the temperature remains constant, and each gram of ice will absorb 80 cal of heat energy in forming 1 g of water.

Once all the ice has melted, the temperature begins to rise again at a uniform rate until the water begins to boil at 100°C. For each degree increase in temperature, each gram will absorb 1 cal of heat energy. During the vaporization process, the temperature remains constant. Each gram of water absorbs 540 cal of heat energy in forming 1 g of water vapor at 100°C. If the resulting water vapor is contained and the heating is continued until all the water is gone, the temperature will again start to rise. The specific heat of steam is 0.48 cal/g · C°.

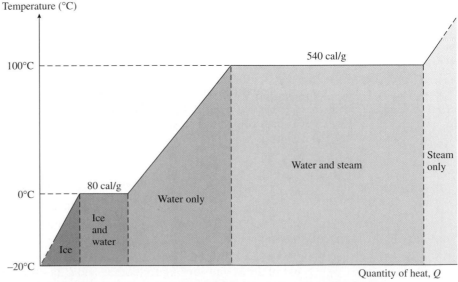

Figure 17.7   The variation of temperature with a change in thermal energy for water.

Example 17.4

What quantity of heat is required to change 20 g of ice at $-25°C$ to steam at $120°C$. Use SI units and take the constants from the tables.

Plan: We will need to separate this problem into five parts: (1) the heat required to bring the ice from $-25°C$ to the melting temperature ($0°C$), (2) the heat required to melt all of this ice, (3) the heat to bring the resulting water from $0°C$ to the steam point ($100°C$), (4) the heat to vaporize all of the water, and (5) the heat to raise the resulting steam to $120°C$. Throughout the entire process, the mass (20 g) does not change. The total heat needed will be the sum of these quantities. With the exception of temperature, we will use SI units for all quantities in this example.

Solution: The mass is converted to kilograms, giving $m = 0.020$ kg, and the necessary constants are taken from the tables:

$$c_w = 4186 \text{ J/kg} \cdot C°, \ c_{ice} = 2090 \text{ J/kg} \cdot C°, \ c_{steam} = 2000 \text{ J/kg} \cdot C°,$$
$$L_f = 3.34 \times 10^5 \text{ J/kg}; \ L_v = 2.26 \times 10^6 \text{ J/kg}$$

The heat to raise the temperature of the ice from $-25°C$ to $0°C$ is

$$Q_1 = mc_w \, \Delta t = (0.020 \text{ kg})(2090 \text{ J/kg} \cdot C°)(25°C)$$
$$= (0.020 \text{ kg})(2090 \text{ J/kg} \cdot C°)(20 \ C°) = 1045 \text{ J}$$

The heat to melt all 20 g of ice is

$$Q_2 = mL_f = (0.020 \text{ kg})(3.34 \times 10^5 \text{ J/kg}) = 6680 \text{ J}$$

The heat to raise the temperature of 20 g of water to $100°C$ is

$$Q_3 = mc_w \, \Delta t = (0.020 \text{ kg})(4186 \text{ J/kg} \cdot C°)(100°C - 0°C) = 8372 \text{ J}$$

The heat to melt all 20 g of ice is

$$Q_4 = mL_v = (0.020 \text{ kg})(2.26 \times 10^6 \text{ J/kg}) = 45,200 \text{ J}$$

Finally, we must increase the steam temperature to $120°C$. We will assume the steam is contained in some fashion because it is in vapor form, and it must be allowed to superheat.

$$Q_5 = mc_s \, \Delta t = (0.020 \text{ kg})(2000 \text{ J/kg} \cdot C°)(120°C - 100°C)$$
$$= (0.020 \text{ kg})(2000 \text{ J/kg} \cdot C°)(20 \ C°) = 800 \text{ J}$$

The total heat required is the sum of these five processes:

$$Q_T = \sum Q = 1045 \text{ J} + 6680 \text{ J} + 8372 \text{ J} + 45,200 \text{ J} + 800 \text{ J}$$
$$Q_T = 62,097 \text{ J} \quad \text{or} \quad Q_T = 62.1 \text{ kJ}$$

When heat is removed from a gas, its temperature drops until it reaches the temperature at which it boiled. As more heat is removed, the vapor returns to the liquid phase. This process is referred to as *condensation*. In condensing, a vapor gives up an amount of heat equivalent to the heat required to vaporize it. Thus, the *heat of condensation* is equivalent to the heat of vaporization. The difference lies only in the direction of heat transfer.

Similarly, when heat is removed from a liquid, its temperature will drop until it reaches the temperature at which it melted. As more heat is removed, the liquid returns to its solid phase. This process is called *freezing* or *solidification*. The heat of solidification is exactly equal to the heat of fusion. Thus, the only distinction between freezing and melting lies in whether heat is being released or absorbed.

Under the proper conditions of temperature and pressure, it is possible for a substance to change from the solid phase directly to the gaseous phase without passing through the liquid phase. This process is referred to as *sublimation*. Solid carbon dioxide (dry ice), iodine, and camphor (mothballs) are examples of substances that are known to sublime at normal temperatures. The quantity of heat absorbed per unit mass in changing from a solid to a vapor is called the ***heat of sublimation.***

Before we leave the subject of fusion and vaporization, it will be instructive to offer examples of their measurement. In any given mixture, the quantity of heat absorbed must equal the quantity of heat released. This principle holds even if a change of phase occurs. The procedure is demonstrated in Examples 17.5 and 17.6.

Example 17.5

Twelve grams of crushed ice at $-10°C$ are dropped into a 50-g aluminum calorimeter cup containing 100 g of water at 50°C. The system is quickly sealed and allowed to reach thermal equilibrium. What is the resulting temperature?

Plan: The heat lost by the calorimeter cup and water must equal the heat gained by the ice, including any changes of phase that takes place. There are three possibilities for the equilibrium temperature: (1) 0°C, with both ice and water remaining; (2) above 0°C, in which case all ice is melted; and (3) below 0°C, if none of the ice melts. Given the initial temperature and amount of water, it seems more likely that all of the ice melts and that the equilibrium temperature t_e is above 0°C. We will make that assumption, and the result will tell us if we are correct.

Solution: Calculate the total heat lost and the total heat gained separately based on our assumptions. To simplify the calculations, we will occasionally omit the units.

$$Heat\ lost = heat\ lost\ by\ cup\ +\ heat\ lost\ by\ water$$
$$= m_c c_c (50°C - t_e) + m_w c_w (50°C - t_e)$$
$$= (50\ g)(0.22\ cal/g \cdot C°)(50°C - t_e) + (100\ g)(1\ cal/g \cdot C°)(50°C - t_e)$$
$$= 550 - 11t_e + 5000 - 100t_e$$
$$= 5550 - 111t_e$$

$$Heat\ gained = Q\ by\ ice\ +\ Q\ by\ melting\ +\ Q\ to\ reach\ t_e$$
$$= m_i c_i (10\ C°) + m_i L_f + m_i c_w (t_e - 0°C)$$
$$= (12\ g)(0.48\ cal/g \cdot C°)(10\ C°) + (12\ g)(80\ cal/g)$$
$$+\ (12\ g)(1\ cal/g \cdot C°)t_e - 0$$
$$= 1018 - 12t_e$$

Now, we set the total heat lost equal to the total heat gained and solve for the final temperature.

$$5550 - 111t_e = 1018 + 12t_e$$
$$123t_e = 4532$$
$$t_e = 36.8°C$$

Example 17.6

Ten grams of steam at 100°C are mixed with 200 g of water and 120 g of ice. Find the final temperature of the system and the composition of the mixture.

Plan: The rather small amount of steam in comparison with ice and water makes us wonder if enough heat may be released by the steam to melt all of the ice. To test this suspicion, we will find the heat required to melt *all* of the ice and then compare that with the maximum heat that could come from the steam (taking the condensed

water down to 0°C.) Then we will be able to apply the conservation laws to find the final temperature and composition of the mixture. Any mixture of ice and water in equilibrium must have a temperature of 0°C.

Solution: The quantity of heat required to melt all of the ice is

$$Q_1 = m_i L_f = (120 \text{ g})(80 \text{ cal/g}) = 9600 \text{ cal}$$

The maximum heat we can expect the steam to give up is

$$Q_2 = m_s L_v + m_s c_w (100°C - 0°C)$$
$$= (10)(540) + (10)(1)(100) = 6400 \text{ cal}$$

Since 9600 cal was needed to melt all the ice and only 6400 cal can be delivered by the steam, the final mixture must consist of ice and water of 0°C.

To determine the final composition of the mixture, note that 3200 additional calories would be required to melt the remaining ice. Hence,

$$m_i L_f = 3200 \text{ cal}$$
$$m_i = \frac{3200 \text{ cal}}{80 \text{ cal/g}} = 40 \text{ g}$$

Therefore, there must be 40 g of ice in the final mixture. The amount of water remaining is

Water remaining = initial water + melted ice + condensed steam
$$= 200 \text{ g} + 80 \text{ g} + 10 \text{ g} = 290 \text{ g}$$

The final composition consists of a mixture of 40 g of ice in 290 g of water at 0°C.

Suppose we had assumed in Example 17.6 that all the ice melted and attempted to solve for t_e, as in Example 17.5. In this case, we would have obtained a value for the equilibrium temperature that was below the freezing point (0°C). Clearly, this kind of answer could result only from a false assumption.

An alternative procedure for solving Example 17.6 would be to solve directly for the number of grams of ice that must have melted to balance the 6400 cal of heat energy released by the steam. It is left as an exercise for you to show that the same results are obtained.

Problem-Solving Strategy

Quantity of Heat and Calorimetry

1. Read the problem carefully and then draw a rough sketch, labeling it with the given information and stating what is to be found. Be careful to include units for all physical quantities.

2. If a heat loss or gain results in a change of temperature, you will need to decide which units are appropriate for the specific heat capacity. The need for the use of consistent units is demonstrated here:

$$Q = mc \, \Delta t \qquad J = (\text{kg})\left(\frac{J}{\text{kg} \cdot \text{C°}}\right)(\text{C°})$$

It is wise to write the units of specific heat on substitution so that the choice of the correct units for other quantities is obvious.

3. If there is a change of phase, the latent heat of fusion or vaporization will be needed. Here again we must be careful to use those units consistent with *joules per kilogram, calories per gram,* or *Btu per pound.* The temperature remains constant during a phase change.

4. Conservation of energy demands that the total of the heat losses be equal to the total of the heat gains. Note

that a decrease of temperature, condensation, or freezing of a liquid indicates a *loss* of heat. A rise in temperature, melting, or vaporizing occurs when there is a *gain* of heat. We might add the absolute values of the losses on the left side and set them equal to the total gains on the right side. As an example, consider a mass m_s of steam at 100°C mixed with a mass m_i of ice at 0°C. The result is water at the equilibrium temperature t_e.

$$m_s L_v + m_s c_w (100 - t_e) = m_i L_f + m_i c_w (t_e - 0)$$

Note that the temperature differences are taken as *high minus low* in each case to yield the *absolute* values lost or gained.

17.6 Heat of Combustion

Whenever a substance is burned, it releases a definite quantity of heat. The quantity of heat per unit mass, or per unit volume, when the substance is burned completely is called the **heat of combustion.** Commonly used units are Btu per pound mass, Btu per cubic foot, calories per gram, and kilocalories per cubic meter. For example, the heat of combustion of coal is approximately 13,000 Btu/lb$_m$. This means that each pound of coal, when completely burned, should release 13,000 Btu of heat energy.

Summary and Review

Summary

In this chapter, you have studied the quantity of heat as a measurable quantity that is based on a standard change. The British thermal unit and the calorie are measures of the heat required to raise the temperature of a unit mass of water by a unit degree. By applying these standard units to experiments with a variety of materials, we have learned to predict heat losses or heat gains in a constructive fashion. The essential concepts presented in this chapter are as follows.

- The *British thermal unit* (Btu) is the heat required to change the temperature of 1 pound-mass of water 1 Fahrenheit degree.
- The *calorie* is the heat required to raise the temperature of 1 gram of water by 1 Celsius degree.
- Several conversion factors may be useful for problems involving thermal energy:

$$1 \text{ Btu} = 252 \text{ cal} = 0.252 \text{ kcal} \qquad 1 \text{ cal} = 4.186 \text{ J}$$

$$1 \text{ Btu} = 778 \text{ ft} \cdot \text{lb} \qquad 1 \text{ kcal} = 4186 \text{ J}$$

- The *specific heat capacity* c is used to determine the quantity of heat Q absorbed or released by a unit mass m as the temperature changes by an interval Δi.

$$c = \frac{Q}{m \, \Delta t} \qquad Q = mc \, \Delta t \qquad \begin{array}{c} \textit{Specific} \\ \textit{Heat Capacity} \end{array}$$

- Conservation of thermal energy requires that in any exchange of thermal energy the heat lost must equal the heat gained.

$$\boxed{Heat \; lost = heat \; gained}$$

$$\boxed{\sum (mc \, \Delta t)\text{loss} = \sum (mc \, \Delta t)\text{gain}}$$

As an example, suppose body 1 transfers heat to bodies 2 and 3 as the system reaches an equilibrium temperature t_e:

$$m_1 c_1 (t_1 - t_e) = m_2 c_2 (t_e - t_2) + m_3 c_3 (t_e - t_3)$$

- The *latent heat of fusion* L_f and the *latent heat of vaporization* L_v are heat losses or gains by a unit mass m during a phase change. There is no change in temperature.

$$L_f = \frac{Q}{m} \qquad Q = mL_f \qquad \begin{array}{c} \textit{Latent Heat} \\ \textit{of Fusion} \end{array}$$

$$L_v = \frac{Q}{m} \qquad Q = mL_v \qquad \begin{array}{c} \textit{Latent Heat} \\ \textit{of Vaporization} \end{array}$$

If a change of phase occurs, the above relationships must be added to the calorimetry equation as appropriate.

Key Terms

boiling point 359
British thermal unit (Btu) 351
calorie 351
calorimeter 356
condensation 361
freezing 361
fusion 359

heat 351
heat capacity 354
heat of combustion 364
heat of sublimation 362
kilocalorie 351
latent heat of fusion 359
latent heat of vaporization 359

mechanical equivalent of heat 352
melting point 359
solidification 361
specific heat capacity 354
sublimation 361
vaporization 359
water equivalent 358

Review Questions

17.1. Discuss the caloric theory of heat. In what ways is it successful in explaining heat phenomena? Where does it fail?

17.2. Blocks of five different metals—aluminum, copper, zinc, iron, and lead—are constructed with the same mass and the same cross-sectional area. Each block is heated to a temperature of 100°C and placed on a block of ice. Which will melt the ice to the greatest depth? List the remaining four metals in the order of decreasing penetration depths.

17.3. On a winter day, the snow is observed to melt from the concrete sidewalk before it melts from the law. Which has the higher heat capacity?

17.4. If two objects have the same heat capacity, are they necessarily constructed of the same material? What if they have the same specific heat capacities?

17.5. Why is temperature considered a *fundamental quantity?*

17.6. A mechanical analogy to the concept of thermal equilibrium is given in Fig. 17.8. When the valve is opened, the water will flow until it has the same level in each tube. What are the analogies to temperature and thermal energy?

17.7. In a mixture of ice and water, the temperature of both ice and water is 0°C. Why, then, does the ice feel colder to the touch?

17.8. Why does steam at 100°C produce a far worse burn than water at 100°C?

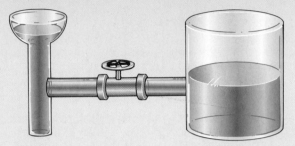

Figure 17.8 A mechanical analogy to the equalization of temperature.

17.9. The temperature of 1 g of iron is raised by 1 C°. How much more heat would be required to raise the temperature of 1 lb of iron by 1 F°?

Problems

Note: Refer to Tables 17.1 and 17.2 for accepted values for specific heat, heat of vaporization, and heat of fusion for the substances in the following problems.

Section 17.2 The Quantity of Heat and Section 17.3 Specific Heat Capacity

17.1. What quantity of heat is required to change the temperature of 200 g of lead from 20 to 100°C?
Ans. 496 cal

17.2. A certain process requires 500 J of heat. Express this energy in calories and in Btu.

17.3. An oven applies 400 kJ of heat to a 4 kg of a substance, causing its temperature to increase by 80 C°. What is the specific heat capacity?
Ans. 1250 J/kg C°

17.4. What quantity of heat will be released when 40 lb of copper cools from 78 to 32°F?

17.5. A 900-kg car traveling initially at 20 m/s is brought to a stop. The work required to stop the car is equal to its change in kinetic energy. If all of this work were converted to heat, what is the equivalent loss in kilocalories? Ans. 43 kcal

17.6. An air conditioner is rated at 15,000 Btu/h. Express this power in kilowatts and in calories per second.

17.7. Hot coffee with a specific heat of 4186 J/kg C° is poured into a 0.5-kg ceramic cup. How much heat is absorbed by the cup if its temperature increases from 20 to 80°C? Ans. 126 kJ

17.8. A 2-kW electric motor is 80 percent efficient. How much heat is lost in 1 hr?

17.9. An 8-kg copper sleeve must be heated from 25 to 140°C so it will expand to fit over a shaft. How much heat is required? Ans. 359 kJ

17.10. How many grams of iron at 20°C must be heated to 100°C so that it releases 1800 cal of heat as it returns to its original temperature?

17.11. A 4-kg chunk of metal (c = 320 J/kg C°) is initially at 300°C. What will its final temperature be if it loses 50 kJ of heat energy? Ans. 261°C

17.12. In a heat treatment, a hot copper part is quenched with water, cooling it from 400 to 30°C. What was the mass of the part if it loses 80 kcal of heat?

Section 17.4 The Measurement of Heat

17.13. A 400-g copper pipe initially at 200°C is dropped into a container filled with 3 kg of water at 20°C. Ignoring other heat exchanges, what is the equilibrium temperature of the mixture?
Ans. 22.2°C

17.14. How much aluminum (c = 0.22 cal/g C°) at 20°C must be added to 400 g of hot water at 80°C in order that the equilibrium temperature be 30°C?

17.15. A 450-g chunk of metal is heated to 100°C and then dropped into a 50-g aluminum calorimeter cup containing 100 g of water. The initial temperature of the cup and water is 10°C, and the equilibrium temperature is 21.1°C. Find the specific heat of the metal. Ans. 0.0347 cal/g C°

17.16. What mass of water initially at 20°C must be mixed with 2 kg of iron to bring the iron from 250°C to an equilibrium temperature of 25°C?

17.17. A worker removes a 2-kg piece of iron from an oven and places it into a 1-kg aluminum container partially filled with 2 kg of water. If the temperature of the water rises from 21 to 50°C, what was the initial temperature of the iron?
Ans. 337°C

17.18. How much iron at 212°F must be mixed with 10 lb of water at 68°F in order to have an equilibrium temperature of 100°F?

17.19. A 1.3-kg copper block is heated to 200°C and then dropped into an insulated container partially filled with 2 kg of water at 20°C. What is the equilibrium temperature?   Ans. 30.3°C

17.20. Fifty grams of brass shot are heated to 200°C and then dropped into a 50-g aluminum cup containing 160 g of water. The temperature of the cup and water is initially 20°C. What is the equilibrium temperature?

Section 17.5 Change of Phase

17.21. A foundry has an electric furnace that can completely melt 540 kg of copper. If the temperature of the copper was initially 20°C, what total heat is needed to melt the copper?
Ans. 2.96×10^8 J

17.22. How much heat is needed to completely melt 20 g of silver at its melting temperature?

17.23. What quantity of heat is needed to convert 2 kg of ice at −25°C to steam at 100°C?
Ans. 6.13×10^6 J

17.24. If 7.57×10^6 J of heat is absorbed in the process of completely melting a 1.60-kg chunk of an unknown metal, what is the latent heat of fusion and what is the metal?

17.25. How many grams of steam at 100°C must be mixed with 200 g of water at 20°C in order for the equilibrium temperature to be 50°C?
Ans. 10.2 g

17.26. What total heat is released when 0.500 lb of steam at 212°F changes to ice at 10°F?

17.27. One hundred grams of ice at 0°C are mixed with 600 g of water at 25°C. What will be the equilibrium temperature for the mixture?
Ans. 10.0°C

17.28. A certain grade of gasoline has a heat of combustion of 4.6×10^7 J/kg. Assuming 100 percent efficiency, how much gasoline must be burned to completely melt 2 kg of copper at its melting temperature?

Additional Problems

17.29. If 1600 J of heat are applied to a brass ball, its temperature rises from 20 to 70°C. What is the mass of the ball?   Ans. 82.1 g

17.30. How much heat does an electric freezer absorb in lowering the temperature of 2 kg of water from 80 to 20°C?

17.31. A heating element supplies an output power of 12 kW. How much time is needed to melt completely a 2-kg silver block? Assume that no power is wasted.   Ans. 14.7 s

17.32. How much ice at −10°C must be added to 200 g of water at 50°C to bring the equilibrium temperature to 40°C?

17.33. Assume that 5 g of steam at 100°C are mixed with 20 g of ice at 0°C. What will be the equilibrium temperature?   Ans. 64.0°C

17.34. How much heat is developed by the brakes of a 4000-lb truck to bring it to a stop from a speed of 60 mi/h?

17.35. Two hundred grams of copper at 300°C are dropped into a 310-g copper calorimeter cup partially filled with 300 g of water. If the initial temperature of the cup and water was 15°C, what is the equilibrium temperature?   Ans. 30.3°C

17.36. How many pounds of coal must be burned to melt completely 50 lb of ice in a heater that is 60 percent efficient?

17.37. If 80 g of molten lead at 327.3°C are poured into a 260-g iron casting initially at 20°C, what will be the equilibrium temperature neglecting other losses?   Ans. 58.9°C

17.38. What equilibrium temperature is reached when 2 lb of ice at 0°F are dropped into a 3-lb aluminum cup containing 7.5 lb of water? The temperature of the cup and water is initially 200°F.

17.39. A solar collector has an area of 5 m², and the power of sunlight is delivered at 550 W/m². This power is used to increase the temperature of 200 g of water from 20 to 50°C. How much time is required?   Ans. 9.13 s

17.40. If 10 g of milk at 12°C are added to 180 g of coffee at 95°C, what is the equilibrium temperature? Assume that milk and coffee are essentially water.

*17.41. How many grams of steam at 100°C must be added to 30 g of ice at 0°C in order to produce an equilibrium temperature of 40°C?   Ans. 6.00 g

*17.42. A 5-g lead bullet moving at 200 m/s embeds itself into a wooden block. Half of its initial energy is absorbed by the bullet. What is the increase in temperature of the bullet?

*17.43. If 4 g of steam at 100°C are mixed with 20 g of ice at −5°C, find the final temperature of the mixture.   Ans. 37.9°C

Critical Thinking Questions

17.44. A large, insulated container holds 120 g of coffee at 85°C. How much ice at 0°C must be added to cool the coffee to 50°C? Now, how much coffee at 100°C must be added to return the contents to 85°C? How many grams are finally in the container?   Ans. 32.3 g, 355 g, 508 g

17.45. Four 200-g blocks are constructed of copper, aluminum, silver, and lead so that they have the same mass and the same base area (although of different heights). The temperature of each block is raised from 20 to 100°C by applying heat at the rate of 200 J/s. Find the time required for each block to reach 100°C.

17.46. Each of the blocks in Question 17.45 is placed on a large block of ice. Find out how much ice is melted by each block when all blocks reach equilibrium at 0°C. Which block sinks deepest, and which sinks the least?

Ans. Pb = 7.78 g, Ag = 13.8 g,
Cu = 23.4 g, Al = 55.1 g, Al, Pb

17.47. In an experiment to determine the latent heat of vaporization for water, a student measures the mass of an aluminum calorimeter cup to be 50 g. After a quantity of water is added, the combined mass of the water and cup is 120 g. The initial temperature of the cup and water is 18°C. A quantity of steam at 100°C is passed into the calorimeter, and the system is observed to reach equilibrium at 47.4°C. The total mass of the final mixture is 124 g. What value will the student obtain for the heat of vaporization?

*17.48. Equal masses of ice at 0°C, water at 50°C, and steam at 100°C are mixed and allowed to reach equilibrium. Will all the steam condense? What will be the temperature of the final mixture? What percentage of the final mixture will be water, and what percentage will be steam?

Ans. no, 100°C, 19.1 percent steam and
80.9 percent water

*17.49. If 100 g of water at 20°C are mixed with 100 g of ice at 0°C and 4 g of steam at 100°C, find the equilibrium temperature and the composition of the mixture.

*17.50. Ten grams of ice at −5°C are mixed with 6 g of steam at 100°C. Find the final temperature and the composition of the mixture.

Ans. 100°C, 13.4 g of water, 2.62 g of steam

18

Transfer of Heat

Transfer of heat is demonstrated by a glass blower. Molten glass is held inside a furnace at the end of a long rod. The heat felt by the hands is due to conduction through the rod; the heat felt by the face is due to radiation. If one could place a hand above the molten glass, the heat felt would be due to both radiation and convection currents. Each of these methods of heat transfer is discussed in this chapter.
(Photo by Paul E. Tippens.)

Objectives

After completing this chapter, you should be able to

1. Demonstrate by definition and example your understanding of *thermal conductivity*, *convection*, and *radiation*.

2. Solve *thermal conductivity* problems involving parameters such as quantity of heat Q, surface area A, surface temperature t, time τ, and material thickness L.

3. Solve problems involving the transfer of heat by convection and discuss the meaning of the *convection coefficient*.

4. Define the *rate of radiation* and *emissivity* and use these concepts to solve problems involving thermal radiation.

We have referred to heat as a form of energy in transit. Whenever there is a difference of temperature between two bodies or between two portions of the same body, heat is said to *flow* in a direction from higher to lower temperature. There are three principal methods by which this heat exchange occurs: *conduction, convection,* and *radiation.* Examples of all three are shown in Fig. 18.1.

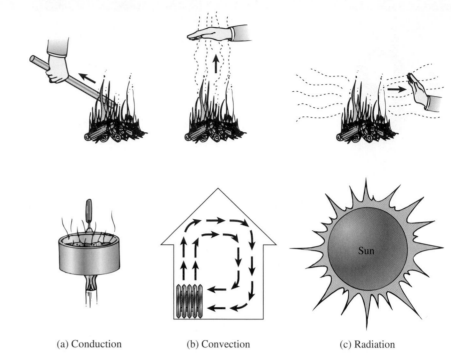

(a) Conduction   (b) Convection   (c) Radiation

Figure 18.1 The three methods of heat transfer: (a) conduction, (b) convection, and (c) radiation.

18.1 Methods of Heat Transfer

Most of our discussion has involved heat transfer by *conduction:* by molecular collisions between neighboring molecules. For example, if we hold one end of an iron rod in a fire, the heat will eventually reach our hand through the process of conduction. The increased molecular activity at the heated end is passed on from molecule to molecule until it reaches the hand. The process will continue as long as there is a difference in temperature along the rod.

> Conduction is the process in which heat energy is transferred by adjacent molecular collisions throughout a material medium. The medium itself does not move.

The most common application of the principle of conduction is probably cooking.

On the other hand, if we place our hand above a fire, as shown in Fig. 18.1b, we can feel the transfer of heat by the rising hot air. This process, called *convection,* differs from conduction because the material medium is moving. Heat is transferred by moving masses instead of being passed along by neighboring molecules.

> Convection is the process by which heat is transferred by the actual mass motion of a fluid.

Convection currents form the basis for heating and cooling most houses.

When we hold our hand to the side of a fire, the primary source of heat is through *thermal radiation.* Radiation involves the emission or absorption of electromagnetic waves originating at the atomic level. These waves travel at the speed of light $(3 \times 10^8$ m/s) and require no material medium for their passage.

> Radiation is the process by which heat is transferred by electromagnetic waves.

The most obvious source of radiant energy is our own Sun. Neither conduction nor convection can play a role in transferring heat energy through space to the Earth. The enormous heat energy received on the Earth is transferred by electromagnetic radiation. Where a material medium is involved, however, the transfer of heat due to radiation is usually small in comparison with that transferred by conduction and convection.

Unfortunately, there are many factors that affect the transfer of heat energy by all three methods. The computation of the quantity of thermal energy transferred in a given process is complicated. The relationships discussed in the following sections are based on empirical observations and depend on ideal conditions. The extent to which these conditions can be met usually determines the accuracy of their predictions.

18.2 Conduction

When two parts of a material are maintained at different temperatures, energy is transferred by molecular collisions from higher to lower temperature. This process of conduction is also aided by the motion of free electrons within the substance. These electrons have become disassociated from their parent atoms and are free to move about when stimulated either thermally or electrically. Most metals are efficient conductors of heat because they have a number of free electrons that can distribute heat in addition to that propagated by molecular agitation. In general, a good conductor of electricity is also a good conductor of heat.

The fundamental law of heat conduction is a generalization of experimental results concerning the flow of heat through a slab of material. Consider the slab of thickness L and area A in Fig. 18.2. One face is maintained at a temperature t and the other at a temperature t'. The quantity of heat Q that flows perpendicular to the face during a time τ is measured. Repeating the experiment for many different materials of different thicknesses and face areas, we are able to make some general observations concerning the conduction of heat:

1. The quantity of heat transferred per unit of time is directly proportional to the temperature difference ($\Delta t = t' - t$) between the two faces.

2. The quantity of heat transferred per unit of time is directly proportional to the area A of the slab.

3. The quantity of heat transferred per unit of time is inversely proportional to the thickness L of the slab.

These results can be expressed in equation form by introducing the proportionality constant k. We write

$$H = \frac{Q}{\tau} = kA\frac{\Delta t}{L} \tag{18.1}$$

where H represents the rate at which heat is transferred. Although the equation was introduced for a slab of material, it also holds for a rod of cross section A and length L.

The proportionality constant k is a property of the material called its ***thermal conductivity.*** From the defining equation, one can see that substances with a large thermal conductivity are good conductors of heat, whereas substances of low conductivity are poor conductors, or *insulators.*

The thermal conductivity of a substance is a measure of its ability to conduct heat and is defined by the relation

$$k = \frac{QL}{\tau A \Delta t} \tag{18.2}$$

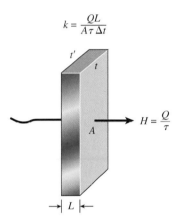

Figure 18.2 Measurement of thermal conductivity.

The numerical value for the thermal conductivity depends on the units chosen for heat, thickness, area, time, and temperature. A substitution of SI units for each of these quantities yields the following accepted units:

$$\text{SI units: J/s} \cdot \text{m} \cdot \text{C}° \quad \text{or} \quad \text{W/m} \cdot \text{K}$$

You will remember that the joule per second (J/s) is the power in watts (W) and that kelvin and Celsius temperature intervals are equal.

Unfortunately, the SI units are rarely used for conductivity in industry today. The choice of units is more often made on the basis of the convenience of measurement. For example, in the USCS, heat is measured in Btu, thickness in inches, area in square feet, time in hours, and the temperature interval in Fahrenheit degrees. Consequently, the units for thermal conductivity, from Eq. (18.2), are as follows:

$$\text{USCS: } k = \text{Btu} \cdot \text{in./ft}^2 \cdot \text{h} \cdot \text{F}°$$

In the metric system, heat transfer in calories is often encountered more than heat transfer expressed in joules. Therefore, the following units are frequently used:

$$\text{Metric units: } k = \text{kcal/m} \cdot \text{s} \cdot \text{C}°$$

The following conversion factors will be helpful:

$$1 \text{ kcal/s} \cdot \text{m} \cdot \text{C}° = 4186 \text{ W/m} \cdot \text{K}$$
$$1 \text{ Btu} \cdot \text{in./ft}^2 \cdot \text{h} \cdot \text{F}° = 3.445 \times 10^{-5} \text{ kcal/m} \cdot \text{s} \cdot \text{C}°$$

The thermal conductivities for various materials are listed in Table 18.1.

Table 18.1

Thermal Conductivities and *R*-values

Substance	Conductivity, *k*			*R*-values*
	W/m · K	kcal/m · s · C°	Btu · in./ft² · h · F°	ft² · h · F°/Btu
Aluminum	205	5.0×10^{-2}	1451	0.00069
Brass	109	2.6×10^{-2}	750	0.0013
Copper	385	9.2×10^{-2}	2660	0.00038
Silver	406	9.7×10^{-2}	2870	0.00035
Steel	50.2	1.2×10^{-2}	320	0.0031
Brick	0.7	1.7×10^{-4}	5.0	0.20
Concrete	0.8	1.9×10^{-4}	5.6	0.18
Corkboard	0.04	1.0×10^{-5}	0.3	3.3
Drywall	0.16	3.8×10^{-5}	1.1	0.9
Fiberglass	0.04	1.0×10^{-5}	0.3	3.3
Glass	0.8	1.9×10^{-4}	5.6	0.18
Polyurethane	0.024	5.7×10^{-6}	0.17	5.9
Sheathing	0.55	1.3×10^{-5}	0.38	2.64
Air	0.024	5.7×10^{-6}	0.17	5.9
Water	0.6	1.4×10^{-4}	4.2	0.24

*R-values based on 1-in. thickness.

Example 18.1

The outside wall of a brick barbeque pit is 6 cm thick. The inside surface is at 150°C, and the outside surface is at 30°C. How much heat is lost in 1 h through an area of 1 m²?

Plan: The rate of heat flow is given by Eq. (18.1). The choice of units for these quantities is determined by the choice for conductivity in Table 18.1. We will use the

SI units, so for brick, $k = 0.7$ W/m · K. Thus, length must be in m, area in m², time in s, and temperature in K, so the heat will be in joules (J).

Given: $A = 1$ m², $L = 0.060$ m, $\tau = 1$ h $= 3600$ s, $\Delta t = (150°C - 30°C)$
$$= 120\ C° = 120\ K$$

Solution: Solving for Q in Eq. (18.1) yields

$$Q = ka\tau \frac{\Delta t}{L}$$

$$Q = \frac{(0.7\ \text{W/mK})(1\ \text{m}^2)(3600\ \text{s})(120\ \text{K})}{0.060\ \text{m}} = 5.04 \times 10^6\ \text{J}$$

Therefore, 5.04 MJ of heat flows each hour from inside the brick wall to the outside.

It is always wise to carry the units of each quantity throughout the entire solution of a problem. This practice will save many needless errors. For example, it is sometimes easy to forget that in USCS units, the thickness must be expressed in *inches* and the area in *square feet*. If the units of thermal conductivity are written with their numerical value during substitution, many such errors can be avoided.

When two materials of different conductivities are connected, the rate at which heat is conducted through each material must be constant over time. There are no sources or sinks of heat energy within the materials, and the beginning and ending points for the heat path are maintained at constant temperatures. A steady flow will eventually be reached since the heat cannot "pile up" or "speed up" at any point. The conductivities of the materials do not change, and the thicknesses are fixed. This means that the temperature intervals for each material must adjust to produce the steady flow of heat through the composite structure.

Example 18.2

The composite wall of a freezing plant (Fig. 18.3) consists of a 10-cm layer of corkboard on the inside and a 14-cm layer of concrete on the outside. The temperature on the inside corkboard surface is $-20°C$, and the temperature of the outside concrete surface is $24°C$. (a) Find the temperature at the interface of the two materials. (b) What is the rate of heat loss in watts per square meter?

Plan: For a steady state, the rate of heat flow through the corkboard is equal to the rate of heat flow through the concrete. The areas are the same, so we can set the heat per unit area (H/A) for flow in the corkboard to the heat per unit area for the concrete. From this

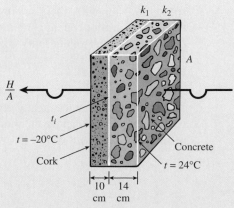

Figure 18.3 Heat conduction through a compound wall.

equation, we can solve for the interface temperature and then use that result to find the rate of heat flow through *either* the cork or the concrete because the rates are the same.

Solution (a): The thermal conductivities for cork and concrete are taken from the tables. We will use the subscript 1 to refer to the corkboard and the subscript 2 to refer to the concrete. The letter t_i is used to represent the temperature at the interface of the two materials. Since the H/A is the same in each material, we may write

$$\frac{H_1}{A_1}(\text{corkboard}) = \frac{H_2}{A_2}(\text{concrete})$$

$$\frac{k_1[t_i - (-20°C)]}{0.10 \text{ m}} = \frac{k_2(24°C - t_i)}{0.14 \text{ m}}$$

$$\frac{(0.04 \text{ W/m} \cdot \text{K})(t_i + 20°C)}{0.10 \text{ m}} = \frac{(0.8 \text{ W/m} \cdot \text{K})(24°C - t_i)}{0.14 \text{ m}}$$

Now, we simplify by multiplying each term by 14 to obtain

$$5.6(t_i + 20°C) = 80(24°C - t_i)$$

$$t_i = 21.1°C$$

Solution (b): The heat flow per unit area per unit time can now be found from Eq. (18.1) applied to either the corkboard or the concrete. For the concrete, we have

$$\frac{H_2}{A_2} = \frac{k_2(24°C - t_i)}{0.14 \text{ m}}$$

$$= \frac{(0.8 \text{ W/m} \cdot \text{K})(24°C - 21.1°C)}{0.14 \text{ m}}$$

$$= 16.4 \text{ W/m}^2$$

Notice that the kelvin interval and the Celsius interval cancel one another because they are equal. The same rate would be calculated through the corkboard. The difference in temperature between the end points of the corkboard is 41.1 C°, whereas the temperature difference in the concrete is only 2.9 C°. The different temperature intervals result primarily from the difference in the thermal conductivities of the walls.

18.3 Insulation: The *R*-Value

Heat losses in homes and in industry are often based on the insulating properties of various composite walls. For example, one might wish to know the effects of replacing dead air spaces in walls with fiberglass insulation. In such cases, the concept of **thermal resistance** *R* has been introduced into engineering applications. The *R*-value of a material of thickness *L* and of thermal conductivity *k* is defined as

$$R = \frac{L}{k}$$

If we recognize that the steady-state flow of heat through a composite wall is constant (see Example 18.2) and apply the thermal conductivity equation to a number of thicknesses of different materials, it can be shown that

$$\frac{Q}{\tau} = \frac{A \, \Delta t}{\sum_i (L_i/k_i)} = \frac{A \, \Delta t}{\sum_i R_i} \tag{18.3}$$

The quantity of heat flowing per unit of time (Q/τ) through a number of thicknesses of different materials is equal to the product of the area A and the temperature difference Δt

divided by the sum of the *R*-values of the various materials. The *R*-values of common building materials are most often given in USCS units, as shown in Table 18.1. For example, fiberglass ceiling insulation that is 6 in. thick has an *R*-value of 18.8 ft² · F° · h/Btu. A 4-in. brick has an *R*-value of 4.00 ft² · F° · h/Btu. These materials placed side by side would have a total *R*-value of 22.8 ft² · F° · h/Btu.

18.4 Convection

Convection has been defined as the process in which heat is transferred by the actual mass motion of a material medium. A current of liquid or gas that absorbs energy at one place and then moves to another place, where it releases heat to a cooler portion of the fluid, is called a ***convection current.*** A laboratory demonstration of a convection current is illustrated in Fig. 18.4. A rectangular section of glass tubing is filled with water and heated at one of the lower corners. The water near the flame is heated and expands, becoming less dense than the cooler water above it. As the heated water rises, it is replaced by cooler water from the lower tube. This process continues until a counterclockwise convection current circulates throughout the tubing. The existence of such a current is demonstrated vividly by dropping ink into the opening at the top. The ink will be carried along by the convection current until it finally returns to the top from the right section of the tube.

If the motion of a fluid is caused by a difference in density that accompanies a change in temperature, the current produced is referred to as ***natural convection.*** The water flowing through the glass tubing in the previous example represents a natural-convection current. When a fluid is called to move by the action of a pump or fan, the current produced is referred to as ***forced convection.***

Both forced- and natural-convection currents occur in the process of heating a home with a conventional gas furnace, like the one shown in Fig. 18.5. A heat exchanger transfers

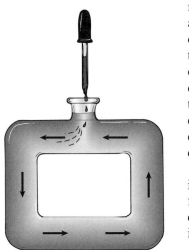
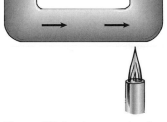

Figure 18.4 An example of natural convection.

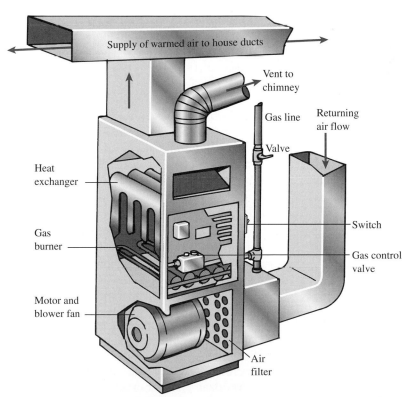

Figure 18.5 Forced-convection currents from a central furnace heat a home. The heat exchanger transfers heat from the burners through the ductwork. Products of combustion are vented to the outside and are never mixed with the air used to heat the home.

Figure 18.6 When a heated slab is placed in a cool fluid, convection currents transfer heat away from the slab at a rate proportional to the difference in temperatures and to the area of the slab.

heat from the gas flames into a metal container. The blower forces air past the heated container and then into the ductwork of a home. The air returns through another set of ducts and reenters the furnace through a filter. A thermostat measures the temperature in the home and regulates the blower and the fuel source to provide the desired quantity of heat. Early gas furnaces wasted about 40 percent of the energy supplied. However, modern units have been able to achieve efficiencies as high as 90 percent by using such techniques as condensed steam or sealed combustion chambers.

The calculation of heat transferred by convection is an enormously difficult task. So many physical properties of a fluid depend upon temperature and pressure that we can hope for only an estimate in most situations. For example, consider the conducting slab of material A and temperature t_s that is submerged completely in a cooler fluid at a temperature t_f, as illustrated in Fig. 18.6. The fluid that comes in contact with the slab will rise and displace the cooler fluid. Experimental observation shows that the rate H at which heat is transferred by convection is proportional to the area A and to the difference in temperature Δt between the slab and the fluid.

Unlike thermal conductivity, convection is not a property of the solid or fluid but depends on many parameters of the system. It is known to vary with the geometry of the solid and its surface finish, the velocity of the fluid, the density of the fluid, and thermal conductivity. Differences in the pressure also affect the transfer of heat by convection. To understand how geometry affects convection, one only has to consider the obvious differences imposed by a floor facing upward or a ceiling facing downward. Several models have been developed to produce mathematical estimates of heat transfer by convection, but none are reliable enough for inclusion in this treatment.

18.5 Radiation

The term *radiation* refers to the continuous emission of energy in the form of electromagnetic waves originating at the atomic level. Gamma rays, x rays, light waves, infrared rays, radio waves, and radar waves are all examples of electromagnetic radiation; they differ only in their wavelength. In this section, we will be concerned with *thermal radiation*.

> Thermal radiation consists of electromagnetic waves emitted or absorbed by a solid, liquid, or gas by virtue of its temperature.

All objects above absolute zero emit radiant energy. At low temperatures, the rate of emission is small, and the radiation is predominantly of long wavelengths. As the temperature is increased, the rate of emission increases rapidly, and the predominant radiation shifts to shorter wavelengths. If an iron rod is heated continuously, it will eventually give off radiation in the visible region; hence, the terms *red hot* and *white hot*.

Experimental measurements have shown that the rate at which thermal energy is radiated from a surface *varies directly with the fourth power of the absolute temperature of the radiating body*. Thus, if the temperature of an object is doubled, the rate at which it emits thermal energy will be increased 16-fold.

An additional factor that must be considered in computing the rate of heat transfer by radiation is the nature of the exposed surfaces. Objects that are efficient emitters of thermal radiation are also efficient absorbers of radiation. An object that absorbs all the radiation incident on its surface is called an **ideal absorber.** Such an object will also be an **ideal radiator.** There is no such thing as an *ideal* absorber; but, in general, the blacker a surface, the better it absorbs thermal energy. For example, a black shirt absorbs more of the Sun's radiant energy than a lighter shirt. Since the black shirt is also a good emitter, its external temperature will be higher than our body temperature, making us uncomfortable.

An ideal absorber or an ideal radiator is sometimes referred to as a **blackbody** for the reasons mentioned previously. The radiation emitted from a blackbody is called *blackbody radiation*. Although such bodies do not actually exist, the concept is useful as a standard for comparing the **emissivity** of various surfaces.

Emissivity *e* is a measure of a body's ability to absorb or emit thermal radiation.

The emissivity is a unitless quantity that has a numerical value between 0 and 1, depending on the nature of the surface. For a blackbody, the emissivity is equal to unity. For a highly polished silver surface, it is near zero.

The *rate of radiation R* of a body is formally defined as the radiant energy emitted per unit area per unit time, in other words, the power per unit area. Symbolically,

$$R = \frac{E}{\tau A} = \frac{P}{A} \tag{18.4}$$

If the radiant power *P* is expressed in watts and the surface area *A* in square meters, the rate of radiation will be in watts per square meter. As we have discussed earlier, this rate depends on two factors, the absolute temperature *T* and the emissivity *e* of the radiating body. The formal statement of this dependence, known as the **Stefan–Boltzmann law,** can be written

$$R = \frac{P}{A} = e\sigma T^4 \tag{18.5}$$

The proportionality constant σ is a universal constant completely independent of the nature of the radiation. If the radiant power is expressed in watts and the surface in square meters, σ has the value of 5.67×10^{-8} W/m² · K⁴. The emissivity *e* has values from 0 to 1, depending on the nature of the radiating surface. A summary of the symbols and their definitions is given in Table 18.2.

Table 18.2

Definition of Symbols in the Stefan–Boltzmann law ($R = e\sigma T^4$)

Symbol	Definition	Comment
R	Energy radiated per unit time per unit area	$\frac{E}{\tau A}$ or $\frac{P}{A}$
e	Emissivity of the surface	0–1
σ	Stefan's constant	5.67×10^{-8} W/m² · K⁴
T^4	The fourth power of the absolute temperature	K⁴

Example 18.3 What power will be radiated from a spherical silver surface 10 cm in diameter if its temperature is 527°C? The emissivity of the surface is 0.04.

Plan: Units are very important. We must convert 527°C to kelvins and determine the area of the spherical surface in square meters (m²). The power radiated from the surface can then be found by solving Eq. (18.5) for *P*.

Solution: The radius is half the diameter, so R = 0.05 m. The area is then

$$A = 4\pi R^2 = 4\pi(0.05 \text{ m})^2; \quad A = 0.0314 \text{ m}^2$$

The absolute temperature is

$$T = 527 + 273 = 800 \text{ K}$$

Solving for P in Eq. (18.5), we obtain

$$P = e\sigma A T^4$$
$$= (0.04)(5.67 \times 10^{-8} \text{ W/m}^2 \cdot \text{K}^4)(0.0314 \text{ m}^2)(800 \text{ K})^4$$
$$= 29.2 \text{ W}$$

We have said that all objects continuously emit radiation, regardless of their temperature. If this is true, why don't the objects eventually run out of fuel? The answer is that they *would* run down if no energy were supplied to them. The filament in an electric light bulb cools rather quickly to room temperature when the supply of electric energy is shut off. It does not cool further because, at this point, the filament is absorbing radiant energy at the same rate that it is emitting radiant energy. The law covering this phenomenon is known as *Prevost's law of heat exchange:*

> A body at the same temperature as its surroundings radiates and absorbs heat at the same rates.

Figure 18.7 shows an isolated object in thermal equilibrium with the walls of its container.

The rate at which energy is absorbed by a body is also given by the Stefan–Boltzmann law [Eq. (18.5)]. Thus, we can figure the net transfer of radiant energy by an object surrounded by walls at a different temperature. For example, consider a thin wire filament in a lamp that is covered with an envelope, as shown in Fig. 18.8. Let the temperature of the filament be denoted by T_1 and the temperature of the surrounding envelope be denoted by T_2. The emissivity of the filament is e, and only radiative processes are considered. In this example, we note that

Net rate of radiation = rate of energy emission − rate of energy absorption
$$R = e\sigma T_1^4 - e\sigma T_2^4$$

$$R = e\sigma(T_1^4 - T_2^4) \qquad \textbf{(18.6)}$$

Equation (18.6) can be applied to any system for computing the net energy emitted by a radiator of temperature T_1 and emissivity e in the presence of surroundings at a temperature T_2.

Figure 18.7 When an object and its surroundings are at the same temperature, the radiant energy emitted is the same as that absorbed.

Figure 18.8 The net energy emitted by a radiator in surroundings of a different temperature.

Summary and Review

Summary

Heat is the transfer of thermal energy from one place to another. We have seen that the rate of transfer by conduction, convection, and radiation can be predicted from experimental formulas. The effects of material, surface areas, and temperature differences must be understood for many industrial applications of heat transfer. The major concepts presented in this chapter are listed as follows.

- In the transfer of heat by conduction, the quantity of heat Q transferred per unit of time τ through a wall or rod of length L is given by

$$H = \frac{Q}{\tau} = kA\frac{\Delta t}{L} \qquad \text{Conduction}$$

where A is the area and Δt is the difference in surface temperatures. The SI unit for H is the *watt* (W). Other commonly used units are kcal/s and Btu/h. From this relation, the thermal conductivity is

$$k = \frac{QL}{\tau A\,\Delta t} \qquad \text{Thermal Conductivity}$$

The SI units for k are W/m · K. Useful conversions can be made from the following definitions:

$$1 \text{ kcal/m} \cdot \text{s} \cdot \text{C}° = 4186 \text{ W/m} \cdot \text{K}$$
$$1 \text{ W/m} \cdot \text{K} = 6.94 \text{ Btu} \cdot \text{in./ft}^2 \cdot \text{h} \cdot \text{F}°$$
$$1 \text{ Btu} \cdot \text{in./ft}^2 \cdot \text{h} \cdot \text{F}° = 3.445 \times 10^{-5} \text{ kcal/m} \cdot \text{s} \cdot \text{C}°$$

- The R-value is an engineering term to measure the thermal resistance offered to the conduction of heat. It is defined as follows:

$$R = \frac{L}{k} \qquad \text{R-Value}$$

This concept applied to a number of thicknesses of different materials yields the following useful equation:

$$\frac{Q}{\tau} = \frac{A\,\Delta t}{\sum_i (L_i/k_i)} = \frac{A\,\Delta t}{\sum_i R_i}$$

The quantity of heat flowing per unit of time (Q/τ) through a number of thicknesses of different materials is equal to the product of the area A and the temperature difference Δt divided by the sum of the R-values of the various materials. In keeping with current engineering practice, the units for the R-value are ft² · F° · h/Btu.

- For heat transfer by radiation, we define the rate of radiation as the energy emitted per unit area per unit time (or simply the power per unit area):

$$R = \frac{E}{\tau A} = \frac{P}{A} \qquad \text{Rate of Radiation, W/m}^2$$

According to the *Stefan–Boltzmann law*, this rate is given by

$$R = \frac{P}{A} = e\sigma T^4 \qquad \sigma = 5.67 \times 10^{-8} \text{ W/m}^2 \cdot \text{K}^4$$

- Prevost's law of heat exchange states that *a body at the same temperature as its surroundings radiates and absorbs heat at the same rates.*

Key Terms

blackbody 376	forced convection 375	rate of radiation 377
conduction 370	ideal absorber 376	Stefan–Boltzmann law 377
convection 370	ideal radiator 376	thermal conductivity 371
convection current 375	natural convection 375	thermal radiation 370
emissivity 376	Prevost's law of heat exchange 378	thermal resistance (*R*-value) 374

Review Questions

18.1. Discuss the vacuum bottle and explain how it minimizes transfer of heat by conduction, convection, and radiation.

18.2. What determines the *direction* of heat transfer?

18.3. Heat flows both by conduction and by radiation. In what ways are they different? In what ways are they similar?

18.4. A hot chunk of iron is suspended centrally within an evacuated calorimeter. Can we determine the specific heat of iron by the techniques introduced in Chapter 18? Discuss.

18.5. Discuss the analogies that exist between steady-state heat flow and the flow of an incompressible fluid.

18.6. A pan of water is placed over a gas burner on a kitchen stove until the water boils vigorously. Discuss the heat transfers that take place. How would you explain the fact that the bubbles forming in the water are carried to the surface in the form of a pyramid instead of rising directly to the surface?

18.7. By placing a flame underneath a paper cup filled with water, it is possible to bring the water to a boil without burning the bottom of the cup. Explain.

18.8. When a piece of paper is wrapped around a stick of wood and the system is heated with a flame, the paper will begin to burn. But if the paper is wrapped tight around a copper rod and heated in the same manner, it does not burn. Why?

18.9. On a cold day, a piece of iron feels colder to the touch than a piece of wood. Explain.

18.10. Copper has about twice the thermal conductivity of aluminum, but its specific heat is a little less than half that of aluminum. A rectangular block is made from each material so that they have identical masses and the same surface area at their bases. Each block is heated to 300°C and placed on the top of a large cube of ice. Which block will stop sinking first? Which will sink deeper?

18.11. Distinguish between thermal conductivity and specific heat as they relate to heat transfer.

18.12. The term *absorptivity* is sometimes used in place of the term *emissivity*. Can you justify this practice?

18.13. Why is more air conditioning required to cool the inside of a navy-blue car than a white car of the same size?

18.14. If a house is to be designed for maximum comfort in both summer and winter, would you prefer a light roof or a dark roof? Explain.

18.15. If you are interested in the number of kilocalories transferred by radiation in a unit of time, the Stefan–Boltzmann law can be written in the form

$$\frac{Q}{\tau} = e\sigma A T^4$$

For this form of the law, show that Stefan's constant σ is equal to

$$1.35 \times 10^{-11} \text{ kcal/m}^2 \cdot \text{s} \cdot \text{K}^4$$

18.16. When a liquid is heated in a glass beaker, a wire gauze is usually placed between the flame and the bottom of the beaker. Why is this a wise practice?

18.17. Does the warm air over a burning fire rise, or is it forced upward by the flames?

18.18. Should a hot-water or steam radiator be painted with an efficient emitter or a poor one? If it is painted black, will it be more efficient? Why?

18.19. Which is a faster process, conduction or convection? Give an illustration to justify your conclusion.

Problems

Note: Refer to Tables 18.1 and 18.2 for thermal conductivities and other constants needed in the solution of the problems in this section.

Section 18.2 Conduction

18.1. A block of copper has a cross section of 20 cm^2 and a length of 50 cm. The left end is maintained at 0°C, and the right end is at 100°C. What is the rate of heat flow in watts? *Ans. 154 W*

18.2. In Prob. 18.1, what is the heat flow if the copper block is replaced with a block of glass having the same dimensions?

18.3. A 50-cm-long brass rod has a diameter of 3 mm. The temperature of one end is 76 C° higher than the other end. How much heat is conducted in 1 min? *Ans. 7.03 J*

18.4. A pane of window glass is 10 in. wide, 16 in. long, and $\frac{1}{8}$ in. thick. The inside surface is at 60°F, and the outside surface is at 20°F. How many Btu are transferred to the outside in a time of 2 h?

18.5. One end of an iron rod 30 cm long and 4 cm^2 in cross section is placed in a bath of ice and water. The other end is placed in a steam bath. How many minutes are needed to transfer 1.0 kcal of heat? In what direction does the heat flow?
Ans. 10.4 min, toward the ice

18.6. A steel plate of thickness 20 mm has a cross section of 600 cm^2. One side is at 170°C, and the other is at 120°C. What is the rate of heat transfer?

18.7. How much heat is lost in 12 h through a 3-in. brick firewall that has an area of 10 ft^2 if one side is at 330°F and the other is at 78°F?
Ans. 50,400 Btu

*18.8. A composite wall 6 m long and 3 m high consists of 12 cm of concrete joined with 10 cm of corkboard. The inside temperature is 10°C, and the outside temperature is 40°C. Find the temperature at the interface of the two materials.

*18.9. What is the steady-state rate of heat flow through the composite wall of Prob. 18.8?

Ans. 4.83 J/s

Section 18.5 Radiation

18.10. What is the power radiated from a spherical blackbody of surface area 20 cm^2 if its temperature is 250°C?

18.11. What is the rate of radiation for a spherical blackbody at a temperature of 327°C? Will this rate

change if the radius is doubled for the same temperature?   Ans. 7.35 kW/m^2, no

18.12. The emissivity of a metallic sphere is 0.3, and at a temperature of 500 K it radiates a power of 800 W. What is the radius of the sphere?

*18.13. If a certain body absorbs 20 percent of the incident thermal radiation, what is its emissivity? What energy will be emitted by this body in 1 min if its surface area is 1 m^2 and its temperature is 727°C?   Ans. 680 kJ

*18.14. The operating temperature of the filament in a 25-W lamp is 2727°C. If the emissivity is 0.3, what is the surface area of the filament?

Additional Problems

18.15. A glass window pane is 60 cm wide, 1.8 m high, and 3 mm thick. The inside temperature is 20°C, and the outside temperature is −10°C. How much heat leaves the house through this window in 1 hr?   Ans. 31 MJ

18.16. A 20-cm thickness of fiberglass insulation covers a 20-m by 15-m attic floor. How many calories of heat are lost to the attic if the temperatures on the sides of the insulation are −10°C and +24°C?

18.17. The bottom of an aluminum pan is 3 mm thick and has a surface area of 120 cm^2. How many calories per minute are conducted through the bottom of the pan if the temperature of the outer surface is 114°C and the temperature of the inner surface is 117°C?   Ans. 0.600 kcal

18.18. A solid wall of concrete is 80 ft high, 100 ft wide, and 6 in. thick. The surface temperatures on the sides of the concrete are 30 and 100°F. What time is required for 400,000 Btu of heat to be transferred?

18.19. The bottom of a hot metal pan has an area of 86 cm^2 and a temperature of 98°C. The pan is placed on top of a corkboard 5 mm thick. The

Formica tabletop under the corkboard is maintained at a constant temperature of 20°C. How much heat is conducted through the cork in 2 min?   Ans. 644 J

*18.20. What thickness of copper is required to have the same insulating value as 2 in. of corkboard?

*18.21. What thickness of concrete is required to have the same insulating value as 6 cm of fiberglass?

Ans. 1.20 m

18.22. A plate-glass window in an office building measures 2 by 6 m and is 1.2 cm thick. Its outer surface is at 23°C, and its inner surface is at 25°C. How many joules of heat pass through the glass in an hour?

18.23. What must be the temperature of a blackbody if its rate of radiation is 860 W/m^2?   Ans. 351 K

18.24. A gray steel ball has an emissivity of 0.75, and when its temperature is 570°C, the power radiated is 800 W. What is the surface area of the ball?

18.25. A 25-W lamp has a filament of area 0.212 cm^2. If the emissivity is 0.35, what is the operating temperature of the filament?   Ans. 2776 K

Critical Thinking Questions

*18.26. The wall of a freezing plant consists of 6 in. of concrete and 4 in. of corkboard. The temperature of the inside cork surface is −15°F, and the temperature of the outside surface is 70°F. What is

the temperature at the interface between the cork and concrete? How much heat is conducted through each square foot of wall in 1 hour?

Ans. 63.7°F, 5.92 Btu/ft^2 h

*18.27. A wooden icebox has walls that are 4 cm thick, and the overall effective surface area is 2 m². How many grams of ice will be melted in 1 min if the inside temperatures of the walls are 4°C an the outside temperatures are 20°C?

*18.28. The filament in a lamp operates at a temperature of 727°C and is surrounded by an envelope at 227°C. If the filament has an emissivity of 0.25 and a surface area of 0.30 cm², what is the operating power of the lamp?   Ans. 0.399 W

*18.29. The thermal conductivity of material A is twice that for material B. Which has the greater R value? When 6 cm of material A is between two walls, the rate of conduction is 400 W/m². What thickness of material B is needed to provide the same rate of heat transfer if the other factors remain constant?

19

Thermal Properties of Matter

Astronaut Mark Lee performs tests during a space walk. The thermal state inside the suit and in space is described by conditions of pressure, temperature, and volume. Changes of any one of these parameters inside the suit can be disastrous. (*Photo by NASA.*)

Objectives

After completing this chapter, you should be able to

1. Write and apply the relationship between the volume and the pressure of a gas at constant temperature (*Boyle's law*).

2. Write and apply the relationship between the volume and the temperature of a gas under conditions of constant pressure (*Charles's law*).

3. Write and apply the relationship between the temperature and pressure of a gas under conditions of constant volume (*Gay-Lussac's law*).

4. Apply the general gas law to the solution of problems involving changes in mass, volume, pressure, and temperature of gases.

5. Define *vapor pressure, dew point,* and *relative humidity* and apply these concepts to the solution of problems.

Now that we have an understanding of the concepts of heat and temperature, we proceed to study the thermal behavior of matter. Four measurable quantities are of interest: the pressure, volume, temperature, and mass of a sample. Together, these variables determine the state of a given sample of matter. Depending on its state, matter may exist in the liquid, solid, or gaseous phase. Thus, it is important to distinguish between the terms *state* and *phase*. We begin by studying the thermal behavior of gases.

19.1 Ideal Gases, Boyle's Law, and Charles's Law

In a gas, the individual molecules are so far apart that the cohesive forces between them are usually small. Even though the molecular structure of different gases may vary considerably, their behavior is affected little by the size of the individual molecules. It is usually safe to say that when a large quantity of gas is confined in a rather small volume, the volume occupied by the molecules is still a tiny fraction of the total volume.

One of the most useful generalizations about gases is the concept of an **ideal gas,** whose behavior is completely unaffected by cohesive forces or molecular volumes. Of course, no real gas is *ideal,* but under ordinary conditions of temperature and pressure, the behavior of any gas conforms closely to the behavior of an ideal gas. Therefore, experimental observations of many real gases can lead to the derivation of general physical laws governing their thermal behavior. The degree to which any real gas obeys these relations is determined by how closely it approximates an ideal gas.

The first experimental measurements of the thermal behavior of gases were made by Robert Boyle (1627–1691). He made an exhaustive study of the changes in the volume of gases as a result of changes in pressure. All other variables, such as mass and temperature, were kept constant. In 1660, Boyle demonstrated that the volume of a gas is inversely proportional to its pressure. In other words, doubling the volume decreases the pressure to one-half its original value. This finding is now know as **Boyle's law.**

(a)   (b)

Figure 19.1 When a gas is compressed at constant temperature, the product of its pressure and its volume is always constant; that is $P_1V_1 = P_2V_2$.

Boyle's Law: Provided that the mass and temperature of a sample of gas are held constant, the volume of the gas is inversely proportional to its absolute pressure.

Another way of stating Boyle's law is to say that the product of the pressure *P* of a gas and its volume *V* will be constant as long as the temperature does not change. Consider, for example, a closed cylinder equipped with a movable piston, as shown in Fig. 19.1. In Fig. 19.1a, the initial state of the gas is described by its pressure P_1 and its volume V_1. If the piston is pressed downward until it reaches the new position shown in Fig. 19.1b, its pressure will increase to P_2 and its volume will decrease to V_2. This process is shown graphically in Fig. 19.2. If the process occurs without a change in temperature, Boyle's law reveals that

$$P_1V_1 = P_2V_2 \qquad \textit{With Constant m and T} \quad \textbf{(19.1)}$$

In other words, the product of pressure and volume in the initial state is equal to the product of pressure and volume in the final state. Equation (19.1) is a mathematical statement of Boyle's law. The pressure *P* must be the *absolute* pressure and not *gauge* pressure.

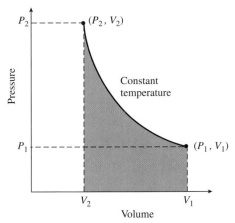

Figure 19.2 A *P-V* diagram illustrating that the pressure of an ideal gas varies inversely with its volume.

Example 19.1

What volume of hydrogen gas at atmospheric pressure is required to fill a 5000-cm^3 tank under a gauge pressure of 530 kPa?

Plan: One atmosphere of pressure is 101.3 kPa. The final absolute pressure is 530 kPa (gauge pressure) plus 101.3 kPa. We will apply Boyle's law to find the volume of hydrogen at 1 atm needed to produce an *inside* pressure of 631 kPa. There will be no need to convert the volume to SI units if the same volume units are acceptable for the answer.

Solution: The absolute initial and final pressures are

$$P_1 = 101.3 \text{ kPa} \qquad p_2 = 530 \text{ kPa} + 101.3 \text{ kPa} = 631 \text{ kPa}$$

The final volume V_2 is 5000 cm^3. Applying Eq. (19.1), we have

$$P_1V_1 = P_2V_2$$
$$(101.3 \text{ kPa})V_1 = (631 \text{ kPa})(5000 \text{ cm}^3)$$
$$V_1 = 31{,}100 \text{ cm}^3$$

Notice that it was not necessary for the units for pressure to be consistent with the units for volume. Since *P* and *V* appear on each side of the equation, it is necessary only to choose the same units for pressure. The units for volume will then be the units substituted for V_2.

In Chapter 16, we used the fact that the volume of a gas increases directly with its temperature to help us define absolute zero. We found the result ($-273°$C) by extending the line on the graph in Fig. 19.3. Of course, any real gas will become a liquid before its volume reaches zero. But the direct relationship is a valid approximation for most gases that are not subjected to extreme conditions of temperature and pressure.

This direct proportionality between volume and temperature was first experimentally tested by Jacques Charles in 1787. *Charles's law* may be stated as follows:

Charles's Law: Provided that the mass and pressure of a gas are held constant, the volume of the gas is directly proportional to its absolute temperature.

If we use the subscript 1 to refer to an initial state of a gas and the subscript 2 to refer to the final state, a mathematical statement of Charles's law is obtained.

$$\frac{V_1}{T_1} = \frac{V_2}{T_2} \qquad \textit{With Constant m and P} \qquad \textbf{(19.2)}$$

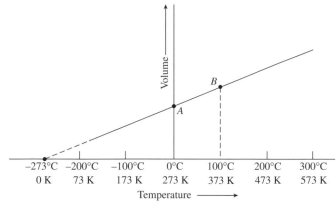

Figure 19.3 The variation of volume as a function of temperature. When the volume is extrapolated to zero, the temperature of a gas is at absolute zero (0 K).

In this equation, V_1 refers to the volume of a gas at the *absolute* temperature T_1, and V_2 is the later volume of the same sample of gas when its absolute temperature is T_2.

The SI unit for volume is the cubic meter (m^3), and, of course, that is the preferred unit. However, it is quite common to find the liter (L) used as a unit for volume, especially when working with gases. The liter is the volume contained by a cube that is 10 cm on a side.

$$1 \text{ L} = 1000 \text{ cm}^3 = 1 \times 10^{-6} \text{ m}^3$$

We will use the liter in some of our examples because it is such a popular unit. As always, you must be careful when using any unit other than the SI units in physical formulas.

Example 19.2

A frictionless cylinder is filled with 2 L of an ideal gas at 23°C. One end of the cylinder is fixed with a movable piston, and the gas is allowed to expand at constant pressure until its volume reaches 2.5 L. What is the new temperature of the gas?

Plan: The mass and pressure of the gas remain constant, so the change in temperature must be proportional to the change in volume, and Charles's law can be applied to find the new temperature. We must remember to use *absolute temperatures*.

Solution: We organize the given information as follows:

Given: $T_1 = 23° + 273° = 296$ K, $V_1 = 2$ L, $V_2 = 2.5$ L; Find: $T_2 = ?$

Now, we solve Charles's law for T_2 as follows:

$$\frac{V_1}{T_1} = \frac{V_2}{T_2} \quad \text{and} \quad T_2 = \frac{V_2 T_1}{V_1}$$

$$T_2 = \frac{(2.5 \text{ L})(296 \text{ K})}{2 \text{ L}} = 370 \text{ K}$$

The final temperature of the gas is 370 K or 97°C.

19.2 ## Gay-Lussac's Law

The three quantities that determine the state of a given mass of gas are its pressure, volume, and temperature. Boyle's law deals with changes in pressure and volume under constant temperature, and Charles's law applies for volume and temperature under constant pressure. The variation in pressure as a function of temperature is described in a law attributed to Gay-Lussac.

Gay-Lussac's Law: If the volume of a sample of gas remains constant, the absolute pressure of the gas is directly proportional to its absolute temperature.

This means that doubling the pressure applied to a gas will cause its absolute temperature to double also. In equation form, *Gay-Lussac's law* may be written as

$$\frac{P_1}{T_1} = \frac{P_2}{T_2} \qquad \textit{With Constant m and V} \quad \textbf{(19.3)}$$

Example 19.3

An automobile tire is inflated to a gauge pressure of 207 kPa (30 lb/in.^{2}) at a time when the surrounding pressure is 1 atm (101.3 kPa) and the temperature is 25°C. After the car is driven, the temperature of the air in the tire increases to 40°C. Assuming the volume changes only slightly, what will be the new gauge pressure in the tire?

Plan: Because the volume and mass are constant, the pressure must increase in the same proportion as the temperature, and we will apply Gay-Lussac's law to find the final absolute pressure. The gauge pressure is then obtained by subtracting the ambient pressure (101.3 kPa).

Solution: First, we will find the absolute temperatures and the absolute pressure.

$$P_1 = 207 \text{ kPa} + 101.3 \text{ kPa} = 308 \text{ kPa}$$

$$T_1 = 25 + 273 = 298 \text{ K}; \qquad T_2 = 40 + 273 = 313 \text{ K}$$

The new pressure is found from Gay-Lussac's law

$$\frac{P_1}{T_1} = \frac{P_2}{T_2} \qquad \text{or} \qquad P_2 = \frac{P_1 T_1}{T_2}$$

$$P_2 = \frac{(308 \text{ kPa})(313 \text{ K})}{298 \text{ K}}; \qquad P_2 = 324 \text{ kPa}$$

The gauge pressure is found by subtracting the ambient air pressure (101.3 kPa).

$$\text{Gauge pressure} = 324 \text{ kPa} - 101 \text{ kPa} = 223 \text{ kPa}$$

A gauge would read this pressure as 223 kPa or about 32.3 lb/in.^{2}.

19.3 General Gas Laws

Thus far, we have discussed three laws that can be used to describe the thermal behavior of gases. Boyle's law, as given by Eq. (19.1), applies for a sample of gas whose temperature is unchanged. Charles's law, as given by Eq. (19.2), applies for a gas sample under a constant pressure. Gay-Lussac's law, in Eq. (19.3), is for a gas sample under constant volume. Unfortunately, none of these conditions is usually satisfied. Normally, a system undergoes changes in volume, temperature, and pressure as a result of a thermal process. A more general relation combines the three laws as follows:

$$\frac{P_1 V_1}{T_1} = \frac{P_2 V_2}{T_2} \qquad \textit{With Constant m} \quad \textbf{(19.4)}$$

where (P_1, V_1, T_1) may be considered the initial coordinates of the initial state and (P_2, V_2, T_2) the coordinates of the final state. In other words, for a given mass, the ratio PV/T is constant for any ideal gas. Equation (19.4) can be remembered by the phrase "a private (PV/T) is always a private."

Example 19.4 An oxygen tank with an internal volume of 20 L is filled with oxygen under an absolute pressure of 6 MPa at 20°C. The oxygen is to be used in a high-flying aircraft, where the absolute pressure is only 70 kPa and the temperature is −20°C. What volume of oxygen can be supplied by the tank under these conditions?

Plan: The given pressures are already absolute pressures, so we convert to absolute temperatures and apply Eq. (19.4).

Solution: After adding 273° to the two Celsius temperatures, we solve for the V_2.

$$\frac{P_1 V_1}{T_1} = \frac{P_2 V_2}{T_2} \quad \text{or} \quad V_2 = \frac{P_1 V_1 T_2}{P_2 T_1}$$

$$V_2 = \frac{(6 \times 10^6 \text{ Pa})(20 \text{ L})(253 \text{ K})}{(70 \times 10^3 \text{ Pa})(293 \text{ K})} = 1480 \text{ L}$$

Let us now consider the effect of a change in mass on the behavior of gases. If the temperature and volume of an enclosed gas are held constant, the addition of more gas will result in a proportional increase in pressure. Similarly, if the pressure and temperature are fixed, an increase in the mass will result in a proportional increase in the volume. We can combine these experimental observations with Eq. (19.4) to obtain the general relation

$$\frac{P_1 V_1}{m_1 T_1} = \frac{P_2 V_2}{m_2 T_2} \tag{19.5}$$

where m_1 is the initial mass and m_2 is the final mass. A study of this relation will reveal that Boyle's law, Charles's law, and Gay-Lussac's law, along with Eq. (19.4), are each special cases of the more general equation (19.5).

Example 19.5 The pressure gauge on a helium storage tank reads 2000 lb/in.^{2} when the temperature is 27°C. The container develops a leak overnight, and the gauge pressure the next morning is found to be 1500 lb/in.^{2} at a temperature of 17°C. What percentage of the original mass of helium remains in the tank?

Plan: We can eliminate volume from consideration since it does not change $(V_1 = V_2)$, so we can simplify Eq. (19.5) and solve it for the ratio (m_2/m_1) of the gas remaining to the gas initially contained by the tank. This ratio will then be expressed as a percentage. Since the initial and final pressures are in the same units, there will be no need to convert the pressures to SI units. However, we *will* need to add 1 atm of pressure (14.7 lb/in.^{2}) to each of the gauge pressure values, and the temperatures must be expressed in kelvins.

Solution: Since $V_1 = V_2$, we simplify Eq. (19.5) to obtain

$$\frac{P_1 V_1}{m_1 T_1} = \frac{P_2 V_2}{m_2 T_2} \qquad \text{or} \qquad \frac{P_1}{m_1 T_1} = \frac{P_2}{m_2 T_2}$$

The ratio m_2/m_1 is the fraction of the helium mass remaining. Hence,

$$\frac{m_2}{m_1} = \frac{P_2 T_1}{P_1 T_2}$$

The pressures and temperatures are adjusted to their absolute values as follows:

$$P_1 = 2000 \text{ lb/in.}^2 + 14.7 \text{ lb/in.}^2 = 2014.7 \text{ lb/in.}^2$$
$$P_1 = 1500 \text{ lb/in.}^2 + 14.7 \text{ lb/in.}^2 = 1514.7 \text{ lb/in.}^2$$
$$T_1 = 27 + 273 = 300 \text{ K}$$
$$T_2 = 17 + 273 = 290 \text{ K}$$

Substitution of these values yields

$$\frac{m_2}{m_1} = \frac{(1514.7 \text{ lb/in.}^2)(300 \text{ K})}{(2014.7 \text{ lb/in.}^2)(290 \text{ K})} = 0.778$$

Therefore, 77.8 percent of the helium still remains inside the container.

Equation (19.5) is general because it accounts for variance in the pressure, volume, temperature, and mass of a gas. The quantity that affects pressure and volume is not the mass of a gas, however, but the number of molecules in the gas. According to the kinetic theory of gases, the pressure is due to molecular collisions with the walls of the container. Increasing the number of molecules will increase the number of particles colliding per second, and the pressure of the gas will become greater. If we are considering a thermal process involving quantities of the same gas, it is safe to apply Eq. (19.5) because the mass is proportional to the number of molecules.

When dealing with different kinds of gas, such as hydrogen compared with oxygen, it is necessary to refer to equal numbers of molecules rather than equal masses. When they are placed in similar containers, 6 g of hydrogen will yield a much greater pressure than 6 g of oxygen. There are many more hydrogen molecules in 6 g of H_2 than there are oxygen molecules in 6 g of O_2. To be more general, we must revise Eq. (19.5) to account for differences in the number of gas molecules instead of the difference in mass. First, we must develop methods of relating the quantity of a gas to the number of molecules present.

19.4 | Molecular Mass and the Mole

Although the mass of individual atoms is difficult to determine because of their size, experimental methods have been successful in measuring **atomic mass.** For example, we know that one atom of helium has a mass of 6.65×10^{-24} g. When working with macroscopic quantities, such as volume, pressure, and temperature, it is much more convenient to compare the *relative masses* of individual atoms.

The relative atomic masses are based on the mass of a reference atom known as *carbon 12*. By arbitrarily assigning exactly 12 *atomic mass units* (u) to this atom, we have a standard for comparison of other atomic masses.

The atomic mass of an element is the mass of an atom of that element compared with the mass of an atom of carbon taken as 12 atomic mass units.

On this basis, the atomic mass of hydrogen is approximately 1 u, and the atomic mass of oxygen is approximately 16 u.

A molecule consists of two or more atoms in chemical combination. The definition of **molecular mass** follows from the definition of relative atomic mass.

The molecular mass M is the sum of the atomic masses of all the atoms making up the molecule.

For example, a molecule of oxygen (O_2) contains two atoms of oxygen. Its molecular mass is 16 u $\times$ 2 = 32 u. A molecule of carbon dioxide (CO_2) contains one atom of carbon and two atoms of oxygen. Thus, the molecular mass of CO_2 is 44 u:

$$
\begin{aligned}
1\,C &= 1 \times 12 = 12\,\text{u} \\
2\,O &= 2 \times 16 = 32\,\text{u} \\
\hline
CO_2 &= 44\,\text{u}
\end{aligned}
$$

In dealing with gases, we have noted that it is more meaningful to treat the amount of substance present in terms of the number of molecules present. This is accomplished by establishing a new unit of measure called the **mole** (mol).

The mole is that quantity of a substance that contains the same number of particles as there are atoms in 12 g of carbon 12.

On the basis of this definition, 1 mol of carbon must be equal to 12 g by definition. Since the molecular mass of any substance is based on carbon 12 as a standard, it follows that

One mole is the mass in grams equal numerically to the molecular mass of a substance.

For example, 1 mol of hydrogen (H_2) is 2 g, 1 mol of oxygen (O_2) is 32 g, and 1 mol of carbon dioxide (CO_2) is 44 g. In other words, 2 g of H_2, 32 g of O_2, and 44 g of CO_2 all have the same number of molecules. This number N_A is known as **Avogadro's number.**

The ratio of the number of molecules N to the number of moles n must equal Avogadro's number N_A. Symbolically,

$$N_A = \frac{N}{n} \qquad \textit{Molecules per Mole} \quad \textbf{(19.6)}$$

There are several experimental methods of determining Avogadro's number. The accepted value for N_A is

$$N_A = 6.023 \times 10^{23} \text{ molecules per mole} \qquad \textit{Avogadro's Number} \quad \textbf{(19.7)}$$

The easiest way to determine the number of moles n contained in a gas is to divide its mass m in grams by its molecular mass M in grams per mole.

$$n = \frac{m}{M} \qquad \textit{Number of Moles} \quad \textbf{(19.8)}$$

Example 19.6 (a) How many moles of gas are present in 200 g of CO_2? (b) How many molecules are present?

Plan: First, we need to determine the molecular mass for CO_2, which was calculated as 44 g/mol earlier in this section. Dividing the mass of the gas by its molecular mass gives the number of moles present. The number of molecules is then found from Avogadro's number.

Solution (a): For 200 g of a gas that contains 44 g/mol, we find from Eq. (19.8) that

$$n = \frac{m}{M} = \frac{200 \text{ g}}{44 \text{ g/mol}}; \qquad n = 4.55 \text{ mol}$$

Solution (b): Since Avogadro's number N_A is the number of molecules per mole, we find the number of gas molecules in 4.55 mol of the gas is

$$n = \frac{N}{N_A} \qquad \text{or} \qquad N = nN_A$$

$$N = (4.55 \text{ mol})(6.023 \times 10^{23} \text{ molecules/mol}); \quad N = 2.74 \times 10^{24} \text{ molecules}$$

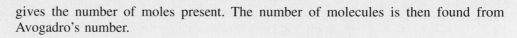

19.5 The Ideal Gas Law

Let us now return to our search for a more general gas law. If we substitute the number of moles n for the mass m in Eq. (19.5), we can write

$$\frac{P_1 V_1}{n_1 T_1} = \frac{P_2 V_2}{n_2 T_2} \tag{19.9}$$

This equation represents the most useful form of a general gas law when all the parameters of an initial state and a final state are known except for a single quantity.

An alternative expression of Eq. (19.9) is

$$\frac{PV}{nT} = R \tag{19.10}$$

where R is known as the ***universal gas constant.*** If we can evaluate R under certain known values of P, V, n, and T, Eq. (19.10) can be used directly without any information concerning initial and final states. The numerical value for R, of course, depends on the choice of units for P, V, n, and T. In SI units, the value has been determined as

$$R = 8.314 \text{ J/mol} \cdot \text{K}$$

Other choices of units lead to the following equivalent values:

$$R = 0.0821 \text{ L} \cdot \text{atm/mol} \cdot \text{K}$$
$$= 1.99 \text{ cal/mol} \cdot \text{K}$$

If the pressure is measured in pascals and the volume in cubic meters, one should use 8.314 J/mol · K for the constant R. Often, however, the pressure is expressed in atmospheres and the volume in liters. Rather than making appropriate conversions, it would probably be simpler to use $R = 0.0821$ L · atm/mol · K.

Equation (19.10) is known as the ***ideal gas law*** and is usually written in the form

$$PV = nRT \tag{19.11}$$

Another useful form of the ideal gas law makes use of the fact that $n = m/M$. Thus,

$$PV = \frac{m}{M}RT \tag{19.12}$$

Provided the density of a real gas is reasonably low, the ideal gas law holds for any gas or even a mixture of several gases. As long as their molecules are far enough apart, we can apply Eq. (19.11), with n being the number of moles.

Example 19.7

Find the volume of 1 mol of any ideal gas at a condition of standard temperature (273 K) and pressure (101.3 kPa).

Plan: Remember that 1 mol of any gas contains the same number of molecules, so as long as we treat the gas as an *ideal* gas, we can use Eq. (19.11) to find its volume. Since 1 mol is at 1 atm pressure, we will use 0.0821 L · atm/mol · K for R.

Solution: Solving for V in Eq. (19.11), we obtain

$$PV = nRT \quad \text{or} \quad V = \frac{nRT}{P}$$

$$V = \frac{(1 \text{ mol})(0.0821 \text{ L} \cdot \text{atm/mol} \cdot \text{K})(273 \text{ K})}{101.3 \times 10^3 \text{ Pa}}$$

$$= 22.4 \text{ L or } 0.0224 \text{ m}^3$$

Thus, 1 mole of any ideal gas at standard temperature and pressure has a volume of 22.4 L.

Example 19.8

How many grams of oxygen will occupy a volume of 1.6 m^3 at a pressure of 200 kPa and a temperature of 27°C?

Plan: We will need to determine the molecular mass of oxygen, which is diatomic; that is, each molecule contains two oxygen atoms. Therefore, there are 32 g/mol ($M = 16\,\text{u} + 16\,\text{u} = 32\,\text{u}$). Using the ideal gas law, we can determine the mass directly from Eq. (19.12).

Solution: The absolute temperature is (27 + 273) or 300 K. Substitution yields

$$PV = \frac{m}{M}RT \quad \text{or} \quad m = \frac{MPV}{RT}$$

$$m = \frac{(32 \text{ g/mol})(200 \times 10^3 \text{ Pa})(1.6 \text{ m}^3)}{(8.314 \text{ J/mol} \cdot \text{K})(300 \text{ K})} = 4110 \text{ g}$$

$$m = 4.11 \text{ kg}$$

19.6 Liquefaction of a Gas

We have defined an ideal gas as one whose thermal behavior is completely unaffected by cohesive forces or molecular volume. Such a gas, compressed at a constant temperature, will remain a gas no matter how great a pressure is applied to it. In other words, it will obey Boyle's law at any temperature. The binding forces necessary for liquefaction are never present.

All real gases experience intermolecular forces. At rather low pressures and high temperatures, however, real gases behave much like an ideal gas. Boyle's law applies because the intermolecular forces under these conditions are practically negligible. A real gas at high temperatures can be compressed in a cylinder, as in Fig. 19.4, to relatively high pressures without liquefying. If the increase in pressure is plotted as a function of the volume, the curve A_1B_1 is obtained. Note the similarity between this curve and that for an ideal gas, as shown in Fig. 19.2.

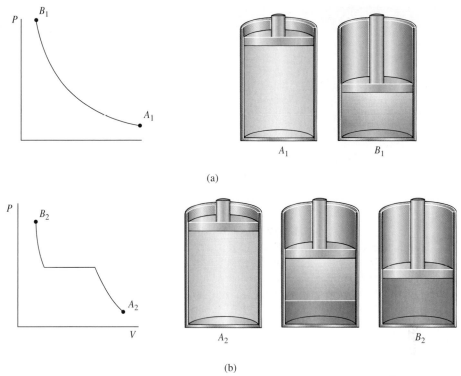

Figure 19.4 (a) Compression of an ideal gas at any temperature or a real gas at high temperature. (b) Liquefaction of a real gas when it is compressed at lower temperatures.

If the same gas is compressed at a much lower temperature, it will begin to condense at a particular pressure and volume. Further compression will continue to liquefy the gas at essentially constant pressure until all the gas has condensed. At that point, a sharp rise in pressure occurs with a slight decrease in volume. The entire process is shown graphically as the curve A_2B_2 in Fig. 19.4.

Let us now begin with the high-temperature compression and perform the experiment at lower and lower constant temperatures. Eventually, a temperature will be reached at which the gas will just begin to liquefy under compression. The highest temperature at which this liquefaction occurs is called the ***critical temperature.***

The critical temperature of a gas is that temperature above which the gas will not liquefy, regardless of the amount of pressure applied to it.

If any gas is to be liquefied, it must first be cooled below its critical temperature. Before this concept was understood, scientists attempted to liquefy oxygen by subjecting it to extreme pressures. Their attempts failed because the critical temperature of oxygen is $-119°C$. After cooling the gas below this temperature, it can be easily liquefied by compression.

19.7 | Vaporization

In Chapter 17, we discussed at length the process of vaporization in which a definite quantity of heat is required to change from the liquid phase to the vapor phase. There are three ways in which vaporization may occur: (1) evaporation, (2) boiling, and (3) sublimation. During evaporation, vaporization occurs at the surface of a liquid as the more energetic molecules leave the surface. In the process of boiling, vaporization occurs within the body of the liquid. Sublimation occurs when a solid vaporizes without passing through the liquid phase. In each case, an amount of energy equal to the latent heat of vaporization or ***sublimation*** must be lost by the liquid or solid.

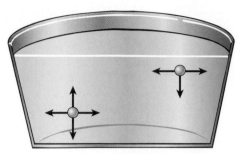

Why Do Zambonis Leave a Steaming Trail?
Zamboni machines, those ice-scraping machines that clean up the ice at ice rinks, make the ice surface clean and fresh. Did you ever wonder why the water they spread leaves a steaming fresh trail? That water is at 180°F (82°C). You might wonder why the machines would spread water that is almost boiling. Why not lay down very cold water? The answer is that evaporation cools the layer of hot water down to freezing very fast. Try this experiment with ice trays: Fill one with boiling water and another with cold tap water. Place both in your freezer. Check them every 15 minutes. Which one freezes faster? The hot water! That is why Zambonis leave steaming trails.

Figure 19.5 A molecule near the surface of a liquid experiences a net downward force. Only the more energetic molecules are able to overcome this force and leave the liquid.

The molecular theory of matter assumes that a liquid consists of molecules crowded fairly close together. These molecules have an average kinetic energy that is related to the temperature of the liquid. Because of random collisions or vibratory motion, however, not all the molecules move at the same rate of speed; some move faster than others.

Because the molecules are so close together, the forces between them are relatively large. As a molecule approaches the surface of a liquid, as in Fig. 19.5, it experiences a resultant downward force. The net force results from the fact that there are no liquid molecules above the surface to offset the downward attraction of these below the surface. Only the *faster-moving* particles can approach the surface with sufficient energy to overcome the retarding forces. These molecules are said to *evaporate* because, on leaving the liquid, they become typical gas particles. They have not changed chemically; the only difference between a liquid and its vapor is the distance between molecules.

Since only the most energetic molecules are able to break away from the surface, the average kinetic energy of the molecules remaining in the liquid is reduced. Hence, *evaporation is a cooling process.* (If you place a few drops of alcohol on the back of your hand, you will feel a cooling sensation.) The rate of evaporation is affected by the temperature of the liquid, the number of molecules above the liquid (the pressure), the exposed surface area, and the extent of ventilation.

19.8 Vapor Pressure

A jar is partially filled with water, as shown in Fig. 19.6. The pressure exerted by the molecules above the surface of the water is measured by an open-tube mercury manometer. In Fig. 19.6a, there are as many molecules of air inside the jar as are contained in an equal

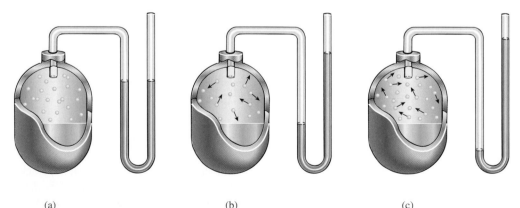

(a)   (b)   (c)

Figure 19.6 Measuring the vapor pressure of a liquid: (a) air pressure only, (b) partial vapor pressure, and (c) saturated vapor pressure.

volume of air outside the jar. In other words, the pressure inside the jar is equal to 1 atm, as indicated by the equal levels of mercury in the manometer.

When a high-energy liquid molecule breaks away from the surface, it becomes a vapor molecule and mixes with the air molecules above the liquid. These vapor molecules collide with air molecules, other vapor molecules, and the walls of the jar. The additional vapor molecules cause a rise in pressure inside the jar, as indicated in Fig. 19.6b. The vapor molecules may also rebound back into the liquid, where they are held as liquid molecules. This process is called ***condensation.*** Eventually the rate of evaporation will equal the rate of condensation, and a condition of equilibrium will exist, as shown in Fig. 19.6c. Under these conditions, the space above the liquid is said to be *saturated.* The pressure exerted by the saturated vapor against the walls of the jar, over and above that exerted by the air molecules, is called the ***saturated vapor pressure.*** It is characteristic of the substance and the temperature but independent of the volume of the vapor.

> The saturated vapor pressure of a substance is the additional pressure exerted by vapor molecules on the substance and its surroundings under a condition of saturation.

Once a condition of saturation is obtained for a substance and its vapor at a particular temperature, the vapor pressure remains essentially constant. If the temperature is increased, the molecules in the liquid will acquire more energy, and evaporation will occur more rapidly. The condition of equilibrium remains upset until once again the rate of condensation has caught up with the rate of evaporation. The saturated vapor pressure of a substance therefore increases with a rise in temperature.

The saturated-vapor-pressure curve for water is plotted in Fig. 19.7. Note that the vapor pressure increases rapidly with temperature. At room temperature (20°C), it is around 17.5 mm of mercury; at 50°C, it has increased to 92.5 mm; and at 100°C, it is equal to 760 mm, or 1 atm. The latter point is important in distinguishing between ***evaporation*** and ***boiling.***

When a liquid boils, bubbles of its vapor can be seen rising toward the surface from within the liquid. The fact that these bubbles are stable and do not collapse indicates that the pressure inside the bubble is equal to the pressure outside the bubble. The pressure inside the bubble is the vapor pressure at that temperature; the pressure on the outside is the pressure at that depth in the liquid. Under this condition of equilibrium, vaporization occurs freely throughout the liquid, causing the liquid to become agitated.

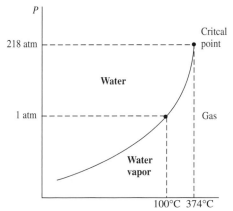

Figure 19.7 The vaporization curve for water. Any point on the curve represents a condition of pressure and temperature that allows water to boil. The curve ends abruptly at the critical temperature because water can exist only as a gas beyond that point.

Boiling is defined as vaporization within the body of a liquid when its vapor pressure equals the pressure in the liquid.

If the pressure on the liquid surface is 1 atm, as it would be in an open container, the temperature at which boiling occurs is called the *normal boiling point* for that liquid. The normal boiling point for water is 100°C because that is the temperature at which the vapor pressure of water is 1 atm (760 mm of mercury). If the pressure on any liquid surface is lower than 1 atm, boiling will occur at a temperature lower than the normal boiling point. If the external pressure is greater than 1 atm, boiling will occur at a higher temperature.

19.9   Triple Point

We have discussed in detail the process of vaporization, and in Fig. 19.7, we constructed a vaporization curve for water. This curve is represented by the line *AB* in the general phase diagram of Fig. 19.8. Any point on this curve represents a temperature and pressure at which water and its vapor can coexist in equilibrium.

A similar curve can be plotted for the temperatures and pressures at which a substance in the solid phase can coexist with its liquid phase. Such a curve is called a ***fusion curve.*** The fusion curve for water is represented by the line *AC* in the phase diagram. At any point on this curve, the rate at which ice is melting is equal to the rate at which water is freezing. Note that, as the pressure increases, the melting temperature (or freezing temperature) is lowered.

A third graph, called the ***sublimation curve,*** can be plotted to show the temperatures and pressures at which a solid may coexist with its vapor. The sublimation curve for water is represented by line *AD* of Fig. 19.8.

Let us now study the phase diagram for water more closely to illustrate the usefulness of such a graph for any substance. The coordinates of any point on the graph represent a particular pressure *P* and a particular temperature *T*. The volume must be considered constant for any thermal change indicated by the graph. For any point that falls in the fork between the vaporization and fusion curves, water will exist in its liquid phase. The vapor and the solid regions are also indicated on the diagram. The point *A*, at which all three curves intersect, is called the ***triple point*** for water. This is the temperature and pressure at which ice, liquid water, and water vapor coexist in equilibrium. Careful measurements have shown that the triple point for water is 0.01°C and 4.62 mm of mercury (Hg).

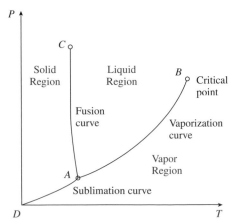

Figure 19.8   Triple-point phase diagram for water or other substance that expands on freezing.

19.10 | Humidity

The air in our atmosphere consists largely of nitrogen and oxygen with small amounts of water vapor and other gases. It is often useful to describe the water-vapor content of the atmosphere in terms of *absolute humidity.*

> The absolute humidity is defined as the mass of water per unit volume of air.

For example, if 7 g of water vapor is contained in every cubic meter of air, the absolute humidity is 7 g/m^3. Other units of absolute humidity are pounds per cubic foot and grains per cubic foot (7000 grains = 1 lb).

A more useful method of expressing the water-vapor content in air is to compare the actual vapor pressure at a particular temperature with the saturated vapor pressure at that temperature. When the atmosphere is holding all the water possible for a certain temperature, it is saturated. The addition of more vapor molecules will simply result in an equal amount of condensation.

> The relative humidity is defined as the ratio of the actual vapor pressure in the air to the saturated vapor pressure at that temperature.

$$Relative\ humidity = \frac{actual\ vapor\ pressure}{saturated\ vapor\ pressure} \tag{19.13}$$

The *relative humidity* is usually expressed as a percentage.

If the air in a room is not already saturated, it can be made so either by adding more water vapor to the air or by lowering the room temperature until the vapor already present is sufficient. The temperature to which the air must be cooled at constant pressure to produce saturation is called the *dew point.* Thus, if ice is placed in a glass of water, moisture will eventually collect on the outside walls of the glass when its temperature reaches the dew point. Given the temperature and the dew point, the relative humidity can be computed from saturated-vapor-pressure tables. Table 19.1 lists the saturated vapor pressure for water at various temperatures.

Table 19.1

Saturated Vapor Pressure for Water

Temperature		Pressure	Temperature		Pressure,
°C	°F	mmHg	°C	°F	mmHg
0	32	4.62	50	122	92.5
5	41	6.5	60	140	149.4
10	50	9.2	70	158	233.7
15	59	12.8	80	176	355.1
17	62.6	14.5	85	185	433.6
19	66.2	16.5	90	194	525.8
20	68	17.5	95	203	633.9
22	71.6	19.8	98	208.4	707.3
24	75.2	22.4	100	212	760.0
26	78.8	25.2	103	217.4	845.1
28	82.4	28.3	105	221	906.1
30	86	31.8	110	230	1074.6
35	95	42.2	120	248	1489.1
40	104	55.3	150	302	3570.5

Example 19.9

On a clear day, the air temperature is 86°F, and the dew point is 50°F. What is the relative humidity?

Plan: First, find the *actual* vapor pressure (at 86°F) and then find the table value for the *saturated* vapor pressure for the dew point (50°F). The relative humidity is the ratio of the saturation pressure for 50°F to the saturation pressure for 86°F.

Solution: From Table 19.1, the saturated pressure at 50°F is 9.2 mm, and the saturated pressure at 86°F is 31.8 mm. Thus, the relative humidity is

$$\frac{9.2 \text{ mm}}{31.8 \text{ mm}} = 0.29$$

The relative humidity is 29 percent.

Summary and Review

Summary

The thermal properties of matter must be understood if we are to apply the many laws discussed in this chapter. The relationships among mass, temperature, volume, and pressure allow us to explain and predict the behavior of gases. The major concepts discussed in this chapter are summarized as follows.

- A useful form of the general gas law that does not involve the use of moles is written on the basis that PV/mT is constant. When a gas in state 1 changes to another state 2, we may write

$$\frac{P_1 V_1}{m_1 T_1} = \frac{P_2 V_2}{m_2 T_2}$$

P = pressure $\quad\quad$ V = volume
m = mass $\quad\quad\quad$ T = absolute temperature

When one or more of the parameters m, P, T, or V is constant, that factor disappears from both sides of the above equation. *Boyle's law, Charles's law,* and *Gay-Lussac's law* are the following special cases:

$$P_1 V_1 = P_2 V_2; \quad \frac{V_1}{T_1} = \frac{V_2}{T_2}; \quad \frac{P_1}{T_1} = \frac{P_2}{T_2}$$

- When applying the general gas law in any of its forms, it must be remembered that the pressure is *absolute pressure* and the temperature is *absolute temperature*.

$$\frac{Absolute}{pressure} = \frac{gauge}{pressure} + \frac{atmospheric}{pressure}$$

$$T_K = t_C + 273 \quad\quad T_R = t_F + 460$$

For example, a pressure measured in an auto tire is 30 lb/in.^{2} at 37°C. These values must be adjusted before substitution into the gas laws:

P = 30 lb/in.^{2} + 14.7 lb/in.^{2} $\quad$ (absolute)
T = 37 + 273 = 310 K

- A more general form of the gas law is obtained by using the concepts of molecular mass M and the number of moles n for a gas. The number of molecules in 1 mol is Avogadro's number N_A.

$$N_A = \frac{N}{n}$$

$$N_A = 6.023 \times 10^{23} \text{ molecules/mol} \quad \begin{array}{l} Avogadro's \\ Number \end{array}$$

The number of moles is found by dividing the mass of a gas (in grams) by its molecular mass M:

$$n = \frac{m}{M} \quad Number\ of\ Moles$$

One often wants to determine the mass, pressure, volume, or temperature of a gas in a single state. The ideal gas law uses the molar concept to arrive at a more specific equation:

$$PV = nRT \quad\quad R = 8.314 \text{ J/mol} \cdot \text{K}$$

It should be noted that use of the constant given above restricts the units of P, V, T, and n to those that are in the constant.

- The *relative humidity* can be computed from saturated-vapor-pressure tables according to the following definition:

$$Relative\ humidity = \frac{actual\ vapor\ pressure}{saturated\ vapor\ pressure}$$

Remember that the *actual* vapor pressure at a particular temperature is the same as the *saturated* vapor pressure for the dew-point temperature. Refer to Example 19.9.

Key Terms

Review Questions

19.1. Distinguish between *state* and *phase*.

19.2. Explain Boyle's law in terms of the molecular theory of matter.

19.3. Explain Charles's law in terms of the molecular theory of matter.

19.4. Why must the absolute temperature be used in Charles's law?

19.5. A closed steel tank is filled with an ideal gas and heated. What happens to the (a) mass, (b) volume, (c) density, and (d) pressure of the enclosed gas?

19.6. Prove the accuracy of the following equations that involve the density ρ of an ideal gas.

(a) $\dfrac{P_1}{\rho_1 T_1} = \dfrac{P_2}{\rho_2 T_2}$

(b) $\rho = \dfrac{PM}{RT}$

19.7. Suppose we wish to express the pressure of an ideal gas in millimeters of mercury and the volume in cubic centimeters. Show that the universal gas constant will be equal to $6.23 \times 10^4 \ \text{mm} \cdot \text{cm}^3/\text{mol} \cdot \text{K}$.

19.8. A mole of any gas occupies 22.4 liters at STP. Can we also say that the same mass of any gas will occupy the same volume? Explain.

19.9. Distinguish between evaporation, boiling, and sublimation.

19.10. From your experience, would you expect alcohol to have a higher vapor pressure than water? Why?

19.11. Explain the principle of operation for the pressure cooker and the vacuum pan in cooking.

19.12. Can a solid have a vapor pressure? Explain.

19.13. If evaporation is a cooling process, is condensation a heating process? Explain.

19.14. Explain the cooling effects of evaporation in terms of the latent heat of vaporization.

19.15. Distinguish between a vapor and a gas by discussing critical temperature.

19.16. Will it take longer to cook an egg by boiling it in water on Mt. Everest or at the seashore? Why?

19.17. Water is brought to a vigorous boil in an open flask. When the flask is removed from the flame and tightly stoppered, the boiling stops. Why? The stoppered flask is then inverted and held under a stream of cold running water. The boiling begins again. Explain. As soon as the flask is removed from the water, boiling stops. If the flask is cooled, boiling begins again. How long can the process of making the water boil by cooling be continued?

19.18. On a cool day, the relative humidity inside a house is the same as the relative humidity outside the house. Are the dew points necessarily the same? Explain.

19.19. Explain what is meant by *critical pressure*.

19.20. Is it possible for ice to exist in equilibrium with boiling water? Explain.

19.21. The formation of moisture on the walls and windows inside a home can cause considerable damage. What causes this moisture? Discuss several ways of reducing or preventing the formation of moisture.

19.22. Discuss the formation of fog and clouds. Why are fog conditions usually worse in the fall and early spring?

Problems

Section 19.3 General Gas Laws

19.1. An ideal gas occupies a volume of $4.00 \ \text{m}^3$ at 200 kPa absolute pressure. What will be the new pressure if the gas is slowly compressed to $2.00 \ \text{m}^3$ at constant temperature? Ans. 400 kPa

19.2. The absolute pressure of a sample of ideal gas is 300 kPa at a volume of $2.6 \ \text{m}^3$. If the pressure decreases to 101 kPa at constant temperature, what is the new volume?

19.3. Two hundred cubic centimeters of an ideal gas at 20°C expands to a volume of $212 \ \text{cm}^3$ at constant pressure. What is the final temperature?

Ans. 37.6°C

19.4. The temperature of a gas sample decreases from 55 to 25°C at constant pressure. If the initial volume was 400 mL, what is the final volume?

19.5. A steel cylinder contains an ideal gas at 27°C. The *gauge* pressure is 140 kPa. If the temperature of the container increases to 79°C, what is the new *gauge* pressure? Ans. 182 kPa

19.6. The absolute pressure of a sample of gas initially at 300 K doubles as the volume remains constant. What is the new temperature?

19.7. A steel cylinder contains 2.00 kg of an ideal gas. Overnight, the temperature and volume remain constant, but the absolute pressure decreases from 500 to 450 kPa. How many grams of the gas leaked out overnight? Ans. 200 g

19.8. Five liters of a gas at 25°C has an absolute pressure of 200 kPa. If the absolute pressure reduces to 120 kPa and the temperature increases to 60°C, what is the final volume?

19.9. An air compressor takes in 2 m³ of air at 20°C and 1 atmosphere (101.3 kPa) pressure. If the compressor discharges into a 0.3-m³ tank at an absolute pressure of 1500 kPa, what is the temperature of the discharged air? Ans. 651 K

19.10. A 6-L tank holds a sample of gas under an absolute pressure of 600 kPa and a temperature of 57°C. What will be the new pressure if the same sample of gas is placed into a 3-L container at 7°C?

19.11. If 0.8 L of a gas at 10°C is heated to 90°C at constant pressure, what will the new volume be? Ans. 1.03 L

19.12. The inside of an automobile tire is under a gauge pressure of 30 lb/in.² at 4°C. After several hours, the inside air temperature rises to 50°C. Assuming constant volume, what is the new gauge pressure?

19.13. A 2-L sample of gas has an absolute pressure of 300 kPa at 300 K. If both pressure and volume experience a two-fold increase, what is the final temperature? Ans. 1200 K

Section 19.4 Molecular Mass and the Mole

19.14. How many moles are contained in 600 g of air? ($M = 29$ g/mol.)

19.15. How many moles of gas are there in 400 g of nitrogen gas? ($M = 28$ g/mol.) How many molecules are in this sample?
Ans. 14.3 mol, 8.60×10^{24} molecules

19.16. What is the mass of a 4-mol sample of air? ($M = 29$ g/mol.)

19.17. How many grams of hydrogen gas ($M = 2$ g/mol) are there in 3.0 moles of hydrogen? How many grams of air ($M = 29$ g/mol) are there in 3.0 moles of air? Ans. 6 g, 87 g

***19.18.** How many molecules of hydrogen gas ($M = 2$ g/mol) are needed to have the same mass as 4 g of oxygen ($M = 32$ g/mol)? How many moles are in each sample?

***19.19.** What is the mass of one molecule of oxygen? ($M = 32$ g/mol). Ans. 5.31×10^{-26} kg

***19.20.** The molecular mass of CO_2 is 44 g/mol. What is the mass of a single molecule of CO_2?

Section 19.5 The Ideal Gas Law

19.21. Three moles of an ideal gas have a volume of 0.026 m³ and a pressure of 300 kPa. What is the temperature of the gas in degrees Celsius?
Ans. 39.7°C

19.22. A 16-L tank contains 200 g of air ($M = 29$ g/mol) at 27°C. What is the absolute pressure of this sample?

19.23. How many kilograms of nitrogen gas ($M = 28$ g/mol) will occupy a volume of 2000 L at an absolute pressure of 202 kPa and a temperature of 80°C? Ans. 3.85 kg

19.24. What volume is occupied by 8 g of nitrogen gas ($M = 28$ g/mol) at standard temperature and pressure (STP)?

19.25. A 2-L flask contains 2×10^{23} molecules of air ($M = 29$ g/mol) at 300 K. What is the absolute gas pressure? Ans. 414 kPa

19.26. A 2-m³ tank holds nitrogen gas ($M = 28$ g/mol) under a gauge pressure of 500 kPa. If the temperature is 27°C, what is the mass of gas in the tank?

19.27. How many moles of gas are contained in a volume of 2000 cm³ at conditions of standard temperature and pressure (STP)? Ans. 0.0893 mol

19.28. A 0.30-cm³ cylinder contains 0.27 g of water vapor ($M = 18$ g/mol) at 340°C. What is its absolute pressure, assuming that the water vapor is an ideal gas?

Section 19.10 Humidity

19.29. If the air temperature is 20°C and the dew point is 12°C, what is the relative humidity?
Ans. 60.8 percent

19.30. The dew point is 20°C. What is the relative humidity when the air temperature is 24°C?

19.31. The relative humidity is 77 percent when the air temperature is 28°C. What is the dew point?
Ans. 23.5°C

19.32. What is the pressure of water vapor in the air on a day when the temperature is 86°F and the relative humidity is 80 percent?

19.33. The air temperature in a room during the winter is 28°C. What is the relative humidity if moisture first starts forming on a window when the temperature of its surface is 20°C?
Ans. 61.8 percent

Additional Problems

19.34. A sample of gas occupies 12 L at 7°C and at an absolute pressure of 102 kPa. Find its temperature when the volume reduces to 10 L and the pressure increases to 230 kPa.

19.35. A tractor tire contains 2.8 ft^3 of air at a gauge pressure of 70 lb/in.^{2}. What volume of air at 1 atm of pressure is required to fill this tire if there is no change in temperature or volume?

Ans. 16.1 ft^3

19.36. A 3-L container is filled with 0.230 mol of an ideal gas at 300 K. What is the pressure of the gas? How many molecules are in this sample of gas?

19.37. How many moles of helium gas (M = 4 g/mol) are there in a 6-L tank when the pressure is 2×10^5 Pa and the temperature is 27°C? What is the mass of the helium?

Ans. 0.481 mol, 1.92 g

19.38. How many grams of air (M = 29 g/mol) must be pumped into an automobile tire if it is to have a gauge pressure of 31 lb/in.^{2}? Assume that the volume of the tire is 5000 cm^3 and its temperature is 27°C.

19.39. The air temperature inside a car is 26°C. The dew point is 24°C. What is the relative humidity inside the car? Ans. 88.9 percent

19.40. The lens in a sensitive camera is clear when the room temperature is 71.6°F and the relative humidity is 88 percent. What is the lowest temperature of the lens if it is not to become foggy from moisture?

*19.41. What is the density of oxygen gas (M = 32 g/mol) at a temperature of 23°C and atmospheric pressure? Ans. 1.32 kg/m^3

*19.42. A 5000-cm^3 tank is filled with carbon dioxide (M = 44 g/mol) at 300 K and 1 atm of pressure. How many grams of CO_2 can be added to the tank if the maximum absolute pressure is 60 atm and there is no change in temperature?

*19.43. The density of an unknown gas at standard temperature and pressure (STP) is 1.25 kg/m^3. What is the density of this gas at 18 atm and 60°C? Ans. 18.4 kg/m^3

Critical Thinking Questions

*19.44. A tank with a capacity of 14 L contains helium gas at 24°C under a gauge pressure of 2700 kPa. (a) What will be the volume of a balloon filled with this gas if the helium expands to an internal absolute pressure of 1 atm and the temperature drops to −35°C? (b) Now suppose the system returns to its original temperature (24°C). What is the final volume of the balloon?

Ans. (a) 310 L, (b) 387 L

*19.45. A steel tank is filled with oxygen. One evening, when the temperature is 27°C, the gauge at the top of the tank indicates a pressure of 400 kPa. During the night, a leak develops in the tank. The next morning, the gauge pressure is only 300 kPa, and the temperature is 15°C. What percentage of the original gas remains in the tank?

*19.46. A 2-L flask is filled with nitrogen (M = 28 g/mol) at 27°C and 1 atm of absolute pressure. A stopcock at the top of the flask is opened to the air, and the system is heated to a temperature of 127°C. The stopcock is then closed, and the system is allowed to return to 27°C. What mass of nitrogen is in the flask? What is the final pressure?

Ans. 1.71 g, 0.750 atm

*19.47. What is the volume of 8 g of sulfur dioxide (M = 64 g/mol) if it has an absolute pressure of 10 atm and a temperature of 300 K? If 10^{20} molecules leak from this volume every second, how long will it take to reduce the pressure by one-half?

*19.48. A flask contains 2 g of helium (M = 4 g/mol) at 57°C and 12 atm absolute pressure. The temperature then decreases to 17°C, and the pressure falls to 7 atm. How many grams of helium have leaked out of the container? Ans. 1.13 L, 0.672 g

*19.49. What must be the temperature of the air in a hot-air balloon in order that the mass of the air is 0.97 times that of an equal volume of air at a temperature of 27°C?

20

Thermodynamics

Heat engines, such as the one driving this locomotive, operate in a cycle to produce output work from input heat. The thermodynamic processes involved in such conversions are the subject of this chapter. (*Photo © vol. 247/Corbis.*)

Objectives

After completing this chapter, you should be able to

1. Demonstrate by definition and examples your understanding of the *first* and *second laws of thermodynamics.*
2. Define and give illustrated examples of *adiabatic, isochoric, isobaric,* and *isothermal* processes.
3. Write and apply a relationship for determining the *ideal efficiency* of a heat engine.
4. Define the *coefficient of performance* for a refrigerator and solve refrigeration problems similar to those discussed in the text.

Thermodynamics treats the transformation of heat energy into mechanical energy and the reverse process, the conversion of work into heat. Since almost all the energy available from raw materials is liberated in the form of heat, it is easy to see why thermodynamics plays such an important role in science and technology.

In this chapter, we will study two basic laws that must be obeyed when heat energy is used to accomplish work. The first law is simply a restatement of the principle of conservation of energy. The second law places restrictions on the efficient use of the available energy.

20.1   Heat and Work

The equivalence of heat and work as two forms of energy has been clearly established. Rumford destroyed the caloric theory of heat by showing that it is possible to remove heat indefinitely from a system as long as external work is supplied. Joule then sealed the case by demonstrating the mechanical equivalence of heat.

Work, like heat, involves a transfer of energy, but there is an important distinction between the two terms. In mechanics, we define **work** as a scalar quantity, equal in magnitude to the product of a force and a displacement. Temperature plays no role in this definition. **Heat,** on the other hand, is energy that flows from one body to another because of a difference in temperature. A temperature difference is a necessary condition for the transfer of heat. *Displacement* is the necessary condition for the performance of work.

The important point in this discussion is to recognize that both heat and work represent changes that occur in a given process. Usually these changes are accompanied by a change in internal energy. Consider the two situations illustrated in Fig. 20.1. In Fig. 20.1a, the internal energy of the water is increased by the performance of mechanical work. In Fig. 20.1b, the internal energy of the water is increased through the flow of heat.

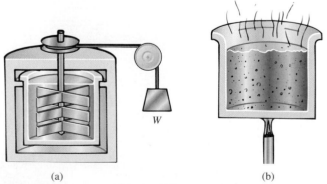

(a)   (b)

Figure 20.1   Increasing the internal energy of a system (a) by the performance of work and (b) by supplying heat to the system.

20.2   The Internal Energy Function

When studying the transformations of heat into work or of work into heat, it is useful to introduce the concept of a thermodynamic *system* and its *surroundings*. By *system,* we mean a collection of molecules or objects on which our attention is focused. The system typically is described by its mass, pressure, volume, and temperature; it is contained in some fashion by its *surroundings*. For example, in a gasoline engine, the system consists of the burning gasoline; the surroundings are the pistons, the cylinder walls, the exhaust system, and other elements.

A system is said to be in ***thermodynamic equilibrium*** if there is no resultant force on the system and if the temperature of the system is the same as its surroundings. This condition requires that no work be done on or by the system and that there be no exchange of heat between the system and its surroundings. Under these conditions, the system has a definite internal energy U. Its *thermodynamic state* can be described by three coordinates: (1) its pressure P, (2) its volume V, and (3) its temperature T. Whenever energy is absorbed or released by such a system in the form of either heat or work, it will reach a new state of equilibrium in such a way that energy is conserved.

In Fig. 20.2, let us consider a general thermodynamic process in which a system is caused to change from an equilibrium state 1 to an equilibrium state 2. In Fig. 20.2a, the system is in thermodynamic equilibrium, with an initial internal energy U_1 and thermodynamic coordinates (P_1, V_1, T_1). In Fig. 20.2b, the system reacts with its surroundings. Heat Q may be absorbed by the system and/or released to its environment. The transfer of heat is considered positive for heat input and negative for heat output. The net heat *absorbed* by

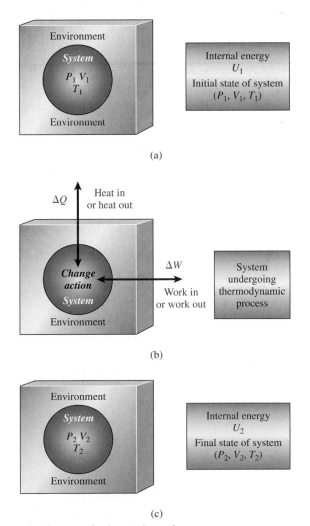

Figure 20.2   A schematic diagram of a thermodynamic process.

the system is represented by ΔQ. Work W may be done *by* the system or *on* the system. Output work is considered positive, and input work is considered negative. Thus, ΔW represents the net work done *by* the system (output work). In Fig. 20.2c, the system has reached its final state 2 and is again in equilibrium, with a final internal energy U_2. Its new thermodynamic coordinates are (P_2, V_2, T_2).

If energy is to be conserved, the change in internal energy

$$\Delta U = U_2 - U_1$$

must represent the difference between the net heat ΔQ absorbed by the system and the net work ΔW done by the system on its surroundings.

$$\Delta U = \Delta Q - \Delta W \qquad (20.1)$$

Thus, the change in internal energy is uniquely defined in terms of the measurable quantities heat and work. Equation (20.1) states the existence of an ***internal energy function U*** *that is determined by the thermodynamic coordinates of a system.* Its value at the final state minus its value at the initial state is equal to the change in energy of the system. Since temperature is associated with the internal energy, it is generally true that an increase or decrease in internal energy also results in an increase or decrease of temperature.

20.3 The First Law of Thermodynamics

The *first law of thermodynamics* is simply a restatement of the principle of conservation of energy:

> Energy cannot be created or destroyed but can change from one form to another.

Applying this law to a thermodynamic process, we note from Eq. (20.1) that

$$\Delta Q = \Delta U + \Delta W \qquad (20.2)$$

This equation represents a mathematical statement of the ***first law of thermodynamics,*** which can be stated as follows:

> **The First Law of Thermodynamics:** In any thermodynamic process, the net heat absorbed by a system is equal to the sum of the net work done by the system and the change in the internal energy of the system.

In applying the first law of thermodynamics, we must recognize that heat Q put into a system is positive, and heat expelled or lost from a system is negative. Work done by a system is positive; work done on a system is negative. An increase in internal energy is positive, and a decrease is negative. These conventions are summarized in Fig. 20.3.

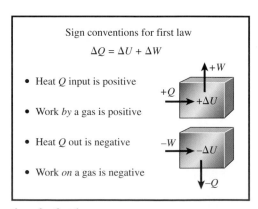

Figure 20.3 Sign Conventions for first law.

For example, if a gas absorbs 800 J of heat and does a net work of 200 J while expelling 300 J of heat in the process, we see that the net heat input is

$$\Delta Q = 800 \text{ J} - 300 \text{ J} = 500 \text{ J}$$

Since $\Delta W = 200$ J, the change in internal energy can be found from Eq. (20.2).

$$\Delta U = \Delta Q - \Delta W = 500 \text{ J} - 200 \text{ J} \qquad \text{or} \qquad \Delta U = +300 \text{ J}$$

The positive value indicates an *increase* in the internal energy of the system.

Example 20.1

A heat engine does 240 J of work during which its internal energy *decreases* by 400 J. What is the net heat exchange for this process?

Plan: The internal energy decreases, so ΔU is negative; the work is done *by* the engine, so ΔW is positive. The magnitude and sign of the thermal energy exchange ΔQ can be found from the first law of thermodynamics.

Solution: Substituting $\Delta U = -400$ J and $\Delta W = +240$ J, we obtain

$$\Delta Q = \Delta U + \Delta W = (-400 \text{ J}) + (240 \text{ J})$$
$$= -400 \text{ J} + 240 \text{ J} = -160 \text{ J}$$

The negative sign for the heat exchange indicates that the net heat is given *out* by the system. If there is no phase change, the temperature of the system will decrease.

20.4 Isobaric Processes and the *P-V* Diagram

It is instructive to look at the energy changes involved with thermodynamic processes by looking at a gas enclosed in a cylinder equipped with a movable, frictionless piston. Let's consider the work done by the expanding gas in Fig. 20.4a. The piston has a cross-sectional area A and rests on a column of gas under a pressure P. Heat can flow in or out of the gas through the cylinder walls. Work can be done on or by the gas by pushing the piston downward, or work can be done by the gas as it expands upward.

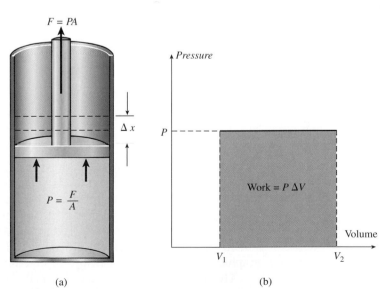

Figure 20.4 (a) Calculating the work done by a gas expanding at constant pressure. (b) The work is equal to the area under the curve on a *P-V* diagram.

Let us first consider the work done by the gas when it expands at a constant pressure ($P = F/A$). The force F exerted by the gas on the piston will be equal to the product of the pressure P and the area A of the piston.

$$F = PA$$

Remember that *work* is equal to the product of force and the parallel displacement. If the piston moves upward through a distance Δx, the work done will be equal to

$$\Delta W = F\,\Delta x = (PA)\Delta x$$

But the increase in volume ΔV of the gas is just $A\,\Delta x$, so we can rearrange the above factors to find that the work done by a gas expanding at constant pressure is given by

$$\Delta W = P\,\Delta V \qquad\qquad (20.3)$$

In other words, the net work is equal to the product of constant pressure and the change in volume. This is an example of what is called an ***isobaric process.*** It should be noted that the change in volume ΔV is the *final value* less the *initial value*, so a decrease in volume yields negative work, and an increase in volume yields positive work.

> An isobaric process is a thermodynamic process that occurs at constant pressure.

Such a process is shown graphically in Fig. 20.4b by plotting the increase in volume as a function of the pressure. This representation, called a ***P-V diagram,*** is extremely useful in thermodynamics. In the previous example, the pressure was constant, so the graph is a straight line. Note that the *area* under the curve (or line in this case) is equal to

$$Area = P(V_2 - V_1) = P\,\Delta V$$

which is also equal to the work done by the expanding gas. This leads us to an important principle:

> When a thermodynamic process involves changes in volume and/or pressure, the work done by the system is equal to the area underneath the curve on a *P-V* diagram.

Example 20.2

Assume the gas inside the cylinder of Fig. 20.4 expands under a constant pressure of 200 kPa, while its volume increases from $2 \times 10^{-3}\ \text{m}^3$ to $5 \times 10^{-3}\ \text{m}^3$. What work is done by this gas?

Solution: The work done is equal to the constant pressure times the change in volume.

$$\text{Work} = (200 \times 10^3\ \text{Pa})(5 \times 10^{-3}\ \text{m}^3 - 2 \times 10^{-3}\ \text{m}^3)$$
$$\text{Work} = 600\ \text{J}$$

In general, the pressure will not be constant during a piston displacement. For example, in the power stroke of a gasoline engine, the fluid is ignited under high pressure, and the pressure decreases as the piston is displaced downward. The *P-V* diagram in this case is a sloping curve, as shown in Fig. 20.5a. The volume increases from V_1 to V_2, while the pressure decreases from P_1 to P_2. To compute the work done in such a process, we must resort to calculus or to a graphical analysis. If the area under the curve can be estimated graphically, then the work can also be determined.

That the area under the curve is equal to the work done when the pressure is not constant can be demonstrated in Fig. 20.5. The area of the narrow shaded rectangle represents the work done by the gas in expanding by an increment ΔV_i under a constant pressure P_i.

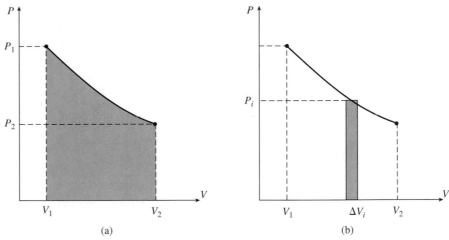

Figure 20.5   Calculating the work done by a gas expanding under a varying pressure.

If the area under the entire curve is split up into many of these rectangles, we can sum all the products $P_i \, \Delta V_i$ to obtain the total work. Thus, the total work is simply the area under the $P\text{-}V$ diagram between the points V_1 and V_2 on the volume axis.

20.5   The General Case for the First Law

The first law of thermodynamics stipulates that energy must be conserved in any thermodynamic process. In the mathematical formulation

$$\Delta Q = \Delta W + \Delta U$$

there are three quantities that may undergo changes. The most general process is one in which all three quantities are involved. For example, the fluid in Fig. 20.6 is expanded while in contact with a hot flame. Considering the gas as a system, there is a net heat transfer ΔQ

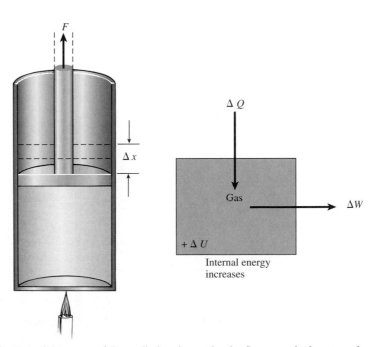

Figure 20.6   Part of the energy ΔQ supplied to the gas by the flame results in external work ΔW. The remainder increases the internal energy ΔU of the gas.

imparted to the gas. This energy is used in two ways: (1) The internal energy ΔU of the gas is increased by a portion of the input thermal energy, and (2) the gas does an amount of work ΔW on the piston that is equivalent to the remainder of the available energy.

Special cases of the first law arise when one or more of the three quantities—ΔQ, ΔW, or ΔU—do not undergo change. In these instances, the first law is considerably simplified. In Sections 20.6 through 20.8, we consider several of these special processes.

20.6   Adiabatic Processes

Suppose a system is completely isolated from its surroundings so that there can be no exchange of thermal energy Q. Any process that occurs completely inside such an isolated chamber is called an ***adiabatic process,*** and the system is said to be surrounded by *adiabatic* walls.

> An adiabatic process is one in which there is no exchange of thermal energy ΔQ between a system and its surroundings.

Applying the first law to a process in which $\Delta Q = 0$, we obtain

$$\Delta W = -\Delta U \qquad\qquad \textit{Adiabatic} \quad \textbf{(20.4)}$$

Equation (20.4) tells us that, in an adiabatic process, the work is done at the *expense* of internal energy. The decrease in thermal energy is usually accompanied by a decrease in temperature.

As an example of an adiabatic process, consider Fig. 20.7, in which a piston is lifted by an expanding gas. If the walls of the cylinder are insulated and the expansion occurs rapidly, the process will be approximately adiabatic. As the gas expands, it does work on the piston but loses internal energy and experiences a drop in temperature. If the process is reversed by forcing the piston back down, work is done *on* the gas $(-\Delta W)$, and there will be an increase in internal energy (ΔU) such that

$$-\Delta W = +\Delta U$$

In this instance, the temperature will rise.

Another example of an adiabatic process that is useful in industrial refrigeration is referred to as a ***throttling process.***

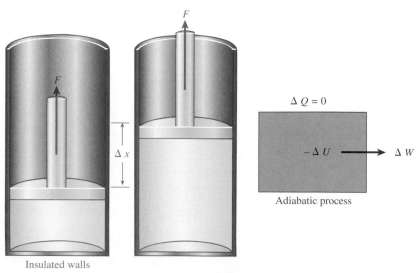

Figure 20.7   In an adiabatic process, there is no transfer of heat, and work is done at the *expense* of internal energy.

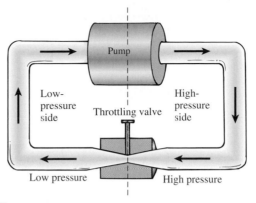

Figure 20.8 The throttling process.

A throttling process is one in which a fluid at high pressure seeps adiabatically through a porous wall or narrow opening into a region of low pressure.

Consider a gas forced by a pump to circulate through the apparatus in Fig. 20.8. Gas from the high-pressure side of the pump is forced through the narrow constriction, called the *throttling valve,* to the low-pressure side. The valve is heavily insulated, so the process is adiabatic, and $\Delta Q = 0$. According to the first law, $\Delta W = -\Delta U$, and the net work done by the gas in passing through the valve is accomplished at the expense of internal energy. In refrigeration, a liquid coolant undergoes a drop in temperature and partial vaporization as a result of the throttling process.

20.7 Isochoric Processes

Another special case for the first law occurs when there is no work done, either *by* the system or *on* the system. This type of process is referred to as an ***isochoric process.*** It is also referred to as an ***isovolumic process*** because there can be no change in volume without the performance of work.

An isochoric process is one in which the volume of the system remains constant.

Applying the first law to a process in which $\Delta W = 0$, we obtain

$$\Delta Q = \Delta U \qquad\qquad \text{Isochoric} \quad \textbf{(20.5)}$$

Therefore, in an isochoric process, all the thermal energy absorbed by a system goes to increase its internal energy. In this instance, there is usually a rise in the temperature of the system.

An isochoric process occurs when water is heated in a container of fixed volume, as shown in Fig. 20.9. As heat is supplied, the increase in internal energy results in a rise in the temperature of the water until it begins to boil. Further increases in internal energy go into the process of vaporization. The volume of the system, consisting of the water and its vapor, remains constant, however, and no external work is done.

When the flame is removed, the process is reversed as heat leaves the system through the bottom of the cylinder. The water vapor will condense, and the temperature of the resulting water will eventually drop to room temperature. This process represents a loss of heat and a corresponding decrease in internal energy, but, once again, no work is done.

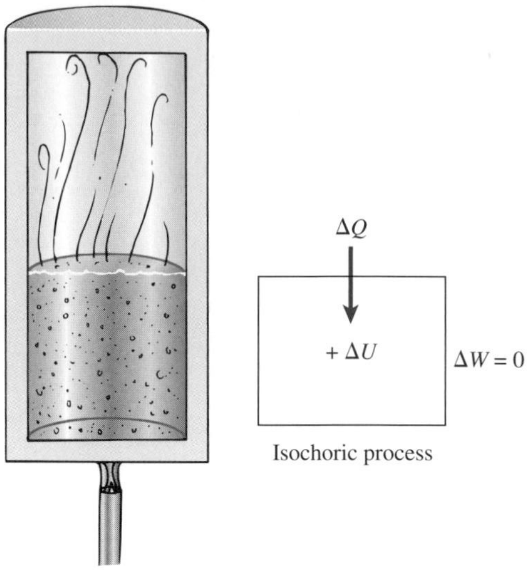

Figure 20.9   In an isochoric process, the volume of the system (water and vapor) remains constant.

20.8   Isothermal Processes

It is possible for the pressure and volume of a gas to vary without a change in temperature. In Chapter 19, we introduced Boyle's law to describe volume and pressure changes during such a process. A gas can be compressed in a cylinder so slowly that it will essentially remain in thermal equilibrium with its surroundings. The pressure increases as the volume decreases, but the temperature is essentially constant.

> An isothermal process is one in which the temperature of the system remains constant.

If there is no change of phase, a constant temperature indicates there is no change in the internal energy of the system. Applying the first law to a process in which $\Delta U = 0$, we obtain

$$\Delta Q = \Delta W \qquad\qquad \textit{Isothermal} \quad \textbf{(20.6)}$$

Thus, in an **isothermal process,** all the energy absorbed by a system is converted into output work.

20.9   The Second Law of Thermodynamics

When we rub our hands together vigorously, the work done against friction increases the internal energy and causes a rise in temperature. The surrounding air forms a large reservoir at a lower temperature, and the heat energy is transferred to the air without changing its temperature appreciably. When we stop rubbing, our hands return to the same state as before. According to the first law of thermodynamics, mechanical energy has been transformed into heat with 100 percent efficiency.

$$\Delta W = \Delta Q$$

Such a transformation can be continued indefinitely as long as work is supplied.

Let us now consider the reverse process. Is it possible to convert heat energy into work with 100 percent efficiency? In the previous example, is it possible to capture all the heat transferred to the air and return it to our hands, causing them to rub together indefinitely of their own accord? On a cold winter day, this process would be a boon to hunters with cold hands. Unfortunately, such a process cannot occur, even though it does not violate the first law. Neither is it possible to retrieve all the heat lost in braking a car to start the wheels rolling again.

We will see that the conversion of heat energy into mechanical work is a losing process. The first law of thermodynamics tells us that we cannot win in such an experiment. In other words, it is impossible to get more work out of a system than the heat put into the system. It does not, however, preclude us from breaking even. Clearly, we need another rule stating that the 100 percent conversion of heat energy into useful work is not possible. This rule forms the basis for the *second law of thermodynamics.*

The Second Law of Thermodynamics: It is impossible to construct an engine that, operating continuously, produces no effect other than the extraction of heat from a reservoir and the performance of an *equivalent* amount of work.

To give more insight and application to this principle, suppose we study the operation and efficiency of heat engines. A particular system might be a gasoline engine, a jet engine, a steam engine, or even the human body. The operation of a heat engine is best described by a diagram similar to that shown in Fig. 20.10. During the operating of such a general engine, three processes occur:

1. A quantity of heat Q_{in} is supplied to the engine from a reservoir at a high temperature T_{in}.

2. Mechanical work W_{out} is done by the engine through the use of a portion of the heat input.

3. A quantity of heat Q_{out} is released to a reservoir at a low temperature T_{out}.

The working substance begins at a specific thermodynamic state described by its temperature, pressure, volume, and number of moles. It then undergoes a series of processes and returns to its initial state. It such a cyclic process, the initial and final internal energies are the same, and $\Delta U = 0$. The first law of thermodynamics, therefore, tells us the net work done in a complete cycle is given by

$$Net\ work = heat\ input - heat\ output$$

$$\Delta W = Q_{in} - Q_{out} \tag{20.7}$$

If a *P-V* diagram is constructed for an engine operating in a complete cycle, a closed loop will be formed, and the area inside this closed loop is equal to the net work done by the engine. All heat engines and refrigerators operate in such a cyclic fashion.

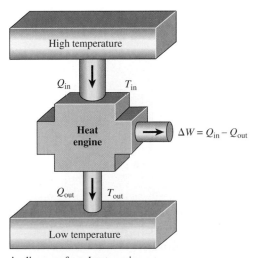

Figure 20.10   A schematic diagram for a heat engine.

The *efficiency* of a **heat engine** is defined as the ratio of the useful work done by the engine to the heat put into the engine, and it is usually expressed as a percentage.

$$Efficiency = \frac{work\ output}{heat\ input}$$

$$e = \frac{Q_{in} - Q_{out}}{Q_{in}} \tag{20.8}$$

For example, an engine that is 25 percent efficient ($e = 0.25$) might absorb 800 J, perform 200 J of work, and reject 600 J as wasted heat. A 100 percent efficient engine is one in which all the input heat is converted to useful work. In this case, no heat would be rejected to the environment ($Q_{out} = 0$). Although such a process would conserve energy, it violates the second law of thermodynamics. The most efficient engine is the one that rejects the *least* possible heat to the environment.

20.10 The Carnot Cycle

All heat engines are subject to many practical difficulties. Friction and the loss of heat through conduction and radiation prevent actual engines from obtaining their maximum efficiency. An idealized engine free of such problems was suggested by Sadi Carnot in 1824. The **Carnot engine** has the maximum possible efficiency for an engine that absorbs heat from a high-temperature reservoir, performs external work, and deposits heat to a low-temperature reservoir. The effectiveness of a given engine can therefore be determined by comparing it with the Carnot engine operating between the same temperatures.

The **Carnot cycle** is illustrated in Fig. 20.11. A gas contained in a cylinder equipped with a movable piston is placed in contact with a reservoir at a high temperature T_{in}. A quantity of heat Q_{in} is absorbed by the gas, which expands isothermally as the pressure decreases. The first stage of a Carnot cycle is shown graphically as the curve *AB* in the *P-V* diagram (Fig. 20.12). The cylinder is next placed on an insulated stand, where it continues to expand adiabatically as the pressure drops to its lowest value. This stage is represented graphically by the curve *BC*. In the third stage, the cylinder is removed from the insulated pad and placed on a reservoir at a low temperature T_{out}. A quantity of heat Q_{out} is exhausted from the gas as it is compressed isothermally from point *C* to point *D* in the *P-V* diagram. Finally, the cylinder is again placed on the insulated pad, where it is compressed adiabatically to its original state along the path *DA*. The engine does external work during the expansion processes and returns to its initial state during the compression processes.

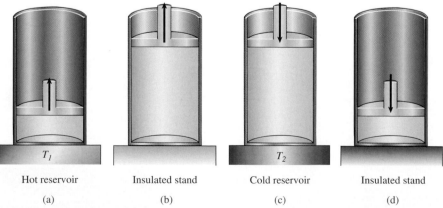

Hot reservoir	Insulated stand	Cold reservoir	Insulated stand
(a)	(b)	(c)	(d)

Figure 20.11 The Carnot cycle: (a) isothermal expansion, (b) adiabatic expansion, (c) isothermal compression, and (d) adiabatic compression.

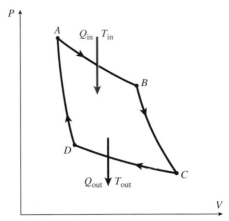

Figure 20.12   The Carnot cycle: A *P-V* diagram for an ideal engine. The net work is equal to the net heat $Q_{in} - Q_{out}$.

20.11 | The Efficiency of an Ideal Engine

The efficiency of a real engine is difficult to predict from Eq. (20.8) because the quantities Q_{in} and Q_{out} are difficult to calculate. Frictional and heat losses through the cylinder walls and around the piston, incomplete burning of the fuel, and even the physical properties of different fuels all frustrate our attempts to measure the efficiency of such engines. We can image an *ideal engine,* however, one that is not restricted by these practical difficulties. The efficiency of such an engine depends only on the quantities of heat absorbed and rejected between two well-defined heat reservoirs. It does not depend on the thermal properties of the working fuel. In other words, regardless of the internal changes in pressure, volume, length, or other factors, all ideal engines have the same efficiency when they are operating between the same two temperatures (T_{in} and T_{out}).

> An ideal engine is one that has the highest possible efficiency for the temperature limits within which it operates.

If we can define the efficiency of an engine in terms of input and output temperatures instead of in terms of the input and output of heat, we will have a more useful formula. For an ideal engine, it can be shown that the ratio of Q_{in}/Q_{out} is the same as the ratio of T_{in}/T_{out}. The actual proof is beyond the scope of this text. The efficiency of an ideal engine can, therefore, be expressed as a function of the absolute temperatures of the input and output reservoirs. Equation (20.8), for an ideal engine, becomes

$$e = \frac{T_{in} - T_{out}}{T_{in}} \tag{20.9}$$

It can be shown that no engine operating between the same two temperatures can be more efficient than would be indicated by Eq. (20.9). This ideal efficiency thus represents an upper limit to the efficiency of any practical engine. The greater the difference in temperature between two reservoirs, the greater the efficiency of any engine.

Example 20.3

An ideal engine, operating between two heat reservoirs at 500 K and 400 K, absorbs 900 J of heat from the high-temperature reservoir during each cycle. What is the efficiency of the engine, and how much heat is rejected to the environment?

Plan: The ideal efficiency is determined from Eq. (20.9) based on absolute temperatures. We can use the efficiency to determine the work output and then subtract that amount from the total input energy to determine how much is wasted.

Solution: The ideal efficiency is

$$e = \frac{T_{in} - T_{out}}{T_{in}} = \frac{500\text{ K} - 400\text{ K}}{500\text{ K}} = 0.200$$

Thus, the ideal efficiency is 20 percent. Now, by definition, the efficiency is the ratio W_{out}/Q_{in}, so we find that

$$e = \frac{W_{out}}{Q_{in}} = 0.200 \qquad \text{or} \qquad W_{out} = (0.200)(900\text{ J}) = 180\text{ J}$$

From the first law of thermodynamics, the net work must equal the net heat exchange.

$$W_{out} = Q_{in} - Q_{out} \qquad \text{or} \qquad Q_{out} = Q_{in} - W_{out}$$
$$Q_{out} = 900\text{ J} - 180\text{ J} \qquad \text{and} \qquad Q_{out} = 720\text{ J}$$

We see an ideal engine having an efficiency of 20 percent takes in 900 J of energy, does 180 J of work, and loses 720 J to the environment.

20.12   Internal Combustion Engines

An internal combustion engine generates the input heat within the engine itself. The most common engine of this variety is the four-stroke gasoline engine, in which a mixture of gasoline and air is ignited by a spark plug in each cylinder. The thermal energy released is converted into useful work by the pressure exerted on a piston by the expanding gases. The four-stroke process is illustrated in Fig. 20.13. During the *intake stroke* (Fig. 20.13a), a mixture of air and gasoline vapor enters the cylinder through the intake valve. Both valves are closed during the *compression stroke* (Fig. 20.13b), as the piston moves upward, causing a rise in pressure. Just before the piston reaches the top, the mixture is ignited, causing a sharp increase in temperature and pressure. In the *power stroke* (Fig. 20.13c), the expanding gases force the piston downward, performing external work. The fourth stroke (Fig. 20.13d) pushes the burned gases out of the cylinder through the exhaust valve. The entire cycle is then repeated for as long as the combustible fuel is supplied to the cylinder.

The ideal cycle used by the engineer to perfect the gasoline engine is shown in Fig. 20.14. It is named the *Otto cycle*, after its inventor. The compression stroke is represented

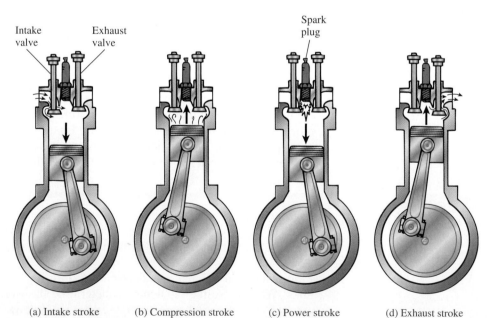

    (a) Intake stroke   (b) Compression stroke   (c) Power stroke   (d) Exhaust stroke
Figure 20.13   The four-stroke gasoline engine: (a) intake stroke, (b) compression stroke, (c) power stroke, and (d) exhaust stroke.

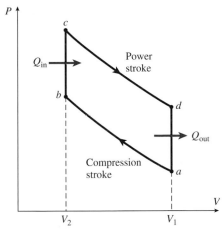

Figure 20.14 The Otto cycle for a four-stroke gasoline engine.

by the curve *ab*. The pressure increases adiabatically as the volume is reduced. At point *b*, the mixture is ignited, supplying a quantity of heat Q_{in} to the system. This causes the sharp rise in pressure indicated by the line *bc*. In the power stroke (*cd*), the gases expand adiabatically, performing external work. The system then cools at constant volume to point *a*, releasing a quantity of heat Q_{out}. The burned gases are exhausted on the next upward stroke, and more fuel is drawn in on the next downward stroke. Then the cycle begins all over again. The volume ratio V_1/V_2, as indicated in the *P-V* diagram, is called the *compression ratio* and is about 8 for most automobile engines.

The efficiency of the ideal Otto cycle can be shown in Eq. (20.10):

$$e = 1 - \frac{1}{(V_1/V_2)^{\gamma - 1}} \qquad \textbf{(20.10)}$$

where γ is the adiabatic constant for the working substance. The adiabatic constant is defined by

$$\gamma = \frac{c_p}{c_v}$$

where c_p is the specific heat of the gas at constant pressure and c_v is the specific heat at constant volume. For monatomic gases, $\gamma = 1.67$, and for diatomic gases, $\gamma = 1.4$. In the gasoline engine, the working substance is mostly air, for which $\gamma = 1.4$. In the ideal case, Eq. (20.10) shows that the higher compression ratios yield higher efficiencies since γ is always greater than 1.

Example 20.4

Compute the efficiency of a gasoline engine for which the compression ratio is 8 and $\gamma = 1.4$.

Solution: From the given information, we note that

$$\frac{V_1}{V_2} = 8 \qquad \text{and} \qquad \gamma - 1 = 1.4 - 1 = 0.4$$

Thus, from Eq. (20.10),

$$e = 1 - \frac{1}{8^{0.4}} = 1 - \frac{1}{2.3} = 57 \text{ percent}$$

In this example, 57 percent represents the maximum possible efficiency of a gasoline engine with the parameters given. Actually, the efficiency of such an engine is normally around 30 percent because of uncontrolled heat losses.

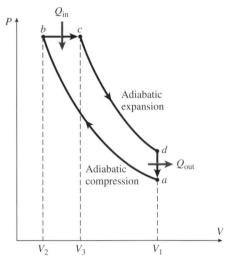

Figure 20.15   The ideal diesel cycle.

A second type of internal combustion engine is the diesel engine. In this engine, the air is compressed to a high temperature and pressure near the top of the cylinder. Diesel fuel, which is injected into the cylinder at this point, ignites and forces the piston downward. The idealized diesel cycle is shown by the *P-V* diagram in Fig. 20.15. Starting at *a,* air is compressed adiabatically to point *b,* where the diesel fuel is injected. The diesel fuel, ignited by the hot air, delivers a quantity of heat Q_{in} at nearly constant pressure (line *bc*). The remainder of the power stroke consists of an adiabatic expansion to point *d,* performing external work. During the exhaust and intake strokes, the gas cools at constant volume to point *a,* losing a quantity of heat Q_{out}. The efficiency of a diesel engine is a function of the compression ratio (V_1/V_2) and the *expansion ratio* (V_1/V_3).

20.13   Refrigeration

A *refrigerator* can be thought of as a heat engine operated in reverse. A schematic diagram of a refrigerator is shown in Fig. 20.16. During every cycle, a compressor or similar device supplies mechanical work *W* to the system, extracting a quantity of heat Q_{cold} from a cold reservoir and depositing a quantity of heat Q_{hot} to a hot reservoir. According to the first law, the input work is given by

$$W = Q_{hot} - Q_{cold}$$

The effectiveness of any refrigerator is determined by the amount of heat Q_{cold} extracted for the least expenditure of mechanical work *W*. The ratio Q_{cold}/W is therefore a measure of the cooling efficiency of a refrigerator and is called its ***coefficient of performance*** K. Symbolically,

$$K = \frac{Q_{cold}}{W} = \frac{Q_{cold}}{Q_{hot} - Q_{cold}} \tag{20.11}$$

For an ideal refrigerator, the maximum efficiency can be expressed in terms of absolute temperatures:

$$K = \frac{T_{cold}}{T_{hot} - T_{cold}} \tag{20.12}$$

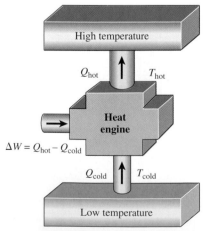

Figure 20.16   A schematic diagram for a refrigerator.

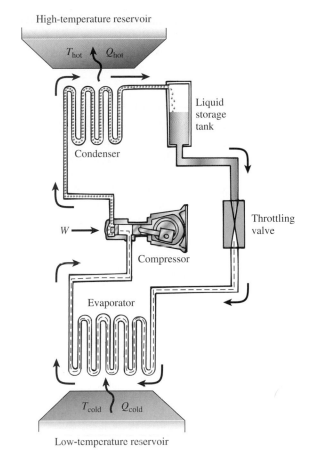

Figure 20.17 The basic components of a refrigeration system.

To understand the refrigeration process better, let us consider the general schematic in Fig. 20.17. This diagram may refer to a number of refrigeration devices, from a commercial plant to a household refrigerator. The working substance, called the ***refrigerant,*** is a fluid that is liquefied easily by an increase in pressure or a drop in temperature. In the liquid phase, it can be vaporized readily by passing it through a throttling process (see Section 20.6) near room temperature. Common refrigerants are ammonia, Freon 12, methyl chloride, and sulfur dioxide. Ammonia, the most common industrial refrigerant, boils at −28°F under a pressure of 1 atm. Freon 12, the most common household refrigerant, boils at −22°F at atmospheric pressure. Variation in pressure radically affects the condensation and evaporation temperatures of all refrigerants.

As shown in the schematic, a typical refrigeration system consists of a ***compressor,*** a ***condenser,*** a ***liquid storage tank,*** a ***throttling valve,*** and an ***evaporator.*** The compressor provides the necessary input work to move the refrigerant through the system. As the piston moves to the right, it sucks in the refrigerant through the intake valve at a little above atmospheric pressure and near room temperature. During the power stroke, the intake valve closes, and the discharge valve opens. The emergent refrigerant, at high temperature and pressure, passes into the condenser, where it is cooled until it liquefies. The condenser may be cooled by running water or by an electric fan. It is during this phase that a quantity of heat Q_{hot} is rejected from the system. The condensed liquid refrigerant, still under a condition of high pressure and temperature, is collected in a liquid reservoir. Then the liquid refrigerant is drawn from the storage tank through a throttling valve, causing a sudden drop in temperature and pressure. As the cold liquid refrigerant flows through the evaporator coils, it absorbs a quantity of heat Q_{cold} from the space and products being cooled. This heat boils the liquid refrigerant and is carried away by the gaseous refrigerant as latent heat of vaporization. This phase is the "payoff" for the entire operation, and all components just contribute to the effective transfer of heat to the evaporator. Finally, the refrigerant vapor leaves the evaporator and is sucked into the compressor to begin another cycle.

Example 20.5

An ideal refrigerator operates between 500 K and 400 K. It extracts 800 J from a cold reservoir during each cycle. How much work is done in each cycle, and how much heat is delivered to the environment?

Plan: First, we will calculate the coefficient of performance K from the given temperatures. Since K is the ratio of the extracted heat Q_{cold} (800 J) to work input, we can next solve for the work done in each cycle. The heat output can then be found since the work is equal to the difference $(Q_{cold} - Q_{hot})$.

Solution: Substitution of the absolute temperatures into Eq. (20.12) gives

$$K = \frac{T_{cold}}{T_{hot} - T_{cold}} = \frac{400 \text{ K}}{500 \text{ K} - 400 \text{ K}}; \qquad K = 4.00$$

The work in a complete cycle is then found as follows:

$$K = \frac{Q_{cold}}{W_{in}} \qquad \text{or} \qquad W_{in} = \frac{Q_{cold}}{K}$$

$$W_{in} = \frac{800 \text{ J}}{4} = 200 \text{ J}$$

Now, since $W_{in} = Q_{hot} - Q_{cold}$, we can find Q_{hot} as follows:

$$Q_{hot} = W_{in} + Q_{cold} = 200 \text{ J} + 800 \text{ J}$$
$$Q_{hot} = 1000 \text{ J}$$

Note that we have used the *maximum* possible coefficient of performance in this example. In an actual refrigerator, a smaller value of K results in the need for more than 200 J of work per cycle.

Summary and Review

Summary

Thermodynamics is the science that treats the conversion of heat into work or the reverse process, the conversion of work into heat. We have seen that, not only must energy be conserved in such processes, but also there are limits on the efficiency. The major concepts presented in this chapter are summarized as follows.

- The *first law of thermodynamics* is a restatement of the conservation-of-energy principle. It says that the net heat ΔQ put into a system is equal to the net work ΔW done by the system plus the net change in internal energy ΔU of the system. Symbolically,

$$\Delta Q = \Delta W + \Delta U \qquad \text{\textit{First Law of Thermodynamics}}$$

- In thermodynamics, work ΔW is often done on a gas. In such cases the work is often represented in terms of pressure and volume. A *P-V* diagram is also useful for measuring ΔW. If the pressure is constant,

$$\Delta W = P \, \Delta V$$

$\Delta W =$ area under *P-V* curve

- Special cases of the first law occur when one of the quantities doesn't undergo a change.

  a. *Adiabatic process:*

  $\Delta Q = 0 \qquad \Delta W = -\Delta U$

  b. *Isochoric process:*

  $\Delta V = 0 \qquad \Delta W = 0 \qquad \Delta Q = \Delta U$

  c. *Isothermal process:*

  $\Delta T = 0 \qquad \Delta U = 0 \qquad \Delta Q = \Delta W$

  d. *Isobaric process:*

  $\Delta P = 0 \qquad \Delta W = P \, \Delta V$

- The *second law of thermodynamics* places restrictions on the possibility of satisfying the *first law.* In short, it points out that in every process, there is some loss of energy due to frictional forces or other dissipative forces. A 100 percent efficient engine, one that converts all input heat to useful output work, is not possible.

- A heat engine is represented generally by Fig. 20.10. The meaning of the symbols used in the following equations can be taken from that figure. The work done by the engine is the difference between input heat and output heat.

$$W = Q_{in} - Q_{out} \qquad \text{\textit{Work} (kcal \textit{or} J)}$$

- The *efficiency e* of an engine is the ratio of the work output to the heat input. It can be calculated for an ideal engine from either of the following relations:

$$e = \frac{Q_{in} - Q_{out}}{Q_{in}}$$

$$e = \frac{T_{in} - T_{out}}{T_{in}} \qquad \text{\textit{Efficiency}}$$

- A *refrigerator* is a heat engine operated in reverse. A measure of the performance of such a device is the amount of cooling you get for the work you must put into the system. Cooling occurs as a result of the extraction of heat Q_{cold} from the cold reservoir. The coefficient of performance K is given by either

$$\text{K} = \frac{Q_{cold}}{Q_{hot} - Q_{cold}} \qquad \text{or} \qquad \text{K} = \frac{T_{cold}}{T_{hot} - T_{cold}}$$

Key Terms

Review Questions

20.1. The latent heat of vaporization of water is 540 cal/g. When 1 g of water is vaporized completely at constant pressure, however, the internal energy of the system increases by only 500 cal. What happened to the remaining 40 cal? Is this an isochoric process? Is it an isothermal process?

20.2. If both heat and work can be expressed in the same units, why is it necessary to distinguish between them?

20.3. Is it necessary to use the concept of molecular energy to describe and use the internal energy function? Explain.

20.4. A gas undergoes an adiabatic expansion. Does it perform external work? If so, what is the source of energy?

20.5. What happens to the internal energy of a gas undergoing (a) adiabatic compression, (b) isothermal expansion, and (c) a throttling process?

20.6. A gas performs external work during an isothermal expansion. What is the source of energy?

20.7. In the text, only one statement was given for the second law of thermodynamics. Discuss each of the following statements, showing them to be equivalent to that given in the text:

a. It is impossible to construct a refrigerator that, working continuously, will extract heat from a cold body and exhaust it to a hot body without the performance of work on the system.

b. The natural direction of heat flow is from a body at high temperature to a body at a low temperature, regardless of the size of each reservoir.

c. All natural spontaneous processes are irreversible.

d. Natural events always proceed in the direction from order to disorder.

20.8. It is energetically possible to extract the heat energy contained in the ocean and use it to power a steamship across the sea. What objections can you offer?

20.9. In an electric refrigerator, heat is transferred from the cool interior to warmer surroundings. Why isn't this in violation of the second law of thermodynamics?

20.10. Consider the performances of external work by the isothermal expansion of an ideal gas. Why isn't this process of converting heat into work in violation of the second law of thermodynamics?

20.11. If natural processes tend to decrease order in the universe, how can you explain the evolution of biological systems to a highly organized state? Does this violate the second law of thermodynamics?

20.12. Will keeping the door of an electric refrigerator open warm or cool a room? Explain.

20.13. What temperature must the cold reservoir have if a Carnot engine is to be 100 percent efficient? Can this ever happen? If it is impossible for a Carnot engine to have a 100 percent efficiency, why is it called the *ideal* engine?

20.14. What determines the efficiency of heat engines? Why is it generally so low?

Problems

Section 20.3 The First Law of Thermodynamics

20.1. In an industrial chemical process, 600 J of heat is supplied to a system, and 200 J of work is done *by* the system. What is the increase in the internal energy of the system?   Ans. 400 J

20.2. Assume that the internal energy of a system decreases by 300 J, while 200 J of work is done by a gas. What is the value of *Q?* Is heat lost or gained by the system?

20.3. In a thermodynamic process, the internal energy of the system increases by 500 J. How much work was done by the gas if 800 J of heat is absorbed?   Ans. 300 J

20.4. A piston does 300 J of work on a gas, which then expands, performing 220 J of work on its surroundings. What is the change in internal energy of the system if the net heat exchange is zero?

20.5. In a chemical laboratory, a technician applies 340 J, of energy to a gas, while the system surrounding the gas does 140 J of work *on* the gas. What is the change in internal energy?   Ans. 480 J

20.6. What is the change in internal energy for Prob. 20.5 if the 140 J of work is done *by* the gas instead of *on* the gas?

20.7. A system absorbs 200 J of heat as the internal energy increases by 150 J. What work is done by the gas?   Ans. 50 J

*20.8. The specific heat of water is 4186 J/kg · C°. How much does the internal energy of 200 g of water change as it is heated from 20 to 30°C? Assume that the volume is constant.

*20.9. At a constant pressure of 101.3 kPa, 1 g of water (1 cm³) is vaporized completely and has a final volume of 1671 cm³ in its vapor form. What work is done by the system against its surroundings? What is the increase in internal energy?
Ans. 169 J, 2090 J

Sections 20.4 Isobaric Processes and the *P-V* Diagram

20.10. The volume of a gas decreases from 5 L to 3 L under a constant pressure of 2 atm. What work is done? Is the work done *by* the gas or *on* the gas? If the internal energy increases by 300 J, what was the net heat exchange? Draw a rough sketch.

20.11. During an isobaric expansion, a steady pressure of 250 kPa causes the volume of a gas to change from 1 L to 3 L. What work was done by this gas?
Ans. 500 J

20.12. A gas confined in the cylinder of an engine has an initial volume of 2×10^{-4} m³. The gas then expands isobarically at 220 kPa. If 350 J of heat is absorbed in the process and the internal energy increases by 150 J, what is the final volume of the gas?

Section 20.5 The General Case for the First Law, Section 20.6 Adiabatic Processes, Section 20.7 Isochoric Processes, and Section 20.8 Isothermal Processes

20.13. An ideal gas expands isothermally while absorbing 4.80 J of heat. The piston has a mass of 3 kg. How high will the piston rise above its initial position?
Ans. 16.3 cm

20.14. The work done on a gas during an adiabatic compression is 140 J. Calculate the increase in internal energy of the system in calories.

20.15. Two kilograms of water, initially at 20°C, is confined to a container, so any changes are isochoric. Then, 9000 J of heat are absorbed by the water, while 1500 J are leaked to the environment due to poor insulation. Find the increase in temperature of the water.
Ans. 0.896 C°

20.16. A gas is confined to a copper can. How much heat must be supplied to increase the internal energy by 59 J? What type of thermodynamic process is involved?

20.17. A gas confined by a piston expands almost isobarically at 100 kPa. When 20,000 J of heat is absorbed by the system, its volume increases from 0.100 m³ to 0.250 m³. What work is done, and what is the change in internal energy?
Ans. 15.0 kJ, 5 kJ

20.18. The specific heat of brass is 390 J/kg C°. A 4-kg piece of brass is heated isochorically, causing the temperature to rise by 10 C°. What is the increase in internal energy?

*20.19. Two liters of an ideal gas have a temperature of 300 K and a pressure of 2 atm. The gas undergoes an isobaric expansion while its temperature is increased to 500 K. What work is done by the gas?
Ans. 270 kJ

*20.20. The diameter of a piston is 6.00 cm, and the length of its stroke is 12 cm. Assume that a constant force of 340 N moves the piston for a full stroke. First, calculate the work based on force and distance. Then, verify your answer by considering pressure and volume.

*20.21. For adiabatic processes, it can be shown that the pressure and volume are related by
$$P_1 V_1^\gamma = P_2 V_2^\gamma$$
where γ is the adiabatic constant, which is 1.40 for diatomic gases and also for the gasoline vapor/air mixture in combustion engines. Use the ideal gas law to prove the companion relationship:
$$T_1 V_1^{\gamma-1} = T_2 V_2^{\gamma-1}$$

*20.22. The compression ratio for a certain diesel engine is 15. The air-fuel mixture ($\gamma = 1.4$) is taken in at 300 K and 1 atm of pressure. Find the pressure and temperature of the gas after adiabatic compression. (Refer to Prob. 20.21).

Section 20.9 The Second Law of Thermodynamics

20.23. What is the efficiency of an engine that does 300 J of work in each cycle while discarding 600 J to the environment?
Ans. 33.3 percent

20.24. During a complete cycle, a system absorbs 600 cal of heat and rejects 200 cal to the environment. How much work is done? What is the efficiency?

20.25. A 37 percent efficient engine loses 400 J of heat during each cycle. What work is done, and how much heat is absorbed in each cycle?
Ans. 235 J, 635 J

20.26. What is the efficiency of an ideal engine that operates between the temperatures of 525 K and 300 K?

20.27. A steam engine takes superheated steam from a boiler at 200°C and rejects it directly into the air at 100°C. What is the ideal efficiency?
Ans. 21.1 percent

20.28. In a Carnot cycle, the isothermal expansion of a gas takes place at 400 K, and 500 cal of heat is absorbed by the gas. How much heat is lost if the system undergoes isothermal compression at 300 K? What is the heat loss, and what work is done?

20.29. A Carnot engine absorbs 1200 cal during each cycle as it operates between 500 K and 300 K. What is the efficiency? How much heat is rejected, and how much work in joules is done during each cycle?   Ans. 40 percent, 720 cal, 2010 J

20.30. The actual efficiency of an engine is 60 percent of its ideal efficiency. The engine operates between temperatures of 460 K and 290 K. How much work is done in each cycle if 1600 J of heat is absorbed?

20.31. A refrigerator extracts 400 J of heat from a box during each cycle and rejects 600 J to a high-temperature reservoir. What is the coefficient of performance?   Ans. 2.00

20.32. The coefficient of performance of a refrigerator is 5.0. How much heat is discarded if the compressor does 200 J of work during each cycle?

20.33. How much heat is extracted from the cold reservoir if the compressor of a refrigerator does 180 J of work during each cycle? The coefficient of performance is 4.0. What heat is rejected to the hot reservoir?   Ans. 720 J, 900 J

20.34. An ideal refrigerator extracts 400 J of heat from a reservoir at 200 K and rejects heat to a reservoir at 500 K. What is the ideal coefficient of performance, and how much work is done in each cycle?

*20.35. A Carnot refrigerator has a coefficient of performance of 2.33. If 600 J of work is done by the compressor in each cycle, how many joules of heat are extracted from the cold reservoir, and how much is ejected to the environment?   Ans. 1400 J, 2000 J

Additional Problems

20.36. In a thermodynamic process, 200 Btu is supplied to produce an isobaric expansion under a pressure of 100 lb/in.^{2}. The internal energy of the system does not change. What is the increase in volume of the gas?

20.37. A 100-cm^3 sample of gas at a pressure of 100 kPa is heated isochorically from point A to point B until its pressure reaches 300 kPa. Then it expands isobarically to point C, where its volume is 400 cm^3. The pressure then returns to 100 kPa at point D, with no change in volume. Finally, it returns to its original state at point A. Draw the P-V diagram for this cycle. What is the net work done for the entire cycle?   Ans. 60 J

20.38. Find the net work done by a gas as it is carried around the cycle shown in Fig. 20.18.

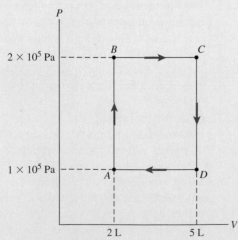

Figure 20.18

20.39. What is the net work done for the process $ABCA$ as described by Fig. 20.19?   Ans. 304 J

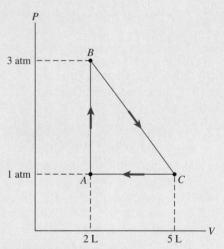

Figure 20.19

*20.40. A real engine operates between 327 and 0°C and has an output power of 8 kW. What is the ideal efficiency for this engine? How much power is wasted if the actual efficiency is only 25 percent?

*20.41. The Otto efficiency for a gasoline engine is 50 percent, and the adiabatic constant is 1.4. Compute the compression ratio.   Ans. 5.66

*20.42. A heat pump takes heat from a water reservoir at 41°F and delivers it to a system of pipes in a house at 78°F. The energy required to operate the heat pump is about twice that required to operate a Carnot pump. How much mechanical work must be supplied by the pump so that it will deliver 1 × 10^6 Btu of heat energy to the house?

20.43. A Carnot engine has an efficiency of 48 percent. If the working substance enters the system at 400°C, what is the exhaust temperature? Ans. 350 K

20.44. During the compression stroke of an automobile engine, the volume of the combustible mixture decreases from 18 to 2 in.^{3}. If the adiabatic constant is 1.4, what is the maximum possible efficiency for the engine?

20.45. How many joules of work must be done by the compressor in a refrigerator to change 1 kg of water at 20°C to ice at −10°C? The coefficient of performance is 3.5. Ans. 126 kJ

20.46. In a mechanical refrigerator, the low-temperature coils of the evaporator are at −30°C, and the condenser has a temperature of 60°C. What is the maximum possible coefficient of performance?

20.47. An engine has a thermal efficiency of 27 percent and an exhaust temperature of 230°C. What is the lowest possible input temperature? Ans. 416 °C

20.48. The coefficient of performance of a refrigerator is 5.0. If the temperature of the room is 28°C, what is the lowest possible temperature that can be obtained inside the refrigerator?

Critical Thinking Questions

20.49. A gas expands against a movable piston, lifting it through 2 in. at constant speed. How much work is done by the gas if the piston weighs 200 lb and has a cross-sectional area of 12 in.^{2}? If the expansion is adiabatic, what is the change in internal energy in Btu? Does ΔU represent an increase or decrease in internal energy?
Ans. 33.3 ft · lb, 0.0428 Btu, decrease

*20.50. Consider the *P-V* diagram shown in Fig. 20.20, where the pressure and volume are indicated for each of the points *A, B, C,* and *D*. Starting at point *A*, a 100 cm^3 sample of gas absorbs 200 J of heat, causing the pressure to increase from 100 to 200 kPa, while the volume increases to 200 cm^3. Next, the gas expands from *B* to *C*, absorbing an additional 400 J of heat while its volume increases to 400 cm^3. (a) Find the net work done and the change in internal energy for each of the processes *AB* and *BC*. (b) What are the net work and the total

change in internal energy for the process *ABC*? (c) What kind of process is illustrated by *BC*?

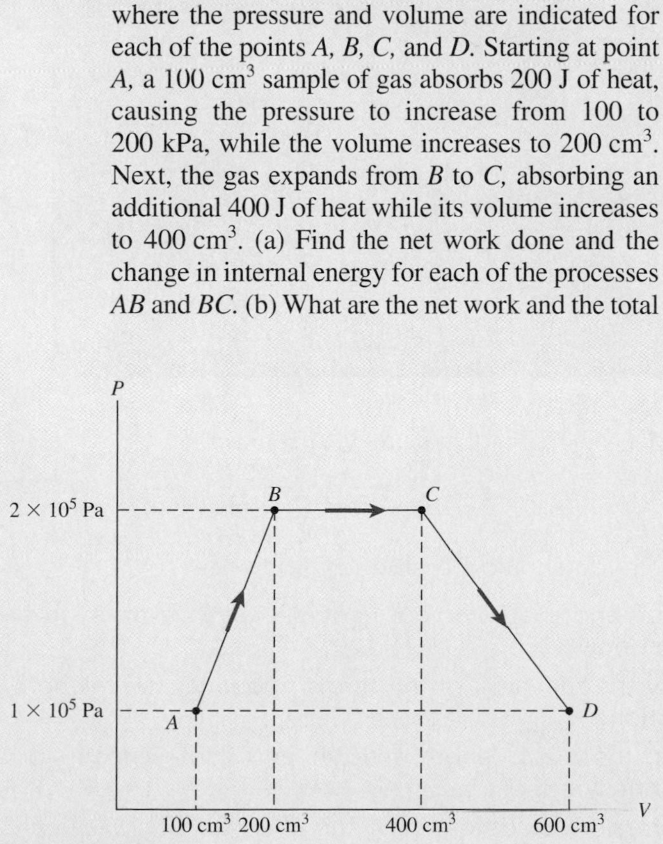

*20.51. The cycle begun in Prob. 20.50 now continues from *C* to *D*, while an additional 200 J of heat is absorbed. (a) Find the net work and the net change in internal energy for the process *CD*. (b) Suppose the system returns to its original state at point *A*. What is the net work for the entire cycle *ABCDA*, and what is the efficiency of the cycle?

*20.52. Consider a specific mass of gas that is forced through an adiabatic throttling process. Before entering the valve, it has internal energy U_1, pressure P_1, and volume V_1. After passing though the valve, it has internal energy U_2, pressure P_2, and volume V_2. The net work done is the work done *by* the gas minus the work done *on* the gas. Show that

$$U_1 + P_1 V_1 = U_2 + P_2 V_2$$

The quantity $U + PV$, called the *enthalpy*, is conserved during a throttling process.

*20.53. A gasoline engine takes in 2000 J of heat and delivers 400 J of work per cycle. The heat is obtained by burning gasoline, which has a heat of combustion of 50 kJ/g. What is the thermal efficiency? How much heat is lost per cycle? How much gasoline is burned in each cycle? If the engine goes through 90 cycles per second, what is the output power?
Ans. 20 percent, 1600 J, 0.040 g, 36 kW

*20.54. Consider a Carnot engine of efficiency *e* and a Carnot refrigerator whose coefficient of performance is *K*. If these devices operate between the same temperatures, derive the following relationship.

$$K = \frac{1 - e}{e}$$

Figure 20.20

21

Mechanical Waves

Mechanical waves formed by a drop of water into a ripple tank. Such waves are able to transmit energy from one place to another as a disturbance, with only localized movements of individual particles. In this chapter, we study the physical properties of mechanical waves. (*Photo © BS15 Photo Disc/Getty.*)

Objectives

After completing this chapter, you should be able to

1. Demonstrate by definition and example your understanding of *transverse* and *longitudinal wave motion.*
2. Define, relate, and apply the meaning of the terms *frequency, wavelength,* and *speed* for wave motion.
3. Solve problems involving the *mass, length, tension,* and *wave velocity* for transverse waves in a string.
4. Write and apply an expression for determining the *characteristic frequencies* for a vibrating string with fixed end points.

Energy can be transferred from one place to another by several means. In driving a nail, the bulk kinetic energy of a hammer is converted into useful work on the nail. Wind, projectiles, and most simple machines also perform work at the expense of material motion. Even the conduction of heat and electricity involves the motion of elementary particles called *electrons*. In this chapter, we study the transfer of energy from one point to another without the physical transfer of material between the points.

21.1 Mechanical Waves

When a stone is dropped into a pool of water, it creates a disturbance that spreads out in concentric circles, eventually reaching all parts of the pool. A small stick, floating on the surface of the water, bobs up and down as the disturbance passes. Energy has been transferred from the point of impact of the stone into the water to the floating stick some distance away. This energy is passed along by the agitation of neighboring water particles. Only the disturbance moves through the water. The actual motion of any particular water particle is comparatively small. Energy propagation by means of a disturbance in a medium instead of the motion of the medium itself is called *wave motion.*

This example is referred to as a *mechanical wave* because its existence depends on a mechanical source and a material medium.

A mechanical wave is a physical disturbance in an elastic medium.

It is important to recognize that all disturbances are not necessarily mechanical. For example, light waves, radio waves, and heat radiation propagate their energy by means of electric and magnetic disturbances. No physical medium is necessary for the transmission of electromagnetic waves. Many of the basic ideas presented in this chapter for mechanical waves are also applicable, however, to electromagnetic waves.

21.2 Types of Waves

Waves are classified according to the motion of a local part of a medium with respect to the direction of wave propagation. One type of wave is the *transverse wave.*

In a transverse wave, the vibration of the individual particles of the medium is perpendicular to the direction of wave propagation.

For example, suppose we fasten one end of a rope to a post and grasp the other end with the hand, as shown in Fig. 21.1. By moving the free end up and down quickly, we send a single

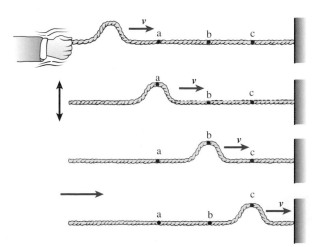

Figure 21.1 In a transverse wave, the individual particles move perpendicular to the direction of wave propagation.

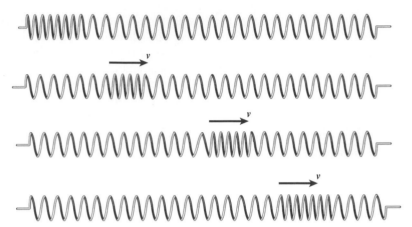

Figure 21.2   In a longitudinal wave, the motion of the individual particles is parallel to the direction of wave propagation. The illustration demonstrates the motion of a condensation pulse.

disturbance called a ***pulse*** down the length of rope. Three equally spaced knots at points *a*, *b*, and *c* demonstrate that the individual particles move up and down while the disturbance moves to the right with velocity **v.**

Another type of wave, which may occur in a coiled spring, is illustrated in Fig. 21.2. The coils near the left end are pinched closely together to form a ***condensation.*** When the distorting force is removed, a condensation pulse is propagated throughout the length of the spring. No part of the spring moves very far from its equilibrium position, but the pulse continues to travel along the spring. Such a wave is called a ***longitudinal wave*** because the spring particles are displaced along the same direction in which the disturbance is traveling.

In a longitudinal wave, the vibration of the individual particles is parallel to the direction of wave propagation.

If the coils of the spring in our example were forced apart at the left, a ***rarefaction*** would be formed, as shown in Fig. 21.3. Upon removal of the disturbing force, a longitudinal rarefaction pulse would be propagated along the spring. In general, a longitudinal wave consists of a series of condensations and rarefactions moving in a determined direction.

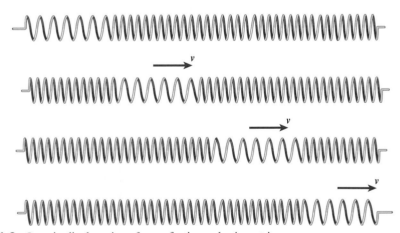

Figure 21.3   Longitudinal motion of a rarefaction pulse in a string.

21.3   Calculating Wave Speed

The speed with which a pulse moves through a medium depends on the elasticity of the medium and the inertia of its particles. The more elastic materials yield greater restoring forces when distorted. The less dense materials offer less resistance to motion. In either

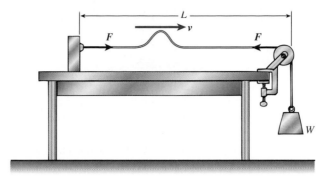

Figure 21.4 Computing the speed of a transverse pulse in a string.

case, the ability of particles to pass on a disturbance to neighboring particles is improved, and a pulse will travel at a greater speed.

Let us consider the motion of a transverse pulse down the string in Fig. 21.4. The string of mass m and length L is maintained under a constant tension F by the suspended weight. When the string is plucked near the left end, a transverse pulse is propagated along the string. The elasticity of the string is measured by the tension F in the string. The inertia of the individual particles is determined by the *mass per unit length* μ of the string. It can be shown that the **wave speed** of a transverse pulse in a string is given by

$$v = \sqrt{\frac{F}{\mu}} = \sqrt{\frac{F}{m/L}} \qquad (21.1)$$

The mass per unit length μ is usually referred to as the **linear density** of the string. If F is expressed in newtons and μ in kilograms per meter, the speed is in meters per second.

Example 21.1

The length L of the string in Fig. 21.4 is 2 m, and the string has a mass of 0.3 g. Find the speed of a transverse pulse in the string if it is under a tension of 20 N.

Plan: First, we will determine the linear density of the string and then calculate the speed from Eq. 21.1. Remember that the SI unit for mass is the kilogram.

Solution:
$$\mu = \frac{m}{L} = \frac{0.3 \times 10^{-3} \text{ kg}}{2 \text{ m}}$$
$$\mu = 1.5 \times 10^{-4} \text{ kg/m}$$

Direct substitution into Eq. (21.1) gives

$$v = \sqrt{\frac{F}{\mu}} = \sqrt{\frac{20 \text{ N}}{1.5 \times 10^{-4} \text{ kg/m}}}$$
$$v = 365 \text{ m/s}$$

Calculation of the speed of a longitudinal pulse will be reserved for Chapter 22, in connection with sound waves.

21.4 **Periodic Wave Motion**

So far, we have been considering single, nonrepeated disturbances called *pulses*. What happens if similar disturbances are repeated periodically? Suppose we attach the left end of a string to the end point of an electromagnetic vibrator, as shown in Fig. 21.5. The end of the

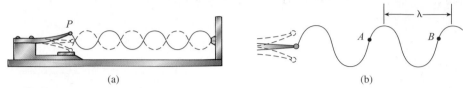

(a)   (b)

Figure 21.5   (a) Production and propagation of a periodic transverse wave. (b) The wavelength λ is the distance between any two particles in phase, such as those at adjacent crests or at points A and B.

metal vibrator is driven with harmonic motion by an oscillating magnetic field. Since the string is attached to the end of the vibrator, a series of periodic transverse pulses is sent down the string. The resulting waves consist of many crests and troughs, which move down the string at a constant rate of speed. The distance between any two adjacent crests or troughs in such a wave train is called the **wavelength**, denoted by λ.

As a wave travels along the string, each particle of the string vibrates about its equilibrium position with the same frequency and amplitude as the vibrating source. The particles of the string are not in corresponding positions, however, at the same times. Two particles are said to be **in phase** if they have the same displacement and if they are moving in the same direction. The particles A and B of Fig. 21.5b are in phase. Since particles at the crests of a given wave train are also in phase, we can provide a more general definition for the wavelength.

> The wavelength λ of a periodic wave train is the distance between any two adjacent particles that are in phase.

Each time the end point P of the vibrator makes a complete oscillation, the wave will move through a distance of one wavelength. The time required to cover this distance is therefore equal to the period T of the vibrating source. Hence, the wave speed v can be related to the wavelength λ and period T by the equation

$$v = \frac{\lambda}{T} \qquad \textbf{(21.2)}$$

The **frequency** f of a wave is the number of waves that pass a particular point in a unit of time. It is the same as the frequency of the vibrating source and is therefore equal to the reciprocal of the period ($f = 1/T$). The units of frequency may be expressed in waves per second, oscillations per second, or cycles per second. The SI unit for frequency is the **hertz** (Hz), which is defined as a cycle per second.

$$1 \text{ Hz} = 1 \text{ cycle/s} = \frac{1}{s}$$

Thus, if 40 waves pass a point every second, the frequency is 40 Hz.

The speed of a wave is more often expressed in terms of its frequency rather than its period. Thus, Eq. (21.2) can be written

$$v = f\lambda \qquad \textbf{(21.3)}$$

Equation (21.3) represents an important physical relationship among the speed, frequency, and wavelength of *any* periodic wave. An illustration of each quantity is given in Fig. 21.6 for a periodic transverse wave.

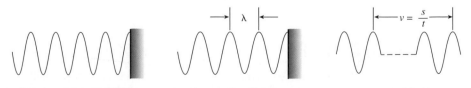

f = waves per second (Hz) λ = wavelength (m) v = speed (m/s)

Figure 21.6   The relationships among the frequency, wavelength, and speed of a transverse wave.

Example 21.2

A man sits near the end of a fishing dock and counts the water waves as they strike a supporting post. In 1 minute, he counts 80 waves. If a particular crest travels 12 m in 8 s, what is the wavelength of the waves?

Plan: We must not confuse frequency, which is waves per second, with velocity, which is the distance a given crest travels per unit of time.

Solution: The frequency and velocity are found from their definitions.

$$f = \frac{80 \text{ waves}}{60 \text{ s}} = 1.33 \text{ Hz}$$

$$v = \frac{x}{t} = \frac{12 \text{ m}}{8 \text{ s}}; \qquad v = 1.50 \text{ m/s}$$

Now, from Eq. (21.3), the wavelength must be

$$\lambda = \frac{v}{f} = \frac{1.50 \text{ m/s}}{1.33 \text{ Hz}}; \qquad \lambda = 1.13 \text{ m}$$

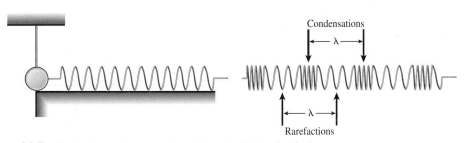

Figure 21.7 Production and propagation of a periodic longitudinal wave.

A longitudinal periodic wave can be generated by the apparatus shown in Fig. 21.7. The left end of a coiled spring is connected to a metal ball supported at the end of a clamped hacksaw blade. When the metal ball is displaced to the left and released, it vibrates with harmonic motion. The resulting condensations and rarefactions are passed along the spring, producing a periodic longitudinal wave. Each particle of the coiled spring oscillates back and forth horizontally with the same frequency and amplitude as the metal ball. The distance between any two adjacent particles that are in phase is the wavelength. As indicated in Fig. 21.7, the distance between adjacent condensations or adjacent rarefactions is a convenient measure of the wavelength. Equation (21.3) also applies for a periodic longitudinal wave.

21.5   Energy of a Periodic Wave

We have seen that each particle in a periodic wave oscillates with simple harmonic motion determined by the source of the wave. The energy content of a wave can be analyzed by considering the harmonic motion of the individual particles. For example, consider a periodic transverse wave in a string at the instant shown in Fig. 21.8. Particle *a* has reached its maximum *amplitude;* its velocity is zero, and it is experiencing its maximum restoring force. Particle *b* is passing through its equilibrium position, where the restoring force is zero. At this instance, particle *b* has its greatest speed and hence its maximum kinetic energy. Particle *c* is at its maximum displacement in the negative direction. As the periodic

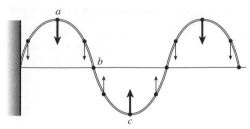

Figure 21.8   The restoring forces that act on particles of a vibrating string.

wave passes along the string, each particle oscillates back and forth through its own equilibrium position.

In Chapter 14 on harmonic motion, we found that the maximum velocity of a particle oscillating with frequency f and amplitude A is given by

$$v_{max} = 2\pi f A$$

When a particle has this speed, it is passing through its equilibrium position, where its potential energy is zero and its kinetic energy is a maximum. Thus, the total energy of the particle is

$$E = U + K = K_{max}$$
$$= \frac{1}{2} m v_{max}^2 = \frac{1}{2} m (2\pi f A)^2$$
$$= 2\pi^2 f^2 A^2 m \tag{21.4}$$

As a periodic wave passes through a medium, each element of the medium is continuously doing work on adjacent elements. Therefore, the energy transmitted along the length of a vibrating string is not confined to one position. Let us apply the result obtained for a single particle to the entire length of a vibrating string. The energy content of the entire string is the sum of the individual energies of its constituent particles. If we let m refer to the entire mass of the string instead of the mass of an individual particle, Eq. (21.4) represents the total wave energy in the string. In a string of length L, the wave energy per unit length is given by

$$\frac{E}{L} = 2\pi^2 f^2 A^2 \frac{m}{L}$$

Substituting μ for the mass per unit length, we can write

$$\frac{E}{L} = 2\pi^2 f^2 A^2 \mu \tag{21.5}$$

The wave energy is proportional to the square of the frequency f, to the square of the amplitude A, and to the linear density μ of the string. It must be recognized that the linear density is not a function of the length of string. This is true because the mass increases in proportion to the length L, so that μ is constant for any length.

Suppose that a wave travels down a length L of a given string with a speed v. The time t required for the wave to travel this length is

$$t = \frac{L}{v}$$

If the energy in this length of string is represented by E, the power P of the wave is given by

$$P = \frac{E}{t} = \frac{E}{L/v} = \frac{E}{L} v \tag{21.6}$$

This represents the *rate* at which energy is propagated down the string. Substitution from Eq. (21.5) yields

$$P = 2\pi^2 f^2 A^2 \mu v \qquad (21.7)$$

The wave power is directly proportional to the energy per unit length and to the wave speed.

The dependence of wave energy and wave power on f^2 and A^2, as found in Eqs. (21.5) and (21.7), is a general conclusion for all kinds of waves. The same ideas will be applied in Chapter 22 when the energy of a sound wave is discussed.

21.6 The Superposition Principle

Until now, we have been considering the motion of a single train of pulses passing through a medium. We now consider what happens when two or more wave trains pass simultaneously through the same medium. Let us consider transverse waves in a vibrating string. The speed of a transverse wave is determined by the tension of the string and its linear density. Since these parameters are a function of the medium and not the source, any transverse wave will have the same speed for a given string under constant tension. However, the frequency and amplitude may vary considerably.

> When two or more wave trains exist simultaneously in the same medium, each wave travels through the medium as though the other were not present.

The resultant wave is a superposition of the component waves. In other words, the resultant displacement of a particular particle on the vibrating string is the algebraic sum of the displacements each wave would produce independently of the other. This is the ***superposition principle:***

> When two or more waves exist simultaneously in the same medium, the resultant displacement at any point and time is the algebraic sum of the displacements of each wave.

It should be noted that the superposition principle, as stated here, applies only for *linear* media—those for which the response is directly proportional to the cause. Also, the sum of displacements is *algebraic* only if the waves have the same plane of polarization. For our purposes, a vibrating string will be assumed to satisfy both these conditions.

The application of this principle is shown graphically in Fig. 21.9. Two waves, indicated by the solid and dashed lines, superpose to form the resultant wave indicated by the heavy line. In Fig. 21.9a, the superposition results in a wave of larger amplitude, referred to as ***constructive interference. Destructive interference*** occurs when the resulting amplitude is smaller, as in Fig. 21.9b.

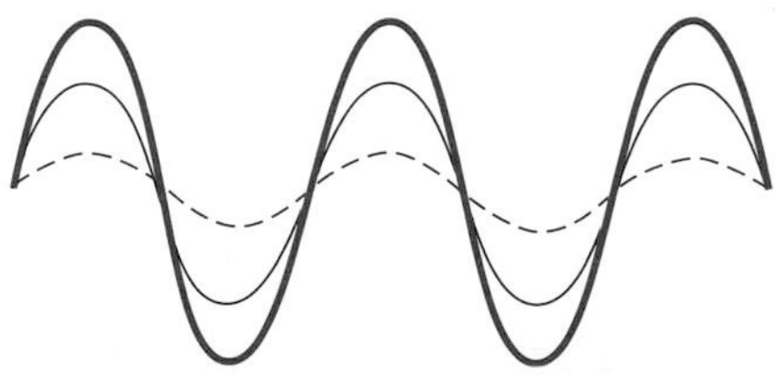

(a) Constructive interference 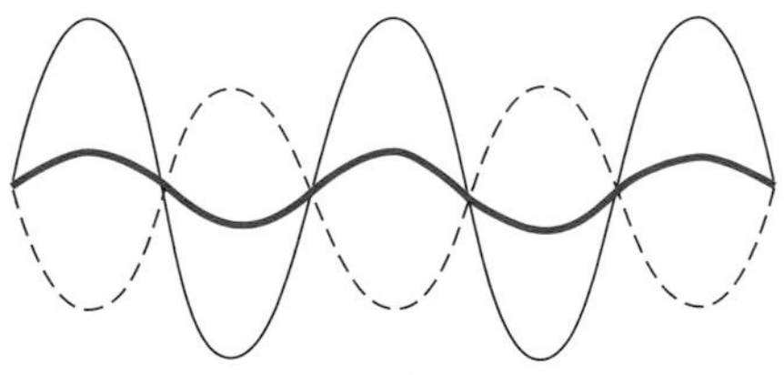   (b) Destructive interference

Figure 21.9   The superposition principle.

21.7 Standing Waves

Let us consider the reflection of a transverse pulse, as shown for the string in Fig. 21.10. When the end of the string is fixed rigidly to a support, the arriving pulse strikes the support, exerting an upward force on it. The reaction force exerted by the support kicks downward on the string, setting up a reflected pulse. Both the displacement and the velocity are reversed in the reflected pulse. In other words, a pulse that arrives as a crest is reflected as a trough with the same speed traveling in the opposite direction, and vice versa.

Suppose we consider the wave set up by a vibrating string whose end points are fixed, as in Fig. 21.11. We can use the superposition principle to analyze the resultant waveform at any instant. In Fig. 21.11a, we consider the incident and reflected waves at a particular time $t = 0$. The incident wave, traveling to the right, is indicated by a light solid line. The reflected wave, traveling to the left, is indicated by a dotted line. The two waves have the same speed and wavelength, but they are oppositely directed. At this instant, all particles of the string lie in a straight horizontal line, as shown by the heavy line. Note that the heavy line is a superposition of the incident and reflected waves at a moment when their displacements add to zero. A snapshot of the string an instant later will show that, with a few exceptions, the particles have all changed positions. This is because the component waves have moved a finite distance.

Let us now consider the resultant wave at time t equal to one-fourth of a period later ($t = \frac{1}{4}T$), as in Fig. 21.11b. The component wave indicated by a solid line will have moved to the *right* a distance of one-fourth wavelength. The component wave indicated by a dashed line will have moved to the *left* a distance of one-fourth wavelength. The resultant wave and hence the shape of the string at this time are indicated on the heavy solid line. Constructive interference has resulted in a wave with twice the amplitude of either of the component waves. When the time t is one-half a period ($t = \frac{1}{2}T$), total destructive interference occurs, and once again the shape of the string is a straight line, as in Fig. 21.11c. At $t = \frac{3}{4}T$, the shape of the string reaches its maximum amplitude in the opposite direction. This constructive interference is shown by the heavy line in Fig. 21.11d.

A series of snapshots of the vibrating string at closely spaced time intervals would reveal a number of loops, as shown in Fig. 21.11e. Such a wave is called a *standing wave.*

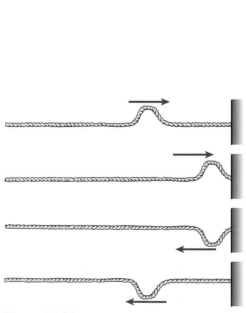

Figure 21.10 The reflection of a transverse pulse at a fixed boundary.

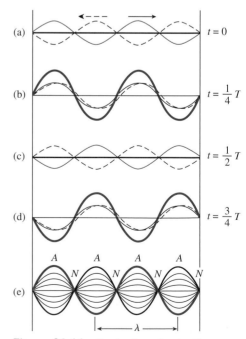

Figure 21.11 Production of a standing wave.

Notice that there are certain points along the string that remain at rest. These positions, called *nodes,* are labeled N in the figure. A flea resting on the vibrating string at these points would not be moved up and down by the wave motion.

Between the nodal points, the particles of the string move up and down with simple harmonic motion. The points of maximum amplitude occur midway between the nodes and are called *antinodes.* A flea resting on the string at any of these points, labeled A, would experience maximum speeds and displacements in the upward and downward oscillation of the string.

The distance between alternate nodes or alternate antinodes in a standing wave is a measure of the wavelength of the component waves.

Longitudinal standing waves may also occur by the continuous reflection of condensation and rarefaction pulses. In this case, the nodes exist where the particles of the medium are stationary, and the antinodes occur where the particles of the medium oscillate with maximum amplitude in the direction of propagation. Longitudinal standing waves will be discussed in Chapter 22 in connection with sound waves.

21.8 Characteristic Frequencies

Let us now consider the possible standing waves that can be set up in a string of length L whose ends are fixed, as in Fig. 21.12. When the string is set into vibration, the incident and reflected wave trains travel in opposite directions with the same wavelength. The fixed end points represent *boundary conditions* that restrict the possible wavelengths that will produce standing waves. These end points must be displacement nodes for any resulting wave patterns.

The simplest possible standing wave occurs when the wavelengths of incident and reflected waves are equal to twice the length of the string. The standing wave consists of a single loop with nodal points at each end, as shown in Fig. 21.12a. This pattern of vibration is referred to as the *fundamental mode of oscillation.* The higher modes of oscillation occur for shorter and shorter wavelengths. From the figure, it is noted that the allowable wavelengths are

$$\frac{2L}{1}, \frac{2L}{2}, \frac{2L}{3}, \frac{2L}{4}, \ldots$$

or, in equation form,

$$\lambda_n = \frac{2L}{n} \qquad n = 1, 2, 3, \ldots \qquad \textbf{(21.8)}$$

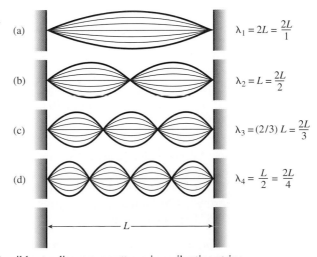

Figure 21.12   Possible standing-wave patterns in a vibrating string.

The corresponding frequencies of vibration are, from $v = f\lambda$,

$$f_n = \frac{nv}{2L} = n\frac{v}{2L} \qquad n = 1, 2, 3, \ldots \qquad \textbf{(21.9)}$$

where v is the speed of the transverse waves. This speed is the same for all wavelengths because it depends only on the characteristics of the vibrating medium. The frequencies given by Eq. (21.9) are called the ***characteristic frequencies of vibration.*** In terms of string tension F and linear density μ, the characteristic frequencies are

$$f_n = \frac{n}{2L}\sqrt{\frac{F}{\mu}} \qquad n = 1, 2, 3, \ldots \qquad \textbf{(21.10)}$$

The lowest possible frequency ($v/2L$) is called the ***fundamental frequency*** f_1. The others, which are integral multiples of the fundamental, are known as the ***overtones.*** The entire series,

$$f_n = nf_1 \qquad n = 1, 2, 3, \ldots \qquad \textbf{(21.11)}$$

consisting of the fundamental and its overtones, is known as the ***harmonic series.*** The fundamental is the first harmonic, the first overtone ($f_2 = 2f_1$) is the second harmonic, the second overtone ($f_3 = 3f_1$) is the third harmonic, and so on.

Example 21.3

A steel piano wire 50 cm long has a mass of 3.05 g and is under a tension of 400 N. What are the frequencies of its fundamental mode of vibration, and what are the first two overtones?

Plan: First, we will calculate the linear density of the string. Again, we must express *length* in meters and *mass* in kilograms. Recall that the fundamental vibration occurs when there is a single loop and $n = 1$ in Eq. (21.10). The first overtone is the second harmonic ($n = 2$), and the second overtone is the third harmonic ($n = 3$).

Solution: The linear density is

$$\mu = \frac{m}{L} = \frac{3.05 \times 10^{-3}\,\text{kg}}{0.500\,\text{m}}; \qquad \mu = 6.10 \times 10^{-3}\,\text{kg/m}$$

The fundamental is found by substituting $n = 1$ into Eq. (21.10).

$$f_1 = \frac{(1)}{2L}\sqrt{\frac{F}{\mu}} = \frac{(1)}{2(0.5\,\text{m})}\sqrt{\frac{400\,\text{N}}{6.10 \times 10^{-3}\,\text{kg/m}}}$$

$$= 265\,\text{Hz}$$

The first and second overtones are

$$f_2 = 2f_1 = 2(256\,\text{Hz}); \qquad f_2 = 512\,\text{Hz}$$
$$f_3 = 3f_1 = 3(256\,\text{Hz}); \qquad f_3 = 768\,\text{Hz}$$

We will see in Chapter 22 that as a string vibrates in one or more of its possible modes, energy is transmitted to the surrounding air in the form of sound waves. These longitudinal waves consist of condensations and rarefactions of the same frequency as the vibrating strings. The human ear interprets these waves as sound. The fundamental frequency of 256 Hz is interpreted as *middle C* on the piano.

Summary and Review

Summary

We have investigated mechanical wave motion in which energy is transferred by a physical disturbance in an elastic medium. The fundamental laws developed in this chapter are important because they also apply for many other types of waves that will be studied later. The essential concepts are summarized as follows.

- The velocity of a transverse wave in a string of mass m and length L is given by

$$v = \sqrt{\frac{F}{\mu}} \qquad \mu = \frac{m}{L} \qquad v = \sqrt{\frac{FL}{m}} \qquad \text{Wave Speed}$$

	Force F	Mass m	Length L	Speed v
SI units	N	kg	m	m/s
USCS units	lb	slug	ft	ft/s

- For any wave of period T or frequency f, the speed v can be expressed in terms of the wavelength λ as follows:

$$v = \frac{\lambda}{T} \qquad v = f\lambda \qquad \text{Frequency is in Hz} = 1/s$$

- The *energy per unit length* and the *power* of wave propagation can be found from

$$\frac{E}{L} = 2\pi^2 f^2 A^2 \mu \qquad P = 2\pi^2 f^2 A^2 \mu v$$

- The characteristic frequencies for the possible modes of vibration in a stretched string are found from

$$f_n = \frac{n}{2L} \sqrt{\frac{F}{\mu}} \qquad n = 1, 2, 3, \ldots$$

Characteristic Frequencies

- The series $f_n = nf_1$ is called the *harmonics*. They are integral multiples of the fundamental f_1. These are mathematical values, and all harmonics may not exist. The actual possibilities beyond the fundamental are called *overtones*. Since all harmonics are possible for the vibrating string, the first overtone is the second harmonic, the second overtone is the third harmonic, and so on.

Key Terms

amplitude 431
antinode 435
characteristic frequencies of vibration 436
condensation 428
constructive interference 433
destructive interference 433
frequency 430
fundamental frequency 436

fundamental mode of oscillation 435
harmonic series 436
hertz 430
in phase 430
linear density 429
longitudinal wave 428
mechanical wave 427
node 435

overtone 436
pulse 428
rarefaction 428
standing wave 434
superposition principle 433
transverse wave 427
wavelength 430
wave motion 427
wave speed 429

Review Questions

21.1. Explain how a water wave is both transverse and longitudinal.

21.2. Describe an experiment to demonstrate that energy is associated with wave motion.

21.3. In a *torsional wave,* the individual particles of the medium vibrate with angular harmonic motion about the axis of propagation. Give a mechanical example of such a wave.

21.4. Discuss the interference of waves. Is there a loss of energy when waves interfere? Explain.

21.5. A transverse pulse is sent down a string of mass m and length L under a tension F. How will the speed of the pulse be affected if (a) the mass of the string is quadrupled, (b) the length of the string is quadrupled, and (c) the tension is reduced by one-fourth?

21.6. Draw graphs of a periodic transverse wave and a periodic longitudinal wave. Indicate on the figures the wavelength and amplitude of each wave.

21.7. Which harmonic is indicated by Fig. 21.12d? Which overtone is present?

21.8. We have seen that boundary conditions determine possible modes of vibration. Draw a diagram of the fundamental and of the first two overtones for a vibrating rod (a) clamped at one end and (b) clamped at its midpoint.

21.9. A vibrating string has a fundamental frequency of 200 Hz. If the length is reduced by one-fourth, what will the new fundamental frequency be? Has the wave speed been altered by shortening the string? Assume constant tension.

21.10. Show graphically the superposition of two waves traveling in the same direction. The second wave has three-fourths the amplitude and one-half the wavelength of the first wave.

21.11. In an experiment with the vibrating string, one end of the string is attached to the tip of a vibrator, and the other end passes over a pulley. Suspended weights are used to produce the fundamental and the first three overtones. What effect will stretching the string have on frequency calculations?

Problems

Section 21.1 Mechanical Waves, Section 21.2 Types of Waves, Section 21.3 Calculating Wave Speed, and Section 21.4 Periodic Wave Motion

21.1. A transverse wave has a wavelength of 30 cm and vibrates with a frequency of 420 Hz. What is the speed of this wave? **Ans. 126 m/s**

21.2. A person on a pier counts the slaps of a wave as the crests hit a post. If 80 slaps are heard in 1 min and a particular crest travels a distance of 8 m in 4 s, what is the length of a single wave?

21.3. A transverse wave is pictured in Fig. 21.13. Find the amplitude, wavelength, period, and speed of the wave if it has a frequency of 12 Hz.
Ans. 12 cm, 28 cm, 83.3 ms, 3.36 m/s

21.4. For the longitudinal wave in Fig. 21.13, find the amplitude, wavelength, period, and speed of the wave if it has a frequency of 8 Hz. If the amplitude were doubled, would any of the other factors change?

21.5. A 500-g metal wire has a length of 50 cm and is under a tension of 80 N. What is the speed of a transverse wave in the wire? **Ans. 8.94 m/s**

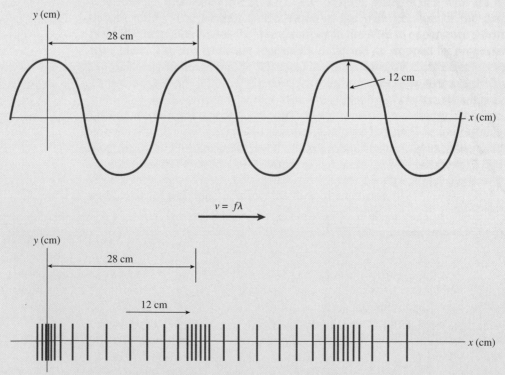

Figure 21.13 The wavelength, velocity, and amplitude for a transverse wave and a longitudinal wave ($A = 12$ cm, and $\lambda = 28$ cm).

21.6. If the wire in Prob. 21.5 is cut in half, what will be its new mass? Show that the speed of the wave is unchanged. Why?

21.7. A 3-m cord under a tension of 200 N sustains a transverse wave speed of 172 m/s. What is the mass of the rope? *Ans. 20.3 g*

21.8. A 200-g cord is stretched over a distance of 5.2 m and placed under a tension of 500 N. Compute the speed of a transverse wave in the cord.

21.9. What tension is required to produce a wave speed of 12.0 m/s in a 900-g string that is 2 m long? *Ans. 64.8 N*

21.10. A wooden float at the end of a fishing line makes eight complete oscillations in 10 s. If it takes 3.60 s for a single wave to travel 11 m, what is the wavelength of the water waves?

21.11. What frequency is required to cause a rope to vibrate with a wavelength of 20 cm when it is under a tension of 200 N? Assume the linear density of the rope to be 0.008 kg/m. *Ans. 791 Hz*

21.12. A tension of 400 N causes a 300-g wire of length 1.6 m to vibrate with a frequency of 40 Hz. What is the wavelength of the transverse waves?

21.13. A horizontal spring is jiggled back and forth at one end by a device that makes 80 oscillations in 12 s. What is the speed of the longitudinal waves if condensations are separated by 15 cm as the wave progresses down the spring? *Ans. 1.00 m/s*

Section 21.5 Energy of a Periodic Wave

21.14. A 2-m length of string has a mass of 300 g and vibrates with a frequency of 2 Hz and an amplitude of 50 mm. If the tension in the rope is 48 N, how much power must be delivered to the string?

21.15. An 80-g string has a length of 40 m and vibrates with a frequency of 8 Hz and an amplitude of 4 cm. Find the energy per unit of length passing along the string. *Ans. 4.04×10^{-3} J/m*

21.16. If the wavelength of the transverse wave in Prob. 21.11 is 1.6 m, what power is supplied by the source?

*21.17. A 300-g string has a length of 2.50 m and vibrates with an amplitude of 8.00 mm. The tension in the string is 46 N. What must be the frequency of the waves in order that the average power be 90.0 W? *Ans. 174 Hz*

Section 21.7 Standing Waves and Section 21.8 Characteristic Frequencies

21.18. A string vibrates with a fundamental frequency of 200 Hz. What is the frequency of the second harmonic and of the third overtone?

21.19. If the fundamental frequency of a wave is 330 Hz, what is the frequency of the fifth harmonic and the second overtone? *Ans. 1650 Hz, 990 Hz*

21.20. The linear density of a string is 0.00086 kg/m. What should be the tension in the rope in order for a 2 m length of this string to vibrate at 600 Hz for its third harmonic?

21.21. A 10-g string, 4 m in length, has a tension of 64 N. What is the frequency of its fundamental mode of vibration? What are the frequencies of the first and second overtones? *Ans. 20, 40, and 60 Hz*

21.22. The second harmonic of a vibrating string is 200 Hz. The length of the string is 3 m, and its tension is 200 N. Compute the linear density of the string.

21.23. A 0.500-g string is 4.3 m long and has a tension of 300 N. If it is fixed at each end and vibrates in three segments, what is the frequency of the standing waves? *Ans. 560 Hz*

21.24. A string vibrates with standing waves in five loops when the frequency is 600 Hz. What frequency will cause the string to vibrate in only two loops?

21.25. A 120-g wire fixed at both ends is 8 m long and has a tension of 100 N. What is the longest possible wavelength for a standing wave? What is the frequency? *Ans. 16 m, 5.10 Hz*

Additional Problems

21.26. The E_5-string on the violin in Fig. 21.14 is to be tuned to a frequency of 660 Hz. From bridge to peg, the length of the string is 33 cm, and its mass is 0.125 g. What must be the tension in the string?

21.27. What is the speed of a transverse wave in a rope of length 2.00 m and mass 80 g under a tension of 400 N? *Ans. 100 m/s*

21.28. A transverse wave travels at a speed of 8.00 m/s. A particular particle on the string moves from its highest point to its lowest point in a time of 0.03 s. What is the wavelength?

21.29. A bass guitar string 750 mm long is stretched with sufficient force to produce a fundamental vibration of 220 Hz. What is the velocity of the transverse waves in this string? *Ans. 330 m/s*

21.30. A 5-kg mass is hung from the ceiling by a 30-g wire that is 1.8 m long. What is the fundamental frequency of vibration for this wire?

Figure 21.14 The length, mass, and tension of a violin string determine the observed frequency. (*Photo by Paul E. Tippens.*)

21.31. A steel guy wire supporting a pole is 18.9 m long and 9.5 mm in diameter. It has a linear density of 0.474 kg/m. When it is struck at one end by a hammer, the pulse returns in 0.3 s. What is the tension in the wire?   Ans. 7530 N

*21.32. A 30-m wire weighing 400 N is stretched with a tension of 1800 N. How much time is required for a pulse to make a round trip if it is struck at one end?

*21.33. Transverse waves have a speed of 20 m/s on a string that has a tension of 8 N. What tension is required to give a wave speed of 30 m/s for the same string?   Ans. 18.0 N

*21.34. The fundamental frequency for a given string is 80 Hz. If the mass of the string is doubled but other factors remain constant, what is the new fundamental frequency?

Critical Thinking Questions

21.35. In a laboratory experiment, an electromagnetic vibrator is used as a source of standing waves in a string. A 1-m length of the string is determined to have a mass of 0.6 g. One end of the string is connected to the tip of the vibrator, and the other passes over a pulley 1 m away and is attached to a weight hanger. A mass of 392 g hanging from the free end causes the string to vibrate in three segments. What is the frequency of the vibrator? What new mass attached to the free end will cause the string to vibrate in four loops? (See Fig. 21.15.) What is the fundamental frequency?
                Ans. 120 Hz, 22.0 g

Vibrator
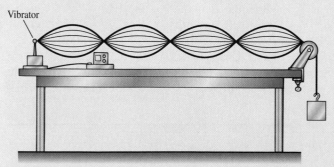

Figure 21.15

*21.36. To understand the parameters that affect wave velocity in a vibrating string, suppose that

$$v = \sqrt{\frac{FL}{m}} = 100 \text{ m/s}$$

What is the new wave speed v' for each of the following changes: (a) $F' = 2F$, (b) $m' = 2 m$, (c) $L' = 2L$?

*21.37. A 2-mW power source generates waves down rope A, and another power source generates waves down an identical rope B. The waves in each rope are of the same frequency f and velocity v. If the amplitude in rope B is twice that of rope A, what power is supplied to rope B?
                Ans. 8.00 mW

*21.38. The fundamental frequency of a steel piano wire is 253 Hz. By what fractional amount must the tension in the wire be increased in order that the frequency be the desired C note (256 Hz)?

*21.39. A variable oscillator allows a laboratory student to adjust the frequency of a source to produce standing waves in a vibrating string. A 1.20-m length of string ($\mu = 0.400$ g/m) is placed under a tension of 200 N. What frequency is necessary to produce three standing loops in the vibrating string? What is the fundamental frequency? What frequency will produce five loops?
                Ans. 884 Hz, 295 Hz, 1470 Hz

22 Sound

Longitudinal sound waves in organ pipes have provided beautiful deep-toned notes in churches for centuries. The length of the pipes and the boundary conditions determine the pitch of the notes; the resonant cavities and the conditions surrounding the pipes determine the quality of the sound.
(*Photo © vol. 32 PhotoDisc/Getty.*)

Objectives

After completing this chapter, you should be able to

1. Define sound and solve problems involving its velocity in metal, in a liquid, and in a gas.

2. Use boundary conditions to derive and apply relationships for calculating the *characteristic frequencies* for an open pipe and for a closed pipe.

3. Compute the intensity level in decibels for a sound whose intensity is given in *watts per square meter*.

4. Use your understanding of the *Doppler effect* to predict the apparent change in sound frequency that occurs as a result of relative motion between a source and an observer.

When a periodic disturbance takes place in air, longitudinal *sound waves* travel out from it. For example, if a tuning fork is struck with a hammer, the vibrating prongs send out longitudinal waves, as shown in Fig. 22.1. An ear, acting as a receiver for these periodic waves, interprets them as sound.

Is the ear necessary for sound to exist? If the tuning fork were struck in the atmosphere of a distant planet, would there be sound, even though no ear could interpret the disturbance? The answer depends on the definition of sound.

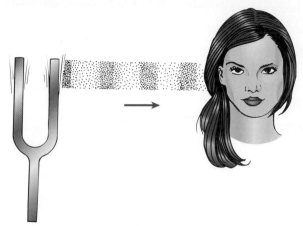

Figure 22.1   A tuning fork acts as a source of longitudinal sound waves.

The term *sound* is used in two different ways. Physiologists define *sound* in terms of the auditory sensations produced by longitudinal disturbances in air. For them, sound does not exist on a distant planet. In physics, on the other hand, we refer to the disturbances themselves rather than the sensations produced.

> Sound is a longitudinal mechanical wave that travels through an elastic medium.

In this case, sound does exist on the planet. In this chapter, *sound* will be used in its physical sense.

22.1   Production of a Sound Wave

Two things must exist to produce a sound wave. There must be a source of mechanical vibration, and there must be an elastic medium through which the disturbance can travel. The source may be a tuning fork, a vibrating string, or a vibrating air column in an organ pipe. *Sounds are produced by vibrating matter.* The requirement of an elastic medium can be demonstrated by placing an electric bell inside an evacuable flask, as shown in Fig. 22.2. When the bell is connected to a battery so it rings continuously, the flask is slowly evacuated. As more and more of the air is pumped from the flask, the sound of the bell becomes fainter and fainter until finally it cannot be heard at all. When

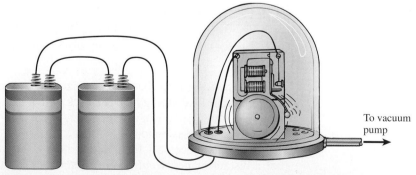

To vacuum pump

Figure 22.2   A bell ringing in a vacuum cannot be heard. A material medium is necessary for the production of sound.

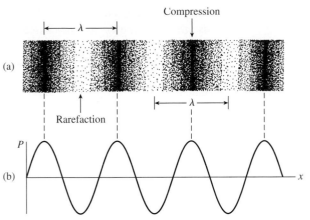

Figure 22.3   (a) Compressions and rarefactions in a sound wave in air at a particular instant. (b) The sinusoidal variation in pressure as a function of displacement.

air is allowed to reenter the flask, the sound of the bell returns. Thus, air is necessary to transmit sound.

Let us now examine more closely the longitudinal sound waves in air as they proceed from a vibrating source. A thin strip of metal clamped tight at its base is pulled to one side and released. As the free end oscillates to and fro with simple harmonic motion, a series of periodic, longitudinal sound waves spread through the air away from the source. The air molecules in the vicinity of the metal strip are alternately compressed and expanded, sending out a wave like that illustrated in Fig. 22.3a. The dense regions where many molecules are packed tightly together are called **compressions.** They are exactly analogous to the **condensations** discussed for longitudinal waves in a coiled spring. The regions with relatively few molecules are referred to as **rarefactions.** The compressions and rarefactions alternate throughout the medium as the individual air particles oscillate to and fro in the direction of wave propagation. Since a compression corresponds to a high-pressure region and a rarefaction corresponds to a low-pressure region, a sound wave can also be represented by plotting the change in pressure P as a function of the distance x. (See Fig. 22.3b.) The distance between two successive compressions or rarefactions is the wavelength.

22.2   The Speed of Sound

Anyone who has seen a weapon being fired at a distance has observed the smoke from the weapon before hearing the report. Similarly, we observe the flash of lightning before hearing the thunder. Even though both light and sound travel with finite speeds, the speed of light is so much greater in comparison that it can be considered instantaneous. The speed of sound can be measured directly by observing the time required for the waves to move through a known distance. In air at 0°C, sound travels at a speed of 331 m/s or 1087 ft/s.

In Chapter 21, we established the idea that wave speed depends on the elasticity of the medium and the inertia of its particles. The more elastic materials sustain greater wave speeds, whereas the denser materials retard wave motion. The following empirical relationships are based on these proportionalities.

For longitudinal sound waves in a wire or rod, the wave speed is given by

$$v = \sqrt{\frac{Y}{\rho}} \qquad\qquad Rod \quad (\mathbf{22.1})$$

where Y is Young's modulus for the solid and ρ is its density. This relation is valid only for rods whose diameters are small in comparison with the longitudinal wavelengths of sound passing through them.

In an *extended solid,* the longitudinal wave speed is a function of the shear modulus S, the bulk modulus B, and the density ρ of the medium. The wave speed can be calculated from

$$v = \sqrt{\frac{B + \frac{4}{3}S}{\rho}} \qquad \text{Extended Solid} \quad \textbf{(22.2)}$$

For longitudinal waves in a fluid, the wave speed is found from

$$v = \sqrt{\frac{B}{\rho}} \qquad \text{Fluid} \quad \textbf{(22.3)}$$

where B is the bulk modulus for the fluid and ρ is its density.

In computing the speed of sound in a gas, the bulk modulus is given by

$$B = \gamma P$$

where γ is the adiabatic constant ($\gamma = 1.4$ for air and diatomic gases) and P is the pressure of the gas. Thus, the speed of longitudinal waves in a gas, from Eq. (22.3), is given by

$$v = \sqrt{\frac{B}{\rho}} = \sqrt{\frac{\gamma P}{\rho}} \qquad \textbf{(22.4)}$$

But for an ideal gas,

$$\frac{P}{\rho} = \frac{RT}{M} \qquad \textbf{(22.5)}$$

where $R = 8.314$ J/mol $\cdot$ kg (universal gas constant)
$T = $ absolute temperature of gas
$M = $ molecular mass of gas

Substitution of Eq. (22.5) into Eq. (22.4) yields

$$v = \sqrt{\frac{\gamma P}{\rho}} = \sqrt{\frac{\gamma RT}{M}} \qquad \text{Gas} \quad \textbf{(22.6)}$$

PHYSICS TODAY

When objects move faster than the speed of sound, they create shock waves. Since the energy from a shock wave is located mostly at the front of the wave, it can create damage, particularly to buildings. For this reason, jets do not fly at supersonic speeds unless they are at high altitudes. Mach number $= v_{object}/v_{sound}$, where $v_{object} > v_{sound}$.

Example 22.1

Compute the speed of sound in an aluminum rod.

Solution: Young's modulus and the density for aluminum are

$$Y = 68{,}900 \text{ MPa} = 6.89 \times 10^{10} \text{ N/m}^2$$
$$\rho = 2.7 \text{ g/cm}^3 = 2.7 \times 10^3 \text{ kg/m}^3$$

From Eq. (22.1),

$$v = \sqrt{\frac{Y}{\rho}} = \sqrt{\frac{6.89 \times 10^{10} \text{ N/m}^2}{2.7 \times 10^3 \text{ kg/m}^3}}$$
$$= \sqrt{2.55 \times 10^7 \text{ m}^2/\text{s}^2} = 5050 \text{ m/s}$$

This speed is approximately 15 times the speed of sound in air.

Example 22.2

Compute the speed of sound in air on a day when the temperature is 27°C. The molecular mass of air is 29.0 g/mol, and the adiabatic constant is 1.4.

Plan: The given information along with the universal gas constant ($R = 8.31$ J/mol · K) can be used to find the speed of sound from Eq. (22.6). Since SI units are required, we must convert the molecular mass to *kilograms* per mole ($M = 29 \times 10^{-3}$ kg/mol).

Solution: The absolute temperature of the air is $T = 27° + 273° = 300$ K.

$$v = \sqrt{\frac{\gamma R T}{M}} = \sqrt{\frac{(1.4)(8.31 \text{ J/mol} \cdot \text{K})(300 \text{ K})}{29 \times 10^{-3} \text{ kg/mol}}}$$

$$v = 347 \text{ m/s}$$

The speed of sound is significantly greater at 27°C than at 0°C. At standard temperature and pressure (273 K, 1 atm), the speed of sound is 331 m/s. It can be seen from Eq. (22.6) that the velocity of sound in air varies directly with the square root of the absolute temperature. Hence, the speed of sound in air can also be approximated by

$$v = (331 \text{ m/s}) \sqrt{\frac{T}{273 \text{ K}}} \tag{22.7}$$

This relation assumes that γ and M do not change and that the velocity of sound is 331 m/s at a temperature of 273 K.

Example 22.3

What is the speed of sound in air at room temperature (20°C)?

Plan: We find the absolute temperature (20° + 273° = 293 K) and substitute directly into Eq. (22.7)

Solution:

$$v = (331 \text{ m/s}) \sqrt{\frac{T}{273 \text{ K}}} = (331 \text{ m/s}) \sqrt{\frac{293 \text{ K}}{273 \text{ K}}} \quad \text{or} \quad v = 343 \text{ m/s}$$

22.3 Vibrating Air Columns

In Chapter 21, we described the possible modes of vibration for a string fixed at both ends. The frequency of the sound waves set up in the air surrounding the string is identical with the frequency of the vibrating string. Thus, the possible frequencies, or the *harmonics,* of sound waves produced by a vibrating string are given by

$$f_n = \frac{nv}{2L} \qquad n = 1, 2, 3, \ldots \tag{22.8}$$

where v is the velocity of transverse waves in the string.

Sound can also be produced by the longitudinal vibrations of an air column in a pipe that is open at both ends, an *open pipe,* or one that is closed at one end, a *closed pipe.* As in the vibrating string, the possible modes of vibration are determined by the boundary conditions. The possible modes of vibration for the air in a closed pipe are illustrated in Fig. 22.4. When a compressional wave is set up in the pipe, the displacement of the air particles at the closed end must be zero.

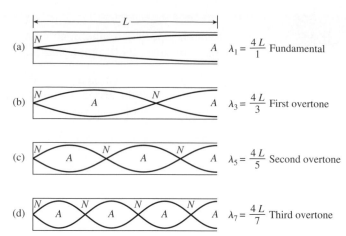

Figure 22.4 Possible standing waves in a closed pipe.

The closed end of a pipe must be a displacement node.

The air at the open end of a pipe has the greatest freedom of motion, so the displacement is free at an open end.

The open end of a pipe must be a displacement antinode.

The sinusoidal curves in Fig. 22.4 represent longitudinal displacements of the air molecules from their equilibrium positions.

The fundamental mode of oscillation for an air column in a closed pipe has a node at the closed end and an antinode at the open end. Thus, the wavelength of the fundamental is four times the length L of the pipe (Fig. 22.4a). The next possible mode, which is the first overtone, occurs when there are two nodes and two antinodes, as shown in Fig. 22.4b. The wavelength of the first overtone is therefore equal to $4L/3$. Similar reasoning will show that the second and third overtones occur for wavelengths equal to $4L/5$ and $4L/7$, respectively. In summary, the possible wavelengths are

$$\lambda_n = \frac{4L}{n} \qquad n = 1, 3, 5, \ldots \tag{22.9}$$

The speed of the sound waves is given by $v = f\lambda$, so the possible frequencies for a *closed pipe* are

$$f_n = \frac{nv}{4L} \qquad n = 1, 3, 5, \ldots \qquad \textit{Closed Pipe} \tag{22.10}$$

Notice that only the *odd harmonics* are allowed for a closed pipe. The first overtone is the third harmonic, the second overtone is the fifth harmonic, and so on.

An air column vibrating in a pipe open at *both* ends must be bounded by displacement antinodes. Figure 22.5 shows the fundamental and first three overtones for an open pipe. Note that the fundamental wavelength is twice the length L of the pipe. When the number of nodes is increased one at a time, the possible wavelengths in an open pipe are

$$\lambda_n = \frac{2L}{n} \qquad n = 1, 2, 3, \ldots \tag{22.11}$$

The possible frequencies are, therefore,

$$f_n = \frac{nv}{2L} \qquad n = 1, 2, 3, \ldots \qquad \textit{Open Pipe} \tag{22.12}$$

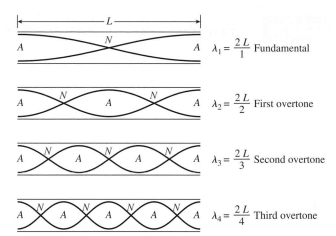

$\lambda_1 = \dfrac{2L}{1}$ Fundamental

$\lambda_2 = \dfrac{2L}{2}$ First overtone

$\lambda_3 = \dfrac{2L}{3}$ Second overtone

$\lambda_4 = \dfrac{2L}{4}$ Third overtone

Figure 22.5   Possible standing waves in an open pipe.

where v is the velocity of the sound waves. Thus, all the harmonics are possible for a vibrating air column in an open pipe. Open pipes of varying lengths are used in many musical instruments, for example, organs, flutes, and trumpets.

Example 22.4

What are the frequencies of the fundamental and first two overtones for a 12-cm closed pipe? Air temperature is 30°C.

Plan: Remember that for a *closed* pipe, only the *odd* harmonics are allowed. The fundamental will be the first harmonic ($n = 1$), the first overtone will occur for $n = 3$, and the second overtone will be for $n = 5$. To find these frequencies, we apply Eq. (22.10) for each case.

Solution: The speed of sound is found from Eq. (20.7), where $T = 30° + 273° = 303$ K.

$$v = (331 \text{ m/s})\sqrt{\dfrac{303 \text{ K}}{273 \text{ K}}} = 349 \text{ m/s}$$

The fundamental occurs for $n = 1$ in Eq. (22.10)

$$f_1 = \dfrac{nv}{4L} = \dfrac{349 \text{ m/s}}{4(0.12 \text{ m})}; \qquad f_1 = 727 \text{ Hz}$$

The first and second overtones occur for $n = 3$ and $n = 5$, respectively.

$$\text{First overtone} = 3f_1 = 2181 \text{ Hz}$$
$$\text{Second overtone} = 5f_1 = 3635 \text{ Hz}$$

Example 22.5

The second overtone for a brass pipe, open at both ends, is 1800 Hz when the speed of sound is 340 m/s. What is the length of the pipe?

Plan: The second overtone occurs for the *third* harmonic ($n = 3$). The length of the pipe will be found by solving for L when $n = 3$.

Solution: From Eq. (22.12) for an open pipe with $n = 3$, we have

$$f_3 = \dfrac{3v}{2L} \quad \text{and} \quad L = \dfrac{3v}{2f_3} = \dfrac{3(340 \text{ m/s})}{2(1800 \text{ Hz})}$$

$$L = 0.283 \text{ m} \quad \text{or} \quad 28.3 \text{ cm}$$

22.4   Forced Vibration and Resonance

When a vibrating body is placed in contact with another body, the second body is forced to vibrate with the same frequency as the original vibrator. For example, if a tuning fork is struck with a hammer and then placed with its base against a wooden tabletop, the intensity of the sound will suddenly be increased. When the tuning fork is removed from the table, the intensity decreases to its original level. The vibrations of the particles in the tabletop in contact with the tuning fork are called *forced vibrations.*

We have seen that elastic bodies have certain natural frequencies of vibration that are characteristic of the material and boundary conditions. A taut string of a particular length can produce sounds of characteristic frequencies. An open or closed pipe also has natural frequencies of vibration. Whenever a body is acted on by a series of periodic impulses having a frequency nearly equal to one of the natural frequencies of the body, the body is set into vibration with a relatively large amplitude. This phenomenon is referred to as *resonance* or *sympathetic vibration.*

An example of resonance is offered by a child sitting in a swing. Experience tells us that the swing can be set into vibration with large amplitude by a series of small pushes at just the right intervals. Such resonance occurs only when the pushes are in phase with the natural frequency of vibration for the swing. A slight variation of the input pulses would result in little or no vibration.

Reinforcement of sound by resonance has many useful applications as well as many unpleasant consequences. The resonance of an air column in an organ pipe amplifies the weak sound of a vibrating air jet. Many musical instruments are designed with resonant cavities to produce varying sounds. Electrical resonance in radio receivers enables the listener to hear weak signals clearly. When tuned to the frequency of a desired station, the signal is amplified by electrical resonance. In poorly designed auditoriums or long hallways, music and voices may have a hollow sound that is unpleasant to the ear. Bridges have been known to collapse because of sympathetic vibrations set up by gusts of wind.

22.5   Audible Sound Waves

We have defined *sound* as a *longitudinal mechanical wave traveling through an elastic medium.* This is a broad definition that makes no restriction whatsoever on the frequencies of sound. The physiologist is concerned primarily with sound waves that are capable of affecting the sense of hearing. Thus, it is useful to divide the sound spectrum into three frequency ranges: *audible* sound, *infrasonic* sound, and *ultrasonic* sound. These ranges are defined as follows:

Audible sound refers to sound waves in the frequency range from 20 to 20,000 Hz.

Sound waves having frequencies below the audible range are termed infrasonic.

Sound waves having frequencies above the audible range are termed ultrasonic.

When studying audible sound, the physiologist uses the terms *loudness, pitch,* and *quality* to describe the sensations produced. Unfortunately, these terms represent sensory magnitudes and are therefore subjective. What is loud to one person is moderate to another. What one person perceives as quality, another considers inferior. As always, the physicist must deal with explicit measurable definitions. The physicist therefore attempts to correlate

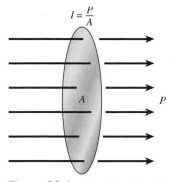

Figure 22.6 The intensity of a sound wave is a measure of the power transmitted per unit of area perpendicular to the direction of wave propagation.

the sensory effects with the physical properties of waves. These correlations can be summarized as follows:

Sensory effects		Physical property
Loudness	$\longleftrightarrow$	Intensity
Pitch	$\longleftrightarrow$	Frequency
Quality	$\longleftrightarrow$	Waveform

The meaning of the terms on the left may vary considerably among individuals. The terms on the right are measurable and objective.

Sound waves constitute a flow of energy through matter. The *intensity* of a given sound wave is a measure of the rate at which energy is propagated through a given volume of space. A convenient method of specifying sound intensity is in terms of the rate at which energy is transferred through a unit area normal to the direction of wave propagation (see Fig. 22.6). Since the rate at which energy flows is the *power* of a wave, the intensity can be related to the power per unit area passing a given point.

Sound intensity is the power transferred by a sound wave through a unit of area normal to the direction of the propagation.

$$I = \frac{P}{A} \tag{22.13}$$

The units for intensity are the ratio of a power unit to an area unit. In SI units, the intensity is in W/m^2, and that is the unit we will use in this text. The rate of energy flow in sound waves is small, however, and industry still uses the $\mu W/cm^2$ in many applications. The conversion factor is

$$1 \ \mu W/cm^2 = 1 \times 10^{-2} \ W/m^2$$

It can be shown by methods similar to those utilized for a vibrating string that the sound intensity varies directly with the square of the frequency f and with the square of the amplitude A of a given sound wave. Symbolically, the intensity I is given by

$$I = 2\pi^2 f^2 A^2 \rho v \tag{22.14}$$

where v is the sound velocity in a medium of density ρ. The symbol A in Eq. (22.14) refers to the amplitude of the sound wave and not the unit area, as in Eq. (22.13).

The intensity I_0 of the faintest audible sound is of the order of 10^{-2} W/m². This intensity, which is referred to as the **hearing threshold,** has been adopted by acoustical experts as the minimum intensity for audible sound.

The hearing threshold represents the standard minimum of intensity for audible sound. Its value at a frequency of 1000 Hz is

$$I_0 = 1 \times 10^{-12} \ W/m^2 = 1 \times 10^{-10} \ \mu W/cm^2 \tag{22.15}$$

The range of intensities over which the human ear is sensitive is enormous. It extends from the hearing threshold I_0 to an intensity 10^{12} times as great. The upper extreme, known as the **pain threshold,** is the intensity that is intolerable for the human ear. The sensation becomes one of feeling pain instead of simply hearing.

The pain threshold represents the maximum intensity the average ear can record without feeling or pain. Its value is

$$I_p = 1 \ W/m^2 = 100 \ \mu W/cm^2 \tag{22.16}$$

In view of the wide range of intensities over which the ear is sensitive, it is more convenient to set up a logarithmic scale for the measurement of sound intensities. Such a scale is established by the following rule:

When the intensity I_1 of one sound is 10 times as great as the intensity I_2 of another, the ratio of intensities is said to be 1 bel (B).

Thus, when comparing the intensities of two sounds, we refer to a difference in *intensity levels* given by

$$\text{B} = \log \frac{I_1}{I_2} \quad \text{bels (B)} \tag{22.17}$$

where I_1 is the intensity of one sound and I_2 is the intensity of the other.

Example 22.6 Two sounds have intensities of 2.5×10^{-8} W/m^2 and 1.2 W/m^2. Compute the difference in intensity levels in bels.

Solution:

$$\text{B} = \log \frac{I_1}{I_2} = \log \frac{1.2 \text{ W/m}^2}{2.5 \times 1.0^{-8} \text{ W/m}^2}$$

$$= \log 4.8 \times 10^7 = 7.68 \text{ B}$$

In practice, the unit of 1 B is too large. To obtain a more useful unit, we define a *decibel* (dB) as one-tenth of a bel. Thus, the answer to Example 22.6 can also be expressed as 76.8 dB.

By using the standard intensity I_0 as a comparison for all intensities, a general scale has been devised for rating any sound. The intensity level in decibels of any sound of intensity I can be found from the general relation

$$\beta = 10 \log \frac{I}{I_0} \quad \textit{decibels (dB)} \tag{22.18}$$

where I_0 is the intensity at the hearing threshold (1×10^{-12} W/m^2). The intensity level for I_0 is zero decibels.

Example 22.7 Compute the intensity level of a sound whose intensity is 1×10^{-4} W/m^2.

Solution:

$$\beta = 10 \log \frac{I}{I_0} = 10 \log \left(\frac{10^{-4} \text{ W/m}^2}{10^{-12} \text{ W/m}^2} \right)$$

$$= 10 \log 10^8 = 10(8) = 80 \text{ dB}$$

Through the logarithmic decibel notation, we have reduced the wide range of intensities to intensity levels from 0 to 120 dB. We must remember, however, that the scale is not linear but logarithmic. A 40-dB sound is much more than twice as intense as a 20-dB sound. A sound that is 100 times as intense as another is only 20 dB larger. Several examples of the intensity levels for common sounds are given in Table 22.1.

Table 22.1

Intensity Levels for Common Sounds

Sound	Intensity Level, dB
Hearing threshold	0
Rustling leaves	10
Whisper	20
Quiet radio	40
Normal conversation	65
Busy street corner	80
Subway car	100
Pain threshold	120
Jet engine	140–160

The intensity of a sound decreases as a person moves away from a source. The change in intensity varies with the square of the distance from the source. For example, a person twice as far away from a source hears a sound that is one-fourth as intense; a person three times as far away hears one that is one-ninth as intense. To see why this occurs, consider that sound radiates outward in all directions from a point source as illustrated in Fig. 22.7. The sound wave is seen as a succession of spherical surfaces. Consider points A and B located at distances r_1 and r_2 from a source that produces a sound of power P. Recalling that $I = P/A$ and that the area of a sphere is $4\pi r^2$, we can write the intensities I_1 and I_2 as follows:

$$I_1 = \frac{P}{4\pi r_1^2} \qquad \text{and} \qquad I_2 = \frac{P}{4\pi r_2^2}$$

The power of the source does not change. Therefore, we can eliminate P from these equations and solve for the ratio of two intensities to obtain

$$\frac{I_1}{I_2} = \frac{r_2^2}{r_1^2} \qquad \text{or} \qquad I_1 r_1^2 = I_2 r_2^2 \tag{22.19}$$

This expression is very useful for determining how the intensity of a sound varies at changes from one location to another.

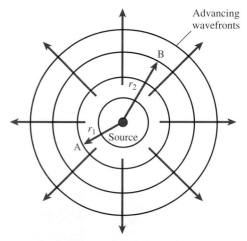

Figure 22.7 Spherical sound waves propagating radially outward from an isometric source. The intensity drops off inversely with the square of the distance from the source.

Example 22.8  A point source emits a sound with an average power of 40 W. What is the intensity at a distance of $r_1 = 3.5$ m from the source? What will be the intensity at a distance of $r_2 = 5$ m?

Plan: The intensity is the power per unit area, and the area surrounding a point source is $4\pi r^2$. The intensity at the first location can be found by substitution of given values. The intensity at the second location is found more easily from Eq. (22.9).

Solution: Given that $r_1 = 3.5$ m and $P = 40$ W, we find the intensity I_1 as follows:

$$I_1 = \frac{P}{4\pi r_1^2} = \frac{40 \text{ W}}{4\pi(3.5 \text{ m})^2}; \qquad \text{or} \qquad I_1 = 0.260 \text{ W/m}^2$$

Next, we apply Eq. (22.9) to find the intensity at $r_2 = 5$ m.

$$I_1 r_1^2 = I_2 r_2^2 \qquad \text{and} \qquad I_2 = \frac{I_1 r_1^2}{r_2^2}$$

$$I_2 = \frac{(0.260 \text{ W/m}^2)(3.5 \text{ m})^2}{(5 \text{ m})^2} = 0.127 \text{ W/m}^2$$

Notice that this inverse square relationship applies to *intensities* and *not* to *intensity levels*.

22.6   Pitch and Quality

The effect of intensity on the human ear manifests itself as *loudness*. In general, sound waves that are more intense are also louder, but the ear is not equally sensitive to sounds of all frequencies. Therefore, a high-frequency sound may not seem as loud as one of lower frequency that has the same intensity.

The *frequency* of a sound determines what the ear judges as the *pitch* of the sound. Musicians designate pitch by letters corresponding to key notes on the piano. For example the C note, D note, and F note each refer to a specific pitch, or frequency. A siren disk, shown in Fig. 22.8, can be used to demonstrate how the pitch is determined by the frequency of a sound. A stream of air is directed against a row of evenly spaced holes. Varying the rate of rotation of the disk causes the pitch of the resulting sound to be increased or decreased.

Two sounds of the same pitch can be distinguished easily. For example, suppose we sound a C note (256 Hz) successively on a piano, a flute, a trumpet, and a violin. Even though each sound has the same pitch, there is a marked difference in the tones. (See Fig. 22.9.) This distinction is said to result from a difference in the *quality* of sound.

Regardless of the source of vibration in musical instruments, several modes of oscillation are usually excited simultaneously. Therefore, the sound produced consists not only of the

Figure 22.8 Demonstrating the relationship between pitch and frequency.

Figure 22.9 A C note played on each of these instruments has the same frequency (pitch), but they give very different sounds due to the different boundary conditions. The quality of a sound is affected by the number and relative intensity of the overtones that are present. (*Photos by Hemera, Inc.*)

fundamental but also of many of the overtones. *The quality of a sound is determined by the number and relative intensities of the overtones present.* The difference in quality between two sounds can be observed objectively by analyzing the complex *waveforms* resulting from each sound. In general, the more complex the wave, the greater the number of harmonics that contribute to it. The pitch (frequency) of the sound from each of the instruments in Fig. 22.9 is the same, and yet the sounds are very different; that is, they differ in *quality.*

22.7 Interference and Beats

In Chapter 21, we discussed the superposition principle as a method for studying interference in transverse waves. Interference also occurs in longitudinal sound waves, and the superposition principle can be applied for them also. A common example of the interference in sound waves occurs when two tuning forks (or other single-frequency sound sources) whose frequencies differ only slightly are struck simultaneously. The sound produced fluctuates in intensity, alternating between loud tones and virtual silence. These regular pulsations are referred to as **beats.** The *vibrato effect* obtained on some organs is an application of this principle. Every vibrato note is produced by two pipes tuned to slightly different frequencies.

To understand the origin of beats, let us examine the interference set up between sound waves proceeding from two tuning forks of slightly different frequency, as shown in Fig. 22.10. The superposition of waves A and B illustrates the origin of beats. The loud tones occur when the waves interfere constructively, and the quiet tones occur when the waves interfere destructively. Observation and calculation show that the two waves interfere constructively $f - f'$ times per second. Thus, we can write

$$\text{Number of beats per second} = |f - f'| \qquad \textbf{(22.20)}$$

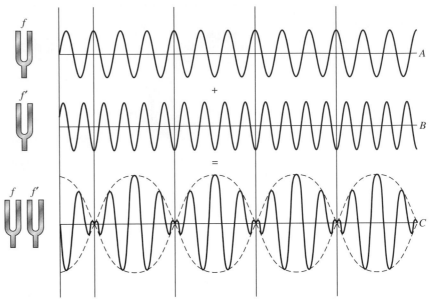

Figure 22.10   Diagram illustrating the origin of beats. The wave *C* is a superposition of waves *A* and *B*.

For example, if tuning forks of frequencies 256 and 259 Hz are struck simultaneously, the resulting sound will pulsate three times every second.

22.8 The Doppler Effect

Whenever a source of sound is moving relative to an observer, the pitch of the sound, as heard by the observer, may not be the same as that perceived when the source is at rest. For example, if we stand near a railway track as a train blowing its whistle approaches, we notice that the pitch of the whistle is *higher* than the normal one when the train is stationary. As the train recedes, the pitch is observed to be *lower* than normal. Similarly, at race tracks, the sound of cars driving toward the stands is considerably higher in pitch than the sound of cars driving away from the stands.

The phenomenon is not restricted to the motion of the source. If the source of sound is stationary, a listener moving toward the source will observe a similar increase in pitch. A listener leaving the source of sound will hear a lower-pitched sound. The change in frequency of sound resulting from relative motion between a source and an observer is called the ***Doppler effect.***

> The Doppler effect refers to the apparent change in frequency of a source of sound when there is relative motion of the source and the listener.

The origin of the Doppler effect can be demonstrated graphically by representing the periodic waves emitted by a source as concentric circles moving radially outward, as in Fig. 22.11. The distance between any two circles represents the wavelength λ of the sound traveling with a velocity *V.* The frequency with which these waves strike the ear determines the pitch of sound heard.

Let us first consider that the source is moving to the right toward a stationary observer *A,* as in Fig. 22.12. As the moving source emits sound waves, each successive wave is emitted from a point closer to the observer than its predecessor. The result is that the distance between successive waves, or the wavelength, is smaller than usual. A smaller wavelength results in a higher frequency of waves, which increases the pitch of the sound heard by observer *A*. Similar reasoning will show that an *increase* in the length of waves reaching observer *B* will cause *B* to hear a *lower*-frequency sound.

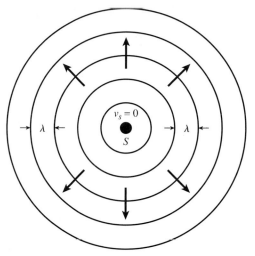

Figure 22.11   Graphic representation of sound waves emitted from a stationary source.

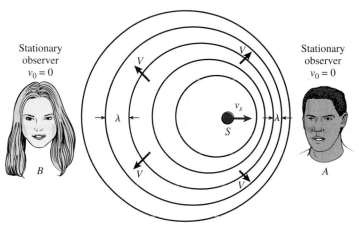

Figure 22.12   Illustration of the Doppler effect. The waves in front of a moving sound are closer together than the waves behind a moving source.

We can now derive a relationship for predicting the change in observed frequency. During one complete vibration of a stationary source (a time equal to the period T), each wave will move through a distance of one wavelength. This distance is represented by λ in Fig. 22.13a and is given by

$$\lambda = VT = \frac{V}{f_s} \qquad \textit{Stationary Source}$$

where V is the velocity of sound and f_s is the frequency of the source. If the source is moving to the right with a velocity v_s, as in Fig. 22.13b, the new wavelength λ' in front of the source will be given by

$$\lambda' = VT - v_sT = (V - v_s)T$$

But $T = 1/f_s$, so we write

$$\lambda' = \frac{V - v_s}{f_s} \qquad \textit{Moving Source} \quad \textbf{(22.21)}$$

This equation will also apply for the wavelength on the left of the moving source if we follow the convention that speeds of approach are considered positive and speeds of recession are considered negative. Thus, if we were computing λ' on the left of our

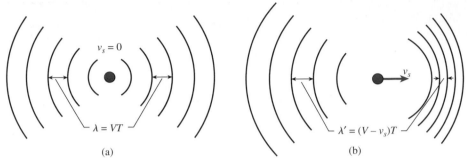

Figure 22.13 Computing the magnitude of sound emitted from a moving source. The source velocity v_s is considered positive for speeds of approach and negative for speeds of recession.

moving source, the negative value would be substituted for v_s, resulting in a larger wavelength.

The velocity of sound in a medium is a function of the properties of the medium and does not depend on the motion of the source. Thus, the frequency f_o heard by a stationary observer from a moving source of frequency f_s is given by

$$f_o = \frac{V}{\lambda'} = \frac{V f_s}{V - v_s} \qquad \textit{Moving Source} \quad \textbf{(22.22)}$$

where V is the speed of sound and v_s is the speed of the source. *The speed v_s is reckoned as positive for speeds of approach and negative for speeds of recession.*

Now we will examine the situation in which a source is stationary and the observer moves toward the source with a velocity v_o. In this case, the wavelength of the sound received does not change, but the number of waves encountered per unit of time (the observed frequency) increases as a result of the observer's speed v_o. *Hence, the observer will hear the frequency*

$$f_o = \frac{f_s(V + v_o)}{V} \qquad \textit{Moving Observer} \quad \textbf{(22.23)}$$

Here, *the speed v_o of the observer should be reckoned as positive for speeds of approach and negative for speeds of recession.*

Often both the source and the observer are moving, so we need a more general relationship that covers all relative motion. Equations (22.22) and (22.23) can be combined to provide such a general formula that works for all instances as long as relative motion of observer and source is along a straight line.

$$f_o = f_s \frac{V + v_o}{V - v_s} \qquad \textit{General Doppler Equation} \quad \textbf{(22.24)}$$

It is seen that Eq. (22.24) will reduce to Eq. (22.22) for a stationary observer ($v_o = 0$) and to Eq. (22.23) for a stationary source ($v_s = 0$).

The sign convention is very important in applying this relationship. The sign of the velocity of sound is always positive. The velocities v_o and v_s are reckoned as positive for speeds of approach and negative for speeds of recession. A velocity of approach means that the source or the observer velocity is directed in such a fashion as to make the two closer together.

Consider an example. A bicyclist traveling to the right at 15 m/s shouts at a runner who is moving ahead of her at 5 m/s in the *same* direction. The source speed v_s is +15 m/s (approach) and the runner (observer) has a speed of −5 m/s (recession). Even though the two are actually getting closer together, the observer's speed v_o is reckoned as *negative* since the direction is *away* from the source.

Example 22.9

On a day when the speed of sound is 340 m/s, a train whistle emits sound at a frequency of 400 Hz. (a) What is the frequency of the sound heard by a stationary observer when the train is moving toward the observer at a velocity of 20 m/s? (b) What is the observed frequency when the train moves at the same speed away from the observer?

Plan: The solution for each of these situations is merely a matter of substitution into the general Doppler equation (22.24). However, the signs substituted for the source velocity must be *positive* in the first instance (approach) and *negative* in the second instance (recession). The observer's velocity is zero in each case.

Solution (a): The source velocity is $v_s = +20$ m/s, $V = +340$ m/s, and $f_s = 400$ Hz.

$$f_o = \frac{f_s(V + v_o)}{V - v_s} = \frac{(400 \text{ Hz})(340 \text{ m/s} + 0)}{340 \text{ m/s} - (+20 \text{ m/s})}$$

$$f_o = \frac{(400 \text{ Hz})(340 \text{ m/s})}{320 \text{ m/s}} = 425 \text{ Hz}$$

Solution (b): The source is receding, so in this case, $v_s = -20$ m/s.

$$f_o = \frac{f_s(V + v_o)}{V - v_s} = \frac{(400 \text{ Hz})(340 \text{ m/s} + 0)}{340 \text{ m/s} - (-20 \text{ m/s})}$$

$$f_o = \frac{(400 \text{ Hz})(340 \text{ m/s})}{360 \text{ m/s}} - 378 \text{ Hz}$$

Example 22.10

A car travels to the left at 20 m/s, with the horn blowing at a frequency of 360 Hz. What frequency is heard by a person in front of the car who is riding a bicycle to the left at 12 m/s? Assume the velocity of sound is 340 m/s.

Plan: Once again, the problem reduces to making the correct choices for the signs for the velocity of the observer and source. The car is *approaching,* so $v_s = +20$ m/s; the bicyclist is *receding,* so $v_o = -12$ m/s.

Solution: The source velocity is $v_s = +20$ m/s, $V = +340$ m/s, and $f_s = 360$ Hz.

$$f_o = \frac{f_s(V + v_o)}{V - v_s} = \frac{(360 \text{ Hz})[340 \text{ m/s} + (-12 \text{ m/s})]}{340 \text{ m/s} - (+20 \text{ m/s})}$$

$$f_o = \frac{(360 \text{ Hz})(328 \text{ m/s})}{320 \text{ m/s}} = 369 \text{ Hz}$$

In working problems involving relative motion of source and observer, it is useful to consider whether the answer you obtain makes sense. For example, if two objects are getting *closer together* as a result of their relative motion, the observed frequency should be higher than that for the source; if they are getting *farther apart,* the frequency should be reduced.

Summary and Review

Summary

We have defined sound as a longitudinal mechanical wave in an elastic medium. Thus, the elasticity and density of a medium will affect the speed of sound as it travels in that medium. Under certain conditions we have seen that standing sound waves can produce characteristic frequencies that we observe as the pitch of the sound. The intensity of sound and the Doppler effect were also discussed in this chapter. The major concepts are summarized as follows.

- Sound is a longitudinal wave traveling through an elastic medium. Its speed in air at 273 K is 331 m/s or 1087 ft/s. At other temperatures, the speed of sound is approximated by

$$v = (331 \text{ m/s}) \sqrt{\frac{T}{273 \text{ K}}} \qquad \textit{Speed of Sound in Air}$$

- The speed of sound in other media can be found from the following:

$$v = \sqrt{\frac{Y}{\rho}} \qquad \textit{Rod}$$

$$v = \sqrt{\frac{\gamma P}{\rho}} = \sqrt{\frac{\gamma RT}{M}} \qquad \textit{Gas}$$

$$v = \sqrt{\frac{B}{\rho}} \qquad \textit{Fluid}$$

$$v = \sqrt{\frac{B + \frac{4}{3}S}{\rho}} \qquad \textit{Extended Solid}$$

- Standing longitudinal sound waves may be set up in a vibrating air column for a pipe that is open at both ends or for one that is closed at one end. The characteristic frequencies are

$$f_n = \frac{nv}{2L} \qquad n = 1, 2, 3, \ldots \qquad \begin{array}{l}\textit{Open Pipe of}\\ \textit{Length L}\end{array}$$

$$f_n = \frac{nv}{4L} \qquad n = 1, 3, 5, \ldots \qquad \begin{array}{l}\textit{Closed Pipe}\\ \textit{of Length L}\end{array}$$

Note that *only the odd harmonics are possible for a closed pipe.* In this case, the first overtone is the third harmonic, the second overtone is the fifth harmonic, and so on.

- The intensity of a sound is the power P per unit area A perpendicular to the direction of propagation.

$$I = \frac{P}{A} = 2\pi^2 f^2 A^2 \rho v \qquad \textit{Intensity, W/m}^2$$

- The intensity level in decibels is given by

$$\beta = 10 \log \frac{I}{I_0} \qquad I_0 = 1 \times 10^{-12} \text{ W/m}^2$$
$$\textit{Intensity Level}$$

- Whenever two waves are nearly the same frequency and exist simultaneously in the same medium, beats are set up such that

$$\text{Number of beats per second} = |f - f'|$$

- The general equation for the Doppler effect is

$$f_o = f_s \frac{V + v_o}{V - v_s} \qquad \textit{General Doppler Equation}$$

where f_o = observed frequency

$\qquad f_s$ = source frequency

$\qquad V$ = velocity of sound

$\qquad v_o$ = velocity of observer

$\qquad v_s$ = velocity of source

Note: Speeds are reckoned as positive for approach and negative for recession.

Key Terms

Review Questions

22.1. What is the physiological definition of sound? What is the meaning of sound in physics?

22.2. Why must astronauts on the surface of the Moon communicate by radio? Can they hear another spacecraft as it lands nearby? Can they hear by touching helmets?

22.3. How is the sound of a person's voice affected by inhaling helium gas? Is the effect one of pitch, loudness, or quality?

22.4. Vocal sounds originate with the vibration of vocal cords. The mouth and nasal openings act as a resonant cavity to amplify and distinguish sounds. Suppose you hum at a constant pitch equal to the C note on a piano. By opening and closing your mouth, what physiological property of the sound is affected?

22.5. The distance in miles to a thunderstorm can be estimated by counting the number of seconds elapsing between the flash of lightning and the arrival of a clap of thunder and dividing the result by 5. Explain why this is a reasonable approximation.

22.6. A store window is broken by an explosion several miles away. A glass of thin crystal shatters when a high note is reached on a violin. Are the causes of damage similar? What physical property of sound was principally responsible in each case?

22.7. Compare the speed of sound in solids, liquids, and gases. Explain the reason for differences in speed.

22.8. Perform a unit analysis of Eq. (22.1) showing that $\sqrt{Y/\rho}$ will yield units of speed.

22.9. How will the speed of sound in a gas be affected if the temperature of the gas is quadrupled?

22.10. An electric bell operates inside an evacuated flask. No sound is heard because of the absence of a medium. Explain what happens if the flask is tilted until the bell touches the walls of the flask.

22.11. Draw diagrams to demonstrate the differences between a progressive longitudinal wave and a standing longitudinal wave.

22.12. A standing wave is set up in a vibrating string. How are the harmonics of the possible sounds related to the number of loops in the string? How are the harmonics related to the number of nodes?

22.13. What effect will closing one end of an open pipe have on the frequency of a vibrating air column?

22.14. Compare the quality of sound produced by a violin with that produced by a tuning fork.

22.15. If the average ear cannot hear sounds of frequencies much in excess of 15,000 Hz, what is the advantage of building stereo music systems that have frequency responses much higher than 15,000 Hz?

22.16. A vibrating tuning fork mounted on a resonating box is moved toward a wall and away from an observer. The resulting sound pulsates in intensity. Explain.

22.17. An instructor attempts to explain the Doppler effect by using baseballs and a bicycle. He proceeds as follows: "Suppose I am at rest, and I release one baseball in the same direction every second at constant speed. Consider me as the source of sound waves and the baseballs as advancing wavefronts. The spacing between the balls at any instant will be constant and analogous to the wavelength of sound waves. Now, suppose I ride a bicycle in the forward direction at a constant speed and continue to release balls in the forward direction and in the backward direction at the same rate and at the same speed. The spacing of the balls in front of me will be closer together because each time I release a ball in that direction, I will have also moved in that direction. Similarly, the balls released in the backward direction will be spaced farther apart than normal." Give a careful analysis of his explanation. In what ways is his analogy correct? In what important aspect does his analogy fail? Why would an equation similar to Eq. (22.21) fail as a means of predicting the spacing of the baseballs? Why does it work for sound waves?

Problems

Section 22.2 The Speed of Sound

22.1. Young's modulus for steel is 2.07×10^{11} Pa, and its density is 7800 kg/m^3. Compute the speed of sound in a steel rod.   Ans. 5150 m/s

22.2. A 3-m length of copper rod has a density of 8800 kg/m^3, and Young's modulus for copper is 1.17×10^{11} Pa. How much time will it take for sound to travel from one end of the rod to the other?

22.3. What is the speed of sound in air ($M = 29$ g/mol and $\gamma = 1.4$) on a day when the temperature is 30°C? Use the approximation formula to check this result.   Ans. 349 m/s

22.4. The speed of longitudinal waves in a certain metal rod of density 7850 kg/m^3 is measured to be 3380 m/s. What is Young's modulus for the metal?

22.5. If the frequency of the waves in Prob. 22.4 is 312 Hz, what is the wavelength?   Ans. 10.8 m

22.6. Compare the theoretical speeds of sound in hydrogen ($M = 2.0$ g/mol, $\gamma = 1.4$) with helium ($M = 4.0$ g/mol, $\gamma = 1.66$) at 0°C.

*22.7. A sound wave is sent from a ship to the ocean floor, where it is reflected and returned. If the round trip takes 0.6 s, how deep is the ocean floor? Consider the bulk modulus for seawater to be 2.1×10^9 Pa and its density to be 1030 kg/m^3.   Ans. 428 m

Section 22.3 Vibrating Air Columns

22.8. Find the fundamental frequency and the first three overtones for a 20-cm pipe at 20°C if the pipe is open at both ends.

22.9. Find the fundamental frequency and the first three overtones for a 20-cm pipe at 20°C if the pipe is closed at one end.   Ans. 429, 1290, 2140, and 3000 Hz

22.10. The auditory canal forms a standing-wave cavity, closed at one end, as shown in Fig. 22.14. Assume the length of this cavity is 2.8 cm. If room temperature is 24°C, what fundamental frequency will be amplified?

22.11. What length of open pipe will produce a fundamental frequency of 356 Hz at 20°C?   Ans. 48.2 cm

22.12. What length of open pipe will produce a frequency of 1200 Hz as its first overtone on a day when the speed of sound is 340 m/s?

22.13. The second overtone of a closed pipe is 1200 Hz at 20°C. What is the length of the pipe?   Ans. 35.7 cm

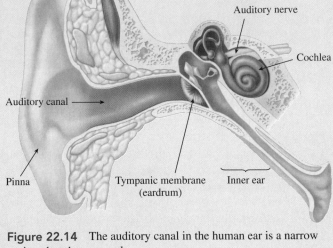

Figure 22.14 The auditory canal in the human ear is a narrow cavity, closed at one end.

22.14. In a resonance experiment, the air in a closed tube of variable length is found to resonate with a tuning fork when the air column is first 6 cm and then 18 cm long. What is the frequency of the tuning fork if the temperature is 20°C?

*22.15. A closed pipe and an open pipe are each 3 m long. Compare the wavelength of the fourth overtone for each pipe at 20°C.   Ans. open = 1.20 m, closed = 1.33 m

Section 22.5 Audible Sound Waves

22.16. What is the intensity level in decibels of a sound that has an intensity of 4×10^{-5} W/m^2?

22.17. The intensity of a sound is 6×10^{-8} W/m^2. What is the intensity level?   Ans. 47.8 dB

22.18. A 60-dB sound is measured at a particular distance from a whistle. What is the intensity of this sound in W/m^2?

22.19. What is the intensity of a 40-dB sound?   Ans. 1×10^{-8} W/m^2

22.20. Compute the intensities for sounds of 10 dB, 20 dB, and 30 dB.

22.21. Compute the intensity levels for sounds of 1×10^{-6} W/m^2, 2×10^{-6} W/m^2, and 3×10^{-6} W/m^2.   Ans. 60.0 dB, 63.0 dB, 64.8 dB

22.22. An isometric source of sound broadcasts a power of 60 W. What are the intensity and the intensity level of a sound heard at a distance of 4 m from this source?

22.23. A 3.0-W sound source is located 6.5 m from an observer. What are the intensity and the intensity level of the sound heard at that distance?
Ans. 5.65×10^{-3} W/m^2, 97.5 dB

22.24. A person located 6 m from a sound source hears an intensity of 2×10^{-4} W/m^2. What intensity is heard by a person located 2.5 m from the source?

*22.25. The intensity level 6 m from a source is 80 dB. What is the intensity level at a distance of 15.6 m from the same source?
Ans. 71.7 dB

Section 22.8 The Doppler Effect

Assume that the speed of sound is 343 m/s for all of these problems.

22.26. A stationary source of sound emits a signal at a frequency of 290 Hz. What are the frequencies heard by an observer (a) moving toward the source at 20 m/s and (b) moving away from the source at 20 m/s?

22.27. A car blowing a 560-Hz horn moves at a speed of 15 m/s as it first approaches a stationary listener and then moves away from a stationary listener at the same speed. What are the frequencies heard by the listener?
Ans. 586 Hz, 537 Hz

22.28. A person stranded in a car blows a 400-Hz horn. What frequencies are heard by the driver of a car passing at a speed of 60 km/h?

22.29. A train moving at 20 m/s blows a 300-Hz whistle as it passes a stationary observer. What are the frequencies heard by the observer as the train passes?
Ans. 319, 300, and 283 Hz

22.30. A child riding a bicycle north at 6 m/s hears a 600-Hz siren from a police car heading south at 15 m/s. What is the frequency heard by the child?

22.31. An ambulance moves northward at 15 m/s. Its siren has a frequency of 600 Hz at rest. A car heads south at 20 m/s toward the ambulance. What frequencies are heard by the car driver before and after the car and ambulance pass one another?
Ans. 664 Hz, 541 Hz

*22.32. A truck traveling at 24 m/s overtakes a car traveling at 10 m/s in the same direction. The trucker blows a 600-Hz horn. What frequency is heard by the car driver?

*22.33. A 500-Hz train whistle is heard by a stationary observer at a frequency of 475 Hz. What is the speed of the train? Is it moving toward the observer or away from the observer?
Ans. 18.1 m/s, away

Additional Problems

22.34. The speed of sound in a certain metal rod is 4600 m/s, and the density of the metal is 5230 kg/m^3. What is Young's modulus for this metal?

22.35. A sonar beam travels in a fluid for a distance of 200 m in 0.12 s. The bulk modulus of elasticity for the fluid is 2600 MPa. What is the density of the fluid?
Ans. 936 kg/m^3

22.36. What is the frequency of the third overtone for a closed pipe of length 60 cm?

22.37. A 40-g string 2 m in length vibrates in three loops. The tension in the string is 270 N. What is the wavelength? What is the frequency?
Ans. 1.33 m, 87.1 Hz

22.38. How many beats per second are heard when two tuning forks of 256 and 259 Hz are sounded together?

22.39. What is the length of a closed pipe if the frequency of its second overtone is 900 Hz on a day when the temperature is 20°C?
Ans. 47.6 cm

22.40. The fundamental frequency for an open pipe is 360 Hz. If one end of this pipe is closed, what will be the new fundamental frequency?

*22.41. A 60-cm steel rod is clamped at one end as shown in Fig. 22.15a. Sketch the fundamental and the first overtone for these boundary conditions. What are the wavelengths in each case?
Ans. 2.40 m, 80.0 cm

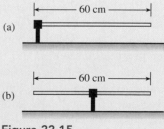

Figure 22.15

*22.42. The 60-cm rod in Fig. 22.15b is now clamped at its midpoint. What are the wavelengths for the fundamental and first overtone?

22.43. The velocity of sound in a steel rod is 5060 m/s. What is the length of a steel rod mounted as shown in Fig. 22.15a if the fundamental frequency of vibration for the rod is 3000 Hz?

Ans. 42.2 cm

*22.44. Find the ratio of the intensities of two sounds if one is 12 dB higher than the other.

22.45. A certain loudspeaker has a circular opening of area 6 cm^2. The power radiated by this speaker is 6×10^{-7} W. What is the intensity of the sound at the opening? What is the intensity level?

Ans. 1 mW/m^2, 90 dB

22.46. The noon whistle at a textile mill has a frequency of 360 Hz. What are the frequencies heard by the driver of a car passing the mill at 25 m/s on a day when sound travels at 343 m/s?

*22.47. What is the difference in intensity levels (dB) for two sounds whose intensities are 2×10^{-5} W/m^2 and 0.90 W/m^2?

Ans. 46.5 dB

Critical Thinking Questions

*22.48. By inhaling helium gas, one can raise the frequency of the voice considerably. For air, $M = 29$ g/mol and $\gamma = 1.4$; for helium, $M = 4.0$ g/mol and $\gamma = 1.66$. At a temperature of 27°C, you sing a C note at 256 Hz. What is the frequency that will be heard if you inhale helium gas and other parameters are unchanged? Notice that both v and f were increased. How do you explain this in view of the fact that $v = f\lambda$? Discuss.

Ans. 751 Hz

*22.49. A toy whistle is made out of a piece of sugarcane that is 8 cm long. It is essentially an open pipe from the air inlet to the far end. Now suppose that a hole is bored at the midpoint so a finger can alternately close and open the hole. If the velocity of sound is 340 m/s, what are the two possible fundamental frequencies obtained by closing and opening the hole at the center of the cane? What is the fundamental frequency if the center hole is covered and the far end is plugged?

*22.50. A tuning fork of frequency 512 Hz is moved away from an observer and toward a flat wall with a speed of 3 m/s. The speed of sound in the air is 340 m/s. What is the apparent frequency of the unreflected sound? What is the apparent frequency of the reflected sound? How many beats are heard each second?

Ans. 508 Hz, 517 Hz, 9 beats/s

*22.51. Using the logarithmic definition of decibels, derive the following expression, which relates the ratio of intensities of two sounds to the difference in decibels for the sounds:

$$\beta_2 - \beta_1 = 10 \log \frac{I_2}{I_1}$$

Use this relationship to work Probs. 22.44 and 22.47.

*22.52. The laboratory apparatus shown in Fig. 22.16 is used to measure the speed of sound in air by the resonance method. A vibrating tuning fork of frequency f is held over the open end of a tube, which is partly filled with water. The length of the air column can be varied by changing the water level. As the water level is gradually lowered from the top of the tube, the sound intensity reaches a maximum at the three levels shown in the figure. The maxima occur whenever the air column resonates with the tuning fork. Thus, the distance between successive resonance positions is the distance between adjacent notes for the standing waves in the air column. The frequency of the fork is 512 Hz, and the resonance positions occur at 17, 51, and 85 cm from the top of the tube. What is the velocity of sound in the air? What is the approximate temperature in the room?

Ans. 348 m/s, 28.3°C

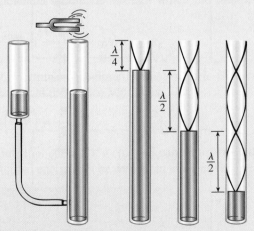

Figure 22.16 Laboratory apparatus for computing the velocity of sound by resonance methods.

22.53. What is the difference in intensity levels of two sounds, one being twice the intensity of the other?

23

The Electric Force

Lighting provides one of the most beautiful displays in nature. With temperatures near that at the surface of the Sun and damaging shock waves, it also represents a significant danger to humans and physical structures. It results from a substantial buildup of charge in clouds, which eventually discharges to the ground along an ionization path created by a column of electrons that extends from the clouds to the ground.

(*Photo © vol. 1 PhotoDisc/Getty.*)

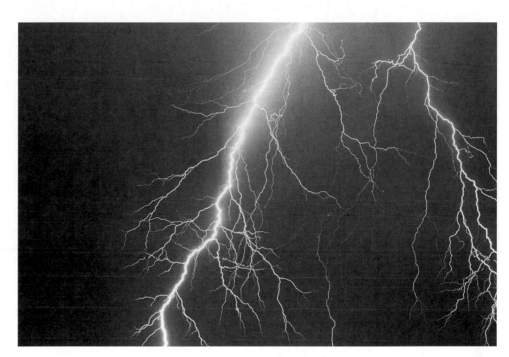

Objectives

After completing this chapter, you should be able to

1. Demonstrate the existence of two kinds of electric charge and verify the *first law of electrostatics* using laboratory materials.

2. Explain and demonstrate the processes of charging by *contact* and by *induction* and use an *electroscope* to determine the nature of an unknown charge.

3. State *Coulomb's law* and apply it to the solution of problems involving electric forces.

4. Define the *electron*, the *coulomb*, and the *microcoulomb* as units of electric charge.

A hard-rubber comb or a plastic rod acquires a strange ability to attract other objects after it is rubbed on a coat sleeve. An annoying *shock* is sometimes experienced when you touch the handle of a car door after sliding across the seat. Stacked sheets of paper tend to resist separation. All these occurrences are examples of *electrification*, a phenomenon that frequently occurs as a result of rubbing objects together. Long ago, the name **charging** was given to the rubbing process, and the electrified object was said to be *charged*. This chapter begins our study of **electrostatics,** the science that treats electric charges at rest.

23.1 The Electric Charge

A good way to begin a study of electrostatics is to experiment with objects that become electrified by rubbing. The materials illustrated in Fig. 23.1 are commonly found in the physics laboratory. In the order of their appearance in the figure, they are a hard-rubber rod resting on a piece of cat's fur, a glass rod resting on a piece of silk, the pith-ball electroscope, suspended pith balls, and the gold-leaf electroscope. A *pith ball* is a light sphere of wood pith painted with metallic paint and usually suspended from a silk thread. An *electroscope* is a sensitive laboratory instrument used to detect the presence of an electric charge.

The pith-ball electroscope can be used to study the effects of electrification. Two metallic-coated pith balls are suspended by silk threads from a common point. We begin by vigorously rubbing the rubber rod with cat's fur (or wool). Then, if the rubber rod is brought near the electroscope, the suspended pith balls will be attracted to the rod, as shown in Fig. 23.2a. After remaining in contact with the rod for an instant, the balls will be repelled from the rod and from each other. When the rod is removed, the pith balls remain separated, as shown in the figure. The repulsion must be due to some property acquired by the pith balls as a result of their contact with the charged rod. We may reasonably assume that some of the *charge* has been transferred from the rod to the pith balls and that all three objects become similarly charged. From these observations, we can state the conclusion that

A force of repulsion exists between two substances that are electrified in the same way.

PHYSICS TODAY

Imagine a newspaper that never needs to be recycled or discarded. The news is transmitted each morning electronically to a specially prepared material similar to paper or to other devices. Moreover, the text and graphics are stable and require no batteries. You can read a book or carry information with you to the beach or any other location. Several companies, such as Xerox and E-Ink, are developing such applications.

One application uses millions of tiny microcapsules that contain positively charged black particles and negatively charged white particles suspended in a clear fluid. Manipulation of an electric field between a transparent top electrode and a bottom electrode can control the position of the white and black particles. Thus, text and graphics can be displayed in the form of pixels, similar to those on a computer monitor.

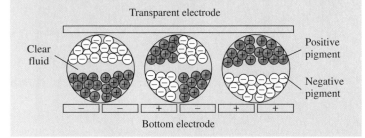

Transparent electrode

Clear fluid

Positive pigment

Negative pigment

Bottom electrode

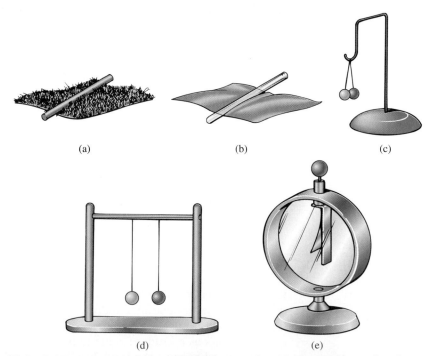

(a)   (b)   (c)

(d)   (e)

Figure 23.1 Laboratory materials for studying electrostatics: (a) a hard-rubber rod resting on a piece of cat's fur, (b) a glass rod resting on a piece of silk, (c) the pith-ball electroscope, (d) suspended pith balls, and (e) the gold-leaf electroscope.

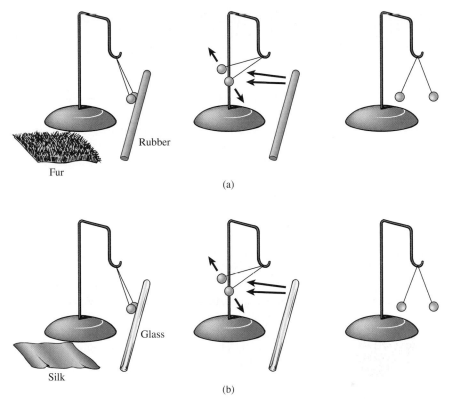

Figure 23.2   (a) Charging the pith-ball electroscope with a rubber rod. (b) Charging the pith balls with a glass rod.

Let us continue our experimentation by picking up the glass rod and rubbing it vigorously on a silk cloth. When the charged rod is brought near the pith balls, the same sequence of events occurs as with the rubber rod. (See Fig. 23.2b.) Does this mean that the nature of the charge is the same on both rods? Our experiment neither proves nor disproves this assumption. In each case, the rod and balls were electrified in the same way, so repulsion occurs in each case.

To test whether the two processes are identical, let us charge one pith ball with a glass rod and the other with a rubber rod. As shown in Fig. 23.3, a force of *attraction* exists between the balls charged in this manner. We conclude that the charges produced on the glass and rubber rods are opposite.

Similar experimentation with many different materials demonstrates that all electrified objects can be divided into two groups: (1) those that have a charge like that produced on glass and (2) those that have a charge like that produced on rubber. According to a convention established by Benjamin Franklin, objects in the former group are said to have a *positive (+) charge,* and objects belonging to the latter group are said to have a *negative (−)*

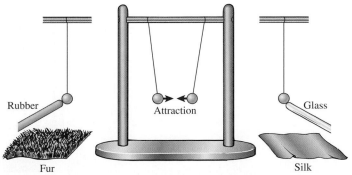

Figure 23.3   A force of attraction exists between two substances that are oppositely charged.

charge. These terms have no mathematical significance; they simply denote the two opposite kinds of electric charge.

We are now in a position to state the ***first law of electrostatics,*** which is based on our experimentation:

Like charges repel, and unlike charges attract.

Two negatively charged objects or two positively charged objects repel each other, as demonstrated by Fig. 23.2a and b, respectively. Figure 23.3 demonstrates that a positively charged object attracts a negatively charged object.

23.2 The Electron

What actually occurs during a rubbing process that causes the phenomenon of electrification? Benjamin Franklin thought that all bodies contained a specified amount of electric fluid that served to keep them in an uncharged state. When two different substances were rubbed together, he postulated that one accumulated an excess of fluid and became positively charged, whereas the other lost fluid and became negatively charged. It is now known that the substance transferred is not a fluid but small amounts of negative electricity called ***electrons.***

The modern atomic theory of matter holds that all substances are made up of atoms and molecules. Each atom has a positively charged central core, called the *nucleus,* which is surrounded by a cloud of negatively charged electrons. The nucleus consists of a number of *protons,* each with a single unit of positive charge, and (except for hydrogen) one or more *neutrons.* As the name suggests, a ***neutron*** is a neutral particle. Normally, an atom of matter is in a *neutral* or *uncharged* state because it contains the same number of protons in its nucleus as there are electrons surrounding the nucleus. A schematic diagram of the neon atom is shown in Fig. 23.4. If, for some reason, a neutral atom loses one or more of its outer electrons, the atom has a net positive charge and is referred to as a positive ***ion.*** A negative ion is an atom that has gained one or more additional charges.

When two particular materials are brought in close contact, some of the loosely held electrons may be transferred from one material to the other. For example, when a hard-rubber rod is rubbed against fur, electrons are transferred from the fur to the rod, leaving an *excess* of electrons on the rod and a *deficiency* of electrons on the fur. Similarly, when a glass rod is rubbed on a silk cloth, electrons are transferred from the glass to the silk. We can now state

An object that has an excess of electrons is negatively charged, and an object that has a deficiency of electrons is positively charged.

A laboratory demonstration of the transfer of charge is illustrated in Fig. 23.5. A hard-rubber rod is rubbed vigorously on a piece of fur. One pith ball is charged negatively with the rod, and the other is touched with the fur. The resulting attraction shows that the fur is oppositely charged. The process of rubbing has left a deficiency of electrons on the fur.

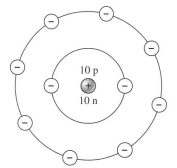

Figure 23.4 The neon atom consists of a tightly packed nucleus containing 10 protons (p) and 10 neutrons (n). The atom is electrically neutral because it is surrounded by 10 electrons.

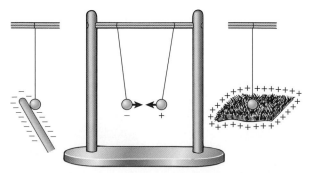

Figure 23.5 Rubbing a hard-rubber rod on a piece of fur transfers electrons from the fur to the rod.

23.3 Insulators and Conductors

A solid piece of matter is composed of many atoms arranged in a manner peculiar to that material. Some materials, primarily metals, have a large number of *free electrons,* which can move about through the material. These materials have the ability to transfer charge from one object to another, and they are called **conductors.**

A conductor is a material through which charge can be easily transferred.

Most metals are good conductors. In Fig. 23.6, a copper rod is supported by a glass stand. The pith balls can be charged by touching the right end of the copper with a charged rubber rod. The electrons are transferred or *conducted* through the rod to the pith balls. Note that none of the charge is transferred to the glass support or to the silk thread. These materials are poor conductors and are referred to as **insulators.**

An insulator is a material that resists the flow of charge.

Other examples of good insulators are rubber, plastic, mica, Bakelite, sulfur, and air.

A semiconductor is a material intermediate in its ability to carry charge.

Examples are silicon, germanium, and gallium arsenide. The ease with which a **semiconductor** carries charge can be varied greatly by the addition of impurities or by a change in temperature.

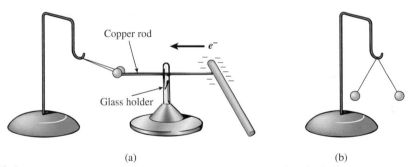

(a)   (b)

Figure 23.6   Electrons are conducted by the copper rod to charge the pith balls.

23.4 The Gold-Leaf Electroscope

The gold-leaf electroscope shown in Fig. 23.7 consists of a strip of gold foil attached to a conducting rod. The rod and the foil are protected from air currents by a cylindrical metal case with glass windows. The rod is fitted with a spherical knob at the top and is insulated from the case by a block of hard rubber or amber. Whenever the knob is given a charge, the repulsion of like charges on the rod and the gold leaf causes the leaf to diverge away from the rod.

Figure 23.8 illustrates charging the electroscope by *contact.* When the knob is touched with the negatively charged rod, electrons flow from the rod to the leaves, leaving an excess of electrons on the electroscope. When the knob is touched with a positively charged rod, electrons are transferred from the knob to the rod, leaving the electroscope deficient in electrons. Note that the residual charge on the electroscope is of the same sign as that of the charging rod.

Once the electroscope is charged, either negatively or positively, it can be used to detect the presence and nature of other charged objects. (See Fig. 23.9.) For example,

Metal knob

Insulator

Metal case

Glass window

Gold leaf

Conducting rod

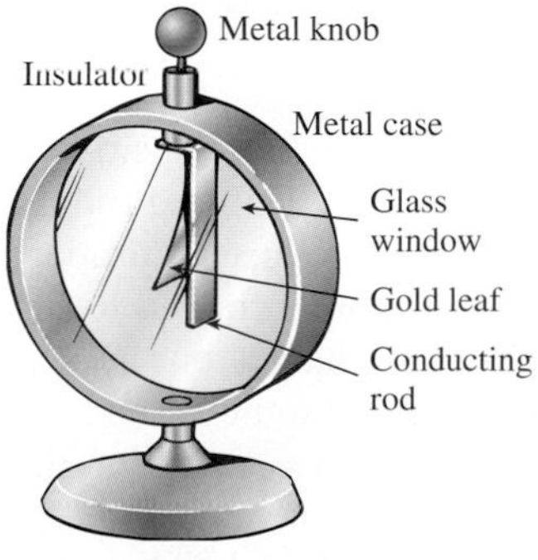

Figure 23.7   The gold-leaf electroscope.

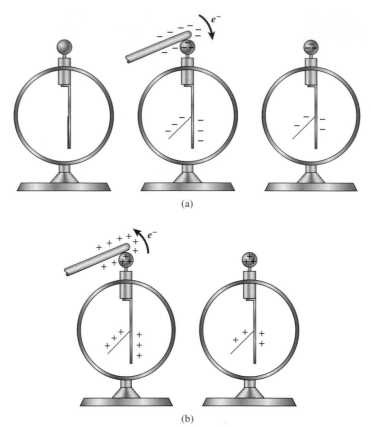

Figure 23.8 Charging the electroscope by contact with (a) a negatively charged rod and (b) a positively charged rod.

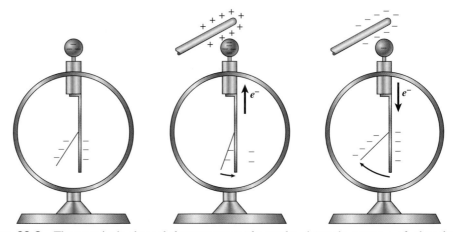

Figure 23.9 The negatively charged electroscope can be used to detect the presence of other charges.

consider what happens to the leaf of a negatively charged electroscope as a positively charged rod is brought near the knob. Some electrons are drawn from the leaf and rod up into the knob. As a result, the leaf converges. Bringing the rod closer produces a proportionately greater convergence of the leaf as more and more electrons are attracted to the knob. There appears to be a direct proportion between the number of charges on the leaf and rod and the force of repulsion between them. Moreover, there must be some *inverse* relation between the separation of the charged rod and the knob and the force attracting the electrons from the leaf and rod of the electroscope. This force becomes stronger as the separation decreases. These observations will help us understand *Coulomb's law,* developed in Section 23.7.

Similar reasoning will show that the leaf of a negatively charged electroscope will be repelled further from the rod when the knob is placed near a negatively charged object. Thus, a charged electroscope can be used to indicate both the polarity and the presence of nearby charges.

23.5 Redistribution of Charge

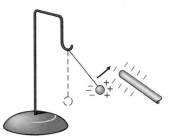

Figure 23.10 Attraction of a neutral body due to a redistribution of charge.

When a negatively charged rod is brought close to an uncharged pith ball, there is an initial attraction, as shown in Fig. 23.10. The attraction of the uncharged object is due to the separation of positive and negative electricity within the neutral body. The proximity of the negatively charged rod repels loosely held electrons to the opposite side of the uncharged object, leaving a deficiency (positive charge) on the near side and an excess (negative charge) on the far side. Since the unlike charge is nearer to the rod, the force of attraction will exceed the force of repulsion and the electrically neutral object will be attracted to the rod. No charge is gained or lost during this process; the charge on the neutral body is simply redistributed.

23.6 Charging by Induction

The redistribution of charge due to the presence of a nearby charged object can be useful in charging objects without contact. This process, called *charging by induction,* can be accomplished without any loss of charge from the charging body. For example, consider the two neutral metal spheres placed in contact as shown in Fig. 23.11. When a negatively charged rod is brought near the left sphere (without touching it), a redistribution of charge occurs. Electrons are forced from the left sphere to the right sphere through the point of contact. Now, if the spheres are separated in the presence of the charging rod, the electrons cannot return to the left sphere. Thus, the left sphere will have a deficiency of electrons (a *positive charge*), and the right sphere will have an excess of electrons (a *negative charge*).

A charge can also be induced on a single sphere. This process is illustrated with the electroscope in Fig. 23.12. A negatively charged rod is placed near the metal knob,

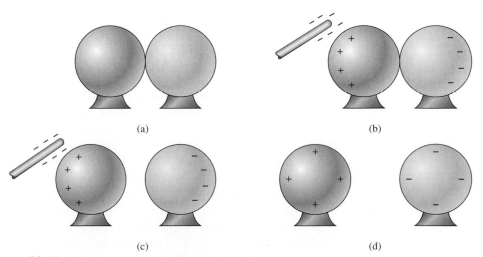

(a)

(b)

(c)

(d)

Figure 23.11   Charging two metal spheres by induction.

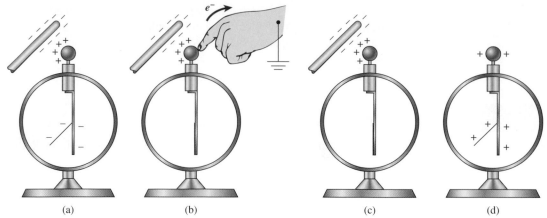

(a)   (b)   (c)   (d)

Figure 23.12 Charging an electroscope by induction. Note that the residual charge is opposite that of the charging body.

causing a redistribution of charge. The repelled electrons cause the leaf to diverge, leaving a deficiency of electrons on the knob. By touching the knob with a finger or by connecting a wire from the knob to the earth, a path is provided for the repelled electrons to leave the electroscope. The body or the ground will acquire a negative charge equal to the positive charge (deficiency) left on the electroscope. When the charging rod is removed, the leaf of the electroscope will again diverge, as shown in the figure. Charging by induction always leaves a residual charge that is opposite that of the charging body.

23.7 Coulomb's Law

As usual, the task of the physicist is to measure the interactions between charged objects in some quantitative fashion. It is not sufficient to state that an electric force exists; we must be able to predict its magnitude.

The first theoretical investigation of the electric forces between charged bodies was accomplished by Charles Augustin de Coulomb in 1784. His studies were made with a torsion balance to measure the variation in force with separation and quantity of charge. The separation r of two charged objects is reckoned as the straight-line distance between their centers. The quantity of charge q can be thought of as the excess number of electrons or protons in the body.

Coulomb found that the force of attraction or repulsion between two charged objects is inversely proportional to the square of their separation distance. In other words, if the distance between two charged objects is reduced by one-half, the force of attraction or repulsion between them will be increased fourfold.

The concept of a quantity of charge was not clearly understood in Coulomb's time. There was no established unit of charge and no means for measuring it, but his experiments clearly showed that the electric force between two charged objects is directly proportional to the product of the quantity of charge on each object. Today, his conclusions are stated in *Coulomb's law:*

> The force of attraction or repulsion between two point charges is directly proportional to the product of the two charges and inversely proportional to the square of the distance between them.

To arrive at a mathematical statement of Coulomb's law, let us consider the charges in Fig. 23.13. The force **F** of attraction between two unlike charges is indicated, and a force of repulsion is shown for like charges. In either case, the magnitude of the force is determined

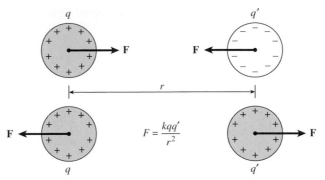

Figure 23.13   Illustrating Coulomb's law.

by the magnitudes of the charges q and q' and by their separation r. From Coulomb's law, we can write

$$F \propto \frac{qq'}{r^2}$$

or

$$F = \frac{kqq'}{r^2} \qquad\qquad\qquad \textbf{(23.1)}$$

The proportionality constant k takes into account the properties of the medium separating the charged bodies and has the dimensions dictated by Coulomb's law.

In SI units, the practical system for the study of electricity, the unit of charge is expressed in **_coulombs_** (C). In this case, the quantity of charge is not defined by Coulomb's law but is related to the flow of charge through a conductor. We will find later that this rate of flow is measured in *amperes*. A formal definition of the coulomb is as follows:

> One coulomb is the charge transferred through any cross section of a conductor in 1 second by a constant current of 1 ampere.

Since current theory is not a part of this chapter, it will suffice to compare the coulomb with the charge of an electron.

$$1 \text{ C} = 6.25 \times 10^{18} \text{ electrons}$$

Obviously, the coulomb is an enormously large unit from the standpoint of most problems in electrostatics. The charge of one electron expressed in coulombs is

$$e^- = -1.6 \times 10^{-19} \text{ C} \qquad\qquad\qquad \textbf{(23.2)}$$

where e^- is the symbol for the electron and the minus sign denotes the nature of the charge.

A more convenient unit for electrostatics is the **_microcoulomb_** (μC), defined by

$$1 \ \mu\text{C} = 10^{-6} \text{ C} \qquad\qquad\qquad \textbf{(23.3)}$$

Since the SI units of force, charge, and distance do not depend on Coulomb's law, the proportionality constant k must be determined by experiment. A large number of experiments have shown that when the force is in newtons, the distance is in meters, and the charge is in coulombs, the proportionality constant is approximately

$$k = 9 \times 10^9 \text{ N} \cdot \text{m}^2/\text{C}^2 \qquad\qquad\qquad \textbf{(23.4)}$$

When applying Coulomb's law in SI units, one must substitute this value for k in Eq. (23.1):

$$F = \frac{(9 \times 10^9 \text{ N} \cdot \text{m}^2/\text{C}^2) \, qq'}{r^2} \qquad\qquad\qquad \textbf{(23.5)}$$

It must be remembered that **F** represents the force on a charged particle and is, therefore, a vector quantity. The *direction* of the force is determined solely by the nature (+ or −) of the charges q and q'. For two charges, each will exert the same force on the other except that they will be in opposite directions (the attraction or repulsion is mutual). Thus, you should first decide which charge is to be addressed and then determine the *direction* of the force on that charge due to the other charge. The direction is determined by the laws for attraction and repulsion; *like charges repel, and unlike charges attract.* The *magnitude* of the force F is found from Coulomb's law by substitution of the absolute values for q, q', and r. The charges must be in *coulombs,* and the distance must be in *meters* if the force is to be in *newtons.*

Example 23.1

Two charges, $q_1 = -8 \ \mu\text{C}$ and $q_2 = +12 \ \mu\text{C}$, are placed 12 cm apart in the air. What is the resultant force on a third charge, $q_3 = -4 \ \mu\text{C}$, placed midway between the other two charges?

Plan: First, we will draw a straight horizontal line and indicate the positions and magnitudes of the three charges, as shown in Fig. 23.14. The focus will be on the central charge q_3, and we will indicate the directions of the forces $\mathbf{F}_1$ and $\mathbf{F}_2$ acting *on* q_3 due to the charges q_1 and q_2. Coulomb's law will give the magnitudes of the forces, and their resultant can be calculated as the vector sum.

Solution: We convert the distance to meters (12 cm = 0.12 m), and one-half of 0.12 m is 0.06 m—the midpoint. The charges are converted to coulombs (1 μC = 1 × 10^{-6} C). The force $\mathbf{F}_1$ on q_3 due to q_1 is found from Coulomb's law. Remember that the sign of the charge is used only to find the direction of forces. We will need only the absolute values for substitution.

$$F_1 = \frac{kq_1q_3}{r^2} = \frac{(9 \times 10^9 \ \text{N} \cdot \text{m}^2/\text{C}^2)(8 \times 10^{-6} \ \text{C})(4 \times 10^{-6} \ \text{C})}{(0.06 \ \text{m})^2}$$

$$F_1 = 80 \ \text{N, repulsion (to the right)}$$

Similarly, the force $\mathbf{F}_2$ on q_3 is equal to

$$F_2 = \frac{kq_2q_3}{r^2} = \frac{(9 \times 10^9 \ \text{N} \cdot \text{m}^2/\text{C}^2)(12 \times 10^{-6} \ \text{C})(4 \times 10^{-6} \ \text{C})}{(0.06 \ \text{m})^2}$$

$$F_2 = 120 \ \text{N, attraction (also to the right)}$$

Finally, the resultant force is the vector sum of $\mathbf{F}_1$ and $\mathbf{F}_2$.

$$F = 80 \ \text{N} + 120 \ \text{N} = 200 \ \text{N, to the right}$$

Note that the signs of the charges were used only to determine the direction of the forces; they were not used in Coulomb's law.

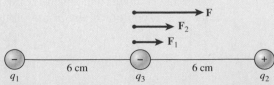

Figure 23.14 Computing the resultant force on a charge placed midway between two other charges.

Problem-Solving Strategy

Electric Forces and Coulomb's Law

1. Read the problem and then draw and label a figure. Indicate positive and negative charges along with given distances. The charges must be in *coulombs,* and the distances must be in *meters.* Remember that $1\ \mu C = 1 \times 10^{-6}$ C and 1 nC $= 1 \times 10^{-9}$ C.

2. Be careful not to confuse the *nature* of the charge ($+$ or $-$) with the sign given to forces and their components. Attraction and/or repulsion determines the direction of electric forces.

3. The *resultant force* on a given charge due to one or more nearby charges is found by vector addition of the force that each charge would exert acting alone. The magnitude of each force is found from Coulomb's law; the direction is determined from the fact that like charges repel and unlike charges attract. Construct a free-body diagram and proceed with vector addition as shown in text examples. You may wish to review vector addition by the component method as discussed in Chapter 3.

4. For charges in equilibrium, recall that the first condition for equilibrium provides that the sum of the x components is zero and the sum of the y components is zero.

Example 23.2

Three charges, $q_1 = +4 \times 10^{-9}$ C, $q_2 = -6 \times 10^{-9}$ C, and $q_3 = -8 \times 10^{-9}$ C, are arranged as shown in Fig. 23.15. What is the resultant force on q_3 due to the other two charges?

Plan: We draw the sketch and free-body diagram, labeling all given information, as shown in Fig. 23.15. The focus must be on charge q_3 as we calculate independently the magnitude and direction of each force due to the other charges. The resultant force is found by the component method. (See Section 3.12 for a review of vector addition.)

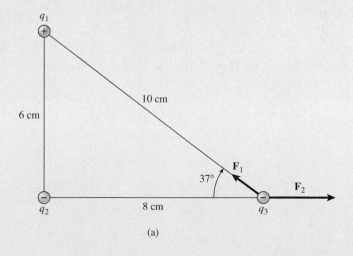

(a)

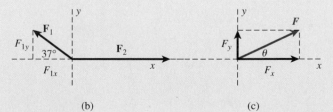

(b)   (c)

Figure 23.15

Solution: Let $\mathbf{F}_1$ be the force on q_3 due to q_1, and let $\mathbf{F}_2$ be the force on q_3 due to q_2. $\mathbf{F}_1$ is a force of attraction (*unlike charges*), and $\mathbf{F}_2$ is a force of repulsion (*like charges*), as shown in Fig. 23.15. The *magnitude* and *direction* of each force are found as follows:

$$F_1 = \frac{kq_1q_3}{r^2} = \frac{(9 \times 10^9 \text{ N} \cdot \text{m}^2/\text{C}^2)(4 \times 10^{-9} \text{ C})(8 \times 10^{-9} \text{ C})}{(0.100 \text{ m})^2}$$

$$= 2.88 \times 10^{-5} \text{ N} = 28.8 \text{ } \mu\text{N} \text{ (}37° \text{ N of W)}$$

$$F_2 = \frac{kq_2q_3}{r^2} = \frac{(9 \times 10^9 \text{ N} \cdot \text{m}^2/\text{C}^2)(6 \times 10^{-9} \text{ C})(8 \times 10^{-9} \text{ C})}{(0.080 \text{ m})^2}$$

$$= 6.75 \times 10^{-5} \text{ N} = 67.5 \text{ } \mu\text{N, east}$$

The resultant force is found using the component method of vector addition. The x and y components of $\mathbf{F}_1$ and $\mathbf{F}_2$ are summarized in Table 23.1.

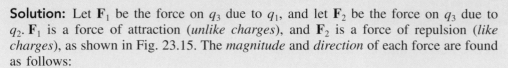

Table 23.1

Vector	Angle ϕ_x	x component	y component
$F_1 = 28.8 \text{ } \mu\text{N}$	37°	$F_{1x} = -(28.8 \text{ } \mu\text{N})(\cos 37°)$ $= -23.0 \text{ } \mu\text{N}$	$F_{1y} = (28.8 \text{ } \mu\text{N})(\sin 37°)$ $= 17.3 \text{ } \mu\text{N}$
$F_2 = 67.5 \text{ } \mu\text{N}$	0°	$F_{2x} = +67.5 \text{ } \mu\text{N}$	$F_{2y} = 0 \text{ } \mu\text{N}$
F	θ	$F_x = \Sigma F_x = +44.5 \text{ } \mu\text{N}$	$F_y = \Sigma F_y = +17.3 \text{ } \mu\text{N}$

From Fig. 23.15c, we apply the Pythagorean theorem to find the magnitude of the resultant force F on q_3:

$$F = \sqrt{F_x^2 + F_y^2}$$

$$= \sqrt{(44.5 \text{ } \mu\text{N})^2 + (17.3 \text{ } \mu\text{N})^2} = 47.7 \text{ } \mu\text{N}$$

Next, the direction is found from the tangent function.

$$\tan\theta = \left|\frac{F_y}{F_x}\right| = \left|\frac{17.3 \text{ } \mu\text{N}}{44.5 \text{ } \mu\text{N}}\right| \quad \text{and} \quad \theta = 21.2° \text{ N of E}$$

Therefore, the resultant force on q_3 is 47.7 μN directed 21° N of E.

Summary and Review

Summary

Electrostatics is the science that treats charges at rest. We have seen that there are two kinds of charge that exist in nature. If an object has an excess of electrons, it is said to be *negatively* charged; if it has a deficiency of electrons, it is *positively* charged. Coulomb's law was introduced to provide a quantitative measure of electrical forces between such charges. The major concepts are listed as follows.

- The first law of electrostatics states that *like charges repel each other and unlike charges attract each other.*
- Coulomb's law states that *the force of attraction or repulsion between two point charges is directly proportional to the product of the two charges and inversely proportional to the separation of the two charges.*

$$F = \frac{kqq'}{r^2} \qquad \text{Coulomb's Law}$$

$$k = 9 \times 10^9 \text{ N} \cdot \text{m}^2/\text{C}^2$$

The force F is in newtons (N) when the separation r is in meters (m) and the charge q is measured in coulombs (C).

- When solving the problems in this chapter, it is important to use the sign of the charges to determine the *direction* of forces and Coulomb's law to determine their *magnitudes*. The resultant force on a particular charge is then found by the methods of vector mechanics.

Key Terms

charging   463
charging by induction   469
conductor   467
coulomb   471
Coulomb's law   470
electron   466

electroscope   464
electrostatics   463
first law of electrostatics   466
insulator   467
ion   466
microcoulomb   471

negative charge   469
neutron   466
pith ball   464
positive charge   469
semiconductor   467

Review Questions

23.1. Discuss several examples of static electricity in addition to those mentioned in the text.

23.2. In the process of rubbing a glass rod with a silk cloth, is charge *created?* Explain.

23.3. What is the nature of the charge on the silk cloth in Question 23.2?

23.4. An insulated stand supports a charged metal ball in the laboratory. Describe several procedures for determining the nature of the charge on the ball.

23.5. During an experiment in the laboratory, two bodies are seen to attract each other. Is this conclusive proof that they are both charged? Explain.

23.6. Two bodies are found to repel each other with an electric force. Is this conclusive proof that they are both charged? Explain.

23.7. One of the fundamental principles of physics is the principle of the conservation of charge, which states that *the total quantity of electric charge in the universe does not change.* Can you offer reasons for accepting this law?

23.8. Describe what happens to the leaf of a positively charged electroscope as (a) a negatively charged rod is brought closer and closer to the knob without touching it, (b) a positively charged rod is brought closer and closer to the knob.

23.9. When charging the leaf electroscope by induction, should the finger be removed before the charging rod is taken away? Explain.

23.10. List the units for each parameter in Coulomb's law for SI units.

23.11. Coulomb's law is valid only when the separation r is large in comparison with the radii of the charge. What accounts for this limitation?

23.12. How many electrons would be required to give a metal sphere a negative charge of (a) 1 C, (b) 1 μC?

Problems

Section 23.7 Coulomb's Law

23.1. Two balls, each having a charge of 3 μC, are separated by 20 mm. What is the force of repulsion between them?   Ans. 202 N

23.2. Two point charges of -3 and $+4$ μC are 12 mm apart in a vacuum. What is the electrostatic force between them?

23.3. An alpha particle consists of two protons ($q_e = 1.6 \times 10^{-19}$ C) and two neutrons (no charge). What is the repulsive force between two alpha particles separated by 2 nm?
                Ans. 2.30×10^{-10} N

23.4. Assume the radius of the electron's orbit around the proton in a hydrogen atom is approximately 5.2×10^{-11} m. What is the electrostatic force of attraction?

23.5. What is the separation of two -4-μC charges if the force of repulsion between them is 200 N?
                Ans. 26.8 mm

23.6. Two identical charges separated by 30 mm experience a repulsive force of 980 N. What is the magnitude of each charge?

*23.7. A 10-μC charge and a -6-μC charge are separated by 40 mm. What is the force between them? The spheres are placed in contact for a few moments and then separated again by 40 mm. What is the new force? Is it attraction or repulsion?
        Ans. 338 N, attraction; 5.62 N, repulsion

*23.8. Two point charges initially attract each other with a force of 600 N. If their separation is reduced to one-third of its original value, what is the new force of attraction?

23.9. A $+60$-μC charge is placed 60 mm to the left of a $+20$-μC charge. What is the resultant force on a -35-μC charge placed midway between the two charges?   Ans. 1.40×10^4 N, left

23.10. A point charge of $+36$ μC is placed 80 mm to the left of a second point charge of -22 μC. What force is exerted on a third charge of $+10$ μC placed at the midpoint?

23.11. For Prob. 23.10, what is the resultant force on a third charge of $+12$ μC placed between the other charges and located 60 mm from the $+36$-μC charge?   Ans. 7020 N, right

23.12. A $+6$-μC charge is 44 mm to the right of a -8-μC charge. What is the resultant force on a -2-μC charge that is 20 mm to the right of the -8-μC charge?

*23.13. A 64-μC charge is located 30 cm to the left of a 16-μC charge. What is the resultant force on a -12-μC charge positioned exactly 50 mm below the 16-μC charge?
                Ans. 2650 N, 113.3°

*23.14. A charge of $+60$ nC is located 80 mm above a -40-nC charge. What is the resultant force on a -50-nC charge located 45 mm horizontally to the right of the -40-nC charge?

*23.15. Three point charges, $q_1 = +8$ μC, $q_2 = -4$ μC, and $q_3 = +2$ μC, are placed at the corners of an equilateral triangle, 80 mm on each side. Assume that the base of the triangle is formed by a line joining the 2- and -4-μC charges. What are the magnitude and direction of the resultant force on the $+8$-μC charge?   Ans. 39 N, 330°

Additional Problems

23.16. What should be the separation of two $+5$-μC charges so that the force of repulsion is 4 N?

23.17. The repulsive force between two pith balls is found to be 60 μN. If each pith ball carries a charge of 8 nC, what is their separation?
                Ans. 98 mm

23.18. Two identical unknown charges experience a mutual repulsive force of 48 N when separated by 60 mm. What is the magnitude of each charge?

23.19. One object contains an excess of 5×10^{14} electrons, and another has a deficiency of 4×10^{14} electrons. What is the force each exerts on the other

if the objects are 30 mm apart? Is it attraction or repulsion?   Ans. 5.12×10^4 N, attraction

23.20. If it were possible to put 1 C of charge on each of two spheres separated by a distance of 1 m, what would be the repulsive force in newtons?

23.21. How many electrons must be placed on each of two spheres separated by 4 mm in order to produce a repulsive force of 400 N?
                Ans. 5.27×10^{12} electrons

23.22. A -40-nC charge is placed 40 mm to the left of a $+6$-nC charge. What is the resultant force on a -12-nC charge placed 8 mm to the right of the $+6$-nC charge?

23.23. A 5-μC charge is placed 6 cm to the right of a 2-μC charge. What is the resultant force on a -9-nC charge placed 2 m to the left of the 2-μC charge? **Ans. 468 mN, right**

23.24. An equal number of electrons are placed on two metal spheres 3.0 cm apart in air. How many electrons are on each sphere if the resultant force is 4500 N?

23.25. A 4-nC charge is placed on a 4-g sphere that is free to move. A fixed 10-μC point charge is 4 cm away. What is the initial acceleration of the 4-μC charge? **Ans. 56.2 m/s^2**

*23.26. Find the resultant force on a $+2$-μC charge that is 60 mm from each of two -4-μC charges that are 80 mm apart in air.

*23.27. Two charges of $+25$ and $+16$ μC are 80 mm apart. A third charge of $+60$ μC is placed between the other charges 30 mm from the $+25$-μC charge. Find the resultant force on the third charge. **Ans. 1.15 $\times$ 10^4 N**

*23.28. A 0.02-g pith ball is suspended freely. The ball is given a charge of $+20$ μC and placed 0.6 m from a charge of $+50$ μC. What will be the initial acceleration of the pith ball?

*23.29. A 4-μC charge is located 6 cm from an 8-μC charge. At what point on a line joining the two charges will the resultant force be zero? **Ans. 2.49 cm from 4-μC charge**

*23.30. A charge of $+8$ nC is placed 40 mm to the left of a -14-nC charge. Where should a third charge be placed if it is to experience a zero resultant force?

*23.31. A $+16$-μC charge is 80 mm to the right of a $+9$-μC charge. Where should a third charge be placed so that the resultant force is zero? **Ans. 34.3 mm right of the 9-μC charge**

*23.32. Two 3-g spheres are suspended from a common point with two 80-mm light silk threads of negligible mass. What charge must be placed on each sphere if their final positions are 50 mm apart?

Critical Thinking Questions

*23.33. A small metal sphere is given a charge of $+40$ μC, and a second sphere 8 cm away is given a charge of -12 μC. What is the force of attraction between them? If the two spheres are allowed to touch and are then again placed 8 cm apart, what new electric force exists between them? Is it attraction or repulsion? **Ans. 675 N, attraction 276 N, repulsion**

*23.34. The total charge on two metal spheres 50 mm apart is 80 μC. If they repel each other with a force of 800 N, what is the charge on each sphere?

*23.35. Four small spheres are each given charges of $q = +20$ μC and placed at the corners of a square with sides of length 6 cm. Show that the resultant force on each charge has a magnitude

equal to 1914 N. What is the direction of the force? What will change if the charges are each $q = -20$ μC? **Ans. 1914 N, 45° away from center**

*23.36. Two charges q_1 and q_2 are separated by a distance r. They experience a force F at this distance. If the initial separation is decreased by only 40 mm, the force between the two charges is doubled. What was the initial separation?

*23.37. Two 8-g pith balls are suspended from silk threads 60 cm long and attached to a common point. When the spheres are given equal amounts of negative charge, the balls come to rest 30 cm apart. Calculate the magnitude of the charge on each pith ball. **Ans. -450 nC**

24

The Electric Field

Smart windows can change a clear, transparent view into a frosty or even dark look with the flip of a wall switch. Tiny particles, known as suspended particle devices (SPD), are placed between two panels of transparent conductive material. When an electric field is established between these panels, these particles line up in a straight line, allowing light to flow through. When the field is removed, they move back to their random orientation and block the light.
(*Courtesy of Switchlite Privacy Glass®, Saint-Gobain Glass Exprover.*)

Objectives

After completing this chapter, you should be able to

1. Define the *electric field* and explain what determines its magnitude and direction.

2. Write and apply an expression that relates the electric field intensity at a point to the distance(s) from the known charge(s).

3. Explain and illustrate the concept of electric field lines and discuss the two rules that must be followed in the construction of such lines.

4. Explain the concept of the *permittivity* of a *medium* and how it affects the field intensity and the construction of field lines.

5. Write and apply *Gauss's law* as it relates to the electric fields surrounding surfaces of known charge density.

In our study of mechanics, we discussed force and motion at great length. Newton's laws of motion were normally used to describe the application and consequences of *contact* forces. A moment's reflection on the universe as a whole convinces us of the enormous number of objects that are *not* in contact.

A projectile experiences a downward force that cannot be explained in terms of its interaction with air particles, planets revolve continuously through the void surrounding the Sun, and the Sun is pulled along an elliptical path by forces that do not touch it. Even at the atomic level, there are no "strings" to hold the electrons in their orbits about the nucleus.

If we are really to understand our universe, we must develop laws to predict the magnitude and direction of forces that are not transmitted by contact. Two such laws have already been discussed:

1. Newton's law of universal gravitation:

$$F_g = G \frac{m_1 m_2}{r^2}$$

(**24.1**)

2. Coulomb's law for electrostatic forces:

$$F_e = k \frac{q_1 q_2}{r^2}$$

(**24.2**)

Newton's law predicts the force that exists between two masses separated by a distance r; Coulomb's law deals with the electrostatic force, as discussed in Chapter 23. In applying such laws, we find it useful to develop certain properties of the space surrounding masses or charges.

24.1 The Concept of a Field

PHYSICS TODAY

Your heart uses an electric potential to make the heart muscle twitch, which causes the blood to be pumped through your body. This potential creates an electric field, which can be monitored through an electrocardiogram (EKG).

Both the electric force and the gravitational force are examples of *action-at-a-distance forces,* which are extremely difficult to visualize. To overcome this fact, early physicists postulated the existence of an invisible material, called *ether,* that was thought to pervade all space. The gravitational force of attraction could then be due to strains in the ether caused by the presence of various masses. Certain optical experiments have now shown the ether theory to be untenable, and we are forced to consider whether space itself possesses properties of interest to the physicist.

It may be postulated that the mere presence of a mass alters the space surrounding it to produce a gravitational force on another nearby mass. We describe this alteration in space by introducing the concept of a *gravitational field* that surrounds all masses. Such a field may be said to exist in any region of space where a test mass will experience a gravitational force. The strength of the field at any point would be proportional to the force a given mass experiences at that point. For example, at every point in the vicinity of the Earth, the gravitational field could be represented quantitatively by

$$\mathbf{g} = \frac{\mathbf{F}}{m}$$

(**24.3**)

where $\mathbf{g}$ = acceleration due to gravity

$\mathbf{F}$ = gravitational force

m = test mass (see Fig. 24.1)

If $\mathbf{g}$ is known at every point above the Earth, the force $\mathbf{F}$ that will act on a given mass m placed at that point can be determined from Eq. (24.3).

The concept of a field can also be applied to electrically charged objects. The space surrounding a charged object is altered by the presence of the charge. We may postulate the existence of an **electric field** in this space.

An electric field is said to exist in a region of space in which an electric charge will experience an electric force.

This definition provides a test for the existence of an electric field. Simply place a charge at the point in question. If an electric force is observed, an electric field exists at that point.

Just as the force per unit mass provides a quantitative definition of a gravitational field, the strength of an electric field can be represented by the force per unit charge. We define the electric field intensity $\mathbf{E}$ at a point in terms of the force $\mathbf{F}$ experienced by a small positive

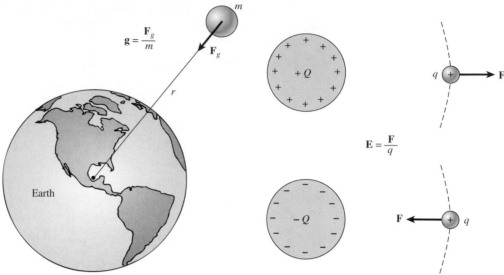

Figure 24.1 The gravitational field at any point above the Earth can be represented by the acceleration **g** that a small mass m would experience if it were placed at that point.

Figure 24.2 The direction of the electric field intensity at a point is the same as the direction in which a positive charge $+q$ would move when placed at that point. Its magnitude is the force per unit charge (F/q).

charge $+q$ when it is placed at that point (see Fig. 24.2). The magnitude of the electric field intensity is given by

$$E = \frac{F}{q} \tag{24.4}$$

In the metric system, a unit of *electric field intensity* is the newton per coulomb (N/C). The usefulness of this definition rests with the fact that, if the field is known at a given point, we can predict the force that will act on any charge placed at that point.

Since the electric field intensity is defined in terms of a *positive* charge, its direction at any point is the same as the electrostatic force on a positive charge at that point.

> The direction of the electric field intensity **E** at a point in space is the same as the direction of the force a positive test charge would experience if it were placed at that point.

On this basis, the electric field in the vicinity of a positive charge $+Q$ would be outward, or away from the charge, as indicated by Fig. 24.3a. In the vicinity of a negative charge $-Q$, the direction of the field would be inward, or toward the charge (Fig. 24.3b).

It must be remembered that the electric field intensity is a property assigned to the *space* that surrounds a charged body. A gravitational field exists above the Earth, whether or not a mass is positioned above the Earth. Similarly, an electric field exists in the neighborhood of a charged body, whether or not a second charge is positioned in the field. If a charge *is* placed in the field, it will experience a force **F** given by

$$\mathbf{F} = q\mathbf{E} \tag{24.5}$$

where **E** = field intensity

q = magnitude of charge placed in field

If q is positive, **E** and **F** will have the same direction; if q is negative, the force **F** will be directed opposite to the field **E**.

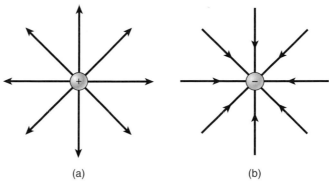

Figure 24.3   (a) The field in the vicinity of a positive charge is directed radially outward at every point. (b) The field is directed inward or toward a negative charge.

Example 24.1

The electric field intensity between the two plates in Fig. 24.4 is constant and directed downward. The magnitude of the electric field intensity is 6×10^4 N/C. What are the magnitude and the direction of the electric force exerted on an electron projected horizontally between the two plates?

Plan: The direction of the field intensity **E** is defined in terms of the force on a positive test charge. The charge of an electron is *negative* ($q_e = -1.6 \times 10^{-19}$ C), meaning that the force on the electron is *upward* (opposite to the field direction). The field intensity is the force per unit charge, so the magnitude of the force will be the product $q_e E$.

Solution: The force, from Eq. (24.5), is

$$F = q_e E = (1.6 \times 10^{-19}\ \text{C})(6 \times 10^4\ \text{N/C})$$
$$= 9.6 \times 10^{-15}\ \text{N}\quad (\textit{upward})$$

Remember that the absolute value of the charge is used. The *direction* of the force **F** on a positive charge is the same as the direction of the field intensity **E;** the force on a negative charge is *opposite* to the field.

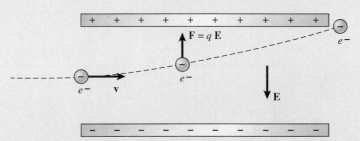

Figure 24.4   An electron projected into an electric field of constant intensity.

Example 24.2

Given that the mass of an electron is 9.1×10^{-31} kg, show that the gravitational force on the electron in Example 24.1 may be neglected.

Plan: The gravitational force is *downward* and is due to the weight ($W = mg$) of the electron. To see what effect this will have on the motion of the electron, we need to see if the weight is significant when compared to the magnitude of the electric field force.

Solution: The weight is the product of the electron mass and gravity.

$$W = mg = (9.1 \times 10^{-31} \text{ kg})(9.8 \text{ m/s}^2)$$
$$W = 8.92 \times 10^{-30} \text{ N}$$

The electric force is larger than the gravitational force by a factor of 1.08×10^{15} and can certainly be neglected in this instance.

24.2 Computing the Electric Intensity

We have discussed one method of measuring the magnitude of the electric field intensity at a point in space. A known charge is placed at the point, and the resultant force is measured. The force per unit charge is then a measure of the electric intensity at that point. The disadvantage of this method is that it bears no obvious relationship to the charge Q that creates the field. Experimentation will quickly show that the magnitude of the electric field surrounding a charged body is directly proportional to the quantity of charge on the body. It can also be demonstrated that at points farther and farther away from a charge Q, a test charge q will experience smaller and smaller forces. The exact relationship is derived from Coulomb's law.

Suppose we wish to calculate the field intensity E at a distance r from a single charge Q, as shown in Fig. 24.5. The force F that Q exerts on a test charge q at the point in question is, from Coulomb's law,

$$F = \frac{kQq}{r^2} \tag{24.6}$$

Substituting this value for F into Eq. (24.4), we obtain

$$E = \frac{F}{q} = \frac{kQq/r^2}{q}$$

$$E = \frac{kQ}{r^2} \tag{24.7}$$

where k is equal to 9×10^9 N $\cdot$ m²/C². The direction of the field is away from Q if Q is positive and toward Q if Q is negative. We now have a relation that allows us to compute the field intensity at a point without having to place a second charge at the point.

Example 24.3

What is the electric field intensity at a distance of 2 cm from a charge of $-12\ \mu$C?

Plan: The charge Q is negative, so the direction of the field will be radially inward, toward the charge. The magnitude is found from Eq. (24.7). Use consistent units.

Solution: Substituting $r = 0.02$ m and $Q = 12 \times 10^{-6}$ C, we obtain

$$E = \frac{kQ}{r^2} = \frac{(9 \times 10^9 \text{ N} \cdot \text{m}^2/\text{C}^2)(12 \times 10^{-6} \text{ C})}{(0.02 \text{ m})^2}$$

$$E = 2.70 \times 10^8 \text{ N/C, toward } Q$$

When more than one charge contributes to the field, as in Fig. 24.6, the resultant field is the vector sum of the contributions due to each charge considered independently.

$$\mathbf{E} = \mathbf{E}_1 + \mathbf{E}_2 + \mathbf{E}_3 + \dots \qquad \textit{Vector sum} \quad \textbf{(24.8)}$$

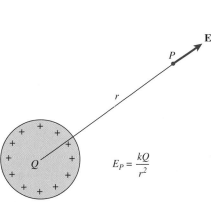

Figure 24.5 Calculating the electric field intensity at a distance r from the center of a single charge Q.

$$E_P = \frac{kQ}{r^2}$$

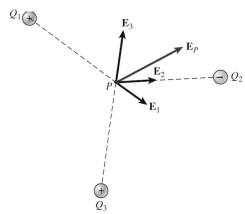

Figure 24.6 The field in the vicinity of a number of charges is equal to the vector sum of the fields due to the individual charges.

The direction of each field is found by considering the force that would be experienced by a positive test charge at the point in question, and the magnitude of the field is found from Eq. (24.7). The resultant electric intensity is then found by the component method of vector addition.

Example 24.4

Two point charges, $q_1 = -6$ nC and $q_2 = +8$ nC, are 12 cm apart, as shown in Figure 24.7. Determine the electric field intensity at point A and at point B, using the information provided in the figure.

Plan: The electric field intensity is a property of *space*. There is no charge at either point A or B in this example. To determine the direction of the field at point A or B, we must imagine a small test positive charge is placed at the point and then recognize the field has the same direction as the force on this test charge. From the figure, we notice the field $\mathbf{E}_1$ at point A due to the -6 nC charge is directed to the left, and the field $\mathbf{E}_2$ due to the $+8$ nC charge is also directed to the left. In each case, this is the way a test positive charge would move when placed at A. The magnitudes of each field are given by applying Eq. (24.7), and the vector sum will give the resultant field intensity at A. Similar reasoning will give the field at point B. However, the math is more complicated due to the $37°$ angle for the $\mathbf{E}_2$ vector.

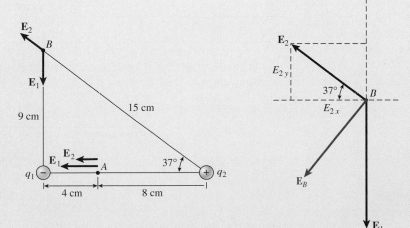

Figure 24.7

Solution (a): The field $\mathbf{E}_1$ at point A due to q_1 is to the *left*. Its magnitude is

$$E_1 = \frac{kq_1}{r_1^2} = \frac{(9 \times 10^9 \text{ N} \cdot \text{m}^2/\text{C}^2)(6 \times 10^{-9} \text{ C})}{(0.04 \text{ m})^2}$$

$$= 3.38 \times 10^4 \text{ N/C, to the left}$$

Remember that the sign of the charge determines the direction of the field, and the negative sign is not used in calculating the *magnitude* of the field.

The electric field $\mathbf{E}_2$ at point A due to q_2 is directed to the *left* and equal to

$$E_2 = \frac{kq_2}{r_2^2} = \frac{(9 \times 10^9 \text{ N} \cdot \text{m}^2/\text{C}^2)(8 \times 10^{-9} \text{ C})}{(0.08 \text{ m})^2}$$

$$= 1.12 \times 10^4 \text{ N/C, to the left}$$

Since the two vectors $\mathbf{E}_1$ and $\mathbf{E}_2$ are each directed to the left, the resultant vector is the simple sum of their magnitudes. Taking the leftward direction as negative, we have

$$\mathbf{E}_1 + \mathbf{E}_2 = -3.38 \times 10^4 \text{ N/C} - 1.12 \times 10^4 \text{ N/C}$$

$$= -4.50 \times 10^4 \text{ N/C (directed to the left)}$$

Solution (b): The field intensity $\mathbf{E}_1$ at B due to q_1 is directed *downward* and is equal to

$$E_1 = \frac{kq_1}{r_1^2} = \frac{(9 \times 10^9 \text{ N} \cdot \text{m}^2/\text{C}^2)(6 \times 10^{-9} \text{ C})}{(0.09 \text{ m})^2}$$

$$= 6.67 \times 10^3 \text{ N/C, downward}$$

The field $\mathbf{E}_2$ at B due to q_2 is directed *away* from q_2 at an angle of 37° N of W and is given by

$$E_2 = \frac{kq_2}{r_2^2} = \frac{(9 \times 10^9 \text{ N} \cdot \text{m}^2/\text{C}^2)(8 \times 10^{-9} \text{ C})}{(0.15 \text{ m})^2}$$

$$= 3.20 \times 10^3 \text{ N/C, 37° N of W}$$

Table 24.1 lists components used to find the resultant field at point B.

Table 24.1

Vector	Angle ϕ_x	x component	y component
$E_1 = 6.67$ kN/C	90°	$E_{1x} = 0$	$E_{1y} = -6.67$ kN/C
$E_2 = 3.20$ kN/C	37°	$E_{2x} = -(3.20 \text{ kN/C})\cos 37°$	$E_{2y} = (3.20 \text{ kN/C}) \sin 37°$
		$= -2.56$ kN/C	$= 1.93$ kN/C
E	θ	$E_x = \Sigma E_x = -2.56$ kN/C	$E_y = \Sigma E_y = -4.74$ kN/C

From Fig. 24.7, we apply the Pythagorean theorem to find the magnitude of the resultant electric field $\mathbf{E}$ at point B.

$$E = \sqrt{E_x^2 + E_y^2}$$

$$= \sqrt{(2.56 \text{ kN/C})^2 + (4.74 \text{ kN/C})^2} = 5.39 \text{ kN/C}$$

Next, the direction is found from the tangent function.

$$\tan \phi = \left| \frac{E_y}{E_x} \right| = \left| \frac{4.74 \text{ kN/C}}{2.56 \text{ kN/C}} \right| \quad \text{and} \quad \phi = 61.6° \text{ S of W}$$

Therefore, the resultant electric field intensity at point B is 5.39 kN/C directed 61.6° S of W.

Problem-Solving Strategy

Electric Fields

1. Read the problem and then draw and label a figure. Indicate positive and negative charges along with given distances. The charges must be in *coulombs,* and the distances must be in *meters.* Remember that $1 \, \mu C = 1 \times 10^{-6} \, C$ and $1 \, nC = 1 \times 10^{-9} \, C$.

2. Remember that the electric field **E** is a property of *space* that allows us to determine the force **F** a unit positive charge q would experience *if* it were placed at a certain point in space. The field exists at a *point* in space independent of whether or not a charge is placed at that point.

3. The magnitude of the electric field due to a single charge is given by

$$E = \frac{kQ}{r^2} \qquad k = 9 \times 10^9 \, N \cdot m^2/C^2$$

4. As discussed for forces, care must be taken not to confuse the nature of a charge (+ or −) with the sign given to electric fields or their components. The direction of the field **E** at a given point is consistent with the direction a *positive* test charge would move if placed at that point.

5. The *resultant electric field* due to a number of charges is found by vector addition of the electric fields due to each charge considered independently. Construct a free-body diagram and proceed with vector addition by the component method.

$$\mathbf{E} = \mathbf{E}_1 + \mathbf{E}_2 + \mathbf{E}_3 + \ldots \qquad \textit{Vector sum}$$

24.3   Electric Field Lines

An ingenious aid to the visualization of electric fields was introduced by Michael Faraday (1791–1867) in his early work in electromagnetism. The method consists of representing both the strength and the direction of an electric field by imaginary lines called **electric field lines.**

> Electric field lines are imaginary lines drawn in such a manner that their direction at any point is the same as the direction of the electric field at that point.

For example, the lines drawn radially outward from the positive charge in Fig. 24.3a represent the direction of the field at any point on the line. The electric lines in the vicinity of a negative charge would be radially inward and directed toward the charge, as in Fig. 24.3b. We shall see later that the density of these lines in any region of space is a measure of the *magnitude* of the field intensity in that region.

In general, the direction of the electric field in a region of space varies from place to place. Thus, the electric lines are normally curved. For instance, let us consider the construction of an electric field line in the region between a positive charge and a negative charge, as illustrated in Fig. 24.8.

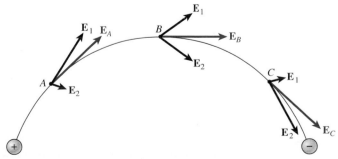

Figure 24.8   The direction of an electric field line at any point is the same as the direction of the resultant electric field intensity at that point.

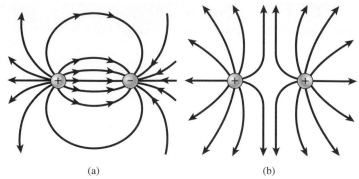

Figure 24.9   (a) A graphical illustration of the electric field lines in the region surrounding two opposite charges. (b) The field lines between two positive charges.

The direction of the electric field line at any point is the same as the direction of the resultant electric field vector at that point. Two rules must be followed when constructing electric field lines:

1. The direction of the field line at any point is the same as the direction in which a positive charge would move if placed at that point.

2. The spacing of the field lines must be such that they are close together where the field is strong and far apart where the field is weak.

Following these general rules, one can construct the electric field lines for the two common cases shown in Fig. 24.9. As a consequence of how electric lines are drawn, *they will always leave positive charges and enter negative charges.* No lines can originate or terminate in space, although one end of an electric line may proceed to infinity.

24.4   Gauss's Law

For any given charge distribution, we can draw an infinite number of electric lines. Clearly, if the spacing of the lines is to be a standardized indication of field strength, we must set a limit on the number of lines drawn in any situation. For example, let us consider the field lines directed radially outward from a positive point charge. (See Fig. 24.10.) We will use the letter N to represent the number of lines drawn. Now let us imagine a spherical surface surrounding the point charge at a distance r from the charge. The field intensity at every point on such a sphere would be given by

$$E = \frac{kq}{r^2} \tag{24.9}$$

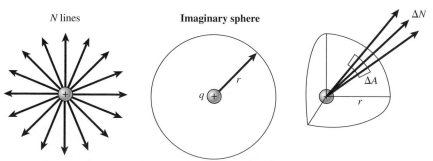

Figure 24.10   Electric field intensity at a distance r from a point charge is directly proportional to the number of lines ΔN penetrating a unit area ΔA of an imaginary spherical surface constructed at that distance.

From the way the field lines are drawn, we might also say that the field at a tiny element of surface area ΔA is proportional to the number of lines ΔN penetrating that area. In other words, the density of lines (lines per unit area) is directly proportional to the field strength. Symbolically,

$$\frac{\Delta N}{\Delta A} \propto E_n \qquad (24.10)$$

The subscript n indicates that the field is everywhere normal to the surface area. This proportionality is true regardless of the total number of lines N that may be drawn. Once we choose a proportionality constant for Eq. (24.10), however, we automatically set a limit to the number of lines drawn for any situation. It has been found that the most convenient choice for this spacing constant is ϵ_0. It is called the ***permittivity of free space*** and is defined by

$$\epsilon_0 = \frac{1}{4\pi k} = 8.85 \times 10^{-12} \text{ C}^2/\text{N} \cdot \text{m}^2 \qquad (24.11)$$

where $k = 9 \times 10^9$ N $\cdot$ m^2/C^2 from Coulomb's law. Hence, Eq. (24.10) can be written

$$\frac{\Delta N}{\Delta A} = \epsilon_0 E_n \qquad (24.12)$$

or

$$\Delta N = \epsilon_0 E_n \Delta A \qquad (24.13)$$

When E_n is constant over the entire surface, the total number of lines radiating outward from the enclosed charge is

$$N = \epsilon_0 E_n A \qquad (24.14)$$

It can be seen that the choice of ϵ_0 is a convenient one by substituting Eq. (24.11) into Eq. (24.9):

$$E_n = \frac{1}{4\pi\epsilon_0} \frac{q}{r^2}$$

Substituting this expression into Eq. (24.14) and recalling that the area of a spherical surface is $A = 4\pi r^2$, we obtain

$$N = \epsilon_0 E_n A$$

$$= \frac{\epsilon_0}{4\pi\epsilon_0} \frac{q}{r^2}(4\pi r^2) = q$$

The choice of ϵ_0 as the proportionality constant has resulted in the fact that *the total number of lines passing normally through a surface is numerically equal to the charge contained within the surface.* Although this result was obtained using a spherical surface, it will apply to any other surface. The more general statement of the result is known as ***Gauss's law:***

The net number of electric lines of force crossing any closed surface in an outward direction is numerically equal to the net total charge within that surface.

$$N = \sum \epsilon_0 E_n A = \sum q \qquad \textit{Gauss's Law} \quad (24.15)$$

Gauss's law can be used to compute the field intensity near surfaces of charge. This represents a distinct advantage over methods developed so far because the previous equations apply only to point charges. The best way to understand the application of Gauss's law is through examples.

24.5   Applications of Gauss's Law

Since most charged conductors have large quantities of charge on them, it is not practical to treat the charges individually. Generally, we speak of the *charge density* σ, defined as the charge per unit area of surface.

$$\sigma = \frac{q}{A} \qquad q = \sigma A \qquad \text{\textit{Charge Density}} \quad \textbf{(24.16)}$$

Example 24.5   Calculate the electric field intensity at a distance r from an infinite sheet of positive charge like the one illustrated in Fig. 24.11.

Plan: Our goal in using Gauss's law is to find an expression relating the electric field to charge density σ. Applications of Gauss's law usually require the construction of an imaginary geometric surface called a *gaussian surface*. The idea is to enclose a net charge inside a surface whose geometry is simple so that its area can be easily determined. The choice of an imaginary surface is dictated by the shape of the charged body. In this example, a wise choice is a cylindrical surface that penetrates the sheet of positive charge so it projects a distance r on either side of the thin sheet. The total charge Σq enclosed by this surface must be equal to $\Sigma \epsilon_0 EA$ according to Gauss's law, and we will use this fact to find an expression for the electric field intensity at the distance r.

Solution: Since the diameter of the cylinder is arbitrary, we will find it convenient to work with the charge density σ, as defined in Eq. (24.16). The area A of each end of the cylinder is the same as the area cut on the sheet of charge. Thus, the total charge contained inside the cylinder is given by

$$\sum q = \sigma A$$

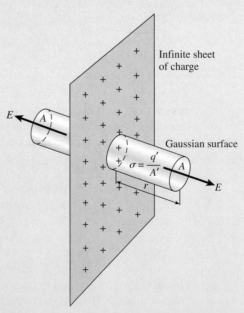

Figure 24.11   Computing the field outside an infinite sheet of positive charge.

Because of the symmetry of the sheet of charge, the resultant field intensity **E** must be directed perpendicular to the sheet at any nearby point. Only the lines of intensity passing through perpendicular to the two surfaces of area *A* need be considered. From Gauss's law, we may write

$$\sum \epsilon_0 EA = \sum q$$
$$\epsilon_0 EA + \epsilon_0 EA = \sigma A$$
$$2\epsilon_0 EA = \sigma A$$

$$E = \frac{\sigma}{2\epsilon_0} \tag{24.17}$$

Note that the field intensity **E** is directed away from the sheet on each side and is independent of the distance *r* from the sheet.

Before you assume that the example of an infinite sheet of charge is impractical, it should be pointed out that *infinity* in a practical sense implies only that the dimensions of the sheet are beyond the point of electrical interaction. In other words, Eq. (24.17) applies when the length and width of the sheet are very large in comparison with the distance *r* from the sheet.

Example 24.6

Show by using Gauss's law that all the excess charge resides on the surface of a charged conductor.

Plan: We will first draw a shape representing an arbitrary charged conductor, such as the one shown in Fig. 24.12. Within such a conductor, charges are free to move if they experience a resultant force. Since like charges repel, we can safely assume that all free charges in the conductor eventually come to rest. Under this condition, the electric field within the conductor must be zero. Otherwise, the charges would be moving. We will construct an imaginary gaussian surface just inside the surface of the conductor, as shown in Fig. 24.12, and apply Gauss's law.

Solution: To show that all the charge is on the surface, we write

$$\sum \epsilon_0 EA = \sum q$$

Substituting $E = 0$, we also find the $\sum q = 0$ or that no charge is enclosed by the surface. Since the gaussian surface can be drawn as close to the outside of the conductor as we wish, we can conclude that all the charge resides on the surface of the conductor. This conclusion holds, even if the conductor is hollow on the inside.

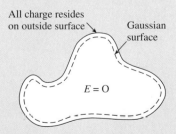

All charge resides on outside surface

Gaussian surface

$E = 0$

Figure 24.12   Gauss's law demonstrating that all the charge resides on the surface of a conductor.

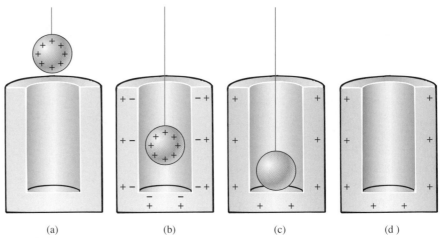

Figure 24.13 Faraday's ice-pail experiment demonstrating surface retention of net charge.

An interesting experiment was devised by Michael Faraday to demonstrate that charge resides on the surface of a hollow conductor. In this experiment, a positively charged ball supported by a silk thread is lowered into a hollow metallic conductor. As shown in Fig. 24.13, a redistribution of charge occurs on the walls of the conductor, drawing the electrons to the inner surface. When the ball makes contact with the bottom of the conductor, the induced charge is neutralized, leaving a net positive charge on the outer surface. Probes with an electroscope will show that no charge resides inside the conductor and that a net positive charge remains on the outer surface.

Example 24.7

A *capacitor* is an electrostatic device consisting of two conductors of area A separated by a distance d. (See Fig. 24.14.) If equal and opposite charges are placed on the conductors, an electric field $\mathbf{E}$ will exist between them. Derive an expression for calculating this electric field in terms of the charge density on the plates.

Plan: We will draw a gaussian cylinder for the inner surface of either plate, as shown in Fig. 24.14. There is no field inside the conducting plate, and the only area penetrated by the field lines is the surface area A' projecting into the space between the plates. We can apply Gauss's law to either plate to find an expression for the electric field intensity.

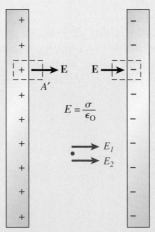

$$E = \frac{\sigma}{\epsilon_0}$$

Figure 24.14 Electric field in the region between two oppositely charged plates is equal to the ratio of the charge density σ to the permittivity ϵ_0.

Solution: Recognizing, that $\Sigma q = \sigma A'$, we solve for E as follows:

$$\sum \epsilon_0 EA = \sum q$$

$$\epsilon_0 EA' = \sigma A'$$

$$E = \frac{\sigma}{\epsilon_0} \tag{24.18}$$

The same result would be obtained if the gaussian cylinder on the right were used. The lines would be directed toward the inside in this case, reflecting the fact that the enclosed charge is *negative*.

Notice that the field between the two plates in Example 24.7 is exactly twice the field due to a thin sheet of charge, as given by Eq. (24.17). One can understand this relationship by considering the field E between the plates as a superposition of the fields due to two oppositely charged sheets.

$$E = E_1 + E_2 = \frac{\sigma}{2\epsilon_0} + \frac{\sigma}{2\epsilon_0} = \frac{\sigma}{\epsilon_0}$$

The field E_1 due to the positive sheet of charge is in the same direction as the field E_2 due to the negative sheet of charge. On the *outside* of the two plates to the left or right E_1 and E_2 are oppositely directed and cancel out, making the resulting field intensity outside the plates equal to zero.

Summary and Review

Summary

The concept of an electric field has been introduced to describe the region surrounding an electric charge. Its magnitude is determined by the force a unit charge will experience at a given location, and its direction is the same as the force on a positive charge at that point. Electric field lines were postulated to give a visual picture of electric fields, and the density of such field lines is an indication of the intensity of the electric field. The major concepts to be remembered are summarized as follows.

- An *electric field* is said to exist in a region of space in which an electric charge will experience an electric force. The *magnitude* of the electric field intensity E is given by the force F per unit of charge q.

$$E = \frac{F}{q} \qquad E = \frac{(9 \times 10^9 \text{ N} \cdot \text{m}^2/\text{C}^2)Q}{r^2}$$

The metric unit for the electric field intensity is the newton per coulomb (N/C). In the previous equation, r is the distance from the charge Q to the point in question.

- The resultant field intensity at a point in the vicinity of a number of charges is the *vector* sum of the contributions from each charge.

$$\mathbf{E} = \mathbf{E}_1 + \mathbf{E}_2 + \mathbf{E}_3 + \cdots \qquad E = \sum \frac{kQ}{r^2}$$

Vector Sum

It must be emphasized that this is a vector sum and not an algebraic sum. Once the magnitude and direction of each vector is determined, the resultant can be found from vector mechanics.

- The permittivity of free space ϵ_0 is a fundamental constant defined as

$$\epsilon_0 = \frac{1}{4\pi k} = 8.85 \times 10^{-12} \text{ C}^2/\text{N} \cdot \text{m}^2$$

Permittivity

- Gauss's law states that the net number of electric field lines crossing any closed surface in an outward direction is numerically equal to the net total charge within that surface.

$$N = \sum \epsilon_0 E_n A = \sum q \qquad \textit{Gauss's Law}$$

- In applications of Gauss's law, the concept of charge density σ as the charge q per unit area A of surface is often utilized:

$$\sigma = \frac{q}{A} \qquad q = \sigma A \qquad \textit{Charge Density}$$

Key Terms

charge density   488
electric field   479
electric field intensity   480

electric field lines   485
gaussian surface   488

Gauss's law   487
permittivity ϵ_0 of free space   487

Review Questions

24.1. Some texts refer to electric field lines as "lines of force." Discuss the advisability of this description.

24.2. Can an electric field exist in a region of space in which an electric charge would not experience a force? Explain.

24.3. Is it necessary that a charge be placed at a point to have an electric field at that point? Explain.

24.4. Using a procedure similar to that for electric fields, show that the gravitational acceleration can be calculated from

$$g = \frac{GM}{r^2}$$

where M = mass of Earth
$ r$ = distance from center of Earth

24.5. Discuss the similarities between electric fields and gravitational fields. In what ways do they differ?

24.6. In Gauss's law, the constant ϵ_0 was chosen as the proportionality factor between line density and field intensity. In a theoretical sense, this is a wise choice because it leads to the conclusion that the total number of lines is equal to the enclosed charge. Is such a choice practical for graphically illustrating field lines? According to Gauss's relation, how many field lines would emanate from a charge of 1 C?

24.7. Justify the following statement: the electric field intensity on the surface of any charged conductor must be directed perpendicular to the surface.

24.8. Electric field lines will never intersect. Explain.

24.9. Suppose you connect an electroscope to the outside surface of Faraday's ice pail. Show graphically what will happen to the gold leaf during each of the steps illustrated in Fig. 24.13.

24.10. Can an electric field line begin and end on the same conductor? Discuss.

24.11. What form would Gauss's law take if we had chosen k for the proportionality constant instead of the permittivity ϵ_0?

24.12. In Gauss's law, demonstrate that the units of $\epsilon_0 EA$ are dimensionally equivalent to the units of charge.

24.13. Show that the field in the region outside the two parallel plates in Fig. 24.14 is equal to zero.

24.14. Why is the field intensity constant in the region between two oppositely charged plates? Draw a vector diagram of the field due to each plate at various points between the plates.

Problems

Section 24.1 The Concept of a Field

24.1. A charge of $+2\ \mu C$ placed at a point P in an electric field experiences a downward force of 8×10^{-4} N. What is the electric field intensity at that point?   Ans. 400 N/C, downward

24.2. A -5-nC charge is placed at point P in Prob. 24.1. What are the magnitude and direction of the force on the -5-nC charge?

24.3. A charge of $-3\ \mu C$ placed at point A experiences a downward force of 6×10^{-5} N. What is the electric field intensity at point A?
          Ans. 20 N/C, upward

24.4. At a certain point, the electric field intensity is 40 N/C, due east. An unknown charge receives a westward force of 5×10^{-5} N. What is the nature and magnitude of the charge?

24.5. What are the magnitude and direction of the force that would act on an electron ($q = -1.6 \times 10^{-19}$ C) if it were placed at (a) point P in Prob. 24.1? Or (b) point A in Prob. 24.3?
          Ans. (a) 6.40×10^{-17} N, up,
          (b) 3.20×10^{-18} N, down

24.6. What must be the magnitude and direction of the electric field intensity between two horizontal plates to produce an upward force of 6×10^{-4} N on a $+60$-μC charge?

24.7. The uniform electric field between two horizontal plates is 8×10^4 C. The top plate is positively charged, and the lower plate has an equal negative charge. What are the magnitude and direction of the electric force acting on an electron as it passes horizontally through the plates?
          Ans. 1.28×10^{-14} N, upward

24.8. Find the electric field intensity at a point P, located 6 mm to the left of an 8-μC charge. What are the magnitude and direction of the force on a -2-nC charge placed at point P?

24.9. Determine the electric field intensity at a point P, located 4 cm above a -12-μC charge. What are the magnitude and direction of the force on a $+3$-nC charge placed at point P?
          Ans. 6.75×10^7 N/c, down, 0.202 N, down

Section 24.2 Computing the Electric Intensity and Section 24.3 Electric Field Lines

24.10. Determine the electric field intensity at the midpoint of a 70-mm line joining a -60-μC charge with a $+40$-μC charge.

24.11. An 8-nC charge is located 80 mm to the right of a $+4$-nC charge. Determine the field intensity at the midpoint of a line joining the two charges.
          Ans. 2.25×10^4 N/C, left

24.12. Find the electric field intensity at a point 30 mm to the right of a 16-nC charge and 40 mm to the left of a 9-nC charge.

24.13. Two equal charges of opposite signs are separated by a horizontal distance of 60 mm. The resultant electric field at the midpoint of the line is 4×10^4 N/C. What is the magnitude of each charge?   Ans. 2 nC

*24.14. A 20-μC charge is 4 cm above an unknown charge q. The electric intensity at a point 1 cm above the 20-μC charge is 2.20×10^9 N/C and is directed upward. What are the magnitude and sign of the unknown charge?

*24.15. A charge of $-20\ \mu C$ is placed 50 mm to the right of a 49 μC charge. What is the resultant field intensity at a point located 24 mm directly above the -20-μC charge?
          Ans. 2.82×10^8 N/C, 297.3°

*24.16. Two charges of $+12$ nC and $+18$ nC are separated horizontally by 28 mm. What is the resultant field intensity at a point 20 mm from each charge and above a line joining the two charges?

*24.17. A +4-nC charge is placed at $x = 0$, and a +6-nC charge is placed at $x = 4$ cm on an x axis. Find the point at which the resultant electric field intensity will be zero.   Ans. $x = 1.80$ cm

Section 24.4 Gauss's Law and Section 24.5 Applications of Gauss's Law

24.18. Use Gauss's law to show that the field outside a solid charged sphere at a distance r from its center is given by

$$E = \frac{1}{4\pi\epsilon_0} \frac{Q}{r^2}$$

where Q is the total charge on the sphere.

*24.19. A charge of +5 nC is placed on the surface of a hollow metal sphere whose radius is 3 cm. Use Gauss's law to find the electric field intensity at a distance of 1 cm from the surface of the sphere. What is the electric field at a point 1 cm inside the surface?   Ans. 2.81×10^4 N/C, zero

*24.20. Two parallel plates, each 2 cm wide and 4 cm long, are stacked vertically so that the field intensity between the two plates is 10,000 N/C directed upward. What is the charge on each plate?

*24.21. A sphere 8 cm in diameter has a charge of 4 μC placed on its surface. What is the electric field intensity at the surface, 2 cm outside the surface, and 2 cm inside the surface?   Ans. 2.25×10^7 N/C, 9.99×10^6 N/C, zero

Additional Problems

24.22. How far from a point charge of 90 nC will the field intensity be 500 N/C?

24.23. The electric field intensity at a point in space is found to be 5×10^5 N/C, directed due west. What are the magnitude and direction of the force on a -4-μC charge placed at that point?   Ans. 2 N, east

24.24. What are the magnitude and direction of the force on an alpha particle ($q = +3.2 \times 10^{-19}$ C) as it passes into an upward electric field of intensity 8×10^4 N/C?

24.25. What is the acceleration of an electron ($e = -1.6 \times 10^{-19}$ C) placed in a constant downward electric field of 4×10^5 N/C? What is the gravitational force on this charge if $m_e = 9.11 \times 10^{-31}$ kg?   Ans. 7.03×10^{16} m/s² N, 8.93×10^{-30} N

24.26. What is the electric field intensity at the midpoint of a 40-mm line between a 6-nC charge and a -9-nC charge? What force will act on a -2-nC charge placed at the midpoint?

*24.27. The charge density on each of two parallel plates is 4 μC/m². What is the electric field intensity between the plates?   Ans. 4.52×10^5 N/C

*24.28. A -2 nC charge is placed at $x = 0$ on the x axis. A +8-nC charge is placed at $x = 4$ cm. At what point will the electric field intensity be equal to zero?

*24.29. Charges of -2 and $+4$ μC are placed at the base corners of an equilateral triangle that has 10-cm sides. What are the magnitude and direction of the electric field intensity at the top corner?   Ans. 3.12×10^6 N/C, 150°

24.30. What are the magnitude and direction of the force that would act on a -2-μC charge

placed at the apex of the triangle described by Prob. 24.29?

*24.31. A 20-mg particle is placed in a uniform downward field of 2000 N/C. How many excess electrons must be placed on the particle for the electric and gravitational forces to balance one another?   Ans. 6.12×10^{11} electrons

*24.32. Use Gauss's law to show that the electric field intensity at a distance R from an infinite line of charge is given by

$$E = \frac{\lambda}{2\pi \epsilon_0 R}$$

where λ is the charge per unit length. Construct a gaussian surface as in Fig. 24.15.

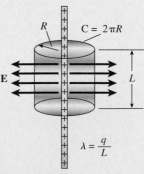

Figure 24.15

*24.33. Use Gauss's law to show that the field just outside any solid conductor is given by

$$E = \frac{\sigma}{\epsilon_0}$$

*24.34. What is the electric field intensity 2 cm from the surface of a sphere 20 cm in diameter having a surface charge density of $+8 \ nC/m^2$?

*24.35. A uniformly charged conducting sphere has a radius of 24 cm and a surface charge density of $+16 \ \mu C/m^2$. What is the total number of electric field lines leaving the sphere?

Ans. 1.16×10^{-5} lines

*24.36. Two charges of $+16 \ \mu C$ and $+8 \ \mu C$ are 200 mm apart in air. At what point on a line joining the two charges will the electric field be zero?

*24.37. Two charges of $+8 \ nC$ and $-5 \ nC$ are 40 mm apart in air. At what point on a line joining the two charges will the electric field intensity be zero?

Ans. 151 mm outside of $-5 \ nC$ charge

Critical Thinking Questions

*24.38. Two equal but opposite charges $+q$ and $-q$ are placed at the base corners of an equilateral triangle whose sides are of length a. Show that the magnitude of the electric field intensity at the apex is the same whether one of the charges is removed or not. What is the angle between the two fields so produced?

*24.39. What are the magnitude and direction of the electric field intensity at the center of the square of Fig. 24.16? Assume that $q = 1 \ \mu C$ and that $d = 4$ cm.   Ans. 3.56×10^7 N, 153.4°

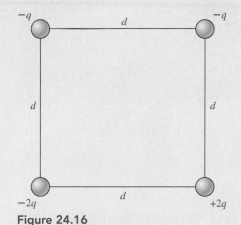

Figure 24.16

*24.40. The electric field intensity between the plates in Fig. 24.17 is 4000 N/C. What is the magnitude of the charge on the suspended pith ball whose mass is 3 mg?   Ans. 4.24 nC

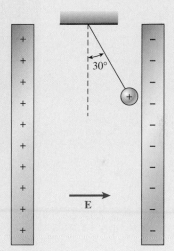

Figure 24.17

*24.41. Two concentric spheres have radii of 20 and 50 cm. The inner sphere has a negative charge of $-4 \ \mu C$, and the outer sphere has a positive charge of $+6 \ \mu C$. Use Gauss's law to find the electric field intensity at distances of 40 and 60 cm from the center of the spheres.

*24.42. The electric field intensity between the two plates in Fig. 24.4 is 2000 N/C. The length of the plates is 4 cm, and their separation is 1 cm. An electron is projected into the field from the left with horizontal velocity of 2×10^7 m/s. What is the upward deflection of the electron at the instant it leaves the plates?   Ans. 0.700 mm

25

Electric Potential

Electrostatic generators, such as the Van de Graff generator shown here, can transfer enormous amounts of charge to a silver dome. Extremely high voltages are produced by these devices. Electrostatic repulsion produces this amazing effect on the woman's hair.
(© *Roger Ressmeyer/Corbis.*)

Objectives

After completing this chapter, you should be able to

1. Demonstrate by definition and example your understanding of *electric potential energy, electric potential,* and *electric potential difference.*

2. Compute the potential energy of a known charge at a given distance from other known charges and state whether the energy is negative or positive.

3. Compute the absolute potential at any point in the vicinity of a number of known charges.

4. Use your knowledge of potential difference to calculate the work required to move a known charge from any point *A* to another point *B* in an electric field created by one or more point charges.

5. Write and apply a relationship between the electric field intensity, the potential difference, and the plate separation for parallel plates of equal and opposite charge.

In our study of mechanics, many problems were simplified by introducing the concepts of energy. The conservation of mechanical energy allowed us to say certain things about the initial and final states of systems without having to analyze the motion between states. The concept of a change of potential energy into kinetic energy avoids the problem of varying forces.

In electricity, many practical problems can be solved by considering the changes in energy experienced by a moving charge. For example, if a certain quantity of work is required to move a charge against electric forces, the charge should have a *potential* for giving up an equivalent amount of energy when it is released. In this chapter, we develop the idea of *electric potential energy.*

25.1   Electric Potential Energy

One of the best ways to understand the concept of ***electric potential energy*** is by comparing it with gravitational potential energy. In the gravitational case, consider that a mass *m* is moved from level *A* in Fig. 25.1 to level *B*. An external force **F** equal to the weight *m***g** must be applied to move the mass against gravity. The work done by this force is the product of *mg* and *h*. When the mass *m* reaches level *B*, it has a potential for doing work relative to level *A*. The system has a *potential energy* (*U*) that is equal to the work done against gravity.

$$U = mgh$$

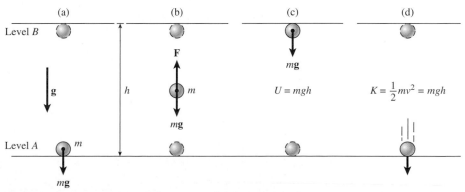

Figure 25.1   A mass *m* lifted against a gravitational field **g** results in a potential energy of *mgh* at level *B*. When released, this energy will be transformed entirely into kinetic energy as it falls to level *A*.

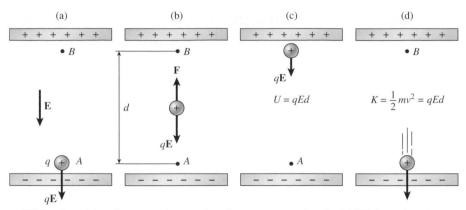

Figure 25.2 A positive charge $+q$ is moved against a constant electric field $\mathbf{E}$ through a distance d. At point B, the potential energy will be qEd relative to point A. When released, the charge will gain an equivalent amount of kinetic energy.

This expression represents the potential for doing work after the mass m is released at level B and falls the distance h. Therefore, the magnitude of the potential energy at B does not depend on the path taken to reach that level.

Now, let us consider a positive charge $+q$ resting at point A in a uniform electric field $\mathbf{E}$ between two oppositely charged plates. (See Fig. 25.2.) An electric force $q\mathbf{E}$ acts downward on the charge. The work done against the electric field in moving the charge from A to B is equal to the product of the force qE and the distance d. Hence, the electric potential energy at point B relative to point A is

$$U = qEd \qquad (25.1)$$

When the charge is released, the electric field will perform this amount of work, and the charge q will have a kinetic energy.

$$K = \frac{1}{2} mv^2 = qEd$$

when it returns to point A.

The previous statements and equations are valid regardless of the path through which the charge is moved. The same work against the gravitational field is required to slide a mass up an inclined plane as would be necessary to lift it vertically. Similarly, the potential energy due to the charge $+q$ at point B is independent of the path. As shown in Fig. 25.3, the potential energy would be the same whether $+q$ were moved along the path L or the path d. The only work that contributes to the potential energy is the work done

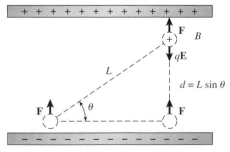

Figure 25.3 Electric potential energy at B is independent of the path taken to reach B.

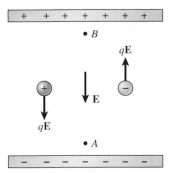

Figure 25.4 A positive charge increases its potential energy when moved from A to B; a negative charge *loses* potential energy when it moves from A to B.

against the electric field force $q\mathbf{E}$. The effective distance moved against this downward electric force is

$$L \sin \theta = d$$

Therefore, the work is the same for either path.

Before we proceed, we should point out an important difference between gravitational potential energy and electric potential energy. In the case of gravity, there is only one kind of mass, and the forces involved are always forces of attraction. Therefore, a mass at higher elevations always has greater potential energy relative to the Earth. This is not true in the electrical case because of the existence of negative charge. For example, in Fig. 25.4, a positive charge has a greater potential energy at point B than at point A. This is true regardless of the reference point chosen for measuring the energy because work has been done *against* the electric field. On the other hand, if a negative charge were moved from point A to point B, work would be done *by* the field. A negative charge would have a *lower* potential energy at B, which is exactly opposite to the situation for a positive charge.

> Whenever a positive charge is moved against an electric field, the potential energy increases; whenever a negative charge moves against an electric field, the potential energy decreases.

This rule is a direct consequence of the fact that the direction of the electric field is defined in terms of a positive charge.

25.2 Calculating Potential Energy

In considering the space between two oppositely charged plates, work computations are fairly simple because the electric field is uniform. The electric force that a charge experiences is constant as long as it remains between the plates. In general, however, the field will not be constant, and we must make allowances for a varying force. For example, consider the electric field in the vicinity of a positive charge Q, as illustrated in Fig. 25.5. The field is directed radially outward, and its intensity falls off inversely with the square of the distance from the center of the charge. The field at points A and B is

$$E_A = \frac{kQ}{r_A^2} \qquad E_B = \frac{kQ}{r_B^2}$$

where r_A and r_B are the respective distances to points A and B.

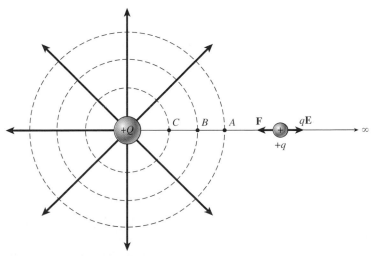

Figure 25.5 The potential energy due to a charge placed in an electric field is equal to the work done *against* electric forces in bringing the charge from infinity to the point in question.

In Fig. 25.5 and in the following discussions, we will use the term infinity ∞ to refer to points that are far away and beyond the point of electrical interaction. Additionally, it will be assumed that there are no other charges present than those specifically indicated in our examples.

The average electric force experienced by a charge $+q$ when it is moved from point A to point B is

$$F = \frac{kQq}{r_A r_B} \tag{25.2}$$

Thus, the work done against the electric field in moving through the distance $r_A - r_B$ is equal to

$$\text{Work}_{A \to B} = \frac{kQq}{r_A r_B}(r_A - r_B)$$

$$= kQq\left(\frac{1}{r_B} - \frac{1}{r_A}\right) \tag{25.3}$$

Note that the work is a function of the distances r_A and r_B. The path traveled is unimportant. The same work would be done against the field in moving a charge from any point on the dashed circle passing through A to any point on the circle passing through B.

Suppose now we compute the work done against electric forces in moving a positive charge from infinity to a point a distance r from the charge Q. From Eq. (25.3), the work is given by

$$\text{Work}_{\infty \to r} = kQq\left(\frac{1}{r} - \frac{1}{\infty}\right)$$

$$= \frac{kQq}{r} \tag{25.4}$$

Since we have shown that the work done against the electric field equals the increase in potential energy, Eq. (25.4) is the potential energy at r relative to infinity. Often potential energy is taken as zero at infinity, so the potential energy of a system composed of a charge q and another charge Q separated by a distance r is

$$U = \frac{kQq}{r} \tag{25.5}$$

The potential energy of the system is equal to the work done against the electric forces in moving the charge $+q$ from infinity to that point.

Example 25.1

A charge of $+2$ nC is 20 cm away from another charge of $+4\ \mu C$. (a) What is the potential energy of the system? (b) What is the change in potential energy if the 2-nC charge is moved to a point 8 cm away from the $+4$-μC charge?

Plan: The potential energy of a system containing two charges q_1 and q_2 is the work required to place them a distance r apart. Equation (25.5) can be used to find the potential energy for $r = 20$ cm and then for $r = 8$ cm. The difference will be the *change* in potential energy.

Solution (a): The potential energy at $r = 20$ cm $= 0.20$ m is

$$U = \frac{kQq}{r} = \frac{(9 \times 10^9\ \text{N} \cdot \text{m}^2/\text{C}^2)(4 \times 10^{-6}\ \text{C})(2 \times 10^{-9}\ \text{C})}{(0.20\ \text{m})}$$

$$= 3.60 \times 10^{-4}\ \text{J}$$

Solution (b): The potential energy at $r = 8$ cm $= 0.08$ m is

$$U = \frac{kQq}{r} = \frac{(9 \times 10^9 \text{ N} \cdot \text{m}^2/\text{C}^2)(4 \times 10^{-6} \text{ C})(2 \times 10^{-9} \text{ C})}{(0.08 \text{ m})}$$

$$= 9.00 \times 10^{-4} \text{ J}$$

The *change* in potential energy is the final energy less the initial energy.

$$\Delta U = 9.00 \times 10^{-4} \text{ J} - 3.6 \times 10^{-4} \text{ J} = 5.4 \times 10^{-4} \text{ J}$$

Notice the difference is positive, indicating an *increase* in potential energy. If the charge Q were negative and all other parameters were unchanged, the potential energy would have *decreased* by this same amount.

25.3 Potential

When we first introduced the concept of an electric field as force per unit charge, we pointed out the primary advantage of such a concept was that it allowed us to assign an electrical property to space. If the field intensity is known at some point, the force on a charge placed at that point can be predicted. It is equally convenient to assign another property to the space surrounding a charge that would allow us to predict the potential energy due to another charge placed at any point. This property of space is called ***potential*** and is defined as follows:

> The potential V at a point a distance r from a charge Q is equal to the work per unit charge done against electric forces in bringing a positive charge $+q$ from infinity to that point.

In other words, the potential at some point *A*, as shown in Fig. 25.6, is equal to *the potential energy per unit charge*. The units of potential are expressed in *joules per coulomb,* defined as a ***volt*** (V).

$$V_A(\text{V}) = \frac{U \text{ (J)}}{q \text{ (C)}} \tag{25.6}$$

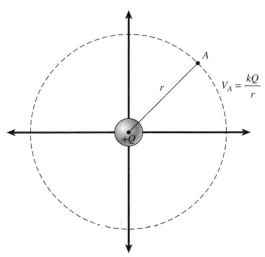

Figure 25.6   Calculating the potential at a distance r from a charge $+Q$.

Thus, a potential of *1 volt* at point *A* means that *if a charge of 1 coulomb were to be placed at A,* the potential energy would be 1 joule. In general, when the potential is known at a point *A,* the potential energy due to a charge *q* at that point can be found from

$$U = qV_A \qquad (25.7)$$

Substitution of Eq. (25.5) into Eq. (25.6) yields an expression for directly computing the potential:

$$V_A = \frac{U}{q} = \frac{kQq/r}{q}$$

$$V_A = \frac{kQ}{r} \qquad \text{\textit{Electric Potential Energy}} \quad (25.8)$$

The symbol V_A refers to the potential at point *A* located at a distance *r* from the charge *Q*.

It is noted now that the potential is the same at equal distances from a spherical charge. For this reason, the dashed lines in Figs. 25.5 and 25.6 are called ***equipotential lines.*** Note that the lines of equal potential are always perpendicular to the electric field lines. If this were not true, work would be done by a resultant force when a charge moved along an equipotential line. Such work would increase or decrease the potential.

Equipotential lines are always perpendicular to electric field lines.

Before offering an example, it must be pointed out that the potential at a point is defined in terms of a positive charge. This means that the potential will be negative at a point in space surrounding a negative charge. As a rule, we must remember that

The potential due to a positive charge is positive, and the potential due to a negative charge is negative.

Using a negative sign for a negative charge *Q* in Eq. (25.8) results in a negative value for the potential.

Example 25.2

(a) Calculate the potential at a point *A* that is 30 cm distant from a charge of $-2\ \mu$C.
(b) What is the potential energy if a $+4$-nC charge is later placed at point *A?*

Plan: At first, there is no potential energy *U* since there is only the *one* charge. However, there is electric potential *V* in the space surrounding the charge. In Part (a), we will use Eq. (25.8) to find the electric potential at a distance of 0.30 m from the -2-μC charge. Then, we will use Eq. (25.7) to find the potential energy when the $+4$-nC charge is placed at *A.*

Solution (a): From Eq. (25.8), we obtain

$$V_A = \frac{kQ}{r} = \frac{(9 \times 10^9\ \text{N} \cdot \text{m}^2/\text{C}^2)(-2 \times 10^{-6}\ \text{C})}{(0.30\ \text{m})}$$

$$= -6.00 \times 10^4\ \text{V}$$

Solution (b): Solving Eq. (25.7) explicitly for *U,* we find the potential energy due to the placement of the $+4$-nC charge

$$U = qV_A = (4 \times 10^{-9}\ \text{C})(-6 \times 10^4\ \text{V})$$

$$= -2.40 \times 10^{-4}\ \text{J}$$

The negative value for potential energy means that work must be done *against* the electric field in moving the charges apart. In this example, 2.40×10^{-4} J of work would have to be done by an external force to remove the charge to infinity.

$$V_A = V_1 + V_2 + V_3$$
$$= \frac{kQ_1}{r_1} + \frac{kQ_2}{r_2} + \frac{kQ_3}{r_3}$$

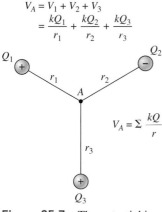

$$V_A = \Sigma \frac{kQ}{r}$$

Figure 25.7 The potential in the vicinity of a number of charges.

Now let us consider the more general case, illustrated by Fig. 25.7, where we wish to calculate the potential at a point in the neighborhood of a number of charges.

The potential in the vicinity of a number of charges is equal to the algebraic sum of the potentials due to each charge.

$$V_A = V_1 + V_2 + V_3 + \ldots$$
$$= \frac{kQ_1}{r_1} + \frac{kQ_2}{r_2} + \frac{kQ_3}{r_3} + \ldots$$

Remember that the potential in the vicinity of a positive charge is positive, and the potential in the vicinity of a negative charge is negative. This means we may insert the sign of the charge into our calculations. In general, the potential at a point in space near other charges is given by

$$V = \Sigma \frac{kQ}{r} \tag{25.9}$$

This equation represents an *algebraic sum* because potential is a scalar quantity and not a vector quantity, as was the case of electric forces and fields.

Example 25.3

Two charges, $Q_1 = +6 \, \mu\text{C}$ and $Q_2 = -6 \, \mu\text{C}$, are separated by 12 cm, as shown in Fig. 25.8. Calculate the potential at points A and B.

Plan: The potential at a particular point is the algebraic sum of the potentials due to each charge, with the distances measured from each charge to that point. The signs of the charge may be used in the addition process to find the total potential.

Solution (a): The potential at A is found from Eq. (25.9).

$$V_A = \frac{kQ_1}{r_1} + \frac{kQ_2}{r_2}$$
$$= \frac{(9 \times 10^9 \, \text{N} \cdot \text{m}^2/\text{C}^2)(6 \times 10^{-6} \, \text{C})}{4 \times 10^{-2} \, \text{m}} + \frac{(9 \times 10^9 \, \text{N} \cdot \text{m}^2/\text{C}^2)(-6 \times 10^{-6} \, \text{C})}{8 \times 10^{-2} \, \text{m}}$$
$$= 13.5 \times 10^5 \, \text{V} - 6.75 \times 10^5 \, \text{V}$$
$$= 6.75 \times 10^5 \, \text{V}$$

This means the electric field will do 6.75×10^5 J of work on each coulomb of positive charge that it moves from A to infinity.

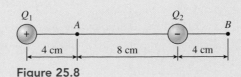

Figure 25.8

Solution (b): The potential at B is

$$V_B = \frac{kQ_1}{r_1} + \frac{kQ_2}{r_2}$$

$$= \frac{(9 \times 10^9 \text{ N} \cdot \text{m}^2/\text{C}^2)(6 \times 10^{-6} \text{ C})}{16 \times 10^{-2} \text{ m}} + \frac{(9 \times 10^9 \text{ N} \cdot \text{m}^2/\text{C}^2)(-6 \times 10^{-6} \text{ C})}{4 \times 10^{-2} \text{ m}}$$

$$= 3.38 \times 10^5 \text{ V} - 13.5 \times 10^5 \text{ V}$$

$$= -10.1 \times 10^5 \text{ V}$$

The negative value indicates the field will hold onto a positive charge. To move 1 C of positive charge from A to infinity, another source of energy must perform 10.1×10^5 J of work. The field would perform negative work in this amount.

25.4 Potential Difference

In practical electricity, we are seldom interested in the work per unit charge to remove a charge to infinity. More often, we want to know the work requirements for moving charges between two points. This leads to the concept of **potential difference.**

> The potential difference between two points is the work per unit positive charge done by electric forces in moving a small test charge from the point of higher potential to the point of lower potential.

Another way of stating this would be to say that the potential difference between two points is the difference in the potentials at those points. For example, if the potential at some point A is 100 V and the potential at another point B is 40 V, the potential difference is

$$V_A - V_B = 100 \text{ V} - 40 \text{ V} = 60 \text{ V}$$

This means that 60 J of work will be done by the field on each coulomb of positive charge moved from A to B. In general, the *work done by the electric field,* or **electric work,** in moving a charge q from point A to point B can be found from

$$\text{Work}_{A \to B} = q(V_A - V_B) \tag{25.10}$$

Example 25.4

(a) What is the potential difference between points A and B in Fig. 25.8? Refer to Example 25.3. (b) How much work is done by the electric field in moving a -2-μC charge from point A to point $B?$

Plan: The difference of potential is simply $V_A - V_B$; the work to move the charge from A to B is the product of q and the difference of potential.

Solution: The potentials at points A and B were calculated in Example 25.3. They are

$$V_A = 6.75 \times 10^5 \text{ V} \qquad V_B = -10.1 \times 10^5 \text{ V}$$

Therefore, the potential difference between points A and B is

$$V_A - V_B = 6.75 \times 10^5 \text{ V} - (-10.1 \times 10^5 \text{ V})$$

$$= 16.9 \times 10^5 \text{ V}$$

Since A is at a higher potential than B, positive work would be done by the field when a *positive* charge is moved from A to B. If a *negative* charge is moved, the work done by the field in moving it from A to B will be negative. In this example, the work is

$$\text{Work}_{A\to B} = q(V_A - V_B)$$
$$= (-2 \times 10^{-9}\,\text{C})(16.9 \times 10^5\,\text{V})$$
$$= -3.37 \times 10^{-3}\,\text{J}$$

Since the work done by this field is negative, another source of energy must supply the work to move the charge.

Problem-Solving Strategy

Electric Potential and Potential Energy

1. Read the problem and then draw and label a figure. Indicate positive and negative charges along with given distances. The charges must be in coulombs, and the distances must be in meters. Remember that $1\,\mu\text{C} = 1 \times 10^{-6}\,\text{C}$ and $1\,\text{nC} = 1 \times 10^{-9}\,\text{C}$.

2. Remember that the electric potential V is a property of *space* that allows us to determine the potential energy U when a charge q is placed at that point. The potential exists at a *point* in space independent of whether a charge is placed at that point.

3. The absolute potential at a point in the vicinity of a number of charges is the algebraic sum of the potentials due to each charge:

$$V = \sum \frac{kQ}{r} \qquad k = 9 \times 10^9\,\text{N} \cdot \text{m}^2/\text{C}^2$$

4. Only *changes* in potential are significant, and the reference point for zero potential can be at infinity or at any other point. Often the point of lowest absolute potential can be chosen as zero.

5. The *work* done by the electric field in moving a charge q from a point A to another point B is simply the product of the charge and the difference of potential:

$$\text{Work}_{AB} = q(V_A - V_B) \qquad \textit{Work by the Electric Field}$$

6. Because the potential in the vicinity of a positive charge is positive and the potential near a negative charge is negative, the signs of the charge and potential can be used algebraically.

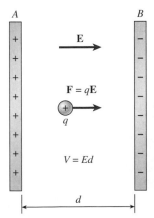

Figure 25.9 The potential between two oppositely charged plates.

Let us now return to the example of a uniform electric field **E** between two oppositely charged plates, as in Fig. 25.9. We shall assume that the plates are separated by a distant d. A charge q placed in the region between the plates A and B will experience a force given by

$$\mathbf{F} = q\mathbf{E}$$

The work done by this force in moving the charge q from plate A to plate B is given by

$$Fd = (qE)d$$

But this work is also equal to the product of the charge q and the potential difference $V_A - V_B$ between the two plates, so we can write

$$q(V_A - V_B) = qEd$$

If we divide through by q and represent the potential difference by the single symbol V, we obtain

$$V = Ed \qquad \qquad \textbf{(25.11)}$$

The potential difference between two oppositely charged plates is equal to the product of the field intensity and the plate separation.

Example 25.5 The potential difference between two plates 5 mm apart is 10 kV. Determine the electric field intensity between the plates.

Solution: Solving for E in Eq. (25.11) gives

$$E = \frac{V}{d} = \frac{10 \times 10^3 \text{ V}}{5 \times 10^{-3} \text{ m}} = 2 \times 10^6 \text{ V/m}$$

As an exercise, the student should show that the *volt per meter* is equivalent to the *newton per coulomb*. The electric field expressed in volts per meter is sometimes referred to as the **potential gradient.**

25.5 Millikan's Oil-Drop Experiment

Now that we have developed the concepts of the electric field and potential difference, we are ready to describe a classic experiment designed to determine the smallest unit of charge. Robert A. Millikan, an American physicist, devised a series of experiments in the early 1900s. A schematic diagram of his apparatus is shown in Fig. 25.10. Tiny oil droplets are sprayed into the region between the two metallic plates. Electrons are freed from air molecules by passing ionizing X rays through the medium. These electrons attach themselves to the oil droplets, giving them a net negative charge.

The downward motion of the oil droplets can be observed by a microscope as they fall slowly under the influence of their weight and the upward viscous force of air resistance. (See Fig. 25.10.) The laws of hydrostatics can be used to calculate the mass m of a particular oil drop from its measured rate of fall.

After the necessary data for determining the mass m have been recorded, an external battery is connected to establish a uniform electric field **E** between the oppositely charged plates.

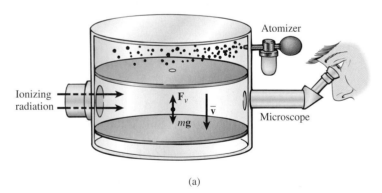

(a)

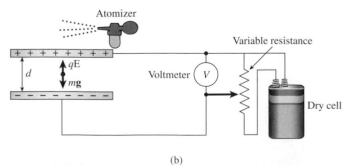

(b)

Figure 25.10 Millikan's oil-drop experiment: (a) The mass m of the droplet is determined from its rate of fall against the viscous force of air resistance. (b) The magnitude of the charge is computed from the equilibrium conditions that suspend the charge between oppositely charged plates.

(See Fig. 25.10.) The magnitude of the field intensity can be controlled by a variable resistor in the electric circuit. The field is adjusted until the upward electric force on the drop is equal to the downward gravitational force so that the oil drop stops moving. Under these conditions,

$$qE = mg \qquad (25.12)$$

where q = net charge of oil drop

m = mass of oil drop

g = acceleration of gravity

The field intensity **E,** as determined from Eq. (25.11), is a function of the applied voltage V and the plate separation d. Therefore, Eq. (25.12) becomes

$$q\frac{V}{d} = mg$$

and the magnitude of the charge on the oil drop is found from

$$q = \frac{mgd}{V} \qquad (25.13)$$

The potential difference V can be read directly from an indicating device called a *voltmeter* attached to the circuit. The other parameters are known.

The charges observed by Millikan were not always the same, but he observed the magnitude of the charge was always an integral multiple of a basic quantity of charge. It was assumed this *least* charge must be the charge of a single electron and the other magnitudes resulted from two or more electrons. Computation of the electronic charge by this method yields

$$e = 1.6065 \times 10^{-19}\,\text{C}$$

which agrees extremely well with the values obtained by other methods.

25.6 The Electronvolt

Let us now consider the energy of a charged particle moving through a potential difference. There are several units in which to measure this energy, but most of the familiar units are inconveniently large. Consider, for example, a charge of 1 C accelerated through a potential difference of 1 V. Its kinetic energy will be

$$K = qEd = qV$$
$$= (1\,\text{C})(1\,\text{V}) = 1\,\text{C} \cdot \text{V}$$

The coulomb-volt, of course, is a joule. But 1 C of charge is inconveniently large when applied to single particles, and the corresponding unit of energy (the joule) is also large. The most convenient unit of energy in atomic and nuclear physics is the **electronvolt** (eV).

The electronvolt is a unit of energy equivalent to the energy acquired by an electron that is accelerated through a potential difference of 1 volt.

The electronvolt differs from the coulomb-volt by the same degree as the difference in the charge of an electron and the charge of 1 C. To compare the two units, suppose we compute the energy in joules acquired by an electron that has been accelerated through a potential difference of 1 V:

$$K = qV$$
$$= (1.6 \times 10^{-19}\,\text{C})(1\,\text{V})$$
$$= 1.6 \times 10^{-19}\,\text{J}$$

Thus, 1 eV is equivalent to an energy of 1.6×10^{-19} J.

Summary and Review

Summary

The concepts of potential energy, potential, and potential difference have been extended to electrical phenomena. The many problems dealing with electrostatic potential have been designed to provide a base for direct current electricity that will be discussed later. The essential elements of this chapter are summarized as follows.

- When a charge q is moved against a constant electric force for a distance d, the potential energy of the system is

$$U = qEd$$

where E is the constant electric field intensity. If the charge is released, it will acquire a kinetic energy

$$K = \frac{1}{2}mv^2 = qEd$$

as it returns for the same distance.

- Because of the existence of positive and negative charges and the opposite effects produced by the same field, we must remember that *the potential energy increases as a positive charge is moved against the electric field, and the potential energy decreases as a negative charge is moved against the same field.*

- In general, the potential energy due to a charge q placed at a distance r from another charge Q is equal to the work done against electric forces in moving the charge $+q$ from infinity.

$$U = \frac{kQq}{r} \qquad \textit{Electric Potential Energy}$$

Note the distance r is not squared, as it was for the electric field intensity.

- The electric *potential V* at a point a distance r from a charge Q is equal to the work per unit charge done against electric forces in bringing a positive charge $+q$ from infinity.

$$V = \frac{kQ}{r} \qquad \textit{Electric Potential}$$

- The unit of electric potential is the joule per coulomb (J/C), which is renamed the volt (V).

$$1\ \text{V} = \frac{1\ \text{J}}{1\ \text{C}}$$

- The potential at a point in the vicinity of a number of charges is equal to the algebraic sum of the potentials due to each charge:

$$V = \sum \frac{kQ}{r} = \frac{kQ_1}{r_1} + \frac{kQ_2}{r_2} + \frac{kQ_3}{r_3} + \dots$$
$$\textit{Algebraic Sum}$$

- The potential difference between two points A and B is the difference in the potentials at those points.

$$V_{AB} = V_A - V_B \qquad \textit{Potential Difference}$$

- The work done by an electric field in moving a charge q from point A to point B can be found from

$$\text{Work}_{AB} = q(V_A - V_B) \qquad \begin{array}{l}\textit{Work and Potential}\\\textit{Difference}\end{array}$$

- The potential difference between two oppositely charged plates is equal to the product of the field intensity and the plate separation.

$$V = Ed \qquad E = \frac{V}{d}$$

Key Terms

Review Questions

25.1. Distinguish clearly between positive and negative work. Distinguish between positive and negative potential energy.

25.2. It is possible for a mass m to increase the potential energy by moving it to a lower elevation? Is it possible for a charged object to increase the potential energy as it is moved to a position of lower potential? Explain.

25.3. Give an example in which the electric potential is zero at some point where the electric field intensity is not zero.

25.4. The electric field inside an electrostatic conductor is zero. Is the electric potential inside the conductor zero also? Explain.

25.5. If the electric field intensity is known at some point, can one determine the electric potential at that point? What information is needed?

25.6. The surface of any conductor is an equipotential surface. Justify this statement.

25.7. Is the direction of the electric field intensity from higher to lower potential? Illustrate.

25.8. Apply the potential concept to the gravitational field to obtain an expression similar to Eq. (25.9)

for computing the potential energy per unit mass. Discuss the applications of such a formula.

25.9. Show that the volt per meter is dimensionally equivalent to the newton per coulomb.

25.10. Distinguish between potential difference and a difference in potential energy.

25.11. A potential difference of 220 V is maintained between the ends of a long high-resistance wire. If the center of the wire is grounded ($V = 0$), what is the potential difference between the center and the end points?

25.12. The potential due to a negative charge is negative, and the potential due to a positive charge is positive. Why? Is it also true that the potential energy duc to a negative charge is negative? Explain.

25.13. Is potential a property assigned to space or to a charge? What is potential energy assigned to?

Problems

Section 25.1 Electric Potential Energy

25.1. A positively charged plate is 30 mm above a negatively charged plate, and the electric field intensity has a magnitude of 6×10^4 N/C. How much work is done *by* the electric field when a $+4\text{-}\mu\text{C}$ charge is moved from the negative plate to the positive plate? Ans. -7.20 mJ

25.2. In Prob. 25.1, how much work is done *on* or against the electric field? What is the electric potential energy at the positive plate?

25.3. The electric field intensity between two parallel plates separated by 25 mm is 8000 N/C. How much work is done *by* the electric field in moving a $-2\text{-}\mu\text{C}$ charge from the negative plate to the positive plate? What is the work done *by* the field in moving the same charge back to the positive plate?
Ans. $+4.00 \times 10^{-4}$ J, -4.00×10^{-4} J

25.4. In Prob. 25.3, what is the potential energy when the charge is at (a) the positive plate and (b) the negative plate?

25.5. What is the potential energy of a $+6\text{-nC}$ charge located 50 mm away from a $+80\text{-}\mu\text{C}$ charge? What is the potential energy if the same charge is 50 mm from a $-80\text{-}\mu\text{C}$ charge?
Ans. $+86.4$ mJ, -86.4 mJ

25.6. At what distance from a $-7\text{-}\mu\text{C}$ charge will a -3-nC charge have a potential energy of 60 mJ? What initial force will the -3-nC charge experience?

25.7. A $+8\text{-nC}$ charge is placed at a point P, 40 mm from a $+12\text{-}\mu\text{C}$ charge. What is the potential

energy per unit charge at point P in joules per coulomb? Will this change if the 8-nC charge is removed? Ans. 2.70×10^6 J/C, no

25.8. A charge of $+6$ μC is 30 mm away from another charge of 16 μC. What is the potential energy of the system?

25.9. In Prob. 25.8, what is the change in potential energy if the $6\text{-}\mu\text{C}$ charge is moved to a distance of only 5 mm? Is this an increase or decrease in potential energy? Ans. 144 J, increase

25.10. A $-3\text{-}\mu\text{C}$ charge is placed 6 mm away from a $-9\text{-}\mu\text{C}$ charge. What is the potential energy? Is it negative or positive?

25.11. What is the change in potential energy when a $3\text{-}\mu\text{C}$ charge is moved from a point 8 cm away from a $-6\text{-}\mu\text{C}$ charge to a point that is 20 cm away? Is this an increase or decrease of potential energy? Ans. $+1.22$ J, increase

25.12. At what distance from a $-7\text{-}\mu\text{C}$ charge must a charge of -12 nC be placed if the potential energy is to be 9×10^{-5} J?

25.13. The potential energy of a system consisting of two identical charges is 4.50 mJ when their separation is 38 mm. What is the magnitude of each charge? Ans. 139 nC

Section 25.3 Potential and Section 25.4 Potential Difference

25.14. What is the electric potential at a point that is 6 cm from a $8.40\text{-}\mu\text{C}$ charge? What is the potential energy of a 2-nC charge placed at that point?

25.15. Calculate the potential at point A that is 50 mm from a -40-μC charge. What is the potential energy if a $+3$-μC charge is placed at point A?

Ans. -7.20 MV, -21.6 J

25.16. What is the potential at the midpoint of a line joining a -12-μC charge with a $+3$-μC charge located 80 mm away from the first charge?

25.17. A $+45$-nC charge is 68 mm to the left of a -9-nC charge. What is the potential at a point located 40 mm to the left of the -9-nC charge?

Ans. 12.4 kV

*25.18. Points A and B are, respectively, 68 mm and 26 mm away from a 90-μC charge. Calculate the potential difference between points A and B. How much work is done by the electric field as a -5-μC charge moves from A to G?

*25.19. Points A and B are, respectively, 40 and 25 mm away from a $+6$-μC charge. How much work must be done against the electric field (by external forces) in moving a $+5$-μC charge from point A to point B?

Ans. $+4.05$ J

*25.20. A $+6$-μC charge is located at $x = 0$ on the x axis, and a -2-μC charge is located at $x = 8$ cm. How much work is done by the electric field in moving a -3-μC charge from the point $x = 10$ cm to the point $x = 3$ cm?

Additional Problems

25.21. Point A is 40 mm above a -9-μC charge, and point B is located 60 mm below the same charge. A -3-nC charge is moved from point B to point A. What is the change in potential energy?

Ans. $+2.02$ mJ

25.22. Two parallel plates are separated by 50 mm in air. If the electric field intensity between the plates is 2×10^4 N/C, what is the potential difference between the plates?

25.23. The potential difference between two parallel plates 60 mm apart is 4000 V. What is the electric field intensity between the plates?

Ans. 66.7 kV/m

25.24. If an electron is located at the plate of lower potential in Prob. 25.23, what will be its velocity when it reaches the plate of higher potential? What is the energy expressed in electronvolts?

25.25. Show that the potential gradient V/m is equivalent to the unit N/C for electric field.

25.26. What is the difference in potential between two points 30 and 60 cm away from a -50-μC charge?

25.27. The potential gradient between two parallel plates 4 mm apart is 6000 V/m. What is the potential difference between the plates? Ans. 24.0 V

25.28. The electric field between two plates separated by 50 mm is 6×10^5 V/m. What is the potential difference between the plates?

25.29. What must be the separation of two parallel plates if the field intensity is 5×10^4 V/m and the potential difference is 400 V? Ans. 8.00 mm

25.30. The potential difference between two parallel plates is 600 V. A 6-μC charge is accelerated through the entire potential difference. What is the kinetic energy given to the charge?

25.31. Determine the kinetic energy of an alpha particle $(+2e)$ that is accelerated through a potential difference of 800 kV. Give the answer in both electronvolts and joules.

Ans. 1.60 MeV, 2.56×10^{-13} J

25.32. A linear accelerator accelerates an electron through a potential difference of 4 MV. What is the energy of an emergent electron in electronvolts and in joules?

25.33. An electron acquires an energy of 2.8×10^{-15} J as it passes from point A to point B. What is the potential difference between these points in volts? Ans. 17.5 kV

*25.34. Show that the total potential energy of the three charges placed at the corners of the equilateral triangle shown in Fig. 25.11 is given by

$$-\frac{3kq^2}{d}$$

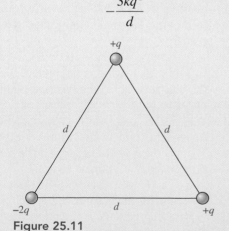

Figure 25.11

*25.35. Assume that $q = 1\ \mu$C and $d = 20$ mm. What is the potential energy of the system of charges in Fig. 25.11? Ans. -1.35 J $\times 10^{-10}$ J

*25.36. The potential at a certain distance from a point charge is 1200 V, and the electric field intensity at that point is 400 N/C. What is the distance to the charge, and what is the magnitude of the charge?

Ans. 3 m, 400 NC

*25.37. Two large plates are 80 mm apart and have a potential difference of 800 kV. What is the magnitude of the force that would act on an electron placed at the midpoint between these plates? What would be the kinetic energy of the electron moving from the low potential plate to the high potential plate?

Ans. 1.60×10^{-12} N, 1.28×10^{-13} J

Critical Thinking Questions

25.38. Plate A has a potential that is 600 V higher than plate B, which is 50 mm below plate A. A $+2$-μC charge moves from plate A to plate B. What is the electric field intensity between the plates? What are the sign and magnitude of the work done by the electric field? Does the potential energy increase or decrease? Now answer the same questions if a -2-μC charge is moved from A to B.

Ans. 12 kV/m, $+1.20$ mJ, decreases, -1.2 mJ, increases

25.39. Point A is a distance $x = +a$ to the right of a $+4$-μC charge. The rightward electric field at point A is 4000 N/C. What is the distance a? What is the potential at point A? What are the electric field and the potential at the point $x = -a$? Find the electric force and the electric potential energy when a -2-nC charge is placed at each point.

*25.40. Points A, B, and C arc at the corners of an equilateral triangle that is 100 mm on each side. At the base of the triangle, a $+8$-μC charge is 100 mm to the left of a -8-μC charge. What is the potential at the apex C? What is the potential at a point D that is 20 mm to the left of the -8-μC charge? How much work is done by the electric field in moving a $+2$-μC charge from point C to point D? Ans. 0, -2.70 MV, $+5.40$ J

*25.41. Two charges, of $+12$ and -6 μC, are separated by 160 mm. What is the potential at the midpoint A of a line joining the two charges? At what point B is the electric potential equal to zero?

*25.42. For the charges and distances shown in Fig. 25.12, find thc potential at points A, B, and C? How much work is done *by* the electric field in moving a $+2$-μC charge from C to A? At what point B is the electric potential equal to zero?

Ans. $V_A = -600$ V, $V_B = +600$ V, $V_C = -300$ V, work $= +0.6$ mJ, work $= -2.4$ mJ

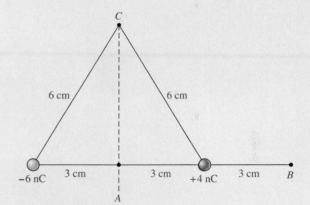

Figure 25.12

*25.43. The horizontal plates in Millikan's oil-drop experiment are 20 mm apart. The diameter of a particular drop of oil is 4 μm, and the density of oil is 900 kg/m^3. Assuming that two electrons attach themselves to the droplet, what potential difference must exist between the plates to establish equilibrium?

26

Capacitance

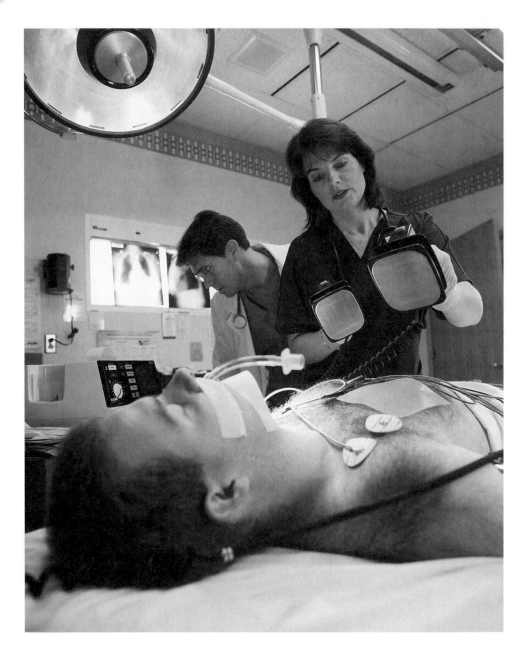

Ventricular defibrillators use large capacitors to shock the heart muscle into establishing its own rhythm. More than 250,000 Americans die each year from sudden cardiac arrest. For every minute that passes without defibrillation, a victim's chances decrease by 7 to 10 percent. In this chapter, we will study the fundamental properties of capacitors. (*Photo © vol. 154/Corbis.*)

Objectives

After completing this chapter, you should be able to

1. Define *capacitance* and apply a relationship among *capacitance*, applied *voltage*, and total *charge*.

2. Compute the capacitance of a *parallel-plate capacitor* when the area of the plates and their separation in a medium of known dielectric constant are given.

3. Write and apply expressions for calculating the *dielectric constant* as a function of the voltage, the electric field, or the capacitance before and after insertion of a dielectric.

4. Calculate the equivalent capacitance of a number of capacitors connected in *series* and in *parallel*.

5. Determine the energy of a charged capacitor, given the appropriate information.

Any charged conductor may be viewed as a reservoir, or source, of electric charge. If a conducting wire is connected to such a reservoir, electric charge can be transferred to perform useful work. In many applications of electricity, large quantities of charge are stored upon a conductor or group of conductors. Any device designed to store electric charge is called a *capacitor.* In this chapter, we discuss the nature and application of these devices.

26.1   Limitations on Charging a Conductor

How much electric charge can be placed on a conductor? Are there practical limits to the number of electrons that can be transferred to or from a conductor? Suppose we connect a large reservoir of positive and negative charges, such as the Earth, to a conducting object, as illustrated in Fig. 26.1a. The energy necessary to transfer electrons from the Earth to the conductor can be provided by an electrical device called a *battery.* Charging the conductor is analogous to pumping air into a hollow steel tank. (See Fig. 26.1b). As more air is pumped into the tank, the pressure opposing the flow of additional air becomes greater. Similarly, as more charge Q is transferred to the conductor, the potential V of the conductor becomes higher, making it increasingly difficult to transfer more charge. We can say that the increase in potential V is directly proportional to the charge Q placed on the conductor. Symbolically,

$$V \propto Q$$

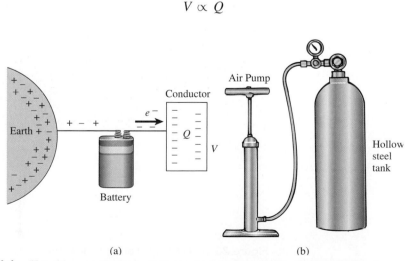

(a)   (b)

Figure 26.1   Charging a conductor is analogous to pumping air into a hollow steel tank.

Therefore, the ratio of the quantity of charge Q to the potential V produced will be a constant for a given conductor. This ratio reflects the ability of a conductor to store charge and is called its **capacitance** C.

$$C = \frac{Q}{V} \tag{26.1}$$

The unit of capacitance is the *coulomb per volt,* which is redefined as a ***farad*** (F). Thus, *if a conductor has a capacitance of 1 farad, a transfer of 1 coulomb of charge to the conductor will raise its potential by 1 volt.*

Let us return to our original question about the limitations placed on charging a conductor. We have said that every conductor has a capacitance C for storing charge. The value of C for a given conductor is not a function of either the charge placed on a conductor or the potential produced. In principle, the ratio Q/V will remain constant as charge is added indefinitely, but the capacitance depends on the *size* and *shape* of a conductor as well as on the nature of the *surrounding medium.*

Suppose we try to place an indefinite quantity of charge Q on a spherical conductor of radius r, as illustrated in Fig. 26.2. The air surrounding the conductor is an insulator, sometimes called a ***dielectric,*** which contains few charges free to move. The electric field intensity E and the potential V at the surface of the sphere are given by

$$E = \frac{kQ}{r^2} \qquad \text{and} \qquad V = \frac{kQ}{r}$$

Since the radius r is constant, both the field intensity and the potential at the surface of the sphere increase in direct proportion to the charge Q. There is a limit, however, to the field intensity that can exist on a conductor without ionizing the surrounding air. When this occurs, the air essentially becomes a conductor, and any additional charge placed on the sphere will "leak off" to the air. This limiting value of electric field intensity for which a material loses its insulation properties is called the ***dielectric strength*** of that material.

> The dielectric strength for a given material is that electric field intensity for which the material ceases to be an insulator and becomes a conductor.

The dielectric strength for dry air at 1 atm pressure is around 3 MN/C. Since the dielectric strength of a material varies considerably with environmental conditions, such as pressure and humidity, it is difficult to compute accurate values.

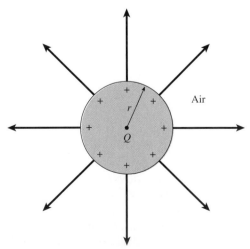

Figure 26.2 The amount of charge that can be placed on a conductor is limited by the dielectric strength of the surrounding medium.

Example 26.1

What is the maximum charge that may be placed on a spherical conductor 1 m in diameter? Assume it is surrounded by air.

Plan: The maximum charge is determined by the field intensity necessary to make the surrounding air a conductor of electrons. We will find the charge necessary to reach the dielectric strength of air for a radius of 0.5 m.

Solution: Given that $r = 0.50$ m and $E_{max} = 3 \times 10^6$ N/C, the maximum charge Q is given by

$$E_{max} = \frac{kQ}{r^2} \quad \text{or} \quad Q = \frac{r^2 E_{max}}{k}$$

$$Q = \frac{(0.5 \text{ m})^2 (3 \times 10^6 \text{ N/C})}{(9 \times 10^9 \text{ N} \cdot \text{m}^2/\text{C}^2)} = 8.33 \times 10^{-5} \text{ C} \quad \text{or} \quad 83.3 \ \mu\text{C}$$

This example illustrates the enormous magnitude of the coulomb when it is applied as a unit of electrostatic charge.

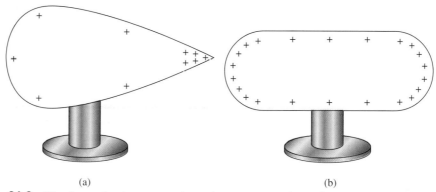

(a)   (b)

Figure 26.3   The charge density on a conductor is greatest at regions of greatest curvature.

Note that the amount of charge that can be placed on a spherical conductor decreases with the radius of the sphere. Thus, smaller conductors can usually hold less charge. But the shape of a conductor also influences its ability to retain charge. Consider the charged conductors illustrated in Fig. 26.3. If these conductors are tested with an electroscope, it will be discovered that the charge on the surface of a conductor is concentrated at points of greatest curvature. Because of the greater charge density in these regions, the electric field intensity is also greater in regions of higher curvature. If the surface is reshaped to a sharp point, the field intensity may become great enough to ionize the surrounding air. A slow leakage of charge sometimes occurs at these locations, producing a ***corona discharge,*** which is often observed as a faint violet glow in the vicinity of the sharply pointed conductor. It is important to remove all sharp edges from electrical equipment to minimize this leakage of charge.

26.2 The Capacitor

When a number of conductors are placed near one another, the potential of each is affected by the presence of the other. For example, suppose we connect a negatively charged plate A to an electroscope, as in Fig. 26.4. The divergence of the gold leaf in the electroscope provides a measure of the potential of the conductor. Now let us suppose that another conductor B is placed parallel to A a short distance away. When the second conductor is grounded, a positive charge will be induced on it as the electrons are forced into the ground. Immediately, the gold leaf will collapse slightly, indicating a drop in the potential of conductor A.

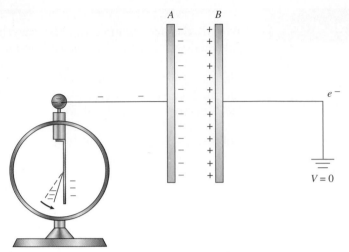

Figure 26.4   A capacitor consists of two closely spaced conductors.

Because of the presence of the induced charge on *B*, less work is required to bring additional units of charge to conductor *A*. In other words, the capacitance of the system for holding charge has been increased by the proximity of the two conductors. Two such conductors in close proximity, carrying equal and opposite charges, constitute a ***capacitor.***

A capacitor consists of two closely spaced conductors carrying equal and opposite charges.

The simplest capacitor is the *parallel-plate capacitor,* illustrated in Fig. 26.4. A potential difference between two such plates can be realized by connecting a battery to them, as shown in Fig. 26.5. Electrons are transferred from plate *A* to plate *B*, producing an equal and opposite charge on the plates. The capacitance of this arrangement is defined as follows:

The capacitance between two conductors having equal and opposite charges is the ratio of the magnitude of the charge on either conductor to the resulting potential difference between the two conductors.

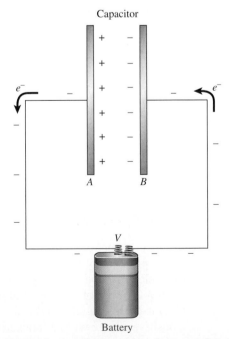

Figure 26.5   Charging a capacitor by transferring charge from one plate to another.

The equation for the capacitance of a capacitor is the same as Eq. (26.1) for a single conductor, except that the symbol V now applies to the *potential difference* and the symbol Q refers to the charge on *either* conductor.

$$C = \frac{Q}{V} \qquad (26.2)$$

$$1 \text{ F} = \frac{1 \text{ C}}{1 \text{ V}}$$

Because of the enormous size of the coulomb as a unit of charge, the farad as a unit of capacitance is usually too large for practical application. Consequently, the following sub-multiples are commonly used:

$$1 \text{ microfarad } (\mu\text{F}) = 10^{-6} \text{ F}$$
$$1 \text{ picofarad } (\text{pF}) = 10^{-12} \text{ F}$$

Capacitances as low as a few picofarads are not uncommon in some electrical communication applications.

Example 26.2

A capacitor having a capacitance of 4 μF is connected to a 60-V battery. What is the charge on the capacitor?

Solution: The charge *on* a capacitor refers to the magnitude of the charge on either plate of the capacitor. From Eq. (26.2),

$$Q = CV = (4 \ \mu\text{F})(60 \text{ V}) = 240 \ \mu\text{C}$$

26.3   Computing the Capacitance

In general, a larger conductor can hold a greater quantity of charge, and a capacitor can store more charge than a single conductor because of the inductive effect of two closely spaced conductors. The closer the spacing of these conductors, the greater the inductive effect and, hence, the easier it becomes to transfer additional charge from one conductor to the other. On the basis of these observations, one might suspect that *the capacitance of a given capacitor will be directly proportional to the area of the plates and inversely proportional to their separation.* The exact relationship can be determined by considering the electric field intensity between the capacitor plates.

The electric field intensity between the plates of the charged capacitor in Fig. 26.6 can be found from

$$E = \frac{V}{d} \qquad (26.3)$$

where V = potential difference between plates, V
$\quad\quad\ d$ = separation of plates, m

An alternative equation for computing the electric field intensity was derived in Chapter 24, using Gauss's law. It relates the field intensity E to the charge density σ as follows:

$$E = \frac{\sigma}{\epsilon_0} = \frac{Q}{A\epsilon_0} \qquad (26.4)$$

where Q = charge on either plate
$\quad\quad\ A$ = area of either plate
$\quad\quad\ \epsilon_0$ = permittivity of vacuum (8.85×10^{-12} C^2/N $\cdot$ m^2)

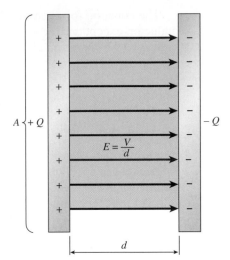

Figure 26.6   The capacitance is directly proportional to the area of either plate and inversely proportional to the plate separation.

For a capacitor with a vacuum between its plates, we combine Eqs. (26.3) and (26.4) to get

$$\frac{V}{d} = \frac{Q}{A\epsilon_0}$$

Realizing that the capacitance C is the ratio of charge to voltage, we can rearrange terms and obtain

$$C_0 = \frac{Q}{V} = \epsilon_0 \frac{A}{d} \qquad\qquad (26.5)$$

The subscript 0 is used to indicate that a vacuum exists between the plates of the capacitor. To a close approximation, Eq. (26.5) can also be used when air is between the capacitor plates.

Example 26.3   Each plate of a parallel capacitor is 2 cm wide and 4 cm long. What should be the separation of the plates of this capacitor in air if the total capacitance is to be 4 pF?

Plan: The required separation can be found by solving for d in Eq. (26.5) after calculating the area of each plate. Recall that 1 pF $= 1 \times 10^{-12}$ F.

Solution: The area of each plate is

$$A = (0.02 \text{ m})(0.04 \text{ m}) = 8 \times 10^{-4} \text{ m}^2$$

$$C_0 = \frac{\epsilon_0 A}{d} \qquad \text{or} \qquad d = \frac{\epsilon_0 A}{C_0}$$

Given that $C = 4 \times 10^{-12}$ F, and $\epsilon_0 = 8.85 \times 10^{-12}$ C^2/N $\cdot$ m^2, we find that

$$d = \frac{(8.85 \times 10^{-12} \text{ C}^2/\text{N} \cdot \text{m}^2)(8 \times 10^{-4} \text{ m}^2)}{4 \times 10^{-12} \text{ F}} = 1.77 \times 10^{-3} \text{ m}$$

The plate separation must be 1.77 mm.

Parallel-plate capacitors are frequently made with a stack of several plates by connecting alternate plates, as shown in Fig. 26.7. By making one of the sets of plates movable, a

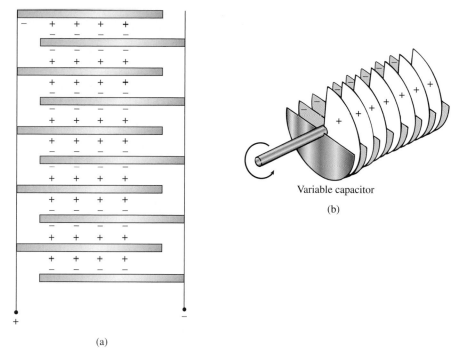

Figure 26.7 (a) A capacitor consisting of a number of stacked plates, alternating with positive and negative charges. (b) A variable capacitor allows one set of plates to be rotated relative to the other, causing a variation in the effective area.

variable capacitor can be constructed. Rotating one set of plates relative to the other set varies the effective area of the capacitor plates, causing a variation in the capacitance. Variable capacitors are sometimes used in the tuning circuits of radios.

26.4 Dielectric Constant; Permittivity

The amount of charge that can be put on a conductor is determined to a large degree by the dielectric strength of the surrounding medium. Similarly, the dielectric strength of the material between the plates of a capacitor limits its ability to store charge. Most capacitors have a nonconducting material, called a *dielectric,* between the plates to provide a *dielectric strength* greater than that for air. The following advantages are realized:

1. A dielectric material provides for a small plate separation without contact.

2. A dielectric increases the capacitance of a capacitor.

3. Higher voltages can be used without danger of dielectric breakdown.

4. A dielectric often provides greater mechanical strength.

Common dielectric materials are mica, paraffined paper, ceramics, and plastics. Alternating sheets of metal foil and paraffin-coated paper can be rolled up to provide a compact capacitor with a capacitance of several microfarads.

To understand the effect of a dielectric, let us consider the insulating material of Fig. 26.8 placed between capacitor plates having a potential difference *V*. The electrons in the dielectric are not free to leave their parent atoms, but they do shift toward the positive plate. The protons and electrons of each atom align themselves as shown in the figure. The material is said to become polarized, and the atoms form *dipoles.* All the positive and negative charges inside the dotted ellipse in Fig. 26.8a neutralize each other. However, a layer of negative charge on one surface and a layer of positive charge on the other are not

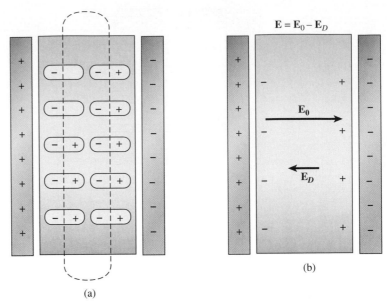

Figure 26.8 (a) The polarization of a dielectric when it is inserted between the plates of a capacitor. (b) The polarization results in an overall reduction in the electric field intensity.

neutralized. An electric field $\mathbf{E}_D$ is set up in the dielectric *opposing* the field $\mathbf{E}_0$, which would exist without the dielectric. The resulting electric field intensity is

$$\mathbf{E} = \mathbf{E}_0 - \mathbf{E}_D \tag{26.6}$$

Therefore, insertion of a dielectric results in a reduction of the field intensity between the capacitor plates.

Since the potential difference V between the plates is proportional to the electric field intensity, $V = Ed$, a reduction in the intensity will cause a drop in potential difference. This fact is illustrated by the example shown in Fig. 26.9. The insertion of a dielectric causes less of a divergence of the gold leaf on the electroscope.

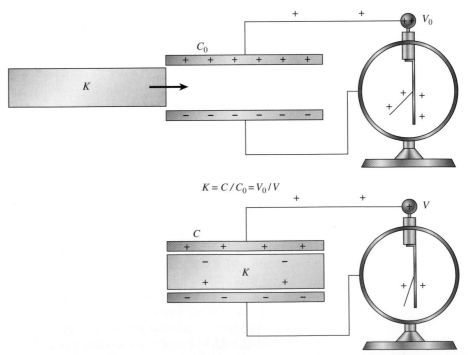

Figure 26.9 The insertion of a dielectric between the plates of a capacitor causes a drop in potential difference, resulting in an increased capacitance.

Table 26.1

Dielectric Constant and Dielectric Strength

Material	Average Dielectric Constant	Average Dielectric Strength, MN/C
Air, dry at 1 atm	1.006	3
Bakelite	7.0	16
Glass	7.5	118
Mica	5.0	200
Nitrocellulose plastics	9.0	250
Paper, paraffined	2.0	51
Rubber	3.0	28
Teflon	2.0	59
Transformer oil	4.0	16

It can be seen from the definition of capacitance, $C = Q/V$, that a drop in voltage will result in an increased capacitance. If we represent the capacitance before insertion of a dielectric by C_0 and the capacitance after insertion by C, the ratio C/C_0 will denote the relative increase in capacitance. Although this ratio varies from material to material, it is constant for a particular dielectric.

The dielectric constant K for a particular material is defined as the ratio of the capacitance C of a capacitor with the material between its plates to the capacitance C_0 for a vacuum.

$$K = \frac{C}{C_0} \qquad (26.7)$$

The *dielectric constant* for various dielectric materials is given in Table 26.1 along with the dielectric strength for the materials. Note that, for air, K is approximately 1.0.

On the basis of proportionalities, it can be shown that the dielectric constant is also given by

$$K = \frac{V_0}{V} = \frac{E_0}{E} \qquad (26.8)$$

where V_0, E_0 = voltage and electric field with vacuum between capacitor plates

V, E = respective values after insertion of dielectric material

The capacitance C of a capacitor having a dielectric between its plates is, from Eq. (26.7),

$$C = KC_0$$

Substituting from Eq. (26.5), we have a relation for computing C directly:

$$C = K\epsilon_0 \frac{A}{d} \qquad (26.9)$$

where A is the area of the plates and d is their separation.

The constant ϵ_0 has been defined earlier as the *permittivity* of a vacuum. Recall from our discussions of Gauss's law that ϵ_0 is actually the proportionality constant that relates the density of electric field lines to the electric field intensity in a vacuum. The permittivity ϵ of a dielectric is greater than ϵ_0 by a factor equal to the dielectric constant K. Thus,

$$\epsilon = K\epsilon_0 \qquad (26.10)$$

On the basis of this relation, we can understand why the dielectric constant, $K = \epsilon/\epsilon_0$, is sometimes referred to as the *relative permittivity*. When we substitute Eq. (26.10) into Eq. (26.9), the capacitance for a capacitor containing a dielectric is simply

$$C = \epsilon\frac{A}{d} \qquad (26.11)$$

This relation is the most general equation for computing capacitance. When a vacuum or air is between the capacitor plates, $\epsilon = \epsilon_0$ and Eq. (26.11) reduce to Eq. (26.5).

Example 26.4

A certain capacitor has a capacitance of 4 μF when its plates are separated by 0.2 mm of vacant space. A battery is used to charge the plates to a potential difference of 500 V and is then disconnected from the system. (a) What will be the potential difference across the plates if a sheet of mica 0.2 mm thick is inserted between the plates? (b) What will the capacitance be after the dielectric is inserted? (c) What is the permittivity of mica?

Plan: When the dielectric is inserted, the capacitor voltage will drop due to induced charge in the dielectric. This results in a decrease of the effective field between the plates. Therefore, the voltage will decrease to an amount determined by the dielectric constant of mica. There will now be a greater capacity for holding charge or an *increased* capacitance. The actual permittivity of the dielectric is the ratio of the new capacitance to the original capacitance.

Solution (a): The dielectric constant for mica is $K = 5$. Thus, Eq. (26.8) gives

$$K = \frac{V_0}{V} \qquad \text{or} \qquad V = \frac{V_0}{K} = \frac{500 \text{ V}}{5} = 100 \text{ V}$$

Solution (b): From Eq. (26.7),

$$K = \frac{C}{C_0} \qquad \text{or} \qquad C = KC_0 = 5(4 \; \mu\text{F}) = 20 \; \mu\text{F}$$

Solution (c): The permittivity is found from Eq. (26.10).

$$K = \frac{\epsilon}{\epsilon_0} \qquad \text{or} \qquad \epsilon = K\epsilon_0 = 5(8.85 \times 10^{-12} \text{ C}^2/\text{N} \cdot \text{m}^2)$$

$$\epsilon = 44.2 \times 10^{-12} \text{ C}^2/\text{N} \cdot \text{m}^2$$

It should be noted that the charge on the capacitor is the same before and after insertion since the voltage source did not stay connected to the capacitor.

Example 26.5

Assume that the source of voltage remains connected to the 4-μF capacitor in Example 26.4. What will be the increase in charge as a result of the insertion of the sheet of mica?

Plan: The voltage will remain at 500 V when the dielectric is inserted. Since the capacitance is increased by the dielectric, an increase in charge will result. The change will be the difference between the final charge and the initial charge.

Solution: The initial charge on the capacitor was

$$Q_0 = C_0 V_0 = (4 \ \mu F)(500 \ V) = 2000 \ \mu C$$

When the mica is inserted, the capacitance increases from $4 \ \mu F$ to $20 \ \mu F$ ($K = 5$).

$$Q = CV = (20 \ \mu F)(500 \ V) = 10{,}000 \ \mu C$$

Thus, the *increase* in charge is

$$\Delta Q = 10{,}000 \ \mu C - 2000 \ \mu C = 8000 \ \mu C$$

This $8000 \ \mu C$ was due to the increased capacitance at the same voltage.

26.5 Capacitors in Parallel and in Series

Electric circuits often contain two or more capacitors grouped together. In considering the effect of such a grouping, it is convenient to resort to the circuit diagram, in which electrical devices are represented by symbols. Four symbols commonly used with capacitors are defined in Fig. 26.10. The high-potential side of a battery is denoted by the longer line. The high-potential side of a capacitor may be represented as a straight line, with a curved line representing the low-potential side. An arrow indicates a variable capacitor. A *ground* is an electrical connection between the wiring of an apparatus and its metal framework or any other large reservoir of positive and negative charges.

First, we consider the effect of a group of capacitors connected along a single path, as shown in Fig. 26.11. Such a connection, in which the positive plate of one capacitor is connected to the negative plate of another, is called a **series connection.** The battery

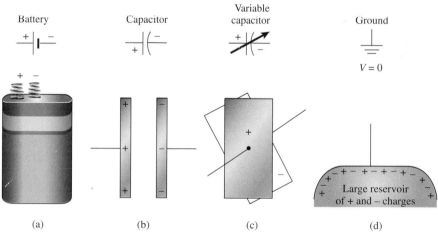

Battery   Capacitor   Variable capacitor   Ground

(a)	(b)	(c)	(d)

Figure 26.10 Definition of symbols frequently used with capacitors.

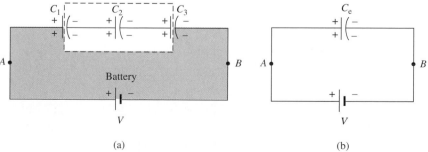

(a)	(b)

Figure 26.11 Computing the equivalent capacitance of a group of capacitors connected in series.

maintains a potential difference V between the positive plate of C_1 and the negative plate of C_3, transferring electrons from one to the other. The charge cannot pass between the plates of a capacitor. Therefore, all the charge inside the dotted circle in Fig. 26.11a is induced charge. For this reason, the charge on each capacitor is identical. We write

$$Q = Q_1 = Q_2 = Q_3$$

where Q is the effective charge transferred by the battery.

All three capacitors can be replaced by an equivalent capacitance C_e without changing the external effect. We now derive an expression for calculating this equivalent capacitance for the series connection. Since the potential difference between A and B is independent of the path, the battery voltage must equal the sum of the potential drops across each capacitor.

$$V = V_1 + V_2 + V_3 \qquad \textbf{(26.12)}$$

If we recall that the capacitance C is defined by the ratio Q/V, Eq. (26.12) becomes

$$\frac{Q}{C_e} = \frac{Q_1}{C_1} + \frac{Q_2}{C_2} + \frac{Q_3}{C_3}$$

For a series connection, $Q = Q_1 = Q_2 = Q_3$, so we can divide out the charge, yielding

$$\frac{1}{C_e} = \frac{1}{C_1} + \frac{1}{C_2} + \frac{1}{C_3} \qquad \textit{Series Connection} \quad \textbf{(26.13)}$$

The total effective capacitance for *two* capacitors in series is

$$C_e = \frac{C_1 C_2}{C_1 + C_2} \qquad \textbf{(26.14)}$$

The derivation of Eq. (26.14) is left as an exercise.

Now, let us consider a group of capacitors connected so that charge can be shared between two or more conductors. When several capacitors are all connected directly to the same source of potential, as in Fig. 26.12, they are said to be connected in *parallel*. From the definition of capacitance, the charge on each parallel capacitor is

$$Q_1 = C_1 V_1 \qquad Q_2 = C_2 V_2 \qquad Q_3 = C_3 V_3$$

The total charge Q is equal to the sum of the individual charges.

$$Q = Q_1 + Q_2 + Q_3 \qquad \textbf{(26.15)}$$

The equivalent capacitance of the entire circuit is $Q = CV$, so Eq. (26.15) becomes

$$CV = C_1 V_1 + C_2 V_2 + C_3 V_3 \qquad \textbf{(26.16)}$$

For a *parallel connection,*

$$V = V_1 = V_2 = V_3$$

since all capacitors are connected to the same potential difference. Hence, the voltages divide out of Eq. (26.16), giving

$$C = C_1 + C_2 + C_3 \qquad \textit{Parallel Connection} \quad \textbf{(26.17)}$$

PHYSICS TODAY

Do you know how much time satellites have in the sunlight to charge their batteries? For low Earth orbit, they have 60 min of sunlight and 35 min of darkness. Geosynchronous Earth orbit (GEO) satellites, which are much farther out, spend less time in Earth's shadow. They spend 22.8 hours in sunlight and 1.2 hours in darkness. The power to run the satellites must come entirely from batteries during the dark period.

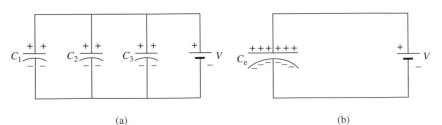

(a)   (b)

Figure 26.12   Equivalent capacitance of a group of capacitors connected in parallel.

Example 26.6

(a) Find the equivalent capacitance of the circuit illustrated in Fig. 26.13a. (b) Determine the charge on each capacitor. (c) What is the voltage across the 4-μF capacitor?

Plan: We will start at the region most distant from the voltage source using the rules for combining capacitors connected in parallel and in series. In this way, we will draw simpler and simpler circuits until we obtain a single equivalent capacitance in series with the source. The charge on the total network and the charge across each capacitor in the network can be determined from the fact that $Q = CV$ and the knowledge of how the voltage is distributed for capacitors in series and in parallel.

Solution (a): The 4- and 2-μF capacitors are in series. Their combined capacitance is found from Eq. (26.14).

$$C_{2,4} = \frac{C_2 C_4}{C_2 + C_4} = \frac{(2\ \mu\mathrm{F})(4\ \mu\mathrm{F})}{2\ \mu\mathrm{F} + 4\ \mu\mathrm{F}}$$
$$= 1.33\ \mu\mathrm{F}$$

These two capacitors can be replaced by their equivalent capacitance, as shown in Fig. 26.13b. The two remaining capacitors are in parallel. Thus, the equivalent capacitance is

$$C_e = C_3 + C_{2,4} = 3\ \mu\mathrm{F} + 1.33\ \mu\mathrm{F}$$
$$= 4.33\ \mu\mathrm{F}$$

Solution (b): The total charge in the network is

$$Q = C_e V = (4.33\ \mu\mathrm{F})(120\ \mathrm{V}) = 520\ \mu\mathrm{C}$$

The charge Q_3 on the 3-μF capacitor is

$$Q_3 = C_3 V = (3\ \mu\mathrm{F})(120\ \mathrm{V}) = 360\ \mu\mathrm{C}$$

The remainder of the charge,

$$Q - Q_3 = 520\ \mu\mathrm{C} - 360\ \mu\mathrm{C} = 160\ \mu\mathrm{C}$$

must be deposited on the series capacitors. Hence,

$$Q_2 = Q_4 = 160\ \mu\mathrm{C}$$

As a check on these values for Q_2 and Q_4, the equivalent capacitance of the two series capacitors can be multiplied by the voltage drop across it:

$$Q_{2,4} = C_{2,4}\,V = (1.33\ \mu\mathrm{F})(120\ \mathrm{V}) = 160\ \mu\mathrm{C}$$

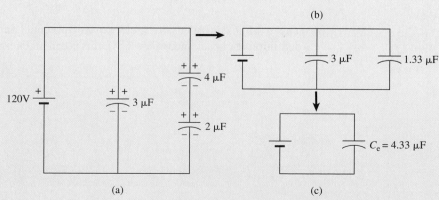

Figure 26.13 Simplifying a problem by substituting equivalent values for capacitance.

Solution (c): The voltage across the 4-μF capacitor is

$$V_4 = \frac{Q_4}{C_4} = \frac{160\ \mu C}{4\ \mu F} = 40\ V$$

The remaining 80 V is across the 2-μF capacitor.

The general facts about capacitors connected in series and in parallel are summarized in Table 26.2.

Table 26.2

Capacitor Circuits

Type of Circuit	Series Circuits	Parallel Circuits
Charge Q	$Q = Q_1 = Q_2 = Q_3$	$Q = Q_1 + Q_2 + Q_3$
Voltage V	$V = V_1 + V_2 + V_3$	$V = V_1 = V_2 = V_3$
Equivalent capacitance	$\dfrac{1}{C_e} = \dfrac{1}{C_1} + \dfrac{1}{C_2} + \dfrac{1}{C_3}$	$C_e = C_1 + C_2 + C_3$
Capacitance for two elements	$C_e = \dfrac{C_1 C_2}{C_1 + C_2}$	$C_e = C_1 + C_2$

26.6 Energy of a Charged Capacitor

Consider a capacitor that is initially uncharged. When a source of potential difference is connected to the capacitor, the potential difference between the plates increases as charge is transferred. As more and more charge builds up on the capacitor, it becomes increasingly difficult to transfer additional charge. Suppose we represent the total charge transferred by Q and the final potential difference by V. The *average* potential difference through which the charge is moved is given by

$$V_{av} = \frac{V_{final} + V_{initial}}{2} = \frac{V + 0}{2} = \frac{1}{2}V$$

Since the total charge transferred is Q, the total work done against electric forces is equal to the product of Q and the average potential difference V_{av}. Thus,

$$\text{Work} = Q\left(\frac{1}{2}V\right) = \frac{1}{2}QV$$

This work is equivalent to the electrostatic potential energy of a charged capacitor. From the definition of capacitance ($Q = CV$), this potential energy can be written in alternative forms:

$$U = \frac{1}{2}QV$$

$$= \frac{1}{2}CV^2$$

$$= \frac{Q^2}{2C} \tag{26.18}$$

When C is in farads, V is in volts, and Q is in coulombs, the potential energy will be in joules. These equations apply equally to all capacitors, regardless of their construction.

Summary and Review

Summary

Stored electric charge is a necessity if large quantities of electrical energy are to be delivered on demand to a modern industrial world. We have studied in this chapter the basic principles that determine the amount of charge that can be stored on capacitors. We have discussed the insertion of capacitors into electric circuits and the factors affecting the distribution of charge in such circuits. The fundamental concepts are summarized as follows.

- Capacitance is the ratio of charge Q to the potential V for a given conductor. For two oppositely charged plates, the Q refers to the charge on either plate and the V refers to the potential difference between the plates.

$$\boxed{C = \frac{Q}{V}} \qquad 1 \text{ farad (F)} = \frac{1 \text{ coulomb (C)}}{1 \text{ volt (V)}}$$
$$\textit{Capacitance}$$

- The dielectric strength is that value for E for which a given material ceases to be an insulator and becomes a conductor. For air, this value is

$$\boxed{E = \frac{kQ}{r^2} = 3 \times 10^6 \text{ N/C}}$$
$$\textit{Dielectric Strength, Air}$$

- For a parallel-plate capacitor, the material between the plates is called the dielectric. The insertion of such a material has an effect on the electric field and the potential between the plates. Consequently, it changes the capacitance. The dielectric constant K for a particular material is the ratio of the capacitance with the dielectric C to the capacitance for a vacuum C_0.

$$\boxed{K = \frac{C}{C_0} \qquad K = \frac{V_0}{V} \qquad K = \frac{E_0}{E}}$$
$$\textit{Dielectric Constant}$$

- The permittivity of a dielectric is greater than the permittivity of a vacuum by a factor equal to the dielectric constant. For this reason, K is sometimes referred to as the *relative permittivity*.

$$\boxed{K = \frac{\epsilon}{\epsilon_0} \qquad \epsilon = K\epsilon_0}$$

$$\boxed{\epsilon_0 = 8.85 \times 10^{-12} \text{ C}^2/\text{N} \cdot \text{m}^2}$$

- The capacitance for a parallel-plate capacitor depends on the surface area A of each plate, the plate separation d, and the permittivity or dielectric constant. The general equation is

$$\boxed{C = \epsilon \frac{A}{d} \qquad C = K\epsilon_0 \frac{A}{d}} \qquad \textit{Capacitance}$$

For a vacuum, $K = 1$ in the above relationship.

- Capacitors may be connected in series, as shown in Fig. 26.11, or in parallel, as shown in Fig. 26.12.

  a. For *series connections,* the charge on each capacitor is the same as the total charge, the potential difference across the battery is equal to the sum of the drops across each capacitor, and the net capacitance is found from

$$\boxed{\begin{aligned} Q_T &= Q_1 = Q_2 = Q_3 \\[6pt] V_T &= V_1 + V_2 + V_3 \\[6pt] \frac{1}{C_e} &= \frac{1}{C_1} + \frac{1}{C_2} + \frac{1}{C_3} \quad \begin{array}{c}\textit{Series}\\\textit{Connections}\end{array} \end{aligned}}$$

  b. For *parallel connections,* the total charge is equal to the sum of the charges across each capacitor, the voltage drop across each capacitor is the same as the drop across the battery, and the effective capacitance is equal to the sum of the individual capacitances.

$$\boxed{\begin{aligned} Q_T &= Q_1 + Q_2 + Q_3 \\[6pt] V_B &= V_1 = V_2 = V_3 \\[6pt] C_e &= C_1 + C_2 + C_3 \quad \begin{array}{c}\textit{Parallel}\\\textit{Connections}\end{array} \end{aligned}}$$

- The potential energy stored in a charged capacitor can be found from any of the following relationships:

$$\boxed{U = \frac{1}{2}QV \qquad U = \frac{1}{2}CV^2 \qquad U = \frac{Q^2}{2C}}$$

When C is in *farads,* V is in *volts,* and Q is in *coulombs,* the potential energy will be in *joules.*

Key Terms

capacitance 514
capacitor 516
corona discharge 515
dielectric 514

dielectric constant 521
dielectric strength 514
farad 514
parallel connection 524

permittivity 521
series connection 523
variable capacitor 519

Review Questions

26.1. Discuss several factors that limit the ability of a conductor to store charge.

26.2. Air is pumped from one metal tank to another, creating a partial vacuum in one tank and high pressure in the other. When the pump is removed, potential energy is stored. The energy is released if the two tanks are reconnected and the pressure in each tank becomes equal. In what ways is this mechanical example analogous to charging and discharging a capacitor?

26.3. Large sparks are often seen jumping from the leather belts driving machinery. Explain.

26.4. The Leyden jar is a capacitor that consists of a glass jar coated inside and out with tinfoil, as shown in Fig. 26.14. Contact with the inside coating is made with a metal chain connected to the central metal rod. From the figure, explain how the capacitor becomes charged. What is the function of the ground wire? What purpose does the glass serve?

26.5. Can lightning be considered a capacitor discharge? Explain.

26.6. Discuss the following statement: the permittivity is a measure of how easily a dielectric will permit the establishment of electric field lines within the dielectric.

26.7. A dielectric with a larger permittivity allows for the storage of greater quantities of charge. Explain.

26.8. Distinguish the dielectric strength of a material from its dielectric constant. What part does each play in the design of a capacitor?

26.9. The term *breakdown voltage* is often used in electronics for capacitors. How would you define such a term? In what ways does it differ from the dielectric strength?

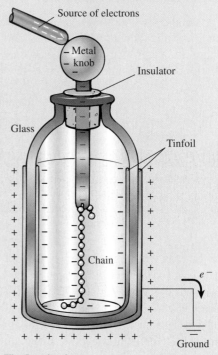

Figure 26.14 The Leyden jar.

26.10. If two point charges are surrounded by a dielectric, will the force each exerts on the other be reduced or increased?

26.11. The unit of permittivity is the $C^2/N \cdot m^2$. Show that the permittivity can be expressed as farads per meter.

26.12. Prove that each of the expressions for potential energy, as given in Eq. (26.18), will yield a proper unit of energy (the joule).

Problems

Section 26.2 The Capacitor

26.1. What is the maximum charge that can be placed on a metal sphere 30 mm in diameter and surrounded by air?   Ans. 75 nC

26.2. How much charge can be placed on a metal sphere of radius 40 mm if it is immersed in transformer oil that has a dielectric strength of 16 MV/m?

26.3. What would be the radius of a metal sphere in air if it could theoretically hold a charge of 1 C?
Ans. 54.8 m

26.4. A 28-μF parallel-plate capacitor is connected to a 120-V source of potential difference. How much charge will be stored on this capacitor?

26.5. A potential difference of 110 V is applied across the plates of a parallel-plate capacitor. If the total charge on each plate is 1200 μC, what is the capacitance?
Ans. 10.9 μF

26.6. Find the capacitance of a parallel-plate capacitor if 1600 μC of charge is on each plate when the potential difference is 80 V.

26.7. What potential difference is required to store a charge of 800 μC on a 40-μF capacitor?
Ans. 20 V

26.8. Write an equation for the potential at the surface of a sphere of radius r in terms of the permittivity of the surrounding medium. Show that the capacitance of such a sphere is given by $C = 4\pi\epsilon r$.

*26.9. A spherical capacitor has a radius of 50 mm and is surrounded by a medium that has a permittivity of 3×10^{-11} C^2/N $\cdot$ m^2. How much charge can be transferred to this sphere by a potential difference of 400 V?
Ans. 4.71×10^{-14} C

Section 26.3 Computing the Capacitance

26.10. A 5-μF capacitor has a plate separation of 0.3 mm of air. What will be the charge on each plate for a potential difference of 400 V? What is the area of each plate?

26.11. The plates of a certain capacitor are 3 mm apart and have an area of 0.04 m^2. What is the capacitance if air is the dielectric?
Ans. 118 pF

26.12. A capacitor has plates of area 0.034 m^2 and a separation of 2 mm in air. The potential difference between the plates is 200 V. What is the capacitance, and what is the electric field intensity between the plates? How much charge is on each plate?

26.13. A capacitor of plate area 0.06 m^2 and plate separation 4 mm has a potential difference of 300 V when air is the dielectric. What is the capacitance for dielectrics of air ($K = 1$) and mica ($K = 5$)?
Ans. 133 pF, 664 pF

26.14. What is the electric field intensity for mica and for air in Prob. 26.13?

26.15. Find the capacitance of a parallel-plate capacitor if the area of each plate is 0.08 m^2, the separation of the plates is 4 mm, and the dielectric is (a) air or (b) paraffined paper ($K = 2$)?
Ans. (a) 177 pF, (b) 354 pF

26.16. Two parallel plates of a capacitor are 4.0 mm apart, and the plate area is 0.03 m^2. Glass ($K = 7.5$) is the dielectric, and the plate voltage is 800 V. What is the charge on each plate, and what is the electric field intensity between the plates?

*26.17. A parallel-plate capacitor with a capacitance of 2.0 nF is to be constructed with mica ($K = 5$) as the dielectric, and it must be able to withstand a maximum potential difference of 3000 V. The dielectric strength of mica is 200 MV/m. What is the minimum area the plates of the capacitor can have?
Ans. 6.78×10^{-4} m^2

Section 26.5 Capacitors in Parallel and in Series

26.18. Find the equivalent capacitance of a 6-μF capacitor and a 12-μF capacitor connected (a) in series and (b) in parallel.

26.19. Find the effective capacitance of a 6-μF capacitor and a 15-μF capacitor connected (a) in series and (b) in parallel.
Ans. 4.29 μF, 21.0 μF

26.20. What is the equivalent capacitance for capacitors of 4, 7, and 12 μF connected (a) in series and (b) in parallel?

26.21. Find the equivalent capacitance for capacitors of 2, 6, and 8 μF connected (a) in series and (b) in parallel.
Ans. 1.26 μF, 16 μF

26.22. A 20- and a 60-μF capacitor are connected in parallel. Then the pair are connected in series with a 40-μF capacitor. What is the equivalent capacitance?

*26.23. If a potential difference of 80 V is placed across the group of capacitors in Prob. 26.22, what is the charge on the 40-μF capacitor? What is the charge on the 20-μF capacitor?
Ans. 2133 μC, 533 μC

*26.24. Find the equivalent capacitance of a circuit in which a 6-μF capacitor is connected in series with two parallel capacitors whose capacitances are 5 and 4 μF.

*26.25. What is the equivalent capacitance for the circuit drawn in Fig. 26.15?
Ans. 6.00 μF

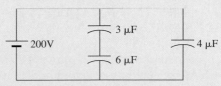

Figure 26.15

*26.26. What is the charge on the 4-μF capacitor in Fig. 26.15? What is the voltage across the 6-μF capacitor?

*26.27. A 6- and a 3-μF capacitor are connected in series with a 24-V battery. What are the charge and voltage across each capacitor?

Ans. $V_3 = 16.0$ V, $Q = 48.0$ μC, $V_6 = 8.00$ V, $Q_6 = 48.0$ μC

*26.28. If the 6- and 3-μF capacitors of Prob. 26.27 are reconnected in parallel with a 24-V battery, what are the charge and voltage across each capacitor?

*26.29. Compute the equivalent capacitance for the entire circuit shown in Fig. 26.16. What is the total charge on the equivalent capacitance?

Ans. 1.74 μF, 20.9 μC

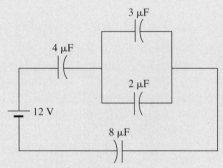

Figure 26.16

*26.30. What are the charge and voltage across each capacitor of Fig. 26.16?

Section 26.6 Energy of a Charged Capacitor

26.31. What is the potential energy stored in the electric field of a 200-μF capacitor when it is charged to a voltage of 2400 V? Ans. 576 J

26.32. What is the energy stored on a 25-μF capacitor when the charge on each plate is 2400 μC? What is the voltage across the capacitor?

26.33. How much work is required to charge a capacitor to a potential difference of 30 kV if 800 μC is on each plate? Ans. 12.0 J

*26.34. A parallel-plate capacitor has a plate area of 4 cm^2 and a separation of 2 mm. A dielectric of constant $K = 4.3$ is placed between the plates, and the capacitor is connected to a 100-V battery. How much energy is stored in the capacitor?

Additional Problems

26.35. What is the breakdown voltage for a capacitor with a dielectric of glass ($K = 7.5$) if the plate separation is 4 mm? The average dielectric strength is 118 MV/m. Ans. 472 kV

26.36. A capacitor has a potential difference of 240 V, a plate area of 5 cm^2, and a plate separation of 3 mm. What are the capacitance and the electric field between the plates? What is the charge on each plate?

26.37. Suppose the capacitor of Prob. 26.36 is disconnected from the 240-V battery and then mica ($K = 5$) is inserted between the plates? What are the new voltage and electric field? If the 240-battery is reconnected, what charge will be on the plates? Ans. 48.0 V, 1.60×10^4 V/m, 1.79 nC

26.38. A 6-μF capacitor is charged with a 24-V battery and then disconnected. When a dielectric is inserted, the voltage drops to 6 V. What is the total charge on the capacitor after the battery has been reconnected?

26.39. A capacitor is formed from 30 parallel plates, each 20 $\times$ 20 cm. If each plate is separated by 2 mm of dry air, what is the total capacitance?

Ans. 5.13 nF

*26.40. Four capacitors, A, B, C, and D, have capacitances of 12, 16, 20, and 26 μF, respectively. Capacitors A and B are connected in parallel. The combination is then connected in series with C and D. What is the effective capacitance?

*26.41. Consider the circuit drawn in Fig. 26.17. What is the equivalent capacitance of the circuit? What are the charge and voltage across the 2-μF capacitor? Ans. 6.00 μF, 18 μC, 9.00 V

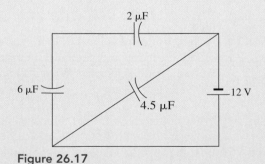

Figure 26.17

*26.42. Two identical 20-μF capacitors, A and B, are connected in parallel with a 12-V battery. What is the charge on each capacitor if a sheet of porcelain

($K = 6$) is inserted between the plates of capacitor B and the battery remains connected?

*26.43. Three capacitors, A, B, and C, have respective capacitances of 2, 4, and 6 μF. Compute the equivalent capacitance if they are connected in series with an 800-V source. What are the charge and voltage across the 4-μF capacitor?

Ans. 1.09 μF, 873 μC, 218 V

*26.44. Suppose the capacitors of Prob. 26.43 are reconnected in parallel with a 12-V source. What is the equivalent capacitance? What are the charge and voltage across the 4-μF capacitor?

*26.45. Show that the total capacitance of a multiple-plate capacitor containing N plates separated by air is given by

$$C_0 = \frac{(N - 1)\epsilon_0 A}{d}$$

where A is the area of each plate and d is the separation of each plate.

*26.46. The energy density u of a capacitor is defined as the potential energy (U) per unit volume (Ad) of the space between the plates. Using this definition and several formulas from this chapter, derive the following relationship for finding the energy density u:

$$u = \frac{1}{2}\epsilon_0 E^2$$

where E is the electric field intensity between the plates.

*26.47. A capacitor with a plate separation of 3.4 mm is connected to a 500-V battery. Use the relation derived in Prob. 26.46 to calculate the energy density between the plates.   Ans. 95.7 mJ/m^3

Critical Thinking Questions

26.48. A certain capacitor has a capacitance of 12 μF when its plates are separated by 0.3 mm of vacant space. A 400-V battery charges the plates and is then disconnected from the capacitor. (a) What is the potential difference across the plates if a sheet of Bakelite ($K = 7$) is inserted between the plates? (b) What is the total charge on the plates? (c) What is the capacitance with the dielectric inserted? (d) What is the permittivity of Bakelite? (e) How much additional charge can be placed on the capacitor if the 400-V battery is reconnected?
Ans. (a) 57.1 V, (b) 4800 μC, (c) 84.0 μF, (d) 6.20 $\times$ 10^{-11} C^2/N m^2, (e) 28.8 mC

*26.49. A medical defibrillator uses a capacitor to revive heart-attack victims. Assume that a 65-μF capacitor in such a device is charged to 1500 V. What is the total energy stored? If 25 percent of this energy passes through a victim in 3 ms, what power is delivered?

*26.50. Consider three capacitors of 10, 20, and 30 μF. Show how these might be connected to produce the maximum and minimum equivalent capaci-

tances and list the values. Draw a diagram of a connection that would result in an equivalent capacitance of 27.5 μF. Show a connection that will result in a combined capacitance of 22.0 μF.
Ans. 60 μF, 5.45 μF

*26.51. A 4-μF air capacitor is connected to a 500-V source of potential difference. The capacitor is then disconnected from the source and a sheet of mica ($K = 5$) is inserted between the plates. What is the new voltage on the capacitor? Now reconnect the 500-V battery and calculate the final charge on the capacitor. By what percentage did the total energy on the capacitor increase due to the dielectric?

*26.52. A 3-μF capacitor and a 6-μF capacitor are connected in series with a 12-V battery. What is the total stored energy of the system? What is the total energy if they are connected in parallel? What is the total energy for each of these connections if mica ($K = 5$) is used as a dielectric for each capacitor?   Ans. 0.144 mJ, 0.648 mJ, 0.720 mJ, 3.24 mJ

27

Current and Resistance

Distribution of Electric power remains a difficult, but very important task in modern times. Power generated by water from dams, the combustion of coal, or other sources must be continually distributed over long distances for use in industry and in our homes. To overcome the electric resistance in long wires, high voltages, from 100,000 to 700,000 volts, are required until transformers reduce the voltage to the 120 or 240-V available at the points of use. In this chapter, we begin the discussion of electric current, resistance, and power for direct current, and later we apply some of the same relationships to alternating current.
(*Photo © vol. 14 PhotoDisc/Getty.*)

Objectives

After completing this chapter, you should be able to

1. Demonstrate by definition and example your understanding of *electric current* and *electromotive force.*

2. Write and apply *Ohm's law* to the solution of problems involving electric resistance.

3. Compute *power losses* as a function of voltage, current, and resistance.

4. Define the *resistivity* of a material and solve problems similar to those in the text.

5. Define the *temperature coefficient of resistance* and calculate the change in resistance that occurs with a change in temperature.

We now leave electrostatics and enter a discussion of charges in motion. We have been concerned with forces, electric fields, and potential energies as they relate to charged conductors. For example, excess electrons, evenly distributed over an insulated spherical surface, will remain at rest. If a wire is connected from the sphere to ground, however, the electrons will flow through the wire to the ground. The flow of charge constitutes an *electric current*. In this chapter, the foundation is lain for a study of direct currents and electric resistance.

27.1 The Motion of Electric Charge

Let us begin our discussion of moving charges by considering the discharge of a capacitor. The potential difference V between the two capacitor plates in Fig. 27.1a is indicated by the electroscope. The total charge Q on either plate is given by

$$Q = CV$$

where C is the capacitance. If a path is provided, electrons on one plate will travel to the other, decreasing the net charge and causing a drop in the potential difference. Thus, a drop in potential, as indicated by the collapsing leaf of the electroscope, means that charge has been transferred. Any conductor used to connect the plates of a capacitor will cause it to discharge. The rate of discharge varies considerably, however, with the size, shape, material, and temperature of the conductor.

If a short, thick wire is connected between the plates of the capacitor, as shown in Fig. 27.1b, the electroscope leaf collapses instantly, indicating a rapid transfer of charge. This current, which exists for a short time, is called a ***transient current.*** If we replace the short, thick wire with a long, thin wire of the same material, we will observe a gradual collapse of the electroscope leaf (Fig. 27.1c). Such opposition to the flow of electricity is called *electric resistance.* A quantitative description of electric resistance will be presented in Section 27.6. It is introduced here to illustrate that the rate at which charge flows through a conductor varies. This rate is referred to as the ***electric current.***

The electric current *I* is the rate of the flow of charge *Q* past a given point *P* on an electric conductor.

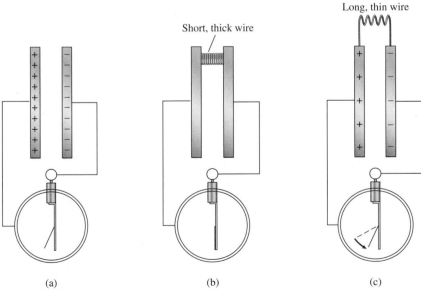

(a)   (b)   (c)

Figure 27.1   (a) A charged capacitor is a source of current. (b) If the capacitor plates are joined by a short, thick wire, the capacitor will discharge instantly. (c) A long, thin wire allows for a gradual discharge.

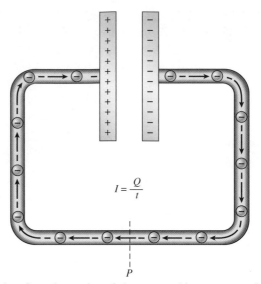

Figure 27.2   Current arises from the motion of electrons and is a measure of the quantity of charge passing a given point in a unit of time.

$$I = \frac{Q}{t} \tag{27.1}$$

The unit of electric current is the *ampere*. One ***ampere*** (A) represents a flow of charge at the rate of *1 coulomb per second* past any point.

$$1\ \text{A} = \frac{1\ \text{C}}{1\ \text{s}}$$

In the example of a discharging capacitor, the current arises from the motion of electrons, as illustrated in Fig. 27.2. The positive charges in a wire are tightly bound and cannot move. The electric field created in the wire because of the potential difference between the plates causes the free electrons in the wire to experience a drift toward the positive plate. The electrons are repeatedly deflected or stopped by processes relating to impurities and thermal motions of the atoms. Consequently, the motion of the electrons is not an accelerated one but a drifting or diffusion process. The average drift velocity of electrons is typically of the order of 4 m/h. This velocity of charge, which is a *distance* per unit of time, should not be confused with current, which is a *quantity* of charge per unit of time.

An analogy to water flowing through a pipe is useful in understanding current flow. The rate of flow of the water in gallons per minute is analogous to the rate of flow of charge in coulombs per second. For a current of 1 A, 6.25×10^{18} electrons (1 C) flow past a given point every second. Just as the size and length of a pipe affect the flow of water, the size and length of a conductor affect the flow of electrons.

Example 27.1   How many electrons pass a point in 5 s if a constant current of 8 A is maintained in a conductor?

Solution: From Eq. (27.1),

$$Q = It = (8\ \text{A})(5\ \text{s})$$
$$= (8\ \text{C/s})(5\ \text{s}) = 40\ \text{C}$$
$$= (40\ \text{C})(6.25 \times 10^{18}\ \text{electrons/C})$$
$$= 2.50 \times 10^{20}\ \text{electrons}$$

27.2 The Direction of Electric Current

Thus far, we have discussed only the magnitude of electric current. The choice of direction is purely arbitrary as long as we apply our definition consistently. The flow of charge caused by an electric field in a gas or a liquid consists of a flow of positive ions in the direction of the field or a flow of electrons opposite to the field direction. As we have seen, the current in a metallic material consists of electrons flowing against the field direction. However, a current that consists of negative particles moving in one direction is electrically the same as a current consisting of positive charges moving in the opposite direction.

There are a number of reasons for preferring the motion of positive charge as an indicator of direction. In the first place, all the concepts introduced in electrostatics—for example, the electric field, potential energy, and potential difference—were defined in terms of positive charges. An electron flows contrary to the electric field and "up a potential hill" from the negative plate to the positive plate. If we define current as a flow of *positive* charge, the loss in energy as charge encounters resistance will be from + to − or "down a potential hill." By convention, we consider all currents as consisting of a flow of positive charge.

> The direction of conventional current is always the same as the direction in which positive charges would move, even if the actual current consists of a flow of electrons.

For a metallic conducting wire, both electron flow and conventional current are indicated in Fig. 27.3. The zigzag line is used to indicate the electric resistance R. Note that the conventional current flows from the positive plate of the capacitor, neutralizing negative charge on the other plate. Conventional current is in the same direction as the electric field $\mathbf{E}$ producing the current.

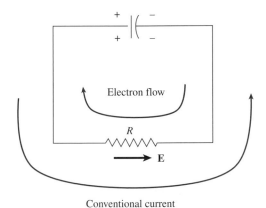

Conventional current

Figure 27.3   In a metallic conductor, the conventional current is in a direction opposite to the actual flow of electrons.

27.3 Electromotive Force

The currents discussed in Sections 27.1 and 27.2 were called *transient currents* because they exist only for a short time. Once the capacitor has been completely discharged, there will no longer be a potential difference to promote the flow of additional charge. If some means were available to keep the capacitor continually charged, a continuous current could be maintained. This would require that electrons be supplied continuously to the negative plate to replace those leaving. In other words, energy must be supplied to replace the energy lost by the charge in the external circuit. In this manner, the potential difference

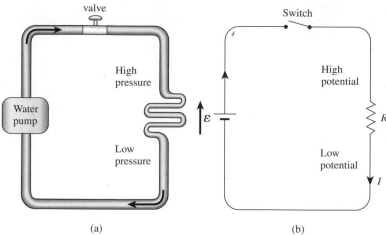

Figure 27.4 The mechanical analogy of a water pump can be used to explain the function of a source of emf in an electric circuit.

between the plates could be maintained, allowing for a continuous flow of charge. A device with the ability to maintain potential difference between two points is called a ***source of electromotive force*** (emf).

The most familiar sources of emf are batteries and generators. Batteries convert chemical energy into electric energy, and generators transform mechanical energy into electric energy.

> A source of electromotive force (emf) is a device that converts chemical, mechanical, or other forms of energy into the electric energy necessary to maintain a continuous flow of electric charge.

In an electric circuit, the source of emf is usually represented by the symbol $\mathcal{E}$.

The function of a source of emf in an electric circuit is similar to the function of a water pump in maintaining the continuous flow of water through a system of pipes. In Fig. 27.4a, the water pump must perform the work on each unit volume of water necessary to replace the energy lost by each unit volume flowing through the pipes. In Fig. 27.4b, the source of emf must do work on each unit of charge that passes through it to raise it to a higher potential. This work must be supplied at a rate equal to the rate at which energy is lost in flowing through the circuit.

By convention, we have assumed the current consists of a flow of positive charge, even though in most cases it is negative electrons. Therefore, the charge loses energy in passing through the resistor from a high potential to a low potential. In the hydraulic analogy, water passes from high pressure to low pressure. When the shut-off valve is closed, pressure exists, but there is no water flow. Similarly, when the electric switch is *open*, there is voltage but no current.

Since ***emf*** is work per unit charge, it is expressed in the same unit as potential difference: the *joule per coulomb,* or *volt.*

> A source of emf of 1 volt will perform 1 joule of work on each coulomb of charge that passes through it.

For example, a 12-V battery performs 12 J of work on each coulomb of charge transferred from the low-potential side ($-$ terminal) to the high-potential side ($+$ terminal). An arrow ($\uparrow$) is usually drawn next to the symbol $\mathcal{E}$ for an emf to indicate the direction in which the source, acting alone, would cause a positive charge to move through the external circuit. The conventional current is directed away from the $+$ terminal of a battery, and the hypothetical positive charge flows "downhill" through external resistance of the $-$ terminal of the battery.

In the following sections, circuit diagrams, like that in Fig. 27.4b, will frequently be used to describe electric systems. Many of the symbols we will use are defined in Fig. 27.5.

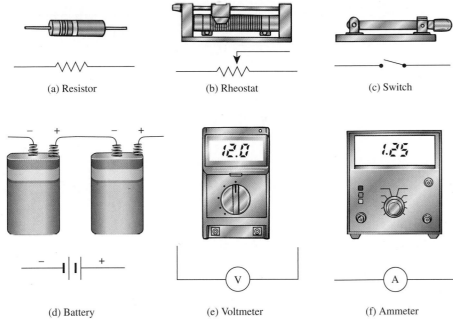

(a) Resistor   (b) Rheostat   (c) Switch

(d) Battery   (e) Voltmeter   (f) Ammeter

Figure 27.5 Conventional symbols used in electric circuit diagrams.

27.4 Ohm's Law; Resistance

Resistance (*R*) is defined as the opposition to the flow of electric charge. Although most metals are good conductors of electricity, all offer some opposition to the surge of electric charge through them. This electric resistance is fixed for many specific materials of known size, shape, and temperature. It is independent of the applied emf and the current passing through it.

The effects of resistance in limiting the flow of charge were first studied quantitatively by Georg Simon Ohm in 1826. He discovered that *for a given resistor at a particular temperature, the current is directly proportional to the applied voltage.* Just as the rate of flow of water between two points depends on the difference of height between them, the rate of flow of electric charge between two points depends on the difference in potential between them. This proportionality is usually stated as ***Ohm's law:***

> The current produced in a given conductor is directly proportional to the difference of potential between its end points.

The current *I* that is observed for a given voltage *V* is therefore an indication of resistance. Mathematically, the resistance *R* of a given conductor can be calculated from

$$R = \frac{V}{I} \qquad V = IR \qquad \qquad Ohm's\ Law \quad (27.2)$$

The greater the resistance *R*, the smaller the current *I* for a given voltage *V*. The unit of measurement of resistance is the ***ohm,*** for which the symbol is the Greek capital letter *omega* (Ω). From Eq. (27.2),

$$1\ \Omega = \frac{1\ V}{1\ A}$$

A resistance of *1 ohm* will allow a current of *1 ampere* when a potential difference of *1 volt* is impressed across its terminals.

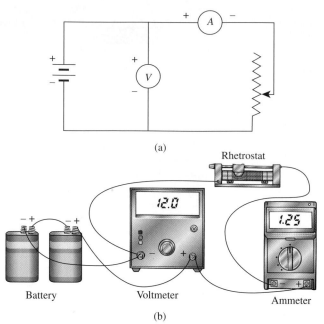

Figure 27.6   (a) A circuit diagram for studying Ohm's law. (b) A pictorial diagram showing how the various elements are connected in the laboratory.

Four devices commonly used in the laboratory to study Ohm's law are a power supply or battery, a voltmeter, an ammeter, and a rheostat. As their names imply, the **voltmeter** and **ammeter** are devices to measure voltage and current. The **rheostat** is simply a variable resistor. A sliding contact changes the number of resistance coils through which charge can flow. A laboratory collection of such devices is illustrated in Fig. 27.6. In this example, a battery provides the necessary direct current. You should study the example in Fig. 27.6a and justify the electrical connections shown pictorially in Fig. 27.6b. Note the voltmeter is connected in parallel with the battery—positive to positive and negative to negative. However, the ammeter, which must read the current through the circuit, is connected in series—positive to negative to positive to negative.

Example 27.2

The slide on the rheostat is at the position shown in Fig. 27.6. The voltmeter indicates a reading of 6.00 V and the ammeter reads 400 mA. (a) What is the resistance across the rheostat? (b) What will the ammeter read if the resistance is doubled?

Plan: We know the current I and the voltage V, so we can apply Ohm's law to find the electric resistance. Remember to use the SI base units of volts and amperes. We will neglect any other resistances that may be encountered.

Solution (a): We solve Ohm's law for resistance R and substitute the given values

$$R = \frac{V}{I} = \frac{6.00 \text{ V}}{0.400 \text{ A}}; \qquad R = 15.0 \text{ }\Omega$$

Solution (b): Doubling the resistance, we substitute $R = 30.0 \text{ }\Omega$ to obtain

$$I = \frac{V}{R} = \frac{6.00 \text{ V}}{30.0 \text{ }\Omega}; \qquad R = 0.200 \text{ A} = 200 \text{ mA}$$

27.5 | Electric Power and Heat Loss

Robots and Ohm's Law
In 1854, the year that Georg Simon Ohm died, Lord Kelvin discovered that the resistance of a wire changes when it is strained (stretched).

Engineers first used this phenomenon to analyze stresses in structural surfaces. For example, by monitoring the current through a fine metal wire mounted alongside a bridge, machines can indicate strain on the bridge based on the wire's varying resistance. Such a device is called a *strain gauge*.

Researchers seeking ways to imitate the complex capabilities of the human hand have used a version of a strain gauge to give robots a sense of touch. Robotic hands, equipped with superhuman strength, have the dexterity to gently crack an egg into a mixing bowl. However, a robot still needs some kind of tactile sensor in order to "see" with its hands.

One of the most promising sensor designs consists of an electronic chip printed with a fine metallic network and covered by a thin sheet of conducting rubber. Pressure on the rubber changes the resistance of the chip, transmitting a "picture" of the strain on the robotic skin.

We have seen that electric charge gains energy within a generating source of emf and loses energy in passing through external resistance. Inside the source of emf, work is done *by the source* in raising the potential energy of charge. As the charge passes through the external circuit, work is done *by* the charge on the components of the circuit. In the case of a pure resistor, the energy is dissipated in the form of heat. If a motor is attached to the circuit, the energy loss is divided between heat and useful work. In any case, the energy gained in the source of emf must equal the energy lost in the entire circuit.

Let us examine the work accomplished inside a source of emf more closely. By definition, *1 joule* of work is accomplished for each *coulomb* of charge moved through a potential difference of *1 volt*. Thus,

$$\text{Work} = Vq \qquad (27.3)$$

where q is the quantity of charge transferred during a time t. But $q = It$, so Eq. (27.3) becomes

$$\text{Work} = VIt \qquad (27.4)$$

where I is the current in *coulombs per second*. This work represents the energy gained by a charge in passing through the source of emf during the time t. An equivalent amount of energy will be dissipated in the form of heat as the charge moves through an external resistance.

The rate at which heat is dissipated in an electric circuit is referred to as the *power loss*. When charge is flowing continuously through a circuit, this power loss is given by

$$P = \frac{\text{work}}{t} = \frac{VIt}{t} = VI \qquad (27.5)$$

When V is in volts and I is in amperes, the power loss is measured in watts. That the product of voltage and current will give a unit of power is shown as follows:

$$(\text{V})(\text{A}) = \frac{\text{J}}{\text{C}}\frac{\text{C}}{\text{s}} = \frac{\text{J}}{\text{s}} = \text{W}$$

Equation (27.5) can be expressed in alternative forms by using Ohm's law ($V = IR$). Substituting for V, we can write

$$P = VI = I^2R \qquad (27.6)$$

Substitution for I in Eq. (27.6) gives another variation:

$$P = VI = \frac{V^2}{R} \qquad (27.7)$$

The relation expressed by Eq. (27.6) is so often used in electrical work that heat loss in electrical wiring is often referred to as an "I-squared-R" loss.

Example 27.3

A small office fan has a label on its base that reads 120 V, 55 W. What is the operating current of this fan, and what is its electric resistance? If the fan is left running overnight for 8 h, how much energy is lost? Assume the laws presented for direct current in this chapter also apply for the office circuit.

Plan: We will recognize that power loss is equal to the product of voltage and current. Thus, we can solve for the amperage and substitute the known values. The resistance can be found by applying Ohm's law or by using another form of the power equation. The energy loss is based on the definition of power as energy expended per unit of time.

Solution: Given that $V = 120$ V and $P = 55$ W, we solve for the current I as follows:

$$P = VI \quad \text{or} \quad I = \frac{P}{V} = \frac{55 \text{ W}}{120 \text{ V}}$$

$$I = 0.458 \text{ A} = 458 \text{ mA}$$

The resistance can be found from Ohm's law, but we will use Eq. (27.7).

$$P = \frac{V^2}{R} \quad \text{or} \quad R = \frac{V^2}{P} = \frac{(120 \text{ V})^2}{55 \text{ W}}$$

$$R = 262 \text{ } \Omega$$

Since 1 h = 3600 s, we convert the time from 8 h to 2.88×10^4 s. Recognizing that power is equal to the work (energy) per unit time, we can solve for the energy loss as follows:

$$\text{Work} = Pt = (55 \text{ W})(2.88 \times 10^4 \text{ s})$$

$$\text{Work} = 1.58 \times 10^6 \text{ J}$$

This energy loss could be expressed as (0.055 kW)(8 h) or 0.440 kW · h. At typical rates, that might cost about a nickel.

27.6 Resistivity

Just as capacitance is independent of the voltage and quantity of charge, the resistance of a conductor is independent of current and voltage. Both capacitance and resistance are inherent properties of a conductor. The resistance of a wire of uniform cross-sectional area, like the one shown in Fig. 27.7, is determined by the following four factors:

1. The kind of material

2. The length

3. The cross-sectional area

4. The temperature

Ohm, the German physicist who discovered the law that now bears his name, also reported that *the resistance of a conductor at a given temperature is directly proportional to its length, inversely proportional to its cross-sectional area, and dependent upon the material from which it is made.* For a given conductor at a given temperature, the resistance can be computed from

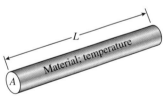

Figure 27.7 The resistance of a wire depends on the kind of material, the length, the cross-sectional area, and the temperature of the wire.

$$R = \rho \frac{L}{A} \tag{27.8}$$

where R = resistance
L = length
A = area

The proportionality constant ρ is a property of the material called its *resistivity,* given by

$$\rho = \frac{RA}{L} \tag{27.9}$$

PHYSICS TODAY

A polygraph, or lie detector test, measures the resistivity of skin. Skin becomes less resistive when a person sweats, which happens unconsciously when a lie is told.

Table 27.1		

Resistivities and Temperature Coefficients at 20°C

Material/Properties	Resistivity $\Omega \cdot m$	Temperature Coefficient of Resistance $1/C°$
Aluminum	2.8×10^{-8}	3.9×10^{-3}
Constantan	49×10^{-8}	—
Copper	1.72×10^{-8}	3.9×10^{-3}
Gold	2.4×10^{-8}	3.4×10^{-3}
Iron	9.5×10^{-8}	5.0×10^{-3}
Nichrome	100×10^{-8}	0.4×10^{-3}
Lead	10×10^{-8}	3.9×10^{-3}
Silver	1.6×10^{-8}	3.8×10^{-3}
Tungsten	5.5×10^{-8}	4.5×10^{-3}

It varies considerably with different materials and is also affected by changes in temperature. When R is in ohms, A is in square meters, and L is in meters, the unit of resistivity is the ohm-meter ($\Omega \cdot m$):

$$\frac{(\Omega)(m^2)}{m} = \Omega \cdot m$$

Table 27.1 lists the resistivities of several common metals.

Example 27.4

A 20-m length of copper wire has a diameter of 0.8 mm. The ends of the wire are placed across the terminals of a 1.5-V battery. What current passes through the wire?

Plan: We will calculate the area of the wire and then calculate the resistance from the length and resistivity of copper. Ohm's law will then give us the current.

Solution: The area of the wire is

$$A = \frac{\pi D^2}{4} = \frac{\pi(8 \times 10^{-4})^2}{4}; \qquad A = 5.03 \times 10^{-7} \, m^2$$

We solve for the resistance using Eq. (27.8).

$$R = \frac{\rho L}{A} = \frac{(1.72 \times 10^{-8} \, \Omega \cdot m)(20 \, m)}{5.03 \times 10^{-7} \, m^2}; \qquad R = 0.684 \, \Omega$$

Finally, from Ohm's law,

$$I = \frac{V}{R} = \frac{1.5 \, V}{0.684 \, \Omega}; \qquad I = 2.19 \, A$$

27.7   Temperature Coefficient of Resistance

For most metallic conductors, the resistance tends to increase as the temperature increases. The increased atomic and molecular movement in the conductor hinders the flow of charge. The increase in resistance for most metals is approximately linear when compared with temperature changes. Experiments have shown that the increase in resistance ΔR is proportional to the initial resistance R_0 and the change in temperature Δt. We can write

$$\Delta R = \alpha R_0 \Delta t \qquad (27.10)$$

The constant α is a characteristic of the material known as the ***temperature coefficient of resistance.*** The defining equation for α can be found by solving Eq. (27.10):

$$\alpha = \frac{\Delta R}{R_0 \Delta t} \qquad (27.11)$$

The temperature coefficient of resistance is the change in resistance per unit resistance per degree change in temperature.

Since the units of ΔR and R_0 are the same, the unit for the coefficient α is inverse Celsius degrees (1/C°). The coefficients for many common materials are given in Table 27.1.

Example 27.5

An iron wire has a resistance of 200 Ω at 20°C. What will be its resistance when heated to a temperature of 80°C?

Plan: First, we will calculate the change in resistance based on the change of temperature and the temperature coefficient for iron, taken from Table 27.1.

Solution: The change in resistance ΔR is found from Eq. (27.10).

$$\Delta R = \alpha R_0 \Delta t$$
$$= (0.005/\text{C}°)(200\ \Omega)(80°\text{C} - 20°\text{C})$$
$$= 60\ \Omega$$

Therefore, the resistance at 80°C is

$$R = R_0 + \Delta R = 200\ \Omega + 72\ \Omega = 260\ \Omega$$

The increase in resistance of a conductor with temperature is large enough to be measured easily. This fact is used in resistance thermometers to measure temperatures accurately. Because of the high melting point of some metals, resistance thermometers can be used to measure extremely high temperatures.

27.8   Superconductivity

In 1911, the Dutch physicist Heike Kamerlingh-Onnes was experimenting with the resistivity of metals at low temperatures. Using liquid helium as the coolant, he was able to cool the metals below 4.2 K (-269°C). While some metals such as platinum and gold reached a constant resistivity at low temperatures, other metals such as mercury exhibited no resistance below a *critical* or *transition temperature,* as shown in Fig. 27.8. This phenomenon of zero resistance or infinite conductivity is called ***superconductivity.*** After many years of study, over 26 elements and a variety of alloys, compounds, and semiconductors have been classified as superconductors, each having a characteristic transition temperature (T_c).

In addition to their electrical properties, superconductors also exhibit extraordinary magnetic properties below their transition temperature. Superconductors placed in a magnetic field will expel all magnetic flux from its interior; in other words, they have perfect diamagnetism. A consequence of this phenomenon, known as the *Meissner-Ochsenfeld effect,* is shown in Fig. 27.9. As the magnet is brought close to the superconductor, the perfect diamagnetism (see Chapter 29) repels the magnet, levitating the magnet above the superconductor.

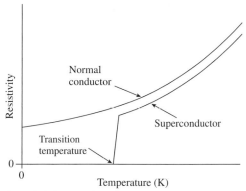

Figure 27.8 The change in resistivity as a function of temperature is shown for normal metallic conductors and for superconductors. Note that the resistivity of a superconductor drops rather abruptly to zero at its critical transition temperature.

Figure 27.9 The levitation of a small piece of metal is accomplished by a superconductor cooled to the temperature of liquid nitrogen. (*Courtesy Science Kit & Boreal Laboratories.*)

Until 1986, the highest recorded transition temperature was 23 K (−250°C) for a niobium-germanium alloy. Experiments done that year show that ceramic materials, usually having insulator properties, exhibit transition temperatures of about 90 K (−183°C). These new high-temperature superconductors have immense technological importance, primarily because liquid nitrogen, with a boiling temperature of 77 K (−196°C), can be used as the coolant. Liquid nitrogen is readily available, much less expensive, and easier to handle than

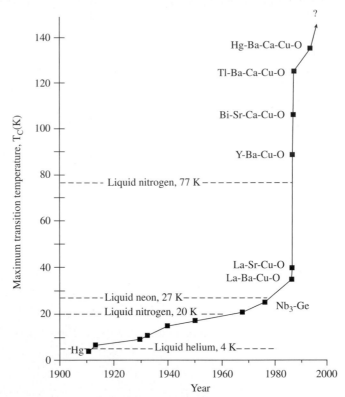

Figure 27.10 A graph depicting the rapid progress with experimentation on superconductors since the work of Bednorz and Müller in 1986. The most recent point indicates the transition temperature of 135 K, which was achieved in 1993 with a compound of mercury, barium, calcium, copper, and oxygen.

liquid helium or liquid hydrogen. An experimental group from Switzerland was able to achieve transition temperatures near 135 K (−138°C). Andreas Schilling, Marco Cantoni, J. D. Guo, and Hans Ott accomplished this feat in 1993 using a compound containing mercury, calcium, barium, copper, and oxygen. Superconductors with room-temperature transitions remain a possibility. The illustration in Fig. 27.10 shows the rapid advance of research since 1986, when J. Georg Bednorz and Karl Alex Müller reported superconductivity at a transitional temperature of 30 K.

Practical applications of superconductors are already realized and under development. Magnets utilizing superconductor coils can achieve higher fields and have cheaper operating costs than conventional magnets because the zero resistivity of superconductors means no loss of energy due to resistive heating. In fact, a persistent current can exist in a superconductor without an applied potential difference. These magnets are now part of diverse instruments, such as supercolliders, magnetic energy storage, and medical diagnostic systems—for example, magnetic resonance imaging (MRI). Superconductor coils are used for windings in motors and generators, and potential applications include underground power transmission lines.

Another application is the magnetic levitated (MAGLEV) vehicle. These vehicles utilize the principle of levitation described earlier in Fig. 27.9. Superconducting magnets on the moving vehicle are located above normal sheet metal. The moving magnets create induced currents (eddy currents) in the metal sheet. These small, induced currents create, in turn, a magnetic field, which repels the moving magnets. A prototype train has already been built in Japan using liquid helium as the coolant.

It should be noted that most of the applications described thus far do not use high-temperature (Type II) superconductors. The older (Type I) superconductors are more flexible and are able to carry more current then the new ceramic superconductors. Small devices called *SQUIDs* (*S*uperconducting *Q*uantum *I*nterference *D*evices), made from high-temperature superconductors, are used, however, in sensitive, electronic measuring devices and computer components.

Summary and Review

Summary

In this chapter, we introduced the *ampere* as a unit of electric current, and we discussed the various quantities that affect its magnitude. Ohm's law described mathematically the relationship among current, resistance, and applied voltage. We also learned the factors that affect electric resistance and applied these concepts to the solution of basic problems in elementary electricity. The major points are summarized as follows.

- Electric current I is the rate of flow of charge Q past a given point on a conductor:

$$I = \frac{Q}{t} \qquad 1 \text{ ampere (A)} = \frac{1 \text{ coulomb (C)}}{1 \text{ second (s)}}$$

- By convention, the *direction* of electric current is the same as the direction in which *positive* charges would move, even if the actual current consists of a flow of negatively charged electrons.
- Ohm's law states that *the current produced in a given conductor is directly proportional to the difference of potential between its end points:*

$$R = \frac{V}{I} \qquad V = IR \qquad \textit{Ohm's Law}$$

The symbol R represents the resistance in ohms (Ω) defined as

$$1 \text{ ohm } (\Omega) = \frac{1 \text{ volt (V)}}{1 \text{ ampere (A)}}$$

- The electric power in watts is given by any of the following:

$$P = VI \qquad P = I^2 R \qquad P = \frac{V^2}{R} \qquad \textit{Power}$$

- The resistance of a wire depends on four factors: (a) the kind of *material,* (b) the *length,* (c) the cross-sectional *area,* and (d) the *temperature.* By introducing a property for the material called its *resistivity ρ,* we can write

$$R = \rho\frac{L}{A} \qquad \rho = \frac{RA}{L} \qquad \textit{SI unit for } \rho\text{: } \Omega \cdot \text{m}$$

- The *temperature coefficient of resistance α* is the change in resistance per unit resistance per degree change in temperature.

$$\alpha = \frac{\Delta R}{R_0 \Delta t} \qquad \Delta R = \alpha R_0 \Delta t$$

Key Terms

Review Questions

27.1. Distinguish clearly between electron flow and conventional current. What are some reasons for preferring the conventional current?

27.2. Use the mechanical analogy of water flowing through pipes to describe the flow of charge through conductors of various lengths and cross-sectional areas.

27.3. A rheostat is connected across the terminals of a battery. What determines the positive and negative terminals on the rheostat?

27.4. Is the electromotive force really a *force?* What is the function of a source of emf?

27.5. What is wrong with the following statement? The resistivity of a material is directly proportional to its length.

27.6. Use Ohm's law to verify Eqs. (27.6) and (27.7).

Problems

Section 27.1 The Motion of Electric Charge,
Section 27.2 The Direction of Electric Current,
Section 27.3 Electromotive Force, and
Section 27.4 Ohm's Law; Resistance

27.1. How many electrons pass a point every second in a wire carrying a current of 20 A? How much time is needed to transport 40 C of charge past this point? **Ans. 1.25×10^{20} electrons, 2 s**

27.2. If 600 C of charge passes a given point in 3 s, what is the electric current in amperes?

27.3. Find the current in amperes when 690 C of charge passes a given point in 2 min. **Ans. 5.75 A**

27.4. If a current of 24 A exists for 50 s, how many coulombs of charge have passed through the wire?

27.5. What is the potential drop across a 4-Ω resistor with a current of 8 A passing through it? **Ans. 32.0 V**

27.6. Find the resistance of a rheostat if the drop in potential is 48 V and the current is 4 A.

27.7. Determine the current through a 5-Ω resistor that has a 40-V drop in potential across it. **Ans. 8.00 A**

27.8. A 2-A fuse is placed in a circuit with a battery having a terminal voltage of 12 V. What is the minimum resistance for a circuit containing this fuse?

27.9. What emf is required to pass 60 mA through a resistance of 20 kΩ? If this same emf is applied to a resistance of 300 Ω, what will be the new current? **Ans. 1200 V, 4 A**

Section 27.5 Electric Power and Heat Loss

27.10. A soldering iron draws 0.75 A at 120 V. How much energy will it use in 15 min?

27.11. An electric lamp has an 80-Ω filament connected to a 100-V dc line. What is the current through the filament? What is the power loss in watts? **Ans. 1.38 A, 151 W**

27.12. Assume the cost of energy in a home is 8 cents per kilowatt-hour. A family goes on a 2-week vacation leaving a single 80-W light bulb burning. What is the cost?

27.13. A 120-V dc generator delivers 2.4 kW to an electric furnace. What current is supplied? What is the resistance? **Ans. 20 A, 6 Ω**

27.14. A resistor develops heat at the rate of 250 W when the potential difference across its ends is 120 V. What is its resistance?

27.15. A 120-V motor draws a current of 4.0 A. How many joules of electrical energy are used in 1 h? How many kilowatt-hours? **Ans. 1.73 MJ, 0.48 kWh**

27.16. A household hair dryer is rated at 2000 W and is designed to operate on a 120-V outlet. What is the resistance of the device?

Section 27.6 Resistivity

27.17. What length of copper ($\rho = 1.78 \times 10^{-8}\ \Omega$ m) wire 1.2 mm in diameter is needed to make a 20-Ω resistor at 20°C? What length of nichrome wire is needed? ($\rho = 100 \times 10^{-8}\ \Omega \cdot$ m) **Ans. 12.7 m, 0.0226 m**

27.18. A 3.0-m length of copper wire ($\rho = 1.78 \times 10^{-8}\ \Omega \cdot$ m) at 20°C has a cross section of 4 mm^2. What is the electric resistance of this wire?

27.19. Find the resistance of 40 m of tungsten wire having a diameter of 0.8 mm at 20°C. ($\rho = 5.5 \times 10^{-8}\ \Omega \cdot$ m) **Ans. 4.37 Ω**

27.20. A certain wire has a diameter of 3 mm and a length of 150 m. It has a resistance of 3.00 Ω at 20°C. What is the resistivity?

27.21. What is the resistance of 200 ft of iron ($\rho = 9.5 \times 10^{-8}\ \Omega \cdot$ m) wire with a diameter of 0.002 in. at 20°C? **Ans. 2860 Ω**

*27.22. A nichrome wire has a length of 40 m at 20°C. What is the diameter if the total resistance is 5 Ω? ($\rho = 100 \times 10^{-8}\ \Omega \cdot$ m)

*27.23. A 115-V source of emf is attached to a heating element that is a coil of nichrome wire ($\rho = 100 \times 10^{-8}\ \Omega \cdot$ m) of cross section 1.20 mm^2. What must be the length of the wire if the resistive power loss is to be 800 W? **Ans. 19.8 m**

Section 27.7 Temperature Coefficient of Resistance

27.24. The resistance of a length of wire ($\alpha = 0.0065/$C°) is 4.00 Ω at 20°C. What is the resistance at 80°C?

27.25. If the resistance of a conductor is 100 Ω at 20°C and 116 Ω at 60°C, what is its temperature coefficient of resistivity? **Ans. $4.00 \times 10^{-3}/$C°**

27.26. A length of copper ($\alpha = 0.0043/$C°) wire has a resistance of 8 Ω at 20°C. What is the resistance at 90°C? At −30°C?

*27.27. The copper windings of a motor experience a 20 percent increase in resistance over their value at 20°C. What is the operating temperature? ($\alpha = 0.0043/C°$).   Ans. 71.3°C

*27.28. What temperature will produce a 25 percent increase in resistance for copper at 20°C ($\alpha = 0.0039/C°$)?

Additional Problems

27.29. A water turbine delivers 2000 kW to an electric generator that is 80 percent efficient and has an output terminal voltage of 1200 V. What current is delivered, and what is the electric resistance?
                              Ans. 1.33 kA, 0.900 Ω

27.30. A 110-V radiant heater draws a current of 6.0 A. How much heat energy in joules is delivered in 1 h?

27.31. A power line has a total resistance of 4 kΩ. What is the power loss through the wire if the current is reduced to 6.0 mA?   Ans. 0.144 W

27.32. A certain wire has a resistivity of 2×10^{-8} Ω · m at 20°C. If its length is 200 m and its cross section is 4 mm^2, what will be its electric resistance at 100°C? Assume that $\alpha = 0.005/C°$ for this material.

27.33. Determine the resistivity of a wire made of an unknown alloy if its diameter is 0.7 mm and 30 m of the wire is found to have a resistance of 4.0 Ω.
                              Ans. 5.13×10^{-8} Ω · m

27.34. The resistivity of a certain wire is 1.72×10^{-8} Ω m at 20°C. A 6-V battery is connected to a 20-m coil of this wire, which has a diameter of 0.8 mm. What is the current in the wire?

27.35. A certain resistor is used as a thermometer. Its resistance at 20°C is 26.00 Ω, and its resistance at 40°C is 26.20 Ω. What is the temperature coefficient of resistance for this material?
                              Ans. $3.85 \times 10^{-4}/C°$

*27.36. What length of copper wire at 20°C has the same resistance as 200 m of iron wire at 20°C? Assume the same cross section for each wire.

*27.37. The power loss in a certain wire at 20°C is 400 W. If $\alpha = 0.0036/C°$, by what percentage will the power loss increase when the operating temperature is 68°C?   Ans. 17.3 percent

Critical Thinking Questions

27.38. A 150-Ω resistor at 20°C is rated at 2.0 W maximum power. What is the maximum voltage that can be applied across the resistor without exceeding the maximum allowable power? What is the current at this voltage?

27.39. The current in a home is alternating current, but the same formulas apply. Suppose a fan motor operating a home cooling system is rated at 10 A for a 120-V line. How much energy is required to operate the fan for a 24-h period? At a cost of 9 cents per kilowatt-hour, what is the cost of operating this fan continuously for 30 days?
                              Ans. 95.0 MJ, $71.28

*27.40. The power consumed in an electrical wire ($\alpha = 0.004/C°$) is 40 W at 20°C. If all other factors are held constant, what is the power consumption when (a) the length is doubled,

(b) the diameter is doubled, (c) the resistivity is doubled, and (d) the absolute temperature is doubled?   Ans. (a) 80 W, (b) 10 W, (c) 80 W, (d) 86.9 W

*27.41. What must be the diameter of an aluminum wire if it is to have the same resistance as an equal length of copper wire of diameter 2.0 mm? What length of nichrome wire is needed to have the same resistance as 2 m of iron wire of the same cross section?

*27.42. An iron wire ($\alpha = 0.0065/C°$) has a resistance of 6.00 Ω at 20°C, and a copper wire ($\alpha = 0.0043/C°$) has a resistance of 5.40 Ω at 20°C. At what temperature will the two wires have the same resistance?   Ans. −18.0°C

28

Direct-Current Circuits

Complex electric circuits, instruments, and devices are used to maintain the U.S. legal volt and to provide for the dissemination of an internationally consistent and accurate standard for voltage measurements.
(*Courtesy of National Institute of Standards and Technology.*)

Objectives

After completing this chapter, you should be able to

1. Determine the effective resistance of a number of resistors connected in *series* and in *parallel*.

2. Write and apply equations involving *voltage, current,* and *resistance* for a circuit containing resistors connected in series and in parallel.

3. Solve problems involving the emf of a battery, its *terminal potential difference,* the *internal resistance,* and the *load resistance.*

4. Write and apply *Kirchhoff's laws* for electrical networks similar to those shown in the text.

Two types of current are in use. **Direct current** (dc) is the continuous flow of charge in only one direction. *Alternating current* (ac) is a flow of charge continually changing in both magnitude and direction. In this chapter, we analyze current, voltage, and resistance for dc circuits. Many of the same methods and procedures can also be applied to ac circuits. The variations required for alternating currents build logically from a strong foundation in dc analysis.

28.1 Simple Circuits; Resistors in Series

An electric circuit consists of any number of branches joined together so that at least one closed path is provided for current. The simplest circuit consists of a single source of emf joined to a single external resistance, as shown in Fig. 28.1. If $\mathscr{E}$ represents the emf and R indicates the total resistance, Ohm's law yields

$$\mathscr{E} = IR \tag{28.1}$$

where I is the current through the circuit. All the energy gained by a charge in passing through the source of emf is lost in flowing through the resistance.

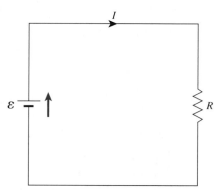

Figure 28.1 A simple electric circuit.

Let us consider the addition of a number of elements to a circuit. Two or more elements are said to be in *series* if they have only *one* point in common that is not connected to some third element. Current can follow only a single path through elements in series. Resistors R_1 and R_2 of Fig. 28.2a are in series because point A is common to both resistors. The resistors in Fig. 28.2b, however, are not in series, because point B is common to three current branches. Electric current entering such a junction may follow two separate paths.

Suppose that three resistors (R_1, R_2, and R_3) are connected in series and enclosed in a box, indicated by the shaded portion of Fig. 28.3. The effective resistance R of the three resistors can be determined from the external voltage V and current I, as recorded by the meters. From Ohm's law,

$$R = \frac{V}{I} \tag{28.2}$$

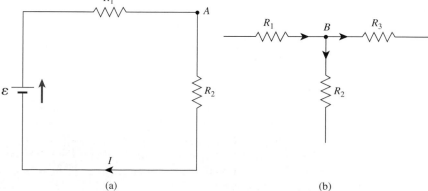

(a) (b)

Figure 28.2 (a) Resistors connected in series. (b) Resistors not connected in series.

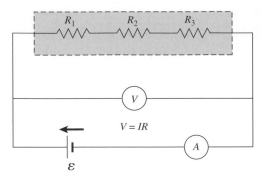

Figure 28.3 The voltmeter–ammeter method of measuring the effective resistance of a number of resistors connected in series.

But what is the relationship of R to the three internal resistances? The current through each resistor must be identical since a single path is provided. Thus,

$$I = I_1 = I_2 = I_3 \qquad \textbf{(28.3)}$$

Utilizing this fact and noting that Ohm's law applies equally well to any part of a circuit, we write

$$V = IR \qquad V_1 = IR_1 \qquad V_2 = IR_2 \qquad V_3 = IR_3 \qquad \textbf{(28.4)}$$

The external voltage V represents the sum of the energies lost per unit of charge in passing through each resistance. Therefore,

$$V = V_1 + V_2 + V_3$$

Finally, if we substitute from Eq. (28.4) and divide out the current, we obtain

$$IR = IR_1 + IR_2 + IR_3$$

$$R = R_1 + R_2 + R_3 \qquad\qquad\qquad \textit{Series} \quad \textbf{(28.5)}$$

To summarize what has been learned about resistors connected in *series:*

1. The current in all parts of a series circuit is the same.

2. The voltage across a number of resistances in series is equal to the sum of the voltage across the individual resistors.

3. The effective resistance of a number of resistors in series is equivalent to the sum of the individual resistances.

Example 28.1

The resistances R_1 and R_2 in Fig. 28.2a are 2 Ω and 4 Ω, respectively. If the source of emf maintains a constant potential difference of 12 V, what is the current delivered to the external circuit? What is the potential drop across each resistor?

Plan: The resistors are connected in series, so each carries the same current determined by the supplied voltage and the sum of the two resistances. Ohm's law applied to each resistor gives the drop across each element.

Solution: For series resistors, the equivalent resistance is

$$R_e = R_1 + R_2 = 2\,\Omega + 4\,\Omega; \qquad R_e = 6\,\Omega$$

The current I through the entire circuit *and* through each resistor is

$$I = \frac{V}{R_e} = \frac{12\text{ V}}{6\,\Omega} \qquad \text{or} \qquad I = 2\text{ A}$$

The voltage drops across each resistor are

$$V_1 = IR_1 = (2 \text{ A})(2 \text{ } \Omega); \qquad V_1 = 4 \text{ V}$$
$$V_2 = IR_2 = (2 \text{ A})(4 \text{ } \Omega); \qquad V_2 = 6 \text{ V}$$

Note that the sum of the voltage drops ($V_1 + V_2$) is equal to 12 V, the total applied voltage.

28.2 Resistors in Parallel

There are several limitations to the operation of series circuits. If a single element in a series circuit fails to provide a conducting path, the entire circuit is opened and current ceases. It would be quite annoying if all electrical devices in a home were to cease functioning whenever one lamp burned out. Moreover, each element in a series circuit adds to the total resistance of the circuit, thereby limiting the total current that can be supplied. These objections can be overcome by providing alternative paths for electric current. Such a connection, in which current can be divided between two or more elements, is called a **parallel connection.**

A **parallel circuit** is one in which two or more components are connected to two common points in the circuit. For example, in Fig. 28.4, the resistors R_2 and R_3 are in parallel because they both have points A and B in common. Note that the current I, provided by the source of emf, is divided between resistors R_2 and R_3.

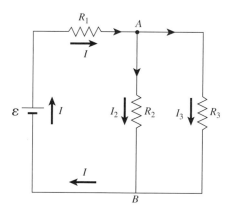

Figure 28.4 The resistors R_2 and R_3 are connected in parallel.

To arrive at an expression for the equivalent resistance R of a number of resistances connected in parallel, we follow a procedure similar to that discussed for a series connection. Assume that three resistors (R_1, R_2, and R_3) are placed inside a box, as shown in Fig. 28.5.

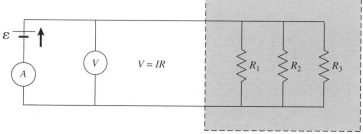

Figure 28.5 Computing the equivalent resistance of a number of resistors connected in parallel.

The total current I delivered to the box is determined by its effective resistance and the applied voltage:

$$I = \frac{V}{R} \tag{28.6}$$

In a parallel connection, the voltage drop across each resistor is the same and equivalent to the total drop in voltage.

$$V = V_1 = V_2 = V_3 \tag{28.7}$$

We see the truth of this statement when we consider that the same energy must be lost by a unit of charge, regardless of the path it travels in the circuit. In this example, charge may flow through any one of the three resistors. Thus, the total current delivered is divided among the resistors.

$$I = I_1 + I_2 + I_3 \tag{28.8}$$

Applying Ohm's law to Eq. (28.8) yields

$$\frac{V}{R} = \frac{V_1}{R_1} + \frac{V_2}{R_2} + \frac{V_3}{R_3}$$

But the voltages are equal, and we can divide them out.

$$\frac{1}{R} = \frac{1}{R_1} + \frac{1}{R_2} + \frac{1}{R_3} \qquad\qquad Parallel \quad (28.9)$$

In summary, for parallel resistors,

1. The total current in a parallel circuit is equal to the sum of the currents in the individual branches.
2. The voltage drops across all branches in a parallel circuit must be of equal magnitude.
3. The reciprocal of the equivalent resistance is equal to the sum of the reciprocals of the individual resistances connected in parallel.

In the case of only two resistors in parallel,

$$\frac{1}{R} = \frac{1}{R_1} + \frac{1}{R_2}$$

Solving this equation algebraically for R, we obtain a simplified formula for computing the equivalent resistance.

$$R = \frac{R_1 R_2}{R_1 + R_2} \tag{28.10}$$

The equivalent resistance of two resistors connected in parallel is equal to their product divided by their sum.

Example 28.2 The total voltage applied to the circuit in Fig. 28.6 is 12 V, and the resistances are $R_1 = 4\ \Omega$, $R_2 = 3\ \Omega$, and $R_3 = 6\ \Omega$. (a) Determine the equivalent resistance of the circuit. (b) Find the current through each resistor.

Plan: The best approach to a problem that contains both series and parallel resistors is to reduce the circuit by steps to its simplest form. This approach is illustrated in

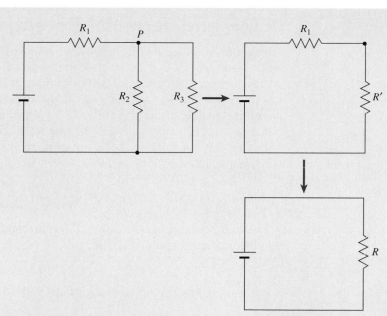

Figure 28.6   Reducing a complex circuit to a simple equivalent circuit.

Fig. 28.6. The two parallel resistors R_2 and R_3 are combined to form an equivalent resistance R', which is then combined in series with R_1 to form a single equivalent resistance R_e for the entire circuit. Ohm's law will then give the current delivered by the source of emf. Finally, by treating the voltages and resistances for each resistor, we will find the current in each element.

Solution (a): The equivalent resistance R' for the parallel resistors is found from Eq. (28.10).

$$R' = \frac{R_2 R_3}{R_2 + R_3} = \frac{(3\ \Omega)(6\ \Omega)}{(3\ \Omega + 6\ \Omega)}; \qquad R' = 2\ \Omega$$

This equivalent resistance R' is in series with R_1, so Eq. (28.5) gives the equivalent resistance for the entire circuit.

$$R_e = R_1 + R' = 4\ \Omega + 2\ \Omega; \qquad R_e = 6\ \Omega$$

Solution (b): The total current delivered by the source of emf is

$$I = \frac{V}{R} = \frac{12\ \text{V}}{6\ \Omega}; \qquad I = 2\ \text{A}$$

Since the resistance R_1 and R' are in *series,* they have the same current as that coming from the source of emf (2 A).

$$I_1 = 2\ \text{A} \qquad \text{and} \qquad I' = 2\ \text{A}$$

When the total current (2 A) reaches point P, it splits, part of it passing through R_2 and the remainder through R_3. These currents are found from Ohm's law.

$$I_2 = \frac{V'}{R_2} = \frac{4\ \text{V}}{3\ \Omega}; \qquad I_2 = 1.33\ \text{A}$$

$$I_3 = \frac{V'}{R_3} = \frac{4\ \text{V}}{6\ \Omega}; \qquad I_3 = 0.667\ \text{A}$$

Note that $I_2 + I_3 = 2$ A, which is the total circuit current.

28.3 EMF and Terminal Potential Difference

In all the preceding problems, we have assumed all resistance to current flow is due to elements of a circuit that are external to the source of emf. This is not strictly true, however, because there is an inherent resistance within every source of emf. This **internal resistance** is represented by the symbol r and is shown schematically as a small resistance in series with the source of emf. (See Fig. 28.7.) When a current I is flowing through the circuit, there is a loss of energy through the external load R_L and also there is a heat loss due to the internal resistance. Thus, the actual terminal voltage V_T across a source of emf $\mathscr{E}$ with an internal resistance r is given by

$$V_T = \mathscr{E} - Ir \tag{28.11}$$

The voltage applied to the external load is therefore less than the emf by an amount equal to the internal potential drop. Since $V_T = IR_L$, Eq. (28.11) can be rewritten

$$V_T = IR_L = \mathscr{E} - Ir \tag{28.12}$$

Solving Eq. (28.12) for the current I, we have

$$I = \frac{\mathscr{E}}{R_L + r} \tag{28.13}$$

The current in a simple circuit containing a single source of emf is equal to the emf voltage divided by the total resistance in the circuit (including internal resistance).

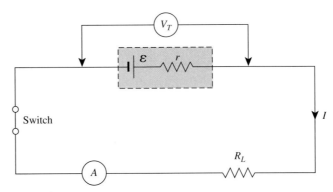

Figure 28.7 Internal resistance.

Example 28.3 A load resistance of 8 Ω is connected to a 12-V battery whose internal resistance is 0.20 Ω. (a) What current is delivered to the load? (b) What will be the reading of a voltmeter placed across the battery terminals while the load is attached?

Plan: The current delivered to the circuit is the ratio of the emf to the total resistance, including the internal resistance of the battery. Once we establish the current, we can find the voltage drop through the internal resistance and subtract it from the emf to find the terminal voltage of the battery.

Solution (a): The current delivered is found from Eq. (28.13).

$$I = \frac{\mathscr{E}}{R_L + r} = \frac{12\text{ V}}{8\ \Omega + 0.2\ \Omega} = 1.46\text{ A}$$

Solution (b): The terminal voltage is

$$V_T = \mathcal{E} - Ir = 12 \text{ V} - (1.46 \text{ A})(0.2 \text{ }\Omega)$$
$$= 12 \text{ V} - 0.292 \text{ V} = 11.7 \text{ V}$$

As a check, we can find the voltage drop across the load R_L:

$$V_T = IR_L = (1.46 \text{ A})(8 \text{ }\Omega) = 11.7 \text{ V}$$

28.4 Measuring Internal Resistance

The internal resistance of a battery can be measured in the laboratory by using a voltmeter, an ammeter, and a known resistance. A voltmeter is an instrument that has an extremely high resistance. When a voltmeter is attached directly to the terminals of a battery, negligible current is drawn from the battery. We can see from Eq. (28.11) that, for zero current, this terminal voltage is equal to the emf ($V_T = \mathcal{E}$). In fact, the emf of a battery is sometimes referred to as its "open-circuit" potential difference. Thus, the emf can be measured with a voltmeter. By connecting a known resistance to the circuit, we can determine the internal resistance by measuring the current delivered to the circuit.

Example 28.4

A battery gives an open-circuit reading of 1.5 V when a voltmeter is placed across its terminals. A small light having a resistance of 3.5 Ω is connected to this battery, and a current of 400 mA is measured. What is the internal resistance of the battery? As the battery ages, its internal resistance increases, and the light grows dimmer. If a much later reading shows a current of only 350 mA, what is the increase Δr of the internal resistance?

Plan: First, we will find the initial internal resistance from Eq. (28.13) by using the known emf, the original current, and the external resistance. The final external resistance can be found from the same relationship by substituting the lower current after aging. The difference in these values represents the increase of internal resistance.

Solution: From Eq. (28.13), we solve for the original resistance r_1 as follows:

$$I = \frac{\mathcal{E}}{R_L + r_1} \quad \text{or} \quad r_1 = \frac{\mathcal{E} - IR_L}{I}$$

$$r_1 = \frac{1.5 \text{ V} - (0.400 \text{ A})(3.5 \text{ }\Omega)}{0.400 \text{ A}}; \quad r_1 = 0.250 \text{ }\Omega$$

Now, the resistance after aging is given when $I = 0.350$ A.

$$r_2 = \frac{1.5 \text{ V} - (0.350 \text{ A})(3.5 \text{ }\Omega)}{0.350 \text{ A}}; \quad r_2 = 0.786 \text{ }\Omega$$

The increase in resistance is the difference ($r_2 - r_1$).

$$\Delta r = 0.786 \text{ }\Omega - 0.250 \text{ }\Omega = 0.536 \text{ }\Omega$$

28.5   Reversing the Current Through a Source of EMF

In a battery, chemical energy is converted into electric energy to maintain current flow in an electric circuit. A generator performs a similar function by converting mechanical energy into electric energy. In either case, the process is reversible. If a source of higher emf is connected in direct opposition to a source of lower emf, the current will pass through the latter from its positive terminal to its negative terminal. Reversing the flow of charge in this manner results in a loss of energy as electric energy is converted into chemical or mechanical energy.

Let us consider the process of charging a battery, as illustrated in Fig. 28.8. As charge flows through the higher source of emf $\mathcal{E}_1$, it gains energy. The terminal voltage for $\mathcal{E}_1$ is given by

$$V_1 = \mathcal{E}_1 - Ir_1$$

in accordance with Eq. (28.12). The output voltage is reduced because of the internal resistance r_1.

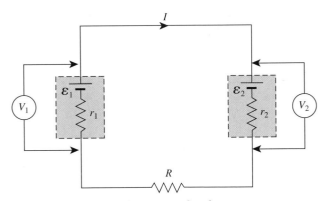

Figure 28.8   Reversing the current through a source of emf.

Energy is lost in two ways as charge is forced through the battery against its normal output direction:

1. Electric energy in the amount equal to $\mathcal{E}_2$ is stored as chemical energy in the battery.

2. Energy is lost to the internal resistance of the battery.

Therefore, the terminal voltage V_2, which represents the total drop in potential across the battery, is given by

$$V_2 = \mathcal{E}_2 + Ir_2 \tag{28.14}$$

where r_2 is the internal resistance. Note that in this case the terminal voltage is *greater* than the emf of the battery. The remainder of the potential supplied by the higher source of emf is lost through the external resistance R.

Throughout the entire circuit, the energy lost must equal the energy gained. Thus, we can write

Energy gained per unit charge = energy lost per unit charge

$$\mathcal{E}_1 = \mathcal{E}_2 + Ir_1 + Ir_2 + IR$$

Solving for the current I yields

$$I = \frac{\mathcal{E}_1 - \mathcal{E}_2}{r_1 + r_2 + R}$$

The current supplied to a continuous electric circuit is equal to the net emf divided by the total resistance of the circuit, including internal resistance.

$$I = \frac{\sum \mathscr{E}}{\sum R} \qquad (28.15)$$

For the purposes of applying Eq. (28.15), an emf is considered negative when the current flows against its normal output direction.

| Example 28.5 | Assume the following values for the parameters of the circuit drawn in Fig. 28.8: $\mathscr{E}_1 = 12$ V, $\mathscr{E}_2 = 6$ V, $r_1 = 0.2\ \Omega$, $r_2 = 0.1\ \Omega$, and $R = 4\ \Omega$. (a) What is the current in the circuit? (b) What is the terminal voltage across the 6-V battery? |

Plan: The current delivered must be equal to the ratio of the net voltage drop around the circuit ($\mathscr{E}_1 - \mathscr{E}_2$) and the total resistance encountered ($r_1 + r_2 + R$). Once the current is known, we can isolate the 6-V battery to find the terminal voltage. Energy is lost through this battery due to both the internal resistance and to the energy required to overcome the natural emf output. Thus, the battery $\mathscr{E}_2$ is being charged, and Eq. (28.15) applies.

Solution (a): From Eq. (28.15), the current is

$$I = \frac{\mathscr{E}_1 - \mathscr{E}_2}{r_1 + r_2 + R} = \frac{12\ \text{V} - 6\ \text{V}}{0.2\ \Omega + 0.1\ \Omega + 4\ \Omega}$$

$$= \frac{6\ \text{V}}{4.3\ \Omega} = 1.40\ \text{A}$$

Solution (b): The terminal voltage of the battery being charged is, from Eq. (28.14),

$$V_2 = \mathscr{E}_2 + Ir_2$$
$$= 6\ \text{V} + (1.4\ \text{A})(0.1\ \Omega)$$
$$= 6.14\ \text{V}$$

28.6 Kirchhoff's Laws

An electrical network is a complex circuit consisting of a number of current loops or meshes. For networks containing several meshes and a number of sources of emf, the application of Ohm's law becomes difficult. A more straightforward procedure for analyzing such circuits was developed in the nineteenth century by Gustav Kirchhoff, a German scientist. His method involves the use of two laws: *Kirchhoff's first law* and *Kirchhoff's second law.*

Kirchhoff's First Law: The sum of the currents entering a junction is equal to the sum of the currents leaving that junction.

$$\sum I_{\text{entering}} = \sum I_{\text{leaving}} \qquad (28.16)$$

Kirchhoff's Second Law: The sum of the emfs around any closed current loop is equal to the sum of all the *IR* drops around that loop.

$$\sum \mathscr{E} = \sum IR \qquad (28.17)$$

A *junction* refers to any point in a circuit at which three or more wires come together. The first law simply states that charge must flow continuously; it cannot pile up at a junction. In

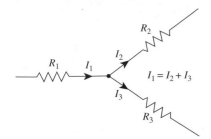

Figure 28.9   The sum of the currents entering a junction must equal the sum of the currents leaving that junction.

Fig. 28.9, if 12 C of charge enters the junction every second, then 12 C must leave it every second. The current delivered to each branch is inversely proportional to the resistance of that branch.

The second law is a restatement of the conservation of energy. If we begin at any point in a circuit and travel around any closed current loop, the energy gained by a unit of charge must equal the energy lost by that charge. Energy is gained through the conversion of chemical or mechanical energy into electric energy by a source of emf. Energy may be lost either in the form of *IR* potential drops or in the process of reversing the current through a source of emf. In the latter case, electric energy is converted into the chemical energy necessary to charge a battery or electric energy is converted to mechanical energy for the operation of a motor.

In applying Kirchhoff's rules, definite procedures must be followed. The steps in the general procedure will be presented by considering the example offered by Fig. 28.10a.

1. Assume a current direction for each loop in the network.

The three loops that may be considered are those illustrated in Fig. 28.10b, c, and d. Considering the entire circuit shown in Fig. 28.10a, the current I_1 is assumed to flow counterclockwise in the top loop, I_2 is assumed to travel to the left in the middle branch, and I_3 is assumed to flow counterclockwise in the lower loop. If we have guessed correctly, the solution to the problem will give a positive value for the current; if we have guessed incorrectly, a negative value will indicate the current is actually in the other direction.

2. Apply Kirchhoff's first law to write a current equation for all but one of the junction points.

Writing the current equation for *every* junction would result in a duplicate equation. In our example, there are two junction points that are labeled *m* and *n*. The current equation for *m* is

$$\sum I_{\text{entering}} = \sum I_{\text{leaving}}$$
$$I_1 + I_2 = I_3 \qquad\qquad\qquad \textbf{(28.18)}$$

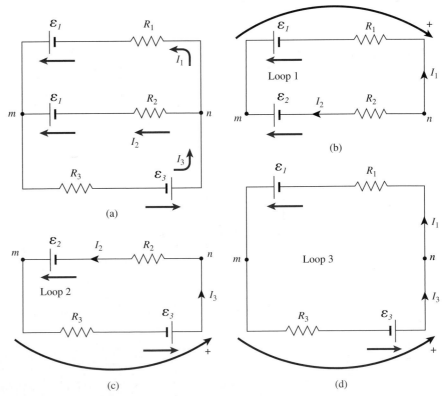

Figure 28.10   Applying Kirchhoff's laws to a complex circuit.

The same equation would result if we considered junction n, and no new information would be given.

3. Indicate by a small arrow, drawn next to the symbol for each emf, the direction in which the source, acting alone, would cause a positive charge to move through the circuit.

In our example, $\mathcal{E}_1$ and $\mathcal{E}_2$ are directed to the left, and $\mathcal{E}_3$ is directed to the right.

4. Apply Kirchhoff's second law ($\Sigma\,\mathcal{E} = \Sigma\,IR$) for one loop at a time. There will be one equation for each loop.

In applying Kirchhoff's second rule, one must begin at a specific point on a loop and *trace* around the loop in a consistent direction back to the starting point. The choice of the *tracing direction* is arbitrary, but, once established, it becomes the positive ($+$) direction for sign conventions. (Tracing directions for the three loops in our example are labeled in Fig. 28.10.) The following sign conventions apply:

1. When the emfs are summed around a loop, the value assigned to the emf is positive if its output (see step 3) is with the tracing direction; it is considered negative if the output is against the tracing direction.

2. An IR drop is considered positive when the assumed current is with the tracing direction and negative when the assumed current opposes the tracing direction.

Let us now apply Kirchhoff's second law to each loop in our example.

Loop 1 Starting at point m and tracing clockwise, we have

$$-\mathcal{E}_1 + \mathcal{E}_2 = -I_1 R_1 + I_2 R_2 \qquad (28.19)$$

Loop 2 Starting at point m and tracing counterclockwise, we have

$$\mathcal{E}_3 + \mathcal{E}_2 = I_3 R_3 + I_2 R_2 \qquad (28.20)$$

Loop 3 Starting at m and tracing counterclockwise we have

$$\mathcal{E}_3 + \mathcal{E}_1 = I_3 R_3 + I_1 R_1 \qquad (28.21)$$

If the equation for loop 1 is subtracted from the equation for loop 2, the equation for loop 3 is obtained, showing that the last loop equation gives no new information.

We now have three independent equations involving only three unknowns. They can be solved simultaneously to find the unknown, and the third loop equation can be used to check the results.

Example 28.6 Use Kirchhoff's laws to solve for the unknown currents in Fig. 28.11.

Plan: It is essential to draw and label a schematic diagram, as shown in Fig. 28.11, labeling all given information, indicating the normal output direction for each emf and the assumed directions for current flow in each circuit. We will choose the junction labeled m and apply Kirchhoff's first law to obtain an equation involving the three unknown currents. At least two other independent equations can be written by applying Kirchhoff's second law to selected current loops. These three equations can be solved simultaneously to find the unknown currents.

Solution: The sum of the currents entering junction m must equal the sum of those leaving that junction. Therefore,

$$\sum I_{\text{entering}} = \sum I_{\text{leaving}}$$
$$I_2 = I_1 + I_3 \qquad (28.22)$$

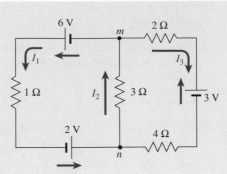

Figure 28.11

Next, the direction of positive output is indicated in the figure, adjacent to each source of emf. Since there are three unknowns, we need at least two more equations from the application of Kirchhoff's second law. Starting at m and tracing counterclockwise around the left loop, we write the voltage equation

$$\sum \mathscr{E} = \sum IR$$
$$6 \text{ V} + 2 \text{ V} = I_1(1 \text{ }\Omega) + I_2(3 \text{ }\Omega)$$
$$8 \text{ V} = (1 \text{ }\Omega)I_1 + (3 \text{ }\Omega)I_2$$

Dividing through by $1 \text{ }\Omega$ and transposing, we obtain

$$I_1 + 3I_2 = 8 \text{ A} \qquad \text{(28.23)}$$

The unit *ampere* arises from the fact that

$$1 \text{ V}/\Omega = 1 \text{ A}$$

Another voltage equation can be written by starting at m and tracing clockwise around the right loop:

$$-3 \text{ V} = I_3(2 \text{ }\Omega) + I_3(4 \text{ }\Omega) + I_2(3 \text{ }\Omega)$$

The negative sign arises from the fact that the output of the source opposes the tracing direction. Simplifying, we have

$$2I_3 + 4I_3 + 3I_2 = -3 \text{ A}$$
$$6I_3 + 3I_2 = -3 \text{ A}$$
$$I_2 + 2I_3 = -1 \text{ A} \qquad \text{(28.24)}$$

The three equations that must be solved simultaneously for I_1, I_2, and I_3 are

$$I_1 - I_2 + I_3 = 0 \qquad \text{(Eq. 28.22)}$$
$$I_1 + 3I_2 = 8 \text{ A} \qquad \text{(Eq. 28.23)}$$
$$I_2 + 2I_3 = -1 \text{ A} \qquad \text{(Eq. 28.24)}$$

From Eq. (28.22), we note

$$I_1 = I_2 - I_3$$

which, substituted into Eq. (28.23), yields

$$(I_2 - I_3) + 3I_2 = 8 \text{ A}$$
$$4I_2 - I_3 = 8 \text{ A} \qquad \text{(28.25)}$$

Now we can solve Eqs. (28.25) and (28.24) simultaneously by eliminating I_3 from the two equations by addition:

$$(28.24): \qquad I_2 + 2I_3 = -1 \text{ A}$$
$$2 \times (28.25): \qquad \underline{8I_2 - 2I_3 = 16 \text{ A}}$$
$$9I_2 = 15 \text{ A}$$
$$I_2 = 1.67 \text{ A}$$

Substituting $I_2 = 1.67$ A into Eqs. (28.23) and (28.24) gives values for the other currents:

$$I_1 = 3 \text{ A} \qquad I_3 = -1.33 \text{ A}$$

The negative value obtained for I_3 indicates our assumed current direction was incorrect. Actually, the current flows opposite the assumed direction. In working problems, however, the minus sign should be retained until all unknowns have been determined.

As a check on these results, we can write one more voltage equation by applying Kirchhoff's second law to the outside loop. Starting at m and tracing counterclockwise, we obtain

$$(6 + 2 + 3) \text{ V} = I_1(1 \, \Omega) - I_3(4 \, \Omega) - I_3(2 \, \Omega)$$
$$I_1 - 6I_3 = 11 \text{ A}$$

Substituting for I_1 and I_3, we obtain

$$3 \text{ A} - (6)(-1.33 \text{ A}) = 11 \text{ A}$$
$$11 \text{ A} = 11 \text{ A} \qquad\qquad \textit{Check}$$

Note again that the negative value for I_3 was used in the mathematics, even though it indicates an incorrect assumption.

Summary and Review

Summary

An understanding of direct-current is essential as an introduction to electrical technology. Most of the advanced study builds on the ideas presented in this chapter. The following summary presents the important facts to be remembered.

- Resistors may be connected in series or in parallel. The general facts about current, voltage, and equivalent resistances are summarized here.

Simple Circuits

Type of Circuit	Series Circuits	Parallel Circuits
	o—W-W-W—o	
Current I	$I = I_1 = I_2 = I_3$	$I = I_1 + I_2 + I_3$
Voltage V	$V = V_1 + V_2 + V_3$	$V = V_1 = V_2 = V_3$
Equivalent resistance	$R = R_1 + R_2 + R_3$	$\dfrac{1}{R} = \dfrac{1}{R_1} + \dfrac{1}{R_2} + \dfrac{1}{R_3}$
Resistance for two elements	$R = R_1 + R_2$	$R = \dfrac{R_1 R_2}{R_1 + R_2}$

- The current supplied to an electric circuit is equal to the *net* emf divided by the total resistance of the circuit, including internal resistances.

$$I = \frac{\sum \mathscr{E}}{\sum R}$$

for example,

$$I = \frac{\mathscr{E}_1 - \mathscr{E}_2}{r_1 + r_2 + R_L}$$

The example is for two opposing batteries of internal resistances r_1 and r_2 when the circuit load resistance is R_L.

- According to Kirchhoff's laws, the current entering a junction must equal the current leaving the junction, and the net emf around any loop must equal the sum of the *IR* drops. Symbolically,

$$\sum I_{\text{entering}} = \sum I_{\text{leaving}}$$
$$\sum \mathscr{E} = \sum IR \qquad \textit{Kirchhoff's Laws}$$

- The following steps should be applied to solving circuits with Kirchhoff's law (see Fig. 28.10):

Step 1 *Assume a current direction for each loop in the network.*

Step 2 *Apply Kirchhoff's first law to write a current equation for all but one of the junction points ($\sum I_{\text{in}} = \sum I_{\text{out}}$).*

Step 3 *Indicate by a small arrow the direction in which each emf acting alone would cause a positive charge to move.*

Step 4 *Apply Kirchhoff's second law ($\sum \mathscr{E} = \sum IR$) to write an equation for all possible current loops. Choose an arbitrary positive tracing direction. An emf is considered positive if its output direction is the same as your tracing direction. An IR drop is considered positive when the assumed current direction is the same as your tracing direction.*

Step 5 *Solve the equations simultaneously to determine the unknown quantities.*

Key Terms

direct current 548
internal resistance 554
junction 557

Kirchhoff's first law 557
Kirchhoff's second law 557
parallel circuit 551

parallel connection 551
series connection 549

Review Questions

28.1. Defend the following statement: the effective resistance of a group of resistors connected in parallel will be less than any of the individual resistances.

28.2. Discuss the advantages and disadvantages of connecting Christmas-tree lights (a) in series, (b) in parallel.

28.3. What is meant by the "open-circuit" potential difference of a battery?

28.4. Distinguish clearly between terminal potential difference and emf.

28.5. Many electrical devices and appliances are designed to operate at the same voltage. How should such devices be connected in an electric circuit?

28.6. Should elements connected in series be designed to function at a constant current or at a constant voltage?

28.7. In an electric circuit, it is desired to decrease the effective resistance by adding resistors. Should these resistors be connected in parallel or in series?

28.8. Describe a method for measuring the resistance of a spool of wire by using a voltmeter, an ammeter, a rheostat, and a source of emf. Draw the circuit diagram. (The rheostat is used to adjust the current to the range required for the ammeter.)

28.9. Given the emf of a battery, describe a laboratory procedure for determining its internal resistance.

28.10. Compare the formulas for computing equivalent capacitance with the formulas developed in this chapter for resistances in series and in parallel.

28.11. Can the terminal voltage of a battery ever be greater than its emf? Explain.

28.12. Solve Eq. (28.9) explicitly for the equivalent resistance R.

28.13. Distinguish between emf (open-circuit potential) and terminal voltage of a battery. What is meant by electrode potential? How is the electrode potential affected by an external discharge current?

28.14. The state of charge of a lead-acid storage battery can be determined by a hydrometer, a device that indicates relative density (in comparison with pure water) of the electrolytic solution. Explain how this information is indicative of the state of charge of a storage battery.

28.15. The emf of a lead-acid storage battery varies with the state of charge. At one instant, the density of the electrolytic solution is found to be 1.29 g/cm^3; at another time, the density is only 1.11 g/cm^3. Compare the electromotive forces of the battery for these two situations.

28.16. Explain why a car battery "goes dead" faster on a cold morning than on a normal day.

28.17. Lead-acid batteries should never be left in a discharged condition for long periods of time. Why is this a bad practice?

28.18. As a part of routine maintenance, distilled water is sometimes added to a lead-acid storage battery. Why is this necessary? Why should the water be distilled? If some acid is spilled from a battery, what precautions should be taken in replacing the acid?

Problems

Section 28.2 Resistors in Parallel

Ignore internal resistances for batteries in this section.

28.1. A 5-Ω resistor is connected in series with a 3-Ω resistor and a 16-V battery. What is the effective resistance, and what is the current in the circuit?
Ans. 8.00 Ω, 2.00 A

28.2. A 15-Ω resistor is connected in parallel with a 30-Ω resistor and a 30-V source of emf. What is the effective resistance, and what total current is delivered?

28.3. In Prob. 28.2, what is the current in 15- and 30-Ω resistors? Ans. 2.00 A, 1.00 A

28.4. What is the equivalent resistance of 2-, 4-, and 6-Ω resistors connected first in series and then in parallel?

28.5. An 18-Ω resistor and a 9-Ω resistor are first connected in parallel and then in series with a 24-V battery. What is the effective resistance for each connection? Neglecting internal resistance, what

is the total current delivered by the battery in each case?
Ans. 6.00 Ω, 27.0 Ω; 4.00 A, 0.899 A

28.6. A 12-Ω resistor and an 8-Ω resistor are first connected in parallel and then in series with a 28-V source of emf. What is the effective resistance and total current in each case?

28.7. An 8-Ω resistor and a 3-Ω resistor are first connected in parallel and then in series with a 12-V source. Find the effective resistance and total current for each connection.
Ans. 2.18 Ω, 5.50 A; 11.0 Ω, 1.09 A

28.8. Given three resistors of 80, 60, and 40 Ω, find their effective resistance when connected in series and when connected in parallel.

28.9. Three resistances of 4, 9, and 11 Ω are connected first in series and then in parallel. Find the effective resistance for each connection.
Ans. 24.0 Ω, 2.21 Ω

*28.10. A 9-Ω resistor is connected in series with two parallel resistors of 6 and 12 Ω. What is the terminal potential difference if the total current from the battery is 4 A?

*28.11. For the circuit described in Prob. 28.10, what is the voltage across the 9-Ω resistor, and what is the current through the 6-Ω resistor?

Ans. 36.0 V, 2.67 A

*28.12. Find the equivalent resistance of the circuit drawn in Fig. 28.12.

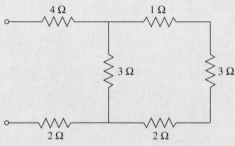

Figure 28.12

*28.13. Find the equivalent resistance of the circuit shown in Fig. 28.13. Ans. 2.22 Ω

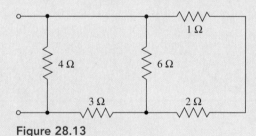

Figure 28.13

*28.14. If a potential difference of 24 V is applied to the circuit in Fig. 28.12, what are the current and voltage across the 1-Ω resistor?

*28.15. If a potential difference of 12-V is applied to the free ends in Fig. 28.13, what are the current and voltage across the 2-Ω resistor?

Ans. 1.60 A, 3.20 V

Section 28.3 EMF and Terminal Potential Difference

28.16. A load resistance of 8 Ω is connected in series with a 18-V battery with an internal resistance of 1.0 Ω. What current is delivered, and what is the terminal voltage?

28.17. A resistance of 6 Ω is placed across a 12-V battery with an internal resistance of 0.3 Ω. What is the current delivered to the circuit? What is the terminal potential difference? Ans. 1.90 A, 11.4 V

28.18. Two resistors, of 7 and 14 Ω, are connected in parallel with a 16-V battery with an internal resistance of 0.25 Ω. What is the terminal potential difference, and what is the current delivered to the circuit?

28.19. The open-circuit potential difference of a battery is 6 V. The current delivered to a 4-Ω resistor is 1.40 A. What is the internal resistance? Ans. 0.286 Ω

28.20. A dc motor draws 20 A from a 120-V dc line. If the internal resistance is 0.2 Ω, what is the emf of the motor?

28.21. For the motor in Prob. 28.20, what is the electric power drawn from the line? What portion of this power is dissipated because of heat losses? What power is delivered by the motor?

Ans. 2400 W, 2320 W, 80 W

28.22. A 2- and a 6-Ω resistor are connected in series with a 24-V battery of internal resistance 0.5 Ω. What is the terminal voltage and the power lost to internal resistance?

*28.23. Determine the total current and the current through each resistor for Fig. 28.14 when $\mathcal{E} = 24$ V, $R_1 = 6$ Ω, $R_2 = 3$ Ω, $R_3 = 1$ Ω, $R_4 = 2$ Ω, and $r = 0.4$ Ω. Ans. $I = 15$ A, $I_1 = 2$ A, $I_2 = 4$ Ω, $I_3 = 6$ A, $I_4 = 9$ A

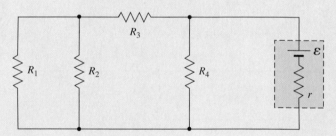

Figure 28.14

*28.24. Find the total current and the current through each resistor for Fig. 28.14 when $\mathcal{E} = 50$ V, $R_1 = 12$ Ω, $R_2 = 6$ Ω, $R_3 = 6$ Ω, $R_4 = 8$ Ω, and $r = 0.4$ Ω.

Section 28.6 Kirchhoff's Laws

28.25. Apply Kirchhoff's second rule to the current loop in Fig. 28.15. What is the net voltage around the loop? What is the net IR drop? What is the current in the loop? Ans. 16 V, 16 V, 2.00 A

28.26. Answer the same questions for Prob. 28.25 when the polarity of the 20-V battery is changed; that is, its output direction is now to the left.

*28.27. Use Kirchhoff's laws to solve for the currents through the circuit shown in Fig. 28.16.

Ans. 190 mA, 23.8 mA, 214 mA

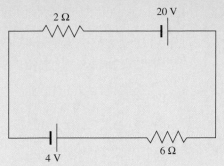

Figure 28.15

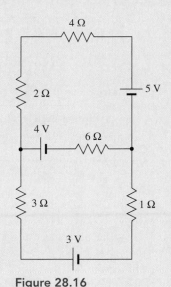

Figure 28.16

*28.28. Use Kirchhoff's laws to solve for the currents in Fig. 28.17.

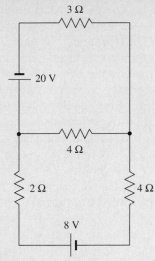

Figure 28.17

*28.29. Apply Kirchhoff's laws to the circuit of Fig. 28.18. Find the currents in each branch.
Ans. 536 mA, 732 mA, 439 mA, 634 mA

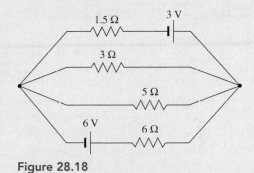

Figure 28.18

Additional Problems

28.30. The current in a single-loop circuit is 6.0 A when the total resistance in the circuit is R. When a 2-Ω resistor is added to the circuit in series with R, the current drops to 4 A. What is the resistance R?

28.31. Resistances of 3, 6, and 9 Ω are first connected in series and then in parallel with a 36-V source of potential difference. Neglecting internal resistance, what is the current leaving the positive terminal of the battery? Ans. 2.00 A, 22.0 A

28.32. Three 3-Ω resistors are connected in parallel. This combination is then placed in series with another 3-Ω resistor. What is the equivalent resistance?

*28.33. Three resistors of 4, 8, and 12 Ω are connected in series with a battery. A switch allows the battery to be connected or disconnected from the circuit. When the switch is open, a voltmeter across the terminals of the battery reads 50 V. When the switch is closed, the voltmeter reads 48 V. What is the internal resistance in the battery? Ans. 1.00 Ω

*28.34. The generator in Fig. 28.19 develops an emf of $\mathcal{E}_1 = 24$ V and has an internal resistance of

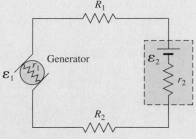

Figure 28.19

0.2 Ω. The generator is used to charge a battery $\mathcal{E}_2 = 12$ V that has an internal resistance of 0.3 Ω. Assume $R_1 = 4$ Ω and $R_2 = 6$ Ω. What is the terminal voltage across the generator? What is the terminal voltage across the battery?

*28.35. What is the power consumed in charging the battery for Prob. 28.34? Show that the power delivered by the generator is equal to the power loss due to resistance and the power consumed in charging the battery.

Ans. 27.4 W = 13.7 W +13.7 W

*28.36. Assume the following values for the parameters of the circuit illustrated in Fig. 28.8: $\mathcal{E}_1 = 100$ V, $\mathcal{E}_2 = 20$ V, $r_1 = 0.3$ Ω, $r_2 = 0.4$ Ω, and $R = 4$ Ω. What are the terminal voltages V_1 and V_2? What is the power lost through the 4-Ω resistor?

*28.37. Solve for the currents in each branch for Fig. 28.20. Ans. 1.20 A, 0.600, 0.600 A

*28.38. If the current in the 6-Ω resistor of Fig. 28.21 is 2 A, what is the emf of the battery? Neglect internal resistance. What is the power loss through the 1-Ω resistor?

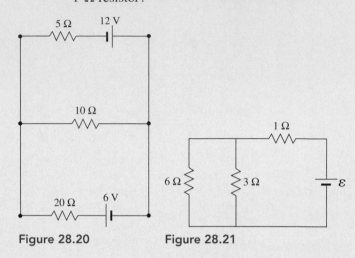

Figure 28.20   Figure 28.21

Critical Thinking Questions

*28.39. A three-way light bulb uses two resistors, a 50-W filament, and a 100-W filament. A three-way switch allows each to be connected in series and provides a third possibility by connecting the two filaments in parallel. Draw a possible arrangement of switches that will accomplish these tasks. Assume that the household voltage is 120 V. What are the resistances of each filament? What is the power of the parallel combination?

Ans. 288 Ω, 144 Ω,150 W

*28.40. The circuit illustrated in Fig. 28.7 consists of a 12-V battery, a 4-Ω resistor, and a switch. When the battery is new, its internal resistance is 0.4 Ω, and a voltmeter is placed across the terminals of the battery. What will be the reading of the voltmeter when the switch is open and when it is closed? After a long period of time, the experiment is repeated, and it is noted that open circuit reading is unchanged, but the terminal voltage has reduced by 10 percent. How do you explain the lower terminal voltage? What is the internal resistance of the old battery?

*28.41. Given three resistors of 3, 9, and 18 Ω, list all the possible equivalent resistances that can be obtained through various connections.

Ans. 2.00 Ω, 2.70 Ω, 6.30 Ω, 9.00 Ω, 11.6 Ω, 20.2 Ω, and 30.0 Ω

*28.42. Referring to Fig. 28.14, assume $\mathcal{E} = 24$ V, $R_1 = 8$ Ω, $R_2 = 3$ Ω, $R_3 = 2$ Ω, $R_4 = 4$ Ω, and $r = 0.5$ Ω. What current is delivered to the circuit by the 24-V battery? What are the voltage and current for the 8-Ω resistor?

*28.43. What is the effective resistance of the external circuit for Fig. 28.22 if internal resistance is neglected? What is the current through the 1-Ω resistor? Ans. 6.08 Ω, 2.58 A

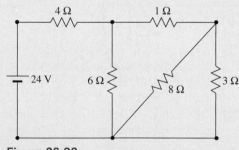

Figure 28.22

29 Magnetism and the Magnetic Field

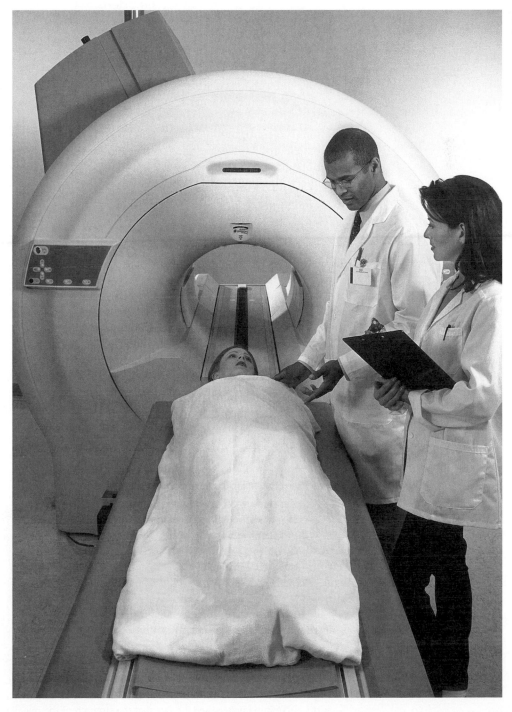

Magnetic resonance imaging (MRI) instruments use strong magnetic fields combined with radio frequency RF pulses to diagnose many medical problems, such as multiple sclerosis, tumors, and infections of the brain, spine, or joints. (*Photo © vol. 275/Corbis.*)

Objectives

After completing this chapter, you should be able to

1. Demonstrate by definition and example your understanding of *magnetism, induction, retentivity, saturation,* and *permeability.*

2. Write and apply an equation relating the magnetic force on a moving charge to its velocity, its charge, and its direction in a field of known magnetic flux density.

3. Determine the magnetic force on a current-carrying wire placed in a known *B* field.

4. Calculate the magnetic flux density (a) at a known distance from a current-carrying wire, (b) at the center of a current loop or coil, and (c) at the interior of a solenoid.

In previous chapters, we saw that electric charges exert forces on one another. In this chapter, we will study magnetic forces. A magnetic force may be generated by electric charges in motion, and an electric force may be generated by a magnetic field in motion. The operation of electric motors, generators, transformers, circuit breakers, televisions, radios, and most electric meters depends on the relationship between electric and magnetic forces. We will begin this chapter by studying the magnetic effects associated with materials and conclude with a discussion of the magnetic effects of charges in motion.

29.1   Magnetism

The first magnetic phenomena to be observed were associated with rough fragments of lodestone (an oxide of iron) found near the ancient city of Magnesia some 2000 years ago. These *natural magnets* were observed to attract bits and pieces of unmagnetized iron. This force of attraction is referred to as **magnetism,** and the device that exerts a magnetic force is called a **magnet.**

If a bar magnet is placed underneath a very thin sheet of glass and we sprinkle iron filings over the entire surface, we notice that the filings arrange themselves into a distinct pattern. Tiny pieces of iron are observed to cling most strongly to the ends of the magnet, as shown in Fig. 29.1. These regions where the magnet's strength appears to be concentrated are called **magnetic poles.**

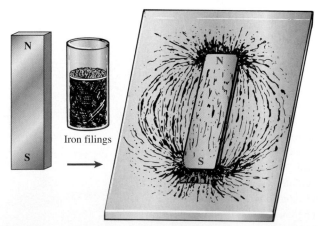

Iron filings

Figure 29.1   A bar magnet is placed under a sheet of thin glass, and iron filings are sprinkled on the surface. Note the pattern and the fact that the strength of the magnet is near the ends.

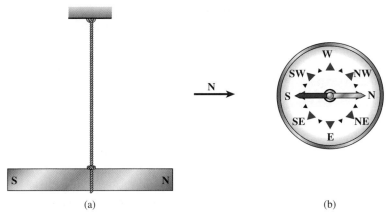

Figure 29.2 (a) A suspended bar magnet will come to rest in a north–south direction. (b) The top view of a magnetic compass.

When any magnetic material is suspended from a string, it turns about a vertical axis. As illustrated in Fig. 29.2, the magnet will align itself in a north–south direction. The end pointing toward the north is called the *north-seeking pole* or the north (N) pole of the magnet. The opposite, *south-seeking,* end is referred to as the south (S) pole of the magnet. It is the polarization of magnetic material that accounts for its usefulness as a compass for navigation. The compass consists of a light magnetized needle pivoted on a low-friction support.

That the north and south poles of a magnet are different can be demonstrated easily. When another bar magnet is brought near a suspended magnet, as in Fig. 29.3, two north poles or two south poles repel each other, whereas the north pole of one end and the south pole of another attract each other. The ***law of magnetic force*** states

Like magnetic poles repel each other; unlike magnetic poles attract each other.

Isolated poles do not exist. No matter how many times a magnet is broken in half, each piece will become a magnet, having both a north and a south pole. We know of no single particle that can create a magnetic field in the same way that a proton or electron can create an electric field.

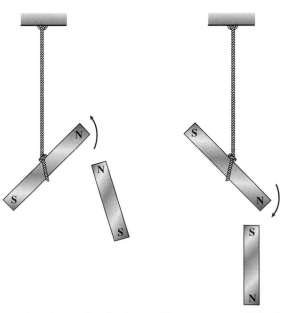

Figure 29.3 Like magnetic poles repel each other; unlike poles attract each other.

The attraction of a magnet for unmagnetized iron and the interacting forces between magnetic poles act through all substances. In industry, ferrous materials in trash are separated by magnets for recycling.

29.2 Magnetic Fields

Every magnet is surrounded by a space in which its magnetic effects are present. Such regions are called **magnetic fields.** Just as electric field lines were useful in describing electric fields, magnetic field lines, called **magnetic flux lines,** are useful for visualizing magnetic fields. The direction of a flux line at any point is indicated by the north pointer of a tiny imaginary compass placed at that point (see Fig. 29.4a). Accordingly, such lines are seen to *leave* the north pole of a magnet and *enter* the south pole. Unlike electric field lines, magnetic flux lines do not have beginning or ending points. They form continuous loops, passing through the metallic bar, shown in Fig. 29.4b. The flux lines in the region between two like or unlike poles are illustrated in Fig. 29.5.

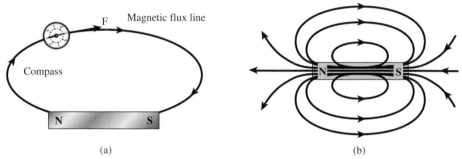

(a)   (b)

Figure 29.4 (a) Magnetic flux lines are in the direction indicated by the north blade of an imaginary compass. (b) The flux lines in the vicinity of a bar magnet.

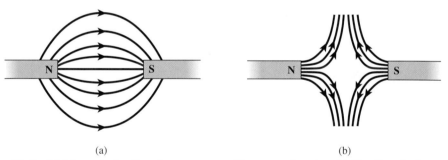

(a)   (b)

Figure 29.5 (a) Magnetic flux lines between two unlike magnetic poles. (b) Flux lines in the space between two like poles.

29.3 The Modern Theory of Magnetism

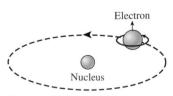

Figure 29.6 Two kinds of electron motion responsible for magnetic properties.

Magnetism in matter is currently believed to result from the movement of electrons in the atoms of substances. If this is true, magnetism is a property of *charge in motion* and is closely related to electric phenomena. According to classical theory, individual atoms of a magnetic substance are, in effect, tiny magnets with north and south poles. The magnetic polarity of atoms is thought to arise primarily from the spin of electrons about their own axes and partially due to their orbital motions around the nucleus. Figure 29.6 illustrates the two kinds of electron motion. Diagrams such as this should not be taken too seriously because there is still much we do not know about the motion of electrons. But we do firmly believe that magnetic fields of all particles must be caused by charge in motion, and such models help us to describe the phenomena.

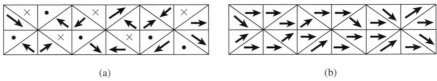

Figure 29.7 (a) Magnetic domains are randomly oriented in an unmagnetized material. (b) The preferred orientation of domains in a magnetized material.

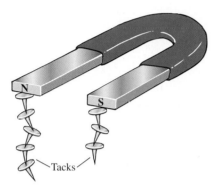

Figure 29.8 Magnetic induction.

The atoms in a magnetic material are grouped into microscopic magnetic regions called **domains.** All the atoms within a domain are believed to be magnetically polarized along a crystal axis. In an unmagnetized material, these domains are oriented in random directions, as indicated by the arrows in Fig. 29.7a. A dot is used to indicate an arrow directed out of the paper, and a cross is used to denote a direction into the paper. If a large number of domains become oriented in the same direction, as in Fig. 29.7b, the material will exhibit strong magnetic properties.

This theory of magnetism is highly plausible because it offers an explanation for many of the observed magnetic effects of matter. For example, an unmagnetized iron bar can be made into a magnet simply by holding another magnet near it or in contact with it. This process, called **magnetic induction,** is illustrated by Fig. 29.8. The tacks become temporary magnets by induction. Note that the tacks on the right become magnetized, even though they do not actually touch the magnet. Magnetic induction is explained by the domain theory. The introduction of a magnetic field aligns the domains, resulting in magnetization.

Induced magnetism is often only temporary, and when the field is removed, the domains gradually become disoriented. If the domains remain aligned to some degree after the field has been removed, the material is said to be *permanently* magnetized. The ability to retain magnetism is referred to as **retentivity.**

Another property of magnetic materials that is explained easily by the domain theory is **magnetic saturation.** There appears to be a limit to the degree of magnetization experienced by a material. Once this limit has been reached, no greater strength of an external field can increase the magnetization. It is believed that all its domains have been aligned.

29.4 · Flux Density and Permeability

In Chapter 24, we stated that electric field lines are drawn so their spacing at any point will determine the strength of the electric field at that point (see Fig. 29.9). The number of lines ΔN drawn through a unit of area ΔA is directly proportional to the electric field intensity E.

$$\frac{\Delta N}{\Delta A} = \epsilon E \qquad (29.1)$$

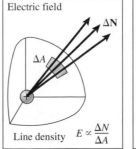
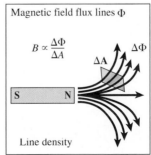

Figure 29.9 Just as the electric field is proportional to the electric line density, the magnetic field is proportional to the magnetic flux density.

The proportionality constant ϵ, which determines the number of lines drawn, is the permittivity of the medium through which the lines pass.

An analogous description of a magnetic field can be presented by considering the magnetic flux Φ passing through a unit of perpendicular area $A_\perp$. This ratio B is called the **magnetic flux density.**

> The magnetic flux density in a region of a magnetic field is the number of flux lines that pass through a unit of perpendicular area in that region.

$$B = \frac{\Phi \; (\text{flux})}{A_\perp \; (\text{area})}$$ (29.2)

The SI unit of magnetic flux is the **weber** (Wb). The unit of flux density would then be webers per square meter, which is redefined as the **tesla** (T). An older unit that remains in use is the *gauss* (G). In summary,

$$1\,\text{T} = 1\,\text{Wb/m}^2 = 10^4\,\text{G}$$ (29.3)

Example 29.1

A rectangular loop 10 cm wide and 20 cm long makes an angle of 30° with respect to the magnetic flux in Fig. 29.10. If the flux density is 0.3 T, compute the magnetic flux Φ penetrating the loop.

Plan: The effective area penetrated by the flux is that component of the area that is perpendicular to the flux. If θ is chosen as the angle that the plane of the loop makes with the field **B,** this component is simply $A \sin \theta$. The definition of the **B** field as flux density will be used to find the flux Φ penetrating that area component.

Solution: The area of the rectangular loop is

$$A = (0.10\,\text{m})(0.20\,\text{m}) = 0.020\,\text{m}^2$$

From Eq. (29.2), the magnitude of the **B** field is defined as the flux per unit area perpendicular to the field. Thus, we write

$$B = \frac{\Phi}{A \sin \theta} \quad \text{or} \quad \Phi = BA \sin \theta$$

The magnetic flux in webers is found by substitution.

$$\Phi = (0.3\,\text{T})(0.02\,\text{m}^2) \sin 30°$$
$$= 3 \times 10^{-3}\,\text{Wb} = 3\,\text{mWb}$$

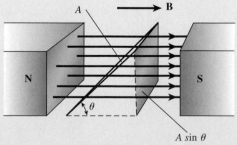

Figure 29.10   Computing the magnetic flux through a rectangular conductor.

The flux density at any point in a magnetic field is affected strongly by the nature of the medium or by the nature of a material placed in the medium. For this reason, it is convenient to define a new magnetic field vector, the *magnetic field intensity* **H,** which does not depend on the nature of a medium. In any case, the number of lines established per unit area is directly proportional to the magnetic field intensity **H.** We can write

$$B = \frac{\Phi}{A_\perp} = \mu H \tag{29.4}$$

where the proportionality constant μ is the **permeability** of the medium through which the flux lines pass. Equation (29.4) is exactly analogous to Eq. (29.1), which was developed for electric fields. The permeability of a medium can thus be thought of as a measure of its ability to establish magnetic flux lines. The greater the permeability of a medium, the more flux lines will pass through a unit of area.

The permeability of free space (vacuum) is denoted by μ_0 and has the following magnitude for SI units:

$$\mu_0 = 4\pi \times 10^{-7} \text{ Wb/A} \cdot \text{m} = 4\pi \times 10^{-7} \text{ T} \cdot \text{m/A}$$

The full meaning of the unit webers per ampere-meter will come later. It is determined by the units of Φ, A, and H of Eq. (29.4), which, for a vacuum, can be written

$$B = \mu_0 H \qquad\qquad\qquad Vacuum \quad (29.5)$$

If a nonmagnetic material, such as glass, is placed in a magnetic field like that in Fig. 29.11, the flux distribution will not vary appreciably from that established for a vacuum. However, when a highly permeable material, such as soft iron, is placed in the field, the flux distribution will be altered considerably. The permeable material becomes magnetized by induction, resulting in a greater field strength for that region. For this reason, the flux density B is also referred to as the *magnetic induction.*

Magnetic materials are classified according to their permeability compared with that of free space. The ratio of the permeability of a material to that of a vacuum is called its *relative permeability* and is given by

$$\mu_r = \frac{\mu}{\mu_0} \tag{29.6}$$

Consideration of Eqs. (29.5) and (29.6) shows the relative permeability of a material is a measure of its ability to change the flux density of a field from its value in a vacuum.

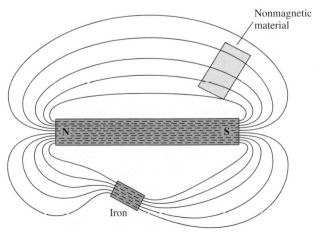

Figure 29.11 A permeable material becomes magnetized by induction, resulting in a greater flux density in that region.

Materials with a relative permeability slightly less than unity have the property of being repelled by a strong magnet. Such materials are said to be *diamagnetic,* and the property is referred to as *diamagnetism.* On the other hand, materials with only slightly greater permeability than that of a vacuum are said to be *paramagnetic.* These materials are feebly attracted by a strong magnet.

A few materials, for example, iron, cobalt, nickel, steel, and alloys of these metals, have extremely high permeabilities, ranging from a few hundred to thousands of times that for free space. Such materials are strongly attracted by a magnet and are said to be *ferromagnetic.*

29.5 Magnetic Field and Electric Current

Although the modern theory of magnetism holds that a magnetic field results from the motion of charges, science has not always accepted this proposition. It is fairly easy to show that a powerful magnet exerts no force on a static charge. In the course of a lecture demonstration in 1820, Hans Oersted set up an experiment to show his students that

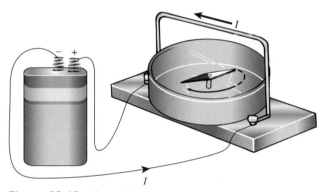

Figure 29.12 Oersted's experiment.

moving charges and magnets also do not interact. He placed the magnetic needle of a compass near a conductor, as illustrated in Fig. 29.12. To his surprise, when a current was sent through the wire, a twisting force was exerted on the compass needle until it pointed almost perpendicular to the wire. Further, the magnitude of the force depended on the relative orientation of the compass needle and the current direction. The maximum twisting force occurred when the wire and compass needle were parallel before the current was established. If they were initially perpendicular, no force was experienced. Evidently, a magnetic field is set up by the charge in motion through the conductor.

In the same year that Oersted made his discovery, Ampère found that forces exist between two current-carrying conductors. Two wires with current in the same direction were found to attract each other, whereas oppositely directed currents caused a force of repulsion. A few years later, Faraday found that the motion of a magnet toward or away from an electric circuit produces a current in the circuit. The relationship between magnetic and electric phenomena could no longer be doubted. Today, all magnetic phenomena can be explained in terms of electric charges in motion.

29.6 The Force on a Moving Charge

Let us investigate the effects of a magnetic field by observing the magnetic force exerted on a charge that passes through the field. In studying these effects, it is useful to imagine a positive-ion tube like that in Fig. 29.13. Such a tube allows us to inject a positive ion of constant charge and velocity into a field of magnetic flux density **B.** By pointing the tube in various directions, we can observe the force exerted on the moving charge. The most striking observation is that the charge experiences a force that is perpendicular to both the magnetic flux density **B** and to the velocity **v** of the moving charge. Note that when the magnetic flux is directed from the left to the right and the charge is moving toward the reader, the charge is deflected upward. Reversing the polarity of the magnets causes the charge to be deflected downward.

The direction of the magnetic force **F** on a positive charge moving with a velocity **v** in a **B** field can be reckoned by applying the *right-hand rule* (see Fig. 29.14), which is stated as follows:

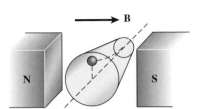

Figure 29.13   The magnetic force **F** on a moving charge is perpendicular both to the flux density **B** and to the charge velocity **v.**

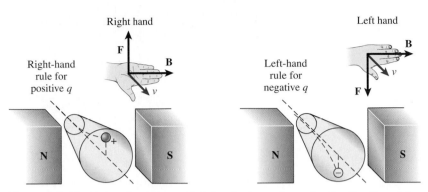

Figure 29.14   Using the *right-* and *left-hand rules* to determine the direction of the magnetic force on a moving charge. Point the fingers in the direction of the **B** field and the thumb in the direction of the moving charge. The open palm will face in the direction of the magnetic force. The right hand is used for positive charges, and the left hand is used for negative charges.

The Right-Hand Rule: Extend the right hand with the fingers pointing in the direction of the **B** field and the thumb pointing in the direction of the velocity **v** of the moving charge. The open palm will be facing in the direction of the magnetic force **F** on a positive charge.

If the moving charge is *negative,* the direction of the force can be determined by using the *left* hand and following the same procedure. In this manner, the direction of the magnetic force is seen to be *opposite* to the direction for a moving positive charge.

Let us now consider the magnitude of the force on a moving charge. Experimentation has shown the magnitude of the magnetic force is directly proportional to the magnitude of the charge q and to its velocity **v**. Greater deflections will be indicated by our positive-ion tube if either of these parameters is increased.

An unexpected variation in the magnetic force will be observed if the ion tube is rotated slowly with respect to the magnetic flux density **B**. As indicated by Fig. 29.15, for a given charge of constant velocity **v,** the magnitude of the force varies with the angle the

Force and Angle of Path

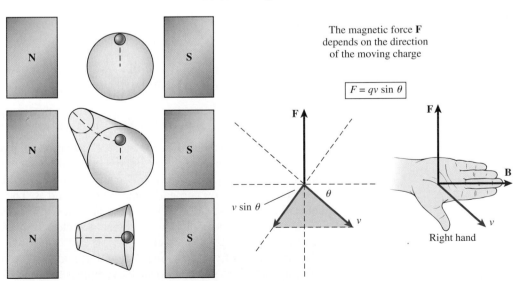

$$F = qv \sin \theta$$

Figure 29.15   The magnetic force is greatest when the path is perpendicular to the field and least when it is parallel to it.

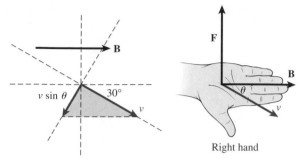

Figure 29.16   The magnetic force on a positive charge moving at 30° with respect to a **B** field.

tube makes with the field. Particle deflection is a maximum when the charge velocity is perpendicular to the field. As the tube is rotated slowly toward **B,** the particle deflection becomes less and less. Finally, when the charge velocity is directed parallel to **B,** no deflection occurs, indicating that the magnetic force has dropped to zero. Evidently, the magnitude of the force is a function not only of the magnitude of the charge and its velocity but also varies with the angle θ between **v** and **B.** This variation is accounted for by stating that the magnetic force is proportional to the component of the velocity, **v** $\sin \theta$, perpendicular to the field direction. (See Fig. 29.16.)

The above observations are summarized by the proportionality

$$F \propto qv \sin \theta \qquad \textbf{(29.7)}$$

If the proper units are chosen, the proportionality constant can be equated to the magnetic flux density B of the field causing the force. In fact, this proportionality is often used to *define* magnetic flux density as the constant ratio:

$$B = \frac{F}{qv \sin \theta} \qquad \textbf{(29.8)}$$

A magnetic field having a flux density of 1 tesla (1 weber per square meter) will exert a force of 1 newton on a charge of 1 coulomb moving perpendicular to the field with a velocity of 1 meter per second.

As a consequence of Eq. (29.8), it can be noted that

$$1 \text{ T} = 1 \text{ N}/(\text{C} \cdot \text{m/s}) = 1 \text{ N/A} \cdot \text{m} \qquad \textbf{(29.9)}$$

These unit relationships are useful in solving problems involving magnetic forces. If we solve for the force F in Eq. (29.8), we obtain

$$F = qvB \sin \theta \qquad \textbf{(29.10)}$$

which is a more useful form for direct calculation of magnetic forces. The force F is in newtons when the charge q is expressed in coulombs, the velocity v is in meters per second, and the flux density **B** is in teslas. The angle θ indicates the direction of **v** with respect to **B.** The force **F** is *always* perpendicular to both **v** and **B.** The direction of these vectors can be determined by application of the right-hand rule.

When representing three-dimensional vectors graphically, it is useful to use the convention that crosses (X) indicate a direction *into* the paper. These symbols might be thought of as the "tails" of vector arrows. We will use dots (·) to indicate the tips of vector arrows pointing *out* of the paper. Two such examples are shown in Fig. 29.17. To test your understanding, you should verify that the force on the positive charge is upward and that the force on the negative charge is to the right.

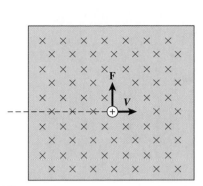 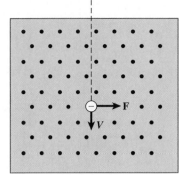

Figure 29.17 The direction of the **B** field is indicated by crosses (into the paper) and dots (out of the paper). Verify that the direction of the force on the positive charge is upward and that the direction of the force on the negative charge is to the right.

Example 29.2

An electron is projected from left to right into a magnetic field directed vertically downward. The velocity of the electron is 2×10^6 m/s, and the magnetic flux density of the field is 0.3 T. Find the magnitude and direction of the magnetic force on the electron.

Plan: The charge on the electron is 1.6×10^{-19} C, the magnitude of the force on the electron is found from Eq. (29.10), and the direction is found by applying the *left-hand rule*. We use the left hand, since the charge on an electron is *negative.*

Solution: The electron is moving in a direction perpendicular to **B.** Thus, $\sin \theta = 1$, and we solve for the force F as follows:

$$F = qvB \sin 90° = (1.6 \times 10^{-19} \text{ C})(2 \times 10^6 \text{ m/s})(0.3 \text{ T})(1)$$
$$F = 9.60 \times 10^{-14} \text{ N}$$

Application of the *left-hand* rule for an electron shows that the direction of the force will be *out of the page,* or toward the reader. (It would be into the page for a positive charge, for example, for a proton or an alpha particle.)

29.7 Force on a Current-Carrying Wire

When an electric current I passes through a conductor lying in a magnetic field **B,** as illustrated in Fig. 29.18, each charge q flowing through the conductor experiences a magnetic force **F.** These forces are transmitted to the conductor as a whole, causing each unit of length to experience a force. If a total quantity of charge q passes with average velocity v through the length L of the wire, we may write

$$F = q\bar{v}B$$

But the average velocity v for each charge passing through the length L in the time t is L/t. Therefore, the net force on the entire length becomes

$$F = q\frac{L}{t}B$$

Now, since $I = q/t$, we rearrange and obtain

$$F = ILB$$

where I represents the current in the wire.

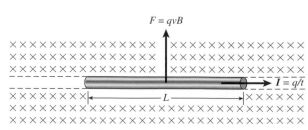

Figure 29.18   Magnetic force on a current-carrying conductor.

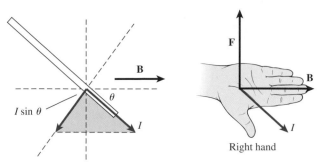

Figure 29.19   The magnetic force on a current-carrying conductor. The current is directed at an angle θ with the **B** field.

Just as the magnitude of the force on a moving charge varies with the velocity direction, the force **F** on current-carrying conductor depends on the angle θ the current makes with the **B** field. As illustrated in Fig. 29.19, it will experience a force **F** in *newtons* given by

$$F = ILB \sin \theta \qquad (29.11)$$

where I is the current in *amperes,* B is the magnetic field in *teslas,* L is the length of the wire in *meters,* and θ is the angle the wire makes with the **B** field.

The direction of the magnetic force on a current-carrying conductor can be determined by the right-hand rule in the same way as for a moving charge (since a current *is* moving charge). As shown in Fig. 29.19, when the thumb points in the direction of the current **I** and the fingers point in the direction of the magnetic field **B,** the palm of the hand will face in the direction of the magnetic force **F.** The direction of the force is *always* perpendicular to both **I** and **B.**

Example 29.3   The wire in Fig. 29.19 makes an angle of 30° with respect to a 0.2-T **B** field. If the length of the wire is 8 cm and a current of 4 A passes through it, determine the magnitude and direction of the resultant force on the wire.

Plan: The magnitude of the force is found by direct substitution into Eq. (29.11), and the direction of the force is given by applying the right-hand rule.

Solution: The length is converted to meters ($L = 8$ cm $= 0.08$ m).

$$F = ILB \sin \theta = (4 \text{ A})(0.08 \text{ m})(0.2 \text{ T}) \sin 30°$$
$$F = 0.032 \text{ N}$$

The direction of the force will be *upward* by applying the right-hand rule. If the current direction were to be reversed, the force would be downward.

29.8 Magnetic Field of a Long, Straight Wire

Oersted's experiment demonstrated that electric charge in motion, or a current, sets up a magnetic field in the space surrounding it. Up to this point, we have been discussing the force that such a field will exert on a second current-carrying conductor or on a charge moving in the field. We will now begin to calculate the magnetic fields produced by electric currents.

Let us first examine the flux density surrounding a long, straight wire carrying a constant current. If iron filings are sprinkled on the paper surrounding the wire in Fig. 29.20,

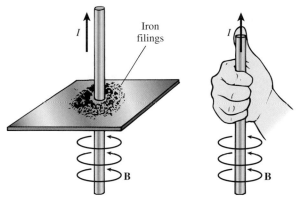

Figure 29.20 *The right-hand-thumb rule:* Grasp wire with right hand; point thumb in direction of *I*. Fingers wrap wire in direction of the circular **B** field.

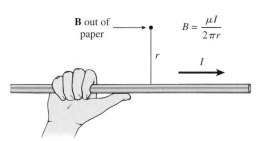

Figure 29.21 The magnetic field **B** at a perpendicular distance *r* from a long, current-carrying conductor.

they will become aligned in concentric circles around the wire. Similar investigation of the area surrounding the wire with a magnetic compass will confirm that the magnetic field is circular and directed in a clockwise fashion, as viewed along the direction of conventional (positive) current. A convenient method devised by Ampère to determine the direction of the field surrounding a straight wire is called the ***right-hand-thumb rule*** (see Fig. 29.20).

If the wire is grasped with the right hand so the thumb points in the direction of the conventional current, the curled fingers of that hand will point in the direction of the magnetic field.

The magnetic induction, or flux density, at a perpendicular distance *d* from a long, straight wire carrying a current *I*, as shown in Fig. 29.21, can be calculated from

$$B = \frac{\mu I}{2\pi r} \qquad\qquad Long\ Wire \quad \textbf{(29.12)}$$

where μ is the permeability of the medium surrounding the wire. In the special cases of a vacuum, air, and nonmagnetic media, the permeability μ_0 is

$$\mu_0 = 4\pi \times 10^{-7}\ \text{T} \cdot \text{m/A} \qquad\qquad \textbf{(29.13)}$$

When using this constant with Eq. (29.12), it is necessary that the current be in amperes, the field be in teslas, and the distance from the wire be in meters.

Example 29.4

Determine the magnetic field **B** in teslas at a distance of 5 cm from a long wire carrying a current of 8 A. Assume the wire is surrounded by air.

Plan: The magnitude of the field is found from Eq. (29.12), and the direction is determined by the right-hand-thumb rule.

Solution: Substitution $r = 5$ cm $= 0.05$ m and $I = 8$ A, gives

$$B = \frac{\mu_0 I}{2\pi r} = \frac{(4\pi \times 10^{-7}\ \text{T} \cdot \text{m/A})(8\ \text{A})}{2\pi(0.05\ \text{m})}$$

$$B = 3.2 \times 10^{-5}\ \text{T}$$

If the surrounding medium is not air or a vacuum, one must consider that the permeability differs from μ_0.

29.9 Other Magnetic Fields

If a wire is bent into a circular loop and connected to a source of current, as shown in Fig. 29.22a, a magnetic field similar to that for a bar magnet will be set up. The right-hand-thumb rule will still serve to give the field direction in a rough manner, but now the flux lines are no longer circular. The magnetic flux density varies considerably from point to point.

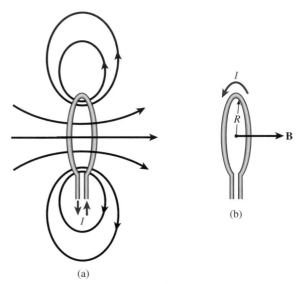

(a)

(b)

Figure 29.22 The magnetic field at the center of a circular loop.

The magnetic induction at the center of a circular loop of radius R carrying a current I is given by Eq. (29.14).

$$B = \frac{\mu I}{2R} \qquad \textit{Center of Loop} \quad \textbf{(29.14)}$$

The direction of B is perpendicular to the plane of the loop. If the wire consists of a coil having N turns, Eq. (29.14) becomes

$$B = \frac{\mu N I}{2R} \qquad \textit{Center of Coil} \quad \textbf{(29.15)}$$

A *solenoid* consists of many circular turns of wire wound in the form of a helix, as illustrated in Fig. 29.23. The magnetic field produced is similar to that of a bar magnet. The

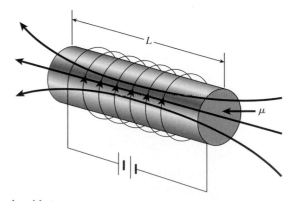

Figure 29.23 The solenoid.

magnetic induction in the interior of a solenoid is given by

$$B = \frac{\mu NI}{L} \qquad \text{Solenoid} \quad (29.16)$$

where N is the number of turns, I is the current in amperes, and L is the length of the solenoid in meters.

Example 29.5

A solenoid is constructed by winding 400 turns of wire on a 20-cm iron core. The relative permeability of the iron is 13,000. What current is required to produce a magnetic induction of 0.5 T in the center of the coil?

Plan: Since we are given the relative permeability, we will need to multiply by μ_0 to find the value of μ to use in Eq. (29.16), which will be used to solve for the current I.

Solution: The relative permeability is 13,000, so, from Eq. (29.6), we have

$$\mu_r = \frac{\mu}{\mu_0} \qquad \text{or} \qquad \mu = \mu_r \mu_0 = (13{,}000)(4\pi \times 10^{-7}\,\text{T} \cdot \text{m/A})$$

$$\mu = 1.63 \times 10^{-2}\,\text{T} \cdot \text{m/A}$$

Given that $N = 400$ turns, $L = 0.20$ m, and $B = 0.5$ T, we solve Eq. (29.16) for the current I.

$$B = \frac{\mu NI}{L} \qquad \text{or} \qquad I = \frac{BL}{\mu N}$$

$$I = \frac{(0.5\,\text{T})(0.20\,\text{m})}{(1.63 \times 10^{-2}\,\text{T} \cdot \text{m/A})(400)}; \qquad I = 0.0153\,\text{A}$$

One type of solenoid, called a *toroid,* is often used in studying magnetic effects. As will be seen in Section 29.10, the toroid consists of a tightly wound coil of wire in the shape of a doughnut. The magnetic flux density in the core of a toroid is also given by Eq. (29.16).

29.10   ## Hysteresis

We have seen that the lines of magnetic flux are more numerous for a solenoid that has an iron core than for a solenoid in air. The flux density is related to the permeability μ of the material serving as a core for the solenoid. Recall the field intensity H and flux density B are related to each other according to the equation

$$B = \mu H$$

Comparison of this relationship with Eq. (29.16) shows that, for a solenoid,

$$H = \frac{NI}{L} \qquad\qquad (29.17)$$

Note that the magnetic intensity is independent of the permeability of the core. It is a function only of the number of turns N, the current I, and the solenoid length L. The magnetic intensity is expressed in *amperes per meter.*

We can study the magnetic properties of matter by observing the flux density B produced as a function of the magnetizing current or as a function of the magnetic intensity H. This can be done more easily when a substance is fashioned into the form of a toroid, as in Fig. 29.24. The magnetic field set up by a current in the magnetizing windings is confined wholly to the toroid. Such a device is often called a *Rowland ring* after J. H. Rowland, who used it to study the magnetic properties of many materials.

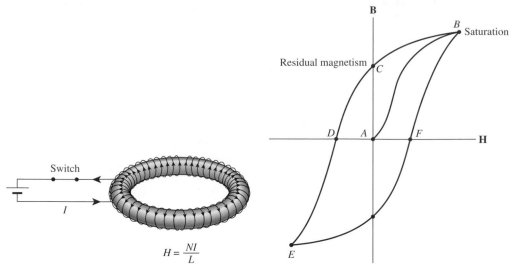

$$H = \frac{NI}{L}$$

Figure 29.24 Rowland's ring.   **Figure 29.25** The hysteresis loop.

Suppose we begin by studying the magnetic properties of a material with an unmagnetized Rowland ring fashioned out of the substance. Initially, $B = 0$ and $H = 0$. The switch is closed, and the magnetizing current I is gradually increased, producing a magnetic intensity given by

$$H = \frac{NI}{L}$$

where L is the circumference of the ring. As the material is subjected to an increasing magnetic intensity H, *the flux density B* increases until the material is *saturated*. Refer to the curve AB in Fig. 29.25. Now, if the current is gradually reduced to zero, the flux density B throughout the core does not return to zero but lags behind the magnetic intensity, as illustrated by the curve BC. (This refers essentially to residual magnetism.) The lack of retraceability is known as ***hysteresis.***

Hysteresis is the lagging of the magnetization behind the magnetic intensity.

The only way to bring the flux density B within the ring back to zero is to reverse the direction of the current through the windings. This procedure builds up the magnetic intensity H in the opposite direction, as shown by curve CD. If the magnetization continues to increase in the negative direction, the material will eventually become saturated again with a reversed polarity. Refer to curve DE. Reducing the current to zero again and then increasing it in the positive direction will yield the curve EFB. The entire curve is called the ***hysteresis loop.***

The area enclosed by a hysteresis loop is an indication of the quantity of energy that is lost (in the form of heat) by carrying a given material through a complete magnetization cycle. The efficiency of many electromagnetic devices depends on the selection of magnetic materials with low hysteresis. On the other hand, materials that are to remain well magnetized should have large hysteresis.

Summary and Review

Summary

We have seen that magnetic fields are created by charge in motion. This basic principle will underlie much of what follows in your study of electromagnetism. The operation of electric motors, generators, transformers, and an endless variety of industrial instruments requires an understanding of the magnetic field. The major concepts are summarized as follows.

- The magnetic flux density **B** in a region of a magnetic field is the number of flux lines that pass through a unit of area perpendicular to the flux.

$$B = \frac{\Phi}{A_\perp} = \frac{\Phi}{A \sin \theta} \qquad \textit{Magnetic Flux Density}$$

where Φ = flux, Wb

A = unit area, m^2

θ = angle that plane of area makes with flux

B = magnetic flux density, T (1 T = 1 Wb/m^2)

- The magnetic flux density **B** is proportional to the magnetic field intensity **H**. The constant of proportionality is the permeability of the medium in which the field exists.

$$B = \frac{\Phi}{A_\perp} = \mu H \qquad \begin{array}{l}\textit{For a Vacuum}\\ \mu_0 = 4\pi \times 10^{-7}\ \text{T} \cdot \text{m/A}\end{array}$$

- The *relative permeability* μ_r is the ratio of μ/μ_0. We can write

$$B = \mu_0 \mu_r H \qquad \text{where} \qquad \mu_r = \frac{\mu}{\mu_0} \qquad \begin{array}{l}\textit{Relative}\\ \textit{Permeability}\end{array}$$

- A magnetic field of flux density equal to 1 T will exert a force of 1 N on a charge of 1 C moving perpendicular to the field with a velocity of 1 m/s. The general case is described by Fig. 29.16, in which the charge moves at an angle θ with the field.

$$F = qvB \sin \theta \quad B = \frac{F}{qv \sin \theta} \qquad \begin{array}{l}\textit{Magnetic Force on}\\ \textit{a Moving Charge}\end{array}$$

The direction of the magnetic force is given by the right-hand rule, as illustrated in Fig. 29.14.

- The force **F** on a wire carrying a current I at an angle θ with a flux density B is given by

$$F = ILB \sin \theta \qquad \begin{array}{l}\textit{Magnetic Force}\\ \textit{on a Conductor}\end{array}$$

where L is the length of the conductor.

- Equations for many common magnetic fields are given as follows:

$$B - \frac{\mu I}{2\pi r} \quad \begin{array}{l}\textit{Long}\\ \textit{Wire}\end{array} \qquad B = \frac{\mu I}{2R} \quad \begin{array}{l}\textit{Center}\\ \textit{of Loop}\end{array}$$

$$B = \frac{\mu NI}{2R} \quad \begin{array}{l}\textit{Center}\\ \textit{of Coil}\end{array} \qquad B = \frac{\mu NI}{L} \quad \textit{Solenoid}$$

Key Terms

Review Questions

29.1. How can you positively determine whether a piece of steel is magnetized? How would you determine its polarity if magnetized?

29.2. In general, magnetic materials with high permeability have low retentivity. Why do you think this is true?

29.3. The Earth acts like a huge magnet with one pole in the Arctic circle and the other in the Antarctic region. Can you justify the following statement: The geographic North Pole is actually near a magnetic south pole? Explain.

29.4. If an iron bar is placed parallel to a north–south direction and is hammered on one end, the bar becomes a temporary magnet. Explain.

29.5. When a bar magnet is broken into several pieces, each part becomes a magnet with a north and a south pole. Apparently, an isolated pole cannot exist. Explain this, using the domain theory of magnetism.

29.6. Heating magnets or passing electric currents through them will cause a reduction in field strength. Explain.

29.7. The strength of a U magnet will be preserved much longer if an iron plate, called a *keeper*, is placed across the north and south poles. Explain.

29.8. A wire lying along a north–south direction supports an electric current from south to north. What happens to the needle of a compass if the compass is placed (a) above the wire, (b) below the wire, and (c) on the right side of the wire?

29.9. Use the right-hand-thumb rule and the right-hand rule for forces to explain why two adjacent wires experience a force of attraction when the currents are in the same direction. Illustrate your point by drawings.

29.10. Explain with the use of diagrams the repulsion of two adjacent wires carrying oppositely directed currents.

29.11. A circular coil in the plane of the paper supports an electric current. Determine the direction of the magnetic flux near the center of the coil when the current is counterclockwise.

29.12. When an electron beam is projected from left to right into a **B** field directed into the paper, the beam is deflected into a circular path. Do the electrons travel clockwise or counterclockwise? Why is the path circular? What if the beam consisted of protons?

29.13. A proton passes through a region of space without being deflected. Can we say positively that no magnetic field exists in that region? Discuss.

29.14. Many sensitive electrical instruments are shielded from magnetic effects by surrounding the device with ferromagnetic material. Explain.

29.15. If B is expressed in teslas and μ is in tesla-meters per ampere, what is the SI unit of H?

29.16. Hardened steel has a thick hysteresis loop, whereas soft iron has a thin loop. Which should be used to produce a permanent magnet? Which should be used if strong temporary magnetization is desired?

Problems

Section 29.1 Magnetism

29.1. The area of a rectangular loop is 200 cm², and the plane of the loop makes an angle of 41° with a 0.28-T magnetic field. What is the magnetic flux penetrating the loop?   Ans. 3.67×10^{-3} Wb

29.2. A coil of wire 30 cm in diameter is perpendicular to a 0.6-T magnetic field. If the coil turns so that it makes an angle of 60° with the field, what is the change in flux?

29.3. A constant horizontal field of 0.5 T pierces a rectangular loop 120 mm long and 70 mm wide. Determine the magnetic flux through the loop when its plane makes the following angles with the **B** field: 0°, 30°, 60°, and 90°.
     Ans. 0, 2.10 mWb, 3.64 mWb, 4.20 mWb

29.4. A flux of 13.6 mWb penetrates a coil of wire 240 mm in diameter. Find the magnitude of the magnetic flux density if the plane of the coil is perpendicular to the field.

29.5. A magnetic flux of 50 μWb passes through a perpendicular loop of wire having an area of 0.78 m². What is the magnetic flux density?
     Ans. 64.1 μT

29.6. A rectangular loop 25 × 15 cm is oriented so that its plane makes an angle θ with a 0.6-T **B** field. What is the angle θ if the magnetic flux linking the loop is 0.015 Wb?

Section 29.6 The Force on a Moving Charge

29.7. A proton ($q = +1.6 \times 10^{-19}$ C) is injected to the right into a **B** field of 0.4 T directed upward. If the velocity of the proton is 2×10^6 m/s, what are the magnitude and direction of the magnetic force on the proton? Ans. 1.28×10^{-13} N, into paper

29.8. An alpha particle ($+2e$) is projected with a velocity of 3.6×10^6 m/s into a 0.12-T magnetic field. What is the magnetic force on the charge at the instant its velocity is directed at an angle of $35°$ with the magnetic flux?

29.9. An electron moves with a velocity of 5×10^5 m/s at an angle of $60°$ with an eastward **B** field. The electron experiences a force of 3.2×10^{-18} N directed into the paper. What are the magnitude of **B** and direction of the velocity?
    Ans. 46.2 μT, v is 60° S of E

29.10. A proton ($+1e$) is moving vertically upward with a velocity of 4×10^6 m/s. It passes through a 0.4-T magnetic field directed to the right. What are the magnitude and direction of the magnetic force?

29.11. If an electron replaces the proton in Prob. 29.10, what will be the magnitude and direction of the magnetic force?
    Ans. 2.56×10^{-13} N, out of paper

*29.12. A particle having a charge q and a mass m is projected into a **B** field directed into the paper. If the particle has a velocity v, show that it will be deflected into a circular path of radius:

$$R = \frac{mv}{qB}$$

Draw a diagram of the motion, assuming a positive charge entering the **B** field from left to right. *Hint:* The magnetic force provides the necessary centripetal force for the circular motion.

*29.13. A deuteron is a nuclear particle consisting of a proton and a neutron bound together by nuclear forces. The mass of a deuteron is 3.347×10^{-27} kg, and its charge is $+1e$. A deuteron projected into a magnetic field of flux density 1.2 T is observed to travel in a circular path of radius 300 mm. What is the velocity of the deuteron? See Prob. 29.12.   Ans. 1.72×10^7 m/s

Section 29.7 Force on a Current-Carrying Wire

29.14. A wire 1 mm in length supports a current of 5.00 A and is perpendicular to a **B** field of 0.034 T. What is the magnetic force on the wire?

29.15. A long wire carries a current of 6 A in a direction $35°$ north of an easterly 0.04-T magnetic field. What are the magnitude and direction of the force on each centimeter of wire
    Ans. 1.38 mN, into paper

29.16. A 12-cm segment of wire carries a current of 4.0 A directed at an angle of $41°$ north of an easterly **B** field. What must be the magnitude of the **B** field if it is to produce a 5-N force on this segment of wire? What is the direction of the force?

29.17. An 80-mm segment of wire is at an angle of $53°$ south of a westward, 2.3-T **B** field. What are the magnitude and direction of the current in this wire if it experiences a force of 2 N directed out of the paper?   Ans. 13.6 A

*29.18. The linear density of a certain wire is 50.0 g/m. A segment of this wire carries a current of 30 A in a direction perpendicular to the **B** field. What must be the magnitude of the magnetic field required to suspend the wire by balancing its weight?

Section 29.9 Other Magnetic Fields

29.19. What is the magnetic induction **B** in air at a point 4 cm from a long wire carrying a current of 6 A?
    Ans. 30 μT

29.20. Find the magnetic induction in air 8 mm from a long wire carrying a current of 14.0 A.

29.21. A circular coil having 40 turns of wire in air has a radius of 6 cm and is in the plane of the paper. What current must exist in the coil to produce a flux density of 2 mT at its center? Ans. 4.77 A

29.22. If the direction of the current in the coil of Prob. 29.21 is clockwise, what is the direction of the magnetic field at the center of the loop?

29.23. A solenoid of length 30 cm and diameter 4 cm is closely wound with 400 turns of wire around a nonmagnetic material. If the current in the wire is 6 A, determine the magnetic induction along the center of the solenoid.   Ans. 10.1 mT

29.24. A circular coil having 60 turns has a radius of 75 mm. What current must exist in the coil to produce a flux density of 300 μT at the center of the coil?

*29.25. A circular loop 240 mm in diameter supports a current of 7.8 A. If it is submerged in a medium of relative permeability 2.0, what is the magnetic induction at the center?   Ans. 81.7 μT

*29.26. A circular loop of radius 50 mm in the plane of the paper carries a counterclockwise current of 15 A. It is submerged in a medium that has a relative permeability of 3.0. What are the magnitude and direction of the magnetic induction at the center of the loop?

Additional Problems

29.27. A $+3\text{-}\mu C$ charge is projected with a velocity of 5×10^5 m/s along the positive x axis perpendicular to a **B** field. If the charge experiences an upward force of 6.0×10^{-3} N, what must be the magnitude and direction of the **B** field?

Ans. 4.00 mT, into paper

29.28. An unknown charge is projected with a velocity of 4×10^5 m/s from right to left into a 0.4-T **B** field directed out of the paper. The perpendicular force of 5×10^{-3} N causes the particle to move in a clockwise circle. What are the magnitude and sign of the charge?

29.29. A -8-nC charge is projected upward at 4×10^5 m/s into a 0.60-T **B** field directed into the paper. The field produces a force ($F = qvB$) that is also a centripetal force (mv^2/R). This force causes the negative charge to move in a circle of radius 20 cm. What is the mass of the charge, and does it move clockwise or counterclockwise?

Ans. 2.40×10^{-15}, clockwise

29.30. What are the magnitude and direction of the **B** field 6 cm above a long segment of wire carrying a 9-A current directed out of the paper? What are the magnitude and direction of the **B** field 6 cm *below* the segment?

29.31. A 24-cm length of wire makes an angle of 32° above a horizontal **B** field of 0.44 T along the positive x axis. What are the magnitude and direction of the current required to produce a force of 4 mN directed out of the paper?

Ans. 71.5 mA, 212°

*29.32. A velocity selector is a device (Fig. 29.26) that utilizes crossed **E** and **B** fields to select ions of only one velocity v. Positive ions of charge q are projected into the perpendicular fields at varying speeds. Ions with velocities sufficient to make the magnetic force equal and opposite to the electric force pass through the bottom of the slit undeflected. Show that the speed of these ions can be found from

$$v = \frac{E}{B}$$

29.33. What is the velocity of protons ($+1e$) injected through a velocity selector (see Prob. 29.32) if $E = 3 \times 10^5$ V/m and $B = 0.25$ T?

Ans. 1.20×10^6 m/s

*29.34. A singly charged Li^7 ion ($+1e$) is accelerated through a potential difference of 500 V and then enters at right angles to a magnetic field of 0.4 T.

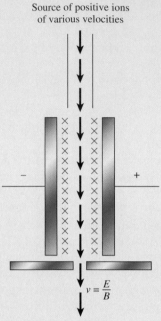

Source of positive ions of various velocities

$$v = \frac{E}{B}$$

Figure 29.26 The velocity selector.

The radius of the resulting circular path is 2.13 cm. What is the mass of the lithium ion?

*29.35. A singly charged sodium ion ($+1e$) moves through a **B** field with a velocity of 4×10^4 m/s. What must be the magnitude of the **B** field if the ion is to follow a circular path of radius 200 mm? (The mass of the ion is 3.818×10^{-27} kg.)

Ans. 4.77 mT

*29.36. The cross sections of two parallel wires are shown in Fig. 29.27 located 8 cm apart in air. The left wire carries a current of 6 A out of the paper, and the right wire carries a current of 4 A into the paper. What is the resultant magnetic induction

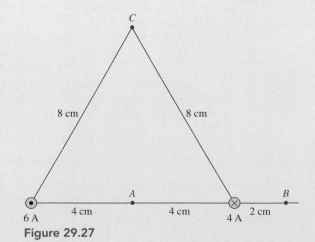

Figure 29.27

at the midpoint A due to both wires? (*Hint:* Superimpose the fields due to each wire to find the resultant field.)

*29.37. What is the resultant magnetic field at point B located 2 cm to the right of the 4-A wire?

Ans. 28.0 μT, downward

*29.38. Two parallel wires (see Fig. 29.28) carrying currents I_1 and I_2 are separated by a distance d. Show that the force per unit length F/L each wire exerts on the other is given by

$$\frac{F}{L} = \frac{\mu I_1 I_2}{2\pi d}$$

*29.39. Two wires lying in a horizontal plane carry parallel currents of 15 A each and are 200 mm apart in air. If both currents are directed to the right, what are the magnitude and direction of the flux density at a point midway between the wires? *Ans. 0*

*29.40. What is the force per unit length that each wire in Prob. 29.39 exerts on the other? Is it attraction or repulsion?

*29.41. A solenoid of length 20 cm and 220 turns carries a coil current of 5 A. What should be the relative permeability of the core to produce a magnetic induction of 0.2 T at the center of the coil?

Ans. 28.9

*29.42. A 1-m segment of wire is fixed so it cannot move, and it carries a current of 6 A directed east. Another 1-m wire segment is located 2 cm above the fixed wire. If the upper wire has a mass of 40 g, what must be the magnitude and direction of the current in the upper wire if its weight is to be balanced by the magnetic force due to the field of the fixed wire?

*29.43. What is the resultant magnetic field at point C in Fig. 29.27? *Ans. 21.8 μT, 96.6°*

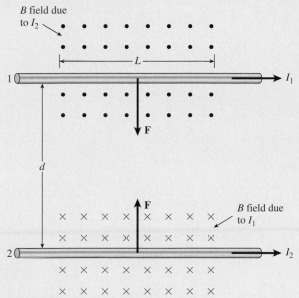

Figure 29.28 Two parallel wires carrying a steady current exert a force on each other.

Critical Thinking Questions

*29.44. A magnetic field of 0.4 T is directed into the paper. Three particles are injected into the field in an upward direction, each with a velocity of 5×10^5 m/s. Particle 1 is observed to move in a clockwise circle of radius 30 cm, particle 2 continues to travel in a straight line, and particle 3 is observed to move counterclockwise in a circle of radius 40 cm. What are the magnitude and sign of the charge per unit mass (q/m) for each of the particles?

Ans. -4.17 MC/kg, zero, $+3.12$ MC/kg

*29.45. A 4.0-A current flows through the circular coils of a solenoid in a counterclockwise direction as viewed along the positive x axis, which is aligned with the air core of the solenoid. What is the direction of the **B** field along the central axis? How many turns per meter of length are required to produce a **B** field of 0.28 T? If the air core is

replaced by a material that has a relative permeability of 150, what current would be needed to produce the same 0.28-T field as before?

*29.46. The plane of a current loop 50 cm long and 25 cm wide is parallel to a 0.3 T **B** field directed along the positive x axis. The 50-cm segments are parallel with the field, and the 25 cm segments are perpendicular to the field. When viewed from the top, the 6-A current is clockwise around the loop. Draw a sketch to show the directions of the **B** field and the directions of the currents in each wire segment. (a) What are the magnitude and direction of the magnetic force acting on each wire segment? (b) What is the resultant torque on the current loop? *Ans. (a) 0.450 N (up),*
-0.450 N (down); (b) 0.225 N m.,
counterclockwise about the $+z$ axis.

*29.47. Consider the two wires in Fig. 29.29, in which the dot indicates current out of the page and the cross indicates current into the page. What is the resultant flux density at points *A*, *B*, and *C?*

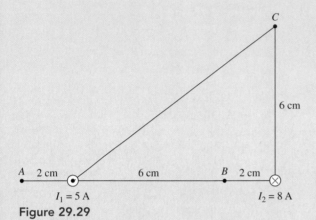

Figure 29.29

*29.48. Two long, fixed, parallel wires *A* and *B* are 10 cm apart in air and carry currents of 6 A and 4 A, respectively, in opposite directions. (a) Determine the net flux density at a point midway between the wires. (b) What is the magnetic force per unit length on a third wire placed midway between *A* and *B* and carrying a current of 2 A in the same direction as A?

Ans. (a) 40 μT; (b) 80 μN/m, toward A

30

Forces and Torques in a Magnetic Field

Airbus A-320 cockpit simulator: A large variety of moving-scale galvanometer gauges tell the pilots about many important physical measurements. These instruments are based on the principle that magnetic torque is proportional to electric current.
(*Photo © Taxi/Getty.*)

Objectives

After completing this chapter, you should be able to

1. Determine the direction of the magnetic force on a current-carrying conductor in a known *B* field.

2. Write and apply equations for calculating the magnetic torque on a coil or a solenoid of known area, number of turns, and current when located in a magnetic field of known flux density.

3. Explain with drawings the function of each part of a laboratory *galvanometer;* describe how it may be converted to an ammeter and to a voltmeter.

4. Calculate the *multiplier resistance* necessary to increase the range of a dc voltmeter that contains a galvanometer of fixed sensitivity.

5. Calculate the *shunt resistance* necessary to increase the range of a galvanometer or ammeter of constant sensitivity.

6. Explain the operation of a simple *dc motor,* discussing the function of each of its parts, with particular emphasis on the *split-ring commutator.*

We have seen that a current-carrying conductor placed in a magnetic field will experience a force that is perpendicular both to the current I and to the magnetic induction B. A coil suspended in a magnetic field will experience a *torque* due to equal and opposite magnetic forces on the sides of the coil. Such forces and torques provide the operating principle of many useful devices. In this chapter, we discuss the galvanometer, the voltmeter, the ammeter, and the dc motor as applications of electromagnetic forces.

30.1   Force and Torque on a Loop

A current-carrying conductor suspended in a magnetic field, as illustrated in Fig. 30.1, will experience a magnetic force given by

$$F = BIL \sin \theta = BI_\perp L \qquad (30.1)$$

where $I_\perp$ refers to the current perpendicular to the **B** field and L is the length of the conductor. The direction of the force is determined from the right-hand rule.

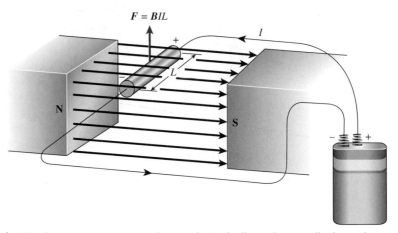

Figure 30.1   The force on a current-carrying conductor is directed perpendicular to the magnetic field, with its direction given by the right-hand rule.

Now let us examine the forces acting on a rectangular current-carrying loop suspended in a magnetic field, shown in Fig. 30.2. The lengths of the sides are a and b, and a current I passes around the loop, as indicated. (The source of emf and method of leading current into the loop are not shown for simplicity.) Sides mn and op of the loop are each of length a and perpendicular to the magnetic induction **B.** Thus, there are exerted on the sides equal and opposite forces of magnitude.

$$F = BIa \qquad (30.2)$$

The force is directed upward for the segment mn and downward for the segment op.

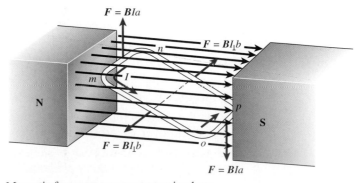

Figure 30.2   Magnetic forces on a current-carrying loop.

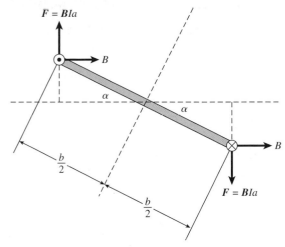

Figure 30.3   Calculating the torque on a current loop.

Similar reasoning will show that equal and opposite forces are also exerted on the other two sides. These forces have a magnitude of

$$F = BIb \sin \alpha$$

where α is the angle that the sides np and mo make with the magnetic field.

Evidently, the loop is in translation equilibrium since the resultant force on the loop is zero. However, the nonconcurrent forces on the sides of length a produce a resultant torque that tends to rotate the coil clockwise. As can be seen from Fig. 30.3, each force produces a torque equal to

$$\tau = BIa\frac{b}{2} \cos \alpha$$

Because the total torque is equal to twice this value, the resultant torque can be found from

$$\tau = BI(a \times b) \cos \alpha \qquad \textbf{(30.3)}$$

Since $a \times b$ is the area A of the loop, Eq. (30.3) can be written

$$\tau = BIA \cos \alpha \qquad \textbf{(30.4)}$$

Note that the torque is a maximum when $\alpha = 0°$, that is, when the plane of the loop is parallel with the magnetic field. As the coil turns about its axis, the angle α increases, reducing the rotational effect of the magnetic forces. When the plane of the loop is perpendicular to the field, the angle $\alpha = 90°$, and the resultant torque is zero. The momentum of the coil will cause it to pass this point slightly, but the direction of the magnetic forces ensures that it will oscillate until it reaches equilibrium with the plane of the loop perpendicular to the field.

If the loop is replaced with a closely wound coil having N turns of wire, the general equation for computing the resultant torque is

$$\tau = NIBA \cos \alpha \qquad \textbf{(30.5)}$$

This equation applies to any complete circuit of area A, and its use need not be restricted to rectangular loops. Any plane loop obeys the same relationships.

Example 30.1

A rectangular coil consisting of 100 turns of wire has a width of 16 cm and a length of 20 cm. The coil is mounted in a uniform magnetic field of flux density 8 mT, and a current of 20 A is sent through the windings. When the coil makes an angle of 30° with the magnetic field, what is the torque tending to rotate the coil?

Solution: Substituting in Eq. (30.5), we obtain

$$\tau = NBIA \cos \alpha$$
$$= (100 \text{ turns})(8 \times 10^{-3} \text{ T})(20 \text{ A})(0.16 \text{ m} \times 0.20 \text{ m})(\cos 30°)$$
$$= 0.443 \text{ N} \cdot \text{m}$$

30.2   Magnetic Torque on a Solenoid

The relationship expressed by Eq. (30.5) applies for computing the torque on a solenoid of area A having N turns of wire. In applying the relation, however, we must remember that the angle α is the angle that each turn of wire makes with the field. It is the *complement* of the angle θ between the solenoid axis and the magnetic field (see Fig. 30.4). An alternative equation for computing the torque on a solenoid is therefore

$$\tau = NBIA \sin \theta \qquad\qquad \textit{Solenoid} \qquad \textbf{(30.6)}$$

You should verify that $\sin \theta$ is equal to $\cos \alpha$ by looking at the figure.

The action of the solenoid in Fig. 30.4 can also be explained in terms of magnetic poles. Applying the right-hand-thumb rule to each turn of wire shows that the solenoid will act as an electromagnet, with north and south poles as indicated.

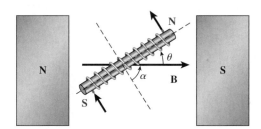

Figure 30.4   Magnetic torque on a solenoid. The angle α is the angle that each turn of wire makes with the **B** field. The angle θ is the angle between the solenoid axis and the **B** field. Remember that $\theta + \alpha = 90°$.

30.3   The Galvanometer

A *galvanometer* is a laboratory device that can be used to measure electric current. It works on the principle that **magnetic torque** is proportional to the current. The essential elements are shown in Fig. 30.5a. A coil of wire, wrapped around a soft iron core, is pivoted on jeweled bearings between the poles of a permanent magnet. Its rotational motion is restrained by a pair of spiral springs, which also serve as current leads to the coil. Depending on the direction of the current being measured, the coil and pointer will rotate either in a clockwise or counterclockwise direction. The permanent magnets are shaped to provide a uniform, *radial field* so that the torque will be directly proportional to the current in the coil. The *sensitivity* of a galvanometer is determined by the spring torque, the friction of the bearings, and the strength of the magnetic field. A typical sensitivity might be 50 μA per scale division. Typically, the zero position on the scale is positioned at the center, as shown in Fig. 30.5b.

The galvanometer can be used as a dc voltmeter and as a dc ammeter in some applications, and they are excellent for studying laboratory circuits that reinforce our understanding of the relationships for current, voltage, and resistance. Sections 30.4 and 30.5 provide a detailed understanding of such applications, along with a discussion of how one can change the range of these instruments.

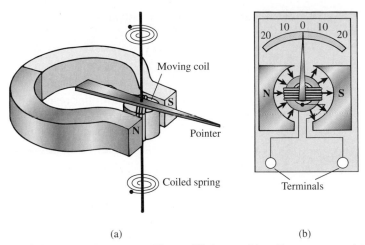

Figure 30.5   The laboratory galvanometer. The equilibrium position (for zero current) is such that the pointer is centrally located so deflection can occur either way.

However, it should be noted that digital devices are much more accurate and less intrusive than the older analog devices. There are several advantages to these instruments. Digital meters contain a battery to power the display, so they draw negligible current from the circuit as compared to galvanometer-based devices. When used to measure voltage, the digital device can have an enormous resistance, usually on the order of 10 MΩ, meaning that the circuit is not altered significantly. Additionally, the problem of polarity that occurs when a galvanometer pegs the wrong way is eliminated. The digital device merely indicates a negative value for one polarity and a positive value for the other polarity.

30.4 The DC Voltmeter

A *voltmeter* is an instrument that measures potential difference between two points in an electric circuit. In this section, we will show how an analog galvanometer can be used for this purpose. The potential difference across a galvanometer is very small, even when a large-scale deflection occurs. Thus, if a galvanometer is to measure voltage, it must be converted to a high-resistance instrument. Suppose, for example, one wants to measure the voltage drop across the battery in Fig. 30.6. This voltage must be measured without appreciably disturbing the current through the circuit. In other words, the voltmeter must draw negligible current. To accomplish this, a large *multiplier resistance* R_m is placed in series with the galvanometer as an integral part of a dc voltmeter.

Note that the galvanometer used in a voltmeter is adjusted so that its equilibrium position is to the extreme left on the scale. This allows for a greater range of measurement but, unfortunately, requires that the current pass through the coils in one direction only. The *sensitivity* of the galvanometer is determined by the current I_g required for *full-scale deflection* (maximum pointer deflection), as indicated in Fig. 30.6.

Suppose the galvanometer coil has a resistance R_g and the meter is designed to yield full-scale deflection for the current I_g. Such a galvanometer, acting alone, could be calibrated to record voltages from zero up to a maximum value given by

$$V_g = I_g R_g \tag{30.7}$$

By properly choosing the multiplier resistance R_m, we can calibrate the meter to read any desired voltage.

Suppose, for example, we want full-scale deflection of the galvanometer for the voltage V_B in Fig. 30.6. The multiplier resistance R_m must be chosen so that only the small current I_g passes through the galvanometer. Under these conditions,

$$V_B = I_g R_g + I_g R_m$$

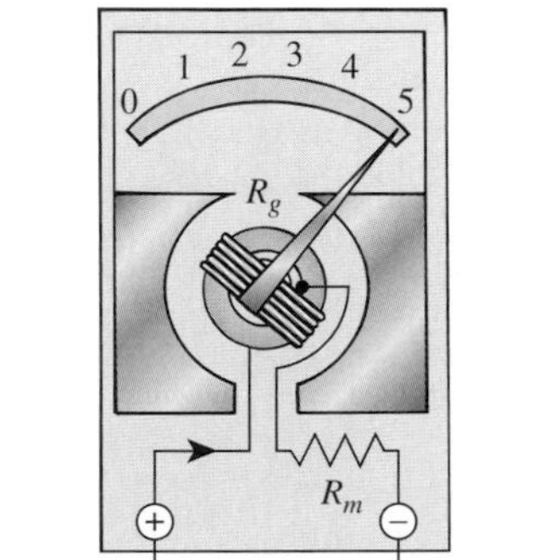

Figure 30.6   The dc voltmeter.

Solving for R_m, we obtain

$$R_m = \frac{V_B}{I_g} - R_g \qquad (30.8)$$

Thus, we see that the multiplier resistance R_m is equal to the total resistance of the device V_B/I_g less the galvanometer resistance R_g.

Example 30.2

A certain galvanometer has an internal resistance of 30 Ω and gives full-scale deflection for a current of 1 mA. Calculate the multiplier resistance necessary to convert this galvanometer into a voltmeter that has a maximum range of 50 V.

Plan: The multiplier resistance R_m must be such that the total drop in voltage through R_g and R_m is 50 V.

Solution: From Eq. (30.8), we obtain

$$R_m = \frac{50\ V}{1 \times 10^{-3}\ A} - 30\ \Omega = 49{,}970\ \Omega$$

Note the *total resistance* of the voltmeter ($R_m + R_g$) is 50 kΩ.

A voltmeter must be connected in *parallel* with the part of the circuit whose potential difference is to be measured. This is necessary so that the large resistance of the voltmeter will not greatly alter the circuit.

30.5 The DC Ammeter

An **ammeter** is a device that, through calibrated scales, gives an indication of the electric current without appreciably altering it. A *galvanometer* is an ammeter, but its range is limited by the extreme sensitivity of the moving coil. The range of the galvanometer can be extended simply by adding a low resistance, called a **shunt,** in parallel with the galvanometer coil (see Fig. 30.7). Placing the shunt in parallel ensures that the ammeter as a whole will have a low resistance, which is necessary if the current is to be essentially unaltered. The major portion of the current will pass through the shunt. Only the small current I_g required for galvanometer deflection will be drawn from the circuit. For example, if 10 A goes through an ammeter, 9.99 A may go through the shunt and only 0.01 A through the coil itself.

Suppose the range of a galvanometer is to be extended to measure a maximum current I of the circuit in Fig. 30.7. A **shunt resistance** R_s must be chosen such that only the current I_g, required for full-scale deflection, passes through the galvanometer coil. The remainder of the current I_s must pass through the shunt. Since R_g and R_s are in parallel, the IR drop across each resistance must be identical:

$$I_s R_s = I_g R_g \qquad (30.9)$$

The shunt current I_s is the difference between the circuit current I and the galvanometer current I_g. Thus, Eq. (30.9) becomes

$$(I - I_g)R_s = I_g R_g$$

Solving for the shunt resistance R_s, we obtain the following useful relation:

$$R_s = \frac{I_g R_g}{I - I_g} \qquad (30.10)$$

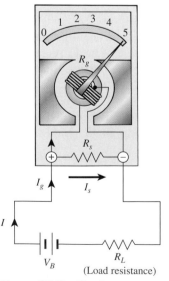

Figure 30.7 The dc ammeter.

Example 30.3 A certain galvanometer has an internal coil resistance of 46 Ω, and a current of 200 mA is required for full-scale deflection. What shunt resistance must be used to convert the galvanometer into an ammeter whose maximum range is 10 A?

Solution: Equation (30.10) gives

$$R_s = \frac{(0.2 \text{ A})(46 \text{ }\Omega)}{10 \text{ A} - 0.2 \text{ A}} = \frac{9.2 \text{ V}}{9.8 \text{ A}} = 0.939 \text{ }\Omega$$

It is important to remember that an ammeter must be connected in *series* with that portion of a circuit through which the current is to be measured. The circuit must be opened at some convenient point and the ammeter inserted. If, by mistake, the ammeter were placed in parallel, the circuit would be shorted across the ammeter because of its extremely low resistance.

30.6   The DC Motor

An *electric motor* is a device that transforms electric energy into mechanical energy. The *dc motor,* like the moving coil of a galvanometer, consists of a current-carrying coil in a magnetic field. However, the motion of the coil in a motor is unrestrained by springs. The design is that the coil will rotate continuously under the influence of magnetic torque.

A simple dc motor, consisting of a single current-carrying loop suspended between two magnetic poles, is illustrated in Fig. 30.8. Normally, the torque exerted on a current-carrying loop would diminish to zero when its plane becomes perpendicular to the field. To provide for continuous rotation of the loop, the current in the loop must be reversed automatically each time it turns through 180°.

The current reversal is accomplished by using a *split-ring commutator,* as shown in Fig. 30.8. The commutator consists of two metal half rings fused to each end of the conducting loop and insulated from each other. As the loop rotates, each brush touches first one half ring and then the other. Thus, the electrical connections are reversed every half revolution at times when the loop is perpendicular to the magnetic field. In this manner, the torque acting on the loop is always in the same direction, and the loop will rotate continuously.

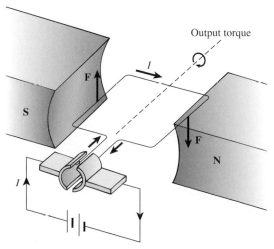

Figure 30.8 The dc motor.

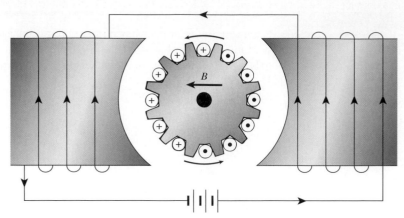

Figure 30.9 Greater, more uniform torque is possible in commercial motors with many armature coils.

Although actual dc motors operate on the principle described for Fig. 30.8, there are a number of designs that increase the available torque and make it more uniform. One such design is shown in Fig. 30.9. A greater magnetic field is established by replacing the permanent magnets with electromagnets. Additionally, the torque can be increased and made more uniform by adding a number of different coils, each having a large number of turns around a slotted iron core called the **_armature._** The commutator is an automatic switching arrangement that maintains the currents in the directions shown in the figure, regardless of the orientation of the armature. More will be said about the dc motor in Chapter 31.

Summary and Review

Summary

Magnetic torques on current loops form the basis for so many applications that a strong foundation is essential. The operation of generators, motors, ammeters, voltmeters, and many industrial instruments is directly affected by magnetic forces and torques. The major concepts to be remembered are summarized as follows.

- The magnetic torque on a current-carrying coil of wire having N turns is given by

$$\tau = NBIA \cos \alpha \qquad \textit{Magnetic Torque}$$

where N = number of turns of wire
B = flux density, T
I = current, A
A = area of the coil of wire, m^2
α = angle plane of coil makes with field

- The same equation applies for a solenoid, except that the angle α is generally replaced with θ, the angle the solenoid axis makes with the field.

$$\tau = NBIA \sin \theta \qquad \textit{Torque on a Solenoid}$$

- The multiplier resistance R_m that must be placed in series with a voltmeter to give full-scale deflection for V_B is found from

$$R_m = \frac{V_B}{I_g} - R_g \qquad \textit{Multiplier Resistance}$$

I_g is the galvanometer current, and R_g is its resistance.
- The *shunt resistance* R_s that must be placed in parallel with an ammeter to give full-scale deflection for a current I is

$$R_s = \frac{I_g R_g}{I - I_g} \qquad \textit{Shunt Resistance}$$

Key Terms

ammeter   594
armature   596
dc motor   595
electric motor   595
full-scale deflection   593

galvanometer   592
magnetic torque   592
multiplier resistance   593
sensitivity   592
shunt   594

shunt resistance   594
split-ring commutator   595
voltmeter   593

Review Questions

30.1. The equal and opposite forces acting on a current loop in a magnetic field form what is called a *couple*. Show that the resultant torque on such a couple is the product of one of the forces and the perpendicular distance between their lines of action.

30.2. Why is it necessary to provide a *radial* magnetic field for the coil of a galvanometer?

30.3. A coil of wire is suspended by a thread with the plane of the loop coinciding with the plane of the paper. If the coil is placed in a magnetic field directed from left to right and if a clockwise current is sent through the coil, describe its motion.

30.4. How are the actions of a galvanometer and a motor similar? How are they different?

30.5. How does the core of a galvanometer coil affect the sensitivity of the instrument?

30.6. Explain the torque exerted on a bar magnet suspended in a magnetic field without referring to magnetic poles. Discuss, from an atomic standpoint, how the observed torque may arise from the same cause as the torque on a current loop.

30.7. Suppose the range of an ammeter is to be increased N-fold. Show that the shunt resistance that must be placed across the terminals of the ammeter is given by

$$R_s = \frac{R_a}{N - 1}$$

where R_a is the ammeter resistance.

30.8. Suppose the range of a given voltmeter is to be increased N-fold. Show that the multiplier resistance that must be placed in series with the voltmeter is given by

$$R_m = (N - 1)R_v$$

where R_v is the voltmeter resistance.

30.9. Show by diagrams how an ammeter and a voltmeter should be connected in a circuit. Compare the resistances of the two devices.

30.10. Discuss the error caused by the insertion of an ammeter into an electric circuit. How is this error minimized?

30.11. Discuss the error caused by the insertion of a voltmeter into a circuit. How is the error minimized?

30.12. A voltmeter is connected to a battery, and the reading is taken. An accurate resistance box is then placed in the circuit and adjusted until the voltmeter reading is one-half of its previous value. Show that the voltmeter resistance must be just equal to the added resistance. (This is called the *half-deflection method* for determining voltmeter resistance.)

30.13. Explain what happens when a voltmeter is erroneously placed in series in a circuit. What happens if an ammeter is mistakenly placed in parallel?

30.14. Prepare a short paper on the following topics:

a. ballistic galvanometer

b. ohmmeter

c. dynamometer

d. voltmeter

30.15. Plot a graph of the torque as a function of time for a single-loop dc motor.

Problems

Section 30.1 Force and Torque on a Loop

30.1. A rectangular loop of wire has an area of 30 cm^2 and is placed with its plane parallel to a 0.56-T magnetic field. What is the magnitude of the resultant torque if the loop carries a current of 15 A?   Ans. 0.0252 N · m

30.2. A coil of wire has 100 turns, each of area 20 cm^2. The coil is free to turn in a 4.0-T field. What current is required to produce a maximum torque of 2.30 N · m?

30.3. A rectangular loop of wire 6 cm wide and 10 cm long is placed with its plane parallel to a magnetic field of 0.08 T. What is the magnitude of the resultant torque on the loop if it carries a current of 14.0 A?   Ans. 6.72 × 10^{-3} N · m

30.4. A rectangular loop of wire has an area of 0.30 m^2. The plane of the loop makes an angle of 30° with a 0.75-T magnetic field. What is the torque on the loop if the current is 7.0 A?

30.5. Calculate the magnetic flux density required to give a coil of 100 turns a torque of 0.5 N · m when its plane is parallel to the field. The dimension of each turn is 84 cm^2, and the current is 9.0 A.   Ans. 66.1 mT

30.6. What current is required to produce a maximum torque of 0.8 N · m on a solenoid having 800 turns of area 0.4 m^2? The flux density is 3.0 mT. What is the position of the solenoid in the field?

30.7. The axis of a solenoid that has 750 turns of wire makes an angle of 34° with a 5-mT field. What is the current if the torque is 4.0 N · m at that angle? The area of each turn of wire is 0.25 m^2.   Ans. 7.63 A

Section 30.3 The Galvanometer, Section 30.4 The DC Voltmeter, and Section 30.5 The DC Ammeter

30.8. A galvanometer coil 50 mm × 120 mm is mounted in a constant radial 0.2-T magnetic field. If the coil has 600 turns, what current is required to develop a torque of 3.6 × 10^{-5} N m?

30.9. A galvanometer has a sensitivity of 20 μA per scale division. What current is required to give full-scale deflection with 25 divisions on either side of the central equilibrium position?   Ans. 500 μA

30.10. A galvanometer has a sensitivity of 15 μA per scale division. How many scale divisions will be covered by the deflecting needle when the current is 60 μA?

30.11. A certain voltmeter draws 0.02 mA for full-scale deflection at 50 V. (a) What is the resistance of the voltmeter? (b) What is the resistance per volt?   Ans. 2.50 MΩ, 50 kΩ/V

*30.12. For the voltmeter in Prob. 30.11, what multiplier resistance must be used to convert this voltmeter to an instrument that reads 150 V full scale?

*30.13. The coil of a galvanometer will burn out if a current of more than 40 mA is sent through it. If the coil resistance is 0.5 Ω, what shunt resistance should be added to permit the measurement of 4.00 A? **Ans. 5.05 mΩ**

*30.14. A current of only 90 μA will produce full-scale deflection of a voltmeter that is designed to read 50 mV full scale. (a) What is the resistance of the meter? (b) What multiplier resistance is required to permit the measurement of 100 mV full scale?

*30.15. An ammeter that has a resistance of 0.10 Ω is connected in a circuit and indicates a current of 10 A at full scale. A shunt having a resistance of 0.01 Ω is then connected across the terminals of the meter. What new circuit current is needed to produce full-scale deflection of the ammeter? **Ans. 110 A**

Additional Problems

30.16. A wire of length 12 cm is made into a loop and placed into a 3.0-T magnetic field. What is the largest torque the wire can experience for a current of 6 A?

30.17. A solenoid has 600 turns of area 20 cm^2 and carries a current of 3.8 A. The axis of the solenoid makes an angle of 30° with an unknown magnetic field. If the resultant torque on the solenoid is 1.25 N $\cdot$ m, what is the magnetic flux density? **Ans. 548 mT**

30.18. A flat coil of wire has 150 turns of radius 4.0 cm, and the current is 5.0 A. What angle does the plane of the coil make with a 1.2-T magnetic field if the resultant torque is 0.60 N $\cdot$ m?

30.19. A circular loop consisting of 500 turns carries a current of 10 A in a 0.25-T magnetic field. The area of each turn is 0.2 m^2. Calculate the torque when the loop makes the following angles with the field: 0°, 30°, 45°, 60°, and 90°. **Ans. 250, 217, 177, 125, and 0 N $\cdot$ m**

30.20. A solenoid of 100 turns has a cross-sectional area of 0.25 m^2 and carries a current of 10 A. What torque is required to hold the solenoid at an angle of 30° with a 40-mT magnetic field?

30.21. A solenoid consists of 400 turns of wire, each of radius 60 mm. What angle does the axis of the solenoid make with the magnetic flux if the current through the wire is 6 A, the flux density is 46 mT, and the resulting torque is 0.80 N m? **Ans. 39.8°**

*30.22. A galvanometer having an internal resistance of 35 Ω requires 1.0 mA for full-scale deflection.

What multiplier resistance is needed to convert this device to a voltmeter that reads a maximum of 30 V?

*30.23. The internal resistance of a galvanometer is 20 Ω, and it reads full scale for a current of 10 mA. Calculate the multiplier resistance required to convert this galvanometer into a voltmeter with a range of 50 V. What is the total resistance of the resulting meter? **Ans. 4980 Ω, 5000 Ω**

*30.24. What shunt resistance is needed to convert the galvanometer of Prob. 30.22 to an ammeter reading 10 mA full scale?

*30.25. A certain voltmeter reads 150 V full scale. The galvanometer coil has a resistance of 50 Ω and produces a full-scale deflection on 20 mV. Find the multiplier resistance of the voltmeter. **Ans. 374,950 Ω**

*30.26. A galvanometer has a coil resistance of 50 Ω and a current sensitivity of 1 mA (full scale). What shunt resistance is needed to convert this galvanometer to an ammeter reading 2.0 A full scale?

*30.27. A laboratory ammeter has a resistance of 0.01 Ω and reads 5.0 A full scale. What shunt resistance is needed to increase the range of the ammeter tenfold? **Ans. 0.00111 Ω**

*30.28. A commercial 3-V voltmeter requires a current of 0.02 mA to produce full-scale deflection. How can it be converted to an instrument with a range of 150 V?

Critical Thinking Questions

30.29. Consider the rectangular, 4- by 6-cm loop of 60 turns of wire, in Fig. 30.10. The plane of the loop makes an angle of 40° with a 1.6-T magnetic field **B** directed along the x axis. The loop is free to rotate about the y axis, and the clockwise current in the coil is 6.0 A. What is the torque, and will it turn toward the x axis or toward the z axis? **Ans. 1.06 N $\cdot$ m, toward z axis**

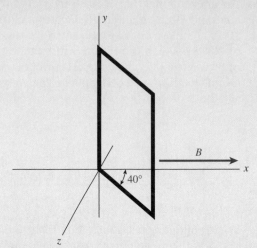

Figure 30.10

B field, show that the torque has a magnitude given by:

$$\tau = \mu B \sin \theta$$

This is sometimes written as a vector (cross) product $\boldsymbol{\tau} = \boldsymbol{\mu} \times \mathbf{B}$.

*30.32. Verify the answer obtained for Prob. 30.19 by applying the formula derived in the previous question. Do not confuse the angle θ with the angle α that the plane of the loop makes with the **B** field.

*30.33. The internal wiring of a three-scale voltmeter is shown in Fig. 30.11. The galvanometer has an internal resistance of 40 Ω, and a current of 1.00 mA will produce full-scale deflection. Find the resistances R_1, R_2, and R_3 for the use of the voltmeter for 10, 50, and 100 V.

Ans. 9960 Ω, 49,960 Ω, 99,960 Ω

*30.30. A voltmeter of range 150 V and total resistance 15,000 Ω is connected in series with another voltmeter of range 100 V and total resistance 20,000 Ω. What will each meter read when they are connected across a 120-V battery of negligible internal resistance?

*30.31. The magnetic moment μ is a vector quantity whose magnitude for a coil of N turns of area A is given by NIA, where I is the current in the coil. The direction of the magnetic moment is perpendicular to the plane of the coil in the direction given by the right-hand-thumb rule (see Fig. 29.20). If such a coil is placed into a uniform

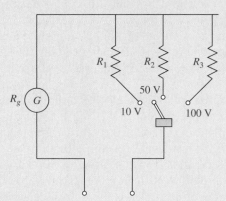

Figure 30.11

31 Electromagnetic Induction

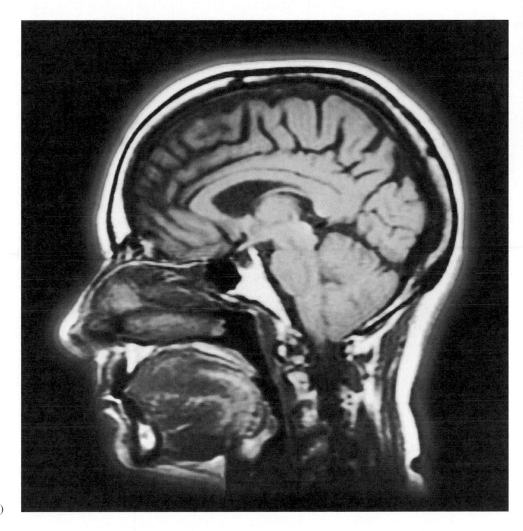

Magnetic Resonance Imaging (MRI) combines strong magnetic fields with radio frequency RF pulses to form a modern medical tool that helps in diagnosing multiple sclerosis, tumors, and infections in the brain, spine, or joints. Magnetic induction, discussed in this chapter, plays a large role in producing the central magnetic field in the core of this device. (*Photo © vol. 54 PhotoDisc/Getty.*)

Objectives

After completing this chapter, you should be able to

1. Explain and calculate the current or emf induced by a conductor moving through a magnetic field.

2. Write and apply an equation relating the induced emf in a length of wire moving with a velocity *v* directed at an angle *θ* with a known *B* field.

3. State Lenz's law and use it or the *right-hand rule* to determine the direction of induced emf or current.

4. Explain the operation of simple ac and dc generators; calculate the instantaneous and maximum emf or current generated by a simple ac generator.

5. Demonstrate with diagrams your knowledge of *series-wound* and *shunt-wound* motors and solve the starting current and operating voltage in electrical problems.

6. Explain the operation of a *transformer* and solve problems involving change in current, voltage, or power.

We have seen that an electric field can produce a magnetic field. In this chapter, you will learn that the reverse is also true: a magnetic field can give rise to an electric field. An electric current is *generated* by a conductor that moves relative to a magnetic field. A rotating coil in a magnetic field *induces* an alternating emf, which produces an *alternating current* (ac). This process is called **electromagnetic induction,** and it is the operating principle behind many electrical devices. For example, electric ac generators and transformers use electromagnetic induction to produce and distribute electric power economically.

| 31.1 | Faraday's Law |

Faraday discovered that when magnetic flux lines are cut by a conductor, an emf is produced between the end points of the conductor. For example, an electric current is induced in the conductor of Fig. 31.1a as it moved downward across the flux lines. (The lowercase symbol i will be used for induced currents and for varying currents.) The faster the movement, the more pronounced the galvanometer deflection. When the conductor is moved upward across the flux lines, a similar observation is made, except that the current is reversed (see Fig. 31.1b). If no flux lines are crossed—for example, the conductor is moved parallel to the field—no current is induced.

Suppose a number of conductors are moved through a magnetic field, as illustrated by dropping a coil of N turns across the flux lines in Fig. 31.2. The magnitude of the induced current is directly proportional to the number of coils and to the rate of motion. Evidently, *an induced emf is produced by the relative motion between the conductor and the magnetic field.* The same effect is observed when the coil is held stationary and the magnet is moved upward.

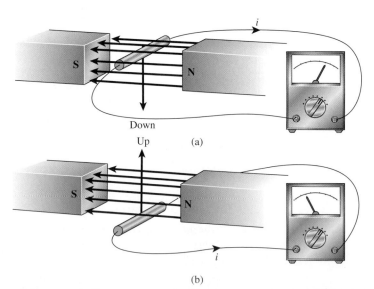

Figure 31.1 When magnetic flux lines are cut by a conductor, an electric current is induced.

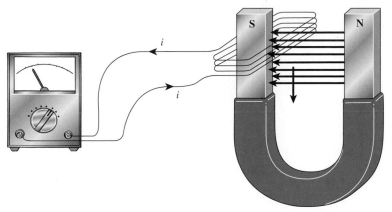

Figure 31.2   Induced emf in a coil is proportional to the number of turns of wire passing through the field.

Summarizing what we have learned from these experiments, we can state that

1. Relative motion between a conductor and a magnetic field induces an emf in the conductor.

2. The direction of the **induced emf** depends on the direction of motion of the conductor with respect to the field.

3. The magnitude of the emf is directly proportional to the rate at which magnetic flux lines are cut by the conductor.

4. The magnitude of the emf is directly proportional to the number of turns of the conductor crossing the flux lines.

A quantitative relationship for computing the induced emf in a coil of N turns is given by

$$\mathscr{E} = -N\frac{\Delta\Phi}{\Delta t} \tag{31.1}$$

where $\mathscr{E}$ = average induced emf

$\Delta\Phi$ = change in magnetic flux occurring during time interval Δt

A magnetic flux changing at the rate of 1 weber per second will induce an emf of 1 volt for each turn of the conductor. The negative sign in Eq. (31.1) means the induced emf is in such a direction as to oppose the change that produced it, as will be explained in Section 31.3.

Now let us discuss how magnetic flux Φ linking a conductor may change. In the simple case of a straight wire moving through lines of flux, $\Delta\Phi/\Delta t$ represents the rate at which the flux linked by the conductor changes. A continuous circuit is necessary for an induced current to exist, however, and more often we are interested in the emf induced in a loop or coil of wire.

Recall that the magnetic flux Φ passing through a loop of effective area A is given by

$$\Phi = BA \tag{31.2}$$

where B is the magnetic flux density. When B is in *teslas* (*webers per square meter*) and A is in *square meters*, Φ is expressed in webers.

A *change in flux* Φ can occur in two principle ways:

1. By changing the flux density **B** going through a constant loop area A:

$$\Delta\Phi = (\Delta B)A \tag{31.3}$$

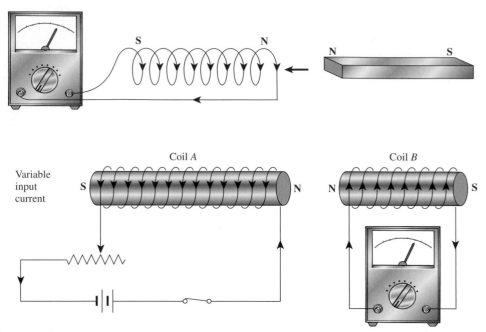

Figure 31.3   (a) Inducing a current by moving a magnet into a coil. (b) A changing current in coil *A* induces a current in coil *B*.

2. By changing the effective area *A* in a magnetic field of constant flux density **B:**

$$\Delta\Phi = B(\Delta A) \tag{31.4}$$

Two examples of changing flux density through a constant, stationary coil area are given in Fig. 31.3. In Fig. 31.3a, the north pole of a magnet is moved through a circular coil. The changing flux density induces a current in the coil, as indicated by the galvanometer. In Fig. 31.3b, no current is induced in coil *B* as long as the current in coil *A* is constant. By quickly varying the resistance in the left circuit, however, the magnetic flux density reaching coil *B* can be increased or decreased. While the flux density is changing, a current is induced in the coil on the right.

Note that when the north (N) pole of the magnet is moved into the coil in Fig. 31.3a, the current flows in a clockwise direction as viewed toward the magnet. Therefore, the end of the *coil* near the N pole of the magnet becomes an N pole also (from the right-hand-thumb rule of Chapter 30). The magnet and the coil will experience a force of repulsion, making it necessary to exert a force to bring them together. If the magnet is removed from the coil, a force of attraction will exist that makes it necessary to exert a force to separate them. We will see in Section 31.3 that such forces are a natural consequence of the conservation of energy.

Example 31.1   A coil of wire having an area of 2×10^{-3} m² is placed in a region of constant flux density equal to 0.65 T. In a time interval of 0.003 s, the flux density is increased to 1.4 T. If the coil consists of 20 turns of wire, what is the induced emf?

Plan: In this case, the area the flux penetrates does not change, and the entire induced emf will be produced by a changing **B** field. Recognizing that the change in flux is the product of the area and the change in **B,** we can find the change in flux and use it to determine the induced emf from Faraday's law.

Solution: First, we will find the change in flux.

$$\Delta\Phi = (\Delta B)A = (B_f - B_0)A$$
$$= (1.4\ \text{T} - 0.65\ \text{T})(2 \times 10^{-3}\ \text{m}^2)$$
$$= 1.50 \times 10^{-3}\ \text{Wb}$$

To find the induced emf, we substitute this change into Eq. (31.1).

$$\mathscr{E} = -N\frac{\Delta\Phi}{\Delta t} = \frac{-N\,\Delta\Phi}{\Delta t}$$
$$= \frac{-(20\ \text{turns})(1.5 \times 10^{-3}\ \text{Wb})}{0.003\ \text{s}} = -10\ \text{V}$$

The negative emf indicates opposition to the *increasing* flux.

The second general way in which the flux linking a conductor may change is by varying the effective area penetrated by the flux. Example 31.2 illustrates this point.

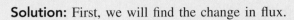

Example 31.2 A square coil 20 cm on a side has 16 turns of wire and is placed perpendicular to a **B** field of flux density 0.8 T. What is the average induced emf if the coil is flipped until its plane is parallel to the field in a time of 0.2 s?

Plan: We will calculate the area of the coil and note that the 0.8-T **B** field remains constant. The change in flux is the product of the change in area (from its original value to zero) and the constant **B** field. The emf can then be found as before.

Solution: The area of the square loop is the square of any side. Thus,

$$A = (0.2\ \text{m})^2 = 0.04\ \text{m}^2$$

This time, the change in flux is due to the changing area.

$$\Delta\Phi = B(\Delta A) = B(A_f - A_0)$$
$$= (0 - 0.04\ \text{m}^2)(0.8\ \text{T})$$
$$= -0.032\ \text{Wb}$$

The negative sign indicates the flux was *decreasing*. The induced emf is

$$\mathscr{E} = -N\frac{\Delta\Phi}{\Delta t} = \frac{-N\,\Delta\Phi}{\Delta t}$$
$$= \frac{-(16\ \text{turns})(-0.032\ \text{Wb})}{0.2\ \text{s}} = 2.56\ \text{V}$$

Note that a *decreasing* flux has resulted in a *positive* emf. This is necessary to conserve energy, as we will see later.

31.2 EMF Induced by a Moving Wire

Another example of a changing area in a constant **B** field is illustrated in Fig. 31.4. Imagine that a moving conductor of length L slides along a stationary U-shaped conductor with a velocity v. The magnetic flux penetrating the loop increases as the area of the loop

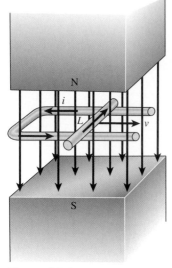

Figure 31.4 The emf induced in a wire of length L moving with a velocity v perpendicular to a magnetic field **B**.

increases. Consequently, an emf is induced in the moving wire, and a current passes around the loop.

The origin of the emf can be understood by recalling that a moving charge in a magnetic field experiences a force given by

$$F = qvB$$

For example, in Fig. 31.4, free charges on the conductor are moved to the right through a magnetic field directed downward. The magnetic force **F** acting on the charges moves them through the length L of wire in a direction given by the right-hand rule (away from the reader for conventional current). The work per unit of charge represents the induced emf, which is given by

$$\mathcal{E} = \frac{\text{work}}{q} = \frac{FL}{q} = \frac{qvBL}{q}$$
$$= BLv \qquad (31.5)$$

If the velocity v of the moving wire is directed at an angle θ with the **B** field, a more general form is needed for Eq. (31.5):

$$\mathcal{E} = BLv \sin \theta \qquad (31.6)$$

Example 31.3

A 0.2-m length of wire moves at a constant velocity of 4 m/s in a direction that is 40° with respect to a magnetic flux density of 0.5 T. Calculate the induced emf.

Solution: Direct substitution into Eq. (31.6) yields

$$\mathcal{E} = (0.5 \text{ T})(0.2 \text{ m})(4 \text{ m/s})(\sin 40°)$$
$$= 0.257 \text{ V}$$

The minus sign does not appear in Eq. (31.6) because the direction of the induced emf is the same as the direction of the magnetic force performing work on the moving charge.

31.3 Lenz's Law

Throughout the discussions of all physical phenomena, one guiding principle stands out above all the rest: the *principle of conservation of energy.* An emf cannot exist without a cause. Whenever an induced current produces heat or performs mechanical work, the necessary energy must come from the work done in inducing the current.

Recall the example discussed in Fig. 31.3a. The north pole of a magnet pushed into a coil induces a current that itself gives rise to another magnetic field. The second field produces a force that opposes the original force. Withdrawing the magnet creates a force that opposes the removal of the magnet. This is an illustration of *Lenz's law:*

> **Lenz's Law:** An induced current will flow in such a direction that it will oppose by its magnetic field the motion of the magnetic field that is producing it.

The more work that is done in moving the magnet into the coil, the greater will be the induced current and, hence, the greater the resisting force. We might have expected this result from the law of conservation of energy. To produce a larger current, we must perform a greater amount of work.

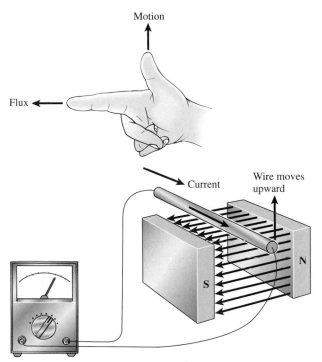

Figure 31.5 The right-hand rule for determining the direction of induced current. This rule is sometimes called *Fleming's Rule*.

The direction of the current induced in a straight conductor moving through a magnetic field can be determined from Lenz's law. However, it it easier to use a modification of the right-hand rule, introduced in Chapter 29, to find the force on a moving charge. This approach, called *Fleming's rule,* is illustrated in Fig. 31.5.

> **Fleming's Rule:** If the thumb, forefinger, and middle finger of the right hand are held at right angles to each other, with the thumb pointing in the direction in which the wire is moving and the forefinger pointing in the field direction (N to S), the middle finger will point in the direction of induced conventional current.

Fleming's rule is easy to apply and useful for studying the currents induced by a simple generator. Students sometimes remember the rule by memorizing *motion–flux–current.* These are the directions given by the thumb, forefinger, and middle finger, respectively.

31.4 The AC Generator

An electric generator converts mechanical energy into electric energy. We have seen that an emf is induced in a conductor when it experiences a change in flux linkage. When the conductor forms a complete circuit, an induced current can be detected. In a generator, a coil of wire is rotated in a magnetic field, and the induced current is transmitted by wires for long distances from its origin.

The construction of a simple generator is shown in Fig. 31.6. Essentially, there are three components: a *field magnet,* an *armature,* and *slip rings* with *brushes.* The field magnet may be a permanent magnet or an electromagnet. The armature for the generator in Fig. 31.6 consists of a single loop of wire suspended between the poles of the field magnet. A pair of slip rings is fused to each end of the loop; they rotate with the loop as it is turned in the magnetic field. Induced current is led away from the system by graphite brushes that ride on each slip ring. Mechanical energy is supplied to the generator by turning the armature in the magnetic field. Electric energy is generated in the form of an induced current.

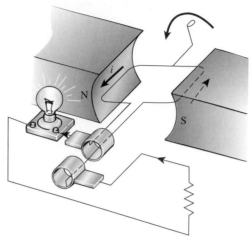

Figure 31.6 The ac generator.

The direction of the induced current must obey Fleming's rule of *motion–flux–current.* In Fig. 31.6, the downward motion of the left wire segment crosses a magnetic flux directed left to right. The induced current is, therefore, toward the slip rings. Similar reasoning shows the current in the right segment of the loop, which is moving upward, will be away from the slip rings.

To understand the operation of an ***ac generator,*** let us follow the loop through a complete rotation, observing the current generated throughout the rotation. Figure 31.7 shows four positions of the rotating coil and the direction of the current delivered to the brushes in each case. Suppose the loop is turned mechanically in a counterclockwise direction. In Fig. 31.7a,

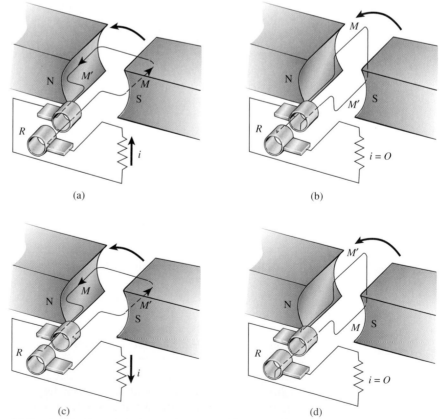

Figure 31.7 As the wire segment *M* moves upward, the current is away from the rings; as it moves downward, the current is toward the rings. In this manner, the rotating loop produces an alternating current.

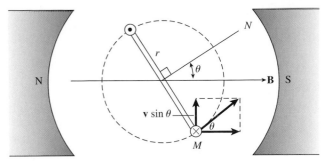

Figure 31.8 Calculating the induced emf.

the loop is horizontal, with side *M* facing the south (S) pole of the magnet. At this point, a maximum current is delivered in the direction shown. In Fig. 31.7b, the loop is vertical, with side *M* facing upward. At this point, no flux lines are being cut, and the induced current drops to zero. When the loop becomes horizontal again, as in Fig. 31.7c, side *M* is now facing the north (N) pole of the magnet. Therefore, the current delivered to the slip ring *R* has changed direction. An induced current flows through the external resistor in a direction opposite to that experienced earlier. In Fig. 31.7d, the loop is vertical again, but now side *M* faces downward. No flux lines are cut, and the induced current again drops to zero. The loop next returns to horizontal as in Fig. 31.7a, and the cycle repeats itself. Thus, the current delivered by such a generator alternates periodically, the direction changing twice each rotation.

The emf generated in each segment of a rotating loop must obey the relation as given in Eq. (31.6):

$$\mathcal{E} = BLv \sin \theta$$

where *v* is the velocity of a moving wire segment of length *L* in a magnetic field of flux density **B**. The direction of the velocity *v* with respect to the **B** field at any instant is denoted by the angle θ. Let us consider the segment *M* of our rotating current loop when it reaches the position shown in Fig. 31.8. The *instantaneous* emf at that position is given by Eq. (31.6). If the loop rotates in a circle of radius *r*, the instantaneous velocity *v* can be found from

$$v = \omega r$$

where ω is the angular velocity in radians per second. Substituting into Eq. (31.6) gives the instantaneous emf

$$\mathcal{E} = BL\omega r \sin \theta \qquad (31.7)$$

An identical emf is induced in the segment of wire opposite *M*, and no *net* emf is generated in the other segments. Hence, the total instantaneous emf is twice the value given by Eq. (31.7), or

$$\mathcal{E}_{inst} = 2BL\omega r \sin \theta \qquad (31.8)$$

But the area *A* of the loop is

$$A = L \times 2r$$

and Eq. (31.8) can be further simplified to

$$\mathcal{E}_{inst} = NBA\omega \sin \theta \qquad (31.9)$$

where *N* is the number of turns of wire.

Equation (31.9) expresses an important principle relating to the study of alternating currents:

If the armature is rotating with a constant angular velocity in a constant magnetic field, the magnitude of the induced emf varies sinusoidally with respect to time.

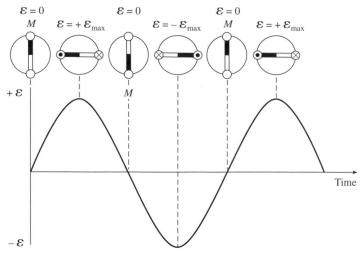

Figure 31.9   Sinusoidal variation of induced emf with time.

This fact is illustrated by Fig. 31.9. The emf varies from a maximum value when $\theta = 90°$ to a zero value when $\theta = 0°$. The maximum instantaneous emf is therefore

$$\mathcal{E}_{max} = NBA\omega \qquad (31.10)$$

since $\sin 90° = 1$. Stating Eq. (31.9) in terms of the maximum emf, we write

$$\mathcal{E}_{inst} = \mathcal{E}_{max} \sin \theta \qquad (31.11)$$

To see the explicit variation of generated emf with time, we should recall that

$$\theta = \omega t = 2\pi f t$$

where f is the number of rotations per second made by the loop. Thus, we can express Eq. (31.11) in the following form:

$$\mathcal{E}_{inst} = \mathcal{E}_{max} \sin 2\pi f t \qquad (31.12)$$

Example 31.4   The armature of a simple ac generator consists of 90 turns of wire, each having an area of 0.2 m². The armature is turned with a frequency of 60 rev/s in a constant magnetic field of flux density 3×10^{-3} T. What is the maximum emf generated?

Plan: The maximum emf occurs when the sine function is equal to 1 in Eq. (31.9). We will convert the frequency from rev/s to rad/s and substitute to find the generated emf.

Solution: Recalling that $\omega = 2\pi f$, we find the angular frequency as follows:

$$\omega = 2\pi f = (2\pi \text{ rad})(60 \text{ rev/s})$$
$$= 377 \text{ rad/s}$$

Substituting this value and other known values into Eq. (31.10), we obtain

$$\mathcal{E}_{max} = NBA\omega$$
$$= (90 \text{ turns})(3 \times 10^{-3} \text{ T})(0.2 \text{ m}^2)(377 \text{ rad})$$
$$= 20.4 \text{ V}$$

Since the induced current is proportional to the induced emf, from Ohm's law, the induced current will also vary sinusoidally according to

$$i_{\text{inst}} = i_{\text{max}} \sin 2\pi ft \qquad \textbf{(31.13)}$$

The maximum current occurs when the induced emf is a maximum. The sinusoidal variation is similar to that plotted in Fig. 31.9.

The SI unit for frequency is the *hertz* (Hz), which is defined as a cycle per second.

$$1 \text{ Hz} = 1 \text{ cycle/s} = 1 \text{ s}^{-1}$$

Thus, a 60-cycle-per-second alternating current has a frequency of 60 Hz.

31.5 The DC Generator

A simple ac generator can be converted easily to a dc generator by substituting a split-ring commutator for the slip rings, as illustrated in Fig. 31.10. The operation is just the reverse of that discussed earlier for a dc motor (Chapter 30). In the motor, an electric current gives rise to an external torque. In the *dc generator,* an external torque generates an electric current. The *commutator* reverses the connections to the brushes twice per revolution. As a result, the current pulsates but never reverses direction. The emf of such a generator varies with time, as shown in Fig. 31.11. Note that the emf is always in the positive direction, but it rises to a maximum and falls to zero twice per complete rotation. Practical dc generators are designed with many coils in several planes so that the generated emf is large and nearly constant.

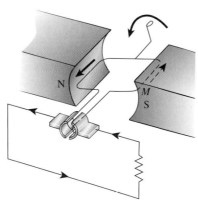

Figure 31.10   An example of a simple dc generator. Verify the direction of the induced current as the segment *M* moves upward in the **B** field directed to the right.

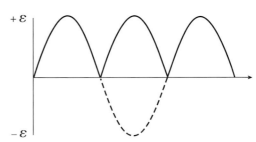

Figure 31.11   Pulsating emf produced by a dc generator.

31.6 Back EMF in a Motor

In an electric motor, a magnetic torque turns a current-carrying loop in a constant magnetic field. We have just seen that a coil rotating in a magnetic field will induce an emf that opposes the cause that gave rise to it. This is true even if a current already exists in the loop. Thus, *every motor is also a generator.* According to Lenz's law, such an induced emf must oppose the current delivered to the motor. For this reason, the emf induced in a motor is called *back emf,* or *counter emf.*

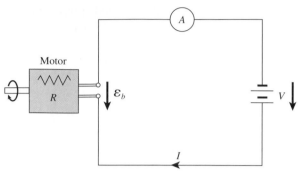

Figure 31.12   Back emf in a dc motor.

The effect of a back emf is to reduce the net voltage delivered to the armature coils of the motor. Consider the circuit illustrated in Fig. 31.12. The net voltage delivered to the armature coils is equal to the applied voltage V less the induced voltage $\mathcal{E}_b$.

Applied voltage − induced voltage = net voltage

According to Ohm's law, the net voltage across the armature coils is equal to the product of the coil resistance R and the current I. Symbolically, we write

$$V - \mathcal{E}_b = IR \tag{31.14}$$

Equation (31.14) tells us that the current through a circuit containing a motor is determined by the magnitude of the back emf. The magnitude of this induced emf, of course, depends on the speed of rotation of the armature. We can show this experimentally by connecting a motor, an ammeter, and a battery in series, as shown in Fig. 31.13. When the armature is rotating, a low current is indicated. The back emf reduces the effective voltage. If the motor is stalled by holding the armature stationary, the back emf will drop to zero. The increased net voltage results in a larger current and can cause the motor to overheat and even burn out.

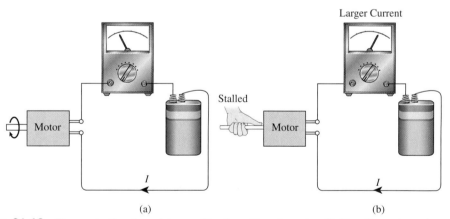

Figure 31.13   Demonstrating the existence of back emf in a dc motor. Stalling the motor reduces the back emf to zero, and this increases the circuit current.

31.7   Types of Motors

DC motors are classified according to how the field coils and the armature are connected. When the armature coils and the field coils are connected in series, as shown in Fig. 31.14, the motor is said to be **series-wound.** In this type of motor, the current energizes both the field windings and the armature windings. When the armature turns slowly, the back emf is small and the current is large. Consequently, a large torque is developed at low speeds.

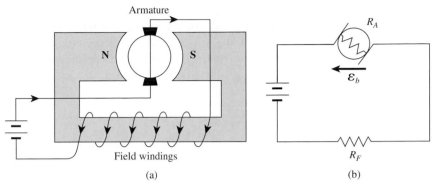

(a)   (b)

Figure 31.14 (a) The series-wound dc motor. (b) Schematic diagram showing how the armature and its resistance are connected in series with the resistance of the field windings. Note the direction of the induced back emf.

In a **shunt-wound** motor, the field windings and the armature windings are connected in parallel, as illustrated by Fig. 31.15. The entire voltage is applied across both windings. The primary advantage of a shunt-wound motor is that it produces more constant torque over a range of speeds. The starting torque is usually lower, however, than a similar series-wound motor.

In some applications, the field windings are in two parts, one connected in series with the armature and the other in parallel with it. Such a motor is called a **compound motor.** The torque produced by a compound motor varies between that of the series and shunt motors.

In *permanent magnet motors,* no field current is necessary. These motors have torque characteristics similar to those of shunt-wound motors.

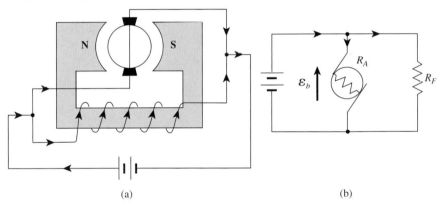

(a)   (b)

Figure 31.15 (a) The shunt-wound dc motor. (b) Schematic diagram showing how the armature and its resistance are connected in parallel with the resistance of the field windings. Note the direction of the induced back emf.

Example 31.5

A 120-V dc shunt motor has an armature resistance of 3 Ω and a field resistance of 260 Ω. When the motor is operating at full speed, the total current is 3 A. (a) What is the back emf of the motor at full speed? (b) Find the current in the motor at the moment the switch is turned on (the starting current).

Plan: When we study the schematic diagram (Fig. 31.15b) for the shunt-wound motor, we see that the back emf occurs in the armature windings. Thus, we need to see how the total current is divided between the field windings and the armature windings to obtain the current I_A. Then, we can write the voltage equations to determine the back emf for full speed operation. The starting current is easier to find since the back emf is zero at that instant.

Solution (a): To determine the back emf, we must see how the current is divided between the armature and field windings. The current I_F in the field windings can be found by writing the voltage equation for the outside loop in Fig. 31.15b:

$$V = I_F R_F$$

from which

$$I_F = \frac{V}{R_F} = \frac{120 \text{ V}}{260 \text{ }\Omega} = 0.46 \text{ A}$$

The current in the armature windings is therefore

$$I_A = 3 \text{ A} - 0.46 \text{ A} = 2.54 \text{ A}$$

Now the voltage equation for the loop containing the battery and the armature is

$$V - \mathscr{E}_b = I_A R_A$$

from which

$$\begin{aligned}
\mathscr{E}_b &= V - I_A R_A \\
&= 120 \text{ V} - (2.54 \text{ A})(3 \text{ }\Omega) \\
&= 120 \text{ V} - 7.62 \text{ V} = 112.4 \text{ V}
\end{aligned}$$

Solution (b): At the instant the switch is turned on, the armature is not yet turning, and consequently $\mathscr{E}_b = 0$. In this case, the armature current is

$$I_A = \frac{V}{R_A} = \frac{120 \text{ V}}{3 \text{ }\Omega} = 40 \text{ A}$$

The field current is still 0.46 A, and the total starting current is

$$\begin{aligned}
I &= I_A + I_F \\
&= 40 \text{ A} + 0.46 \text{ A} = 40.5 \text{ A}
\end{aligned}$$

31.8 The Transformer

It was noted earlier that a changing current in one wire loop will induce a current in a nearby loop. The induced current arises from the changing magnetic field associated with a changing current. Alternating current has a distinct advantage over direct current because of the inductive effect of a current that varies constantly in magnitude and direction. The most common application of this principle is offered by the **transformer**, a device that increases or decreases the voltage in an ac circuit.

A simple transformer is illustrated in Fig. 31.16. There are three essential parts: (1) a primary coil connected to an ac source, (2) a secondary coil, and (3) a soft iron core. As an alternating current is sent through the primary coil, magnetic flux lines move back and forth through the iron core, inducing an alternating current in the secondary coil.

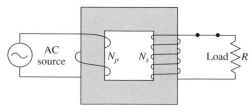

Figure 31.16 Transformer.

The constantly changing magnetic flux is established throughout the core of the transformer and passes through both primary and secondary coils. The emf $\mathcal{E}_p$ induced in the primary coil is given by

$$\mathcal{E}_p = -N_p \frac{\Delta\Phi}{\Delta t} \qquad (31.15)$$

where N_p = number of primary turns

$\Delta\Phi/\Delta t$ = rate at which flux changes

Similarly, the emf $\mathcal{E}_s$ induced in the secondary coil is

$$\mathcal{E}_s = -N_s \frac{\Delta\Phi}{\Delta t} \qquad (31.16)$$

where N_s is the number of secondary turns. Since the same flux changes at the same rate through each coil, we can divide Eq. (31.15) by Eq. (31.16) to obtain

$$\frac{\mathcal{E}_p}{\mathcal{E}_s} = \frac{N_p}{N_s} \qquad (31.17)$$

$$\frac{Primary\ voltage}{Secondary\ voltage} = \frac{primary\ turns}{secondary\ turns}$$

The induced voltage is in direct proportion to the number of turns. If the ratio of secondary turns N_s to primary turns N_p is varied, an input (primary) voltage can provide any desired output (secondary) voltage. For example, if there are 40 times as many turns in the secondary coil, an input voltage of 120 V will be increased to $40 \times 120 = 4800$ V in the secondary coil. A transformer that produces a larger output voltage is called a ***step-up transformer.***

A ***step-down transformer*** can be constructed by making the number of primary turns greater than the number of secondary turns. Using a step-down transformer gives a lower output voltage.

Transformer efficiency is defined as the ratio of the power output to the power input. Recalling that electric power is equal to the product of voltage and current, we can write the efficiency E of a transformer as

$$E = \frac{power\ output}{power\ input} = \frac{\mathcal{E}_s i_s}{\mathcal{E}_p i_p} \qquad (31.18)$$

where i_p and i_s are the currents in the primary coil and the secondary coil, respectively. Most electric transformers are carefully designed for extremely high efficiencies, normally above 90 percent.

It is important to recognize that there is no power gain as a result of transformer action. When the voltage is stepped up, the current must be stepped down so that the product $\mathcal{E}i$ does not increase. To see this more clearly, let us assume that a given transformer is 100 percent efficient. For this perfect transformer, Eq. (31.18) becomes

$$\mathcal{E}_s i_s = \mathcal{E}_p i_p$$

or

$$\frac{i_p}{i_s} = \frac{\mathcal{E}_s}{\mathcal{E}_p} \qquad (31.19)$$

This equation shows clearly the inverse relationship between current and induced voltage.

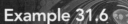

Example 31.6

An ac generator that delivers 20 A at 6000 V is connected to a step-up transformer. What is the output current at 120,000 V if the transformer efficiency is 100 percent?

Plan: For 100 percent efficiency, the ratio of primary current to secondary current is the same as the ratio of *secondary* voltage to *primary* voltage. Substitution of known values allows us to find the output, or secondary current.

Solution: From Eq. (31.19), we write

$$\frac{i_p}{i_s} = \frac{\mathscr{E}_s}{\mathscr{E}_p} \qquad \text{or} \qquad i_s = \frac{\mathscr{E}_p i_p}{\mathscr{E}_s}$$

$$i_s = \frac{(6000 \text{ V})(20 \text{ A})}{(120,000 \text{ V})} = 1.0 \text{ A}$$

Note in Example 31.6 that the current was *reduced* from 20 A to 1 A, while the voltage was *increased* by a factor of 20. Since heat losses in transmission lines vary directly with the square of the current (i^2R), this means electric power can be transmitted for large distances without significant loss. When the electric power reaches its destination, step-down transformers are used to provide the desired current at lower voltages.

Summary and Review

Summary

Electromagnetic induction allows for the production of an electric current in a conducting wire. This is the basic operating principle behind many electrical devices. An understanding of the concepts summarized as follows is necessary for most applications involving the use of alternating current.

- A magnetic flux changing at the rate of 1 Wb/s will induce an emf of 1 V for each turn of a conductor. Symbolically,

$$\mathcal{E} = -N\frac{\Delta\Phi}{\Delta t} \qquad \textit{Induced EMF}$$

- Two principal ways in which the flux changes are

$$\Delta\Phi = \Delta BA \qquad \Delta\Phi = B\,\Delta A$$

- The induced emf due to a wire of length L moving with a velocity v at an angle θ with a field **B** is given by

$$\mathcal{E} = BLv\sin\theta \qquad \textit{EMF Due to Moving Wire}$$

- According to *Lenz's law*, the induced current must be in such a direction that it produces a magnetic force that opposes the force causing the motion.
- *Fleming's rule:* If the thumb, forefinger, and middle finger of the right hand are held at right angles to each other, with the thumb pointing in the direction in which the wire is moving and the forefinger pointing in the field direction (N to S), the middle finger will point in the direction of induced conventional current (*motion–flux–current*).

- The instantaneous emf generated by a coil of N turns moving with an angular velocity ω or frequency f is

$$\mathcal{E}_{\text{inst}} = NBA\omega\sin\omega t \qquad \mathcal{E}_{\text{inst}} = 2\pi fNBA\sin 2\pi ft$$

- The maximum emf occurs when the sine is equal to 1. Thus,

$$\mathcal{E}_{\text{max}} = NBA\omega \qquad \mathcal{E}_{\text{inst}} = \mathcal{E}_{\text{max}}\sin 2\pi ft$$

- Since the induced current is proportional to $\mathcal{E}$, we also have

$$i_{\text{inst}} = i_{\text{max}}\sin 2\pi ft \qquad \textit{Instantaneous Current}$$

- The back emf in a motor is the induced voltage that causes a reduction in the net voltage delivered to a circuit.

$$\textit{Applied voltage} - \textit{induced back emf} = \textit{net voltage}$$

$$V - \mathcal{E}_b = IR \qquad \mathcal{E}_b = V - IR$$

- For a transformer having N_p primary and N_s secondary turns,

$$\frac{\textit{Primary voltage}}{\textit{Secondary voltage}} = \frac{\textit{primary turns}}{\textit{secondary turns}} \qquad \frac{\mathcal{E}_p}{\mathcal{E}_s} = \frac{N_p}{N_s}$$

- The efficiency of a transformer is

$$E = \frac{\text{power output}}{\text{power input}} = \frac{\mathcal{E}_p i_p}{\mathcal{E}_s i_s} \qquad \begin{array}{l}\textit{Transformer}\\\textit{Efficiency}\end{array}$$

Key Terms

ac generator   608
armature   607
back cmf   611
commutator   611
compound motor   613
dc generator   611

electromagnetic induction   602
field magnet   607
Fleming's rule   607
induced emf   603
Lenz's law   606
series-wound motor   612

shunt-wound motor   613
slip rings   607
step-down transformer   615
step-up transformer   615
transformer   614
transformer efficiency   615

Review Questions

31.1. Discuss the various factors that influence the magnitude of an induced emf in a length of wire moving in a magnetic field.

31.2. A bar magnet is held in a vertical position with the north pole facing upward. If a closed-loop coil is dropped over the north end of the magnet, what is the direction of the induced current viewed from the top of the magnet?

31.3. A circular loop is suspended with its plane perpendicular to a magnetic field directed from left to right. The loop is removed from the field by moving it upward quickly. What is the direction of the induced current viewed along the field direction? Is a force required to remove the loop from the field?

31.4. An induction coil is essentially a transformer that operates on direct current. As shown in Fig. 31.17, the induction coil consists of a few primary turns wound around an iron core with a large number of secondary turns surrounding the primary. A battery current magnetizes the core so that it attracts the armature of the interruptor and opens the circuit periodically. When the circuit is opened, the field collapses, and a large emf is induced in the secondary coil, producing a spark at the output terminals. What is the function of the capacitor C connected in parallel with the interruptor? Explain how an induction coil is used in the ignition system of an automobile.

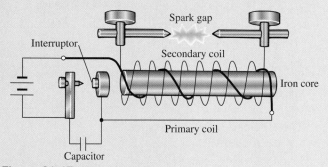

Interruptor

Spark gap

Secondary coil

Iron core

Primary coil

Capacitor

Figure 31.17   Induction coil.

31.5. Explain clearly how an ac generator can be converted to a dc generator. How would you proceed to convert an ac generator into an ac motor?

31.6. When the electric motor in a plant is starting, a worker notices the lights are momentarily dimmed. Explain.

31.7. What type of dc motor should be purchased to operate a winch used to lift heavy objects? Why?

31.8. What type of motor should be used to operate an electric fan where uniform torque is desired at high speeds?

31.9. Explain how the existence of back emf in a motor helps keep its speed constant. *Hint:* What happens to $\mathcal{E}_b$ and i when the armature speed increases or decreases?

31.10. There are three primary ways in which power is lost through the operation of a transformer: (a) wire-resistance losses, (b) hysteresis losses, and (c) eddy-current losses. (*Eddy currents* are induced current loops that occur in the mass of a magnetic material resulting from a changing flux.) Explain how energy is wasted by these three processes.

31.11. Explain with the use of diagrams how transformers make it possible to transmit current economically from power installations to homes many miles away.

31.12. An ac generator produces 60 Hz alternating voltage. How many degrees of armature rotation will correspond to one-fourth of a cycle?

31.13. Why is it more economical for power companies to provide alternating current than direct current?

31.14. Prepare a brief report on the following topics and explain the part played by electromagnetic induction.

a. the betatron

b. induction coil

c. the telephone

d. eddy currents

e. magnetohydrodynamic generator

f. electric power transmission

g. the universal motor

h. induction motor

i. synchronous motor

Problems

Section 31.2 EMF Induced by a Moving Wire

31.1. A coil of wire 8 cm in diameter has 50 turns and is placed in a B field of 1.8 T. If the B field is reduced to 0.6 T in 0.002 s, what is the induced emf?   Ans. -151 V

31.2. A square coil of wire having 100 turns of area 0.044 m^2 is placed with its plane perpendicular to a constant B field of 4 mT. The coil is flipped to a position parallel with the field in a time of 0.3 s. What is the induced emf?

31.3. A coil of 300 turns moving perpendicular to the flux in a uniform magnetic field experiences a flux linkage of 0.23 mWb in 0.002 s. What is the induced emf?   Ans. -34.5 V

31.4. The magnetic flux linking a loop of wire changes from 5 to 2 mWb in 0.1 s. What is the average induced emf?

31.5. A coil of 120 turns is 90 mm in diameter and has its plane perpendicular to a 60-mT magnetic field produced by a nearby electromagnet. The current in the electromagnet is cut off, and as the field collapses, an emf of 6 V is induced in the coil. How long does it take for the field to disappear?
Ans. 7.63 ms

31.6. A coil of 56 turns has an area of 0.3 m². Its plane is perpendicular to a 7-mT magnetic field. If this field collapses to zero in 6 ms, what is the induced emf?

31.7. A wire 0.15 m long moves at a constant velocity of 4 m/s in a direction that is 36° with respect to a 0.4-T magnetic field. The axis of the wire is perpendicular to the magnetic flux lines. What is the induced emf?
Ans. 0.141 V

31.8. A 0.2-m wire moves at an angle of 28° with an 8-mT magnetic field. The wire length is perpendicular to the flux. What velocity v is required to induce an emf of 60 mV?

Section 31.4 The AC Generator and Section 31.5 The DC Generator

31.9. The magnetic field in the air gap between the magnetic poles and the armature of an electric generator has a flux density of 0.7 T. The length of the wires on the armature is 0.5 m. How fast must these wires move to generate a maximum emf of 1.00 V in each armature wire?
Ans. 2.86 m/s

31.10. A single loop of wire has a diameter of 60 mm and makes 200 rpm in a constant 4-mT magnetic field. What is the maximum emf generated?

31.11. The armature of a simple generator has 300 loops of diameter 20 cm in a constant 6-mT magnetic field. What must be the frequency of rotation in revolutions per second in order to induce a maximum emf of 7.00 V?
Ans. 19.7 rev/s

31.12. An armature in an ac generator consists of 500 turns, each of area 60 cm². The armature is rotated at a frequency of 3600 rpm in a uniform 2-mT magnetic field. What is the frequency of the alternating emf? What is the maximum emf generated?

31.13. In Prob. 31.12, what is the instantaneous emf at the time when the plane of the coil makes an angle of 60° with the magnetic flux?
Ans. 1.13 V

31.14. The armature of a simple ac generator has 100 turns of wire, each having a radius of 5.00 cm.

The armature turns in a constant 0.06-T magnetic field. What must be the rotational frequency in rpm to generate a maximum voltage of 2.00 V?

31.15. A circular coil has 70 turns, each 50 mm in diameter. Assume the coil rotates about an axis that is perpendicular to a magnetic field of 0.8 T. How many revolutions per second must the coil make to generate a maximum emf of 110 V?
Ans. 159 rev/s

*31.16. The armature of an ac generator has 800 turns, each of area 0.25 m². The coil rotates at a constant 600 rpm in a 3-mT field. What is the maximum induced emf? What is the instantaneous emf 0.43 s after the coil passes a position of zero emf?

*31.17. A 300-Ω resistor is connected in series with an ac generator of negligible internal resistance. The armature of the generator has 200 turns of wire 30 cm in diameter, and it turns at 300 rpm in a constant 5-mT field. What is the instantaneous current through the resistor 0.377 s after the coil passes a position of zero emf?
Ans. 4.90 mA

Section 31.6 Back EMF in a Motor

31.18. A 120-V dc motor draws a current of 3.00 A in operation and has a resistance of 8.00 Ω. What is the back emf when the motor is operating, and what is the starting current?

31.19. The armature coil of the starting motor in an automobile has a resistance of 0.05 Ω. The motor is driven by a 12-V battery, and the back emf at operating speed is 6.00 V. What is the starting current? What is the current at full speed?
Ans. 240 A, 120 A

31.20. A 220-V dc motor draws a current of 10 A in operation and has an armature resistance of 0.4 Ω. What is the back emf when the motor is operating, and what is the starting current?

*31.21. A 120-V series-wound dc motor has a field resistance of 90 Ω and an armature resistance of 10 Ω. When operating at full speed, a back emf of 80 V is generated. What is the total resistance of the motor? What is the starting current? What is the operating current?
Ans. 100 Ω, 1.20 A, 0.400 A

*31.22. The efficiency of the motor in Prob. 31.21 is the ratio of the power output to the power input. Determine the efficiency based on the known data.

Section 31.8 The Transformer

31.23. A step-up transformer has 400 secondary turns and only 100 primary turns. A 120-V alternating voltage is connected to the primary coil. What is the output voltage?
Ans. 480 V

31.24. A step-down transformer is used to drop an alternating voltage from 10,000 to 500 V. What must be the ratio of secondary turns to primary turns? If the input current is 1.00 A and the transformer is 100 percent efficient, what is the output current?

31.25. A step-up transformer is 95 percent efficient and has 80 primary turns and 720 secondary turns. If the primary draws a current of 20 A at 120 V, what are the current and voltage for the secondary?

Ans. 2.11 A, 1080 V

31.26. A 25-W lightbulb has a resistance of 8.0 Ω while burning. The light is powered from the secondary of a small transformer connected to a 120-V circuit. What must be the ratio of secondary turns to primary turns in this application? Assume 100 percent efficiency.

Additional Problems

31.27. A 70-turn coil of wire has an area of 0.06 m^2 and is placed perpendicular to a constant 8-mT magnetic field. Calculate the induced emf if the coil flips 90° in 0.02 s.   Ans. 1.68 mV

31.28. A coil of area 0.2 m^2 has 80 turns of wire and is suspended with its plane perpendicular to a uniform magnetic field. What must be the flux density to produce an average emf of 2 V as the coil is flipped parallel to the field in 0.5 s?

31.29. The flux through a 200-turn coil changes from 0.06 to 0.025 Wb in 0.5 s. The coil is connected to an electric light, and the combined resistance is 2 Ω. What is the average induced emf, and what average current is delivered to the light filament?   Ans. 14.0 V, 7.00 A

31.30. A 90-mm length of wire moves with an upward velocity of 35 m/s between the poles of a magnet. The magnetic field is 80 mT directed to the right. If the resistance in the wire is 5.00 mΩ, what are the magnitude and direction of the induced current?

31.31. A generator develops an emf of 120 V and has a terminal potential difference of 115 V when the armature current is 25.0 A. What is the resistance of the armature?   Ans. 0.200 Ω

31.32. The coil of an ac generator rotates at a frequency of 60 Hz and develops a maximum emf of 170 V. The coil has 500 turns, each of area 4 × 10^{-3} m^2.

What is the magnitude of the magnetic field in which the coil rotates?

31.33. A generator produces a maximum emf of 24 V when the armature rotates at 600 rpm. Assuming nothing else changes, what is the maximum emf when the armature rotates at 1800 rpm?

Ans. 72.0 V

*31.34. A shunt-wound motor connected across a 117-V line generates a back emf of 112 V when the armature current is 10 A. What is the armature resistance?

*31.35. A 110-V shunt-wound motor has a field resistance of 200 Ω connected in parallel with an armature resistance of 10 Ω. When the motor is operating at full speed, the back emf is 90 V. What is the starting current, and what is the operating current?   Ans. 11.6 A, 2.55 A

*31.36. A 120-V shunt motor has a field resistance of 160 Ω and an armature resistance of 1.00 Ω. When the motor is operating at full speed, it draws a current of 8.00 A. Find the starting current. What series resistance must be added to reduce the starting current to 30 A?

*31.37. A shunt generator has a field resistance of 400 Ω and an armature resistance of 2.00 Ω. The generator delivers a power of 4000 W to an external line at 120 V. What is the emf of the generator?

Ans. 187 V

Critical Thinking Questions

31.38. A coil of wire has 10 loops, each of diameter D, placed inside a B field that varies at the rate of 2.5 mWb/s. If the induced emf is 4 mV, what is the diameter of the coil? What will the induced emf be if the diameter is doubled? What is the

induced emf if the rate of change in the B field is doubled?   Ans. 45.2 cm, 16 mV, 8 mV

31.39. In Fig. 31.18a, the single loop of area 0.024 m^2 is connected to a 4-mΩ resistor. The magnet moves to the left through the center of the loop,

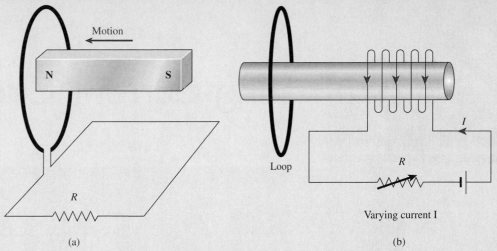

Motion

N   S

Loop

R

(a)

R

Varying current I

(b)

Figure 31.18 Using Lenz's law to determine the direction of the induced currents.

causing the magnetic flux to increase at the rate of 2 mWb/s. What are the magnitude and direction of the current through the resistor? What if the magnet is then pulled back out of the loop with the same speed?

31.40. In Fig. 31.18b, a changing B field is produced first by an increasing current through the loops and then by a decreasing current through the loops. In each case, will the induced current be downward or upward on the near side of the single loop? Ans. up, down

*31.41. A coil of 50 turns is rotated clockwise with a frequency of 60 Hz in a constant 3.0-mT magnetic field. The direction of the B field is along the

positive x axis, and the coil of area 0.070 m² rotates clockwise in the x–y plane. When the plane of the loop is parallel to the field, what is the direction of the current as viewed from the top (clockwise or counterclockwise)? What is the maximum induced emf? Starting time $t = 0$ when the emf is zero. At what later time will its emf first rise to 2.00 V?

*31.42. When the motor on a heat pump is first turned on, it momentarily draws 40.0 A. The current then immediately drops to a steady value of 12.0 A. If the motor operates on a 120-V power source, what is the back emf generated while the motor is running? Ans. 84 V

32

Alternating-Current Circuits

Giant generators at Hoover Dam produce over 2000 megawatts of power. The alternating current enters step-up transformers to provide the high voltage necessary to transfer energy to distant locations. In this chapter, we will study the elements of alternating-current circuits that include resistance, capacitance, and inductance connected in series. (*Photo by U.S. Bureau of Reclamation.*)

Objectives

After completing this chapter, you should be able to

1. Determine the instantaneous current for charging and discharging a *capacitor* and for the growth and decay of current in an *inductor*.

2. Write and apply equations for calculating the *inductance* and *capacitance* for inductors and capacitors in an ac circuit.

3. Explain with diagrams the phase relationships for a circuit with (a) pure resistance, (b) pure capacitance, and (c) pure inductance.

4. Write and apply equations for calculating the *impedance*, the *phase angle*, and the *effective current* for a series ac circuit containing resistance, capacitance, and inductance.

5. Write and apply an equation for calculating the resonant frequency for an ac circuit.

6. Define and be able to determine the *power factor* for a series ac circuit.

About 99 percent of the energy generated in the United States is in ac form. There are reasons for the predominant use of ac circuits. A rotating coil in a magnetic field induces an alternating emf in an extremely efficient manner. Besides, the transformer provides a convenient method of transmitting ac currents over long distances with a minimal power loss.

The only element of importance in the dc circuit (besides a source of emf) was the resistor. Since alternating currents behave differently from direct currents, additional circuit elements become important. In addition to the normal resistance, electromagnetic induction and capacitance play important roles. In this chapter, we present a few elementary aspects of alternating current in electric circuits.

32.1 The Capacitor

In Chapter 26, we discussed the capacitor as an electrostatic device able to store charge. Charging and discharging capacitors in an ac circuit provide an effective means of regulating and controlling the flow of charge. Before discussing the effects of **capacitance** in an ac circuit, however, it will be useful to describe the growth and decay of charge on a capacitor.

Consider the circuit illustrated in Fig. 32.1, containing only a capacitor and a resistor. When the switch is moved to S_1, the capacitor begins to be charged rapidly by the current i. As the potential difference Q/C between the capacitor plates rises, however, the rate of flow of charge to the capacitor decreases. At any instant, the iR drop through the resistor must equal the difference between the terminal voltage V_B of the battery and the back emf of the capacitor. Symbolically,

$$V_B - \frac{Q}{C} = iR \tag{32.1}$$

where i = instantaneous current

Q = instantaneous charge on capacitor

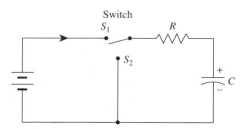

Figure 32.1 Circuit diagram illustrating a method for charging and discharging a capacitor.

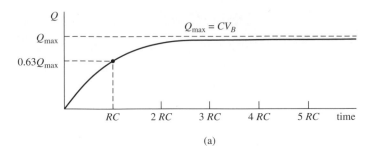

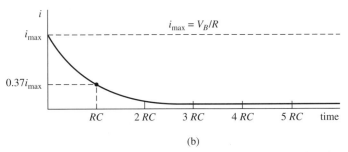

Figure 32.2 (a) The charge on a capacitor rises, approaching, but never reaching, its maximum value. (b) The current decreases, approaching zero as the charge builds to its maximum value.

Initially, the charge Q is zero, and the current i is a maximum. Thus, at time $t = 0$.

$$Q = 0 \quad \text{and} \quad i = \frac{V_B}{R} \tag{32.2}$$

As the charge on the capacitor builds up, it produces a back emf Q/C opposing the flow of additional charge; the current i decreases. Both the increase in charge and the decrease in current are exponential functions, as shown by the curves in Fig. 32.2. If it were possible to continue charging indefinitely, the limits at $t = \infty$ would be

$$Q = CV_B \quad \text{and} \quad i = 0 \tag{32.3}$$

The methods of calculus applied to Eq. (32.1) show that the instantaneous charge is given by

$$Q = CV_B(1 - e^{-t/RC}) \tag{32.4}$$

and that the instantaneous current is given by

$$i = \frac{V_B}{R} e^{-t/RC} \tag{32.5}$$

where t is the time. The logarithmic constant e is 2.71828 to six significant figures. Substitution of $t = 0$ and $t = \infty$ into the above relations will yield Eqs. (32.2) and (32.3), respectively.

The equations for computing instantaneous charge and current are simplified at the particular instant when $t = RC$. This time, usually denoted by τ, is called the **time constant** of the circuit.

$$\tau = RC \qquad \qquad \textit{Time Constant} \tag{32.6}$$

We see from Eq. (32.4) that the charge Q rises to $(1 - 1/e)$ times its final value in one time constant:

$$Q = CV_B\left(1 - \frac{1}{e}\right) = CV_B(0.63)$$

$$Q = 0.63 \, CV_B \tag{32.7}$$

In a capacitance circuit, the charge on a capacitor will rise to 63 percent of its maximum value after charging for a period of one time constant.

Substituting $\tau = RC$ into Eq. (32.5) shows the current delivered to the capacitor decreases to $1/e$ times its initial value in one time constant:

$$i = \frac{V_B}{R}\frac{1}{e} = 0.37\frac{V_B}{R} \tag{32.8}$$

In a capacitive circuit, the current delivered to a capacitor will decrease to 37 percent of its initial value after being charged for a period of one time constant.

Now let us consider the problem of a discharging capacitor. For practical reasons, *a capacitor is considered to be fully charged after a period of time equal to five time constants* ($5RC$). If the switch in Fig. 32.1 has been in position S_1 for at least this long, it can be assumed that the maximum charge CV_B is on the capacitor. By moving the switch to position S_2, the voltage source is removed from the circuit, and a path is provided for discharge. In this case, the voltage Eq. (32.1) reduces to

$$-\frac{Q}{C} = iR \tag{32.9}$$

Both the charge and the current decay along curves similar to that shown for the charging current in Fig. 32.2b. The instantaneous charge is found from

$$Q = CV_B e^{-t/RC} \tag{32.10}$$

and the instantaneous current is given by

$$i = \frac{-V_B}{R}e^{-t/RC} \tag{32.11}$$

The negative sign in the current equation indicates that the direction of i in the circuit has been reversed.

After discharging for one time constant, the charge and the current will have decayed to $1/e$ times their initial values. This can be shown by substituting τ into Eqs. (32.10) and (32.11).

In a capacitance circuit, the charge and the current will decay to 37 percent of their initial values after the capacitor has discharged for a length of time equal to one time constant.

The capacitor is considered to be fully discharged after five time constants ($5RC$).

Example 32.1

A 12-V battery having an internal resistance of 1.5 Ω is connected to a 4-μF capacitor through leads having a resistance of 0.5 Ω. (a) What is the initial current delivered to the capacitor? (b) How long will it take to charge the capacitor fully? (c) What is the value of the current after one time constant?

Plan: Initially, there is no back emf coming from the capacitor because no charge has built up on it. Therefore, the initial current can be found from Ohm's law using the battery voltage and the total circuit resistance. The time constant τ for the circuit is found as the product RC, and the time to fully charge the capacitor is approximately five time constants. After one time constant, the current will have decayed to 37 percent of its initial value as the back emf builds on the capacitor.

Solution (a): The starting current i_0 is

$$i_0 = \frac{\mathcal{E}_B}{R + r} = \frac{12 \text{ V}}{1.5 \text{ }\Omega + 0.5 \text{ }\Omega}; \qquad i_0 = 6.0 \text{ A}$$

Solution (b): Since $\tau = RC$ and the time for full charge is $T = 5\tau$, we can say the capacitor is fully charged after a time

$$T = 5RC = 5(1.5 \text{ }\Omega + 0.5 \text{ }\Omega)(4 \text{ }\mu\text{F})$$
$$T = 5(2 \text{ }\Omega)(4 \text{ }\mu\text{F}) = 40 \text{ }\mu\text{s}$$

Solution (c): After one time constant, we find the current, from Eq. (32.8), will be equal to 37 percent of its initial value (6 A).

$$i_\tau = (0.37)(6.0 \text{ A}) = 2.22 \text{ A}$$

In the preceding discussions, we simplified the approach by using direct currents. When an alternating voltage is impressed upon a capacitor, there are surges of charge into and out of the capacitor plates. Therefore, an alternating current is maintained in a circuit, even though there is no path between the capacitor plates. The effect of capacitance in an ac circuit will be discussed in Section 32.4.

32.2   The Inductor

Another important element in an ac circuit is the *inductor*, which consists of a continuous loop or coil of wire. (See Fig. 32.3.) In Chapter 31, we showed that a change in magnetic flux in the region enclosed by such a coil will induce an emf in the coil. Until now, the flux changes originated from sources outside the coil itself. We now consider the emf induced in a coil as a result of changes in its *own* current. Regardless of how the flux change occurs, the induced emf must be given by

$$\mathcal{E} = -N\frac{\Delta\Phi}{\Delta t} \tag{32.12}$$

where N = number of turns

$\Delta\Phi/\Delta t$ = rate at which flux changes

When the current through an inductor increases or decreases, a *self-induced* emf arises in the circuit that *opposes* the change. Consider the circuit illustrated in Fig. 32.3. When the switch is closed, the current rises from zero to its maximum value $i = V_B/R$. The inductor responds to this increasing current by setting up an induced back emf. Since the geometry of the inductor is fixed, the rate of change in flux, $\Delta\Phi/\Delta t$, or the induced emf $\mathcal{E}$, is

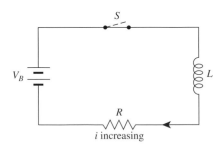

Figure 32.3   The inductor.

proportional to the rate of change in current, $\Delta i/\Delta t$. This proportionality is expressed in the equation

$$\mathcal{E} = -L\frac{\Delta i}{\Delta t} \qquad (32.13)$$

The proportionality constant L is called the **inductance** of the circuit. Solving explicitly for the inductance in Eq. (32.13), we write

$$L = -\frac{\mathcal{E}}{\Delta i/\Delta t} \qquad (32.14)$$

The unit of inductance is the **henry** (H).

> A given inductance has an inductance of 1 henry (H) if an emf of 1 volt is induced by a current changing at the rate of 1 ampere per second.

$$1\ \text{H} = 1\ \text{V} \cdot \text{s/A}$$

The inductance of a coil depends on its geometry, the number of turns, the spacing of the turns, and the permeability of its core but not on voltage and current values. In this respect, the inductor is similar to capacitors and resistors.

We shall now consider the growth and decay of current in an inductive circuit. The circuit illustrated in Fig. 32.4 contains an inductor L, a resistor R, and a battery V_B. The switch is positioned so that the battery can be alternately connected and disconnected from the circuit. When the switch is moved to position S_1, a current begins to grow in the circuit. As the current rises, the induced emf $-L(\Delta i/\Delta t)$ is established in opposition to the battery voltage V_B. The net emf must equal the iR drop through the resistor. Thus,

$$V_B - L\frac{\Delta i}{\Delta t} = iR \qquad (32.15)$$

A mathematical analysis of Eq. (32.15) will show that the rise in current as a function of time is given by

$$i = \frac{V_B}{R}(1 - e^{-(R/L)t}) \qquad (32.16)$$

This equation shows that the current i is zero when $t = 0$ and has a maximum of V_B/R when $t = \infty$. The effect of inductance in a circuit is to delay the establishment of this maximum current. The rise and decay of current in an inductive current are shown in Fig. 32.5.

The time constant for an inductive circuit is

$$\tau = \frac{L}{R} \qquad (32.17)$$

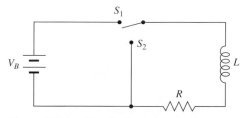

Figure 32.4 Circuit for studying inductance.

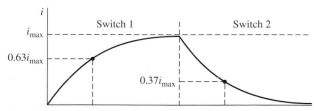

Figure 32.5 Rise and decay of current in an inductor.

τ is in *seconds* when L is in *henrys* and R is in *ohms*. Insertion of this value into Eq. (32.16) shows that

> In an inductive circuit, the current will rise to 63 percent of its final value in one time constant (L/R).

After the current in Fig. 32.4 has attained a steady value, if the switch is moved to position S_2, the current will decay exponentially, as shown in Fig. 32.5. The equation for the decay is

$$i = \frac{V_B}{R}e^{-(R/L)t} \tag{32.18}$$

Substitution of L/R into Eq. (32.18) shows that

> In an inductive circuit, the current decays to 37 percent of its initial value in one time constant (L/R).

Once again, for practical reasons the rise or decay time for an inductor is considered to be five time constants ($5L/R$).

32.3   Alternating Currents

Now that we are familiar with the basic elements in an ac circuit, it is necessary to understand more about alternating currents. The quantitative description of an alternating current is much more complicated than that for direct currents, whose magnitude and direction are constant. An alternating current flows back and forth in a circuit and has no "direction" in the sense direct current has. In addition, the magnitude varies sinusoidally with time, as we learned in our discussions of the ac generator.

The variation in emf or current for an ac circuit can be represented by a rotating vector or by a sine wave. These representations are compared in Fig. 32.6. The vertical component of the rotating vector at any instant is the instantaneous magnitude of the voltage or the current. One complete revolution of the rotating vector or one complete sine wave on the curve represents one *cycle*. The number of complete cycles per second experienced by an alternating current is called its ***frequency*** and provides an important description of the current. The relationship between the instantaneous emf $\mathscr{E}$ or the instantaneous current i and

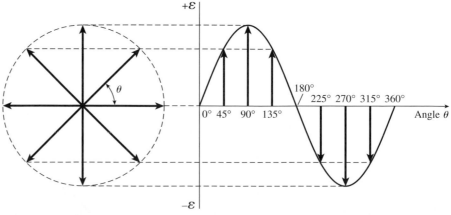

Figure 32.6 Rotating vector and its corresponding sine wave can be used to represent ac current or voltage.

the frequency was established in Chapter 31:

$$\mathcal{E} = \mathcal{E}_{max} \sin 2\pi ft \qquad (32.19)$$

$$i = i_{max} \sin 2\pi ft \qquad (32.20)$$

Note that the average value for the current in an ac circuit is zero since the magnitude alternates between i_{max} and $-i_{max}$. Even though there is no *net* current, however, charge is in motion, and electric energy can be released in the form of heat or useful work. The best method of measuring the effective strength of alternating currents is to find the dc value that will produce the same *heating* effect or develop the same *power* as the alternating current. This current value, called the **effective current** i_{eff}, is found to be 0.707 times the maximum current. A similar relation holds for the effective emf or voltage in an ac circuit. Thus,

$$i_{eff} = 0.707 i_{max} \qquad (32.21)$$

$$\mathcal{E}_{eff} = 0.707 \mathcal{E}_{max} \qquad (32.22)$$

One effective ampere is that alternating current that will develop the same power as 1 ampere of direct current.

One effective volt is that alternating voltage that will produce an effective current of 1 ampere through a resistance of 1 ohm.

AC meters are calibrated to show effective values. For example, if an ac meter measures household voltage to be 120 V at 10 A, Eqs. (32.12) and (32.22) will show that the maximum values of current and voltage are

$$i_{max} = \frac{10 \text{ A}}{0.707} = 14.14 \text{ A}$$

$$V_{max} = \frac{120 \text{ V}}{0.707} = 170 \text{ V}$$

Therefore, the house voltage actually varies between $+170$ and -170 V, and the current ranges from $+14.14$ to -14.14 A. The usual frequency of voltage variation is 60 Hz.

32.4 Phase Relation in AC Circuits

In all dc circuits, the voltage and the current reach maximum and zero values at the same time and are said to be *in phase*. The effects of inductance and capacitance in ac circuits prevent the voltage and current from reaching maxima and minima at the same time. In other words, the current and voltage in most ac circuits are *out of phase*.

To understand phase relations in an ac circuit, suppose we first consider a circuit containing a *pure* resistor in series with an ac generator, as in Fig. 32.7. This is an idealized circuit in which the inductive and capacitive effects are negligible. Many household devices, such as lights, heaters, and toasters, approximate a condition of pure resistance. In such devices, the instantaneous voltage V and current i are in phase. That is, variations in voltage will result in simultaneous variations of current. When the voltage is a maximum, the current is also a maximum. When the voltage is zero, the current is zero.

Next, we consider the phase relation between current and voltage across a *pure inductor*. The circuit illustrated in Fig. 32.8 contains only an inductor in series with the ac generator. We have seen that the presence of inductance in a circuit whose current is changing at the rate $\Delta i/\Delta t$ results in a back emf

$$\mathcal{E} = -L\frac{\Delta i}{\Delta t}$$

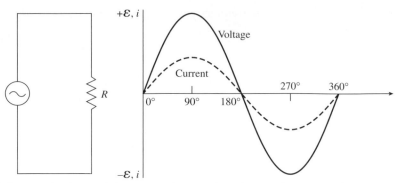

Figure 32.7 In a circuit containing pure resistance, the voltage and current are in phase.

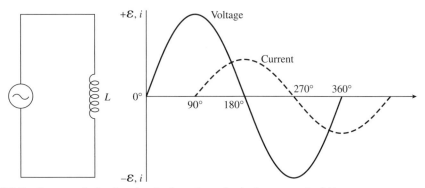

Figure 32.8 In a pure inductive circuit, the voltage leads the current by 90°.

which delays the current in reaching its maximum. The voltage reaches a maximum, while the current is still at zero. When the voltage reaches a minimum, the current is at a maximum. In a circuit containing only inductance, the voltage is said to lead (occur before) the current by one-fourth of a cycle (or 90°). See the curve in Fig. 32.8.

> In a circuit containing pure inductance, the voltage leads the current by 90°.

The effect of capacitance in an ac circuit is opposite to that of inductance. For the circuit shown in Fig. 32.9, the voltage must *lag behind* the current since the flow of charge to the capacitor is necessary to build up an opposing emf. When the applied voltage is decreasing, charge flows from the capacitor. The rate of flow of this charge reaches a maximum when the applied voltage is zero.

> In a circuit containing pure capacitance, the voltage lags behind the current by 90°.

This means that the variations in voltage occur one-fourth of a cycle *later* than the corresponding variations in current.

A useful mnemonic device for remembering the phase relationships for capacitive and inductive circuits is

"$\mathscr{E}LI$ the $IC\mathscr{E}$ man"

Remember that $\mathscr{E}$ represents emf, I represents current, C represents capacitance, and L represents inductance. $\mathscr{E}LI$ the $IC\mathscr{E}$ man tells us that the voltage $\mathscr{E}$ leads the current I in an inductor L, and that the current I leads the voltage $\mathscr{E}$ in a capacitor C.

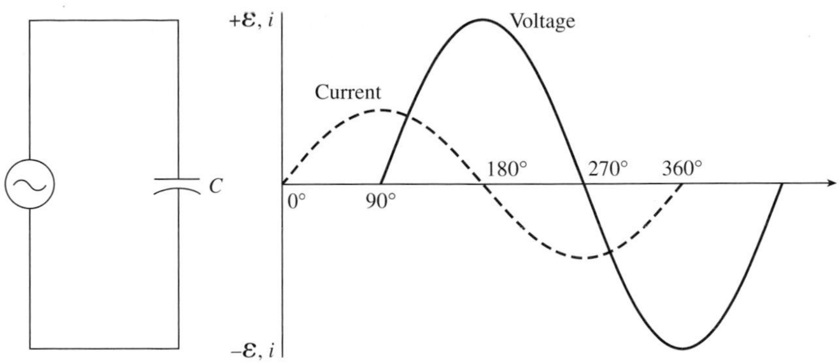

Figure 32.9 In a circuit containing only capacitance, the voltage lags the current by 90°.

32.5   Reactance

In a dc circuit, the only opposition to current results from the material through which charge passes. Relative heat losses, which obey Ohm's law, also occur in ac circuits. However, for alternating currents, we must also contend with inductance and capacitance. Both inductors and capacitors *impede* the flow of an alternating current, and their effects must be considered along with the opposition of normal resistance.

> The reactance of an ac circuit may be defined as its nonresistive opposition to the flow of alternating current.

We first consider the opposition to the flow of an alternating current through an inductor. Such opposition, called **inductive reactance,** arises from the self-induced back emf produced by a changing current. The magnitude of the inductive reactance X_L is determined by the inductance L of the inductor and by the frequency f of the alternating current and can be found from the formula

$$X_L = 2\pi f L \tag{32.23}$$

The inductive reactance is measured in *ohms* when the inductance is in *henrys* and the frequency is in *hertz*.

The effective current i in an inductor is determined from its inductive reactance X_L and the effective voltage V by an equation analogous to Ohm's law:

$$V = iX_L \tag{32.24}$$

Example 32.2   A coil having an inductance of 0.5 H is connected to a 120-V, 60-Hz power source. If the resistance of the coil is neglected, what is the effective current through the coil?

Solution: The inductive reactance is

$$X_L = 2\pi f L = (2\pi)(60 \text{ Hz})(0.5 \text{ H}) = 188.4 \ \Omega$$

The current is found from Eq. (32.24):

$$i = \frac{V}{X_L} = \frac{120 \ \Omega}{188.4 \ \Omega} = 0.637 \text{ A}$$

Opposition to alternating current is also experienced because of the capacitance in a circuit. The ***capacitive reactance*** X_C is found from

$$X_C = \frac{1}{2\pi fC} \tag{32.25}$$

where C = capacitance

f = frequency of alternating current

Capacitive reactance is expressed in *ohms* when C is in *farads* and f is in *hertz*. Once the capacitive reactance X_C of a capacitor is known, the effective current i can be found from

$$V = iX_C \tag{32.26}$$

where V is the applied voltage.

32.6 The Series AC Circuit

In general, an ac circuit contains resistance, capacitance, and inductance in varying amounts. A series combination of these parameters is illustrated in Fig. 32.10. The total voltage drop in a dc circuit is the simple sum of the drop across each element in the circuit. In the ac circuit, however, the voltage and current are not in phase with each other. Recall that V_R is always in phase with the current but V_L leads the current by 90° and V_C lags the current by 90°. Clearly, if we are to determine the effective voltage V of the entire circuit, we must develop a means of treating phase differences.

This can best be accomplished by using a vector diagram, called the ***phase diagram.*** (See Fig. 32.11.) In this method, the effective values of V_R, V_L, and V_C are plotted as rotating vectors. The phase relationship is expressed in terms of the ***phase angle*** Φ, which is a measure of how much the voltage leads the current in a particular circuit element. For example, in a pure resistor, the voltage and the current are in phase, and $\Phi = 0$. For an inductor, $\Phi = +90°$, and in a capacitor $\Phi = -90°$. The negative phase angle occurs when the voltage lags behind the current. Following this scheme, V_R appears as a vector along the x axis, V_L is represented by a vector pointing vertically upward, and V_C is directed downward.

The effective voltage V in an ac circuit can be defined as the vector sum of V_R, V_L, and V_C as they exist on the phase diagram. It can be seen from Fig. 32.11 that the magnitude of **V** is

$$V = \sqrt{V_R^2 + (V_L - V_C)^2} \tag{32.27}$$

You should verify this equation by applying Pythagoras's theorem to the vector diagram.

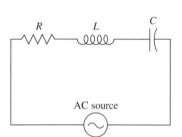

Figure 32.10 Series ac circuit containing resistance, inductance, and capacitance.

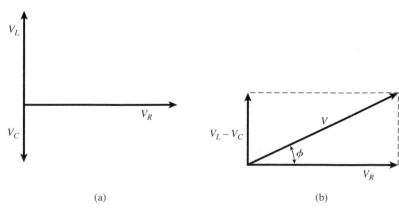

(a)   (b)

Figure 32.11 Phase diagram.

Note from the phase diagram that a value of V_L that is greater than V_C results in a positive phase angle. In other words, if the circuit is inductive, the voltage leads the current. In a capacitive circuit, $X_C > X_L$, and a negative phase angle will result, indicating that the voltage lags the current. In any case, the magnitude of the phase angle can be found from

$$\tan \Phi = \frac{V_L - V_C}{V_R} \qquad (32.28)$$

A more useful form of Eq. (32.27) can be found by recalling that

$$V_R = iR \qquad V_L = iX_L \qquad V_C = iX_C$$

Upon substitution, we find that

$$V = i\sqrt{R^2 + (X_L - X_C)^2} \qquad (32.29)$$

The quantity multiplied by the current in Eq. (32.29) is a measure of the combined opposition the circuit offers to alternating current. It is called the ***impedance*** and is denoted by the symbol Z.

$$Z = \sqrt{R^2 + (X_L - X_C)^2} \qquad (32.30)$$

The higher the impedance in a circuit, the lower the current for a given voltage. Since R, X_L, and X_C are measured in *ohms*, the impedance is also expressed in ohms.

Therefore, the effective current i in an ac circuit is given by

$$i = \frac{V}{Z} \qquad (32.31)$$

where V = applied voltage

Z = impedance in circuit

It must be remembered that Z depends on the frequency of the alternating current as well as on the resistance, inductance, and capacitance.

Since the voltage across each element depends directly on resistance or reactance, an alternative phase diagram can be constructed by treating R, X_L, and X_C as vector quantities. Such a diagram can be used to compute the impedance, as indicated in Fig. 32.12. The phase angle Φ in this representation can be found from

$$\tan \Phi = \frac{X_L - X_C}{R} \qquad (32.32)$$

Of course, this angle is the same as that given by Eq. (32.28).

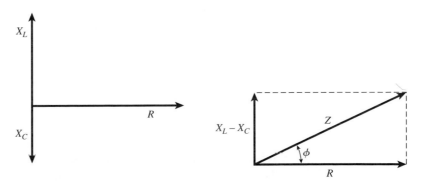

Figure 32.12   Impedance diagram.

Example 32.3

A 40-Ω resistor, a 0.4-H inductor, and a 10-μF capacitor are connected in series with an ac source that generates 120-V, 60 Hz, alternating current. (a) Find the impedance of the circuit. (b) What is the phase angle? (c) Determine the effective current in the circuit.

Plan: We will first calculate the inductive reactance X_L and capacitive reactance X_C based on their definitions. Then, we will find the impedance by combining the reactance with the circuit resistance, using Eq. (32.30). The effective current is the ratio of the applied voltage to the impedance, and the phase angle is found from the tangent function.

Solution (a): The inductive reactance and capacitive reactance are found as follows:

$$X_L = 2\pi f L = (2\pi)(60 \text{ Hz})(0.4 \text{ H})$$
$$= 151 \ \Omega$$

$$X_C = \frac{1}{2\pi f C} = \frac{1}{(2\pi)(60 \text{ Hz})(10 \times 10^{-6} \text{ F})}$$
$$= 265 \ \Omega$$

The impedance of the circuit is

$$Z = \sqrt{R^2 + (X_L - X_C)^2}$$
$$= \sqrt{(40 \ \Omega^2) + (151 \ \Omega - 265 \ \Omega)^2} = 121 \ \Omega$$

Solution (b): From Eq. (32.32), the phase angle is

$$\tan \Phi = \frac{151 \ \Omega - 265 \ \Omega}{40 \ \Omega} = -2.85$$

$$\Phi = -71°$$

The negative sign is used to indicate the phase angle is in the fourth quadrant.

Solution (c): Finally, we can determine the effective current from the known impedance.

$$i = \frac{V}{Z} = \frac{120 \text{ V}}{121 \ \Omega} = 0.992 \text{ A}$$

The negative phase angle indicates the voltage will lag behind this effective current. The circuit is more capacitive than it is inductive.

32.7   Resonance

Since inductance causes the current to lag behind the voltage and capacitance causes the current to lead the voltage, their combined effect is to cancel each other. The total reactance is given by $X_L - X_C$, and the impedance in a circuit is a minimum when $X_L = X_C$. When this is true, only the resistance R remains, and the current will be a maximum. Setting $X_L = X_C$, we can write

$$2\pi f_r L = \frac{1}{2\pi f_r C}$$

and

$$f_r = \frac{1}{2\pi \sqrt{LC}} \qquad\qquad (32.33)$$

When the applied voltage has this frequency, called the *resonant frequency,* the current in the circuit will be a maximum. Additionally, it should be pointed out that since the current is limited only by resistance, it will be in phase with the voltage.

The antenna circuit in a radio receiver contains a variable capacitor that acts as a tuner. The capacitance is varied until the resonant frequency is equal to a particular signal frequency. The current peaks when this happens, and the receiver responds to the incoming signal.

32.8 The Power Factor

In ac circuits, no power is consumed because of capacitance or inductance. Energy is merely stored at one instant and released at another, causing the current and voltage to be out of phase. Whenever the current and voltage are in phase, the power P delivered is a maximum given by

$$P = iV$$

where i = effective current

V = *effective voltage*

This condition is satisfied when the ac circuit contains only resistance R or when the circuit is in resonance ($X_L = X_C$).

Normally, however, an ac circuit contains sufficient reactance to limit the effective power. In any case, the power delivered to the circuit is a function only of the component of the voltage V that is in phase with the current. From Fig. 32.11, this component is V_R, and we can write

$$V_R = V \cos \Phi$$

where Φ is the phase angle. Thus, the effective power consumed in an ac circuit is

$$P = iV \cos \Phi \tag{32.34}$$

The quantity $\cos \Phi$ is called the ***power factor*** of the circuit. Note that $\cos \Phi$ can vary from zero in a circuit containing pure reactance ($\Phi = 90°$) to unity in a circuit containing only resistance ($\Phi = 0°$).

Equation (32.30) and Fig. 32.12 show that the power factor can also be found from

$$\cos \Phi = \frac{R}{Z} = \frac{R}{\sqrt{R^2 + (X_L - X_C)^2}} \tag{32.35}$$

Example 32.4

(a) What is the power factor for the circuit described in Example 32.3? (b) What power is absorbed in the circuit?

Solution (a): The power factor is found in Eq. (32.35):

$$\cos \Phi = \frac{R}{Z} = \frac{40 \ \Omega}{121 \ \Omega} = 0.33$$

Solution (b): The power absorbed in the circuit is

$$P = iV \cos \Phi = (0.992 \text{ A})(120 \text{ V})(0.33)$$
$$= 39.3 \text{ W}$$

The power factor is sometimes expressed as a percentage instead of as a decimal. For example, the power factor of 0.33 in Example 32.4 could be expressed as 33 percent. Most commercial ac circuits have power factors from 80 to 90 percent because they usually contain more inductance than capacitance. Since this requires the electric-power companies to furnish more current for a given power, the power companies extend a lower rate to users with power factors above 90 percent. Commercial users can improve their inductive power factors by adding capacitors, for instance.

Summary and Review

Summary

There are three principal elements in ac circuits: the *resistor,* the *capacitor,* and the *inductor.* A resistor is affected by ac current in the same manner as for dc circuits, and the current is determined by Ohm's law. The capacitor regulates and controls the flow of charge in an ac circuit; its opposition to the flow of electrons is called *capacitive reactance.* The inductor experiences a self-induced emf that adds *inductive reactance* to the circuit. The combined effect of all three elements in opposing electric current is called *impedance.* The major points to remember are summarized as follows.

- When a capacitor is being charged, the instantaneous values of the charge Q and the current i are found from

$$Q = CV_B(1 - e^{-t/RC}) \qquad i = \frac{V_B}{R}e^{-t/RC}$$

- The charge on the capacitor will rise to 63 percent of its maximum value as the current delivered to the capacitor decreases to 37 percent of its initial value during a period of one time constant τ.

$$\tau = RC \qquad \textit{Time Constant}$$

- When a capacitor is discharging, the instantaneous values of the charge and current are given by

$$Q = CV_B e^{-t/RC} \qquad i = \frac{-V_B}{R}e^{-t/RC}$$

Both the charge and the current decay to 37 percent of their initial values after discharging for one time constant.

- When alternating current passes through a coil of wire, an inductor, a self-induced emf arises to oppose the change. This emf is given by

$$\mathscr{E} = -L\frac{\Delta i}{\Delta t} \qquad L = -\frac{\mathscr{E}}{\Delta i/\Delta t}$$

This constant L is called the *inductance.* An inductance of one henry (H) exists if an emf of 1 V is induced by a current changing at the rate of 1 A/s.

- The rise and decay of current in an inductor are found from

$$i = \frac{V_B}{R}(1 - e^{-(R/L)t}) \qquad \textit{Current Rise}$$

$$i = \frac{V_B}{R}e^{-(R/L)t} \qquad \textit{Current Decay}$$

- In an inductive circuit, the current will rise to 63 percent of its maximum value or decay to 37 percent of its maximum in a period of one time constant. For an inductor, the time constant is

$$\tau = \frac{L}{R} \qquad \textit{Time Constant}$$

- Since alternating currents and voltages vary continuously, we speak of an *effective ampere* and an *effective volt* that are defined in terms of their maximum values as follows:

$$i_{\text{eff}} = 0.707 i_{\text{max}} \qquad \mathscr{E}_{\text{eff}} = 0.707 \mathscr{E}_{\text{max}}$$

- Both capacitors and inductors offer resistance to the flow of alternating current (called *reactance*), calculated from

$$X_L = 2\pi f L \qquad \textit{Inductive Reactance } X_L$$

$$X_C = \frac{1}{2\pi f C} \qquad \textit{Capacitive Reactance } X_C$$

The symbol f refers to the frequency of the alternating current in hertz. One hertz is one cycle per second.

- The voltage, current, and resistance in a series ac circuit are studied with the use of phasor diagrams. Figure 32.12 illustrates such a diagram for X_C, X_L, and R. The resultant of these vectors is the effective resistance of the entire circuit jcalled the *impedance* Z.

$$Z = \sqrt{R^2 + (X_L - X_C)^2} \qquad \textit{Impedance}$$

- If we apply Ohm's law to each part of the circuit and then to the entire circuit, we obtain the following useful equations. First, the total voltage is given by

$$V = \sqrt{V_R^2 + (V_L - V_C)^2} \qquad \textit{Voltage}$$

$$V_R = iR \qquad V_L = iX_L$$

$$V_C = iX_C \qquad V = iZ$$

$$V = i\sqrt{R^2 + (X_L - X_C)^2} \qquad \textit{Ohm's Law}$$

- Because the voltage leads the current in an inductive circuit and lags the current in a capacitive circuit, the voltage and current maxima and minima usually do not coincide. The phase angle Φ is given by

$$\tan \Phi = \frac{V_L - V_C}{V_R}$$

or

$$\tan \Phi = \frac{X_L - X_C}{R}$$

- The resonant frequency occurs when the net reactance is zero ($X_L = X_C$):

$$f_r = \frac{1}{2\pi\sqrt{LC}} \qquad \textit{Resonant Frequency}$$

- No power is consumed because of capacitance or inductance. Since power is a function of the component of the impedance along the resistance axis, we can write

$$P = iV \cos \Phi \qquad \textit{Power Factor} = \cos \Phi$$

$$\cos \Phi = \frac{R}{Z}$$

$$\cos \Phi = \frac{R}{\sqrt{R^2 + (X_L - X_C)^2}}$$

Key Terms

Review Questions

32.1. Inductance in a circuit depends on its geometry and on the proximity of magnetic materials. Which of the following coils have higher inductances, and why?

  a. closely or widely spaced turns

  b. long coil or short coil

  c. large cross section or small cross section

  d. iron core or air core

32.2. Show that the unit for the time constant L/R and RC for inductive and capacitive circuits is the second.

32.3. Sketch a curve of current versus time for (a) a circuit of high inductance and (b) a circuit of lower inductance.

32.4. Plot a curve of voltage versus time for (a) a charging capacitor and (b) a discharging capacitor. Compare these curves with those in the text for the current.

32.5. By plotting a curve of voltage as a function of time, show how the voltage changes in an inductive circuit. Compare these curves with those in the text for the current.

32.6. Should someone interested in establishing the breakdown voltage of a capacitor in an ac circuit be concerned with the average voltage, maximum voltage, or effective voltage?

32.7. A coil of wire is connected across the terminals of a 110-V dc battery. An ammeter connected in series with the coil indicates a current of 5 A. What happens to the current if an iron core is inserted into the coil? Now, disconnect the dc source, remove the iron core, and reconnect the system to a 110-V ac generator. The ammeter is adjusted to read ac effective amperes. Has the current decreased, increased, or stayed the same? What happens to the current if the iron core is inserted? How do you account for your observations?

32.8. An incandescent lamp is connected in series with a 110-V ac generator and a variable capacitor. If the capacitance is increased, will the lamp glow more brightly, or will it be dimmed? Explain. What would happen if the generator is replaced by a battery?

32.9. As the capacitance of a circuit increases, what happens to the resonant frequency of the circuit?

32.10. As the frequency is increased in an inductive circuit, what happens to the current in the circuit?

32.11. Inductive reactance depends *directly* on the frequency of the alternating current, whereas capacitive reactance varies *inversely* with the frequency. Both oppose the flow of charge in an ac circuit. Explain why their relationship to the frequency differs.

32.12. When a circuit is tuned to its resonant frequency, what is the power factor?

32.13. What is the power factor of a circuit containing (a) pure resistance, (b) pure inductance, (c) pure capacitance?

Problems

Section 32.1 The Capacitor

32.1. A series dc circuit consists of a 4-μF capacitor, a 5000-Ω resistor, and a 12-V battery. What is the time constant for this circuit?   Ans. 20 ms

32.2. What is the time constant for a series dc circuit containing a 6-μF capacitor and a 400-Ω resistor connected to a 20-V battery?

32.3. For the circuit described in Prob. 32.1, what are the initial current and the final current? How much time is needed to be assured that the capacitor is fully charged?   Ans. 2.40 mA, 0, 100 ms

32.4. For the circuit in Prob. 32.2, what is the maximum charge for the capacitor, and how much time is required to fully charge the capacitor?

*32.5. An 8-μF capacitor is connected in series with a 600-Ω resistor and a 24-V battery. After a time equal to one time constant, what are the charge on the capacitor and the current in the circuit?
          Ans. 121 μC, 14.7 mA

*32.6. Assume the fully charged capacitor of Prob. 32.5 is allowed to discharge. After one time constant, what are the current in the circuit and the charge on the capacitor?

*32.7. Assume the 8-μF capacitor in Prob. 32.5 is fully charged and is then allowed to decay. Find the time when the current in the circuit will have decayed to 20 percent of its initial value.
          Ans. 7.73 ms

*32.8. A 5-μF capacitor is connected in series with a 12-V source of emf and a 4000-Ω resistor. How much time is required to place 40 μC of charge on the capacitor?

Section 32.2 The Inductor

32.9. A series dc circuit contains a 4-mH inductor and an 80-Ω resistor in series with a 12-V battery. What is the time constant for the circuit? What are the initial and final currents?
          Ans. 50 μs, 0, 150 mA

32.10. A 5-mH inductor, a 160-Ω resistor, and a 50-V battery are connected in series. What time is required for the current in the inductor to reach 63 percent of its steady-state value? What is the current at that instant?

32.11. What is the current in Prob. 32.9 after a time of one time constant?   Ans. 94.8 mA

32.12. Assume the inductor in Prob. 32.10 has reached its steady-state value. If the circuit is broken, how much time must pass before we can be sure that the current in the inductor is zero?

32.13. The current in a 25-mH solenoid increases from 0 to 2 A in a time of 0.1 s. What is the magnitude of the self-induced emf?   Ans. −500 mV

32.14. A series dc circuit contains a 0.05-H inductor and a 40-Ω resistor in series with a 90-V source of emf. What are the maximum current in the circuit and the time constant?

*32.15. What is the instantaneous current in the inductor of Prob. 32.14 after a decay time of 1.0 ms?
          Ans. 1.24 A

*32.16. A 6-mH inductor, a 50-Ω resistor, and a 38-V battery are connected in series. At what time after the connection will the current reach an instantaneous value of 600 mA?

Section 32.3 Alternating Currents

32.17. An ac voltmeter, when placed across a 12-Ω resistor, reads 117 V. What are the maximum values for the voltage and current?
          Ans. 165 V, 13.8 A

32.18. In an ac circuit, the voltage and current reach maximum values of 120 V and 6.00 A. What are the effective ac values?

32.19. A capacitor has a maximum voltage rating of 500 V. What is the highest effective ac voltage that can be supplied to it without breakdown?

Ans. 354 V

32.20. A certain appliance is supplied with an effective voltage of 220 V under an effective current of 20 A. What are the maximum and minimum values?

Section 32.5 Reactance

32.21. A 6-μF capacitor is connected to a 40-V 60-Hz ac line. What is the reactance? What is the effective ac current in the circuit containing pure capacitance?

Ans. 442 Ω, 90.5 mA

32.22. A 2-H inductor of negligible resistance is connected to a 50-V 50-Hz ac line. What is the reactance? What is the effective ac current in the coil?

32.23. A 50-mH inductor of negligible resistance is connected to a 120-V 60-Hz ac line. What is the inductive reactance? What is the effective ac current in the circuit?

Ans. 18.8 Ω, 6.37 A

32.24. A 6-μF capacitor is connected to a 24-V 50-Hz ac source. What is the current in the circuit?

32.25. A 3-μF capacitor connected to a 120-V ac line draws an effective current of 0.5 A. What is the frequency of the source?

Ans. 221 Hz

32.26. Find the reactance of a 60-μF capacitor in a 600-Hz ac circuit. What is the reactance if the frequency is reduced to 200 Hz?

32.27. The frequency of an alternating current is 200 Hz, and the inductive reactance for a single inductor is 100 Ω. What is the inductance?

Ans. 79.6 mH

*32.28. A 20-Ω resistor, a 2-μF capacitor, and a 0.70-H inductor are available. Each of these, in turn, is connected to a 120-V 60-Hz ac source as the only circuit element. What is the effective ac current in each case?

Section 32.6 The Series AC Circuit

*32.29. A 300-Ω resistor, a 3-μF capacitor, and a 4-H inductor are connected in series with a 90-V 50-Hz ac source. What is the net reactance of the circuit? What is the impedance?

Ans. 196 Ω, 358 Ω

*32.30. What is the effective ac current delivered to the ac series circuit described in Prob. 32.29? What is the peak value for this current?

*32.31. A series ac circuit consists of a 100-Ω resistor, a 0.2-H inductor, and a 3-μF capacitor connected to a 110-V 60-Hz ac source. What are the inductive reactance, the capacitative reactance, and the impedance for the circuit?

Ans. 75.4 Ω, 884 Ω, 815 Ω

*32.32. What are the phase angle and the power factor for the circuit described in Prob. 32.31?

*32.33. Assume all the elements of Prob. 32.28 are connected in series with the given source. What is the effective current delivered to the circuit?

Ans. 113 mA

*32.34. A series ac circuit contains a 12-mH inductor, an 8-μF capacitor, and a 40-Ω resistor connected to a 110-V 200-Hz ac line. What is the effective ac current in the circuit?

*32.35. When a 6-Ω resistor and a pure inductor are connected to a 100-V 60-Hz ac line, the effective current is 10 A. What is the inductance? What is the power loss through the resistor, and what is the power loss through the inductor?

Ans. 24.5 mH, 600 W, 0

*32.36. A capacitor is in series with a resistance of 35 Ω and connected to a 220-V ac line. The reactance of the capacitor is 45 Ω. What is the effective ac current? What is the phase angle? What is the power factor?

*32.37. A coil having an inductance of 0.15 H and a resistance of 12 Ω is connected to a 110-V 25-Hz line. What is the effective current in the circuit? What is the power factor? What power is lost in the circuit?

Ans. 4.16 A, 45.4 percent, 208 W

*32.38. What is the resonant frequency for the circuit described in Prob. 32.29?

*32.39. What is the resonant frequency for the circuit described in Prob. 32.34?

Ans. 514 Hz

Additional Problems

32.40. For the circuit drawn in Fig. 32.13, find the current leaving the battery after one time constant: (a) when the switch is closed to S_1. (b) when the switch is closed to S_2.

32.41. If Switch 2 in Fig. 32.13 is closed for several minutes and then suddenly opened, what will be the instantaneous current after one time constant?

Ans. 88.3 mA

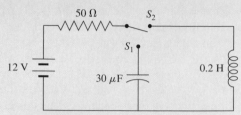

Figure 32.13

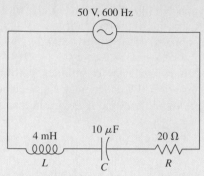

Figure 32.14

32.42. A resonant circuit has an inductance of 400 μH and a capacitance of 100 pF. What is the resonant frequency?

32.43. An *LR* dc circuit has a time constant of 2 ms. What is the inductance if the resistance is 2 kΩ? What is the instantaneous current 2 ms after the circuit is connected to a 12-V battery?
Ans. 4.00 H, 3.79 mA

32.44. A series dc circuit consists of a 12-V battery, a 20-Ω resistor, and an unknown capacitor. The time constant is 40 ms. What is the capacitance? What is the maximum charge on the capacitor?

*32.45. A 2-H inductor having a resistance of 120 Ω is connected to a 30-V battery. How much time is required for the current to reach 63 percent of its maximum value? What is the initial rate of current increase in amperes per second? What is the final current? Ans. 16.7 ms, 15 A/s, 250 mA

*32.46. Consider the circuit shown in Fig. 32.14. What is the impedance? What is the effective current? What is the power loss in the circuit?

*32.47. A tuner circuit contains a 4-mH inductor and a variable capacitor. What must be the capacitance if the circuit is to resonate at a frequency of 800 Hz? Ans. 9.89 μF

*32.48. An 8-μF capacitor is in series with a 40-Ω resistor and connected to a 117-V 60-Hz ac source. What is the impedance? What is the power factor? How much power is lost in the circuit?

*32.49. Someone wants to construct a circuit that has a resonant frequency of 950 kHz. If a coil in the circuit has an inductance of 3 mH, what capacitance should be added to the circuit? Ans. 9.36 pF

*32.50. A 50-μF capacitor and a 70-Ω resistor are connected in series across a 120-V 60-Hz ac line. Determine the current in the circuit, the phase angle, and the total power loss.

*32.51. Refer to Prob. 32.50. What is the voltage across the resistor? What is the voltage across the capacitor? What inductance should be added to the circuit to reach resonance?
Ans. 95.6 V, 72.5 V, 141 mH

Critical Thinking Questions

*32.52. The resistance of an 8-H inductor is 200 Ω. If this inductor is suddenly connected across a potential difference of 50 V, what is the initial rate of increase of the current in A/s? [See Eq. (32.15).] What is the final steady current? At what rate is the current increasing after one time constant. At what time after connecting the source of emf, does the current equal one-half of its final value?
Ans. 250 mA, 6.25 A/s, 2.31 A/s, 27.7 ms

*32.53. In Chapter 29, we found that the magnetic flux density within a solenoid was given by:

$$B = \frac{\Phi}{A} = \frac{\mu_0 NI}{l}$$

Here we have used *l* for the length of the solenoid in order to avoid confusion with the symbol *L* used for inductance. We also know that induced emf is found in two ways:

$$\mathcal{E} = -N\frac{\Delta\Phi}{\Delta t} \quad \text{and} \quad \mathcal{E} = -L\frac{\Delta i}{\Delta t}$$

Prove the following relationships for a solenoid of length *l* having *N* turns of area *A*:

$$L = \frac{N\Phi}{I} \quad \text{and} \quad L = \frac{\mu N^2 A}{l}$$

*32.54. An inductor consists of a coil 30 cm long with 300 turns of area 0.004 m^2. Find the inductance (see previous Prob. 32.53). A battery is connected, and the current builds from zero to 2.00 A in 0.001 s. What is the average induced emf in the coil?   Ans. 1.51 mH, −3.02 V

*32.55. An inductor, a resistor, and a capacitor are connected in series with a 60-Hz ac line. A voltmeter connected to each element in the circuit gives the following readings: $V_R = 60$ V, $V_L = 100$ V, and $V_C = 160$ V. What is the total voltage drop in the circuit? What is the phase angle?

*32.56. The antenna circuit in a radio receiver consists of a variable capacitor and a 9-mH coil. The resistance of the circuit is 40 Ω. A 980-kHz radio wave produces a potential difference of 0.2 mV across the circuit. Determine the capacitance required for resonance. What is the current at resonance?   Ans. 2.93 pF, 5 mA

*32.57. A series RLC circuit has elements as follows: $L = 0.6$ H, $C = 3.5$ μF, and $R = 250$ Ω. The circuit is driven by a generator that produces a *maximum* emf of 150 V at 60 Hz. What is the effective ac current? What is the phase angle? What is the average power loss in the circuit?

33

Light and Illumination

The Hubble telescope captured this view of the spiral galaxy, NGC 4414. Based on careful and accurate brightness measurements, astronomers were able to determine light coming from this galaxy required more than 60,000,000 years to reach the Earth. In this chapter, we will study light as the visible region in a spectrum of electromagnetic waves. (*Photo by Hubble Heritage Team [AURA/STS CI/NASA].*)

Objectives

After completing this chapter, you should be able to

1. Discuss the historical investigation into the nature of light and explain how light sometimes behaves as a wave and sometimes as particles.

2. Describe the broad classifications in the electromagnetic spectrum on the basis of frequency, wavelength, or energy.

3. Write and apply formulas for the relationship among velocity, wavelength, and frequency and between energy and frequency for electromagnetic radiation.

4. Describe experiments that will result in a reasonable estimation of the speed of light.

5. Illustrate with drawings your understanding of the formation of shadows, labeling the *umbra* and *penumbra*.

6. Demonstrate your understanding of the concepts of *luminous flux, luminous intensity*, and *illumination* and solve problems similar to those in the text.

An iron bar resting on a table is in thermal equilibrium with its surroundings. From its outward appearance, one would never guess that it is active internally. All objects are continuously emitting radiant heat energy that is related to their temperature. The bar is in thermal equilibrium only because it is radiating and absorbing energy at the same rates. If the balance is upset by placing one end of the bar in a hot flame, the bar becomes more active internally and emits heat energy at a greater rate. As the heating continues to around 600°C, some of the radiation emitted from the bar becomes *visible;* that is, it affects our sense of sight. The color of the bar becomes a dull red, which turns brighter as more heat is supplied.

The radiant energy emitted by the object before this effect is visible consists of ***electromagnetic waves*** of longer wavelengths than red light. Such waves are referred to as ***infrared rays,*** meaning "beyond the red." If the temperature of the bar is increased to around 1500°C, it becomes white hot, indicating a further extension of the radiant energy into the ***visible region.***

This example sets the stage for our discussion of ***light.*** The nature of light is no different fundamentally from the nature of other electromagnetic radiation, for example, heat, radio waves, or ultraviolet radiation. The characteristic that distinguishes light from the other radiation is its energy.

Light is electromagnetic radiation that is capable of affecting the sense of sight.

The energy content of a visible light photon varies from about 2.8×10^{-19} J to around 5.0×10^{-19} J.

33.1   What Is Light?

The answer to that question has been extremely elusive throughout the history of science. The long search for an answer provides an inspiring example of the scientific approach to the solution of a problem. Every hypothesis put forth to explain the nature of light was tested both by logic and by experimentation. The ancient philosophers' contention that visual rays are emitted from the eye to the seen object failed the test of both logic and experience.

By the latter part of the seventeenth century, two theories were advanced to explain the nature of light: the particle (corpuscular) theory and the wave theory. The principal advocate of the corpuscular theory was Sir Isaac Newton. The wave theory was upheld by Christian Huygens (1629–1695), a Dutch mathematician and scientist who was 13 years older than Newton. Each theory set out to explain the characteristics of light observed at that time. Three important characteristics can be summarized as follows:

1. *Rectilinear propagation:* Light travels in straight lines.
2. *Reflection:* When light is incident on a smooth surface, it turns back into the original medium.
3. *Refraction:* The path of light changes when it enters a transparent medium.

According to the corpuscular theory, tiny particles of unsubstantial mass were emitted by light sources such as the Sun or a flame. These particles traveled outward from the source in straight lines at enormous speeds. When the particles entered the eye, the sense of sight was stimulated. Rectilinear propagation was easily explained in terms of particles. In fact, one of the strongest arguments for the corpuscular theory was based on this property. It was reasoned that particles cast sharp shadows, as illustrated in Fig. 33.1a, whereas waves can bend around edges. Such bending of waves, as shown in Fig. 33.1b, is called ***diffraction.***

The sharp shadows formed by light beams indicated to Newton that light must consist of particles. Huygens, on the other hand, explained that the bending of water waves and sound waves around obstacles is easily noticed because of the long wavelengths. He

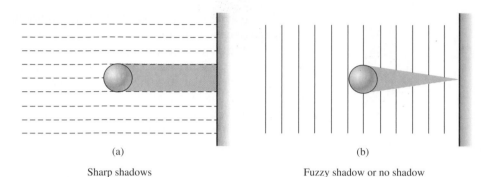

Sharp shadows   Fuzzy shadow or no shadow

Figure 33.1 A strong argument for the particle theory of matter is the formation of sharp shadows. Waves were known to bend around obstacles in their path.

reasoned that if light were a wave with a short wavelength, it would appear to cast a sharp shadow because the amount of bending would be small.

It was also difficult to explain why particles traveling in straight lines from many directions could cross without impeding one another. In a paper published in 1690, Huygens wrote the following:

> If, furthermore, we pay attention to, and weigh up, the extraordinary speed with which light spreads in all directions, and also the fact that coming, as it does, from quite different, indeed from opposite, directions, the rays interpenetrate without impeding one another, then we may well understand that whenever we see a luminous object, this cannot be due to the transmission of matter which reaches us from the object, as for instance a projectile or an arrow flies through the air.

Huygens explained the propagation of light in terms of the motion of a disturbance through the distance between a source and the eye. He based his argument on a simple principle that is still useful today in describing the propagation of light. Suppose we drop a stone into a quiet pool of water. A disturbance is created that moves outward from the point of impact in a series of concentric waves. The disturbance continues even after the stone has struck the bottom of the pool. Such an example prompted Huygens to postulate that disturbances existing at all points along a moving wavefront at one instant can be considered sources for the wavefront at the next instant. Huygens's principle states

> Every point on an advancing wavefront can be considered a source of secondary waves called wavelets. The new position of the wavefront is the envelope of the wavelets emitted from all points of the wavefront in its previous position.

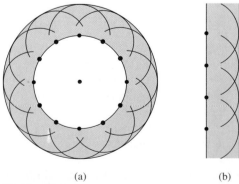

(a)   (b)

Figure 33.2 Illustration of Huygens's principle (a) for a spherical wave and (b) for a plane wave.

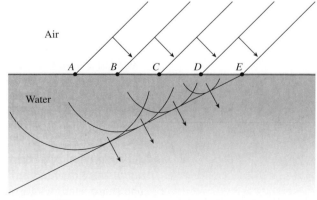

Figure 33.3 Huygens's explanation of refraction in terms of the wave theory.

The application of this principle is illustrated in Fig. 33.2 for the common cases of a plane wave and a circular wave.

Huygens's principle was particularly successful in explaining reflection and refraction. Figure 33.3 shows how the principle can be used to explain the bending of light as it passes from air to water. When the plane waves strike the water surface at an angle, points *A, C,* and *E* become the sources of new wavelets. The envelope of these secondary wavelets indicates a change in direction. A similar construction can be made to explain reflection.

Reflection and refraction were also easily explained in terms of the particle theory. Figures 33.4 and 33.5 illustrate models that can be used to explain reflection and refraction on the basis of tiny corpuscles. Perfectly elastic particles of unsubstantial mass rebounding from an elastic surface could explain the regular reflection of light from smooth surfaces. Refraction could be analogous to the change in direction of a rolling ball as it encounters an incline. This explanation required that the particles of light travel faster in the refracted medium, whereas the wave theory required light to travel more slowly in the refracted medium. Newton recognized that if it could ever be shown that light travels more slowly in a material medium than it does in air, he would have to abandon the particle theory. It was not until the middle of the nineteenth century that Jean Foucault successfully demonstrated that light travels more slowly in water than it does in air.

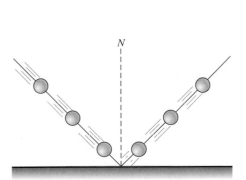

Figure 33.4 Explanation of reflection in terms of the particle theory of light.

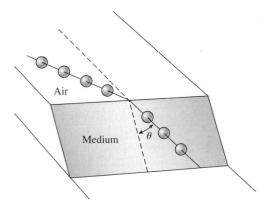

Figure 33.5 Refraction of light as it passes from air to another medium was explained by this mechanical example.

33.2 The Propagation of Light

The discovery of interference and diffraction in 1801 and 1816 swung the debate solidly toward Huygens's wave theory. Clearly, interference and diffraction could be explained only in terms of a wave theory. However, there still remained one problem. All wave phenomena were thought to require the existence of a medium. How, for example, could light waves travel through a vacuum if there were nothing to "vibrate"? Indeed, how could light reach the Earth from the Sun or other stars through millions of miles of empty space? To avoid this contradiction, physicists postulated the existence of a "light-carrying ether." This all-penetrating universal medium was thought to fill all the space between and within all material bodies. But what was the nature of this ether? Certainly it could not be a gas, solid, or liquid that obeyed the physical laws known at that time. And yet the wave theory could not be denied in light of the evidence of interference and diffraction. No other choice seemed possible except for the definition of ether as "that which carries light."

In 1865, a Scottish physicist, James Clerk Maxwell, set out to determine the properties of a medium that would carry light and also account for the transmission of heat and electric energy. His work demonstrated that *an accelerated charge can radiate electromagnetic waves into space.* Maxwell explained that the energy in an electromagnetic wave is equally

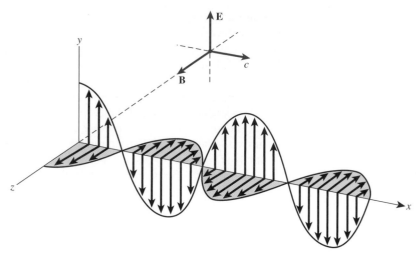

Figure 33.6 Electromagnetic theory holds that light is propagated as oscillating transverse fields. The energy is divided equally between electric **E** and magnetic **B** fields, which are mutually perpendicular.

divided between electric and magnetic fields that are mutually perpendicular. Both fields oscillate perpendicular to the direction of wave propagation, as shown in Fig. 33.6. Thus, a light wave would not have to depend on the vibration of matter. It could be propagated by oscillating transverse fields. Such a wave could "break off" from the region around an accelerating charge and fly off into space with the velocity of light. Maxwell's equations predicted that heat and electric action, as well as light, were propagated at the speed of light as electromagnetic disturbances.

Experimental confirmation of Maxwell's theory was achieved in 1885 by H. R. Hertz, who proved that radiation of electromagnetic energy can occur *at any frequency.* In other words, light, heat radiation, and radio waves are of the same nature, and they all travel at the speed of light (3×10^8 m/s). All types of radiation can be reflected, focused by lenses, polarized, and so forth. It seemed that the wave nature of light could no longer be doubted.

Confirmation of the electromagnetic theory paved the way for the eventual downfall of the "light-carrying ether" postulate. In 1887, A. A. Michelson, an American physicist, showed conclusively that the velocity of light is a constant, independent of the motion of the source. He could not establish any difference between the speed of light traveling in the direction of the Earth's motion and traveling opposite the Earth's motion. Those who are interested in a thorough discussion of the subject should look up the *Michelson–Morley experiment.* Einstein later interpreted Michelson's results to mean that the notion of ether must be abandoned in favor of a completely empty space.

Maxwell was able to show that the speed of any electromagnetic wave is related to the permeability μ and the permittivity ϵ of the medium in which it travels. For free space, we have shown these values to be

$$\mu_0 = 4\pi \times 10^{-7} \, \text{N} \cdot \text{s}^2/\text{C}^2 \qquad \text{and} \qquad \epsilon_0 = 8.85 \times 10^{-12} \, \text{C}^2/\text{N} \cdot \text{m}^2$$

The velocity of light can be found from

$$c = \frac{1}{\sqrt{\mu_0 \epsilon_0}} \tag{33.1}$$

Substitution of the constants reveals the velocity of light is

$$c = 2.99792458 \times 10^8 \, \text{m/s} \tag{33.2}$$

Because electromagnetic waves travel at the same speed as measured for light, the treatment of light as an electromagnetic wave is supported.

With the redefinition of the meter in 1983, the international standard for the measurement of length became the distance traveled by light in a time of 1/299792458 s as measured by a cesium clock. This definition established the exact velocity of light as a universal constant equal to the value given in Eq. (33.2). Useful approximations are 3×10^8 m/s and 186,000 mi/s. These values may be used in most physical calculations without fear of significant error.

33.3 The Electromagnetic Spectrum

Today, the electromagnetic spectrum is known to spread over a tremendous range of frequencies. A chart of the electromagnetic spectrum is presented in Fig. 33.7. The wavelength λ of electromagnetic radiation is related to its frequency f by the general equation

$$c = f\lambda \tag{33.3}$$

where c is the velocity of light (3×10^8 m/s). In terms of wavelengths, the tiny segment of the electromagnetic spectrum referred to as the *visible region* lies between 0.00004 and 0.00007 cm.

Because of the small wavelengths of light radiation, it is more convenient to define smaller units of measure. A common unit is the *nanometer* (nm).

One nanometer (1 nm) is defined as one-billionth of a meter.

$$1 \text{ nm} = 10^{-9} \text{ m} = 10^{-7} \text{ cm}$$

The visible region of the electromagnetic spectrum extends from 400 nm for violet light to approximately 700 nm for red light. Other older units are the *millimicron* (mμ), which is the same as a nanometer, and the *angstrom* (Å), which is 0.1 nm.

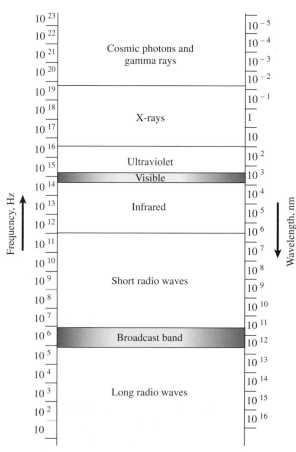

Figure 33.7 Electromagnetic spectrum.

Example 33.1 The wavelength of yellow light from a sodium flame is 589 nm. Compute its frequency.

Solution: The frequency is found from Eq. (33.3).

$$f = \frac{c}{\lambda} = \frac{3 \times 10^8 \text{ m/s}}{589 \times 10^{-9} \text{ m}}$$
$$= 5.09 \times 10^{14} \text{ Hz}$$

Newton was the first to make detailed studies of the visible region by dispersing "white light" through a prism. In order of increasing wavelength, the spectral colors are violet (450 nm), blue (480 nm), green (520 nm), yellow (580 nm), orange (600 nm), and red (640 nm). Anyone who has seen a rainbow has seen the effects that different wavelengths of light have on the human eye.

The electromagnetic spectrum is continuous; there are no gaps between one form of radiation and another. The boundaries we set are purely arbitrary, depending upon our ability to sense one small portion directly and to discover and measure those portions outside the visible region.

The first discovery of radiation of wavelengths longer than those of red light was accomplished in 1800 by William Herschel. These waves are now known as thermal radiation and are referred to as *infrared waves.*

Shortly after the discovery of infrared waves, radiation of wavelengths shorter than visible light were noticed. These waves, now known as *ultraviolet waves,* were discovered in connection with their effect on certain chemical reactions.

How far the infrared region might extend in the direction of longer wavelengths was not known throughout most of the nineteenth century. Fortunately, Maxwell's electromagnetic theory opened the door for the discovery of many other classifications of radiation. The spectrum of electromagnetic waves is now conveniently divided into the eight major regions shown in Fig. 33.7: (1) long radio waves, (2) short radio waves, (3) the infrared region, (4) the visible region, (5) the ultraviolet region, (6) X-rays, (7) gamma rays, and (8) cosmic photons.

33.4 The Quantum Theory

The work of Maxwell and of Hertz in establishing the electromagnetic nature of light waves was truly one of the most important events in the history of science. Not only was the wave nature of light explained, but the door was opened to an enormous range of electromagnetic waves. Remarkably, only two years after Hertz's verficiation of Maxwell's wave equations, the wave theory of light was again challenged. In 1887, Hertz noticed an electric spark would jump more readily between two charged spheres when their surfaces were illuminated by the light from another spark. This phenomenon, known as the *photoelectric effect,* is demonstrated by the apparatus in Fig. 33.8. A beam of light strikes a metal surface A in an evacuated tube. Electrons ejected by the light are drawn to the collector B by external batteries. The flow of electrons is indicated by a device called an *ammeter.* The photoelectric effect defied explanation in terms of the wave theory. In fact, the ejection of electrons could be accounted for more easily in terms of the old particle theory. Still, there could be no doubt of the wave properties either. Science faced a remarkable paradox.

The photoelectric effect, along with several other experiments involving the emission and absorption of radiant energy, could not be accounted for purely in terms of Maxwell's electromagnetic wave theory. In an attempt to bring experimental observation into agreement with theory. Max Planck, a German physicist, published his *quantum hypothesis* in 1901. He found that the problems with the radiation theory lay with the assumption that energy is radiated continuously. It was postulated that electromagnetic energy is absorbed or emitted in

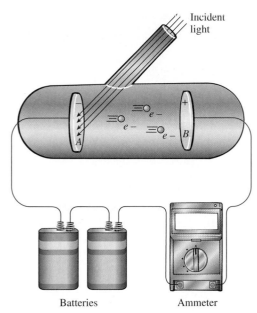

Figure 33.8   The photoelectric effect.

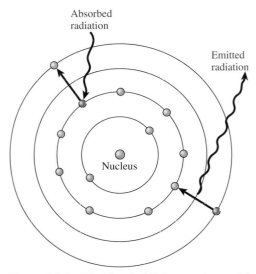

Figure 33.9   The historical Bohr atom is a useful way to visualize transitions among energy levels.

discrete packets, or *quanta*. The energy content of these quanta, or **photons** as they were called, is proportional to the frequency of the radiation. Planck's equation can be written

$$E = hf \tag{33.4}$$

where E = energy of photon

f = frequency of photon

h = proportionality factor called *Planck's constant* $(6.626 \times 10^{-34} \, \text{J} \cdot \text{s})$

In 1905, Einstein extended the idea proposed by Planck and postulated that the energy in a light beam does not spread continuously through space. By assuming that light energy is concentrated in small packets (photons) whose energy content is given by Planck's equation, Einstein was able to predict the photoelectric effect mathematically. At last, theory was reconciled with experimental observation.

Thus, it appears that light is *dualistic.* The wave theory is retained by considering the photon to have a frequency and an energy proportional to the frequency. The modern practice is to use the wave theory when studying the propagation of light. The corpuscular theory, on the other hand, is necessary to describe the interaction of light with matter. We may think of light as radiant energy transported in photons carried along by a wave field.

The origin of light photons was not understood until Niels Bohr in 1913 devised a model of the atom based on **quantum theory.** Bohr postulated that electrons can move about the nucleus of an atom only in certain orbits, or *discrete energy levels,* as shown in Fig. 33.9. Atoms were said to be *quantized.* If the atoms are somehow energized, as by heat, the orbital electrons may jump into a higher orbit. At some later time, these excited electrons will fall back to their original level, releasing as photons the energy that was originally absorbed. Although Bohr's model was not strictly correct, it provided a basis for understanding the emission and absorption of electromagnetic radiation in quantum units.

33.5   Light Rays and Shadows

One of the first properties of light to be studied was rectilinear propagation and the formation of shadows. Instinctively, we rely quite heavily on this property for estimating distances, directions, and shapes. The formation of sharp shadows on a sundial is used to estimate time. In this section, we discuss how we can predict the formation of shadows.

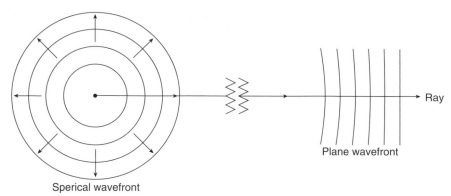

Sperical wavefront

Figure 33.10   A ray is an imaginary line, drawn perpendicular to advancing wavefronts, that indicates the direction of light propagation.

According to Huygens's principle, every point on a moving wavefront can be considered a source for secondary wavelets. The wavefront at any time is the envelope of these wavelets. Thus, the light emitted in all directions by the point source of light in Fig. 33.10 can be represented by a series of spherical wavefronts moving away from the source at the speed of light. For our purposes, a ***point source*** of light is one whose dimensions are small in comparison with the distances studied. Note that the spherical wavefronts become essentially plane wavefronts in any specific direction at a long distance away from the source. An imaginary straight line drawn perpendicular to the wavefronts in the direction of the moving wavefronts is called a ***ray.*** There are, of course, an infinite number of rays starting from the point source.

Any dark-colored object absorbs light, but a black one absorbs nearly all the light it receives. Light that is not absorbed upon striking an object is either reflected or transmitted. If all the light incident upon an object is reflected or absorbed, the object is said to be *opaque.* Since light cannot pass through an opaque body, a shadow will be produced in the space behind the object. The shadow formed by a point source of light is illustrated in Fig. 33.11. Since light is propagated in straight lines, rays drawn from the source past the edges of the opaque object form a sharp shadow proportional to the shape of the object. Such a region in which no light has entered is called an ***umbra.***

If the source of light is an extended one rather than a point, the shadow will consist of two portions, as shown in Fig. 33.12. The inner portion receives no light from the source and is therefore the *umbra.* The outer portion is called the ***penumbra.*** An observer within the penumbra would see a portion of the source but not all the source. An observer located outside both regions would see all the source. Solar and lunar eclipses can be studied by similar construction of shadows.

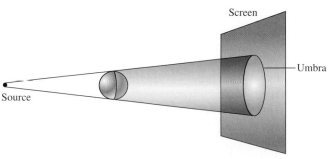

Figure 33.11   Shadow formed by a point source of light.

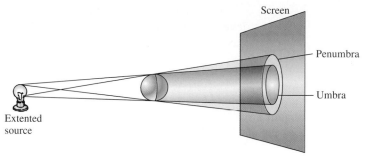

Figure 33.12   Shadows formed by an extended source of light.

33.6 Luminous Flux

Most sources of light emit electromagnetic energy distributed over many wavelengths. Electric power is supplied to a lamp, and radiation is emitted. The radiant energy emitted per unit of time by the lamp is called the radiant power or the ***radiant flux.*** Only a small portion of this radiant power is in the visible region—the region between 400 and 700 nm—and is called the ***luminous flux.*** The sensation of sight depends only on the visible, or *luminous,* energy radiated per unit of time.

> The luminous flux is that part of the total radiant power emitted from a light source that is capable of affecting the sense of sight.

In a common incandescent lightbulb, only about 10 percent of the energy radiated is luminous flux. The bulk of the radiant power is nonluminous.

The human eye is not equally sensitive to all colors. In other words, equal radiant power of different wavelengths does not produce equal brightness. A 40-W green lightbulb appears brighter than a 40-W blue lightbulb. A graph portraying the response of the eye to various wavelengths is shown in Fig. 33.13. Note that the sensitivity curve is bell-shaped around the center of the visible spectrum. Under normal conditions, the eye is most sensitive to yellow-green light of wavelength 555 nm. The sensitivity falls off rapidly for longer and shorter wavelengths.

If the unit chosen for luminous flux is to correspond to the sensual response of the human eye, a new unit must be defined. The watt (W) is not sufficient because the visual sensations are not the same for different colors. What is needed is a unit that measures *brightness.* Such a unit is the *lumen* (lm), which is determined by comparison with a standard source.

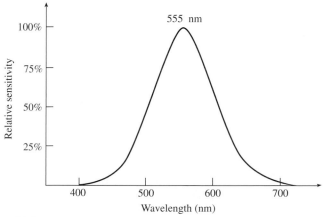

Figure 33.13   Sensitivity curve.

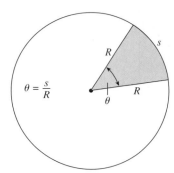

Figure 33.14 Definition of a plane angle θ expressed in radians.

To understand the definition of a lumen in terms of the standard source, we must first develop the concept of a solid angle. A solid angle in ***steradians*** (sr) is defined the same way a plane angle is defined in radians. In Fig. 33.14 the angle θ in radians is

$$\theta = \frac{s}{R} \qquad \text{rad} \tag{33.5}$$

where s is the arc length and R is the radius. The solid angle Ω is similarly defined by Fig. 33.15. It may be thought of as the opening from the tip of a cone that is subtended by a segment of area on the spherical surface.

> One steradian (sr) is the solid angle subtended at the center of a sphere by an area A on its surface that is equal to the square of its radius R.

In general, the solid angle in steradians is given by

$$\Omega = \frac{A}{R^2} \qquad \text{sr} \tag{33.6}$$

The steradian, like the radian, is a unitless quantity.

Just as there are 2π rad in a complete circle, it can be shown by Eq. (33.6) that there are 4π sr in a complete sphere.

$$\Omega = \frac{A}{R^2} = \frac{4\pi R^2}{R^2} = 4\pi \text{ sr}$$

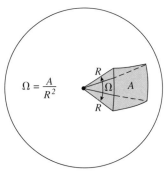

Figure 33.15 Definition of a solid angle Ω in steradians.

Notice the solid angle is independent of the distance from the source. There are 4π sr in a sphere, regardless of the length of its radius.

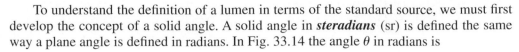

Example 33.2   What solid angle is subtended at the center of an 8-m-diameter sphere by a 1.5-m^2 area on its surface?

Solution: Equation (33.6) yields

$$\Omega = \frac{A}{R^2} = \frac{1.5 \text{ m}^2}{(4 \text{ m})^2}; \qquad \Omega = 0.0938 \text{ sr}$$

We are now in a position to clarify the definition of a unit that measures luminous flux. The ***lumen*** is defined by comparison with an internationally recognized standard source.

> One lumen (lm) is the luminous flux (or visible radiant power) emitted from a $\frac{1}{60}$-cm^2 opening in a standard source and included within a solid angle of 1 sr.

The standard source consists of a hollow enclosure maintained at the temperature of solidification of platinum, about 1773°C. In practice, it is more convenient to use standard incandescent lamps that have been rated by comparison with the standard.

Another convenient definition of the lumen utilizes the sensitivity curve (Fig. 33.13) as a basis for establishing luminous flux. By referring to the standard source, 1 lm is defined in terms of the radiant power of yellow-green light.

> One lumen is equivalent to $\frac{1}{680}$ W of yellow-green light of wavelength 555 nm.

To determine the luminous flux emitted by light of a different wavelength, the luminosity curve must be used to compensate for visual sensitivity.

Example 33.3

A source of monochromatic red light (600 nm) produces a visible radiant power of 4 W. What is the luminous flux in lumens?

Plan: The luminous flux can be estimated by studying the luminosity curve. We know the curve peaks at 680 lumens per watt for yellow-green light (555 nm.) The percent drop in relative intensity for the red light can be estimated from the luminosity curve (see Fig. 33.13) and used to find the luminous flux.

Solution: If the light were yellow-green (555 nm) instead of red, it would have a luminous flux F given by

$$F = (680 \text{ lm/W})(4 \text{ W}) = 2720 \text{ lm}$$

From the sensitivity curve, red light of wavelength 600 nm evokes about 59 percent of the response obtained with yellow-green light. Hence, the luminous flux issuing from the red light source is

$$F = (0.59)(2720 \text{ lm}) = 1600 \text{ lm}$$

The luminous flux is often calculated in the laboratory by determining the illumination it produces on a known surface area.

33.7 Luminous Intensity

Light travels radially outward in straight lines from a source that is small in comparison with its surroundings. For such a source of light, the luminous flux included in a solid angle Ω remains the same at all distances from the source. Therefore, it is frequently more useful to speak of the *flux per unit solid angle* than simply to express the total flux. The physical quantity that expressed this relationship is called the ***luminous intensity.***

The luminous intensity *I* of a source of light is the luminous flux *F* emitted per unit solid angle Ω.

$$I = \frac{F}{\Omega} \tag{33.7}$$

The unit for intensity is the *lumen per steradian* (lm/sr), called a *candela* (cd). The candela, or ***candle*** as it was sometimes called, originated when the international standard was defined in terms of the quantity of light emitted by the flame of a certain make of candle. This standard was found unsatisfactory, and eventually it was replaced by the platinum standard.

Most light sources project light in a particular direction. An ***isotropic source*** is one that emits light uniformly in all directions. We can find an expression for calculating the total flux emitted by an isotropic source by recalling the solid angle for such a source would be 4π sr. The flux F, from Eq. (33.7), would be

$$F = \Omega I = 4\pi I \qquad \textit{Isotropic Source} \tag{33.8}$$

For example, if the luminous intensity of a light source is *1 candle* (1 lm/sr), the luminous flux leaving the source would be 4π *lumens*.

Example 33.4

A spotlight is equipped with a 40-cd bulb that concentrates a beam onto a vertical wall. The beam covers an area of 9 m² on the wall, and the spotlight is located 20 m from the wall. Calculate the luminous intensity for the spotlight.

Plan: We will first calculate the total flux F leaving an isotropic force, assuming the beam is not concentrated at all. Then, we assume all of this flux is concentrated into the solid angle subtended by the 9-m² area on the wall. The intensity I will be the flux for that solid angle.

Solution: The total flux emitted by the 40-cd bulb is found from Eq. (33.7).

$$F = 4\pi I = (4\pi)(40 \text{ cd}) = 160\pi \text{ lm}$$

This total flux is concentrated by reflectors and lenses into a solid angle given by

$$\Omega = \frac{A}{R^2} = \frac{9 \text{ m}^2}{(20 \text{ m})^2} = 0.0225 \text{ sr}$$

The intensity of the beam is now found from Eq. (33.7).

$$I = \frac{F}{\Omega} = \frac{160 \pi \text{ lm}}{0.0225 \text{ sr}} = 2.23 \times 10^4 \text{ cd}$$

Note the units of intensity (cd) and the units of flux (lm) are the same dimensionally. This is true because the solid angle in steradians is dimensionless.

33.8   Illumination

If the intensity of a source is increased, the luminous flux transmitted to each unit of surface area in the vicinity of the source is also increased. The surface appears brighter. In the measurement of light efficiency, the engineer is concerned with the density of luminous flux falling on a surface. We are therefore led to a discussion of the ***illumination*** of a surface.

> The illumination E of a surface A is defined as the luminous flux F per unit area.

$$E = \frac{F}{A} \tag{33.9}$$

When the flux F is measured in lumens and the area A in square meters, the illumination E has the units of *lumens per square meter,* or *lux* (lx). When A is expressed in square feet, E is expressed in *lumens per square foot*. The lumen per square foot is sometimes loosely referred to as the *footcandle*.

Direct application of Eq. 33.9 requires a knowledge of the luminous flux falling on a given surface. Unfortunately, the flux of common light sources is difficult to determine. For this reason, Eq. 33.9 is most often used to calculate the flux when A is known and E is computed from the measured intensity.

To see the relationship between intensity and illumination, let us consider a surface A at a distance R from a point source of intensity I, as shown in Fig. 33.16. The solid angle Ω subtended by the surface at the source is

$$\Omega = \frac{A}{R^2}$$

where the area A is perpendicular to the emitted light. If the luminous flux makes an angle θ with the normal to the surface, as shown in Fig. 33.17, we must consider the projected area $A \cos \theta$. This represents the effective area the flux "sees." Therefore, the solid angle, in general, can be found from

$$\Omega = \frac{A \cos \theta}{R^2}$$

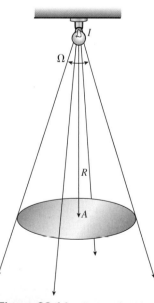

Figure 33.16   Computing the illumination of a surface perpendicular to the incident flux.

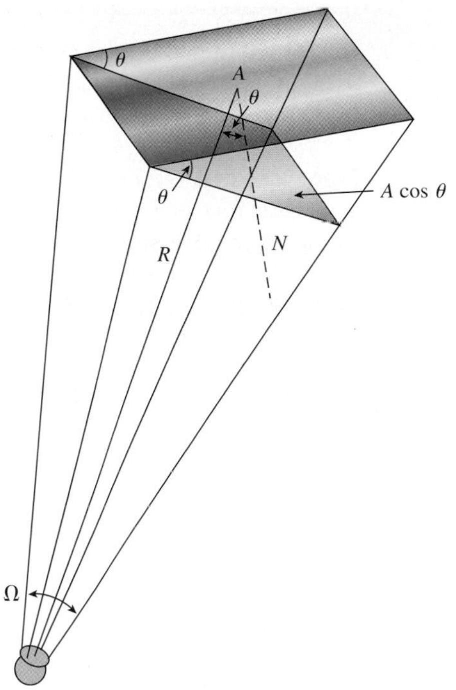

Figure 33.17 When a surface makes an angle θ with the incident flux, the illumination E is proportional to the component $A\ \cos\theta$ of the surface perpendicular to the flux.

Solving for the luminous flux F in Eq. (33.7), we obtain

$$F = I\Omega = \frac{IA\ \cos\theta}{R^2} \tag{33.10}$$

We are now ready to express the illumination as a function of intensity. Substituting Eq. (33.10) into the defining equation for illumination gives the following:

$$E = \frac{F}{A} = \frac{IA\ \cos\theta}{AR^2}$$

or

$$E = \frac{I\ \cos\theta}{R^2} \tag{33.11}$$

For the special case in which the surface is normal to the flux, $\theta = 0°$, and Eq. (33.11) is simplified to

$$E = \frac{I}{R^2} \qquad\qquad \textit{Normal Surface} \quad \textbf{(33.12)}$$

You should verify that the units of *candela per square meter* are equivalent dimensionally to the units of *lumens per square meter,* or *lux.*

Example 33.5

A 100-W incandescent lamp has a luminous intensity of 125 cd. What is the illumination of a surface located 3 ft below the lamp?

Solution: By direct substitution into Eq. (33.12), we obtain

$$E = \frac{I}{R^2} = \frac{125\ \text{cd}}{(3\ \text{ft})^2} = 13.9\ \text{lm/ft}^2$$

Example 33.6 A 300-cd tungsten-filament lamp is located 2.0 m from a surface whose area is 0.25 m². The luminous flux makes an angle of 30° with the normal to the surface. (a) What is the illumination? (b) What is the luminous flux striking the surface?

Plan: We will refer to Fig. 33.17 and note that the illumination decreases with the square of distance. Multiplying by the cosine of the given angle will give the effective area penetrated by the flux. We can then calculate the illumination and the flux.

Solution (a): The illumination is found directly from Eq. (33.11).

$$E = \frac{I \cos \theta}{R^2} = \frac{(300 \text{ cd})(\cos 30°)}{(2 \text{ m})^2} = 65 \text{ lx}$$

Solution (b): The flux falling on the surface is found by solving for F in Eq. (33.9). Thus,

$$F = EA = (65 \text{ lx})(0.25 \text{ m}^2)$$
$$= 16.2 \text{ lm}$$

The equations involving illumination and luminous intensity are mathematical formulations of the *inverse-square law,* which can be stated as follows:

> The illumination of a surface is proportional to the luminous intensity of a point light source and is inversely proportional to the square of the distance.

If a light that is illuminating a surface is raised to twice the original height, the illumination will be only one-fourth as great. If the lamp distance is tripled, the illumination is reduced by one-ninth. This inverse-square relationship is illustrated in Fig. 33.18.

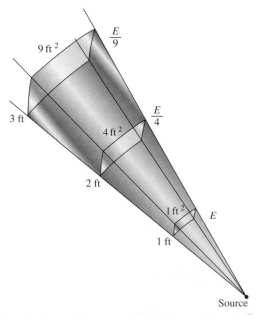

Figure 33.18   The luminous intensity I is constant for a given solid angle Ω. However, the illumination E (flux per unit area) decreases with the square of the distance from a point source.

Summary and Review

Summary

The investigation of the nature of light is still continuing, but experiments show it sometimes behaves as particles and sometimes as a wave. Modern theory holds that light is electromagnetic radiation and that its radiant energy is transported in photons carried along by a wave field. The main ideas and formulas presented in this chapter are summarized as follows.

- The wavelength λ of electromagnetic radiation is related to its frequency f by the general equation

$$c = f\lambda \qquad c = 3 \times 10^8 \text{ m/s}$$

- The range of wavelengths for visible light goes from 400 nm for violet to 700 nm for red.

$$1 \text{ nm} = 10^{-9} \text{ m} \qquad \textit{The nanometer is} \\ \textit{used for wavelengths.}$$

- The energy of light photons is proportional to the frequency.

$$E = hf \qquad E = \frac{hc}{\lambda} \qquad h = 6.626 \times 10^{-34} \text{ J} \cdot \text{s}$$

The constant h is *Planck's constant*.
- *Luminous flux* has been defined as that portion of the total radiant power emitted from a light source that is capable of affecting the sense of sight. Since visual brightness and sensitivity varies for different individuals, it was necessary to define luminous intensity in terms of a standard source and a well-defined solid angle (the steradian). By comparison to such standards, we are able to treat the illumination of surfaces, which is so important for the design of industrial workplaces.

- The *luminous intensity* of a light source is the luminous flux F per unit solid angle Ω. *Luminous flux* is the radiant power in the visible region. It is measured in *lumens*.

$$1 \text{ lm} = \frac{1}{680} \text{ W} \quad \text{for 555-nm light} \quad \textit{The Lumen}$$

$$\Omega = \frac{A}{R^2} \qquad \textit{Solid Angle in Steradians}$$

$$I = \frac{F}{\Omega} \qquad \begin{array}{c} \textit{Luminous Intensity} \\ (1 \text{ cd} = 1 \text{ lm/sr}) \end{array}$$

- For an isotropic source, one emitting light in all directions, the luminous flux is

$$F = 4\pi I \qquad \textit{Isotropic Source}$$

- The illumination E of a surface A is defined as the luminous flux per unit area.

$$E = \frac{F}{A} \qquad E = \frac{I \cos \theta}{R^2} \qquad \begin{array}{c} \textit{Illumination} \\ (\text{lm/m}^2), \text{lx} \end{array}$$

Key Terms

candle 653
diffraction 643
electromagnetic wave 643
illumination 654
infrared ray 643
infrared wave 648
isotropic source 653
light 643

lumen 652
luminous flux 651
luminous intensity 653
nanometer 647
penumbra 650
photoelectric effect 648
photons 649
point source 650

quantum theory 649
radiant flux 651
ray 650
steradian 652
ultraviolet wave 648
umbra 650
visible region 643

Review Questions

33.1. What is meant by the dualistic nature of light? In what ways does light behave as particles? In what ways does light behave as a wave?

33.2. Explain how the energy of an electromagnetic wave depends on its frequency and how it depends on the wavelength.

33.3. When light enters glass from air, its energy in glass is the same as its energy in air. Is its frequency the same? What about its wavelength? Explain.

33.4. Microwave ovens, television, and radar utilize electromagnetic waves between infrared and radio waves. Compare the energy, frequency, and wavelengths of these waves with the energy, frequency, and wavelengths of visible radiation.

33.5. Review the definition of a radian and discuss how the steradian for solid angles is similar to the radian for plane angles. How many radians are in a complete circle? How many steradians are in a complete sphere?

33.6. Draw a diagram illustrating a solar eclipse, labeling the umbra and penumbra regions. If you view a partial eclipse of the Sun, are you standing in the umbra or the penumbra region?

33.7. Can you justify the following definition for a lumen? *A lumen is equal to the luminous flux falling on a surface on 1 square meter, all points of which are 1 meter from a uniform point source of 1 candela.*

33.8. An older unit for illumination was the *foot-candle,* that illumination E received by a surface 1 ft^2 located at a distance of 1 ft from a 1-cd source of light. Explain how this definition is the same as that given in this text.

33.9. Describe the distribution of luminous flux from an incandescent lamp. Why is the lamp not an isotropic source?

33.10. Discuss the factors that will affect the illumination of a table in a machine shop.

33.11. Illumination is sometimes referred to as *flux density.* Explain why such a term might be appropriate.

33.12. *Photometry* is the science of measuring light. The intensity of a light source can be determined with the photometer, shown in Fig. 33.19. The luminous intensity I_x of an unknown source is found by visually comparing the unknown with a standard source of known intensity I_s. If the distances from each source are adjusted so that the grease spot is illuminated equally by each source, the unknown intensity I_x can be calculated from the inverse-square law. Derive the photometry equation

$$\frac{I_x}{r_x^2} = \frac{I_s}{r_s^2} \quad \textit{Photometry Equation} \quad \textbf{(33.13)}$$

where r_s is the distance to the standard and r_x is the distance to the unknown source.

33.13. If two 40-W light bulbs are compared by using the photometer, will they necessarily be located at equal distances from the grease spot?

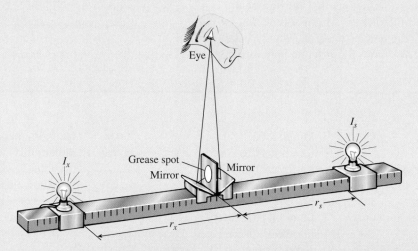

Figure 33.19 The grease-spot photometer is used to measure the intensity of an unknown source of light by comparing it to a standard source.

Problems

Section 33.2 The Propagation of Light and Section 33.3 The Electromagnetic Spectrum

33.1. An infrared spectrophotometer scans the wavelengths from 1 to 16 μm. Express this range in terms of the frequencies of the infrared rays.

Ans. 30.0 $\times$ 10^{13} to 1.88 $\times$ 10^{13} Hz

33.2. What is the frequency of violet light of wavelength 410 nm?

33.3. A microwave radiator used in measuring automobile speeds emits a radiation of frequency 1.2×10^9 Hz. What is the wavelength?

Ans. 250 mm

33.4. What is the range of frequencies for visible light?

33.5. If Plank's constant h is equal to 6.626×10^{-34} J·s, what is the energy of light of wavelength 600 nm?

Ans. 3.31×10^{-19} J

33.6. What is the frequency of light that has an energy of 5×10^{-19} J?

33.7. The frequency of yellow-green light is 5.41×10^{-14} Hz. Express the wavelength of this light in nanometers and in angstroms.

Ans. 555 nm, 5550 A

33.8. What is the wavelength of light that has an energy of 7×10^{-19} J?

33.9. The Sun is approximately 93 million miles from Earth. How much time is required for the light emitted by the Sun to reach us on Earth?

Ans. 8.33 min

33.10. A helium–neon laser beam has a frequency of 4.74×10^{14} Hz and a power of 1 mW. What is the average number of photons per second propagated by this beam?

33.11. The light reaching us from the nearest star, Alpha Centauri, requires 4.3 years to reach us. How far is this in miles? In kilometers?

Ans. 2.53×10^{13} mi, 4.07×10^{13} km

33.12. A spacecraft circling the Moon at a distance of 384,000 km from the Earth communicates by radio with a base on Earth. How much time elapses between the sending and receiving of a signal?

33.13. A spacecraft sends a signal that requires 20 min to reach the Earth. How far away is the spacecraft from Earth?

Ans. 3.60×10^{11} m

Section 33.5 Light Rays and Shadows

33.14. The shadow formed on a screen 4 m away from a point source of light is 60 cm tall. What is the height of the object located 1 m from the source and 3 m from the shadow?

33.15. A point source of light is placed 15 cm from an upright 6-cm ruler. Calculate the length of the shadow formed by the ruler on a wall 40 cm from the ruler.

Ans. 22.0 cm

33.16. How far must an 80-mm-diameter plate be placed in front of a point source of light if it is to form a shadow 400 mm in diameter at a distance of 2 m from the light source?

33.17. A source of light 40 mm in diameter shines through a pinhole in the tip of a cardboard box 2 m from the source. What is the diameter of the image formed on the bottom of the box if the height of the box is 60 mm?

Ans. 1.20 mm

*33.18. A lamp is covered with a box, and a 20-mm-long narrow slit is cut in the box so that light shines through. An object 30 mm tall blocks the light from the slit at a distance of 500 mm. Calculate the length of the umbra and penumbra formed on a screen located 1.50 m from the slit.

Section 33.8 Illumination

33.19. What is the solid angle subtended at the center of a 3.20-m-diameter sphere by a 0.5-m^2 area on its surface?

Ans. 0.195 sr

33.20. A solid angle of 0.080 sr is subtended at the center of a 9.00-cm-diameter sphere by surface area A on the sphere. What is this area?

33.21. An $8\frac{1}{2} \times 11$-cm sheet of metal is illuminated by a source of light located 1.3 m directly above it. What is the luminous flux falling on the metal if the source has an intensity of 200 cd. What is the total luminous flux emitted by the light source?

Ans. 1.11 lm, 2510 lm

33.22. A 40-W monochromatic source of yellow-green light (555 nm) illuminates a 0.5-m^2 surface at a distance of 1.0 m. What is the luminous intensity of the source, and how many lumens fall on the surface?

33.23. What is the illumination produced by a 200-cd source on a small surface 4.0 m away?

Ans. 12.5 lx

33.24. A lamp 2 m from a small surface produces an illumination of 100 lx on the surface. What is the intensity of the source?

33.25. A tabletop 1 m wide and 2 m long is located 4.0 m from a lamp. If 40 lm of flux fall on this surface, what is the illumination E of the surface?

Ans. 20.0 lx

33.26. What should be the location of the lamp in Prob. 33.25 in order to produce twice the illumination?

*33.27. A point source of light is placed at the center of a sphere 70 mm in diameter. A hole is cut in the surface of the sphere, allowing the flux to pass through a solid angle of 0.12 sr. What is the diameter of the opening?

Ans. 13.7 mm

Additional Problems

33.28. When light of wavelength 550 nm passes from air into a thin glass plate and out again into the air, the frequency remains constant, but the speed through the glass is reduced to 2×10^8 m/s. What is the wavelength inside the glass?

33.29. A 30-cd standard light source is compared with a lamp of unknown intensity using a grease-spot photometer (see Fig. 33.19). The two light sources are placed 1 m apart, and the grease spot is moved toward the standard light. When the grease spot is 25 cm from the standard light source, the illumination is equal on both sides. Compute the unknown intensity.   Ans. 270 cd

33.30. Where should the grease spot in Prob. 33.29 be placed for the illumination by the unknown light source to be exactly twice the illumination of the standard source?

33.31. The illumination of a given surface is 80 lx when it is 3 m away from the light source. At what distance will the illumination be 20 lx?   Ans. 600 m

33.32. A light is suspended 9 m above a street and provides an illumination of 35 lx at a point directly below it. Determine the luminous intensity of the light.

*33.33. A 60-W monochromatic source of yellow-green light (555 nm) illuminates a 0.6 m² surface at a distance of 1.0 m. What is the solid angle subtended at the source? What is the luminous intensity of the source?   Ans. 0.60 sr, 68,000 cd

*33.34. At what distance from a wall will a 35-cd lamp provide the same illumination as an 80-cd lamp located 4.0 m from the wall?

*33.35. How much must a small lamp be lowered to double the illumination on an object that is 80 cm directly under it?   Ans. 23.4 cm

*33.36. Compute the illumination of a given surface 140 cm from a 74-cd light source if the normal to the surface makes an angle of 38° with the flux.

*33.37. A circular tabletop is located 4 m below and 3 m to the left of a lamp that emits 1800 lm. What illumination is provided on the surface of the table? What is the area of the tabletop if 3 lm of flux falls on its surface?   Ans. 4.58 lx, 0.655 m²

*33.38. What angle θ between the flux and a line drawn normal to a surface will cause the illumination of that surface to be reduced by one-half when the distance to the surface has not changed?

*33.39. All the light from a spotlight is collected and focused on a screen of area 0.30 m². What must be the luminous intensity of the light for an illumination of 500 lx to be achieved?   Ans. 150 cd

*33.40. A 300-cd light is suspended 5 m above the left edge of a table. Find the illumination of a small piece of paper located a horizontal distance of 2.5 m from the edge of the table.   Ans. 8.59 lx

Critical Thinking Questions

33.41. A certain radio station broadcasts at a frequency of 1150 kHz; a red beam of light has a frequency of 4.70×10^{14} Hz; and an ultraviolet ray has a frequency of 2.4×10^{16} Hz. Which has the highest wavelength? Which has the greatest energy? What are the wavelengths of each electromagnetic wave?
Ans. radio, ultraviolet, 261 m, 639 nm, 12.5 nm

*33.42. An unknown light source A located 80 cm from a screen produces the same illumination as a standard 30-cd light source at point B located 30 cm from the screen. What is the luminous intensity of the unknown light source?

*33.43. The illumination of a surface 3.40 m directly below a light source is 20 lx. Find the intensity of the light source. At what distance below the light source will the illumination be doubled? Is the luminous flux also doubled at this location?
Ans. 231 cd, 2.40 m, no

*33.44. The illumination of an isotropic source is E_A at a point A on a table 30 cm directly below the source. At what horizontal distance from A on the tabletop will the illumination be reduced by one-half?

34

Reflection and Mirrors

Laser light is used to study reflection properties of light. Note that the angle of reflection is equal to the angle of incidence as the direction of the light beam is changed several times. (*Photo © Firefly Productions/Corbis.*)

Objectives

After completing this chapter, you should be able to

1. Define and illustrate with drawings your understanding of the following terms: *virtual images, real images, converging mirror, diverging mirror, magnification, focal length,* and *spherical aberration.*
2. Use ray-tracing techniques to construct images formed by spherical mirrors.
3. Predict mathematically the nature, size, and location of images formed by spherical mirrors.
4. Determine the magnification and/or the *focal length* of spherical mirrors by mathematical and experimental methods.

The eye responds to light. Every object viewed is seen with light—either the light emitted by the object or light that is reflected from it. We now have a general understanding of the nature of light, and we have studied luminous objects and methods of measuring the light emitted from them.

Although all light can be traced to sources of energy, for example, the Sun, an electric lightbulb, or a burning candle, most of what we see in the physical world is a result of reflected light. In this chapter, we treat the laws describing how light is turned back into its original medium as a result of striking a surface. Although this phenomenon, called *reflection,* can be interpreted in terms of Maxwell's electromagnetic wave theory, it is much simpler to describe it by tracing *rays.*

The ray treatment, generally referred to as **geometrical optics,** is based on the application of Huygens's principle. Remember that light rays are imaginary lines drawn perpendicular to advancing wavefronts in the direction of light propagation.

34.1 The Laws of Reflection

When light strikes the boundary between two media, such as air and glass, one or more of three things can happen. As illustrated in Fig. 34.1, some of the light incident on a glass surface is reflected, and some passes into the glass. The light that enters the glass is partially absorbed and partially transmitted. The transmitted light usually undergoes a change in direction, called **refraction.** In this chapter, we will concern ourselves only with the phenomenon of reflection.

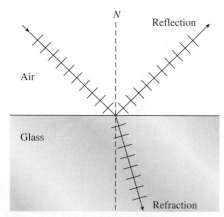

Figure 34.1   When light strikes the boundary between two media, it may be reflected, refracted, or absorbed.

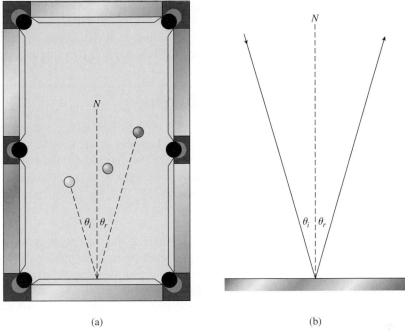

(a)   (b)

Figure 34.2   Reflection of light follows the same path as that expected for a bouncing billiard ball. The angle of incidence is equal to the angle of reflection.

The reflection of light obeys the same general law of mechanics that governs other bouncing phenomena; that is, the angle of incidence equals the angle of reflection. For example, let us consider the pool table in Fig. 34.2a. To hit the black ball on the right, we must aim at a point on the rail such that the incident angle θ_i is equal to the reflected angle θ_r. Similarly, light reflected from a smooth surface, as in Fig. 34.2b, has equal angles of incidence and reflection. The angles θ_i and θ_r are measured with respect to the normal to the surface. Two basic laws of reflection can be stated:

The angle of incidence is equal to the angle of reflection.

The incident ray, the reflected ray, and the normal to the surface all lie in the same plane.

Light reflection from a smooth surface, in Fig. 34.3a, is called ***regular,*** or ***specular reflection.*** Light striking the surface of a mirror or glass is specularly reflected. If all the incident light that strikes a surface were reflected in this manner, we could not see the

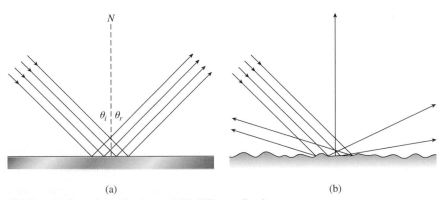

(a)   (b)

Figure 34.3   (a) Specular reflection and (b) diffuse reflection.

surface. We would see only images of other objects. It is **_diffuse reflection_** (Fig. 34.3b) that enables us to see a surface. A rough or irregular surface will spread out and scatter the incident light, resulting in illumination of the surface. Reflection of light from brick, concrete, or newsprint provides an example of diffuse reflection.

34.2 Plane Mirrors

A highly polished surface that forms images by specular reflection of light is called a _mirror._ The mirrors hanging on the walls of our homes are usually flat, or _plane,_ and we all have a certain familiarity with the images framed by **_plane mirrors._** In every case, the image appears to be as far behind the mirror as the actual object is in front of it. As shown in Fig. 34.4, the images also appear right–left reversed. Anyone who has learned to tie a necktie or apply makeup by looking in a mirror is well aware of these effects.

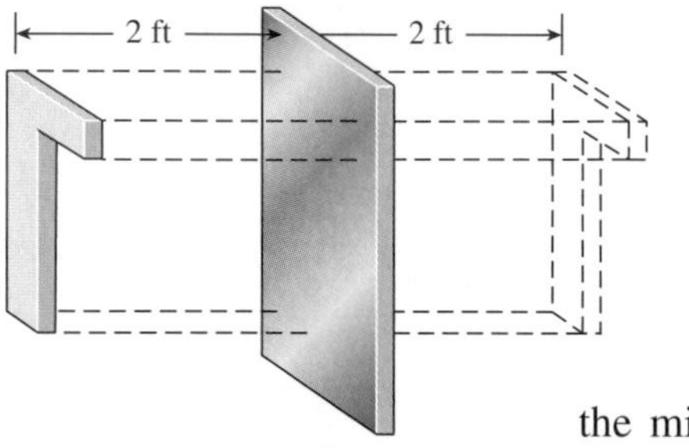

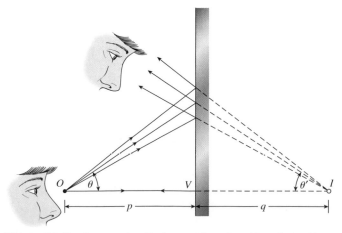

Figure 34.4 Images formed by mirrors are right–left reversed.

To understand the formation of images by a plane mirror, let us first consider image I formed by rays emitted from a point O in Fig. 34.5. Four light rays are traced from the point source of light. The light ray OV is reflected back on itself at the mirror. Since the reflected light appears to have traveled the same distance as the incident light, the image will appear to be an equal distance behind the mirror when viewed along the normal to the reflecting surface. When the reflected light is viewed at an angle to the mirror, the same conclusion results: the image distance q is equal to the object distance p. This is true because the angle θ is equal to the angle θ' in the figure. Thus, we can say that

> The object distance is equal in magnitude to the image distance for a plane mirror.

$$p = q \qquad\qquad \textit{Plane Mirror} \quad (\textbf{34.1})$$

We now consider the image formed by an extended object, as shown in Fig. 34.6. An extended object may be thought of as consisting of many point objects arranged according to the shape and size of the object. Each point on the object will have an image point located an equal distance behind the mirror. It follows that the image will have the same size and shape as the object. Right and left will be reversed, however, as discussed earlier.

Notice that the images formed by a plane mirror are, in truth, reflections of real objects. The images themselves are not real because no light passes through them. These images that _appear_ to the eye to be formed by rays of light but that in truth do not exist are called **_virtual images._** A **_real image_** is an image formed by actual light rays.

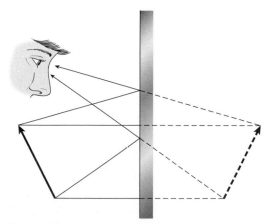

Figure 34.5 Constructing the image of a point object formed by a plane mirror.

Figure 34.6 Image of an extended object.

A virtual image is one that seems to be formed by light coming from the image, but no light rays actually pass through it.

A real image is formed by actual light rays that pass through it. Real images can be projected onto a screen.

Since virtual images are not formed by real light rays, they cannot be projected onto a screen.

Real images cannot be formed by a plane mirror because the light reflected at a plane surface diverges. But if a plane mirror forms virtual images that do not physically exist, how can we see them? The full answer to this question must wait for a discussion of refraction and lenses. A preliminary answer is illustrated by Fig. 34.7, which also serves to demonstrate the two types of images. The eye makes use of the principle of refraction to reconverge the reflected light that *seems* to come from the virtual image. A *real* image is therefore projected on the retina of the eye. This image, which is formed by real, reflected light rays, is interpreted by the brain to have originated from a point behind the mirror. The brain is conditioned to the rectilinear propagation of light. It is fooled when light is somehow caused to change directions. Men who do not believe the brain can be conditioned to interpret images should try to tie someone's necktie without looking in a mirror. In this case, the real object seems less natural than its virtual image.

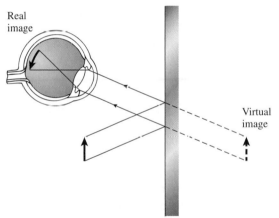

Figure 34.7 Image formed by a plane mirror is virtual. Such images appear to the eye to be located behind the mirror.

34.3 Spherical Mirrors

The same geometrical methods applied for the reflection of light from a plane mirror can be used for a curved mirror. The angle of incidence is still equal to the angle of reflection, but the normal to the surface changes at every point along the surface. A complicated relation between the object and its image results.

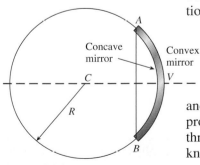

Figure 34.8 Definition of terms for spherical mirrors.

Most curved mirrors used in practical application are spherical. A **spherical mirror** is a mirror that may be thought of as a portion of a reflecting sphere. The two kinds of spherical mirrors are illustrated in Fig. 34.8. If the inside of the spherical surface is the reflecting surface, the mirror is said to be **concave.** If the outside portion is the reflecting surface, the mirror is **convex.** In either case, R is the **radius of curvature,** and C is the *center of curvature* for the mirrors. The segment AB, often useful in optical problems, is called the **linear aperture** of the mirror. The dashed line CV that passes through the center of curvature and the topographical center, or *vertex,* of the mirror is known as the *axis* of the mirror.

Let us now examine the reflection of light from a spherical surface. As a simple case, suppose a beam of parallel light rays to be incident on a concave surface, as shown in

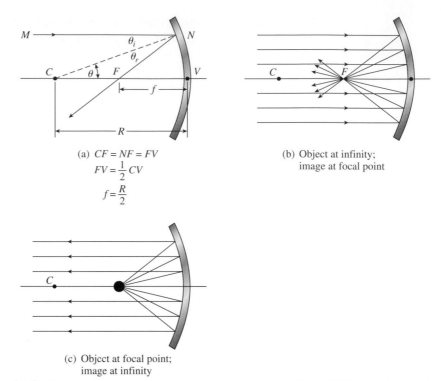

(a) $CF = NF = FV$
$$FV = \frac{1}{2}\,CV$$
$$f = \frac{R}{2}$$

(b) Object at infinity;
image at focal point

(c) Object at focal point;
image at infinity

Figure 34.9   The focal point of a converging mirror (concave surface). (a) The focal length is one-half the radius of curvature; (b) the object at infinity and the image at the focal point; and (c) the object at the focal point and the image at infinity.

Fig. 34.9. Since the mirror is perpendicular to the axis at its vertex V, a light ray CV is reflected back on itself. In fact, any light ray that proceeds along a radius of the mirror will be reflected back along itself. The parallel light ray MN is reflected so that the angle of incidence θ_i is equal to the angle of reflection θ_r. Both angles are measured with respect to the radius CN. The geometry of the reflection is such that the reflected ray passes through a point F on the axis halfway between the center of curvature C and the vertex V. The point F, to which parallel light rays converge, is called the *focal point* of the mirror. The distance from F to V is called the **focal length** f. It is left as an exercise to show from Fig. 34.9a that

$$f = \frac{R}{2} \tag{34.2}$$

The focal length f of a concave mirror is one-half of its radius of curvature R.

All light rays from a distant object, such as the Sun, will converge at the focal point F, as shown in Fig. 34.9b. For this reason, concave mirrors are frequently called **converging mirrors.** The focal point can be found experimentally by converging sunlight to a point on a piece of paper. The point along the axis of the mirror where the image formed on the paper is brightest will correspond to the focal point of the mirror.

Since light rays are reversible, a source of light placed at the focal point of a converging mirror will form its image at infinity. In other words, the emerging light beam will be parallel to the axis of the mirror, as shown in Fig. 34.9c.

A similar principle holds for a convex mirror, as illustrated in Fig. 34.10. Note that a parallel light beam incident on a convex surface diverges. The reflected light rays *appear* to come from a point F located behind the mirror, but no light rays actually pass through it. Even though the focal point is virtual, the distance VF is still called the *focal length* of the convex mirror. Since the actual light rays diverge when striking such a surface, a convex mirror is called a **diverging mirror.** Equation (34.2) also applies for a convex mirror. To be

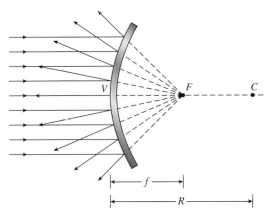

Figure 34.10 The focal point of a diverging mirror (convex surface).

consistent with theory (to be developed later), however, the focal length f and the radius R must be reckoned as negative for diverging mirrors.

34.4 Images Formed by Spherical Mirrors

The best method for understanding the formation of images by mirrors is through geometrical optics, or *ray tracing*. This method consists of considering the reflection of a few rays diverging from some point of an object O that is *not* on the mirror axis. The point at which all these reflected rays will intersect determines the image location. We will discuss three rays whose path can be traced easily. Each ray is illustrated for a converging (concave) mirror in Fig. 34.11 and for a diverging (convex) mirror in Fig. 34.12.

Ray 1: A ray parallel to the mirror axis passes through the focal point of a concave mirror or seems to come from the focal point of a convex mirror.

Ray 2: A ray that passes through the focal point of a concave mirror or proceeds toward the focal point of a convex mirror is reflected parallel to the mirror axis.

Ray 3: A ray that proceeds along a radius of the mirror is reflected back along its original path.

In any given situation, only two of these rays are necessary to locate the image of a point. By choosing rays from an extreme point of the object, the remainder of the image

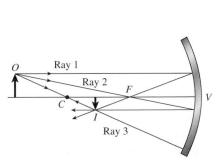

Figure 34.11 The principal rays for constructing the images formed by converging (concave) mirrors.

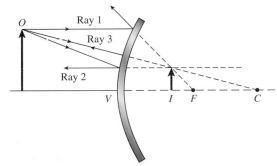

Figure 34.12 The principal rays for constructing the images formed by diverging (convex) mirrors.

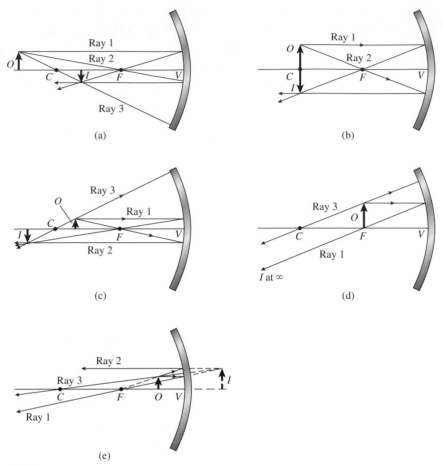

Figure 34.13   Images formed by a converging mirror for the following object distances: (a) beyond the center of curvature *C*, (b) at *C*, (c) between *C* and the focal length *F*, (d) at *F*, and (e) inside *F*.

can usually be filled in by symmetry. In the figures, dashed lines are used to identify virtual rays and virtual images.

To illustrate the graphical method and at the same time visualize some of the possible images, let us consider several images formed by a concave mirror. Figure 34.13a illustrates the image formed by an object *O* that is located outside the center of curvature of the mirror. Note that the image is between the focal point *F* and the center of curvature *C*. The image is *real, inverted,* and *smaller* than the object.

In Fig. 34.13b, the object *O* is located at the center of curvature *C*. The concave mirror forms an image at the center of curvature that is *real, inverted,* and the *same size* as the object.

In Fig. 34.13c, the object *O* is located between *C* and *F*. Ray tracing shows that the image is located beyond the center of curvature. It is *real, inverted,* and *larger* than the object.

When the object is at the focal point *F*, all reflected rays are parallel (see Fig. 34.13d). Since the reflected rays will never intersect when extended in either direction, no image will be formed. (Some people prefer to say that the image distance is infinite.)

When the object is located inside the focal point *F*, as shown in Fig. 34.13e, the image *appears* to be behind the mirror. This can be seen by extending the reflected rays to a point behind the mirror. Thus, the image is *virtual.* Notice also that the image is *enlarged* and erect (right side up). The **magnification** in this instance is the principle behind shaving mirrors and other mirrors that form enlarged virtual images.

On the other hand, all images formed by *convex* mirrors have the same characteristics. As was illustrated in Fig. 34.12, such images are *virtual, erect,* and *reduced in size.* This results in a wider field of view and accounts for many of the uses of convex mirrors. Automobile rear-view mirrors are usually convex to give maximum viewing capability. Some stores have large convex mirrors conveniently located to give them a panoramic view to help detect shoplifters.

34.5 | The Mirror Equation

Now that we have a feel for the characteristics and formation of images, it will be useful to develop an analytical approach to image formation. Consider the reflection of light from a point object O, as illustrated in Fig. 34.14 for a concave mirror. The ray OV is incident along the axis of the mirror and is reflected back on itself. Ray OM is selected arbitrarily and proceeds toward the mirror at an angle α with the axis of the mirror. This ray is incident at an angle θ_i and reflected at an equal angle θ_r. The light rays reflected at M and V cross at the point I, forming an image of the object. The object distance p and image distance q are measured from the vertex of the mirror and indicated in the figure. The image at I is a *real* image since it is formed by actual light rays that pass through it.

Now, let us consider the image formed by an extended object OA, as shown in Fig. 34.15. The image of the point O is found to be I, as before. By tracing rays from the top of the arrow, we are able to draw the image of A at B. The ray AM passes through the center of curvature and is reflected back on itself. A ray AV that strikes the vertex of the mirror forms equal angles θ_i and θ_r. Rays VB and AM cross at B, forming an image of the top of the arrow at that point. The rest of the image IB can be constructed by tracing similar rays for corresponding points on the object OA. Notice that the image is *real* and *inverted*.

The following quantities are identified in Fig. 34.15.

$$\text{Object distance} = OV = p$$
$$\text{Image distance} = IV = q$$
$$\text{Radius of curvature} = CV = R$$
$$\text{Object size} = OA = y$$
$$\text{Image size} = IB = y'$$

We now attempt to relate these quantities. From the figure, it is noted that angles OCA and VCM are equal. Labeling this angle by α, we can write

$$\tan \alpha = \frac{y}{p - R} = \frac{-y'}{R - q}$$

from which

$$\frac{-y'}{y} = \frac{R - q}{p - R} \tag{34.3}$$

The image size y' is negative because it is inverted in the figure. Similarly, angles θ_i and θ_r in the figure are equal, so

$$\tan \theta_i = \tan \theta_r \qquad \frac{y}{p} = \frac{-y'}{q} \tag{34.4}$$

Combining Eqs. (34.3) and (34.4), we have

$$\frac{-y'}{y} = \frac{q}{p} = \frac{R - q}{p - R} \tag{34.5}$$

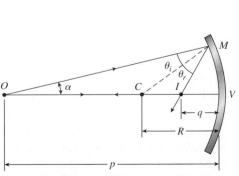

Figure 34.14 Converging mirror forms a point image of a point object.

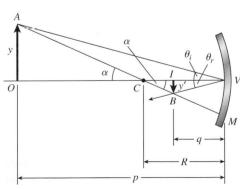

Figure 34.15 Deriving the mirror equation.

Rearranging terms, we obtain the important relation

$$\frac{1}{p} + \frac{1}{q} = \frac{2}{R} \tag{34.6}$$

This relationship is known as the ***mirror equation.*** It is more often written in terms of the focal length f of the mirror, instead of the radius of curvature. Recalling that $f = R/2$, we can rewrite Eq. (34.6) as

$$\frac{1}{p} + \frac{1}{q} = \frac{1}{f} \tag{34.7}$$

A similar derivation can be made for a convex mirror, and the same equations apply if the proper sign convention is adopted. Object and image distances, p and q, must be considered *positive* for real objects and images and *negative* for virtual objects and images. The radius of curvature R and focal length f must be considered *positive* for converging (concave) mirrors and *negative* for diverging (convex) mirrors.

Example 34.1

What is the focal length of a converging mirror whose radius of curvature is 20 cm? What are the nature and location of an image formed by the mirror if an object is placed 15 cm from the vertex of the mirror?

Plan: The *nature* of an image is determined by answering three basic questions: (1) is it *erect* or *inverted;* (2) is it *enlarged* or *diminished;* (3) and is it *real* or *virtual?* Each of these questions can be answered graphically by geometrical optics or mathematically by sign conventions. First, we will construct a rough diagram for the problem similar to that shown in Fig. 34.13c. The focal length of the converging mirror is one-half the given radius of curvature. Using this information and the given object distance, we can solve for the image distance as the only unknown in the mirror equation.

Solution: From the ray-tracing diagram for this example (Fig. 34.13c), we can see visually that the image is *real* (formed in front of the mirror), *inverted,* and *enlarged.* The focal length is positive for a *converging* mirror and given by

$$f = \frac{R}{2} = \frac{+20 \text{ cm}}{2}; \qquad f = +10 \text{ cm}$$

The object distance p is $+15$ cm, and the image distance can be found by solving the mirror equation explicitly for q.

$$\frac{1}{p} + \frac{1}{q} = \frac{1}{f}; \qquad q = \frac{pf}{p - f}$$

Substituting $f = +10$ cm and $p = +15$ cm, we obtain

$$q = \frac{(15 \text{ cm})(10 \text{ cm})}{15 \text{ cm} - 10 \text{ cm}} = +30 \text{ cm}$$

The positive sign for q verifies that the image is *real.* Note that it was not necessary to convert the length to SI units as long as all units of length are the same.

Although direct substitution into the mirror equation is easy enough, it is usually better to solve the equation algebraically for the unknown quantity and then substitute the given information. The signs of the substituted values are one of the main sources of error.

You may find the following forms useful in most mirror problems:

$$p = \frac{qf}{q - f} \qquad q = \frac{pf}{p - f} \qquad f = \frac{pq}{p + q} \qquad \textbf{(34.8)}$$

The sign convention that must be followed for the mirror equation requires a clear understanding of whether objects and images are real or virtual. Remember that *real* images are formed by real rays of light, and they must be *in front* of the mirror. Virtual images appear to be *behind* the mirror. A *virtual object* occurs sometimes when a virtual image, formed by one mirror, becomes the object for another in problems involving multiple reflections. These are rarely encountered, and the object distance is nearly always positive. A summary of the sign conventions is given as follows:

1. The object distance p is positive for real objects and negative for virtual objects.
2. The image distance q is positive for real images and negative for virtual images.
3. The radius of curvature R and the focal length f is positive for converging mirrors and negative for diverging mirrors.

This convention applies only to the numerical values substituted into the mirror equation. The quantities q, p, and f should maintain their signs unchanged until the substitution is made.

Example 34.2

A convex mirror of focal length 6 cm is placed 4 cm from a coin. Locate and describe the image formed by this mirror.

Plan: For a diverging mirror, all images are virtual, erect, and diminished. The appropriate ray diagram will be similar to Fig. 34.12. We can find the location of the image by applying the mirror equation.

Solution: Recalling that the focal length is negative for diverging mirrors, we can substitute $f = -6$ cm and $p = 4$ cm into the mirror equation.

$$q = \frac{pf}{p - f} = \frac{(4 \text{ cm})(-6 \text{ cm})}{4 \text{ cm} - (-6 \text{ cm})}$$

$$= \frac{-24 \text{ cm}^2}{10 \text{ cm}} = -2.4 \text{ cm}$$

Since the image distance is negative, we have confirmed the image is *virtual*.

34.6

Magnification

The images formed by spherical mirrors may be larger, smaller, or equal in size to the objects. The ratio of the image size to the object size is the *magnification M* of the mirror.

$$\textit{Magnification} = \frac{\textit{image size}}{\textit{object size}} = \frac{y'}{y} \qquad \textbf{(34.9)}$$

The *size* refers to *any* linear dimension—height, width, diameter, or even a mark on the object. Referring to Eq. (34.5) and to Fig. 34.15, we obtain the useful relation

$$M = \frac{y'}{y} = \frac{-q}{p} \qquad \textit{Magnification Equation} \quad \textbf{(34.10)}$$

where q is the image distance and p is the object distance. A convenient feature of Eq. (34.10) is that an *inverted image will always have a negative magnification and an erect image will have a positive magnification.*

Example 34.3

A source of light 6 cm high is located 60 cm from a concave mirror whose focal length is 20 cm. Find the nature, size, and location of the image.

Plan: We will draw a rough ray diagram, similar to Fig. 34.13a, with the object located beyond the center of curvature. The sketch will show that the image is real, inverted, and enlarged. These conclusions must be verified with appropriate signs in our solution. Next, we will organize the given information and use the mirror equation and the magnification equation to find the location and size of the image.

Solution: Given that $f = +20$ cm and $p = 60$ cm, we find that

$$q = \frac{pf}{p - f} = \frac{(60 \text{ cm})(20 \text{ cm})}{60 \text{ cm} - 20 \text{ cm}}$$

$$= \frac{1200 \text{ cm}^2}{40 \text{ cm}} = +30 \text{ cm}$$

The image distance is positive, confirming the image is real. Now, the image size is found by solving for y' in the magnification equation.

$$M = \frac{y'}{y} = \frac{-q}{p} \qquad \text{or} \qquad y' = \frac{-qy}{p}$$

$$y' = \frac{(-30 \text{ cm})(6 \text{ cm})}{60 \text{ cm}} = -3 \text{ cm}$$

The negative sign verifies the image is inverted. Note that the magnification is $-1/2$.

Problem-Solving Strategy

Reflection and Mirrors

1. Read the problem carefully and draw a horizontal line representing the mirror axis. Indicate, by points on the mirror axis, the location of the radius R and focal length f of the mirror. (Remember that $f = R/2$.) Draw the concave or convex mirror as a curved line and position the object as an erect arrow at its approximate location in front of the mirror.

2. Construct a geometric ray-tracing diagram to get a visual representation of the problem. A rough sketch should suffice, unless the problem requires a graphical solution.

3. Make a list of the given quantities, taking care to give the appropriate sign to each value. The radius and focal length are *positive* for *converging mirrors* and *negative* for *diverging mirrors*. Image distances q are *positive*

when measured to *real images* and *negative* when measured to *virtual images*. The image size y' is *positive* for *erect images* and *negative* for *inverted images*.

4. Use the following equations to substitute and solve for the unknown quantities. Do not confuse the signs of operation (addition or subtraction) with the signs of substitution.

$$\frac{1}{p} + \frac{1}{q} = \frac{1}{f} \qquad M = \frac{y'}{y} = \frac{-q}{p}$$

$$f = \frac{R}{2}$$

5. It may be necessary to eliminate an unknown by solving both the mirror equation and the magnification equation simultaneously.

Example 34.4

In a laboratory experiment, it is desired to form an image that is one-half as large as an object. How far must the object be held from a diverging mirror of radius 40 cm?

Plan: The focal length is one-half the radius, so $f = -20$ cm. The negative sign is necessary for a diverging mirror. The problem is that we do not know either p or q. However, these two unknowns each appear in the magnification equation and in the mirror equation. Simultaneous solution of these two equations will allow us to eliminate the unknown image distance q in order to solve for the object distance p.

Solution: As usual, we will draw a ray diagram for the problem similar to that for any diverging mirror. (See Fig. 34.12.) The image must be virtual, erect, and diminished. The erect image means that the magnification is positive $(+1/2)$, and the magnification equation gives

$$M = \frac{-q}{p} = +\frac{1}{2} \quad \text{or} \quad q = \frac{-p}{2}$$

Now, we can find another expression for q from the mirror equation.

$$q = \frac{pf}{p - f}$$

The two expressions for q must be equal, so we may write

$$\frac{\cancel{p}f}{p - f} = \frac{-\cancel{p}}{2} \quad \text{or} \quad \frac{f}{p - f} = \frac{-1}{2}$$

The solution for q is

$$2f = -1(p - f)$$
$$2f = -p + f$$
$$p = f - 2f$$
$$p = -f$$

Thus, the object distance is

$$p = -f = -(-20 \text{ cm}); \quad p = +20 \text{ cm}$$

When an image is held at a distance from a diverging mirror that is equal to the focal length, the image size will be one-half of the object size.

34.7 | Spherical Aberration

In practice, spherical mirrors form reasonably sharp images as long as their apertures are small compared with their focal lengths. When large mirrors are used, however, some of the rays from objects strike near the outer edges and are focused to different points on the axis. This focusing defect, illustrated in Fig. 34.16, is known as *spherical aberration.*

A *parabolic mirror* does not exhibit this defect. Theoretically, parallel light rays incident on a parabolic reflector will focus at a single point on the mirror axis, as shown in Fig. 34.17. A small source of light located at the focal point of a parabolic reflector is the principle used in many spotlights and search lights. The beam emitted from such a device is parallel to the axis of the reflector.

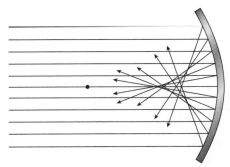

Figure 34.16 Spherical aberration.

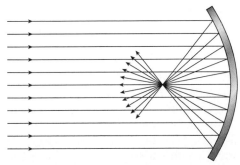

Figure 34.17 Parabolic reflector focuses all incident parallel light to the same point.

Summary and Review

Summary

In this chapter, we have studied the reflective properties of converging and diverging spherical mirrors. The focal length and radius of curvature of such mirrors determine the nature and size of the images they form. Application of the formulas and ideas given in this chapter are necessary for understanding the operation and use of many technical instruments. The main concepts are summarized as follows.

- The formation of images by spherical mirrors can be visualized more easily with ray-tracing techniques. The three principal rays are listed in the following. You should refer to Fig. 34.18a for converging mirrors and to Fig. 34.18b for diverging mirrors.

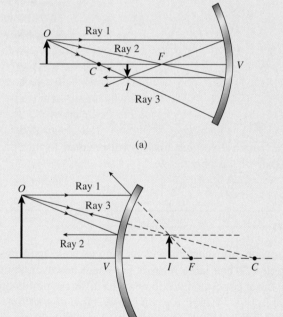

(a)

(b)

Figure 34.18 Ray tracing for (a) converging mirror and (b) a diverging mirror.

Ray 1: A ray parallel to the mirror axis passes through the focal point of a concave mirror or seems to come from the focal point of a convex mirror.

Ray 2: A ray that passes through the focal point of a concave mirror or proceeds toward the focal point of a convex mirror is reflected parallel to the mirror axis.

Ray 3: A ray that proceeds along a radius of the mirror is reflected back along its original path.

- Before listing the *mirror equations,* you should review what the symbols mean and the sign conventions.

R = radius of curvature,
     + for converging, − for diverging

f = focal length, + for converging, − for diverging

p = object distance, + for real object, − for virtual

q = image distance, + for real images, − for virtual

y = object size, + if erect, − if inverted

y' = image size, + if erect, − if inverted

M = magnification, + if both erect or both inverted

- The mirror equations can be applied to either converging (concave) or diverging (convex) spherical mirrors:

$$f = \frac{R}{2} \qquad M = \frac{y'}{y} = \frac{-q}{p}$$

$$\frac{1}{p} + \frac{1}{q} = \frac{1}{f} \qquad \textit{Mirror Equation}$$

- Alternative forms for the last equation are

$$p = \frac{qf}{q - f} \qquad q = \frac{pf}{p - f} \qquad f = \frac{pq}{p + q}$$

Key Terms

Review Questions

34.1. Discuss the statement: One cannot "see" the surface of a perfect mirror.

34.2. Prove by a diagram that rays diverging from a point source of light appear to diverge from a virtual point after reflection from a plane surface.

34.3. Can an image of a real object be projected on a screen by a plane mirror? By a convex mirror? By a concave mirror?

34.4. State the laws of reflection and show how they may be demonstrated in a laboratory.

34.5. Use the mirror equation to show that the image of an infinitely distant object is formed at the focal point of a spherical mirror.

34.6. Use the mirror equation to show that the image of an object placed at the focal point of a concave mirror is located at infinity.

34.7. Use the mirror equation to show that, for a plane mirror, the image distance is equal in magnitude to the object distance. What is the magnification of a plane mirror?

34.8. In a concave shaving mirror, will greater magnification be achieved when the object is closer to the focal point or when it is closer to the vertex? Use diagrams to verify your conclusion.

34.9. Do objects moving closer to the vertex of a convex mirror form larger or smaller virtual images? Explain with diagrams.

34.10. Without looking at Fig. 34.13 in the text, construct the images formed by a concave mirror when the object is (a) beyond C, (b) at C, (c) between C and F, (d) at F, and (e) between F and V. Discuss the nature and relative size of each image.

34.11. Several small spherical mirrors are lying on a laboratory table. Describe how you would distinguish the diverging mirrors from the converging mirrors without touching them.

34.12. For real objects, is it possible to construct an inverted image by using a diverging mirror? What can you say about the magnification of diverging mirrors?

34.13. You wish to choose a shaving mirror that will give maximum magnification with an erect image. Does the focal length of the mirror play a part in determining its magnification? Explain.

34.14. Two concave spherical mirrors have the same focal length, but one has a larger linear aperture. Which forms the sharper image? Why?

34.15. Show how it is possible for a plane mirror to form a real image if light from the object is first converged by a concave mirror.

Problems

Section 34.2 Plane Mirrors

34.1. A man 1.80 m tall stands 1.2 m from a large plane mirror. How tall is his image? How far is he from his image?   Ans. 1.80 m, 2.40 m

34.2. What is the shortest mirror length required to enable a 1.68-m-tall woman to see her entire image?

*34.3. A plane mirror moves at a speed of 30 km/h away from a stationary person. How fast does this person's image appear to be moving in the opposite direction?   Ans. 60 km/h

*34.4. The optical lever is a sensitive measuring device that utilizes minute rotations of a plane mirror to measure small deflections. The device is illustrated in Fig. 34.19. When the mirror is in position 1, the light ray follows the path IVR_1. If the mirror is rotated through an angle θ to position 2, the ray will follow the path IVR_2. Show that the reflected beam turns through an angle 2θ, which is twice the angle through which the mirror itself turns.

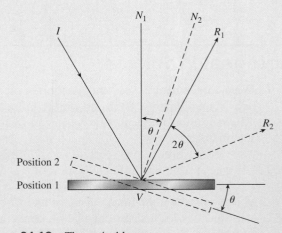

Figure 34.19   The optical lever.

Section 34.3 Spherical Mirrors

34.5. A lightbulb 3 cm high is placed 20 cm in front of a concave mirror with a radius of curvature of 15 cm. Determine the nature, size, and location of the image formed. Sketch the ray-tracing diagram.

       Ans. Real, $y' = -1.8$ cm, $q = +12$ cm

34.6. A spherical concave mirror has a focal length of 20 cm. What are the nature, size, and location of the image formed when a 6-cm-tall object is located 15 cm from this mirror?

34.7. A 8-cm-long pencil is placed 10 cm from a diverging mirror of radius 30 cm. Determine the nature, size, and location of the image formed. Sketch the ray-tracing diagram.

       Ans. Virtual, $y' = +4.80$ cm, $q' = -6.00$ cm

34.8. A spherical convex mirror has a focal length 25 cm. What are the nature, size, and location of the image formed of a 5-cm-tall object located 30 cm from the mirror?

34.9. An object 5 cm tall is placed halfway between the focal point and the center of curvature of a concave spherical mirror of radius 30 cm. Determine the location and magnification of the image.

       Ans. $q = +45$ cm, $M = -2.00$

34.10. A 4-cm-high source of light is placed in front of a spherical concave mirror that has a radius of 40 cm. Determine the nature, size, and location of the images formed for the following object distances: (a) 60 cm, (b) 40 cm, (c) 30 cm, (d) 20 cm, and (e) 10 cm. Draw the appropriate ray-tracing diagrams.

*34.11. At what distance from a concave spherical mirror of radius 30 cm must an object be placed to form an enlarged, inverted image located 60 cm from the mirror?   Ans. $p = 20$ cm

Section 34.6 Magnification

34.12. What is the magnification of an object if it is located 10 cm from a mirror and its image is erect and seems to be located 40 cm behind the mirror? Is this mirror diverging or converging?

34.13. A Christmas tree ornament has a silvered surface and a diameter of 3 in. What is the magnification of an object placed 6 in. from the surface of this ornament?   Ans. $+0.111$

34.14. What type of mirror is required to form an image on a screen 2 m away from the mirror when an object is placed 12 cm in front of the mirror? What is the magnification?

*34.15. A concave shaving mirror has a focal length of 520 mm. How far away from it should an object be placed for the image to be erect and twice its actual size?   Ans. 260 mm

*34.16. If a magnification of $+3$ is desired, how far should the mirror of Prob. 34.15 be placed from the face?

*34.17. An object is placed 12 cm from the surface of a spherical mirror. If an erect image is formed that is one-third the size of the object, what is the radius of the mirror? Is it converging or diverging?

       Ans. -6.00 cm, diverging

*34.18. A concave spherical mirror has a radius of 30 cm and forms an inverted image on a wall 90 cm away. What is the magnification?

Additional Problems

34.19. What are the nature, size, and location of the image formed when a 6-cm-tall object is located 15 cm from a spherical concave mirror of focal length 20 cm?

       Ans. Virtual, erect, $q = -60$ cm, $y' = +24$ cm

34.20. An erect image has a magnification of $+0.6$. Is the mirror diverging or converging? What is the object distance if the image distance is -12 cm?

34.21. An object is located 50 cm from a converging mirror that has a radius of 40 cm. What are the image distance and the magnification?

       Ans. $q = +33.3$ cm, $M = -0.667$

34.22. What is the focal length of a diverging mirror if the image of an object located 200 mm from the mirror appears to be a distance of 120 mm behind the mirror?

34.23. A silver ball is 4.0 cm in diameter. Locate the image of a 6-cm-long object located 9 cm from the surface of the ball. What is the magnification?

       Ans. -9.0 mm, $+0.100$

34.24. An object 80 mm tall is placed 400 mm in front of a diverging mirror of radius -600 mm. Determine the nature, size, and location of the image.

*34.25. An object 10 cm tall is located 20 cm from a spherical mirror. If an erect image 5 cm tall is formed, what is the focal length of the mirror?

       Ans. $f = -20$ cm

*34.26. What is the magnification if the image of an object is located 15 cm from a diverging mirror of focal length -20 cm?

*34.27. An object is placed 200 mm from the vertex of a convex spherical mirror that has a radius of 400 mm. What is the magnification of the mirror? *Ans. M = +1/2*

*34.28. A convex spherical mirror has a radius of −60 cm. How far away should an object be held if the image is to be one-third the size of the object?

*34.29. What should be the radius of curvature of a convex spherical mirror if it is to produce an image one-fourth as large as an object that is located 40 in. from the mirror? *Ans. R = −26.7 in.*

*34.30. A convex mirror has a focal length of −500 mm. If an object is placed 400 mm from the vertex, what is the magnification?

*34.31. A spherical mirror forms a real image 18 cm from the surface. The image is twice as large as the object. Find the location of the object and the focal length of the mirror.
Ans. p = 9.00 cm, f = 6.00 cm

*34.32. A certain mirror placed 2 m from an object produces an erect image enlarged three times. Is the mirror diverging or converging? What is the radius of the mirror?

*34.33. The magnification of a mirror is −0.333. Where is the object located if its image is formed on a card 540 mm from the mirror? What is the focal length? *Ans. p = 1.62 m, f = +405 mm*

*34.34. What should be the radius of curvature of a concave mirror if it is to produce an image one-fourth as large as an object 50 cm away from the mirror?

*34.35. A spherical shaving mirror has a magnification of +2.5 when an object is located 15 cm from the surface. What is the focal length of the mirror?
Ans. f = +25 cm

Critical Thinking Questions

34.36. A baseball player is 6 ft tall and stands 30 ft in front of a plane mirror. The distance from the top of his cap to his eyes is 8 in. Draw a diagram showing location of the images formed of his feet and of the top of his cap. What is the minimum length of mirror required for him to see his entire image? If he walks 10 m closer to the mirror, what is the new separation of object and image?
Ans. 36 in., 40 ft

*34.37. The diameter of the Moon is 3480 km, and it is 3.84×10^8 m away from Earth. A telescope on Earth utilizes a spherical mirror that has a radius of 8.00 m to form an image of the Moon. What is the diameter of the image formed? What is the magnification of the mirror?

*34.38. An image 60 mm long is formed on a wall located 2.3 m away from a source of light 20 mm high. What is the focal length of this mirror? Is it diverging or converging? What is the magnification?
Ans. 862 mm, converging, −3.00

*34.39. Derive an expression for calculating the focal length of a mirror in terms of the object distance p and the magnification M. Apply it to Prob. 34.25. Derive a similar relation for calculating the image distance q in terms of M and p. Apply it to Prob. 34.33.

*34.40. A concave mirror of radius 800 mm is placed 600 mm from a plane mirror that faces it. A source of light placed midway between the mirrors is shielded so that the light is first reflected from the concave surface. What are the position and magnification of the image formed after reflection from the plane mirror? (*Hint:* Treat the image formed by the first mirror as the object for the second mirror.)
Ans. 1.8 m behind plane mirror, +4.00

35

Refraction

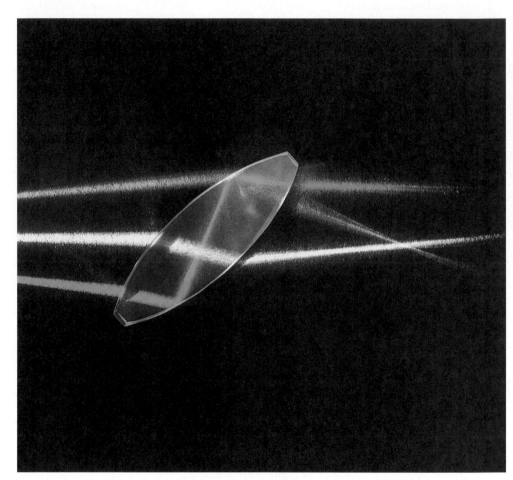

Refraction and total internal reflection of light beams are observed in this photograph as three rays of light pass through a biconvex lens. In this chapter, we will study the laws of refraction and the conditions for total internal reflection. (*Photo © Richard Megna/ Fundamental Photographs.*)

Objectives

After completing this chapter, you should be able to

1. Define the *index of refraction* and state three laws that describe the behavior of refracted light.
2. Apply Snell's law to the solution of problems involving the transmission of light in two or more media.
3. Determine the change in velocity or wavelength of light as it moves from one medium into another.
4. Explain the concepts of *total internal reflection* and the *critical angle* and use these ideas to solve problems similar to those in the text.

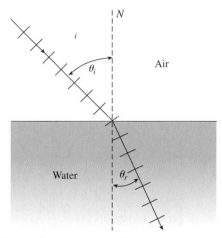

Figure 35.1 Refraction of a wavefront at the boundary between two media.

Light travels in straight lines at a constant speed in a uniform medium. If the medium changes, the speed will also change, and the light will travel in a straight line along a new path. The bending of a light ray as it passes obliquely from one medium to another is known as **refraction.** The principle of refraction is illustrated in Fig. 35.1 for a light wave entering water from the air. The angle θ_i that the incident beam makes with the normal to the surface is referred to as the *angle of incidence.* The angle θ_r between the refracted beam and the normal is called the *angle of refraction.*

Refraction explains familiar phenomena such as the apparent distortion of objects partially submerged in water. The stick appears to be bent at the surface of the water in Fig. 35.2a, and the fish in Fig. 35.2b appears to be closer to the surface than it really is. In this chapter, we study the properties of refractive media and develop equations to predict their effect on incident light rays.

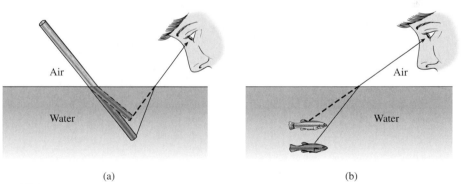

(a)   (b)

Figure 35.2 Refraction is responsible for the distortion of images. (a) The stick appears to be bent. (b) The fish seems to be closer to the surface than it actually is.

35.1 Index of Refraction

The velocity of light in a material substance is generally less than the free-space velocity of 3×10^8 m/s. In water, the speed of light is almost 2.25×10^8 m/s, which is just about three-fourths of its velocity in air. Light travels about two-thirds as fast in glass, or around 2×10^8 m/s. The ratio of the velocity c of light in a vacuum to the velocity v of light in a particular medium is called the ***index of refraction*** n for that material.

The index of refraction n of a particular material is the ratio of the free-space velocity of light to the velocity of light through the material.

$$n = \frac{c}{v} \qquad \textit{Index of Refraction} \quad \textbf{(35.1)}$$

The index of refraction is a unitless quantity, which is generally greater than unity. For water, $n = 1.33$, and for glass, $n = 1.5$. Table 35.1 lists the index of refraction for several common substances. Note that the values given apply for yellow light (589 nm). The velocity of light in a material is different for different wavelengths. This effect, known as *dispersion,* will be discussed in Section 35.4. When the wavelength of light is not specified, the index is usually assumed to correspond to that for yellow light.

Table 35.1

Index of Refraction for Yellow Light of Wavelength 589 nm

Substance	n	Substance	n
Benzene	1.50	Glycerin	1.47
Carbon disulfide	1.63	Ice	1.31
Diamond	2.42	Quartz	1.54
Ethyl alcohol	1.36	Rock salt	1.54
Fluorite	1.43	Water	1.33
Glass		Zircon	1.92
Crown	1.52		
Flint	1.63		

Example 35.1

Compute the velocity of yellow light in a diamond that has a refractive index of 2.42.

Plan: The refractive index is the ratio of free-space light velocity to the velocity in the medium, so we can solve for that velocity by substitution.

Solution: From Eq. (35.1), we have

$$n = \frac{c}{v} \qquad \text{or} \qquad v = \frac{c}{n}$$

$$v = \frac{3 \times 10^8 \text{ m/s}}{2.42} = 1.24 \times 10^8 \text{ m/s}$$

35.2 The Laws of Refraction

Two basic laws of refraction have been known and observed since ancient times. These laws are stated as follows and are illustrated in Fig. 35.3.

The incident ray, the refracted ray, and the normal to the surface all lie in the same plane.

The path of a ray refracted at the interface between two media is exactly reversible.

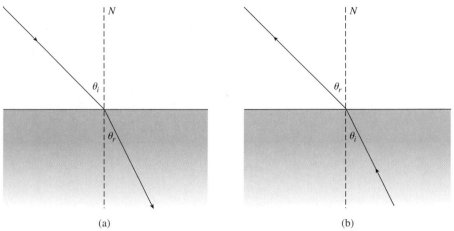

(a)   (b)

Figure 35.3 (a) The incident ray, the refracted ray, and the normal to the surface are in the same plane. (b) Refracted rays are reversible.

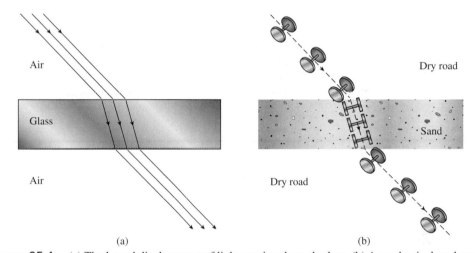

(a)   (b)

Figure 35.4 (a) The lateral displacement of light passing through glass. (b) A mechanical analogy.

These two laws are demonstrated easily by observation and experiment. It is of much more importance, in a practical sense, however, to understand and predict the *degree* of bending that occurs.

To understand how a change in the velocity of light can alter its path through a medium, let us consider the mechanical analogy shown in Fig. 35.4. In Fig. 35.4a, light incident on the glass plate is first bent toward the normal as it passes through the denser medium, and then it is bent away from the normal as it returns to the air. In Fig. 35.4b, the action of wheels encountering a patch of sand resembles the behavior of light. As they approach the sand, one wheel strikes first and slows down. The other wheel continues at the same speed, causing the axle to assume a new angle. When both wheels are in the sand, the wheels again move in a straight line with uniform speed. The first wheel to enter the sand is also the first to leave, and it speeds up as it leaves the patch of sand. Thus, the axle swings around to its original direction. The path of the axle is analogous to the path of a wavefront.

The change in direction of light as it enters another medium can be analyzed with the help of the wavefront diagram in Fig. 35.5. A plane wave in a medium of refractive index n_1 meets at the plane surface of a medium whose index is n_2. The angle of incidence is labeled θ_2. In the figure, it is assumed the second medium has a greater optical density than the first ($n_2 > n_1$). An example is light passing from air ($n_1 = 1$) to water ($n_2 = 1.33$). The line AB represents the wavefront at time $t = 0$ when it just comes into contact with

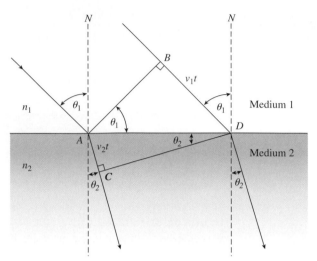

Figure 35.5   Deriving Snell's law.

medium 2. The line *CD* represents the same wavefront after the time *t* required to enter the second medium completely. Light travels from *B* to *D* in medium 1 in the same time *t* required for light to travel from *A* to *C* in medium 2. Assuming the velocity v_2 in the second medium is smaller than the velocity v_1 in the first medium, the distance *AC* will be shorter than the distance *BD*. These lengths are given by

$$AC = v_2 t \qquad BD = v_1 t$$

It can be shown from geometry that angle *BAD* is equal to θ_1 and that angle *ADC* is equal to θ_2, as indicated in Fig. 35.5. The line *AD* forms a hypotenuse that is common to the two triangles *ADB* and *ADC*. From the figure,

$$\sin \theta_1 = \frac{v_1 t}{AD} \qquad \sin \theta_2 = \frac{v_2 t}{AD}$$

Dividing the first equation by the second, we obtain

$$\frac{\sin \theta_1}{\sin \theta_2} = \frac{v_1}{v_2} \tag{35.2}$$

The ratio of the sine of the angle of incidence to the sine of the angle of refraction is equal to the ratio of the velocity of light in the incident medium to the velocity of light in the refracted medium.

This rule was first discovered by the seventeenth-century Dutch astronomer Willebrord Snell and is called **Snell's law** in his honor. An alternative form for the law can be obtained by expressing the velocities v_1 and v_2 in terms of the indexes of refraction for the two media. Recall that

$$v_1 = \frac{c}{n_1} \qquad \text{and} \qquad v_2 = \frac{c}{n_2}$$

Utilizing these relations in Eq. (35.2), we write:

$$n_1 \sin \theta_1 = n_2 \sin \theta_2 \tag{35.3}$$

Since the sine of an angle increases as the angle increases, we see that an increase in the index of refraction results in a decrease in the angle, and vice versa.

Example 35.2

Light passes at an angle of incidence of 35° from water into the air. What is the angle of refraction if the index of refraction for water is 1.33?

Plan: The angle of refraction θ_{air} can be found from Snell's law.

Solution: Given that $n_{air} = 1.0$, $n_{water} = 1.33$, and $\theta_{water} = 35°$, we have

$$n_a \sin \theta_a = n_w \sin \theta_w$$
$$(1.0) \sin \theta_a = (1.33) \sin 35°$$
$$\sin \theta_a = 0.763$$
$$\theta_a = 49.7°$$

The index of refraction in air (1.0) was *less* than that for the water (1.33), so the refracted angle in air is larger than the angle of incidence.

Example 35.3

A ray of light in water ($n_w = 1.33$) is incident at 40° on the glass bottom of a container, as shown in Fig. 35.6. If the refracted ray makes an angle of 33.7° with the normal, what is the index of refraction for the glass?

Plan: Water is the incident medium, and glass is the refracted medium. Given the angles and the index for water, we can apply Snell's law to find the index for glass.

Solution: Substituting $n_w = 1.33$, $\theta_w = 40°$, and $\theta_g = 33.7°$, we obtain

$$n_g \sin \theta_g = n_w \sin \theta_w$$
$$n_g \sin 33.7° = (1.33) \sin 40°$$
$$n_g = \frac{(1.33) \sin 40°}{\sin 33.7°}$$
$$n_g = 1.54$$

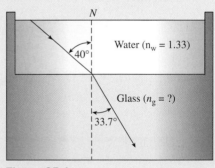

Figure 35.6

35.3

Wavelength and Refraction

We have seen that light slows down when passing into a medium of greater *optical density*. What happens to the wavelength of light entering a new medium? In Fig. 35.7, light traveling in air at a velocity c encounters a medium through which it travels at the reduced speed v_m. Upon returning to the air, it again travels at the speed c of light in air. This does not violate the conservation of energy because the energy of a light wave is proportional to

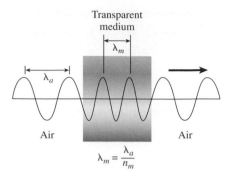

Transparent
medium

Air   Air

$$\lambda_m = \frac{\lambda_a}{n_m}$$

Figure 35.7 The wavelength of light is reduced when it enters a medium of greater optical density.

its frequency. The frequency f is the same inside the medium as it is outside the medium. That this is true will be realized if you consider the frequency is the number of waves passing any point per unit of time. The same number of waves leaves the medium per second as enters it. Thus, the frequency inside the medium cannot change. The velocity is related to the frequency and wavelength by

$$c = f\lambda_a \qquad \text{and} \qquad v_m = f\lambda_m \qquad \qquad \textbf{(35.4)}$$

where c and v_m are the speeds in air and inside the medium and λ_a and λ_m are the respective wavelengths. Since the velocity decreases inside the medium, the wavelength inside the medium must decrease proportionately for the frequency to remain constant. Dividing the first equation by the second in Eq. (35.4) yields

$$\frac{c}{v_m} = \frac{f\lambda_a}{f\lambda_m} = \frac{\lambda_a}{\lambda_m}$$

If we substitute $v_m = c/n_m$, we obtain

$$n_m = \frac{\lambda_a}{\lambda_m}$$

Therefore, the wavelength λ_m inside the medium is reduced by

$$\lambda_m = \frac{\lambda_a}{n_m} \qquad \qquad \textbf{(35.5)}$$

where n_m is the index of refraction of the medium and λ_a is the wavelength of the light in air.

Example 35.4 Monochromatic red light of wavelength 640 nm passes from air into a glass plate of refractive index 1.5. What is the wavelength of this light inside the glass?

Plan: The wavelength will be less in the glass due to the slower speed of the light.

Solution: Direct substitution into Eq. (35.5) yields

$$\lambda_g = \frac{\lambda_a}{n_g} = \frac{640 \text{ nm}}{1.5}; \qquad \lambda_g = 427 \text{ nm}$$

The wavelength in the glass indicates the color is blue. If you are observing this effect, why does the color still appear to be red?

Suppose a monochromatic light ray in medium 1 enters medium 2. We should remember the following four ratios are equal. In other words, we can set any two ratios equal to form an equation for finding a desired unknown.

$$\frac{n_1}{n_2} = \frac{\sin \theta_2}{\sin \theta_1} = \frac{v_2}{v_1} = \frac{\lambda_2}{\lambda_1} \tag{35.6}$$

Note that all ratios except for one are the ratio for medium 2 to medium 1. The order is reversed *only* for the indices of refraction, which is n_1/n_2.

35.4 Dispersion

We have already mentioned that the velocity of light in different substances varies with different wavelengths. We defined the index of refraction as the ratio of the free-space velocity c to the velocity inside a medium.

$$n = \frac{c}{v_m}$$

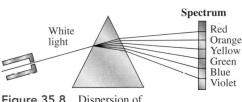

Figure 35.8 Dispersion of light by a prism.

The values given in Table 35.1 are valid strictly for monochromatic yellow light (589 nm). A different wavelength of light, such as blue light or red light, results in a slightly different index of refraction. Red light travels faster through a particular medium than blue light. This can be shown by passing white light through a glass prism, as in Fig. 35.8. Because of the different speeds inside the medium, the beam is *dispersed* into its component colors.

Dispersion is the separation of light into its component wavelengths.

From such an experiment, we conclude white light is actually a mixture of light, consisting of several colors. The projection of a dispersed beam is called a *spectrum.*

35.5 Total Internal Reflection

A fascinating phenomenon, known as *total internal reflection,* can occur when light passes obliquely from one medium to a medium with a lower optical density. To understand this phenomenon, consider a source of light submerged in medium 1, as illustrated in Fig. 35.9. Consider the four rays A, B, C, and D, which diverge from the submerged source. Ray A passes into medium 2 normal to the interface. The angle of incidence and the angle of refraction are both zero for this special case. Ray B is incident at an angle θ_1 and refracted away from the normal at an angle θ_2. The angle θ_2 is greater than θ_1 because the index of refraction for

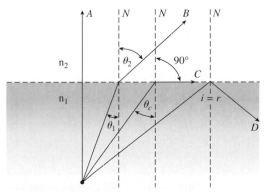

Figure 35.9 Critical angle of incidence.

medium 1 is greater than that for medium 2 ($n_1 > n_2$). As the angle of incidence θ_1 increases, the angle of refraction θ_2 also increases until the refracted ray C emerges tangent to the surface. The angle of incidence θ_c for which this occurs is known as the ***critical angle.***

> The critical angle θ_c is the limiting angle of incidence in a denser medium that results in an angle of refraction of 90°.

A ray approaching the surface at an angle greater than the critical angle is reflected back inside medium 1. The ray D in Fig. 35.9 does not pass into the upper medium at all but is *totally internally reflected* at the interface. This type of reflection follows the same laws as any other reflection; in other words, the angle of incidence is equal to the angle of reflection. Total internal reflection can occur only when light is incident from a denser medium ($n_1 > n_2$).

The critical angle for two given media can be calculated from Snell's law.

$$n_1 \sin \theta_c = n_2 \sin \theta_2$$

where θ_c is the critical angle and $\theta_2 = 90°$. Simplifying, we write

$$n_1 \sin \theta_c = n_2(1)$$

or

$$\sin \theta_c = \frac{n_2}{n_1} \qquad \textit{Critical Angle} \quad \textbf{(35.7)}$$

Since the sine of the critical angle can never be greater than 1, the index of refraction n_1 for the incident medium must be greater than the index n_2 for the refracted medium. If this were not the case, the light would be reflected at the boundary, back into the incident medium.

Example 35.5

What is the critical angle for a glass to air surface if the refractive index of the glass is 1.5?

Solution: Direct substitution into Eq. (35.7) yields

$$\sin \theta_c = \frac{n_a}{n_g} = \frac{1.0}{1.5} = 0.667$$

$$\theta_c = 42°$$

The fact that the critical angle for glass is 42° makes it possible to use 45° prisms in many optical instruments. Two such uses are illustrated in Fig. 35.10. In Fig. 35.10a, a 90° reflection can be obtained with little loss of intensity. In Fig. 35.10b, a 180° deflection is obtained. In each case, total internal reflection occurs because the angles of incidence are all 45° and therefore greater than the critical angle.

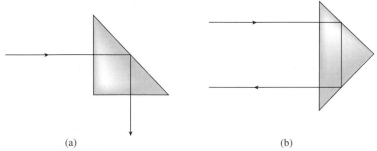

(a)   (b)

Figure 35.10   Right-angle prisms make use of the principle of total internal reflection to deviate the path of light.

35.6   Fiber Optics and Applications

The application of fiber optics in communications has resulted in an information explosion. Optical fiber has a significantly larger bandwidth than copper wire, which means more information can be transmitted during a fixed time period. This increased information-carrying capability provides significant new possibilities, including interactive television and cable channel selections numbering in the thousands.

Although it is easy to see how fiber optics is changing the world around us, it may come as a surprise that its ability to transmit information depends primarily on a single physical phenomenon, *total internal reflection*. As discussed in Section 35.5, total internal reflection results when light passing through one medium encounters a second medium of lower optical density. An optical fiber consists of two such mediums.

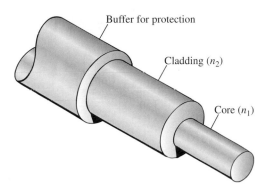

Figure 35.11   The basic structure of an optical fiber. The index of refraction for the core (n_1) must be greater than the index of refraction for the cladding (n_2).

The structure of an optical fiber is illustrated in Fig. 35.11. Notice that the core of the fiber serves as the transmission medium, while the cladding acts to contain the transmitted signal. This means the core must have a higher index of refraction than the cladding ($n_1 > n_2$).

Using the ray model of propagation, we can think of signals propagating through a fiber as distinct rays of light. As illustrated in Fig. 35.12, each ray has a fixed angle of incidence with the core/cladding boundary. As long as the beams maintain an angle of incidence equal to or greater than the fiber's critical angle, the signal will be confined to the fiber. (*Note:* Since an optical fiber is actually a cylindrical wave guide, light can propagate only at specific angles greater than the critical angle. These specific angles can be determined by solving the appropriate wave guide equations.)

As mentioned previously, fiber optics has played a significant role in the expansion of communication. At its basic level, optical fiber communications consist of sending information from a source to a destination by transmitting pulses of light. This is analogous to the historical practice of ship-to-ship communications using Morse code. When one ship wished to communicate with another, a light source was flashed on and off in a sequence that both parties understood. This principle can be extended to fiber-optic communications, where numerous and highly efficient data coding schemes have been created. The encoding scheme chosen depends on the application and cost considerations.

Using optical fiber in communication systems has many advantages. Among the more important are immunity to electromagnetic interference, improved data security, higher transmission speed, and increased signal bandwidth. These advantages help make optical fiber the communications medium of choice for the foreseeable future.

While communication dominates the field of fiber optics, many other applications exist for optical fiber. Figure 35.13 illustrates a fluid level sensor. In this sensor, the behavior of light when traveling between different media is exploited. Notice the fiber end has been

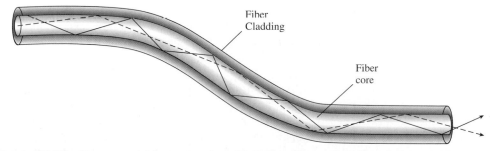

Figure 35.12   The ray model for propagation of light through an optical fiber. As long as curves are gentle enough to keep the angle of incidence greater than the critical angle, the transmitted light will experience total internal reflection.

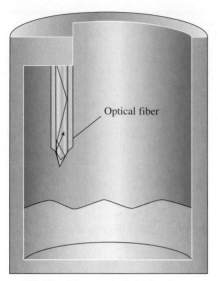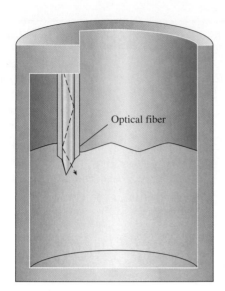

Figure 35.13 Fiber-optic fluid level sensor.

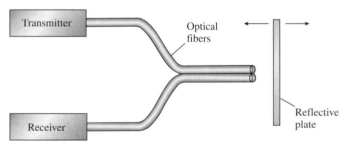

Figure 35.14 Fiber-optic displacement sensor.

specially shaped for this application so that when exposed to air (low fluid level), a majority of light transmitted down the fiber will be reflected back. When the fluid being sensed covers the fiber end, however, the index of the fluid more closely matches the index of the fiber core, thus reducing the amount of reflected light. This variation in reflected light is used to determine fluid level.

Another possible use for optical fiber is as a displacement sensor. This sensor exploits the spreading behavior of light as it leaves the end of a fiber. With this sensor, the distance between the fiber ends and the reflective plate in Fig. 35.14 can be measured. As the plate moves farther away from the fiber ends, less light is reflected from the transmitting fiber into the receiving fiber. Likewise, as the plate moves closer to the fiber ends, more light is reflected into the receiving fiber. This varying level of receiver intensity is used to determine the plate's position.

In medicine, fiber optics is making an impact in both diagnostics and treatment. For diagnostic purposes, a device known as a fiber-optic endoscope allows internal organs to be inspected visually. In this device, two optical fibers are used—one for illuminating the area of interest and the other for transmitting the image being viewed. When connected to a video monitor, this device gives doctors the ability to see what previously required exploratory surgery.

In treatment, blocked arteries can be cleared using the LASTAC (laser enhanced transluminal angioplasty) system. In this system, laser light is transmitted through an optical fiber that has been inserted into a blocked artery. The laser vaporizes the impeding plaque, thus clearing the artery. With this treatment, more serious surgery can often be avoided.

35.7   Is Seeing the Same as Believing?

Because we are accustomed to light traveling in straight lines, refraction and total internal reflection often present us with pictures we do not believe. Atmospheric refraction accounts for many illusions that are referred to as *mirages.* Figure 35.15 provides two examples of such occurrences. In Fig. 35.15a, a layer of hot air in contact with the heated ground is less dense than the cool layers of air above it. Consequently, light from distant objects is refracted upward, making them appear inverted.

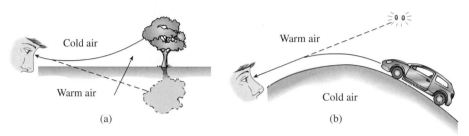

(a)   (b)

Figure 35.15   Atmospheric refraction accounts for the mirage in (a) and explains the phenomenon of *looming* in (b).

At night, the situation is sometimes reversed: The cool layer of air is beneath warmer layers. The headlights of the car in Fig. 35.15b appear to be *looming* in the air. Many scientists believe that some of the unidentified flying objects (UFOs) reported for centuries can be explained in terms of atmospheric refraction.

Combinations of refraction and total internal reflection account for the bizarre photograph in Fig. 35.16. The picture was taken by an underwater camera looking upward at a girl sitting on the edge of a pool with her legs dangling in the water. An explanation is given in Fig. 35.17. The upper portion of the picture results from refraction at the surface. The inverted legs in the middle of the picture are due to total internal reflection at the surface of the water. The bottom of the picture represents the only undistorted image since the legs are viewed directly and in the same medium.

Figure 35.16   A photograph taken by an underwater camera presents a bizarre picture of a person sitting at the edge of a swimming pool. (*Photo from the film* Introduction to Optics, *Education Development Center.*)

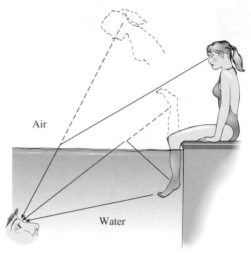

Figure 35.17   Combinations of refraction, reflection, and total internal reflection serve to deceive the eye.

35.8   Apparent Depth

Refraction causes an object submerged in a liquid of higher index of refraction to appear closer to the surface than it actually is. This shallowing effect is illustrated in Fig. 35.18. The object O appears to be at I because of the refraction of light from the object. The ***apparent depth*** is denoted by q, and the actual depth is denoted by p. Snell's law applied at the surface gives

$$\frac{\sin \theta_1}{\sin \theta_2} = \frac{n_2}{n_1} \qquad (35.8)$$

If we can relate the ratio of the indexes of refraction to actual and apparent depths, a useful relation can be obtained for predicting the apparent depths of submerged objects. From Fig. 35.18, it is noted that

$$\angle AOB = \theta_1 \qquad \text{and} \qquad \angle AIB = \theta_2$$

and that

$$\sin \theta_1 = \frac{d}{OA} \qquad \sin \theta_2 = \frac{d}{IA}$$

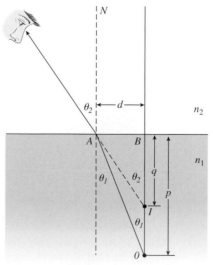

Figure 35.18   Relation between apparent depth and actual depth.

Using this information in Eq. (35.8), we obtain

$$\frac{\sin \theta_1}{\sin \theta_2} = \frac{d/OA}{d/IA} = \frac{IA}{OA} \tag{35.9}$$

If we restrict ourselves to rays that are nearly vertical, the angles θ_1 and θ_2 will be small, so the following approximations apply:

$$OA \approx p \qquad \text{and} \qquad IA \approx q$$

Applying these approximations to Eqs. (35.8) and (35.9), we can write

$$\frac{\sin \theta_1}{\sin \theta_2} = \frac{n_2}{n_1} = \frac{q}{p}$$

$$\frac{Apparent\ depth\ q}{Actual\ depth\ p} = \frac{n_2}{n_1} \tag{35.10}$$

Example 35.6

A coin rests on the bottom of a container filled with water ($n_w = 1.33$). The apparent distance of the coin from the surface is 9 cm. How deep is the container?

Plan: We recognize the apparent depth is the virtual image q of the coin and not the actual depth p. We can find p from the given indices of refraction.

Solution: Solving for p in Eq. (35.10) gives the actual depth of the container.

$$\frac{q}{p} = \frac{n_w}{n_a} \qquad \text{or} \qquad p = \frac{q n_w}{n_a}$$

$$p = \frac{(9\ \text{cm})(1.33)}{(1.0)} = 12\ \text{cm}$$

The apparent depth is approximately three-fourths of the actual depth.

Summary and Review

Summary

Refraction has been defined as the bending of a light ray as it passes obliquely from one medium to another. We have seen that the degree of bending can be predicted based on either the change in velocity or the known index of refraction for each medium. The concepts of refraction, critical angle, dispersion, and internal reflection play important roles in the operation of many instruments. The major concepts covered in this chapter are summarized as follows.

- The index of refraction of a particular material is the ratio of the free-space velocity of light c to the velocity v of light through the medium.

$$n = \frac{c}{v} \qquad c = 3 \times 10^8 \text{ m/s} \qquad \textit{Index of Refraction}$$

- When light enters from medium 1 and is refracted into medium 2, Snell's law can be written in the following two forms (see Fig. 35.19):

$$n_1 \sin \theta_1 = n_2 \sin \theta_2 \qquad \frac{v_1}{v_2} = \frac{\sin \theta_1}{\sin \theta_2} \qquad \textit{Snell's Law}$$

- When light enters medium 2 from medium 1, its wavelength is changed by the fact that the index of refraction is different.

$$\frac{\lambda_1}{\lambda_2} = \frac{n_2}{n_1} \qquad \lambda_2 = \frac{n_1 \lambda_1}{n_2}$$

- The critical angle θ_c is the maximum angle of incidence from one medium that will still produce

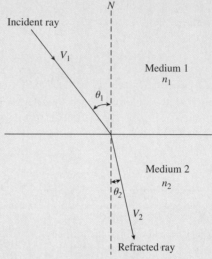

Figure 35.19 Snell's law.

refraction (at 90°) into a bordering medium. From the definition, we obtain

$$\sin \theta_c = \frac{n_2}{n_1} \qquad \textit{Critical Angle}$$

- Refraction causes an object in one medium to be observed at a different depth when viewed from above in another medium.

$$\frac{\textit{Apparent depth } q}{\textit{Actual depth } p} = \frac{n_2}{n_1}$$

Key Terms

apparent depth   690
critical angle   686
dispersion   680

index of refraction   679
optical density   683
refraction   679

Snell's law   682
spectrum   685
total internal reflection   685

Review Questions

35.1. State three laws of refraction and show how they can be demonstrated in the laboratory.

35.2. Is the index of refraction a constant for a particular medium? Explain.

35.3. Explain how the day is lengthened by atmospheric refraction.

35.4. A coin is placed on the bottom of a bucket so that it is just out of sight when viewed at an angle from the top. Show by the use of diagrams why the coin becomes visible if the bucket is filled with water.

35.5. Will objects of higher optical density have greater or smaller critical angles when surrounded by air?

35.6. On the basis of topics discussed in this chapter, explain why a diamond is much more brilliant than a glass replica.

35.7. Explain why right-angle prisms are more efficient reflectors than mirrored surfaces.

35.8. Why are colors observed in the light from a diamond?

35.9. A child stands waist deep in a swimming pool that has a uniform depth throughout. Why does it appear to him that he is standing in the deepest part of the pool?

35.10. The wavelength λ of a certain source of radiation is increased to 2λ. If the index of refraction was measured originally to be 1.5, what will it be when the wavelength is doubled?

35.11. What is the critical angle for an irregularly shaped piece of glass submerged in a liquid of the same index of refraction? Why would the glass be invisible in this case?

Problems

Section 35.1 Index of Refraction

35.1. The speed of light through a certain medium is 1.6×10^8 m/s in a transparent medium. What is the index of refraction in that medium?
                                                                 Ans. 1.88

35.2. If the speed of light is to be reduced by one-third, what must be the index of refraction for the medium through which the light travels?

35.3. Compute the speed of light in (a) crown glass, (b) diamond, (c) water, and (d) ethyl alcohol.
          Ans. 2.00×10^8 m/s, 1.24×10^8 m/s, 2.26×10^8 m/s, 2.21×10^8 m/s

35.4. If light travels at 2.1×10^8 m/s in a transparent medium, what is the index of refraction in that medium?

Section 35.2 The Laws of Refraction

35.5. Light is incident at an angle of 37° from air to flint glass ($n = 1.6$). What is the angle of refraction into the glass?   Ans. 22.1°

35.6. A beam of light makes an angle of 60° with the surface of water. What is the angle of refraction into the water?

35.7. Light passes from water ($n = 1.33$) to air. The beam emerges into air at an angle of 32° with the horizontal water surface. What is the angle of incidence inside the water?   Ans. 39.6°

35.8. Light in air is incident at 60° and is refracted into an unknown medium at an angle of 40°. What is the index of refraction for the unknown medium?

35.9. Light strikes from medium A into medium B at an angle of 35° with the horizontal boundary. If the angle of refraction is also 35°, what is the relative index of refraction between the two media?
                                                                 Ans. 1.43

35.10. Light incident from air at 45° is refracted into a transparent medium at an angle of 34°. What is the index of refraction for the material?

*35.11. A ray of light originating in air (Fig. 35.20) is incident on water ($n = 1.33$) at an angle of 60°. It then passes through the water entering glass ($n = 1.50$) and finally emerges back into air again. Compute the angle of emergence.
                                                                 Ans. 60°

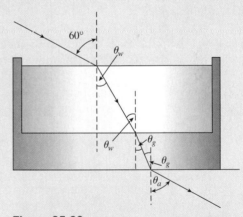

Figure 35.20

*35.12. Prove that, no matter how many parallel layers of different media are traversed by light, the entrance angle and the final emergent angle will be equal as long as the initial and final media are the same.

Section 35.3 Wavelength and Refraction

35.13. The wavelength of sodium light is 589 nm in air. Find its wavelength in glycerine.   Ans. 401 mm

35.14. The wavelength decreases by 25 percent as it goes from air to an unknown medium. What is the index of refraction for that medium?

35.15. A beam of light has a wavelength of 600 nm in air. What is the wavelength of this light as its passes into glass ($n = 1.50$)? Ans. 400 nm

35.16. Red light (520 nm) changes to blue light (478 nm) when it passes into a liquid. What is the index of refraction for the liquid? What is the velocity of the light in the liquid?

*35.17. A ray of monochromatic light of wavelength 400 nm in medium A is incident at 30° at the boundary of another medium B. If the ray is refracted at an angle of 50°, what is its wavelength in medium B? Ans. 613 nm

Section 35.5 Total Internal Reflection

35.18. What is the critical angle for light moving from quartz ($n = 1.54$) to water ($n = 1.33$)?

35.19. The critical angle for a given medium relative to air is 40°. What is the index of refraction for the medium? Ans. 1.56

35.20. If the critical angle of incidence for a liquid to air surface is 46°, what is the index of refraction for the liquid?

35.21. What is the critical angle relative to air for (a) diamond, (b) water, and (c) ethyl alcohol? Ans. 24.4°, 48.8°, 47.3°

35.22. What is the critical angle for flint glass immersed in ethyl alcohol?

*35.23. A right-angle prism like the one shown in Fig. 35.10a is submerged in water. What is the minimum index of refraction needed for the material to achieve total internal reflection? Ans. 1.88

Additional Problems

35.24. The angle of incidence is 30°, and the angle of refraction is 26.3°. If the incident medium is water, what might the refractive medium be?

35.25. The speed of light in an unknown medium is 2.40×10^8 m/s. If the wavelength of light in this unknown medium is 400 nm, what is the wavelength in air? Ans. 500 nm

35.26. A ray of light strikes a pane of glass at an angle of 30° with the glass surface. If the angle of refraction is also 30°, what is the index of refraction for the glass?

35.27. A beam of light is incident on a plane surface separating two media of indexes 1.6 and 1.4. The angle of incidence is 30° in the medium of higher index. What is the angle of refraction? Ans. 34.8°

35.28. In going from glass ($n = 1.50$) to water ($n = 1.33$), what is the critical angle for total internal reflection?

35.29. Light of wavelength 650 nm in a particular glass has a speed of 1.7×10^8 m/s. What is the index of refraction for this glass? What is the wavelength of this light in air? Ans. 1.76, 1146 nm

35.30. The critical angle for a certain substance is 38° when it is surrounded by air. What is the index of refraction of the substance?

35.31. The water in a swimming pool is 2 m deep. How deep does it appear to a person looking vertically down? Ans. 1.50 m

35.32. A plate of glass ($n = 1.50$) is placed over a coin on a table. The coin appears to be 3 cm below the top of the glass plate. What is the thickness of the glass plate?

Critical Thinking Questions

*35.33. Consider a horizontal ray of light striking one edge of an equilateral prism of glass ($n = 1.50$) as shown in Fig. 35.21. At what angle θ will the ray emerge from the other side? Ans. 77.0°

*35.34. What is the minimum angle of incidence at the first face of the prism in Fig. 35.21 such that the beam is refracted into air at the second face? (Larger angles do not produce total internal reflection at the second face.)

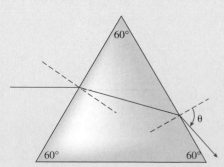

Figure 35.21

*35.35. Light passing through a plate of transparent material of thickness t suffers a lateral displacement d, as shown in Fig. 35.22. Compute the lateral displacement if the light passes through glass surrounded by air. The angle of incidence θ_1 is 40°, and the glass ($n = 1.50$) is 2 cm thick.

Ans. 5.59 mm

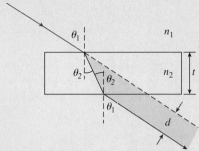

Figure 35.22

*35.36. A rectangularly shaped block of glass ($n = 1.54$) is submerged completely in water ($n = 1.33$). A beam of light traveling in the water strikes a vertical side of the glass block at an angle of incidence θ_1 and is refracted into the glass, where it continues to the top surface of the block. What is the minimum angle θ_1 at the side such that the light does not go out of the glass at the top?

*35.37. Prove that the lateral displacement in Fig. 35.22 can be calculated from

$$d = t \sin \theta_1 \left(1 - \frac{n_1 \cos \theta_1}{n_2 \cos \theta_2} \right)$$

Use this relationship to verify the answer to Question 35.35.

36

Lenses and Optical Instruments

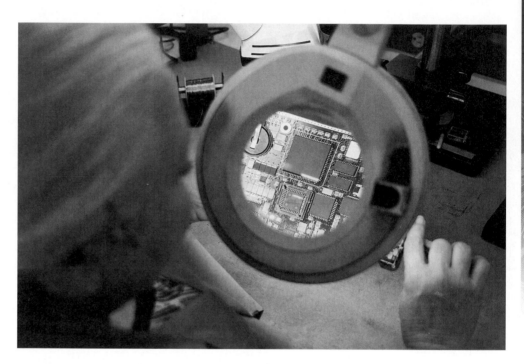

A convex lens is used in this magnifying glass, which allows for close observation of a computer circuit. Converging and diverging lenses can create real or virtual images for a variety of applications, from simple magnifying glasses to complex industrial microscopes. (*Photo © vol. 39 PhotoDisc/Getty.*)

Objectives

After completing this chapter, you should be able to

1. Determine mathematically or experimentally the focal length of a lens and state whether it is converging or diverging.

2. Apply the lensmaker's equation to solve for unknown parameters related to the construction of lenses.

3. Use ray-tracing techniques to construct images formed by diverging and converging lenses for various object locations.

4. Predict mathematically or determine experimentally the nature, size, and location of images formed by converging and diverging lenses.

A *lens* is a transparent object that alters the shape of a wavefront passing through it. Lenses are usually constructed of glass and shaped so that refracted light will form images similar to those discussed for mirrors. Anyone who has examined objects through a magnifying glass, observed distant objects through a telescope, or experimented in photography knows something of the effects lenses have on light. In this chapter, we study the images formed by lenses and discuss their application.

36.1   Simple Lenses

The simplest way of understanding how a lens works is to consider the refraction of light by prisms, as illustrated in Fig. 36.1. When Snell's law is applied to each surface of a prism, light is bent toward the normal when entering a prism and away from the normal on leaving. The effect, in either case, is to cause the light beam to be deviated toward the base of the prism. The light rays remain parallel because both the entrance and emergent surfaces are planes forming equal angles with all rays passing the prism. Thus, a prism merely alters the direction of a wavefront.

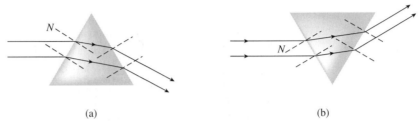

(a)   (b)

Figure 36.1   Parallel rays of light are bent toward the base of a prism and remain parallel.

Suppose we place two prisms base to base, as shown in Fig. 36.2a. Light incident from the left will converge, but it will not come to a focus. To focus the light rays to a point, the extreme rays must be deviated more than the central rays. This is accomplished by grinding the surfaces so that they have a uniformly curved cross section, as indicated in Fig. 36.2b. A lens that brings a parallel beam of light to a point focus in this fashion is called a ***converging lens.***

A converging lens is one that refracts and converges parallel light to a point focus beyond the lens.

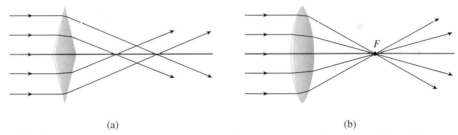

(a)   (b)

Figure 36.2   (a) Two prisms placed base to base will converge rays but will not bring all rays to a common focus. (b) A converging lens can be constructed by curving the surfaces uniformly.

The curved surfaces of lenses may be of any regular shape, such as spherical, cylindrical, or parabolic. Since spherical surfaces are easier to make, most lenses are constructed with two spherical surfaces. The line joining the centers of the two spheres is known as the *axis* of the lens. Three examples of converging lenses are shown in Fig. 36.3: *double convex,*

Converging lenses

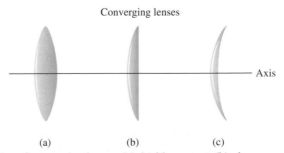

(a)   (b)   (c)

Figure 36.3   Examples of converging lenses: (a) double convex, (b) plano-convex, and (c) converging meniscus.

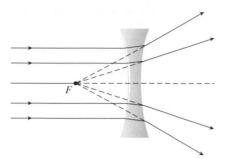

Figure 36.4 A diverging lens refracts light so that it appears to come from a point on the same side of the lens as the incident light.

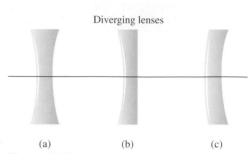

Figure 36.5 Examples of diverging lenses: (a) double concave, (b) plano-concave, and (c) diverging meniscus.

plano-convex, and *converging meniscus.* Note that converging lenses are thicker in the middle than at the edge.

A second type of lens can be constructed by making the edges thicker than the middle, as shown in Fig. 36.4. Parallel light rays passing through such a lens bend toward the thicker part, causing the beam to diverge. Projection of the refracted light rays shows that the light appears to come from a virtual focal point in front of the lens.

> A diverging lens is one that refracts and diverges parallel light from a point located in front of the lens.

Examples of **diverging lenses** are *double concave, plano-concave,* and *diverging meniscus.* See Fig. 36.5.

36.2 Focal Length and the Lensmaker's Equation

A lens is regarded as "thin" if the thickness is small in comparison with the other dimensions involved. As with mirrors, image formation by thin lenses is a function of the focal length; however, there are important differences. One obvious difference is that light may pass *through* a lens in two directions. This results in two focal points for each lens, as shown in Fig. 36.6 for a converging lens and in Fig. 36.7 for a diverging lens. The former has a *real focus F,* and the latter has a **virtual focus F′.** The distance between the optical center of a lens and the focus on either side of the lens is the **focal length f.**

> The focal length f of a lens is reckoned as the distance from the optical center of the lens to either focus.

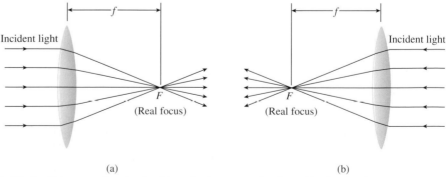

Figure 36.6 Demonstrating the focal length of a converging lens. The focal point is real because actual light rays pass through it.

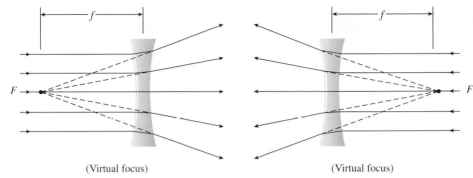

(Virtual focus)   (Virtual focus)

Figure 36.7   Demonstrating the virtual focal points of a diverging lens.

Since light rays are reversible, a source of light placed at either focus of a converging lens results in a parallel light beam. This can be seen by reversing the direction of the rays, illustrated in Fig. 36.6.

The focal length f of a lens is not equal to one-half the radius of curvature, as for spherical mirrors; it depends on the index of refraction n of the material from which it is made. It also is determined by the radii of curvature R_1 and R_2 of its surfaces, as defined in Fig. 36.8a. For thin lenses, these quantities are related by the equation

$$\frac{1}{f} = (n - 1)\left(\frac{1}{R_1} + \frac{1}{R_2}\right)$$ **(36.1)**

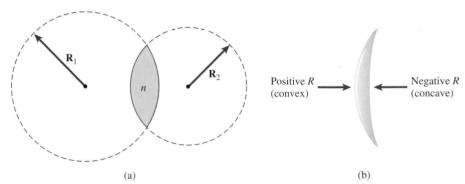

(a)   (b)

Figure 36.8   (a) The focal point of a lens is determined by the radii of its surfaces and the index of refraction. (b) The sign convention for the radius of a lens surface.

Because Eq. (36.1) involves the construction parameters for a lens, it is referred to as the **_lensmaker's equation._** It applies equally to converging and diverging lenses if the following sign convention is utilized:

- The radius of curvature (either R_1 or R_2) is considered positive if the surface is curved outward (convex) and negative if the surface is curved inward (concave). See Fig. 36.8b.

- The focal length f of a converging lens is considered positive, and the focal length of a diverging lens is considered negative.

Example 36.1

A lensmaker plans to construct a plano-concave lens out of glass ($n_g = 1.5$). What should be the radius of its curved surface in order to make a diverging lens with a focal length of -30 cm?

Plan: A *plano-concave* lens is shown in Fig. 36.5b. Remember that the focal length is negative for a diverging lens and that the radius of the plane surface is considered as infinity. The lensmaker's equation will give the radius of the curved surface, which should be a negative quantity. It does not matter which surface is chosen as R_1.

Solution: We are given that $f = -30$ cm, $R_1 = \infty$, and $n = 1.5$. From Eq. (36.1), we solve for R_2 as follows:

$$\frac{1}{f} = (n-1)\left(\frac{1}{\infty} + \frac{1}{R_2}\right) = (n-1)\left(0 + \frac{1}{R_2}\right)$$

$$\frac{1}{f} = \left(\frac{n-1}{R_2}\right) \quad \text{or} \quad R_2 = (n-1)f$$

$$R_2 = (1.5 - 1)(-30 \text{ cm}) = (0.5)(-30 \text{ cm})$$

$$R_2 = -15.0 \text{ cm}$$

The radius of curvature is negative, which is what we expected for a diverging lens.

Example 36.2

A *meniscus lens* has a convex surface of radius 10 cm and a concave surface of radius -15 cm. If the lens is constructed from glass with an index of refraction equal to 1.52, what will be the focal length?

Plan: We don't know until we find the sign of f whether this lens will be converging (Fig. 36.3c) or diverging (Fig. 36.5c). We will assign $R_1 = +10$ cm and $R_2 = -15$ cm and solve the lensmaker's equation for the focal length.

Solution: Substituting the given information, we find that

$$\frac{1}{f} = (n-1)\left(\frac{1}{R_1} + \frac{1}{R_2}\right)$$

$$= (1.52 - 1)\left(\frac{1}{10 \text{ cm}} + \frac{1}{(-15 \text{ cm})}\right)$$

$$= 0.52\left(\frac{1}{10 \text{ cm}} - \frac{1}{15 \text{ cm}}\right)$$

$$= 0.52\left(\frac{15 \text{ cm} - 10 \text{ cm}}{150 \text{ cm}^2}\right) = \frac{0.52}{30 \text{ cm}}$$

Now, we can solve for the focal length as follows:

$$\frac{1}{f} = \frac{0.52}{30 \text{ cm}} \quad \text{or} \quad f = \left(\frac{30 \text{ cm}}{0.52}\right)$$

$$f = 57.7 \text{ cm}$$

The fact that the focal length is positive indicates it is a *converging* meniscus lens.

36.3 | Image Formation by Thin Lenses

To understand how images are formed by lenses, we now introduce ray-tracing methods similar to those discussed for spherical mirrors. The method consists of tracing two or more rays from a chosen point on the object and using the point of intersection as the image of that point. The entire deviation of a ray passing through a thin lens can be considered to take place at a plane through the center of the lens. In Section 36.2, it was noted that a lens has two focal points. We define the *first focal point* F_1 as the one located on the same side of the lens as the incident light. The *second focal point* F_2 is located on the opposite, or far, side of the lens. With these definitions in mind, there are three principal rays that can be traced easily through a lens. These rays are illustrated in Fig. 36.9 for a converging lens and in Fig. 36.10 for a diverging lens:

Ray 1: A ray parallel to the axis passes through the second focal point F_2 of a converging lens or appears to come from the first focal point F_1 of a diverging lens.

Ray 2: A ray that passes through the first focal point F_1 of a converging lens or proceeds toward the second focal point F_2 of a diverging lens is refracted parallel to the lens axis.

Ray 3: A ray that passes through the geometrical center of a lens will not be deviated.

The intersection of any two of these rays (or their extensions) from an object point represents the image of that point. Since a real image produced by a lens is formed by rays of light that actually pass through the lens, *a real image is always formed on the side of the lens opposite the object. A virtual image will appear to be on the same side of the lens as the object.*

To illustrate the graphical method and, at the same time, to understand the various images formed by lenses, we will consider several examples. Images formed by a converging lens are shown in the object locations described in Fig. 36.11a to e.

Notice the images formed by a *convex* lens are similar to those formed by *concave* mirrors. This is true because they both converge light. Since concave lenses diverge light, we would expect them to form images similar to those formed by a diverging mirror (convex mirror). Figure 36.12 demonstrates this similarity.

Images of real objects formed by diverging lenses are always virtual, erect, and diminished in size.

To avoid confusion, one should identify both lenses and mirrors as either converging or diverging. Diverging lenses are often used to reduce or neutralize the effect of converging lenses.

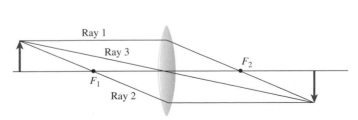

Figure 36.9 Principal rays for image construction using a converging lens.

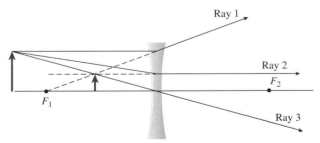

Figure 36.10 Principal rays for constructing images formed by diverging lenses.

(a) Object located at a distance beyond twice the focal length. A real, inverted, and diminished image is formed between F_2 and $2F_2$ on the opposite side of the lens.

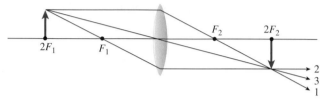

(b) Object at a distance equal to twice the focal length. A real, inverted image the same size as the object is located at $2F_2$ on the opposite side of the lens.

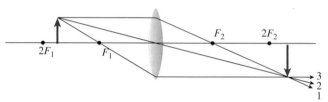

(c) Object located at a distance between one and two focal lengths from the lens. A real, inverted, and enlarged image is formed beyond $2F_2$ on the opposite side of the lens.

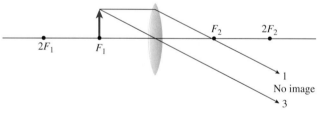

(d) Object at the first focal point F_1. No image is formed. The refracted rays are parallel.

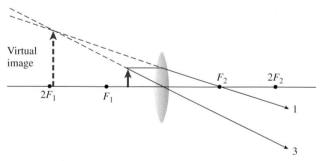

(e) Object located inside the first focal point. A virtual, erect, and enlarged image is formed on the same side of the lens as the object.

Figure 36.11   Image construction is shown for the following object distances: (a) beyond $2F_1$, (b) at $2F_1$, (c) between $2F_1$, and F_1, and (d) at F_1, and (e) inside F_1.

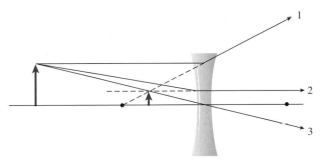

Figure 36.12 Images formed by diverging lenses are always virtual, erect, and diminished in size.

36.4 The Lens Equation and Magnification

The characteristics, size, and location of images can also be determined analytically from the **lens equation.** This important relation can be deduced by applying plane geometry to Fig. 36.13. The derivation is similar to the one used to derive the mirror equation, and the final form is exactly the same. The lens equation can be written

$$\frac{1}{p} + \frac{1}{q} = \frac{1}{f} \tag{36.2}$$

where p = object distance

q = image distance

f = focal length of lens

The same sign conventions established for mirrors can be used in the lens equation if converging and diverging lenses are compared with converging and diverging mirrors. This convention is summarized as follows:

1. The object distance p and the image distance q are considered positive for real objects and images and negative for virtual objects and images.

2. The focal length f is considered positive for converging lenses and negative for diverging lenses.

The following alternative forms of the lens equation are useful in solving optical problems:

$$p = \frac{fq}{q - f} \qquad q = \frac{fp}{p - f}$$

$$f = \frac{qp}{p + q} \tag{36.3}$$

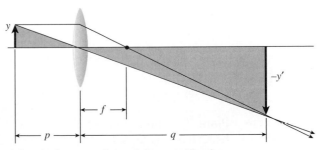

Figure 36.13 Deriving the lens equation and the magnification.

You should verify each of these forms by solving the lens equation explicitly for each parameter in the equation.

The *magnification* of a lens is also derived from Fig. 36.13 and has the same form as discussed for mirrors. Recalling that the magnification M is defined as the ratio of image size y' to the object size y, we can write

$$M = \frac{y'}{y} = -\frac{q}{p} \tag{36.4}$$

where q is the image distance and p is the object distance. A *positive magnification indicates the image is erect, whereas a negative magnification occurs only when the image is inverted.*

Example 36.3

An object 4 cm high is located 10 cm from a thin, converging lens having a focal length of 20 cm. What are the nature, size, and location of the image?

Plan: To get a visual estimate of the nature, size, and location of the image, we will draw a rough ray diagram for an object located inside the focal length. (See Fig. 36.11e.) The quantitative solution for the location and size of the image can be determined from the lens equation and the magnification equation.

Solution: Given that $f = 20$ cm and $p = 10$ cm, we solve for q as follows:

$$q = \frac{pf}{p-f} = \frac{(10 \text{ cm})(20 \text{ cm})}{10 \text{ cm} - 20 \text{ cm}}$$

$$= \frac{200 \text{ cm}^2}{-10 \text{ cm}} = -20 \text{ cm}$$

The negative sign agrees with the ray diagram and shows that the image is virtual. The image size is found by substituting the object size, $y = 4$ cm, into Eq. (36.4).

$$M = \frac{y'}{y} = \frac{-q}{p}$$

$$y' = \frac{-qy}{p} = \frac{-(-20 \text{ cm})(4 \text{ cm})}{10 \text{ cm}}$$

$$y' = +8 \text{ cm}$$

The positive sign indicates the image is *erect.* This example illustrates the principle of the magnifying glass. A converging lens held closer to an object than its focal point produces an erect, enlarged, and virtual image.

Example 36.4

A diverging meniscus lens has a focal length of -16 cm. If the lens is held 10 cm from an object, where is the image located? What is the magnification?

Plan: This time, the ray diagram will be similar to Fig. 36.12. For any diverging lens, the image will always be virtual, erect, and diminished. The location and

magnification of the image will be found from the lens equation and the magnification equation.

Solution: Direct substitution yields

$$q = \frac{pf}{p - f} = \frac{(10 \text{ cm})(-16 \text{ cm})}{10 \text{ cm} - (-16 \text{ cm})}$$

$$= \frac{-160 \text{ cm}^2}{10 \text{ cm} + 16 \text{ cm}} = -6.15 \text{ cm}$$

The negative sign shows that the image is virtual. The magnification is

$$M = \frac{-q}{p} = \frac{-(-6.15 \text{ cm})}{10 \text{ cm}}$$

$$M = +0.615$$

The sign and magnitude of the answer indicate the image is erect and diminished.

Problem-Solving Strategy

Lenses and Optical Instruments

1. Read the problem carefully and draw a horizontal line representing the lens axis. Indicate the location of points equal to f and $2f$ on each side of a vertical line representing the concave or convex lens. Represent the object as an erect arrow at its approximate location to the left of the lens.

2. Construct a geometric ray-tracing diagram to get a visual representation of the problem. A rough sketch should suffice, unless the problem requires a graphical solution. Remember that converging and diverging lenses and mirrors form similar images, except that the virtual images are formed on the *opposite* side for mirrors and on the *same* side for lenses. Real images are formed by the actual light rays, and virtual images are formed where the light only appears to originate, for example, behind the mirror or on the same side of the lens as the incident light.

3. Make a list of the given quantities, taking care to place the appropriate sign to each value. The radius and focal length are *positive* for *converging* mirrors and *negative* for *diverging* mirrors. Image distances q are *positive* when measured to *real* images and *negative* when measured to *virtual* images. The image size y' is *positive* for erect images and *negative* for inverted images.

4. Use the following equations to substitute and solve for the unknown quantities. Do not confuse the signs of operation (addition or subtraction) with the signs of substitution.

$$\frac{1}{p} + \frac{1}{q} = \frac{1}{f} \qquad M = \frac{y'}{y} = \frac{-q}{p}$$

5. It may be necessary to eliminate an unknown by solving both the lens equation and the magnification equation simultaneously.

36.5 Combinations of Lenses

When light passes through two or more lenses, the combined action can be determined by considering the image that would be formed by the first lens as the object for the second lens, and so on. Consider, for example, the lens arrangement illustrated in Fig. 36.14. Lens 1 forms a real, inverted image I_1 of the object O. By considering this intermediate image as a real object for lens 2, the final image I_2 is seen to be real, erect, and enlarged. The lens equation can be applied successively to the two lenses to determine the location of the final image analytically.

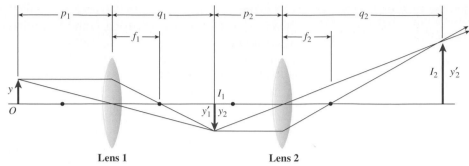

Figure 36.14 The microscope.

The total magnification produced by a system of lenses is the product of the magnifications produced by each lens in the system. The truth of this can be seen from Fig. 36.14. The magnifications in this case are

$$M_1 = \frac{y_1'}{y_1} \qquad M_2 = \frac{y_2'}{y_2}$$

Since $y_1' = y_2$, the product $M_1 M_2$ yields

$$\frac{y_1'}{y_1} \frac{y_2'}{y_2} = \frac{y_2'}{y_1}$$

But y_2'/y_1 is the overall magnification M. In general, we can write

$$M = M_1 M_2 \qquad\qquad (36.5)$$

Applications of the above principles are found for the microscope, the *telescope,* and other optical instruments.

36.6 The Compound Microscope

A compound *microscope* consists of two converging lenses, arranged as shown in Fig. 36.15. The left lens is of short focal length and is called the ***objective lens.*** This lens has a large magnification and forms a real, inverted image of the object being studied. The image is further enlarged by the eyepiece, which forms a virtual final image. The total magnification achieved is the product of the magnifications of the objective lens and the ***eyepiece.***

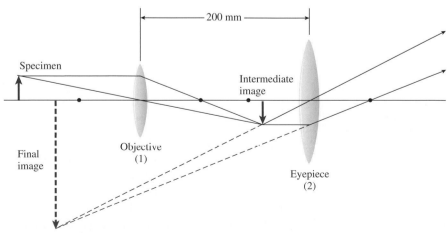

Figure 36.15 Combinations of lenses.

Example 36.5

In a compound microscope, the objective lens has a focal length of 8 mm, and the eyepiece has a focal length of 40 mm. The distance between the two lenses is 200 mm, and the final image appears to be at a distance of 250 mm from the eyepiece. (a) How far is the object from the objective lens? (b) What is the total magnification? A diagram of this arrangement is shown in Fig. 36.15.

Plan: We will begin by labeling the objective lens as "1" and the eyepiece as "2." Since more information is given about parameters that affect the second lens (the *eyepiece*), we will compute the position of the intermediate image first. That information can then be applied to the first lens to determine the location of the specimen. The overall magnification is the product of the magnifications due to each lens.

Solution (a): The distance from lens 2 to the final image is $q_2 = -250$ mm. The distance to the intermediate image is p_2, which is found from the lens equation.

$$p_2 = \frac{f_2 q_2}{q_2 - f_2} = \frac{(40 \text{ mm})(-250 \text{ mm})}{-250 \text{ mm} - 40 \text{ mm}}$$
$$= 34.5 \text{ mm}$$

The negative sign was used for q_2 because of the *virtual* image. Now that the distance p_2 is known, we can find the image distance q_1 for the first image formed by lens 1.

$$q_1 = 200 \text{ mm} - 34.5 \text{ mm}; \qquad q_1 = 165.5 \text{ mm}$$

Using this value in the lens equation for lens 1 allows us to find the object distance p_1.

$$p_1 = \frac{q_1 f_1}{q_1 - f_1} = \frac{(165.5 \text{ mm})(8 \text{ mm})}{165.5 \text{ mm} - 8 \text{ mm}}$$
$$= 8.41 \text{ mm}$$

Therefore, the object is located 8.41 mm in front of the objective lens.

Solution (b): The total magnification is the product of the individual magnifications:

$$M_1 = \frac{-q_1}{p_1} \qquad \text{and} \qquad M_2 = \frac{-q_2}{p_2}$$

The total magnification $M_T = M_1 M_2$, so

$$M_T = \left(\frac{-q_1}{p_1}\right)\left(\frac{-q_2}{p_2}\right) = \frac{q_1 q_2}{p_1 p_2}$$
$$M_T = \frac{(165.5 \text{ mm})(-250 \text{ mm})}{(8.41 \text{ mm})(34.5 \text{ mm})} = -143$$

The negative magnification indicates the final image is inverted. This microscope would be rated 143×, and the object under study should be placed 8.41 mm from the objective lens.

PHYSICS TODAY

Holographic Memory
The CD-ROM and DVD medium is a thin disk of metallized polycarbonate that stores data as pits and ridges. The data can be read by reflecting a laser beam off this changing surface. However, these disks are essentially two-dimensional in that data is stored only on the disk. A new technology could produce optical data in three dimensions—a hologram. In experiments done at CalTech, crystals of lithium niobate doped with iron and manganese atoms could be excited by different kinds of light and could store the energy until exposed to red or ultraviolet light. The stored pattern, created by a coordinated pair of red lasers in the presence of ultraviolet light, was held by the electrons of both iron and manganese atoms. IBM is very interested in this technology because it could potentially lead to 3-D optical read–write memory devices.

36.7 Telescope

The optical system of a refracting *telescope* is essentially the same as that of a microscope. Both instruments use an eyepiece, or *ocular,* to enlarge the image produced by an objective lens, but a telescope is used to examine large, distant objects, and a microscope is used for small, nearby objects.

The refracting telescope is illustrated in Fig. 36.16. The objective lens forms a real, inverted, and diminished image of the distant object. As with the microscope, the eyepiece forms an enlarged and virtual final image of the distant object.

A telescope image is usually smaller than the object being observed. Linear magnification is therefore not a meaningful way to describe the effectiveness of a given telescope. A better measure would be to compare the size of the final image with the size of the object observed without the telescope. If the image seen by the eye is larger than it would be without the telescope, the effect will be to make the object appear closer to the eye than it actually is.

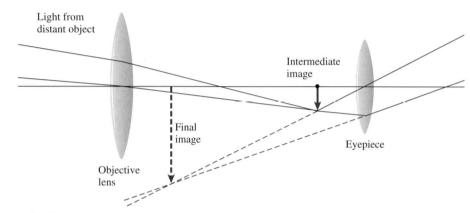

Figure 36.16 The refracting telescope.

36.8 Lens Aberrations

Spherical lenses often fail to produce perfect images because of defects inherent in their construction. Two of the most common defects are known as **spherical aberration** and **chromatic aberration.** Spherical aberration, as discussed earlier for mirrors, is the inability of a lens to focus all parallel rays to the same point. (See Fig. 36.17.)

> Spherical aberration is a lens defect in which the extreme rays are brought to a focus nearer the lens than rays entering near the optical center of the lens.

This effect can be minimized by placing a **diaphragm** in front of the lens. The diaphragm blocks off the extreme rays, producing a sharper image along with a reduction in light intensity.

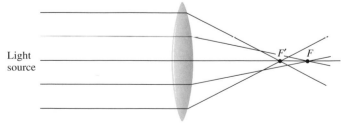

Figure 36.17 Spherical aberration.

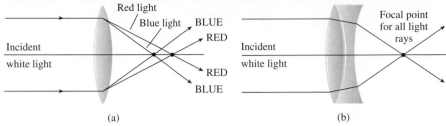

Figure 36.18 (a) Chromatic aberration and (b) an achromatic lens.

In Chapter 35, we discussed the fact that the index of refraction for a given transparent material varies with the wavelength of the light passing through it. Thus, if white light is incident upon a lens, the rays of the component colors are not focused at the same point. The defect, known as *chromatic aberration,* is illustrated in Fig. 36.18a, where blue light is shown to focus nearer the lens than red light.

> Chromatic aberration is a lens defect that reflects its inability to focus light of different colors to the same point.

The remedy for this defect is the **achromatic lens,** illustrated in Fig. 36.18b. Such a lens can be constructed by combining a converging lens of crown glass ($n = 1.52$) with a diverging lens of flint glass ($n = 1.63$). These lenses are chosen and constructed so that the dispersion of one is equal and opposite to that of the other.

Summary and Review

Summary

A lens is a transparent device that converges or diverges light to or from a focal point. Lenses are used extensively in the design of many industrial instruments, and an understanding of how they form images is valuable. A summary of the major concepts discussed in this chapter is provided as follows.

- Image formation by thin lenses can be understood more easily through ray-tracing techniques, as shown in Fig. 36.9 for converging lenses and in Fig. 36.10 for diverging lenses. Remember that the first focal point F_1 is the one on the same side of the lens as the incident light. The second focal point F_2 is on the far side.

  Ray 1: A ray parallel to the axis passes through the second focal point F_2 of a converging lens or appears to come from the first focal point F_1 of a diverging lens.

  Ray 2: A ray that passes through F_1 of a converging lens or proceeds toward F_2 of a diverging lens is refracted parallel to the lens axis.

  Ray 3: A ray that passes through the geometrical center of a lens will not be deviated.

- The *lensmaker's equation* is a relationship among the focal length, the radii of the two lens' surfaces, and the index of refraction of the lens material. The meaning of these parameters is seen from Fig. 36.8.

$$\frac{1}{f} = (n - 1)\left(\frac{1}{R_1} + \frac{1}{R_2}\right) \qquad \textit{Lensmaker's Equation}$$

R_1 or R_2 is positive if the outside surface is convex, negative if concave; f is considered positive for a converging lens and negative for a diverging lens.

- The lens equations for object and image locations and for the magnification are the same as for the mirror equations.

$$\frac{1}{p} + \frac{1}{q} = \frac{1}{f} \qquad p = \frac{qf}{q - f}$$

$$q = \frac{pf}{p - f} \qquad f = \frac{pq}{p + q}$$

$$\textit{Magnification} = \frac{image\ size}{object\ size} \qquad M = \frac{y'}{y} = \frac{-q}{p}$$

p or q is positive for real and negative for virtual; y or y' is positive if erect and negative if inverted.

Key Terms

achromatic lens   709
chromatic aberration   708
converging lens   697
diaphragm   708
diverging lens   698
eyepiece   706
focal length   698
lens   696
lens equation   703
lensmaker's equation   699
magnification   704
meniscus lens   700
microscope   706
objective lens   706
spherical aberration   708
telescope   708
virtual focus   698

Review Questions

36.1. Illustrate by diagrams the effect of a converging lens on a plane wavefront passing through it. What would the effect of a diverging lens be?

36.2. Explain why olives appear to be larger when viewed through their cylindrical glass container. Is the magnification due to the liquid, the glass, or both?

36.3. What happens to the focal length of a converging lens when it is immersed in water? What happens to the focal length of a diverging lens?

36.4. Distinguish between a *real* focus and a *virtual* focus. Which is on the same side of the lens as the incident light?

36.5. Distinguish between the first focal point and the second focal point as defined in the text.

36.6. The focal length of a converging lens is 20 cm. Choose a suitable scale and determine by graphical construction the nature, location, and magnification for the following object distances: (a) 15 cm,

(b) 20 cm, (c) 30 cm, (d) 40 cm, (e) 60 cm, (f) infinity.

36.7. An object is moved from the surface of a lens to the focal point of the lens. Explain what happens to the image if the lens is (a) converging, (b) diverging.

36.8. Describe what happens to the magnification and location of an image as the object moves from infinity to the surface of (a) a converging lens, (b) a diverging lens.

36.9. Discuss the similarities and the differences between lenses and mirrors.

36.10. A camera uses a diaphragm to control the amount of light reaching the film. On a bright day, the diaphragm is almost closed, whereas on a dark day, it must be opened wide to expose the film properly. Discuss the quality of the images produced in each case if the lens is not corrected for aberrations.

36.11. In a microscope, the objective lens has a short focal length, whereas in a telescope, the objective lens has a long focal length. Explain the reason for the difference in focal lengths.

36.12. Derive the lens equation with the help of Fig. 36.13.

36.13. Derive the magnification relation [Eq. (36.4)] with the help of Fig. 36.13.

36.14. Describe two methods you might use to determine the focal length of a double-concave lens.

36.15. Describe an experiment to determine the focal length of a double-concave lens.

36.16. Without referring to the text, write down the various sign conventions that must be applied when working with thin lenses.

36.17. According to convention, the object distance is considered negative when measured to a *virtual object*. Suggest some examples of virtual objects.

Problems

Section 36.2 Focal Length and the Lensmaker's Equation

(Assume $n = 1.50$ unless told otherwise.)

36.1. A plano-convex lens is to be constructed out of glass so that it has a focal length of 40 cm. What is the radius of curvature of the curved surface?
Ans. 20.0 cm

36.2. If one uses a glass double-convex lens to obtain a focal length of 30 cm, what must be the curvature of each convex surface?

36.3. The curved surface of a plano-concave lens has a radius of -12 cm. What is the focal length if the lens is made from a material with a refractive index of 1.54?
Ans. -22.2 cm

36.4. A converging meniscus lens has a concave surface with a radius of -20 cm and a convex surface with a radius of 12 cm. What is the focal length?

36.5. A converging lens such as the one shown in Fig. 36.8a is made of glass. The first surface has a radius of 15 cm, and the second surface has a radius of 10 cm. What is the focal length?
Ans. 12.0 cm

36.6. A meniscus lens has a convex surface with a radius of 20 cm and a concave surface with a radius of -30 cm. What is the focal length if the refractive index is 1.54?

36.7. A plano-convex lens is ground from crown glass ($n = 1.52$). What should be the radius of the curved surface if the desired focal length is to be 400 mm?
Ans. 208 mm

36.8. The magnitudes of the concave and convex surfaces of a glass lens are 200 and 600 mm, respectively. What is the focal length? Is it diverging or converging?

36.9. A plastic lens ($n = 1.54$) has a convex surface of radius 25 cm and a concave surface of -70 cm. What is the focal length? Is it diverging or converging?
Ans. 72.0 cm, converging

Section 36.3 Image Formation by Thin Lenses

36.10. A 7-cm-long pencil is placed 35 cm from a thin converging lens of focal length 25 cm. What are the nature, size, and location of the image formed?

36.11. An object 8 cm high is placed 30 cm from a thin converging lens of focal length 12 cm. What are the nature, size, and location of the image formed?
Ans. real, inverted, $y' = -5.33$ cm, $+20$ cm

36.12. A virtual, erect image appears to be located 40 cm in front of a lens of focal length 15 cm. What is the object distance?

36.13. A 50-mm-tall object is placed 12 cm from a converging lens of focal length 20 cm. What are the nature, size, and location of the image?
Ans. virtual, erect, $y' = 125$ mm, $q = -30$ cm

36.14. An object located 30 cm from a thin lens has a real, inverted image located a distance of 60 cm on the opposite side of the lens. What is the focal length of the lens?

36.15. A light source is 600 mm from a converging lens of focal length 180 mm. Construct the image using ray diagrams. What is the image distance? Is the image real or virtual?   Ans. 257 mm, real

36.16. A plano-convex lens is held 40 mm from a 6-cm object. What are the nature and location of the image formed if the focal length is 60 mm?

36.17. An object 6 cm high is held 4 cm from a diverging meniscus lens of focal length −24 cm. What are the nature, size, and location of the image?
     Ans. virtual, $y' = 5.14$ cm, $q = -3.43$ cm

*36.18. The focal length of a converging lens is 200 mm. An object 60 mm high is mounted on a movable track so that the distance from the lens can be varied. Calculate the nature, size, and location of the image formed for the following object distances: (a) 150 mm, (b) 200 mm, (c) 300 mm, (d) 400 mm, (e) 600 mm.

36.19. An object 450 mm from a converging lens forms a real image 900 mm from the lens. What is the focal length of the lens?   Ans. +30 cm

Section 36.4 The Lens Equation and Magnification

36.20. An object is located 20 cm from a converging lens. If the magnification is −2, what is the image distance?

36.21. A pencil is held 20 cm from a diverging lens of focal length −10 cm. What is the magnification?
     Ans. +0.333

*36.22. A magnifying glass has a focal length of 27 cm. How close must this glass be held to an object to produce an erect image three times the size of the object?

*36.23. A magnifying glass held 40 mm from a specimen produces an erect image that is twice the object size. What is the focal length of the lens?
     Ans. +80 mm

*36.24. What is the magnification of a lens if the focal length is 40 cm and the object distance is 65 cm?

Additional Problems

36.25. The radius of the curved surface in a plano-concave lens is 20 cm. What is the focal length if $n = 1.54$?   Ans. −37.0 cm

36.26. A thin meniscus lens is formed with a concave surface of radius −40 cm and a convex surface of radius +30 cm. If the resulting focal length is 79.0 cm, what was the index of refraction of the transparent material?

36.27. A converging lens has a focal length of 20 cm. An object is placed 15.0 cm from the lens. Find the image distance and the nature of the image.
     Ans. −60 cm, virtual

36.28. How far from a source of light must a lens be placed if it is to form an image 800 mm from the lens? The focal length is 200 mm.

36.29. A source of light 36 cm from a lens projects an image on a screen 18.0 cm from the lens. What is the focal length of the lens? Is it converging or diverging?   Ans. 12.0 cm, converging

36.30. What is the minimum film size needed to project the image of a student who is 2 m tall? Assume the student is located 2.5 m from the camera lens and the focal length is 55.0 mm.

*36.31. When parallel light strikes a lens, the light diverges, apparently coming from a point 80 mm behind the lens. How far from an object should this lens be held to form an image one-fourth the size of the object?   Ans. 240 mm

*36.32. How far from a diverging lens should an object be placed so that its image will be one-fourth the size of the object? The focal length is −35 cm.

*36.33. The first surface of a thin lens has a convex radius of 20 cm. What should be the radius of the second surface to produce a converging lens of focal length 8.00 cm?   Ans. +5.00 cm

*36.34. Two thin, converging lenses are placed 60 cm apart and have the same axis. The first lens has a focal length 10 cm, and the second has a focal length of 15.0 cm. If an object 6.0 cm high is placed 20 cm in front of the first lens, what are the location and size of the final image? Is it real or virtual?

*36.35. A converging lens of focal length 25 cm is placed 50 cm in front of a diverging lens that has a focal length of −25 cm. If an object is placed 75 cm in front of the converging lens, what is the location of the final image? What is the total magnification? Is the image real or virtual?
     Ans. $q_2 = -8.33$ cm in front of diverging lens, $M = -0.333$, virtual

Critical Thinking Questions

*36.36. A camera consists of a converging lens of focal length 50 mm mounted in front of a light-sensitive film as shown in Fig. 36.19. When infinite objects are photographed, how far should the lens be from the film? What is the image distance when an object is photographed 500 mm from the lens? What is the magnification?

Ans. 50 mm, 55.5 mm, −0.111

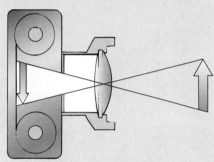

Figure 36.19 The camera.

*36.37. An object is placed 30 cm from a screen. At what points between the object and the screen can a lens of focal length 5 cm be placed to obtain an image on the screen?

*36.38. A simple projector is illustrated in Fig. 36.20. The condenser provides even illumination of the film by the light source. The frame size of regular 8-mm film is 5 × 4 mm. An image is to be projected 600 × 480 mm on a screen located 6 m from the projection lens. What should be the focal length of the projection lens? How far should the film be from the lens?

Ans. f = 49.6 mm, q = 50.0 mm

*36.39. A telescope has an objective lens of focal length 900 mm and an eyepiece of focal length 50 mm. The telescope is used to examine a 30-cm-tall rabbit at a distance of 60 m. What is the distance between the lenses if the final image is 25 cm in front of the eyepiece? What is the apparent height of the rabbit as seen through the telescope?

*36.40. The Galilean telescope consists of a diverging lens as the eyepiece and a converging lens as the objective. The focal length of the objective is 30 cm, and the focal length of the eyepiece is −2.5 cm. An object 40 m away from the objective has a final image located 25 cm in front of the diverging lens. What is the separation of the lenses? What is the total magnification?

Ans. 27.5 cm, −0.068

*36.41. The focal length of the eyepiece of a particular microscope is 3.0 cm, and the focal length of the objective lens is 19 mm. The separation of the two lenses is 26.5 cm, and the final image formed by the eyepiece is at infinity. How far should the objective lens be placed from the specimen being studied?

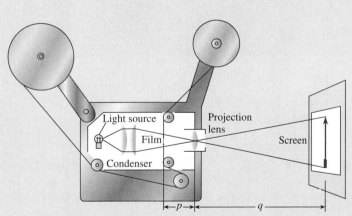

Figure 36.20 The projector.

37

Interference, Diffraction, and Polarization

Interference of light reflected from a CD illustrates the patterns of dark and light fringes that can occur with reflected light. In this chapter, we will study the conditions for the production of light and dark fringes due to the interference and diffraction of light. (*Photo © vol. 88/Corbis.*)

Objectives

After completing this chapter, you should be able to

1. Demonstrate by definition and drawings your understanding of the terms *constructive interference, destructive interference, diffraction, polarization,* and *resolving power.*

2. Describe Young's experiment and be able to use the results to predict the location of bright and dark fringes.

3. Discuss the use of a diffraction grating, derive the grating equation, and apply it to the solution of optical problems.

Light is dualistic in nature, sometimes exhibiting the properties of particles and sometimes those of waves. The demonstrative proof of the wave nature of light came with the discovery of interference and diffraction. Then polarization studies showed that, unlike sound waves, light waves are transverse instead of longitudinal.

In this chapter, we study these phenomena and their significance in physical optics. We will find that the neat geometrical ray approach, which was so helpful in studying mirrors and lenses, must be discarded in favor of a more rigorous wave analysis.

37.1 Diffraction

When light waves pass through an aperture or past the edge of an obstacle, they always bend to some degree into the region not directly exposed to the light source. This phenomenon is called ***diffraction.***

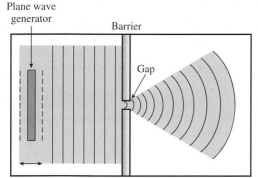

Figure 37.1 Schematic diagram illustrating the diffraction of plane water waves through a narrow gap.

Diffraction is the ability of waves to bend around obstacles placed in their path.

To understand this bending of waves, let us consider what happens when water waves strike a narrow opening. A plane-wave generator can be used in a ripple tank, as shown in Fig. 37.1. The vibrating strip of metal serves as a source of waves at one end of a tray of water. The plane waves strike the barrier, spreading out into the region behind the gap. The diffracted waves appear to originate at the gap in accordance with ***Huygens's wave principle: each point on a wavefront can be regarded as a new source of secondary waves.***

A similar experiment can be performed with light, but for the diffraction to be observable, the slit in the barrier must be narrow. In fact, diffraction is pronounced only when the dimensions of an opening or an obstacle are comparable to the wavelength of the waves striking it. This explains why diffraction of water waves and sound waves is often observed in nature and the diffraction of light is not.

37.2 Young's Experiment: Interference

The first convincing evidence of diffraction was demonstrated by Thomas Young in 1801. A schematic diagram of Young's apparatus is shown in Fig. 37.2. Light from a monochromatic source falls on a slit A, which acts as a source of secondary waves. Two more slits S_1 and S_2 are parallel to A and equidistant from it. Light from A passes through both S_1 and S_2

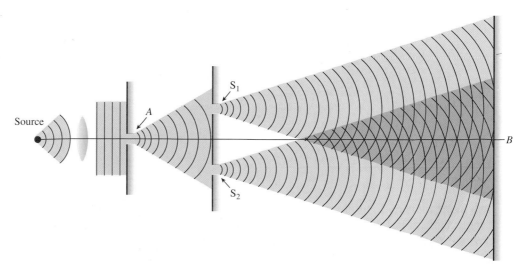

Figure 37.2 Illustration of Young's experiment.

Figure 37.3 Photograph of an interference pattern in Young's experiment. (*From F. A. Jenkins and H. E. White, Fundamentals of Optics, 4th ed., McGraw-Hill Company, New York, 1976. Reprinted by permission.*)

and then on to a viewing screen. The expected pattern is just two noninterfering light bands, with the remainder of the screen dark. Instead, the screen is illuminated, as shown in Fig. 37.3. Even the point B located behind the barrier in direct line with slit A is illuminated. It is easy to see why this experiment caused early physicists to doubt that light consisted of particles traveling in straight lines. The results could be explained only in terms of the wave theory.

The illumination of the screen in alternate bright and dark lines can also be explained in terms of the wave theory. To understand their origin, we recall the *superposition principle*, introduced in Chapter 21 for the study of *interference* in waves:

When two or more waves exist simultaneously in the same medium, the resultant amplitude at any point is the sum of the amplitudes of the composite waves at that point.

Two waves are said to interfere *constructively* when the amplitude of the resultant wave is greater than the amplitudes of either component wave. *Destructive interference* occurs when the resultant amplitude is smaller.

In Young's experiment, light waves reach slits S_1 and S_2 at the same time and originate from a single source of one wavelength. Therefore, the secondary wavelets leaving slits S_1 and S_2 are *in phase*. The sources are said to be *coherent.* Figure 37.4 shows how the light and

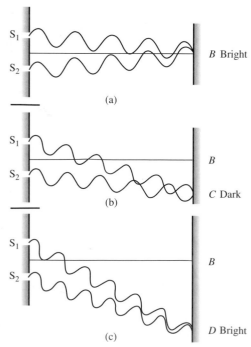

Figure 37.4 Origin of light and dark bands in an interference pattern.

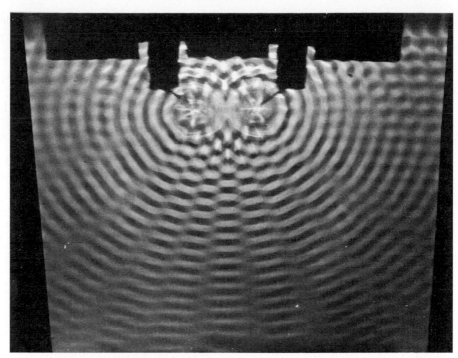

Figure 37.5 Interference pattern set up in water waves by two coherent sources. (*From* PSSC Physics, *D. C. Heath and Company, Lexington, Mass., 1965.*)

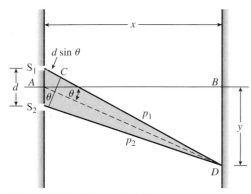

Figure 37.6 Theoretical interpretation of the double-slit experiment.

dark bands are produced. The bright bands occur whenever the waves arriving from the two slits interfere constructively. Dark bands occur when destructive interference takes place. At point B in the center of the screen, light travels the same distances p_1 and p_2 from each slit. The difference in path length $\Delta p - 0$ and constructive interference results in a bright central band. At point C, the difference in path lengths Δp causes the waves to interfere destructively. Another bright band occurs at point D when the waves from each slit again reinforce each other. The overall interference pattern is similar to that produced by the water waves in Fig. 37.5.

Let us now consider the theoretical conditions necessary for the production of bright and dark bands. Consider light reaching the point D at a distance y from the central axis AB, as shown in Fig. 37.6. The separation of the two slits is represented by d, and the screen is located at a distance x from the slits. The point D on the screen makes an angle θ with the axis of the system. The line S_2C is drawn so that the distances CD and p_2 are equal. As long as the distance x to the screen is much greater than the slit separation d, we may consider that p_1, p_2, and DA are all approximately perpendicular to the line S_2C. Therefore, the angle S_1S_2C is equal to the angle θ, and the difference Δp in path lengths of light coming from S_1 and S_2 is given by

$$\Delta p = p_1 - p_2 = d \sin \theta$$

Constructive interference will occur at D when this difference in path length is equal to

$$0, \lambda, 2\lambda, 3\lambda, \ldots, n\lambda$$

where λ is the wavelength of the light. Therefore, the conditions for bright *fringes* are given by

$$d \sin \theta = n\lambda \qquad n = 0, 1, 2, 3, \ldots \qquad (37.1)$$

where d is the slit separation and θ is the angle the fringe makes with the axis.

The conditions required for the formation of dark fringes at D will be satisfied when the path difference is

$$\frac{\lambda}{2}, \frac{3\lambda}{2}, \frac{5\lambda}{2}, \ldots$$

Under these conditions, destructive interference will cancel the waves. Thus, dark fringes will occur when

$$d \sin \theta = n\frac{\lambda}{2} \qquad n = 1, 3, 5, \ldots \qquad (37.2)$$

The preceding equations can be put into a more useful form by expressing them in terms of the measurable distances x and y. For small angles,

$$\sin \theta \approx \tan \theta = \frac{y}{x}$$

Substitution of y/x for $\sin \theta$ in Eqs. (37.1) and (37.2) yields

Bright fringes: $$\frac{yd}{x} = n\lambda \qquad n = 0, 1, 2, \ldots \qquad (37.3)$$

Dark fringes: $$\frac{yd}{x} = n\frac{\lambda}{2} \qquad n = 1, 3, 5, \ldots \qquad (37.4)$$

In experiments designed to measure the wavelength of light, x and d are known initially. The distance y to any particular fringe can be measured and used to determine the wavelength.

Example 37.1

In *Young's experiment,* the two slits are 0.04 mm apart, and the screen is located 2 m away from the slits. The third bright fringe from the center is displaced 8.3 cm from the central fringe. (a) Determine the wavelength of the incident light. (b) Where will the second dark fringe appear?

Plan: The third bright fringe results from the third instance ($n = 3$) of constructive interference beyond the central maximum ($n = 0$). The wavelength of the light must be consistent with the given information for bright fringes when $n = 3$. Dark fringes occur when the path length differs by odd-multiples of half wavelengths. Thus, the second dark fringe occurs when $n = 3$ in Eq. (37.4). The vertical displacement at the screen is found by using the same wavelength as that found in the first part of the problem. (See Fig. 37.6.)

Solution (a): For the third bright fringe, $n = 3$, $y = 8.3$ cm or 0.083 m, and $d = 0.04$ mm or 4×10^{-5} m. The wavelength is found from Eq. (37.3).

$$\frac{yd}{x} = 3\lambda \qquad \text{or} \qquad \lambda = \frac{yd}{3x}$$

$$\lambda = \frac{(0.083 \text{ m})(4 \times 10^{-5} \text{ m})}{3(2 \text{ m})} = 5.53 \times 10^{-7} \text{ m}$$

The wavelength of the light must be 553 nm.

Solution (b): The displacement of the second dark fringe is found by setting $n = 3$ in Eq. (37.4).

$$\frac{yd}{x} = \frac{3\lambda}{2} \quad \text{or} \quad y = \frac{3\lambda x}{2d}$$

$$y = \frac{3(5.53 \times 10^{-7}\,\text{m})(2\,\text{m})}{2(4 \times 10^{-5}\,\text{m})} = 4.15 \times 10^{-2}\,\text{m}$$

The second dark fringe occurs 4.15 cm below the central maximum.

37.3 The Diffraction Grating

If many parallel slits similar to those in Young's experiment are spaced regularly and of the same width, a brighter and sharper diffraction pattern can be obtained. Such an arrangement is known as a ***diffraction grating.*** Gratings are made by ruling thousands of parallel grooves on a glass plate with a diamond point. The grooves act as opaque barriers to light, and the clear spaces form the slits. Most laboratory gratings are ruled from 10,000 to 30,000 lines per inch.

A parallel beam of monochromatic light striking a diffraction grating, as shown in Fig. 37.7, is diffracted in a manner similar to that in Young's experiment. Only a few slits are shown in the figure, each separated by a distance d. Each slit acts as a source of secondary ***Huygens' wavelets,*** producing an interference pattern. A lens is used to focus the light from the slits on a screen.

The diffracted rays leave the slits in many directions. In the forward direction, the paths of all rays will be of the same length, setting up a condition for constructive interference. A central bright image of the source is formed on the screen. In certain other directions, the diffracted rays will also be in phase, giving rise to other bright images. One of these directions θ is illustrated in Fig. 37.7. The first bright line formed on either side of the central image is called the ***first-order image.*** The second bright line on either side of the central maximum is called the *second-order fringe,* and so forth. The condition for the formation of these bright fringes is the same as that derived for Young's experiment (Fig. 37.8). Therefore, we can write the grating equation as

$$d \sin \theta_n = n\lambda \qquad n = 1, 2, 3, \ldots \tag{37.5}$$

where d = spacing of slits

λ = wavelength of incident light

θ_n = deviation angle for nth bright fringe

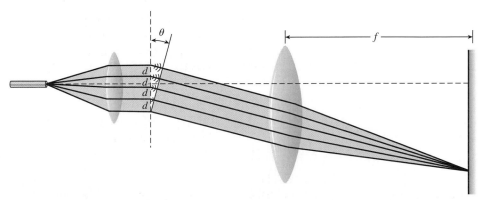

Figure 37.7   The diffraction grating.

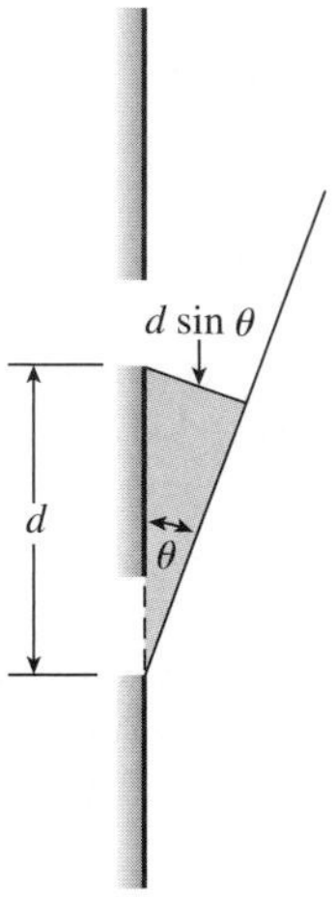

Figure 37.8 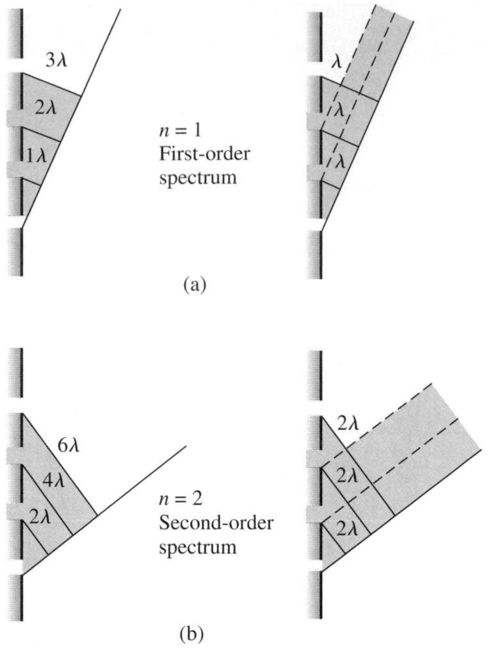   Figure 37.9

The first-order bright fringe occurs when $n = 1$. As illustrated in Fig. 37.9a, this image occurs when the paths of diffracted rays from each slit differ by an amount equal to one wavelength. The second-order image occurs when the paths differ by two wavelengths (Fig. 37.9b).

Example 37.2

A diffraction grating having 20,000 lines per inch is illuminated by parallel light of wavelength 589 nm. What are the angles at which the first- and second-order bright fringes occur?

Plan: We will first recognize that "lines per inch" is really the reciprocal of the slit separation d, which would be "inches per line." Then, we convert d to SI units (m/line) and substitute into Eq. (37.5) to find the first- ($n = 1$) and second- ($n = 2$) order bright fringes.

Solution: The slit spacing d is found as follows:

$$d = \frac{1}{20{,}000 \text{ lines/in.}}\left(\frac{0.0254 \text{ m}}{1 \text{ in.}}\right); \qquad d = 1.27 \times 10^{-6} \text{ m/line}$$

The angle for the *first-order* bright fringe is found by substituting $n = 1$ into Eq. (37.5).

$$\sin \theta_1 = \frac{n\lambda}{d} = \frac{(1)(5.89 \times 10^{-7} \text{ m})}{1.27 \times 10^{-6} \text{ m}} = 0.464$$

$$\theta_1 = 27.6°$$

The angle for the *second-order* fringe occurs when $n = 2$.

$$\sin \theta_2 = \frac{n\lambda}{d} = \frac{(2)(5.89 \times 10^{-7} \text{ m})}{1.27 \times 10^{-6} \text{ m}} = 0.928$$

$$\theta_2 = 68.1°$$

The third-order fringe will not be formed due to the fact that $\sin \theta$ cannot exceed 1.00. The beam will not be deviated by an angle greater than 90°.

37.4 Resolving Power of Instruments

We have learned that light passing through a small opening or past an obstacle is diffracted, so the images formed are fuzzy. Interference fringes near the edges of images sometimes make it difficult to determine the exact shape of the source. Figure 37.10 shows the diffraction pattern formed by passing light through a small circular opening. Notice the large central maximum surrounded by dark and bright interference bands. This diffraction is of extreme importance in optical instruments because it sets the ultimate limit on the possible magnification.

To understand this limitation, consider light from the two sources in Fig. 37.11, passing through a small circular opening in an opaque barrier. In Fig. 37.11a, the images of the sources *A* and *B* are distinguished as separate images. The sources are said to be *resolved*. If they are brought closer together, however, as in Fig. 37.11b, their images overlap, resulting in a confused image. When the sources are so close together (or the opening is so small) that the separate images can no longer be distinguished, the sources are said to be *unresolved*.

The resolving power of an instrument is a measure of its ability to produce well-defined separate images.

A useful method of expressing the ***resolving power*** of an instrument is in terms of the angle θ subtended at the opening by the objects being resolved. (See Fig. 37.11.) The smallest angle θ_0 for which the images can be distinguished separately is a measure of the resolving power. No matter how perfectly a lens is constructed, the image of a point source of light will not be focused at a point. It will appear as only a tiny, bright dot with light and dark

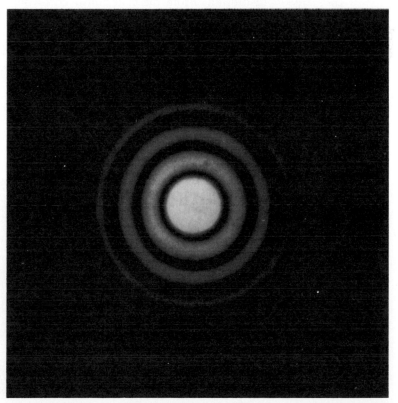

Figure 37.10   Photograph of the interference pattern formed by passing light through a small circular opening. (*Reprinted from Cagnet et al.,* Atlas of Optical Phenomena, *Springer-Verlag. Reprinted by permission.*)

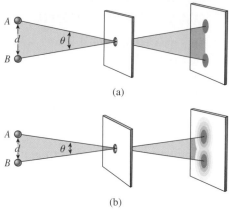

(a)

(b)

Figure 37.11   (a) The images of the sources A and B are easily distinguished. (b) As the sources are brought closer together, the images overlap, resulting in a confused image.

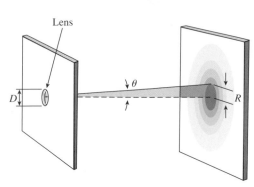

Figure 37.12   The limit of resolution.

fringes around it. ***Resolution*** improves as the diameter of a lens is increased. It can be shown that for a given lens of diameter D (see Fig. 37.12), the optimum resolution occurs for the angular width θ_0, subtended by the radius of the central image in the diffraction pattern. In a telescope this limiting angle is

$$\theta_0 = \frac{1.22\lambda}{D} \qquad \textbf{(37.6)}$$

where λ is the wavelength of the light and D is the diameter of the objective lens.

Figure 37.13 illustrates the resolution of two point sources of light. In Fig. 37.13a the two images can still be distinguished; in Fig. 37.13b, the images are at the limit of resolution. From the criterion used to obtain Eq. (37.6), two such images are just resolved when the central maximum of one pattern coincides with the first dark fringe of the other pattern.

To illustrate the topics we have been discussing in this section, consider a telescope. In Fig. 37.14, two objects are separated by a distance s_0 and are located at a distance p from the objective lens in the telescope. The angle between the objects, in radians, is approximately

$$\theta_0 = \frac{s_0}{p} \qquad \textbf{(37.7)}$$

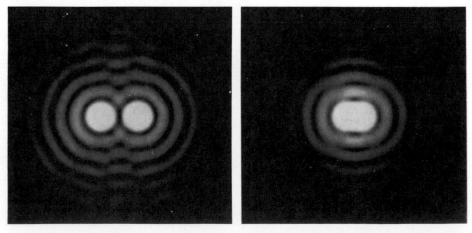

Figure 37.13   (a) Separate images formed from two point sources. (b) The images of two sources at the limit of resolution. (*From Cagnet et al.*, Atlas of Optical Phenomena, *Springer-Verlag. Reprinted by permission.*)

$$\theta_0 = \frac{s_0}{P}$$

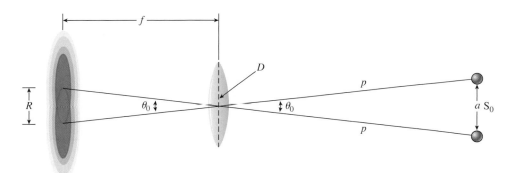

Figure 37.14   Resolution of two distant objects by a spherical lens.

The subscript 0 is used to show that θ_0 and s_0 represent the *minimum conditions* for resolution. Thus, we can rewrite Eq. (37.6) as

$$\theta_0 = \frac{1.22\lambda}{D} = \frac{s_0}{P} \qquad (37.8)$$

Since θ_0 represents the *minimum* separation that can be resolved, this distance is also used to indicate the *resolving power* of the instrument.

Example 37.3

One of the largest refracting telescopes in the world is the 40-in.-diameter instrument at Yerkes Observatory in Wisconsin; its objective lens has a focal length of 19.8 m (65 ft). (a) What is the minimum separation of two features on the Moon's surface such that they are just resolved by this telescope? (b) What is the radius of the central maximum in the diffraction pattern set up by the objective lens? For white light, the central wavelength of 500 nm can be used for computing the resolution. The Moon is 3.84×10^8 m from the Earth.

Plan: The minimum separation s_0 of two images on the Moon's surface is determined at the point where the circular diffraction patterns overlap at their centers. The limiting angle θ_0 for this resolution is given by the ratio s_0/p and also by $1.22\lambda/D$. The equality of these ratios allows us to find the unknown separation s_0. The radius R of the central maximum at minimum resolution is found by realizing that the ratio of the radius to the focal length is equal to the ratio s_0/p.

Solution (a): We must first convert all distances to meters, Thus,

$$p = 3.84 \times 10^5 \text{ km} = 3.84 \times 10^8 \text{ m}$$
$$\lambda = 500 \text{ nm} = 5 \times 10^{-7} \text{ m}$$
$$D = 40 \text{ in. } (2.54 \times 10^{-2} \text{ m/in.}) = 1.02 \text{ m}$$

The minimum separation s_0 is found by solving for s_0 in Eq. (37.8).

$$s_0 = 1.22\frac{\lambda p}{D}$$

$$= \frac{(1.22)(5 \times 10^{-7} \text{ m})(3.84 \times 10^8 \text{ m})}{1.02 \text{ m}} = 230 \text{ m}$$

Solution (b): From Eq. (37.7) and Fig. 37.14, the radius R of the central maximum is

$$R = f\theta_0 = f\frac{s_0}{p}$$

$$= \frac{(19.8 \text{ m})(230 \text{ m})}{3.84 \times 10^8 \text{ m}} = 1.19 \times 10^{-5} \text{ m}$$

The 200-in. reflecting telescope at Mt. Palomar in California can distinguish features on the Moon 46.3 m (or 152 ft) apart.

37.5   Polarization

All the phenomena discussed so far can be explained on the basis of either transverse or longitudinal waves. Interference and diffraction occur in sound waves, which are longitudinal, as well as in water waves, which are transverse. More experimental evidence is needed to determine whether light waves are longitudinal or transverse. In this section, we introduce a property of light waves that can be interpreted only in terms of transverse waves.

Let us first consider a mechanical example of transverse waves in a vibrating string. If the source of the wave causes each particle of the rope to vibrate up and down in a single plane, the waves are *plane-polarized.* If the rope is vibrated in such a manner that each particle moves in a random manner at all possible angles, the waves are *unpolarized.*

> Polarization is the process by which the transverse oscillations of a wave motion are confined to a definite pattern.

As an illustration of the *polarization* of a transverse wave, consider the rope passing through slotted frames, as shown in Fig. 37.15. The unpolarized vibrations pass through slot A and emerge polarized in the vertical plane. This frame is called the *polarizer.* Only waves with vertical vibrations can pass through the slot; all other vibrations will be blocked. The slotted frame B is called the *analyzer* because it can be used to test whether the incoming waves are plane-polarized. If the analyzer is rotated so that the slots in B are perpendicular to those in A, all the incoming waves are stopped. This can happen only if the waves reaching B are polarized in a plane perpendicular to the slots in B.

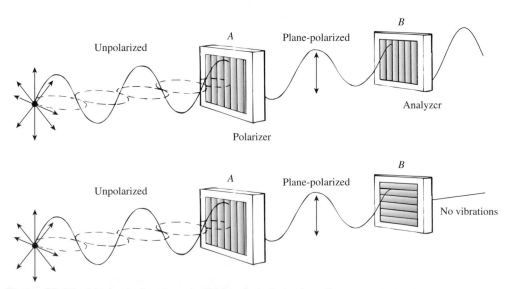

Figure 37.15   Mechanical analogy explaining the polarization of a transverse wave.

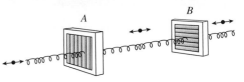

Figure 37.16 A longitudinal wave cannot be polarized.

Polarization is a characteristic of *transverse* waves. Should the rope in our example be replaced by a spring, as in Fig. 37.16, the longitudinal waves would pass on through the slots, regardless of their orientation.

Now, let us consider light waves. In Chapter 33, we discussed the electromagnetic nature of light waves. Recall that such a wave consists of an oscillating electric field and an oscillating magnetic field perpendicular to each other and to the direction of propagation. Therefore, light waves consist of *oscillating fields* rather than of vibrating particles, as was the case for the waves in a rope. If it can be shown that these oscillations can be polarized, we can state conclusively the oscillations are transverse.

A number of substances exhibit different indices of refraction for light with different planes of polarization relative to their crystalline structure. Some examples are calcite, quartz, and tourmaline. Plates can be constructed from these materials that transmit light in only a single plane of oscillation. Thus, they can be used as polarizers for incident light whose oscillations are randomly oriented. Analogous to the vibrating rope passing through slots, two polarizing plates can be used to determine the transverse nature of light waves.

As shown in Fig. 37.17, light emitted by most sources is unpolarized. Upon passing through a tourmaline plate (the polarizer), the light beam emerges plane-polarized but with reduced intensity. Another plate serves as the analyzer. As this plate is rotated with respect to the polarizer, the intensity of the light passing through the system is gradually reduced until relatively no light is transmitted. It is therefore demonstrated that light waves are transverse rather than longitudinal.

Polarization has demonstrated that transverse nature of light in laboratories for many years, but the useful applications of the principle were not realized until the development of ***Polaroid sheets.*** These sheets are constructed by sandwiching a thin layer of iodosulfate crystals between two sheets of plastic. The crystals are aligned by a strong electric field. Two such sheets can be used to control the intensity of light. Photographers use Polaroid filters to vary the intensity and to reduce the glare from reflected light. Complex engineering studies can be made by examining stress patterns in certain plastic tool models. The light and dark fringes set up by polarized light give an indication of the areas of varying stress.

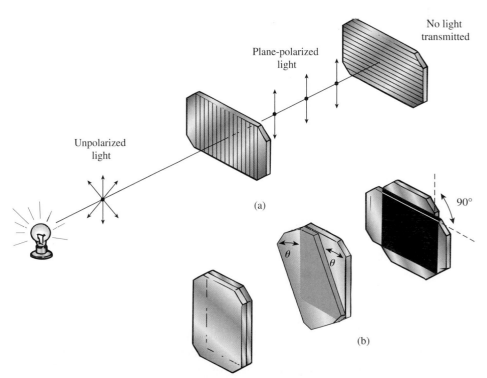

Figure 37.17 (a) Proof that light can be polarized. (b) The reduction of transmitted light intensity as the analyzer is rotated from 0° to 90°.

Summary and Review

Summary

In this chapter, we have discussed several ways in which light behaves as a wave. The bending of light around obstacles placed in its path is called *diffraction*. A combination of diffraction and interference of light waves led to Young's experiment and to the diffraction grating. Modern industrial instruments use these concepts for a variety of applications. The major ideas presented in this chapter are summarized as follows.

- In Young's experiment, interference and diffraction account for the production of bright and dark fringes. The location of these fringes is given by the following equations (see Fig. 37.6):

  Bright fringes: $\dfrac{yd}{x} = n\lambda \qquad n = 0, 1, 2, 3, \ldots$

  Dark fringes: $\dfrac{yd}{x} = n\dfrac{\lambda}{2} \qquad n = 1, 3, 5, 7, \ldots$

- In a diffraction grating of slit separation d, the wavelengths of the nth order fringes are given by

$$d \sin \theta_n = n\lambda \qquad n = 1, 2, 3, \ldots$$

- The resolving power of an instrument is a measure of its ability to produce well-defined, separate images. The minimum conditions for resolution are illustrated in Fig. 37.14. For this situation, the resolution equation is

$$\theta_0 = 1.22\frac{\lambda}{D} = \frac{s_0}{p} \qquad \textit{Resolving Power}$$

Key Terms

analyzer   724
coherent   716
diffraction   715
diffraction grating   719
first-order image   719
fringe   717

Huygens's wave principle   715
Huygen's wavelet   719
interference   716
plane-polarized   724
polarization   724
polarizer   724

Polaroid sheet   725
resolution   722
resolving power   721
superposition principle   716
Young's experiment   718

Review Questions

37.1. Radio waves and light waves are both electromagnetic radiations. Explain why radio waves can be received behind tall buildings, whereas light cannot reach these areas.

37.2. Consider plane water waves striking a gap in a barrier, as in Fig. 37.1. Explain how the diffraction varies as (a) the slit width is decreased and (b) the wavelength of incoming waves is reduced.

37.3. Which effects in Young's experiment are due to diffraction, and which are due to interference?

37.4. In a diffraction grating, how does the spacing of the lines affect the separation of the fringes in the interference pattern?

37.5. If white light were incident upon a diffraction grating, instead of monochromatic light, what would the resulting interference pattern look like?

37.6. In Young's experiment, what effect will reducing the wavelength of incident light have on the interference pattern?

37.7. What effect does increasing the aperture of a lens have on its resolving power? Will a larger wavelength of light result in increased resolution if other conditions are kept constant?

37.8. The resolving power of some microscopes is increased by illuminating the object with ultraviolet light. Explain.

37.9. A polarizing plate absorbs about 50 percent of the intensity of an unpolarized beam. What accounts for this?

37.10. Assume a beam of unpolarized light passes through a polarizer and an analyzer, as shown in Fig. 37.17. The analyzer has been rotated by an angle θ from the position for maximum transmission. Only the component of the amplitude A in the plane-polarized beam that lies along the axis is transmitted by the analyzer. This component is given by $A \cos \theta$. The intensity I of light is proportional to the square of the amplitude A. Show that the intensity I of the beam transmitted by the analyzer is given by

$$I = I_0 \cos^2 \theta \qquad (37.9)$$

where θ is the angle at which the analyzer is placed relative to the position for maximum transmitted intensity I_0.

37.11. Which of the following waves can be polarized: (a) X-rays, (b) water waves, (c) sound waves, (d) radio waves?

37.12. When white light falls on a prism, it is dispersed, forming a spectrum of colors, with the red component receiving the least deviation. Compare this spectrum with that produced by a diffraction grating.

37.13. Give a strong argument for the wave theory of light. Counter this argument with reasons for considering that light consists of particles.

37.14. Suppose that you are using a diffraction grating with a spacing of 3000 lines per centimeter. Discuss the usefulness of this grating for examining (a) the infrared radiation of wavelength 3 nm and (b) the ultraviolet radiation of wavelength 100 nm.

Problems

Section 37.2 Young's Experiment: Interference

37.1. Light from a laser has a wavelength of 632 nm. Two rays from this source follow paths that differ in length. What is the minimum path difference required to cause (a) constructive interference and (b) destructive interference?
Ans. 632 nm, 316 nm

37.2. Find the difference in path length required in Prob. 37.1 to provide the very next instances of constructive and destructive interference.

37.3. Monochromatic light illuminates two parallel slits 0.2 mm apart. On a screen 1.0 m from the slits, the first bright fringe is separated from the central fringe by 2.50 mm. What is the wavelength of the light? Ans. 500 nm

37.4. Monochromatic light from a sodium flame illuminates two slits separated by 1.0 mm. A viewing screen is 1.0 m from the slits, and the distance from the central bright fringe to the bright fringe nearest it is 0.589 mm. What is the frequency of the light?

37.5. Two slits 0.05 mm apart are illuminated by green light of wavelength 520 nm. A diffraction pattern is formed on a viewing screen 2.0 m away. What is the distance from the center of the screen to the first bright fringe? What is the distance to the third dark fringe? Ans. 2.08 cm, 5.20 cm

*37.6. For the situation described in Prob. 37.5, what is the separation of the two first-order bright fringes located on each side of the central band?

*37.7. Young's experiment is performed using monochromatic light of wavelength 500 nm. The slit separation is 1.20 mm, and the screen is 5.00 m away. How far apart are the bright fringes?
Ans. 2.08 mm

*37.8. In Young's experiment, it is noted that the second dark fringe appears at a distance of 2.5 cm from the central bright fringe. Assume the slit separation is 60 μm and the screen is 2.0 m away. What is the wavelength of the incident light?

Section 37.3 The Diffraction Grating

37.9. A diffraction grating having 300 lines per millimeter is illuminated by light of wavelength 589 nm. What are the angles at which the first- and second-order bright fringes are formed?
Ans. 10.0°, 20.7°

37.10. A diffraction grating has 250,000 lines per meter. What is the wavelength of incident light if the second-order bright fringe occurs at 12.6°?

37.11. A small sodium lamp emits light of wavelength 589 nm, which illuminates a grating marked with 6000 lines per centimeter. Calculate the angular deviation of the first- and second-order bright fringes? Ans. 20.7°, 45.0°

37.12. A parallel beam of light illuminates a diffraction grating with 6000 lines per centimeter. The second-order bright fringe is located 32.0 cm from the central image on a screen 50 cm from the grating. Calculate the wavelength of the light.

*37.13. The visible light spectrum ranges in wavelength from 400 to 700 nm. Find the angular width of the first-order spectrum produced by passing a white light through a grating marked with 20,000 lines per inch. Ans. 15.0°

*37.14. An infrared spectrophotometer uses gratings to disperse infrared light. One grating is ruled with 240 lines per millimeter. What is the maximum wavelength that can be studied with this grating?

Section 37.4 Resolving Power of Instruments

37.15. Light of wavelength 600 nm falls on a circular opening of diameter 0.32 mm. A diffraction pattern forms on a screen 80 cm away. What is the distance from the center of the pattern to the first dark fringe? Ans. 1.83 mm

37.16. The limiting angle of resolution for an objective lens in an optical instrument is 3×10^{-4} rad for a 650-nm light source. What is the diameter of the circular opening?

37.17. A certain radio telescope has a parabolic reflector that is 70 m in diameter. Radio waves from outer space have a wavelength of 21 cm. Calculate the theoretical limit of resolution for this telescope. Ans. 3.66×10^{-4} rad

37.18. Using a telescope with a 60-m-diameter lens, find how far apart two objects can be resolved if they are located a distance in space equal to that from the Earth to the Sun (93 million miles).

37.19. What is the angular limit of resolution of a person's eye when the diameter of the opening is 3 mm? Assume the wavelength of the light is 500 nm. Ans. 2.03×10^{-4} rad

37.20. At what distance could the eyes described in Prob. 37.19 resolve wires in a door screen that are separated by 2.5 mm?

Additional Problems

37.21. In Young's experiment, a 600-nm light illuminates a slit located 2.0 m from a screen. The second bright fringe formed on the screen is 5 mm from the central maximum. What is the slit width? Ans. 480 μm

37.22. A transmission grating is ruled with 5000 lines per centimeter. For light of wavelength 550 nm, what is the angular deviation of the third-order bright fringe?

37.23. Monochromatic light passes through two slits separated by 0.24 mm. In the pattern formed on a screen 50 cm away, the distance between the first bright fringe of the left of the central maximum and the first fringe on the right is 2.04 mm. What is the wavelength of the light? Ans. 490 nm

37.24. A transmission grating ruled with 6000 lines per centimeter forms a second-order bright fringe at an angle of 53° from the central fringe. What is the wavelength of the incident light?

*37.25. If the separation of the two slits in Young's experiment is 0.10 mm and the distance to the screen is 50 cm, find the distance between the first dark fringe and the third bright fringe when the slits are illuminated with light of wavelength 600 nm. Ans. 7.50 mm

*37.26. Light from a mercury-arc lamp is incident on a diffraction grating ruled with 7000 lines per inch. The spectrum consists of a yellow line (579 nm) and a blue line (436 nm). Compute the angular separation (in radians) of these lines in the third-order spectrum.

*37.27. A telescope will be used to resolve two points on a mountain 160 km away. If the separation of the points is 2.0 m, what is the minimum diameter for the objective lens? Assume that the light has an average wavelength of 500 nm. Ans. 4.86 cm

Critical Thinking Questions

*37.28. A Michelson interferometer, as shown in Fig. 37.18, can be used to measure a small distance. The beam splitter partially reflects and partially transmits monochromatic light of wavelength λ from the source S. One mirror M_1 is fixed and another M_2 is movable. The light rays reaching the eye from each mirror differ, causing constructive and destructive interference patterns to move across the scope as the mirror M_2 is moved a distance x. Show that this distance is given by

$$x = m\frac{\lambda}{2}$$

where m is the number of dark fringes that cross an indicator line on the scope as the mirror moves a distance x.

37.29. A Michelson interferometer (see Prob. 37.28) is used to measure the advance of a small screw. How far has the screw advanced if krypton-86 light ($\lambda = 606$ nm) is used and 4000 fringes move across the field of view as the screw advances?

*37.30. A diffraction grating has 500 lines per millimeter ruled on its glass surface. White light passes through the grating and forms several spectra on a screen 1.0 m away. Is the deviation of colors

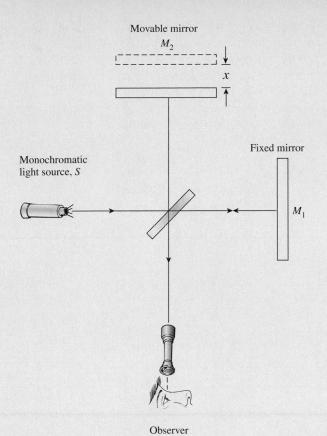

Movable mirror

M_2

x

Monochromatic
light source, S

Fixed mirror

M_1

Observer

Figure 37.18   The Michelson interferometer.

with a grating different from those experienced
for prisms? On the screen, what is the distance
between the first-order blue line (400 nm) and the
first-order red line (680 nm)? How many
complete spectra (400–700 nm) are possible for
these conditions?

Ans. yes,
$\Delta y = 15.9$ cm, two ($n = 2.85$)

*37.31. The taillights of an automobile are 1.25 m apart.
Assume the pupil of a person's eye has a
diameter of 5 mm and the light has an average
wavelength of 604 nm. At night, on a long
straight highway, how far away can the two
taillights be resolved? Suppose you squint your
eyes, forming a slit in which the limiting angle
changes from $\theta_0 = 1.22 \lambda/D$ to $\theta_0 = \lambda/d$. What
is the new distance for resolution of the images?

*37.32. The intensity of unpolarized light is reduced by
one-half when it passes through a polarizer. In
the case of the plane-polarized light reaching the
analyzer, the intensity I of the transmitted beam
is given by

$$I = I_0 \cos^2 \theta$$

where I_0 is the maximum intensity transmitted
and θ is the angle through which the analyzer has
been rotated. Consider three Polaroid plates
stacked so that the axis of each is turned 30° with
respect to the spreading plate. By what percentage will the incident light be reduced in intensity
when it passes through all three plates?

Ans. 28.1 percent

38

Modern Physics and the Atom

Photomicrograph of tungsten crystal shows the kaleidoscope pattern of minute atoms in a nearly perfect crystal. This particular photomicrograph is of a tungsten tip with a radius of 25 nm, and it is magnified 2,700,000 times using a superpowered Muller field ion microscope. In this chapter, we will study the physics and applications that have resulted from a study of the atom. (*Photo © Bettmann/Corbis.*)

Objectives

After completing this chapter, you should be able to

1. Discuss Einstein's postulates and determine the relativistic changes in length, mass, and time.
2. Demonstrate your understanding of the photoelectric effect, the equivalence of mass and energy, the de Broglie wavelength, and the Bohr model of the atom.
3. Demonstrate your understanding of emission and absorption spectra and discuss with drawings the Balmer, Lyman, and Paschen spectral series.
4. Calculate the energy absorbed or emitted by the hydrogen atom when the electron moves to a higher or a lower energy level.

By 1900, physical events had been observed that could not be explained satisfactorily by the laws of classical physics. Light, which was thought to be a wave phenomenon, was found to exhibit properties of particles as well. Newton's concepts of absolute mass, length, and time were found inadequate to describe certain physical events. The light produced by an electric spark in gases did not produce a continuous spectrum on passing through a prism or diffraction grating. These and other unexplained phenomena signaled the beginning of entirely new ways of viewing the world around us.

In 1905, Einstein's first paper on relativity was published; it was followed in 1916 by a second paper. They put classical physics into a new perspective. Einstein's relativity laid the basis for a universal physics that limited classical Newtonian physics to situations involving speeds considerably less than the speed of light. The work of Einstein stimulated a rash of work by others, which has had immense ramifications; it established parameters for maximum energy use, space travel, modern electronics, chemical analysis, X-rays, nuclear weaponry, and many other applications.

38.1 Relativity

Einstein's two papers on relativity altered physics dramatically, but they are not generally understood by many people outside the scientific community. To understand relativity, you must put aside all your preconceived ideas and be willing to view physical events from a new perspective.

The special theory of relativity, published in 1905, is based on two postulates. The first states that every object is in motion compared with something else, that *there is no such thing as absolute rest*. For example, picture a freight car moving along a railroad track at 40 mi/h. The cargo in relation to the freight car is stationary, but in relation to the earth, it is moving at 40 mi/h. According to the first postulate, it is impossible to name anything that is at absolute rest; an object is at rest (or moving) only in relation to some specified reference point.

Einstein's first postulate also tells us that if we see something changing its position with respect to us, we have no way of knowing whether *it* is moving or *we* are moving. If you walk to a neighbor's house, it is just as correct according to this postulate to say that the house came to you. This sounds absurd to us because we are accustomed to using the earth as a frame of reference. Einstein's laws were designed to be completely independent of such a preferred reference frame. From the standpoint of physics, the first postulate is often restated as follows:

> The laws of physics are the same for all frames of reference moving at a constant velocity with respect to one another.

Nineteenth-century physicists had suggested that there was a preferred frame of reference, the so-called luminiferous ether. This was the medium through which they thought electromagnetic waves propagated. However, experiments like the famous Michelson–Morley experiment of 1887 (discussed in Chapter 33) and others were unable to detect the existence of the ether. These experiments are the basis for Einstein's second revolutionary postulate:

> The free-space velocity of light c is constant for all observers, independent of their state of motion.

To see why this second postulate was so revolutionary, let us consider a bus traveling at 50 km/h, as in Fig. 38.1. A person riding in the bus first throws a baseball with a speed (relative to the person) of 20 km/h toward the front of the bus. Then a second ball is thrown with the same speed toward the rear of the bus. To the person who threw the balls, each ball traveled at 20 km/h, one toward the front of the bus and one toward the rear of the bus. But to an observer on the ground, the velocity of the bus adds to the velocity of the first ball, which appears to be traveling at 70 km/h in the same direction as the bus. And when the

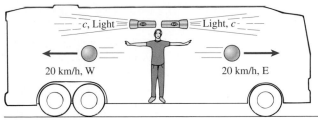

(a) Velocities relative to bus

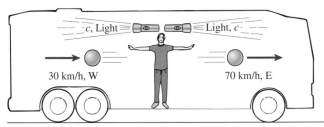

(b) Velocities relative to ground

Figure 38.1   The velocity of the bus is 50 km/h to the right (east). Two balls thrown with equal speeds (20 km/h), one to the right and the other to the left, have different speeds relative to the ground. The speed of light is independent, however, of a frame of reference.

velocity of the bus is added to the second ball, the observer on the ground sees it traveling at 30 km/h *in the same direction as the bus.* But the speed of light does not change and will appear the same to the person on the bus and the observer on the ground. Light always travels at the same constant speed.

$$c = 3 \times 10^8 \text{ m/s}$$

whether it is traveling with the source or away from the source.

38.2 Simultaneous Events: The Relativity of Time

A general understanding of *Einstein's postulates* does not require extensive mathematics. It requires only that we look at things in a fresh and different way. Our understanding of the world around us depends largely on our experience. The time of flight from Atlanta to New York may be listed as 2 h, and we would assume an observer on the ground and another on the plane would each record the same time interval. Certainly, our experience seems to support such a conclusion. We will see, however, that time intervals in one reference frame are in general not the same as those measured from a second frame that is in motion relative to the first. Thus, the observer aboard the plane will judge the trip to require less time than that reckoned by the person on the ground. Of course, the apparent discrepancy would not be noticeable at normal speeds.

Our measurements of time are usually based on the occurrence of simultaneous events. For example, the sweeping second hand on a stopwatch passes the point marked 0 just as a sprinter leaves the starting block. It later passes the point marked 10 just as the sprinter crosses the finish line. The judgment that the time interval was 10 s is based on the measurement of these simultaneous events. But Einstein's postulates force us to ask whether events judged to be simultaneous in one reference frame would also be judged as simultaneous in a moving frame of reference.

Einstein developed a thought experiment to illustrate the relativity of simultaneous measurements. Imagine a boxcar is moving along a railroad track with uniform velocity, as shown in Fig. 38.2. Two bolts of lightning strike the ends of the boxcar, leaving marks A_0 and B_0 on the boxcar and marks A and B on the ground. An observer P_0 is at the midpoint

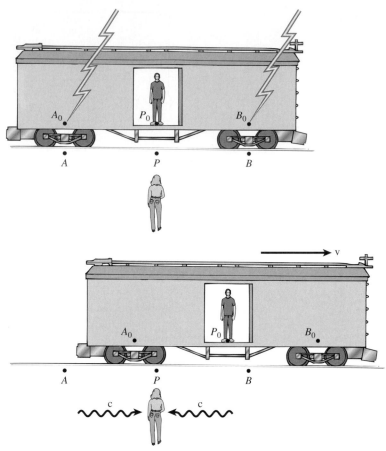

Figure 38.2 Two lightning bolts strike the ends of a boxcar, marking both car and track. The person at point P on the ground observes the events to be simultaneous. The person at point P_0 on the car judges that the event B_0 occurred before the event at A_0.

of the boxcar, and another observer P is on the ground halfway between points A and B. The events seen by each observer are light signals arriving from the lightning bolts. It must also be noted that the speed c of the light is constant for each observer and is not affected by the motion of the boxcar.

The light signals reaching the ground-observer P are judged to arrive at the same time. The signals traveled the same distance, and the events are judged to occur simultaneously. Now consider the same events as observed by the person P_0 aboard the boxcar. By the time the light signal from B_0 has reached the point P on the ground, it has already passed point P_0 on the moving boxcar. The light from A_0 has not yet reached P_0. Thus, the judgment of the person on the boxcar is that events did not occur at the same time. Who is correct? Since according to the principle of relativity, there is no preferred frame of reference, we must conclude that each is correct from his or her own perspective.

38.3   Relativistic Length, Mass, and Time

The relativity of simultaneous measurements has far-reaching consequences when material objects approach the speed of light. All physical measurements must account for relative motion. Time, length, and mass measurements will not be the same for all observers. A comparison of measurements made in different frames of reference is easier if we establish what is called a "proper" frame of reference. It might be noted that two events occurring in the *same* frame of reference are more fundamental than those same events as observed from another frame. We will define a ***proper time interval*** Δt_0 as an interval between two events that occur

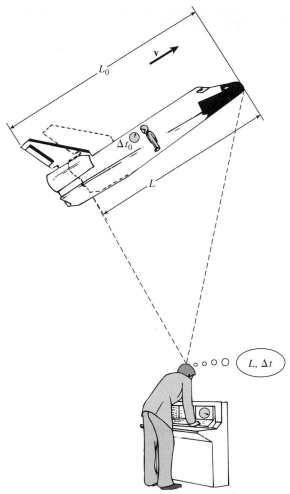

Figure 38.3 The length of objects and the duration of events are affected by relative motion. The person aboard the rocket measures the length L_0 and the time interval Δt_0; the person in the laboratory observes a shorter length L and records a longer time interval Δt.

at the same space point. For example, consider a rocket ship moving in space at a speed v relative to an observer in a laboratory on the Earth. (See Fig. 38.3.) A person aboard this ship measures *proper length L_0, proper mass m_0,* and *proper time intervals Δt_0*. The person on the ground who measures these same events that actually occur on the ship will obtain different values for L, m, and Δt. Each observer is correct from his or her own viewpoint.

A series of relativistic equations can be developed to predict how measurements are affected by relative motion. In each case, the effect becomes more pronounced as the velocity v of objects approaches the limiting velocity c of light. If we find that the proper length of a rocket ship (see Fig. 38.3) is L_0, its length L when it is moving at a relative speed v will be given by

$$L = L_0 \sqrt{1 - \frac{v^2}{c^2}} \qquad \textit{Relativistic Contraction} \qquad \textbf{(38.1)}$$

Such foreshortening of length in the direction of motion is referred to as ***relativistic contraction.***

This means that the length L of a moving object is observed to be shorter by a factor of $\sqrt{1 - v^2/c^2}$ than its length at rest (its proper length). A study of the formula will reveal that the observed length L will be equal to the proper length L_0 when $v = 0$ (the object is at rest). It will begin to shorten as the velocity approaches c.

Example 38.1

When a rocket ship is at rest with respect to us, its length is 100 m. How long do we measure its length to be when it moves by us at a speed of 2.4×10^8 m/s, or $0.8c$?

Solution: In this case, the proper length L_0 is 100 m. Substitution into Eq. (38.1) gives

$$L = L_0 \sqrt{1 - \frac{v^2}{c^2}} = (100 \text{ m}) \sqrt{1 - \frac{(0.8c)^2}{c^2}}$$

$$= (100 \text{ m}) \sqrt{1 - \frac{0.64c^2}{c^2}} = (100 \text{ m}) \sqrt{1 - 0.64}$$

$$= (100 \text{ m}) \sqrt{0.36} = (100 \text{ m})(0.6) = 60 \text{ m}$$

As we have seen, time intervals are also affected by relative motion. A clock held by a person aboard the spaceship in Fig. 38.3 indicates a proper time interval Δt_0 that is shorter than that (Δt) observed from the laboratory on the Earth. The time interval Δt as recorded by the fixed observer is given by

$$\Delta t = \frac{\Delta t_0}{\sqrt{1 - v^2/c^2}} \qquad \textit{Time Dilation} \quad \textbf{(38.2)}$$

This slowing of time (longer time intervals) as a function of velocity is referred to as **time dilation.**

To be sure that you understand this equation, you should recognize that Δt and Δt_0 represent time *intervals,* or the time elapsed from the beginning until the end of an event. Consequently, a clock that is running more slowly will record longer time intervals. We can say that time has stopped when it becomes impossible to measure an event; in other words, the time interval is infinite. This is exactly what is predicted by the time-dilation equation in the limit where $v = c$.

$$\Delta t = \frac{\Delta t_0}{\sqrt{1 - c^2/c^2}} = \frac{\Delta t_0}{\sqrt{1 - 1}} = \frac{\Delta t_0}{0} = \infty$$

Example 38.2

Suppose we observe a rocket ship moving past us at $0.85c$, as in Example 38.1. We measure the time between ticks on the rocket-ship clock to be 1.67 s. What time between ticks is measured by the captain of the rocket ship?

Plan: The *proper time* Δt_0 is the time interval indicated by the moving clock on board the rocket ship. That is where the actual event occurs. Since we measure a time of 1.67 s from a position external to the moving clock, our measurement is the **relativistic time** Δt. Once we have distinguished between relativistic time and proper time, the application of the time-dilation equation will provide the solution.

Solution: Substituting $\Delta t_0 = 1.67$ s into Eq. (38.2) yields

$$\Delta t = \frac{\Delta t_0}{\sqrt{1 - v^2/c^2}} \qquad \text{or} \qquad \Delta t_0 = \Delta t \sqrt{1 - v^2/c^2}$$

$$\Delta t_0 = (1.67 \text{ s}) \sqrt{1 - \frac{(0.85c)^2}{c^2}} = (1.67 \text{ s}) \sqrt{1 - 0.722}$$

$$= (1.67 \text{ s})(0.527) = 0.880 \text{ s}$$

The proper time measured by the captain is only 52.7 percent of the relative time measured by us.

Let us now consider yet another physical quantity that varies with relative velocity. For momentum to be conserved independent of a frame of reference, the mass of a body must vary in the same proportion as length and time. If the *rest mass,* or the *proper mass,* of a body is m_0, the relativistic mass m of a body moving with a speed v will be given by

$$m = \frac{m_0}{\sqrt{1 - v^2/c^2}} \qquad \textit{Relativistic Mass} \qquad (38.3)$$

Remember that the subscript "0" refers to the proper, or rest, mass.

Example 38.3

The rest mass of an electron is 9.1×10^{-31} kg. A lab technician measures the velocity of an electron in a vacuum tube to be 90 percent of the velocity of light. What will be the laboratory measurement of its mass?

Plan: Since the mass m is determined in the laboratory, it will be the **relativistic mass.** The mass $m_0 = 9.1 \times 10^{-31}$ kg is the *proper* mass. We will find m by direct substitution into the equation for the *relativistic* mass.

Solution: The mass measured by the laboratory technician is given by Eq. (38.3).

$$m = \frac{m_0}{\sqrt{1 - v^2/c^2}} = \frac{(9.1 \times 10^{-31} \text{ kg})}{\sqrt{1 - \dfrac{(0.9c)^2}{c^2}}}$$

$$= 20.9 \times 10^{-31} \text{ kg}$$

The relativistic mass of the electron is larger than its rest mass by a factor of 2.30.

It should be clear from Eq. (38.3) that, if m_0 is not equal to zero, the value for the relativistic mass m approaches infinity as v approaches c. This would mean that an infinite force would be needed to accelerate a nonzero mass to the velocity of light. Apparently, the free-space velocity of light represents an upper limit for the speed of such masses. If the rest mass *is* zero, however, as for photons of light, the relativistic-mass equation *does* allow for $v = c$.

As startling as the predictions of Einstein's equations are, they are experimentally verified daily in the laboratory. Their conclusions are amazing to us only because we do not have direct experience with such fantastic speeds. The giant atom smashers, the betatrons, and many other devices for accelerating particles are in use all over the world to accelerate atomic and nuclear particles at speeds close to the speed of light. Protons in the big Brookhaven accelerator have been accelerated to within 99.948 percent of the speed of light. Their mass is shown to be increased precisely as relativity predicts. Electrons can now be accelerated to more than $0.9999999c$, causing their mass to increase by more than 40,000 times the rest value.

38.4   Mass and Energy

Before Einstein, physicists had always considered mass and energy as separate quantities that must be *conserved* separately. Now mass and energy must be considered as different ways of expressing the same quantity. If we say that mass can be converted to energy and energy to mass, we must recognize that mass and energy are the same thing expressed in different units. Einstein found the conversion factor to be equal to the square of the velocity of light.

$$E_0 = m_0 c^2 \qquad (38.4)$$

From the way the equation is written, it is easy to see that a tiny bit of mass corresponds to an enormous amount of energy. For example, an object whose rest mass m_0 is 1 kg has a rest energy E_0 of 9×10^{16} J.

A more general discussion of energy must take the effects of relativity into account. The expression for the total energy of a particle of rest mass m_0 and momentum $p = mv$ can be written

$$E = \sqrt{(m_0c^2)^2 + p^2c^2} \tag{38.5}$$

Now, if we substitute for m_0 from the relativistic-mass relationship, Eq. (38.3), the total energy reduces to

$$E = mc^2 \tag{38.6}$$

where m represents the relativistic mass. This is the most general form for the total energy of a particle.

Note that Eq. (38.5) reduces to $E_0 = m_0c^2$ when the velocity is zero and hence $p = 0$. Furthermore, if we consider velocities considerably less than c, the equation simplifies to

$$E = \frac{1}{2}m_0v^2 + m_0c^2 \tag{38.7}$$

Here we have the usual expression for kinetic energy with a new term added for the *rest energy*.

The more general expression for the kinetic energy of a particle must consider the effects of relativity. Recall that the kinetic energy E_k at speed v is defined as the work that must be done to accelerate a particle from rest to speed v. By the methods of calculus, it can be shown that the relativistic kinetic energy of a particle is given by

$$E_K = (m - m_0)c^2 \tag{38.8}$$

This represents the difference between the total energy of a particle and its rest-mass energy.

Example 38.4

An electron is accelerated to a speed of $0.9c$. Compare its relativistic kinetic energy with the value based on Newtonian mechanics.

Plan: In Example 38.3, it was shown that the relativistic mass of an electron at this speed would be 20.9×10^{-31} kg. Since its rest mass is 9.1×10^{-31} kg, we can use these values to solve for the relativistic kinetic energy based on Eq. (38.8). To make the comparison for Newtonian kinetic energy, we will use the rest mass at the speed of $0.9c$.

Solution: The relativistic kinetic energy is

$$E_k = (m - m_0)c^2$$
$$= (20.9 \times 10^{-31} \text{ kg} - 9.1 \times 10^{-31} \text{ kg})(3 \times 10^8 \text{ m/s})^2$$
$$= (11.8 \times 10^{-31} \text{ kg})(9 \times 10^{16} \text{ m}^2/\text{s}^2)$$
$$= 10.6 \times 10^{-14} \text{ J}$$

The Newtonian value is based on $K = \frac{1}{2}m_0v^2$, where $v = 0.9c$.

$$\frac{1}{2}m_0v^2 = \frac{1}{2}(9.1 \times 10^{-31} \text{ kg})(0.9c)^2$$
$$= (4.55 \times 10^{-31} \text{ kg})(0.81c^2)$$
$$= (4.55 \times 10^{-31} \text{ kg})(0.81)(3 \times 10^8 \text{ m/s})^2$$
$$= 3.32 \times 10^{-14} \text{ J}$$

The relativistic kinetic energy is more than three times that of the Newtonian value.

38.5 Quantum Theory and the Photoelectric Effect

Recall from Chapter 33 that the **photoelectric effect** led to the establishment of a dualistic theory of light. (See Fig. 38.4.) The ejection of electrons as a result of incident light could not be accounted for in terms of the existing electromagnetic theory.

In an attempt to bring experiment into agreement with theory, Max Planck postulated that electromagnetic energy is absorbed or emitted in discrete packets, or *quanta*. The energy of such quanta, or *photons,* is proportional to the frequency of the radiation. **Planck's equation** can be written

$$E = hf \tag{38.9}$$

where h is the proportionality constant known as *Planck's constant.* Its value is

$$h = 6.63 \times 10^{-34} \, \text{J} \cdot \text{s}$$

Einstein used Planck's equation to explain the photoelectric effect. He reasoned that if light is emitted in photons of energy hf, it must also travel as photons. When a quantum of light strikes a metallic surface, it has an energy equal to hf. If all this energy is transferred to a single electron, the electron might be expected to leave the metal with energy hf. At least an amount of energy W is needed, however, to remove the electron from the metal. The term W is called the **work function** of the surface. Thus, the ejected electron leaves with a maximum kinetic energy given by

$$E_K = \frac{1}{2}mv_{\text{max}}^2 = hf - W \tag{38.10}$$

This is *Einstein's photoelectric equation.*

As the frequency of incident light is varied, the maximum energy of the ejected electron varies. The lowest frequency f_0 at which an electron is emitted occurs when $E_K = 0$. In this case,

$$f_0 = \frac{W}{h} \tag{38.11}$$

The quantity f_0 is often called the **threshold frequency.**

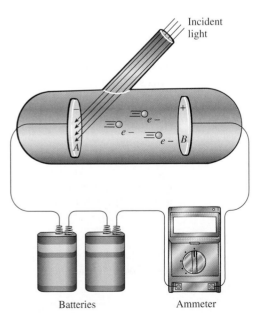

Figure 38.4   The photoelectric effect.

Example 38.5 Light of wavelength 650 nm is required to cause electrons to be ejected from the surface of a particular metal. What is the kinetic energy of the ejected electrons if the surface is bombarded with light of wavelength 450 nm?

Plan: We will first determine the energy in joules provided by the threshold wavelength of 650 nm. That will be the work function W for the surface. At that point, electrons have zero velocity. We will then subtract the work function from the energy provided by light of wavelength equal to 450 nm, giving the kinetic energy available to the ejected electrons. This is essentially an application of the photoelectric equation.

Solution: The work function $W = hf_0$ for the surface is equal to the energy of the threshold wavelength ($\lambda_0 = 650$ nm). Recalling that $f_0 = c/\lambda_0$, we can write

$$W = hf_0 = \frac{hc}{\lambda_0}$$

$$= \frac{(6.63 \times 10^{-34}\,\text{J} \cdot \text{s})(3 \times 10^8\,\text{m/s})}{650 \times 10^{-9}\,\text{m}}$$

$$= 3.06 \times 10^{-18}\,\text{J}$$

The energy of the 450-nm light is

$$hf = \frac{hc}{\lambda} = \frac{(6.63 \times 10^{-34}\,\text{J} \cdot \text{s})(3 \times 10^8\,\text{m/s})}{450 \times 10^{-9}\,\text{m}}$$

$$= 4.42 \times 10^{-19}\,\text{J}$$

From Einstein's photoelectric equation,

$$E_K = hf - W$$

$$= 4.42 \times 10^{-19}\,\text{J} - 3.06 \times 10^{-19}\,\text{J} = 1.36 \times 10^{-19}\,\text{J}$$

38.6 Waves and Particles

Electromagnetic radiation has a dual character in its interaction with matter. Sometimes it exhibits wave properties, as demonstrated by interference and diffraction. At other times, as in the photoelectric effect, it behaves like particles, which we have called *photons*. In 1924, Louis de Broglie was able to demonstrate this duality of matter by deriving a relationship for the wavelength of a particle.

This relationship can be seen by looking at two expressions for the energy of a photon. We have already seen from Planck's work that the energy of a photon can be expressed as a function of its wavelength λ.

$$E = hf = \frac{hc}{\lambda}$$

Another expression for the energy of a particle of rest mass m_0 was given earlier in Section 38.4. Equation (38.5) stated that

$$E = \sqrt{(m_0c^2)^2 + p^2c^2}$$

This equation also shows that photons have momentum $p = mv$ due to their relativistic mass. The rest mass m_0 of a photon is zero, however, so Eq. (38.5) becomes

$$E = \sqrt{p^2c^2} = pc \tag{38.12}$$

Since we also know that $E = hc/\lambda$, we can write

$$\frac{hc}{\lambda} = pc$$

from which the wavelength of a photon is given by

$$\lambda = \frac{h}{p} \qquad (38.13)$$

de Broglie proposed that all objects have wavelengths related to their momentum—whether the objects are wavelike or particlelike. For example, the wavelength of an electron or any other particle is given by de Broglie's equation, which can be rewritten

$$\lambda = \frac{h}{mv} \qquad \textit{de Broglie Wavelength} \quad (38.14)$$

Example 38.6 What is the de Broglie wavelength of an electron that has a kinetic energy of 100 eV?

Plan: To find the de Broglie wavelength for this electron, we need to know its mass and velocity. The mass of an electron is known to be 9.1×10^{-31} kg, and its velocity can be found from its given kinetic energy. We will need to convert 100 eV to joules and apply the Newtonian expression for kinetic energy. If the velocity is reasonable and not approaching relativistic speeds, we can use it to find the de Broglie wavelength.

Solution: We convert the 100 eV to joules by recalling that $1 \text{ eV} = 1.6 \times 10^{-19}$ J.

$$K = 100 \text{ eV}\left(\frac{1.6 \times 10^{-19} \text{ J}}{1 \text{ eV}}\right) = 1.60 \times 10^{-17} \text{ J}$$

The Newtonian kinetic energy is $\frac{1}{2}m_0 v^2$, so we can solve for the velocity as follows:

$$K = \frac{1}{2}m_0 v^2 \qquad \text{or} \qquad v^2 = \frac{2K}{m_0}$$

$$v^2 = \frac{2(1.6 \times 10^{-17} \text{ J})}{9.1 \times 10^{-31} \text{ kg}} = 3.52 \times 10^{13} \text{ m}^2/\text{s}^2$$

The velocity of the electron is, therefore,

$$v = \sqrt{3.52 \times 10^{13} \text{ m}^2/\text{s}^2} = 5.93 \times 10^6 \text{ m/s}$$

Since the velocity is much less than 10 percent of the velocity of light, the speed is not relativistic, and we were justified in using the Newtonian relationship for energy. The de Broglie wavelength can now be calculated from Eq. (38.14).

$$\lambda = \frac{h}{mv} = \frac{6.63 \times 10^{-34} \text{ J} \cdot \text{s}}{(9.1 \times 10^{-31} \text{ kg})(5.93 \times 10^6 \text{ m/s})}$$

$$= 1.23 \times 10^{-10} \text{ m} = 0.123 \text{ nm}$$

Note from de Broglie's equation that the higher the velocity of the particle, the *shorter* its wavelength. Remember that the relativistic mass must be used if the velocity of the given particle is large enough to warrant it. Generally speaking, if the velocity of the particle is greater than one-tenth of the velocity of light, you should consider the relativistic values.

38.7 The Rutherford Atom

Early attempts to explain the structure of the atom recognized that matter was electrically neutral and if charges existed within the atom, they must be neutralized in some fashion, meaning equal numbers of positive and negative charges. In 1911, Ernest Rutherford conducted experiments in which he bombarded a thin metal foil with a stream of alpha particles, as illustrated in Fig. 38.5. An *alpha particle* is a tiny, positively charged particle emitted by a radioactive substance such as radium. Most of the positively charged particles penetrated the foil easily, as indicated by a flash of light when they struck the zinc sulfide screen, and a few were deflected slightly. Much to Rutherford's surprise, however, others were deflected at extreme angles. Some were even deflected backward. (Rutherford himself said that it was like firing a cannonball through tissue paper and having it come back and hit you in the face.) These extreme deflections could not be explained in terms of the Thomson model for the atom. If charge really was scattered through the atom, the electrical forces would be far too weak to repulse the alpha particles at the large angles actually observed.

Rutherford explained these results by assuming all the positive charge of an atom is concentrated into a small region, called the *nucleus* of the atom. The electrons were assumed to be distributed in the space around this positive charge. With the atom consisting largely of empty space, the fact that most of the alpha particles passed right through the foil is explained easily. Furthermore, Rutherford found the large-angle scattering to be a consequence of electrostatic repulsion. Once the alpha particle has penetrated the electrons surrounding the positive charge, it is close to a large positive charge of great mass. Large repulsive electrostatic forces are therefore expected.

Rutherford has been credited with the discovery of the nucleus because he was able to develop formulas to predict the observed scattering of alpha particles. On the basis of his calculations, the diameter of the nucleus was estimated to be approximately one ten-thousandth of the diameter of the atom itself. The positive part of the atom was seen to be concentrated in the nucleus, approximately 10^{-5} nm in diameter. Rutherford believed the electrons were grouped around the nucleus, so the diameter of the whole atom is about 0.1 nm.

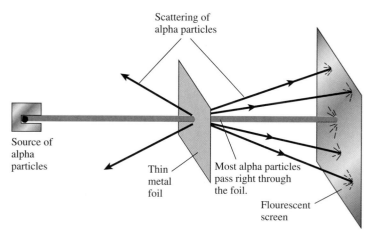

Figure 38.5   Rutherford scattering of alpha particles provided the first evidence for the atomic nucleus.

38.8 Electron Orbits

An immediate difficulty accompanying the Rutherford atom was related to the stability of the atomic electrons. We know from Coulomb's law that the electrons should be attracted to the nucleus. One possible explanation is that the electrons are moving in circles around the nucleus much as planets revolve around the Sun. The necessary centripetal force would be provided by coulomb attraction.

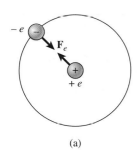

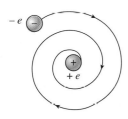

(a)

(b)

Figure 38.6 (a) A stable electron orbit in which the centripetal force is provided by the electrostatic force $\mathbf{F}_e$. (b) Instability due to electromagnetic radiation, which should cause the electron to lose energy and spiral into the nucleus.

Consider, for example, the hydrogen atom, which consists of a single proton and a single electron. We might expect the electron to remain in a constant orbit about the nucleus, as illustrated in Fig. 38.6a. The charge of the electron is labeled $-e$, and the equal but opposite charge of the proton is labeled $+e$. At a distance r from the nucleus, the electrostatic force of attraction on the electron is given by applying Coulomb's law where $k = \frac{1}{4}\pi\epsilon_0$.

$$F_e = \frac{e^2}{4\pi\epsilon_0 r^2} \tag{38.15}$$

where each charge has the magnitude e. For a stable orbit, this force must exactly equal the centripetal force, given by

$$F_c = \frac{mv^2}{r} \tag{38.16}$$

where m is the mass of the electron traveling with a velocity v. Setting $F_e = F_c$, we have

$$\frac{e^2}{4\pi\epsilon_0 r^2} = \frac{mv^2}{r} \tag{38.17}$$

Solving for the radius r, we obtain

$$r = \frac{e^2}{4\pi\epsilon_0 mv^2} \tag{38.18}$$

According to classical theory, Eq. (38.18) should predict the orbital radius r of the electron as a function of its speed v.

The problem with this approach is that the electron must be continually accelerated under the influence of the electrostatic force. According to classical theory, an accelerated electron must radiate energy. The total energy of the electron would therefore decrease gradually, reducing the speed of the electron. As can be seen from Eq. (38.18), gradual reduction in electron speed v results in smaller and smaller orbits. Thus, the electron should spiral into the nucleus, as shown in Fig. 38.6b. The fact that it does not compels us to acknowledge a fundamental inconsistency in Rutherford's atom.

38.9 Atomic Spectra

All substances radiate electromagnetic waves when they are heated. Since each element is different, such emitted radiation can be expected to provide clues to atomic structure. These electromagnetic waves are analyzed by a **spectrometer,** which uses a prism or a diffraction grating to organize the radiation into a pattern called a *spectrum*. For an incandescent source of light, the spectrum is *continuous;* in other words, it contains all wavelengths and is similar to a rainbow. If, however, the source of light is a hot gas under low pressure, the spectrum of the emitted light consists of a series of bright lines separated by dark regions. Such spectra are called **line emission spectra.** The chemical composition of a vaporized material can be determined by comparing its spectrum with known spectra.

A line emission spectrum for hydrogen is shown in Fig. 38.7. The sequence of lines, called a **spectral series,** has a definite order, the lines becoming more and more crowded as the limit of the series is approached. Each line corresponds to a characteristic frequency or wavelength (color). The line of longest wavelength, 656.3 nm, is in the red and is labeled H_α. The others are labeled, in order, as H_β, H_γ, and so on.

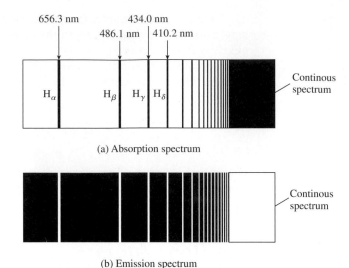

656.3 nm   434.0 nm
486.1 nm   410.2 nm

H_α H_β H_γ H_δ

Continous
spectrum

(a) Absorption spectrum

Continous
spectrum

(b) Emission spectrum

Figure 38.7   Line spectra for the Balmer series of the hydrogen atom.

It is also possible to obtain similar information from a gas or vapor in an unexcited state. When light is passed through a gas, certain discrete wavelengths are *absorbed*. These **absorption spectra** are like those produced by emission except that the characteristic wavelengths appear as *dark* lines on a light background. The absorption spectrum for hydrogen is compared with the emission spectrum in Fig. 38.7.

As early as 1884, Johann Jakob Balmer found a simple mathematical relationship for predicting the characteristic wavelengths of some of the lines in the hydrogen spectrum. His formula is

$$\frac{1}{\lambda} = R\left(\frac{1}{2^2} - \frac{1}{n^2}\right) \tag{38.19}$$

where λ = wavelength
R = Rydberg constant
$n = 3, 4, 5, \ldots$

If λ is measured in meters, the value for R is

$$R = 1.097 \times 10^7 \text{ m}^{-1}$$

The series of wavelengths predicted by Eq. (38.19) is called the *Balmer series*.

Example 38.7

Using Balmer's equation, determine the wavelength of the H_α line in the hydrogen spectrum. (The first line occurs when $n = 3$.)

Solution: Direct substitution yields

$$\frac{1}{\lambda} = 1.097 \times 10^7 \text{ m}^{-1}\left(\frac{1}{2^2} - \frac{1}{3^2}\right) = 1.524 \times 10^6 \text{ m}^{-1}$$

from which

$$\lambda = 656.3 \text{ nm}$$

Other characteristic wavelengths are found by setting $n = 4, 5, 6$, and so on. The limit of the series is found by setting $n = \infty$ in Eq. (38.19).

Since the discovery of Balmer's equation, several other series spectra have been discovered for hydrogen. In general, all these discoveries can be summarized by the single equation

$$\frac{1}{\lambda} = R\left(\frac{1}{l^2} - \frac{1}{n^2}\right) \tag{38.20}$$

where l and n are integers with $n > l$. The series predicted by Balmer corresponds to $l = 2$ and $n = 3, 4, 5, \dots$. The *Lyman series* is in the ultraviolet region and corresponds to $l = 1$, $n = 2, 3, 4, \dots$; the *Paschen series* is in the infrared and corresponds to $l = 3$, $n = 4, 5, 6, \dots$; and the *Brackett series,* also in the infrared, corresponds to $l = 4$, $n = 5$, $6, 7, \dots$.

38.10   The Bohr Atom

Observations of atomic spectra have indicated that atoms emit only a few rather definite frequencies. This fact does not agree with the Rutherford model, which predicts an unstable atom emitting radiant energy of all frequencies. Any theory of atomic structure must account for the regularities observed in atomic spectra.

The first theory to explain the line spectrum of the hydrogen atom satisfactorily was offered by Niels Bohr in 1913. He assumed, like Rutherford, that the electrons were in circular orbits about a dense, positively charged nucleus, but he decided that electromagnetic theory cannot be strictly applied on the atomic level. Thus, he avoided the problem of orbital instability due to emitted radiation. Bohr's first postulate is as follows:

An electron may exist only in those orbits where its angular momentum is an integral multiple of $h/2\pi$.

Thus, contrary to classical prediction, electrons may be in certain, specified orbits without the emission of radiant energy.

The basis for Bohr's first postulate can be seen in terms of de Broglie wavelengths. The stable orbits are those in which an integral number of electron wavelengths can be fitted into the circumference of the Bohr orbit. Such orbits would allow for standing waves, as illustrated in Fig. 38.8 for four wavelengths. The conditions for such standing waves would be given by

$$n\lambda = 2\pi r \qquad n = 1, 2, 3, \dots \tag{38.21}$$

where r is the radius of an electron orbit that contains n wavelengths.

Since $\lambda = h/mv$, we can rewrite Eq. (38.21) as

$$n\frac{h}{mv} = 2\pi r$$

from which it can be shown that the angular momentum mvr is given by

$$mvr = \frac{nh}{2\pi} \tag{38.22}$$

The number n, called the ***principal quantum number,*** may take on the values $n = 1, 2, 3, \dots$.

A *second postulate* given by Bohr places even further restrictions on atomic theory by incorporating the quantum theory.

If an electron changes from one stable orbit to any other, it loses or gains energy in discrete quanta equal to the difference in energy between the initial and final states.

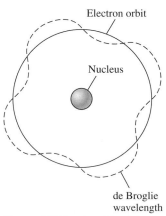

Figure 38.8 A stable electron orbit, showing a circumference equal to four de Broglie wavelengths.

In equation form, Bohr's second postulate can be written

$$hf = E_i - E_f$$

where hf = energy of emitted or absorbed photon

E_i = initial energy

E_f = final energy

Let us return to our discussion of the hydrogen atom to see whether Bohr's postulates will help to bring theory into harmony with observed spectra. Recall that application of Coulomb's law and Newton's law resulted in Eq. (38.18) for the radius r of the orbiting electron.

$$r = \frac{1}{4\pi\epsilon_0} \frac{e^2}{mv^2}$$

According to Bohr's theory,

$$mvr = \frac{nh}{2\pi}$$

Solving these two equations simultaneously for the radius r and for the velocity v, we obtain

$$r = n^2 \frac{\epsilon_0 h^2}{\pi m e^2} \tag{38.23}$$

$$v = \frac{e^2}{2\epsilon_0 n h} \tag{38.24}$$

These equations predict the possible radii and velocities for the electron, where $n = 1, 2, 3, \ldots$.

We now derive an expression for the total energy of a hydrogen atom for any electron orbit.

$$E_T = E_K + E_P$$

The kinetic energy is found by substituting from Eq. (38.24),

$$E_K = \frac{1}{2}mv^2 = \frac{me^4}{8\epsilon_0^2 n^2 h^2} \tag{38.25}$$

The potential energy of the atom for any orbit is

$$E_P = \frac{-1}{4\pi\epsilon_0} \frac{e^2}{r} = -\frac{me^4}{4\epsilon_0^2 n^2 h^2} \tag{38.26}$$

after substitution of r from Eq. (38.23). The potential energy is negative because outside work is necessary to remove the electron from the atom. Adding Eq. (38.25) to Eq. (38.26), we find the total energy to be

$$E_T = -\frac{me^4}{8\epsilon_0^2 n^2 h^2} \tag{38.27}$$

Returning to Bohr's second postulate, we are now in a position to predict the energy of an emitted or absorbed photon. Normally, the electron is in its ground state, corresponding to $n = 1$. If the atom *absorbs* a photon, the electron may jump to one of the outer orbits. From an *excited* state, it will soon fall back to a lower orbit, *emitting* a photon in the process.

Suppose an electron is in an outer orbit of quantum number n_i and then returns to a lower orbit of quantum number n_f. The decrease in energy must be equal to the energy of the emitted photon.

$$E_i - E_f = hf$$

Substituting the total energy for each state from Eq. (38.27), we obtain

$$hf = E_i - E_f$$

$$= -\frac{me^4}{8\epsilon_0^2 n_i^2 h^2} + \frac{me^4}{8\epsilon_0^2 n_f^2 h^2}$$

$$= \frac{me^4}{8\epsilon_0^2 h^2}\left(\frac{1}{n_f^2} - \frac{1}{n_i^2}\right)$$

Dividing both sides of this relation by h and remembering that $f = c/\lambda$, we can write

$$\frac{1}{\lambda} = \frac{me^4}{8\epsilon_0^2 h^3 c}\left(\frac{1}{n_f^2} - \frac{1}{n_i^2}\right) \tag{38.28}$$

Substituting in the values, we obtain

$$\frac{me^4}{8\epsilon_0^2 h^3 c} = 1.097 \times 10^7 \ m^{-1}$$

which is equal to Rydberg's constant R. Therefore, Eq. (38.28) is simplified to

$$\frac{1}{\lambda} = R\left(\frac{1}{n_f^2} - \frac{1}{n_i^2}\right) \tag{38.29}$$

The above relationship is exactly the same in form as Eq. (38.20), which was established by Balmer and others from experimental data. Thus, the **Bohr atom** brings theory into harmony with observation.

Example 38.8

Determine the wavelength of the photon emitted from a hydrogen atom when the electron jumps from the first excited state to the ground state.

Plan: The first excited state corresponds to $n_i = 2$, and the ground state corresponds to $n_f = 1$. Therefore, we can find the energy of the emitted photon from Eq. (38.28).

Solution: Remembering that $R = 1.097 \times 10^7 \ m^{-1}$, we obtain

$$\frac{1}{\lambda} = R\left(\frac{1}{n_f^2} - \frac{1}{n_i^2}\right)$$

$$= (1.097 \times 10^7 \ m^{-1})\left(\frac{1}{1^2} - \frac{1}{2^2}\right) = 8.23 \times 10^6 \ m^{-1}$$

Solving for λ gives the wavelength

$$\lambda = 1.22 \times 10^{-7} \ m = 122 \ nm$$

The energy of the emitted photon could be found from $E = hf = hc/\lambda$.

38.11 Energy Levels

From Bohr's work, we now have a picture of an atom in which orbital electrons can occupy a number of **energy levels.** The total energy at the nth level is given by Eq. (38.27).

$$E_n = -\frac{me^4}{8\epsilon_0^2 n^2 h^2} \tag{38.30}$$

When the hydrogen atom is in its stable *ground* state, the quantum number n is equal to 1. The possible *excited* states are given for $n = 2, 3, 4, \ldots$.

Example 38.9

Determine the energy of an electron in the ground state for a hydrogen atom.

Plan: The ground state for the hydrogen atom is represented by the first energy level, which corresponds to the quantum number $n = 1$. The energy of this state represents the total energy required to remove the electron to infinity, and it can be found from Eq. (38.30).

Solution: The following constants are needed:

$$\epsilon_0 = 8.85 \times 10^{-12} \, \text{C}^2/\text{N} \cdot \text{m}^2 \qquad h = 6.63 \times 10^{-34} \, \text{J} \cdot \text{s}$$
$$m = 9.1 \times 10^{-31} \, \text{kg} \qquad e = 1.6 \times 10^{-19} \, \text{C}$$

Direct substitution into Eq. (38.30), with $n = 1$, yields

$$E_1 = -\frac{me^4}{8\epsilon_0^2 n^2 h^2}$$

$$= -\frac{(9.1 \times 10^{-31} \, \text{kg})(1.6 \times 10^{-19} \, \text{C})^4}{8(8.85 \times 10^{-12} \, \text{C}^2/\text{N} \cdot \text{m}^2)^2(1)^2(6.63 \times 10^{-34} \, \text{J} \cdot \text{s})^2}$$

$$= -2.17 \times 10^{-18} \, \text{J}$$

A more convenient unit for measuring energy at the atomic level is the electronvolt. Recall from Chapter 23 that an electronvolt (eV) is the energy acquired by an electron when it is accelerated through a potential difference of 1 V. Consequently,

$$1 \, \text{eV} = 1.6 \times 10^{-19} \, \text{J}$$

In applications in which a larger unit of energy is required, the megaelectronvolt (MeV) is more appropriate.

$$1 \, \text{MeV} = 10^6 \, \text{eV}$$

From the previous example, the energy of the ground-state electron can be expressed in electronvolts as follows:

$$E_1 = -2.17 \times 10^{-18} \, \text{J} \frac{1 \, \text{eV}}{1.6 \times 10^{-19} \, \text{J}}$$

$$= -13.6 \, \text{eV}$$

This fact can be used to write Eq. (38.30) in a simpler form:

$$E_n = \frac{-13.6 \, \text{eV}}{n^2} \tag{38.31}$$

Similar calculations will yield smaller negative values for the outer orbits. If the electron were to be entirely removed from the atom, a case where $n = \infty$, 13.6 eV of energy would be required. ($E_\infty = 0$.) Similarly, a photon of energy 13.6 eV could be emitted if an electron were to be captured by an ionized hydrogen atom and wind up in the ground state.

The atomic spectra observed for hydrogen is now understood in terms of energy levels. The Lyman series results from electrons returning from some excited state to the ground state, as seen in Fig. 38.9. The Balmer, Paschen, and Brackett series occur when the final states are orbits for which $n = 2$, $n = 3$, and $n = 4$, respectively.

An energy-level diagram for hydrogen is shown in Fig. 38.10. Such diagrams are often used to describe the various energy states of atoms.

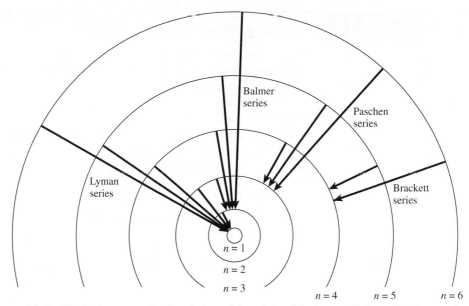

Figure 38.9   The Bohr atom and a description of the origin of the Lyman, Balmer, Paschen, and Brackett spectral series.

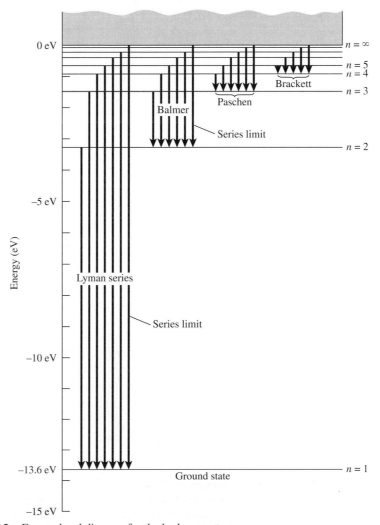

Figure 38.10   Energy-level diagram for the hydrogen atom.

38.12   Lasers and Laser Light

One of the most useful applications based on quantum physics and a study of the atom is the laser. The intense, highly focused, and coherent light emitted by these devices is responsible for many advances in science. In medicine, trained ophthalmologists can repair the retina of an eye with laser spot welds. By combining laser light with fiber optics (see Section 35.6), a revolution is already under way in the fields of electronics and communications. Powerful lasers have even been developed to drill small holes in diamonds.

The principle behind the operation of lasers is relatively easy to understand. It is merely an application of the quantum theory discussed earlier in this chapter for the energy levels of an atom. Basically there are three ways in which photons can interact with matter: (1) *absorption*, (2) *spontaneous emission*, and (3) *stimulated emission*. Each of these three events are described in Fig. 38.11.

Absorption and emission were discussed in Section 38.11, where it was pointed out that the absorption of a photon can excite an atom by raising an electron to a higher energy level, as is illustrated by Fig. 38.11a. Such an electron is in an *excited state*, and eventually it will drop to its original level and undergo what is called spontaneous emission (see Fig. 38.11b). In each case, the absorbed or the emitted photon has an energy given by

$$E_2 - E_1 = hf$$

Spontaneous emission accounts for the light we see from lightbulbs and many other traditional sources of light. Although the light has a definite energy hf for a given photon, the total light emitted consists of many such photons of varying energies. Therefore, spontaneous emission produces light that is not directional, and it cannot be focused sharply.

Stimulated emission provides the key to the operation and effectiveness of lasers. In fact, the word *laser* is an approximate acronym for "light amplification by stimulated emission of radiation." Suppose an atom is initially in an excited state E_2, as in Fig. 38.11c, and a second photon of energy $hf = E_2 - E_1$ is incident on the atom. Since the incident photon energy is the same as the excitation energy of the electron, there is an increased probability that the electron will drop to its lower energy level, thereby emitting a second photon of the same energy. Such stimulated emission, when coupled with the existence of the incident photon, has the effect of producing two photons with only one incident photon. Each photon has the same energy, direction, and polarization. These photons can, in turn, stimulate other atoms to emit similar photons. Thus, a chain reaction occurs in which many photons of light are emitted, forming the intense, coherent light characteristic of lasers.

Other factors are important for the effective operation of lasers. For example, for stimulated emission to occur, the atoms of the working substance must be in an excited state. At room temperature, most of the atoms have electrons in the ground state with only a few in the higher level. Such a normal population is shown in Fig. 38.12a. By introducing external energy in the form of heat, intense light, or electrical discharges, the population of electrons can be inverted. As in Fig. 38.12b, a population inversion places more electrons in the higher energy level than there are in the lower level. Conditions are therefore ripe for stimulating an avalanche of photons, all of the same frequency and highly directional.

The helium–neon laser, illustrated in Fig. 38.13, is commonly found in physics laboratories. A high voltage is placed across a low-pressure mixture of helium and neon gas

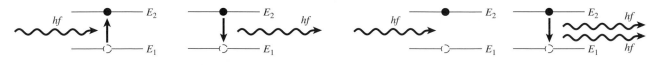

(a) Absorption   (b) Spontaneous emission   (c) Stimulated emission

Figure 38.11   (a) An electron, originally in its ground state E_1, is boosted to the higher level E_2 because of the absorption of a photon. (b) Spontaneous emission occurs and a photo is emitted, $hf = E_2 - E_1$. (c) Stimulated emission occurs when an incident photon of energy hf causes the emission of a second photon of the same energy.

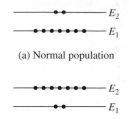

(a) Normal population

(b) Population inversion

Figure 38.12 Population inversion occurs when applied external energy produces atoms with more electrons in the higher energy level than in the lower energy level.

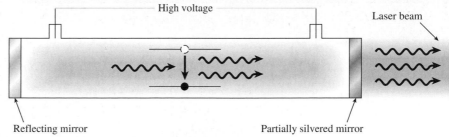

Figure 38.13 A schematic diagram of the helium–neon laser.

contained in a glass tube. Spontaneous emission occurs first, then a chain reaction of stimulated emissions. Both ends of the tube are silvered to form mirrors that reflect the photons back and forth, increasing the opportunities for additional stimulated emissions. One end is only partially silvered so that some of the photons can escape from the tube and form a laser beam.

38.13 Modern Atomic Theory

Although the Bohr atom remains a convenient way to describe the atom, a much more refined theory has been found necessary. The model of the electron as a point particle moving in a perfectly circular orbit does not explain many atomic phenomena. Additional quantum numbers have been established to describe the shape and orientation of the electron cloud about the nucleus, as well as the spin motion of the electrons. It has also been established that no two electrons of the same atom can exist in exactly the same state, even though they may exist at the same Bohr energy level. A more complete description of modern atomic theory can be found in textbooks in the field of atomic physics.

Summary and Review

Summary

The works of Einstein, Bohr, de Broglie, Balmer, and many others have led to a much clearer understanding of nature. We no longer view the world as though all phenomena can be seen, touched, and observed in traditional ways. Greater understanding of the atom has led to many industrial applications based on the principles discussed in this chapter. A summary of the major topics is given as follows.

- According to Einstein's equations of relativity, length, mass, and time are affected by relativistic speeds. The changes become more significant as the ratio of an object's velocity v to the free-space velocity of light c becomes larger.

$$L = L_0 \sqrt{1 - \frac{v^2}{c^2}} \quad \begin{array}{l} Relativistic \\ Contraction \end{array}$$

$$m = \frac{m_0}{\sqrt{1 - v^2/c^2}} \quad Relativistic\ Mass$$

$$\Delta t = \frac{\Delta t_0}{\sqrt{1 - v^2/c^2}} \quad Time\ Dilation$$

In the previous equations, $c = 3 \times 10^8$ m/s.

- The total energy of a particle of rest mass m_0 and speed v can be written in either of the following forms:

$$E = mc^2 \quad E = \sqrt{m_0^2 c^4 + p^2 c^2} \quad Total\ Energy$$

In these equations, m is the relativistic mass as determined by the speed v, and p is the momentum mv.

- The *relativistic kinetic energy* is found from

$$E_K = (m - m_0)c^2 \quad Relativistic\ Kinetic\ Energy$$

- The quantum theory of electromagnetic radiation relates the energy of such radiation to its frequency f or wavelength λ.

$$E = hf \quad E = \frac{hc}{\lambda} \quad h = 6.63 \times 10^{-34}\ \mathrm{J \cdot s}$$

- In the photoelectric effect, the kinetic energy of the ejected electrons is the energy of the incident radiation hf less the work function of the surface W.

$$E_K = \frac{1}{2}mv^2 = hf - W \quad Photoelectric\ Equation$$

- The lowest frequency f_0 at which a photoelectron is ejected is the threshold frequency. It corresponds to the work-function energy W.

$$f_0 = \frac{W}{h} \quad W = hf_0 \quad Threshold\ Frequency$$

- By combining wave theory with particle theory, de Broglie was able to give the following equation for the wavelength of any particle whose mass and velocity are known:

$$\lambda = \frac{h}{mv} \quad h = 6.63 \times 10^{-34}\ \mathrm{J \cdot s} \quad \begin{array}{l} de\ Broglie \\ Wavelength \end{array}$$

- Bohr's first postulate states that the angular momentum of an electron in any orbit must be a multiple of $h/2\pi$. His second postulate states that the energy absorbed or emitted by an atom is in discrete amounts equal to the difference in energy levels of an electron. These concepts are given as equations:

$$mvr = \frac{nh}{2\pi} \quad hf = E_i - E_f \quad Bohr's\ Postulates$$

- Absorption and emission spectra for gases verify the discrete nature of radiation. The wavelength λ or frequency f that corresponds to a change in electron energy levels is given by

$$\frac{1}{\lambda} = R\left(\frac{1}{n_f^2} - \frac{1}{n_i^2}\right) \quad f = Rc\left(\frac{1}{n_f^2} - \frac{1}{n_i^2}\right)$$

$$R = \frac{me^4}{8\epsilon_0^2 h^3 c} = 1.097 \times 10^7\ \mathrm{m}^{-1} \quad \begin{array}{l} Rydberg's \\ Constant \end{array}$$

- The total energy of a particular quantum state n for the hydrogen atom is given by:

$$E_n = -\frac{me^4}{8\epsilon_0^2 n^2 h^2} \quad or \quad E_n = -\frac{13.6\ \mathrm{eV}}{n^2}$$

where $\epsilon_0 = 8.85 \times 10^{-12}\ \mathrm{C^2/N \cdot m^2}$
$e = 1.6 \times 10^{-19}\ \mathrm{C}$
$m = 9.1 \times 10^{-31}\ \mathrm{kg}$
$h = 6.63 \times 10^{-34}\ \mathrm{J \cdot s}$

Key Terms

absorption spectra   744
alpha particle   742
Bohr atom   747
Einstein's postulates   733
energy level   747
excited state   750
line emission spectra   743

photoelectric effect   739
Planck's equation   739
principal quantum number   745
proper time interval   734
relativistic contraction   735
relativistic mass   737
relativistic time   736

spectral series   743
spectrometer   743
threshold frequency   739
time dilation   736
work function   739

Review Questions

38.1. An astronaut holds a clock of mass m and length L. Assume the astronaut passes you at relativistic speed. Compare your measurements of m, L, and Δt with those made by the astronaut for the same clock.

38.2. Recalling Newton's second law of motion, what happens to the thrust requirements for propelling rockets to higher and higher relativistic speeds? Theoretically, what force would be required to achieve the velocity of light?

38.3. You are enclosed in a box with six opaque walls. Are there any experiments you can perform inside the box to prove that you are (a) moving with constant linear velocity, (b) accelerating, or (c) rotating with constant angular velocity?

38.4. Combine Eqs. (38.5) and (38.3) to obtain Einstein's equation for the total relativistic energy, $E = mc^2$.

38.5. Suppose you want photoelectrons to have a kinetic energy E_K and you know the work function W of the surface; how would you determine the required wavelength λ of the incident light?

38.6. Describe an experiment you might perform to determine the work function of a surface, assuming you have a light source of varying wavelength.

38.7. Sketch on a single diagram the Balmer series and the Lyman series for the hydrogen emission spectrum. What is meant by the *series limit?*

38.8. Explain clearly what is meant when we say that the energy of the ground state is -13.6 eV for the hydrogen atom. What is the significance of the minus sign?

38.9. Describe an experiment that will demonstrate (a) a line emission spectrum, (b) a line absorption spectrum, and (c) a continuous spectrum.

38.10. Hydrogen atoms in their ground state are bombarded by electrons that have been accelerated through a potential difference of 12.8 V. Which lines of the Lyman series will be emitted by the hydrogen atoms?

Problems

Section 38.1 Relativity

38.1. A spaceship travels past an observer at a speed of $0.85c$. A person aboard the spacecraft observes that it requires 6.0 s for him to walk the length of his cabin. What time would the observer record for the same event?   Ans. 11.4 s

38.2. A rocket A moves past a lab B at a speed of $0.9c$. A technician in the lab records 3.50 s for the time of an event that occurs on the rocket. What is the time as reckoned by a person aboard the rocket?

38.3. A blinking light on a spacecraft moves past an observer at $0.75c$. The observer records that the light blinks at a frequency of 2.0 Hz. What is the actual frequency of the blinking light?
Ans. 3.02 Hz

38.4. A particle on a table has a diameter of 2 mm when at rest. What must be the speed of an observer who measures the diameter as 1.69 mm?

38.5. A blue meterstick is aboard ship A, and a red meterstick is aboard ship B. If ship A moves past B at $0.85c$, what will be the length of each meterstick as reckoned by a person aboard ship A?
Ans. $L_B = 1.00$ m, $L_R = 52.7$ cm

38.6. Three metersticks travel past an observer at speeds of $0.1c$, $0.6c$, and $0.9c$. What lengths would be recorded by the observer?

38.7. What mass is required to run about 1 million 100-W lightbulbs for 1 year?   Ans. 35.0 g

38.8. Elementary particles called *mu-mesons* rain down through the atmosphere at 2.97×10^8 m/s. At rest the mu-meson would decay on average $2 \ \mu s$ after it came into existence. What is the lifetime of these particles from the viewpoint of an observer on Earth?

Section 38.5 Quantum Theory and the Photoelectric Effect

38.9. The first photoelectrons are emitted from a copper surface when the wavelength of incident radiation is 282 nm. What is the threshold frequency for copper? What is the work function for a copper surface?   Ans. 1.06×10^{15} Hz, 4.40 eV

38.10. If the photoelectric work function of a material is 4.0 eV, what is the minimum frequency of light required to eject photoelectrons? What is the threshold frequency?

38.11. The energy E of a photon in joules is found from the product hf. Often we are given the wavelength of light and need to find its energy in electronvolts. Show that

$$E = \frac{1240}{\lambda}$$

such that if λ is in nanometers, E will be the energy in electronvolts.

38.12. Use the equation derived in Prob. 38.11 to verify that light of wavelength 490 nm has an energy of 2.53 eV. Also show that a photon with an energy of 2.10 eV has a wavelength of 590 nm.

*38.13. The threshold frequency for a certain metal is 2.5×10^{14} Hz. What is the work function? If light of wavelength 400 nm shines on this surface, what is the kinetic energy of ejected photoelectrons?   Ans. 1.04 eV, 2.07 eV

*38.14. When light of frequency 1.6×10^{15} Hz strikes a material surface, electrons just begin to leave the surface. What is the maximum kinetic energy of photoelectrons emitted from this surface when illuminated with light of frequency 2.0×10^{15} Hz?

*38.15. The work function of a nickel surface is 5.01 eV. If a nickel surface is illuminated by light of wavelength 200 nm, what is the kinetic energy of the ejected electrons?   Ans. 1.21 eV

*38.16. The stopping potential is a reverse voltage that just stops the electrons from being emitted in a photoelectric application. The stopping potential is therefore equal to the kinetic energy of ejected photoelectrons. Find the stopping potential for Prob. 38.13.

Section 38.6 Waves and Particles

38.17. What is the de Broglie wavelength of a proton ($m = 1.67 \times 10^{-27}$ kg) when it is moving with a speed of 2×10^7 m/s?   Ans. 1.99×10^{-14} m

38.18. The de Broglie wavelength of a particle is 3×10^{-14} m. What is the momentum of the particle?

38.19. Recalling formulas for kinetic energy and momentum, show that for nonrelativistic speeds, the momentum of a particle can be found from

$$p = \sqrt{2mE_k}$$

where E_k is the kinetic energy and m is the mass of the particle.

*38.20. Determine the kinetic energy of an electron if its de Broglie wavelength is 2×10^{-11} m.

*38.21. What is the de Broglie wavelength of the waves associated with an electron that has been accelerated through a potential difference of 160 V?   Ans. 9.71×10^{-11} m

*38.22. The charge on a proton is $+1.6 \times 10^{-19}$ C, and its rest mass is 1.67×10^{27} kg. What is the de Broglie wavelength of a proton if it is accelerated from rest through a potential difference of 500 V?

Section 38.9 Atomic Spectra

38.23. Determine the wavelength of the first three spectral lines of atomic hydrogen in the Balmer series.   Ans. 656, 486, and 434 nm

38.24. Find the wavelengths of the first three spectral lines of atomic hydrogen in the Paschen series.

38.25. Determine the radius of the $n = 4$ Bohr level of the classical Bohr hydrogen atom.   Ans. 850 nm

38.26. What is the classical radius of the first Bohr orbit in the hydrogen atom?

38.27. Determine the wavelength of the photon emitted from a hydrogen atom when the electron jumps from the $n = 3$ Bohr level to ground level.   Ans. 103 nm

*38.28. What is the maximum wavelength of an incident photon if it can ionize a hydrogen atom originally in its second excited state ($n = 3$)?

*38.29. What are the shortest and longest possible wavelengths in the Balmer series?   Ans. 364 and 656 nm

Additional Problems

38.30. At a cost of 9 cents per kilowatt, what is the cost of the maximum energy to be released from a 1-kg mass?

38.31. An event that occurs on a spaceship traveling at $0.8c$ relative to the Earth is observed by a person on the ship to last for 3 s. What time would be observed by a person on Earth? How far will the person on Earth judge that the spaceship has traveled during this event?

Ans. 5.00 s, 1.2×10^9 km

38.32. When monochromatic light of wavelength 450 nm strikes a cathode, photoelectrons are emitted with a velocity of 4.8×10^5 m/s. What is the work function for the surface in electronvolts? What is the threshold frequency?

38.33. In the hydrogen atom, an electron falls from the $n = 5$ level to the $n = 2$ level and emits a photon in the Balmer series. What are the wavelength and energy of the emitted light?

Ans. 434 nm, 2.86 eV

38.34. Calculate the frequency and the wavelength of the H_β line of the Balmer series. The transition is from the $n = 4$ level of the Bohr atom.

38.35. A spaceship A travels past another ship B with a relative velocity of $0.2c$. Observer B determines that it takes a person on ship A exactly 3.96 s to perform a task. What time will be measured for the same event by observer A? Ans. 3.88 s

*38.36. The rest mass of an electron is 9.1×10^{-31} kg. What is the mass of an electron traveling at a speed of 2×10^8 m/s? What is the total energy of the electron? What is its relativistic kinetic energy?

*38.37. What is the de Broglie wavelength of an electron with a kinetic energy of 50 MeV?

Ans. 0.174 pm

*38.38. The rest mass of a proton is 1.67×10^{-27} kg. What is the total energy of a proton that has been accelerated to a velocity of 2.5×10^8 m/s? What is its relativistic kinetic energy?

*38.39. Compute the mass and the speed of protons having a relativistic kinetic energy of 235 MeV. The rest mass of a proton is 1.67×10^{-27} kg.

Ans. 2.09×10^{-27} kg, 1.8×10^8 m/s

*38.40. How much work is required to accelerate a 1-kg mass from rest to a speed of $0.1c$? How much work is required to accelerate this mass from an initial speed of $0.3c$ to a final speed of $0.9c$? (Use the work-energy theorem.)

*38.41. A particle of mass m is traveling at $0.9c$. By what factor is its relativistic kinetic energy greater than its Newtonian kinetic energy? Ans. 3.20

*38.42. What is the momentum of a 40-eV photon? What is the wavelength of an electron with the same momentum as this photon?

*38.43. When monochromatic light of wavelength 410 nm strikes a cathode, photoelectrons are emitted with a velocity of 4.0×10^5 m/s. What is the work function for the surface, and what is the threshold frequency?

Ans. 2.58 eV, 6.21×10^{14} Hz

*38.44. What is the velocity of a neutron ($m = 1.675 \times 10^{-27}$ kg) that has a de Broglie wavelength of 0.1 nm? What is its kinetic energy in electronvolts?

*38.45. What is the velocity of a particle that has a relativistic kinetic energy of twice its rest mass energy? Ans. $0.943c$

*38.46. Compute the relativistic mass and the speed of electrons that have a relativistic kinetic energy of 1.2 MeV.

Ans. 30.4×10^{-31} kg, 2.86×10^8 m/s

Critical Thinking Questions

38.47. A blue spacecraft is traveling at $0.8c$ relative to a red spacecraft. On the blue ship, a person moves a blue block a distance of 8 m in 3.0 s. On the red ship, a person moves a red block a distance of 4 m in 2.0 s. (a) What are the measurements of these four parameters from the viewpoint of the person on the blue ship? (b) What are the same measurements from the perspective of a person on the red ship?

Ans. Blue: $L_b = 8.00$ m, $t_b = 3.00$ s, $L_r = 2.40$ m, $t_r = 3.33$ s; Red: $L_b = 4.80$ m, $t_b = 5.00$ s, $L_r = 4.00$ m, $t_r = 2.00$ s

*38.48. Use the work-energy theorem to compare the work required to change relativistic speeds with values obtained from Newtonian physics. (a) The speed changes from $0.1c$ to $0.2c$. (b) The speed changes from $0.7c$ to $0.8c$.

*38.49. An electron in the hydrogen atom drops from the $n = 5$ level to the $n = 1$ level. What are the frequency, wavelength, and energy of the emitted photon? In which series does this photon occur?

How much energy must be absorbed by the atom in order to kick the electron back up to the fifth level?   Ans. 95.0 nm, 3.16×10^{15} Hz, 2.09×10^{-18} J, Lyman, 2.09×10^{15} J

*38.50. In a photoelectric experiment shown in Fig. 38.14, a source of emf is connected in series with a galvanometer G. Light falling on the metal cathode produces photoelectrons. The source of emf is biased against the flow of electrons, retarding their motion. The potential difference V_0 just sufficient to stop the most energetic photoelectrons is called the *stopping potential*. Assume that a surface is illuminated with light of wavelength 450 nm, causing electrons to be ejected from the surface at a maximum speed of 6×10^5 m/s. What is the work function for the surface, and what is the stopping potential?

*38.51. In a photoelectric experiment, 400-nm light falls on a certain metal, and photoelectrons are emitted. The potential required to stop the flow of electrons is 0.20 V. What is the energy of the incident photons? What is the work function? What is the threshold frequency?
         Ans. 3.11 eV, 2.91 eV, 7.02×10^{14} Hz

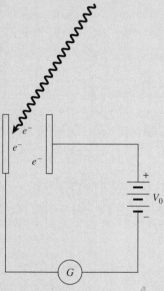

Figure 38.14   The stopping potential V_0 is the potential difference that stops the transition of the most energetic photoelectrons ejected by the incident radiation.

39

Nuclear Physics and the Nucleus

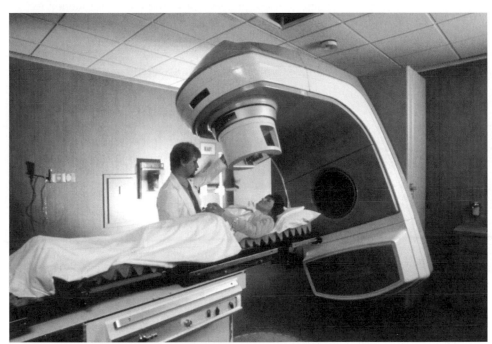

A linear accelerator (LINAC) is the device commonly used for external beam radiation treatments for patients with cancer. The linear accelerator can also be used in stereotactic radiosurgery. It delivers a uniform dose of high-energy X-rays to the region of the patient's tumor. These X-rays can destroy the cancer cells, while sparing the surrounding normal tissue.
(Photo © Larry Mulvehill/Photo Researchers, Inc.)

Objectives

After completing this chapter, you should be able to

1. Define the *mass number* and the *atomic number* and demonstrate your understanding of the nature of fundamental nuclear particles.

2. Define *isotopes* and discuss the use of a mass spectrometer to separate isotopes.

3. Calculate the mass defect and the binding energy per nucleon for a particular isotope.

4. Demonstrate your understanding of radioactive decay and nuclear reactions; describe alpha particles, beta particles, and gamma rays, listing their properties.

5. Calculate the activity and the quantity of radioactive isotope remaining after a period of time if the half-lives and the initial values are given.

6. State the various conservation laws and discuss their application to nuclear reactions.

7. Draw a rough sketch of a nuclear reactor, describing the various components and their functions in the production of nuclear power.

The work of Rutherford and Bohr left us with a picture of the atom as a dense, positively charged nucleus surrounded by a cloud of electrons at distinct energy levels. From this point of view, the nucleus is the center of an atom, containing most of the atom's mass. The behavior of the atom is also affected by the nucleus because in the neutral atom, the total number of positive charges in the nucleus must equal the number of electrons.

In this chapter, we look at the basic internal structure of the nucleus. We will find that classical physics is not adequate to describe interactions at this level. Topics to be discussed include nuclear binding energy, radioactivity, and nuclear energy. The emphasis will be on providing a broad understanding of the atomic nucleus and its behavior.

Nuclear technology has grown enormously since its beginning in the early 1940s. The study of the atomic nucleus was once a subject reserved mainly for physicists, but today there are few people whose lives are not touched by some aspect of nuclear science. As patients, we see the doctor use radioactive materials to diagnose a condition or treat it. As citizens, we are concerned with the promises and dangers of large-scale nuclear power production. More than ever before, technicians and engineers need a better understanding of the atomic nucleus and its potential.

39.1 The Atomic Nucleus

PHYSICS TODAY

A new instrument, the positron emission tomography scanner, or PET scanner, uses isotopes that emit positrons. Such an isotope is included in a solution injected into the patient's body. In the body, the isotope decays, releasing a positron. The positron annihilates an electron, emitting two gamma rays. The PET scanner detects the gamma rays and pinpoints the site of the positron-emitting isotope. A computer is then used to make a three-dimensional map of the isotope distribution. This map can show details such as the use of nutrients in certain areas of the brain. For example, if a person in a PET scanner were solving a physics problem, more nutrients would flow to the part of the brain being used to solve the problem. The decay of the positrons in this part of the brain would increase, and the PET scanner could map this area.

All matter is composed of different combinations of *at least* three fundamental particles: protons, neutrons, and electrons. For example, a beryllium atom (Fig. 39.1) consists of a nucleus that contains four protons and five neutrons. The fact that beryllium is electrically neutral requires that four electrons surround the nucleus. The two inner electrons are at a different energy level ($n = 1$) than the outer two electrons ($n = 2$).

Rutherford's scattering experiments demonstrated that the nucleus contains most of the mass of an atom and that the nucleus is only about one ten-thousandth of the diameter of the atom. Thus, a typical atom with a diameter of 10^{-10} m (100 pm) would have a nucleus about 10^{-14} m (10 fm) in diameter. The prefixes *pico* (10^{-12}) and *femto* (10^{-15}) are useful for nuclear dimensions. Since the diameter of the atom is 10,000 times that of its nucleus, the atom, and therefore matter, consists largely of space that is almost empty.

Let us review what is known about the fundamental particles. The electron has a mass of 9.1×10^{-31} kg and a charge of $e = -1.6 \times 10^{-19}$ C. The proton is the nucleus of a hydrogen atom. It has a mass of 1.673×10^{-27} kg and a positive charge equal in magnitude to the charge of an electron ($+e$). Since the mass of an electron is extremely small, the mass of a proton is approximately the same as the mass of a hydrogen atom, which consists of one proton and one electron. The proton has a diameter of approximately 3 fm.

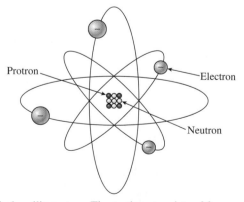

Figure 39.1 A model of a beryllium atom. The nucleus consists of four protons and five neutrons surrounded by four electrons. The positive charge of the protons is balanced exactly by the negatively charged electrons in the neutral electron.

Table 39.1			

Fundamental Particles

Particle	Symbol	Mass, kg	Charge, C
Electron	e	9.1×10^{-31}	-1.6×10^{-19}
Proton	p	1.673×10^{-27}	$+1.6 \times 10^{-19}$
Neutron	n	1.675×10^{-27}	0

The other nuclear particle, the neutron, is present in the nuclei of all elements except hydrogen. It has a mass of 1.675×10^{-27} kg, which is slightly greater than that of the proton, but it has no charge. Thus, although neutrons contribute to the mass of a nucleus, they do not affect the net positive charge of the nucleus, which is due only to protons. The neutron also has a diameter of approximately 3 fm. Table 39.1 summarizes the data we have discussed for three fundamental particles.

From what we now know about the fundamental particles, it is clear that diagrams like Fig. 39.1 cannot be taken too seriously. Distances are not normally presented to scale in such schematic representations. Moreover, classical laws of physics often do not apply for the microworld of the nucleus.

A true understanding of atomic and nuclear events will require a new way of thinking. For example, one might ask what holds the nucleus together. Clearly, if Coulomb's electrostatic repulsion applies in the nucleus, it must be overcome by a much larger force. Both this much larger force and the electrostatic force are immense compared with the gravitational force. This third force is called the **nuclear force.**

The nuclear force is a strong, short-range force. If two **nucleons** (which are protons or neutrons) are separated by approximately 1 fm, a strong attractive force occurs that drops quickly to zero as their separation becomes larger. The force appears to be the same, or nearly the same, between two protons or two neutrons or between a neutron and a proton. If one nucleon is completely surrounded by other nucleons, its nuclear force field will be saturated, and it cannot exert any force on nucleons outside those surrounding it.

39.2 | The Elements

For many centuries, scientists have been studying the various elements found on Earth. A number of attempts have been made to organize the different elements according to their chemical and/or physical properties. The modern grouping of elements is the periodic table. One form of the periodic table is printed in Table 39.2.

Each element is assigned a number that distinguishes it from any other element. For example, the number for hydrogen is 1, the number for helium is 4, and the number for oxygen is 8. These numbers equal the number of protons in the nucleus of that element. The number is given the symbol Z and is called the **atomic number.**

> The atomic number Z of an element is equal to the number of protons in the nucleus of an atom of that element.

The atomic number determines indirectly the chemical properties of an element because Z determines the number of electrons needed to balance the positive charge of the nucleus. The chemical nature of an atom depends on the number of electrons, in particular, the outermost, or valence, electrons.

As the number of protons in a nucleus increases, so does the number of neutrons. In lighter elements, the increase is approximately one to one, but heavier elements may have

Table 39.2

The Periodic Table

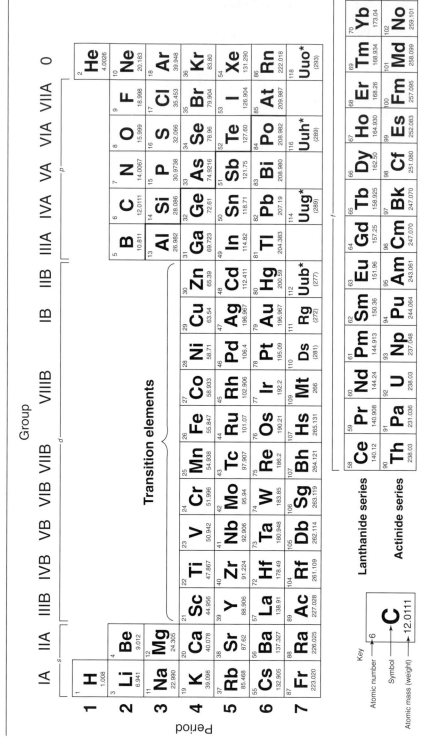

Group

Transition elements

Lanthanide series

Actinide series

Key

Atomic number → 6
Symbol → **C**
Atomic mass (weight) → 12.0111

Atomic weight values listed in parentheses are approximate.
***Indicates unnamed elements.**

more than $1\frac{1}{2}$ times more neutrons that protons. For example, oxygen has 8 protons and 8 neutrons, whereas uranium has 92 protons and 146 neutrons. The total number of nucleons in a nucleus is called the ***mass number A.***

> The mass number A of an element is equal to the total number of protons and neutrons in its nucleus.

If we represent the number of neutrons by N, we can write the mass number A in terms of the atomic number Z and the number of neutrons.

$$A = Z + N \qquad (39.1)$$

Thus, the mass number of uranium is $92 + 146$, or 238.

A general way of describing the nucleus of a particular atom is to write the symbol for the element with its mass number and atomic number shown as follows:

$$\substack{\text{mass number} \\ \text{atomic number}}[\text{symbol}] = {}_{Z}^{A}X$$

For example, the uranium atom has the symbol ${}_{92}^{238}U$.

An alphabetical listing of all the elements is given in Table 39.3. The symbols for the first four elements are given as follows:

Element	Symbol	Protons	Neutrons	Electrons
Hydrogen	${}_{1}^{1}H$	1	1	1
Helium	${}_{2}^{4}He$	2	4	2
Lithium	${}_{3}^{7}Li$	3	7	3
Beryllium	${}_{4}^{9}Be$	4	9	4

Example 39.1

How many neutrons are in the nucleus of an atom of mercury ${}_{80}^{201}Hg$?

Plan: The symbol for mercury shows the atomic number is 80 and the mass number is 201. The atomic number gives the number of protons in the nucleus, and the mass number is the number of nuclear particles including protons and neutrons.

Solution: The number of neutrons is found by solving for N in Eq. (39.1)

$$N = A - Z = 201 - 80; \qquad N = 121 \text{ neutrons}$$

39.3 The Atomic Mass Unit

The small masses of nuclear particles call for an extremely small unit of mass. Scientists normally express atomic and nuclear masses in ***atomic mass units*** (u).

> One atomic mass unit (1 u) is exactly equal to one-twelfth of the mass of the most abundant form of the carbon atom.

In terms of the kilogram, the atomic mass unit is

$$1 \text{ u} = 1.6606 \times 10^{-27} \text{ kg} \qquad (39.2)$$

Table 39.3

International Atomic Weights [Based on Carbon 12]*

Element	Symbol	Atomic Number	Atomic Weight
Actinium	Ac	89	(227)
Aluminum	Al	13	26.9815
Americium	Am	95	(243)
Antimony	Sb	51	121.75
Argon	Ar	18	39.948
Arsenic	As	33	74.9216
Astatine	At	85	(210)
Barium	Ba	56	137.34
Berkelium	Bk	97	(247)
Beryllium	Be	4	9.0122
Bismuth	Bi	83	208.980
Boron	B	5	10.811
Bromine	Br	35	79.909
Cadmium	Cd	48	112.40
Calcium	Ca	20	40.08
Californium	Cf	98	(251)
Carbon	C	6	12.01115
Cerium	Ce	58	140.12
Cesium	Cs	55	132.905
Chlorine	Cl	17	35.453
Chromium	Cr	24	51.996
Cobalt	Co	27	58.9332
Copper	Cu	29	63.546
Curium	Cm	96	(247)
Darmstadium	Ds	110	(281)
Dubnium	Db	105	(262)
Dysprosium	Dy	66	162.50
Einsteinium	Es	99	(254)
Erbium	Er	68	167.26
Europium	Eu	63	151.96
Fermium	Fm	100	(253)
Fluorine	F	9	18.9984
Francium	Fr	87	(223)
Gadolinium	Gd	64	157.25
Gallium	Ga	31	69.72
Germanium	Ge	32	72.59
Gold	Au	79	196.967
Hafnium	Hf	72	178.49
Helium	He	2	4.0026
Holmium	Ho	67	164.930
Hydrogen	H	1	1.00797
Indium	In	49	114.82
Iodine	I	53	126.9044
Iridium	Ir	77	192.2
Iron	Fe	26	55.847
Krypton	Kr	36	83.80
Lanthanum	La	57	138.91
Lawrencium	Lr	103	(262)
Lead	Pb	82	207.19
Lithium	Li	3	6.939
Lutetium	Lu	71	174.97
Magnesium	Mg	12	24.312
Manganese	Mn	25	54.9380
Mendelivium	Md	101	(256)
Mercury	Hg	80	200.59
Molybdenum	Mo	42	95.94
Neodymium	Nd	60	144.24
Neon	Ne	10	20.183
Neptunium	Np	93	(237)
Nickel	Ni	28	58.71
Niobium	Nb	41	92.906
Nitrogen	N	7	14.0067
Nobelium	No	102	(254)
Osmium	Os	76	190.2
Oxygen	O	8	15.9994
Palladium	Pd	46	106.4
Phosphorus	P	15	30.9738
Platinum	Pt	78	195.09
Plutonium	Pu	94	(244)
Polonium	Po	84	(210)
Potassium	K	19	39.102
Praseodymium	Pr	59	140.907
Promethium	Pm	61	(145)
Protactinium	Pa	91	(231)
Radium	Ra	88	(226)
Radon	Rn	86	(222)
Rhenium	Re	75	186.22
Rhodium	Rh	45	102.91
Roentgenium	Rg	111	(272)
Rubidium	Rb	37	85.47
Ruthenium	Ru	44	101.07
Rutherfordium	Rf	104	(261)
Samarium	Sm	62	150.35
Scandium	Sc	21	44.956
Selenium	Se	34	78.96
Silicon	Si	14	28.086
Silver	Ag	47	107.870
Sodium	Na	11	22.9898
Strontium	Sr	38	87.62
Sulfer	S	16	32.064
Tantalum	Ta	73	180.948
Technetium	Tc	43	(99)
Tellurium	Te	52	127.60
Terbium	Tb	65	158.924
Thallium	Tl	81	204.37
Thorium	Th	90	232.038
Thulium	Tm	69	168.934
Tin	Sn	50	118.69
Titanium	Ti	22	47.90
Tungsten (Wolfram)	W	74	183.85
Uranium	U	92	238.03
Vanadium	V	23	50.942
Xenon	Xe	54	131.30
Ytterbium	Yb	70	173.04
Yttrium	Y	39	88.905
Zinc	Zn	30	65.37
Zirconium	Zr	40	91.22

*Values in parentheses are mass numbers of longest-lived or best-known isotopes.

Example 39.2

The periodic table shows the average atomic mass of barium to be 137.34 u. What is the average mass of the barium nucleus?

Plan: We will first convert the mass of an electron from kilograms to atomic mass units. Then, the average mass of the barium nucleus is found by subtracting the mass of the surrounding cloud of electrons from the average mass of the entire atom.

Solution: The mass of an electron is 9.1×10^{-31} kg, so

$$m_e = 9.1 \times 10^{-31} \text{ kg} \left(\frac{1 \text{ u}}{1.6606 \times 10^{-27} \text{ kg}} \right) = 5.5 \times 10^{-4} \text{ u}$$

Since the atomic number Z of barium is 56, there must be the same number of electrons in order that the entire atom be electrically neutral. This total mass is

$$m_T = 56 \text{ electrons} \left(\frac{5.5 \times 10^{-4} \text{ u}}{\text{electron}} \right) = 0.0308 \text{ u}$$

The average atomic mass was given as 137.34 u. Thus, the average nuclear mass is

$$m_{\text{avg}} = 137.34 \text{ u} - 0.0308 \text{ u} = 137.31 \text{ u}$$

In Example 39.2, it must be emphasized that the discussion involved *average* values and does not represent the mass of a particular nucleus of barium. Whereas all barium nuclei must have the same number of protons, some may contain more or less neutrons than others. The periodic table gives the average of the naturally occurring atoms of barium. Moreover, the atomic masses given in the table include the masses of the surrounding electrons.

In our study of nuclear physics, it will be useful to remember the following masses expressed in atomic mass units:

Proton mass: $m_p = 1.007276$ u

Neutron mass: $m_n = 1.008665$ u

Electron mass: $m_e = 0.00055$ u

It should be remembered from Chapter 38 that mass can be equated to units of energy from Einstein's relation

$$E = mc^2$$

Consequently, a mass of 1 u corresponds to an energy given by

$$(1 \text{ u}) \, c^2 = (1.66 \times 10^{-27} \text{ kg})(3 \times 10^8 \text{ m/s})^2 = 1.49 \times 10^{-10} \text{ J}$$

In more convenient units of electronvolts, we can write

$$(1 \text{ u})c^2 = 1.49 \times 10^{-10} \text{ J} \frac{1 \text{ eV}}{1.6 \times 10^{-19} \text{ J}}$$

$$= 9.31 \times 10^8 \text{ eV} = 931 \text{ MeV}$$

The conversion factor from mass units (u) to energy units (MeV) is therefore

$$c^2 = 931 \text{ MeV/u} \tag{39.3}$$

As an exercise, you might verify that the electron and the proton have rest-mass energies of 0.511 and 938 MeV, respectively.

39.4 | ## Isotopes

It is possible for two atoms of the same element to have nuclei containing different numbers of neutrons. Such atoms are called *isotopes.*

> Isotopes are atoms that have the same atomic number Z but different mass numbers A.

For example, naturally occurring carbon is a mixture of two isotopes. The most abundant form, $^{12}_{6}C$, has six protons and six neutrons in its nucleus. Another form, $^{13}_{6}C$, has an extra neutron. Some elements have as many as 10 different isotopic forms.

Experimental verification of the existence of isotopes is accomplished with a *mass spectrometer.* This device, illustrated in Fig. 39.2, is used to separate the isotopes of an element. A source of singly ionized atoms of a particular element is positioned above the velocity selector. These ions are missing one electron and therefore have a charge of $+e$. They are propelled at varying speeds into the crossed **E** and **B** fields of the velocity selector. Ions with velocities sufficient to make the magnetic force $\mathbf{F}_m$ equal and opposite to the electric force $\mathbf{F}_e$ will pass through the bottom slit undeflected. Recalling that $F_e = eE$ and $F_m = evB$, we write

$$evB = eE$$

from which

$$v = \frac{E}{B} \tag{39.4}$$

Only ions with this velocity will pass through the slit at the bottom of the selector.

The fast-moving positive ions next pass into the lower region, where another **B** field causes them to experience a perpendicular magnetic force. The magnitude of this force will be constant and equal to evB, but its direction will always be at right angles to the velocity of the ion. The result is a circular path of radius R. The magnetic force provides the necessary centripetal force. In this case, we have

$$F_{\text{magnetic}} = F_{\text{centripetal}}$$

or

$$evB = \frac{mv^2}{R}$$

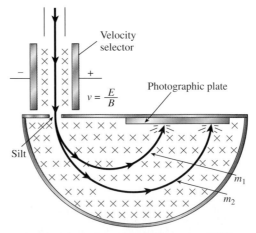

Figure 39.2   The mass spectrometer is used to separate isotopes of different mass; the crosses indicate that the direction of the magnetic field is into the paper.

where m is the mass of the ion of charge e. Solving for R, we find the radius of the semi-circular path is given by

$$R = \frac{mv}{eB} \qquad (39.5)$$

Since v, e, and B are constant, Eq. (39.5) gives the radius as a function of the mass of given ions. Ions of different mass will strike the photographic plate at different positions because their semicircular paths are different. Wherever a beam of ions strikes the plate, a darkened line will be produced. The distance of a particular line from the slit is twice the radius in which that beam of ions moves. In this manner, the mass can be determined from Eq. (39.5).

The mass spectrometer is used to separate and study isotopes. Most elements are mixtures of atoms with different mass numbers. For example, if an ion beam of pure lithium is injected into the mass spectrometer, two types of atoms are observed. The darker bank occurs because around 92 percent of the atoms have a mass of 7.016 u. The remaining 8 percent, producing a lighter band, are atoms with mass 6.015 u. These two isotopes of lithium are written $^{7}_{3}\text{Li}$ and $^{6}_{3}\text{Li}$.

Since some elements, for example, tin, have many different isotopic forms, it is not surprising that the average atomic mass for elements is often not very close to an integer. Average atomic mass is affected by the mass numbers and relative abundance of each isotopic form. For example, chlorine has an average atomic mass of 35.453 u, which results from a mixture of the two isotopes $^{35}_{17}\text{Cl}$ and $^{37}_{17}\text{Cl}$. The lighter chlorine isotope occurs about three times as often as the heavier one.

Example 39.3

When chlorine is studied with the mass spectrometer, it is noted that an intense line occurs 24 cm from the entrance slit. Another lighter line appears at a distance of 25.37 cm. If the mass of the ions that form the first line is 34.980 u, what is the mass of the other isotope?

Plan: The given distances must be converted from diameters to radii for the two ion paths. See Fig. 39.2. Equation (39.5) gives an expression for the radius of an ion path in terms of its mass and velocity. Since the velocity and charge of each ion are the same, the ratio of the two masses must equal the corresponding ratio of diameters. That will allow us to solve for the unknown ion mass.

Solution: The radii for the two paths are

$$R_1 = 12 \text{ cm} \quad \text{and} \quad R_2 = 12.685 \text{ cm}$$

Now, from Eq. (39.5),

$$R_1 = \frac{m_1 v}{eB} \quad \text{and} \quad R_2 = \frac{m_2 v}{eB}$$

Since e, v, and B are constant, we have

$$\frac{m_1}{m_2} = \frac{R_1}{R_2}$$

from which the mass m_2 is found to be

$$m_2 = \frac{m_1 R_2}{R_1} = \frac{(34.980 \text{ u})(12.685 \text{ cm})}{(12 \text{ cm})}$$

$$= 36.977 \text{ u}$$

39.5 The Mass Defect and Binding Energy

One of the startling results that can be demonstrated with the mass spectrometer is that the mass of a nucleus is not exactly equal to the sum of the masses of its nucleons. Let us consider, for example, the helium atom, $^4_2$He, which has two electrons about a nucleus containing two protons and two neutrons. The atomic mass is found from the periodic table to be 4.0026 u.

Now let us compare this value with the mass of all the individual particles that make up the atom:

$$2p = 2(1.007276\ \text{u}) = 2.014552\ \text{u}$$
$$2n = 2(1.008665\ \text{u}) = 2.017330\ \text{u}$$
$$2e = 2(0.00055\ \text{u}) = \underline{0.001100\ \text{u}}$$
$$\text{Total mass} = 4.032982\ \text{u}$$

The mass of the parts (4.0330 u) is apparently greater than the mass of the atom (4.0026 u).

$$m_{\text{parts}} - m_{\text{atom}} = 4.0330\ \text{u} - 4.0026\ \text{u}$$
$$= 0.0304\ \text{u}$$

When protons and neutrons join to form a helium nucleus, the mass is decreased in the process. This difference is called the ***mass defect.*** A mass defect can be shown to exist for atoms of all elements.

The mass defect is defined as the difference between the rest mass of a nucleus and the sum of the rest masses of its constituent nucleons.

We have seen from Einstein's work that mass and energy are equivalent. We might suppose, then, that the mass decrease in joining nucleons together will result in an energy decrease. Since energy is conserved, a decrease in the energy of the system means that energy must be released in joining the system together. In the case of helium, this energy would come from a mass of 0.0304 u and would be equal to

$$E = mc^2 = (0.0304\ \text{u})\frac{931\ \text{MeV}}{1\ \text{u}} = 28.3\ \text{MeV}$$

The total energy that would be released if we could build a nucleus from protons and neutrons is called the *binding energy* of the nucleus. As we have just seen, the binding energy of $^4_2$He is 28.3 MeV, as illustrated in Fig. 39.3a.

We can also reverse the above process and state that the ***binding energy*** is the energy required to break a nucleus apart into its constituent particles.

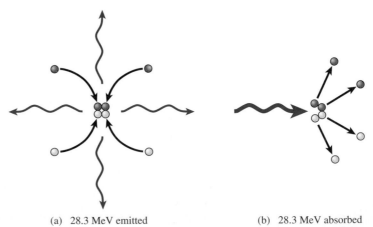

(a) 28.3 MeV emitted   (b) 28.3 MeV absorbed

Figure 39.3   (a) When two protons and two neutrons are fused together to form a helium nucleus, energy is released. (b) The same amount of energy is required to break the nucleus apart into its constituent nucleons.

The binding energy of a nucleus is defined as the energy required to separate a nucleus into its constituent nucleons.

In our example, an energy of 28.3 MeV must be supplied to $_2^4\text{He}$ to separate the nucleus into two protons and two neutrons (Fig. 39.3b).

An isotope of atomic number Z and mass number A consists of Z protons, Z electrons, and $N = (A - Z)$ neutrons. If we neglect the binding energy of the electrons, a neutral isotope would have the same mass as Z neutral hydrogen atoms plus the mass of the neutrons. The masses of $_1^1\text{H}$ and m_n are

$$m_H = 1.007825 \text{ u} \qquad m_n = 1.008665 \text{ u} \qquad (39.6)$$

If we represent the atomic mass by M, the binding energy E_B can be approximated by

$$E_B = [(Zm_H + Nm_n) - M]c^2 \qquad \textit{Binding Energy} \quad (39.7)$$

In applying this equation, we should remember that $N = A - Z$ and that $c^2 = 931$ MeV/u.

Example 39.4

Determine the total binding energy and the binding energy per nucleon for the $_7^{14}\text{N}$ nucleus. Assume the atomic mass for nitrogen-14 is 14.003074 u.

Plan: We will add up the masses of all the protons, neutrons, and electrons inside the nucleus and determine the total mass of the individual parts of the atom. Then, we will subtract from that total the atomic mass of the nitrogen-14 atom. The difference is the mass defect, which can be converted to the binding energy by multiplying by c^2. The entire process can be concluded by substitution into Eq. (39.7). Note the mass of the electrons is taken care of by using the mass of a hydrogen atom, which includes the electrons.

Solution: For nitrogen-14, $Z = 7$, $N = 7$, and $M = 14.003074$ u.

$$E_B = [(Zm_H + Nm_n) - M]c^2$$
$$= \{[7(1.007825 \text{ u}) + 7(1.008665 \text{ u})] - 14.003074 \text{ u}\}(931 \text{ MeV/u})$$
$$= (0.112356 \text{ u})(931 \text{ MeV/u}) = 104.6 \text{ MeV}$$

Since $_7^{14}\text{N}$ contains 14 nucleons, the binding energy per nucleon is

$$\frac{E_B}{A} = \frac{104.6 \text{ MeV}}{14 \text{ nucleons}} = 7.47 \text{ MeV/nucleon}$$

The atomic mass M used in Eq. (39.7) must be taken for the particular isotope of the element, not from the periodic table (Table 39.2) or from Table 39.3. These tables give the atomic masses of the naturally occurring mixture of isotopes for each element. The atomic mass of $_6^{12}\text{C}$, for example, is exactly 12.0000 u by definition. The periodic table gives a value of 12.01115 u because naturally occurring carbon contains small amounts of $_6^{13}\text{C}$ in addition to the more abundant $_6^{12}\text{C}$ atoms. The term **nuclide** is used to refer to a particular isotope that has a particular number of nuclear particles and, hence, a specific mass. The masses of several common nuclides are given in Table 39.4. These are the masses that should be used in determining mass defects and binding energies. Electron masses are included, so the given masses are basically the mass of the nucleus plus Z atomic electrons.

The binding energy per nucleon, as computed in Example 39.4, is an important way of comparing the nuclei of various elements. A plot of the binding energy per nucleon as a function of increasing mass number is shown in Fig. 39.4 for many

Table 39.4

Atomic Masses for Several Nuclides

Nuclide	Atomic Number	Mass Number	Atomic Mass
Hydrogen	1	1	1.007825 u
Deuterium	1	2	2.014102 u
Tritium	1	3	3.016049 u
Helium 3	2	3	3.016030 u
Helium 4	2	4	4.002603 u
Lithium 6	3	6	6.015126 u
Lithium 7	3	7	7.016003 u
Beryllium 9	4	9	9.012186 u
Boron 11	5	11	11.009305 u
Carbon 12	6	12	12.000000 u
Carbon 13	6	13	13.003354 u
Nitrogen 14	7	14	14.003074 u
Oxygen 16	8	16	15.994915 u
Oxygen 17	8	17	16.999132 u
Neon 20	10	20	19.992440 u
Copper 64	29	64	63.929759 u
Tin 120	50	120	119.902108 u
Gold 197	79	197	196.966541 u
Mercury 204	80	204	203.973865 u
Thallium 206	81	206	205.976104 u
Polonium 216	84	216	216.001922 u
Radon 222	86	222	222.017531 u
Radium 224	88	224	224.020218 u
Radium 226	88	226	226.025360 u
Thorium 233	90	233	233.041469 u
Thorium 234	90	234	234.043630 u
Protactinium 233	91	233	233.040130 u
Uranium 238	92	238	238.050786 u

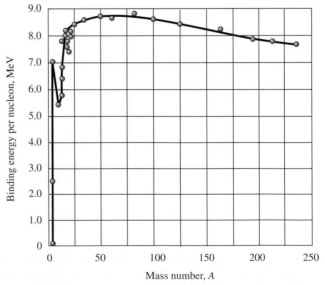

Figure 39.4 The average binding energy per nucleon for the most stable nucleus at each mass number.

stable nuclei. Note that the mass numbers toward the center (50 to 80) yield the highest binding energy per nucleon. Elements ranging from $A = 50$ to $A = 80$ are the most stable.

39.6 Radioactivity

The strong nuclear force holds the nucleons tight in the nucleus, overcoming the coulomb repulsion of protons. The balance of forces is not always maintained, however, and sometimes particles or photons are emitted from the nuclei of atoms. Such unstable nuclei are said to be *radioactive* and have the property of **radioactivity.**

All naturally occurring elements with atomic numbers greater than 83 are radioactive. They are slowly decaying and disappearing from the Earth. Uranium and radium are two of the better-known examples of naturally radioactive elements. A few other naturally occurring, lighter, and less active elements have also been discovered.

Unstable nuclei are also produced artificially as by-products of nuclear reactors, for study in laboratories, or for other purposes. In addition, some elements are made radioactive naturally by bombardment with high-energy photons.

There are three major forms of radioactive emission from atomic nuclei:

1. **Alpha particles** (α) An alpha particle is the nucleus of a helium atom and consists of two protons and two neutrons. It has a charge of $+2e$ and a mass of 4.001506 u. Because of their positive charges and relatively low speeds ($\approx 0.1c$), alpha particles do not have great penetrating power.

2. **Beta particles** (β) There are two kinds of beta particles, a beta minus particle (β^-) and a beta plus particle (β^+). The beta minus particle is simply an electron of charge $-e$ and mass equal to 0.00055 u. A beta plus particle, also called a *positron,* has the same mass as an electron but the opposite charge ($+e$). These particles are generally emitted at speeds near the velocity of light. The beta minus particles are much more penetrating than alpha particles, but beta plus particles easily combine with electrons; then rapid annihilation of both the positrons and electrons occurs, with the emission of gamma rays.

3. **Gamma rays** (γ) A gamma ray is a high-energy electromagnetic wave similar to heat and light but of much higher frequency. These rays have no charge or rest mass and are the most penetrating radiation emitted by radioactive elements.

To understand why these forms of radiation are emitted, it is helpful to look at nuclei that are relatively stable. Plotting the number of neutrons N versus the number of protons Z for these stable nuclei gives a rough graph like that shown in Fig. 39.5. Note that the light elements are stable when the ratio of Z to N is close to 1. More neutrons are required for stability in the heavier elements. The additional nuclear forces of the extra neutrons are needed to balance the higher electric forces that result as more protons are collected. Whenever a nucleus occurs that deviates very much from the line, it is unstable and will emit some form of radiation, thereby achieving stability.

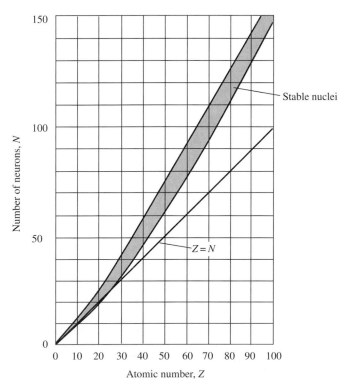

Figure 39.5 A comparison of the number of neutrons as a function of the atomic number. Notice that nuclei of higher Z have the greater proportion of neutrons.

| 39.7 | ## Radioactive Decay |

Let us look at radioactive decay by alpha, beta, and gamma radiation and see what occurs during each process. The emission of an alpha particle $_2^4\alpha$ reduces the number of protons in the parent nucleus by 2 and the number of nucleons by 4. Symbolically, we write

$$_Z^A X \rightarrow \ _{Z-2}^{A-4} Y + \ _2^4\alpha + \text{energy} \tag{39.8}$$

The energy term results from the fact that the rest energy of the products is less than that of the parent atom. The difference in energy is carried away primarily by the kinetic energy imparted to the alpha particle. The recoil kinetic energy of the much more massive daughter atom is small by comparison.

Example 39.5

Write the reaction that occurs when $_{88}^{226}\text{Ra}$ decays by alpha emission.

Solution: Applying Eq. (39.8), we write

$$_{88}^{226}\text{Ra} \rightarrow \ _{86}^{222}\text{Rn} + \ _2^4\alpha + \text{energy}$$

Notice that the unstable element radium has been transformed into a new element, radon, which is closer to the stability line.

Next consider the emission of beta-minus particles from the nucleus. If beta-minus particles are electrons, how can an electron come from a nucleus containing only protons and neutrons? This can be answered, at least in part, by analogy to the Bohr atom. We have seen that photons, which do not exist in the atom, are emitted by atoms when they change from one state to another. Similarly, electrons, which do not exist in nuclei, can be emitted as a form of radiation when the nucleus changes from one state to another. When such a change does occur, the total charge must be conserved. This requires the conversion of a neutron into a proton and an electron.

$$_0^1 n \rightarrow \ _1^1 p + \ _{-1}^0 e$$

Thus, in beta-minus emission, a neutron is replaced by a proton. The atomic number Z increases by 1, and the mass number is unchanged. Symbolically,

$$_Z^A X \rightarrow \ _{Z+1}^A Y + \ _{-1}^0 \beta + \text{energy} \tag{39.9}$$

An example of beta emission is the decay of an isotope of neon into sodium:

$$_{10}^{23}\text{Ne} \rightarrow \ _{11}^{23}\text{Na} + \ _{-1}^0 \beta + \text{energy}$$

The increase in Z is necessary to conserve charge.

Similarly, in positron (beta-plus) emission, a proton in the nucleus decays to a neutron and a positron.

$$_1^1 p \rightarrow \ _0^1 n + \ _{+1}^0 e$$

The atomic number Z decreases by 1, and the mass number A is unchanged. Symbolically,

$$_Z^A X \rightarrow \ _{Z-1}^A Y + \ _{+1}^0 \beta + \text{energy} \tag{39.10}$$

An example of positron emission is the decay of an isotope of nitrogen into an isotope of carbon:

$$_7^{13}\text{N} \rightarrow \ _6^{13}\text{C} + \ _{+1}^0 \beta + \text{energy}$$

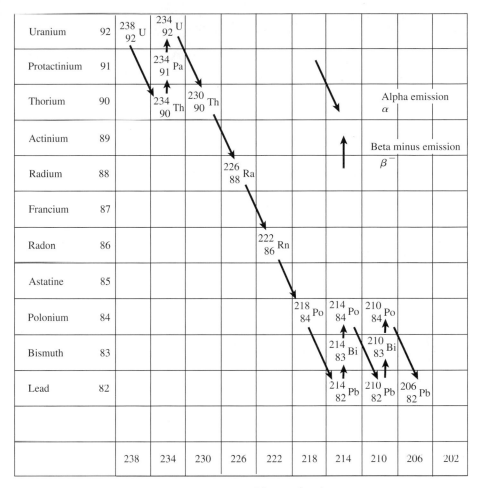

		238	234	230	226	222	218	214	210	206	202

Uranium 92, Protactinium 91, Thorium 90, Actinium 89, Radium 88, Francium 87, Radon 86, Astatine 85, Polonium 84, Bismuth 83, Lead 82

$^{238}_{92}U$ $^{234}_{92}U$ $^{234}_{91}Pa$ $^{234}_{90}Th$ $^{230}_{90}Th$ $^{226}_{88}Ra$ $^{222}_{86}Rn$ $^{218}_{84}Po$ $^{214}_{84}Po$ $^{210}_{84}Po$ $^{214}_{83}Bi$ $^{210}_{83}Bi$ $^{214}_{82}Pb$ $^{210}_{82}Pb$ $^{206}_{82}Pb$

Alpha emission α

Beta minus emission β^-

Mass number, A

Figure 39.6 The uranium series of disintegration. Uranium decays, through a series of alpha- and beta-minus emissions, from ^{238}U to ^{206}Pb.

In both types of beta emission, the kinetic energy is shared mostly by the beta particle and another particle called a *neutrino*. The neutrino has virtually no rest mass and no electric charge, but it can have both energy and momentum.

In gamma emission, the parent nucleus maintains the same atomic number Z and the same mass number A. The gamma photon simply carries away energy from an unstable nucleus. Frequently, a succession of alpha and beta decays is accompanied by gamma decays, which carry off excess energy.

The radioactive disintegration of $^{238}_{92}U$ is shown in Fig. 39.6 as a series of decays through a number of elements until it becomes a stable $^{206}_{82}Pb$ nucleus.

39.8 Half-Life

A radioactive material continues to emit radiation until all the unstable atoms have decayed. The number of unstable nuclei decaying or disintegrating every second for a given isotope can be predicted on the basis of probability. This number is referred to as the **activity R,** given by

$$R = \frac{-\Delta N}{\Delta t} \tag{39.11}$$

where N is the number of undecayed nuclei. The negative sign is included because N is decreasing with time. The units for R are inverse seconds (s^{-1}).

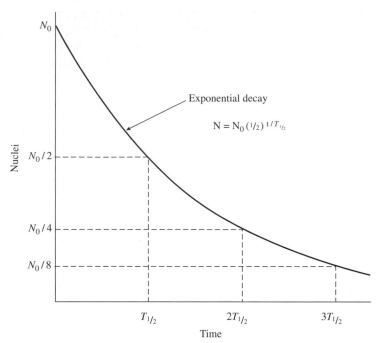

Figure 39.7   The radioactive decay curve, illustrating the half-life as the time $T_{1/2}$ required for one-half of the unstable nuclei, present at time $t = 0$, to decay.

In practice, the activity in disintegrations per second is so large that a more convenient unit, the ***curie*** (Ci), is defined as follows.

> One curie (Ci) is the activity of a radioactive material that decays at the rate of 3.7×10^{10} disintegrations per second.

$$1 \text{ Ci} = 3.7 \times 10^{10} \text{ s}^{-1} \qquad (39.12)$$

The activity of 1 g of radium is slightly less than 1 Ci.

The random nature of nuclear decay means the activity R at any time is directly proportional to the number of nuclei remaining; in other words, as the number of remaining nuclei decreases with time, the activity must also decrease with time. Therefore, if we plot the number of remaining nuclei as a function of time, as illustrated in Fig. 39.7, we see that radioactive decay is not linear. The time it takes for this curve to drop to one-half its original value is different for each radioactive isotope, and it is called the ***half-life.***

> The half-life $T_{1/2}$ of a radioactive isotope is the length of time in which one-half of its unstable nuclei will decay.

For example, the half-life of radium 226 is 1620 yr; 1 g of this isotope will decay to 0.5 g in 1620 yr, to 0.25 g in 2(1620 yr), to 0.125 g in 3(1620 yr), and so on.

We can use this definition of the half-life to determine how many nuclei are present at a time t. If we start out at time $t = 0$ with a number N_0 of unstable nuclei, then after n half-lives have passed, there will be left a number of nuclei N given by

$$N = N_0\left(\frac{1}{2}\right)^n \qquad (39.13)$$

The number n of half-lives in the period of time t is, of course, $t/T_{1/2}$. Thus, a more applicable form of the above relation is

$$N = N_0\left(\frac{1}{2}\right)^{t/T_{1/2}}$$

Since the amount of radioactive material is determined by the number of nuclei present, an equation similar to Eq. (39.13) can be used to compute the mass of remaining radioactive material after a number of half-lives.

The same idea applied to the activity R of a radioactive sample or to the portion of radioactive mass remaining in a given sample yields the following similar equations:

$$R = R_0\left(\frac{1}{2}\right)^{t/T_{1/2}} \quad \text{and} \quad m = m_0\left(\frac{1}{2}\right)^{t/T_{1/2}} \tag{39.14}$$

Example 39.6

The worst by-product of ordinary nuclear reactors is the radioactive isotope plutonium 239, which has a half-life of 24,400 yr. Suppose the initial activity of a sample containing 1.64×10^{20} $^{239}_{94}\text{Pu}$ nuclei is 4 mCi. (a) How many of these nuclei remain after 73,200 yr? (b) What will be the activity at that time?

Plan: After each half-life, the number of radioactive nuclei reduces by one-half. We will determine how many half-lives are contained in 73,200 yr and use it as the exponent of $\frac{1}{2}$. That number is multiplied by the number of beginning nuclei to obtain the number remaining. A similar calculation can be made for the radioactivity.

Solution (a): Substitution into Eq. (39.13) yields

$$N = N_0\left(\frac{1}{2}\right)^{t/T_{1/2}} = 1.64 \times 10^{20}\left(\frac{1}{2}\right)^{73,200\ \text{yr}/24,400\ \text{yr}}$$

$$= 1.64 \times 10^{20}\left(\frac{1}{2}\right)^{3} = 1.64 \times 10^{20}\left(\frac{1}{8}\right)$$

$$= 2.05 \times 10^{19}\ \text{nuclei}$$

Solution (b): We obtain the remaining activity from Eq. (39.14).

$$R = R_0\left(\frac{1}{2}\right)^{t/T_{1/2}} = 4\ \text{mCi}\left(\frac{1}{8}\right) = 0.5\ \text{mCi}$$

Both these calculations assume that no new $^{239}_{94}\text{Pu}$ nuclei are being created by other processes. It is easy to see from this example why disposal of some radioactive materials is such a difficult problem.

39.9 Nuclear Reactions

In a chemical reaction, the atoms of two molecules react to form different molecules. In a *nuclear reaction,* nuclei, radiation, and/or nucleons collide to form different nuclei, radiation, and nucleons. If the colliding objects are charged, at least one of the colliding masses must be accelerated to a relatively high velocity. Normally, the bombarding particle has low rest mass, for example, a proton $^{1}_{1}p$ or an alpha particle $^{4}_{2}\alpha$. These nuclear projectiles are accelerated with many different devices, for instance, Van de Graaff generators, cyclotrons, and linear accelerators.

In the nuclear reactions we will study, several conservation laws must be observed, primarily *conservation of charge, conservation of nucleons,* and *conservation of mass-energy.*

Conservation of Charge: The total charge of a system can neither be increased nor decreased in a nuclear reaction.

Conservation of Nucleons: The total number of nucleons in the interaction must remain unchanged.

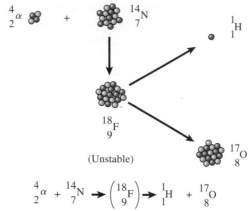

$$\overset{4}{2}\alpha + \overset{14}{7}N \longrightarrow \left(\overset{18}{9}F\right) \longrightarrow \overset{1}{1}H + \overset{17}{8}O$$

Figure 39.8 Striking a nitrogen-14 nucleus with an alpha particle.

Conservation of Mass-Energy: The total mass-energy of a system must remain unchanged in a nuclear reaction.

Now let us observe what happens when an alpha particle $^4_2\alpha$ strikes a nucleus in a sample of nitrogen gas $^{14}_7N$. (See Fig. 39.8.) The first step is the entry of the alpha particle, which adds 2 protons and 2 neutrons to the nucleus. The atomic number Z is increased by 2, and the mass number A is increased by 4. The resulting nucleus is an *unstable* compound nucleus of fluorine $^{18}_9F$. This unstable nucleus quickly disintegrates into the final products, oxygen $^{17}_8O$ and hydrogen ^1_1H. The overall reaction can be written

$$\overset{4}{2}\alpha + \overset{14}{7}N \rightarrow \overset{1}{1}H + \overset{17}{8}O \tag{39.15}$$

Note how charge and nucleons are conserved in these reactions. There was a net charge of $+9e$ before the reaction and a net charge of $+9e$ after the reaction, and there are 18 nucleons before and after the reaction.

39.10 Nuclear Fission

Before the discovery of the neutron in 1932, alpha particles and protons were the primary particles used to bombard atomic nuclei, but as charged particles, they have the disadvantage of being repelled electrostatically by the nucleus. Consequently, large energies are required before nuclear reactions can occur.

Since neutrons have zero electric charge, they can easily penetrate the nucleus of an atom with no coulomb repulsion. Fast neutrons may pass completely through a nucleus or may cause it to disintegrate. Slow neutrons may be captured by a nucleus, creating an unstable isotope, which may disintegrate.

Whenever the absorption of an incoming neutron causes a nucleus to split into two smaller nuclei, the reaction is called ***nuclear fission,*** and the product nuclei are called *fission fragments.*

Nuclear fission is the process by which heavy nuclei are split into two or more nuclei of intermediate mass numbers.

Whenever a slow neutron is captured by a uranium nucleus $^{235}_{92}U$, an unstable nucleus $(^{236}_{92}U)$ is produced that may decay in several ways into smaller product nuclei (Fig. 39.9). Such fission reactions may produce fast neutrons, beta particles, and gamma rays in addition to the product nuclei. For this reason, the products of a fission process, including fallout from a nuclear explosion, are highly radioactive.

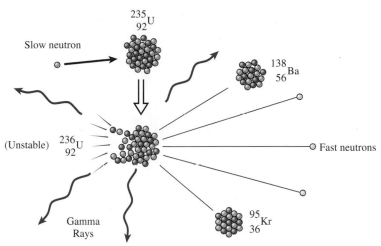

Figure 39.9   Nuclear fission of ^{235}U by capture of a slow neutron.

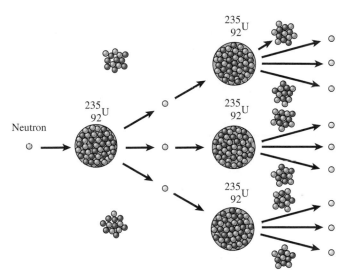

Figure 39.10   Nuclear chain reaction.

The fission fragments have a smaller mass number and therefore about 1 MeV more binding energy per nucleon (see Section 39.5). As a result, fission releases a large amount of energy. In the previous example, approximately 200 MeV per fission is produced.

Because each nuclear fission releases more neutrons, which may lead to additional fission, a **_chain reaction_** is possible. As seen in Fig. 39.10, the three neutrons released from the fission of $^{235}_{92}$U produce three additional fissions. Thus, starting with one neutron, we have liberated nine after only two steps. If such a chain reaction is not controlled, it can lead to an explosion of enormous magnitude.

39.11   Nuclear Reactors

A **_nuclear reactor_** is a device that controls the nuclear fission of radioactive material, producing new radioactive substances and large amounts of energy. These devices are used to furnish heat for electric power generation, propulsion, and industrial processes; to produce new elements or radioactive materials for a multitude of applications; and to supply neutrons for scientific experimentation.

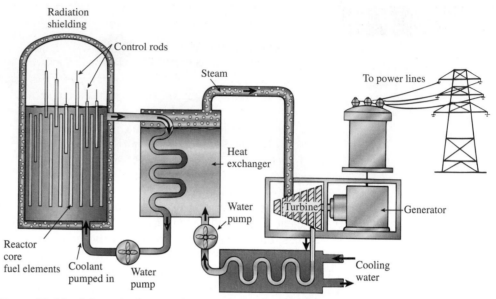

Figure 39.11 Schematic diagram of a nuclear reactor. Water heated under pressure in the reactor core is pumped into a heat exchanger, where it produces steam to operate a turbine.

A schematic diagram of a typical reactor is given in Fig. 39.11. The basic components are (1) a *core* of nuclear fuel, (2) a *moderator* for slowing down fast neutrons, (3) *control rods* or other means for regulating the fission process, (4) a *heat exchanger* for removing heat generated in the core, and (5) *radiation shielding.* Steam produced by the reactor is used to drive a turbine, which generates electricity. The spent steam is changed to water in the condenser and pumped back to the heat exchanger for another cycle.

The essential ingredient in the reactor is the fissionable material, or nuclear fuel. About the only naturally occurring fissionable material is $^{235}_{92}U$, which constitutes about 0.7 percent of naturally available uranium. The remaining 99.3 percent is $^{238}_{92}U$. Fortunately, $^{238}_{92}U$ is a *fertile* material, meaning that it changes to a fissionable material when struck by neutrons. Plutonium $^{239}_{94}Pu$ produced in this manner can provide new fuel for the reactor.

The production of additional fuel as a part of the reactor's operation has led to the design of *breeder reactors,* in which there is a net increase in fissionable material. In other words, the reactor produces more fuel than it consumes. This does not violate the law of conservation of energy. It provides only for the production of fissionable material from fertile materials.

The fissionable fuel in most reactors depends on the availability of slow neutrons, which are more likely to produce fission. Fast neutrons liberated by the fission of nuclei must therefore be slowed down. For this reason, reactor fuel is embedded in a suitable substance called a **moderator.** The function of this substance is to slow neutrons without capturing them.

Neutrons have a mass about the same as that of a hydrogen atom. It might be expected, then, that substances containing hydrogen atoms would be effective as moderators of neutrons. The neutron is analogous to a marble, which can be stopped by a collision with another marble but will merely bounce off a cannonball because of the great mass difference. Water (H_2O) and heavy water, containing $^{2}_{1}H$ instead of $^{1}_{1}H$, are often used as moderators. Other suitable materials are graphite and beryllium.

To control the nuclear furnace, it is necessary to regulate the number of neutrons that initiate the fission process. Substances such as boron and cadmium capture neutrons efficiently and are excellent control materials. A typical reactor has control rods that can be inserted into the reactor at variable distances. By adjusting the position of these rods, the activity of the nuclear furnace is controlled. A supplementary set of rods is available to allow the reactor to be shut down completely in an emergency.

39.12 Nuclear Fusion

In our earlier discussion on mass defect, we calculated that 28.3 MeV of energy is released in the formation of $^4_2$He from its component nucleons. This joining together of light nuclei into a single heavier nucleus is called ***nuclear fusion.*** This process provides the fuel for stars like our own Sun, and it is also the principle behind the hydrogen bomb. Many consider the fusion of hydrogen into helium as the ultimate fuel.

The use of nuclear fusion as a controlled source of energy is not without problems. It is still believed by most physicists that extremely high temperatures will be necessary to sustain nuclear fusion. The fusing nuclei would require millions of electronvolts of kinetic energy to overcome their coulombic repulsion. In the hydrogen bomb, this enormous energy is supplied by an atomic explosion, which then triggers the fusion process. The peaceful production of fusion by this method presents the problem of containment. The nuclear fuel would need to be so hot that it would instantly disintegrate any known substance. Present research methods involve containment by magnetic fields or rapid heating by powerful lasers. It is easy to see why the idea of "cold fusion" through an electrolytic process has generated so much excitement.

If the problems of fusion are ever solved, this energy source could provide a solution to our formidable problem of dwindling resources. The deuterium commonly found in seawater could provide us with an almost inexhaustible supply of fuel. It would represent more than a billion times the energy available in all our coal and oil reserves. In addition, it appears that fusion reactors would have much less of a problem with radioactive residue than that currently experienced with fission reactors.

Summary and Review

Summary

In this chapter, we have studied the fundamental particles that make up the nuclei of atoms. Protons and neutrons are held together in the nucleus by strong nuclear forces that are active only within the nucleus. When such particles are joined together, the resulting mass is less that of the constituent parts. For massive nuclei, it was also pointed out that energy results from tearing these nuclei apart. In either case, there is an enormous potential for useful energy. The major concepts that must be remembered in this chapter are as follows.

- The fundamental nuclear particles discussed in this chapter are summarized in the following table. The masses are given in atomic mass units (u), and the charge is in terms of the electronic charge $+e$ or $-e$, which is 1.6×10^{-19} C.

Fundamental Particles

Particle	Symbol	Mass, u	Charge
Electron	$_{-1}^{0}e, \ _{-1}^{0}\beta$	0.00055	$-e$
Proton	$_{1}^{1}p, \ _{1}^{1}H$	1.007276	$+e$
Neutron	$_{0}^{1}n$	1.008665	0
Positron	$_{+1}^{0}e, \ _{+1}^{0}\beta$	0.00055	$+e$
Alpha particle	$_{2}^{4}\alpha, \ _{2}^{4}He$	4.001506	$+2e$

The atomic masses of the various elements are given in the text.

- The atomic number Z of an element is the number of protons in its nucleus. The mass number A is the sum of the atomic number and the number of neutrons N. These numbers are used to write the nucleus symbol:

$$A = Z + N \qquad Symbol: {}_{Z}^{A}X$$

- One *atomic mass unit* (1 u) is equal to one-twelfth the mass of the most abundant carbon atom. Its value in kilograms is given as follows. Also, since $E = mc^2$, we can write the conversion factor from mass to energy as c^2.

$$1 \text{ u} = 1.6606 \times 10^{-27} \text{ kg} \qquad c^2 = 931 \text{ MeV/u}$$

$$1 \text{ MeV} = 10^6 \text{ eV} = 1.6 \times 10^{-13} \text{ J}$$

In the mass spectrometer, the velocity v and the radius R of the singly ionized particles are

$$v = \frac{E}{B} \qquad R = \frac{mv}{eB} \qquad Mass\ Spectrometer$$

- The *mass defect* is the difference between the rest mass of a nucleus and the sum of the rest masses of its nucleons. The *binding energy* is obtained by multiplying the mass defect by c^2.

$$E_B = [(Zm_H + Nm_n) - M]c^2 \qquad Binding\ Energy$$

where $m_H = 1.007825$ u
$m_n = 1.008665$ u
$c^2 = 931$ MeV/u
$M = $ atomic mass
$N = A - Z$
$Z = $ atomic number

- Several general equations for radioactive decay are

$$_{Z}^{A}X \rightarrow {}_{Z-2}^{A-4}Y + {}_{2}^{4}\alpha + \text{energy} \qquad Alpha\ Decay$$

$$_{Z}^{A}X \rightarrow {}_{Z+1}^{A}Y + {}_{-1}^{0}\beta + \text{energy} \qquad Beta\text{-}Minus\ Decay$$

$$_{Z}^{A}X \rightarrow {}_{Z-1}^{A}Y + {}_{+1}^{0}\beta + \text{energy} \qquad Beta\text{-}Plus\ Decay$$

- The *activity* R of a sample is the rate at which the radioactive nuclei decay. It is generally expressed in curies (Ci).

$$\text{One } curie \text{ (1 Ci)} = 3.7 \times 10^{10}$$
$$\text{disintegrations per second (s}^{-1})$$

- The *half-life* of a sample is the time $T_{1/2}$ in which one-half the unstable nuclei will decay.
- The number of unstable nuclei remaining after a time t depends on the number n of half-lives that have passed. If N_0 nuclei exist at time $t = 0$, then a number N exist at time t. We have

$$N = N_0 \left(\frac{1}{2}\right)^n \qquad \text{where } n = \frac{t}{T_{1/2}}$$

- The activity R and mass m of the radioactive portion of a sample are found from similar relations:

$$R = R_0 \left(\frac{1}{2}\right)^n \qquad m = m_i \left(\frac{1}{2}\right)^n$$

- In any nuclear equation, the number of nucleons on the left side must equal the number of nucleons on the right side. Similarly, the net charge must be the same on each side.

Key Terms

activity 771
alpha particle 769
atomic mass unit 761
atomic number 759
beta particle 769
binding energy 766
chain reaction 775
conservation of charge 773
conservation of mass-energy 773

conservation of nucleons 773
curie 772
gamma rays 769
half-life 772
isotopes 764
mass defect 766
mass number 761
mass spectrometer 764
moderator 776

nuclear fission 774
nuclear force 759
nuclear fusion 777
nuclear reactor 775
nucleon 759
nuclide 767
radioactivity 769

Review Questions

39.1. Write the symbol $_Z^A X$ for the most abundant isotopes of (a) cadmium, (b) silver, (c) gold, (d) polonium, (e) magnesium, and (f) radon.

39.2. From the curve describing the binding energy per nucleon (Fig. 39.4), would you expect the mass defect to be greater for chromium $_{24}^{52}Cr$ or uranium $_{92}^{238}U$? Why?

39.3. The binding energy is greater for the mass numbers in the central part of the periodic table. Discuss the significance of this in relation to nuclear fission and nuclear fusion. How do you explain the release of energy in both fusion and fission in view of the fact that one process brings nuclei together and the other tears them apart?

39.4. How is the stability of an isotope affected by the ratio of the mass number A to the atomic number Z? Does the element whose ratio is closest to 1 always appear to be the more stable?

39.5. Define and compare alpha particles, beta particles, and gamma rays. Which are likely to do the most damage to human tissue?

39.6. Given a source that emits alpha, beta, and gamma radiation, draw a diagram showing how you could demonstrate the charge and penetrating power of each type of radiation. Assume you have at your disposal a source of a magnetic field and several thin sheets of aluminum.

39.7. Describe and explain, step by step, the decay of $_{92}^{238}U$ to the stable isotope of lead, $_{82}^{206}Pb$. (Refer to Fig. 39.6.)

39.8. Write in the missing symbol, in the form $_Z^A X$, for the following nuclear disintegrations:

a. $_{90}^{234}Th \rightarrow _{91}^{234}Pa + \underline{\hspace{1cm}}$
b. $_{15}^{32}P \rightarrow \underline{\hspace{1cm}} + _{+0}^{1}e$

c. $_{94}^{239}Pu \rightarrow _{90}^{234}Th + \underline{\hspace{1cm}}$
d. $_{92}^{238}U \rightarrow \underline{\hspace{1cm}} + _2^4\alpha$

39.9. Write the missing symbol for the following nuclear reactions:

a. $_1^2H + _1^3H \rightarrow _2^4H + \underline{\hspace{1cm}}$
b. $_{12}^{25}Mg + \underline{\hspace{1cm}} \rightarrow _{13}^{28}Al + _1^1H$
c. $_4^9Be + _2^4\alpha \rightarrow _6^{12}C + \underline{\hspace{1cm}}$
d. $_1^1H + \underline{\hspace{1cm}} \rightarrow _6^{12}C + _2^4He$

39.10. Explain the function of the following components of a nuclear reactor: (a) uranium, (b) radiation shielding, (c) moderator, (d) control rods, (e) heat exchanger, and (f) condenser.

39.11. Give examples to show how beta decay and alpha decay tend to bring unstable nuclei closer to the stability curve of Fig. 39.5.

39.12. Radon has a half-life of 3.8 days. Consider a sample of radon having a mass m and an activity R. What mass of radioactive radon remains after 3.8 days? Does this mean the activity is reduced to one-half in the time of one half-life?

39.13. Radioactive carbon $_7^{14}C$ has a half-life of 5570 h. In a living organism, the relative concentration of this isotope is the same as it is in the atmosphere because of the interchange of materials between the organism and the air. When an organism dies, this interchange stops, and radioactive decay begins without replacement from the living organism. Explain how this principle can be used to determine the age of fossil remains.

Problems

Refer to Table 39.4 for nuclidic masses.

Section 39.2 The Elements

39.1. How many neutrons are in the nucleus of $^{208}_{82}$Pb? How many protons? What is the ratio N/Z?

Ans. 126, 82, 1.54

39.2. The nucleus of a certain isotope contains 143 neutrons and 92 protons. Write the symbol for this nucleus.

39.3. From a stability curve, it is determined that the ratio of neutrons to protons for a cesium nucleus is 1.49. What is the mass number for this isotope of cesium?

Ans. 137

39.4. Most nuclei are nearly spherical in shape and have a radius that can be approximated by

$$r = r_0 A^{1/3} \qquad r_0 = 1.2 \times 10^{-15} \text{ m}$$

What is the approximate radius of the nucleus of a gold atom ($^{197}_{79}$Au)?

39.5. Study Table 39.4 for information on the several nuclides. Determine the ratio of N/Z for the following nuclides: beryllium-9, copper-64, and radium 224.

Ans. 1.25, 1.21, 1.55

Section 39.3 The Atomic Mass Unit

39.6. Find the mass in grams of a gold particle containing two million atomic mass units.

39.7. Consider a 2-kg cylinder of copper. What is the mass in atomic mass units? In megaelectronvolts? In joules?

Ans. 1.20×10^{27} u, 1.12×10^{30} MeV, 1.79×10^{17} J

39.8. A certain nuclear reaction releases an energy of 5.5 MeV. How much mass (in atomic mass units) is required to produce this energy?

39.9. The periodic table gives the average mass of a silver atom as 107.842 u. What is the average mass of the silver nucleus?

Ans. 107.816 u

*39.10. Consider the mass spectrometer as illustrated by Fig. 39.2. A uniform magnetic field of 0.6 T is placed across both upper and lower sections of the spectrometer, and the electric field in the velocity selector is 120 V/m. A singly charged neon atom ($+1.6 \times 10^{-19}$ C) of mass 19.992 u passes through the velocity selector and into the spectrometer. What is the velocity of the neon atom as it emerges from the velocity selector?

*39.11. What is the radius of the circular path followed by the neon atom of Prob. 39.10?

Ans. 6.92 cm

Section 39.5 The Mass Defect and Binding Energy

*39.12. Calculate the mass defect and binding energy for the neon-20 atom ($^{20}_{10}$Ne).

*39.13. Calculate the binding energy and the binding energy per nucleon for tritium ($^{3}_{1}$H). How much energy in joules is required to tear the nucleus apart into its constituent nucleons?

Ans. 8.48 MeV, 2.83 MeV/nucleon, 1.36×10^{-12} J

*39.14. Calculate the mass defect of $^{7}_{3}$Li. What is the binding energy per nucleon?

*39.15. Determine the binding energy per nucleon for carbon-12 ($^{12}_{6}$C).

Ans. 7.68 MeV/nucleon

*39.16. What are the mass defect and the binding energy for a gold atom ($^{197}_{79}$Au)?

*39.17. Determine the binding energy per nucleon for tin-120 ($^{120}_{50}$Sn).

Ans. 8.50 MeV/nucleon

Section 39.7 Radioactive Decay

39.18. The activity of a certain sample is rated as 2.8 Ci. How many nuclei will have disintegrated in a time of 1 min?

39.19. The cobalt nucleus ($^{60}_{27}$Co) emits gamma rays of approximately 1.2 MeV. How much mass is lost by the nucleus when it emits a gamma ray of this energy?

Ans. 0.00129 u

39.20. The half-life of the radioactive isotope indium-109 is 4.30 h. If the activity of a sample is 1 mCi at the start, how much activity remains after 4.30, 8.60, and 12.9 h?

39.21. The initial activity of a sample containing 7.7×10^{11} bismuth-212 nuclei is 4.0 mCi. The half-life of this isotope is 60 min. How many bismuth-212 nuclei remain after 30 min? What is the activity at the end of that time?

Ans. 5.44×10^{11} nuclei, 2.83 mCi

*39.22. Strontium-90 is produced in appreciable quantities in the atmosphere during a nuclear explosion. If this isotope has a half-life of 28 years, how long will it take for the initial activity to drop to one-fourth of its original activity?

*39.23. Consider a pure, 4.0-g sample of radioactive gallium-67. If the half-life is 78 h, how much time is required for 2.8 g of this sample to decay?

Ans. 135.5 h

*39.24. If one-fifth of a pure radioactive sample remains after 10 h, what is the half-life?

Section 39.9 Nuclear Reactions

*39.25. Determine the minimum energy released in the nuclear reaction

$$^{19}_{9}F + ^{1}_{1}H \rightarrow ^{4}_{2}He + ^{16}_{8}O + energy$$

The atomic mass of $^{19}_{9}F$ is 18.998403 u.

Ans. 8.11 MeV

*39.26. Determine the approximate kinetic energy imparted to the alpha particle when radium-226 decays to form radon-222. Neglect the energy imparted to the radon nucleus.

*39.27. Find the energy involved in the production of two alpha particles in the reaction

$$^{7}_{3}Li - ^{1}_{1}H \rightarrow ^{4}_{2}He + ^{4}_{2}He + energy$$

Ans. 17.3 MeV

*39.28. Compute the kinetic energy released in the beta-minus decay of thorium-233.

*39.29. What must be the energy of an alpha particle if it bombards a nitrogen-14 nucleus, producing $^{17}_{8}O$ and $^{1}_{1}H$? ($^{17}_{8}O = 16.999130$ u)

$$^{4}_{2}He + ^{14}_{7}N + Energy \rightarrow ^{17}_{8}O + ^{1}_{1}H$$

Ans. 1.19 MeV

Additional Problems

*39.30. What is the average mass in kilograms of the nucleus of a boron-11 atom?

*39.31. What are the mass defect and the binding energy per nucleon for boron-11?

Ans. 0.0818 u, 6.92 megaelectronvolts per nucleon

*39.32. Find the binding energy per nucleon for thallium-206.

*39.33. Calculate the energy required to separate the nucleons in mercury-204. Ans. 1.61 GeV

*39.34. The half-life of a radioactive sample is 6.8 h. How much time passes before the activity drops to one-fifth of its initial value?

*39.35. How much energy is required to tear apart a deuterium atom? Ans. 2.22 MeV

*39.36. Plutonium-232 decays by alpha emission with a half-life of 30 min. How much of this substance remains after 4 h if the original sample had a mass of 4.0 g? Write the equation for the decay.

*39.37. If 32×10^9 atoms of a radioactive isotope are reduced to only 2×10^9 atoms in a time of 48 h, what is the half-life of this material?

Ans. 12.0 h

*39.38. A certain radioactive isotope retains only 10 percent of its original activity after a time of 4 h. What is its half-life?

*39.39. When a $^{6}_{3}Li$ nucleus is struck by a proton, an alpha particle and a produce nucleus are released. Write the equation for this reaction. What is the net energy transfer in this case?

Ans. 4.02 MeV

*39.40. Uranium-238 undergoes alpha decay. Write the equation for the reaction and calculate the disintegration energy.

*39.41. A 9-g sample of radioactive material has an initial activity of 5.0 Ci. Forty minutes later, the activity is only 3.0 Ci. What is the half-life? How much of the pure sample remains?

Ans. 54.3 min, 5.40 g

Critical Thinking Questions

*39.42. Nuclear fusion is a process that can produce enormous energy without the harmful by-products of nuclear fission. Calculate the energy released in the following nuclear fusion reaction:

$$^{3}_{2}H + ^{3}_{2}He \rightarrow ^{4}_{2}He + ^{1}_{1}H + ^{1}_{1}H$$

Ans. 12.9 MeV

*39.43. Carbon-14 decays very slowly with a half-life of 5740 years. Carbon dating can be accomplished by seeing what fraction of carbon-14 remains, assuming that the decay process began with the death of a living organism. What would be the age of a chunk of charcoal if it was determined that the radioactive C-14 remaining was only 40 percent of what would be expected in a living organism?

*39.44. The velocity selector in a mass spectrometer has a magnetic field of 0.2 T perpendicular to an electric field of 50 kV/m. The same magnetic field is across the lower region. What is the velocity of singly charged lithium-7 atoms as they leave the selector? If the radius of the circular path in the spectrometer is 9.10 cm, what is the atomic mass of the lithium atom?

Ans. 2.50×10^5 m/s, 7.014 u

*39.45. A nuclear reactor operates at a power level of 2.0 MW. Assuming approximately 200 MeV of energy is released for a single fission of U-235, how many fission processes are occurring each second in the reactor?

*39.46. Consider an experiment that bombards $^{14}_{7}\text{N}$ with an alpha particle. One of the two product nuclides is $^{1}_{1}\text{H}$. The reaction is

$$^{4}_{2}\text{He} + {}^{14}_{7}\text{N} \rightarrow {}^{A}_{Z}\text{X} + {}^{1}_{1}\text{H}$$

What is the product nuclide indicated by the symbol X? How much kinetic energy must the alpha particle have in order to produce the reaction?   Ans. 1.19 MeV

*39.47. When passing a stream of ionized lithium atoms through a mass spectrometer, the radius of the path followed by $^{7}_{3}\text{Li}$ (7.0169 u) is 14.00 cm. A lighter line is formed by the $^{6}_{3}\text{Li}$ (6.0151 u). What is the radius of the path followed by the $^{6}_{3}\text{Li}$ isotopes?

Index